Lecture Notes in Artificial Intelligence 13281

Subseries of Lecture Notes in Computer Science

More information about this subseries at https://link.springer.com/bookseries/1244

João Gama · Tianrui Li · Yang Yu ·
Enhong Chen · Yu Zheng · Fei Teng (Eds.)

Advances in Knowledge Discovery and Data Mining

26th Pacific-Asia Conference, PAKDD 2022
Chengdu, China, May 16–19, 2022
Proceedings, Part II

Editors
João Gama
Laboratory of Artificial Intelligence and Decision Support
University of Porto
Porto, Portugal

Tianrui Li
School of Computing and Artificial Intelligence
Southwest Jiaotong University
Chengdu, China

Yang Yu
National Key Laboratory for Novel Software Technology
Nanjing University
Nanjing, China

Enhong Chen
School of Computer Science and Technology
University of Science and Technology of China
Hefei, China

Yu Zheng
JD iCity, JD Technology & JD Intelligent Cities Research
Beijing, China

Fei Teng
School of Computing and Artificial Intelligence
Southwest Jiaotong University
Chengdu, China

ISSN 0302-9743 ISSN 1611-3349 (electronic)
Lecture Notes in Artificial Intelligence
ISBN 978-3-031-05935-3 ISBN 978-3-031-05936-0 (eBook)
https://doi.org/10.1007/978-3-031-05936-0

LNCS Sublibrary: SL7 – Artificial Intelligence

This Springer imprint is published by the registered company Springer Nature Switzerland AG
The registered company address is: Gewerbestrasse 11, 6330 Cham, Switzerland

General Chairs' Preface

On behalf of the Organizing Committee, it is our great pleasure to welcome you to the 26th Pacific-Asia Conference on Knowledge Discovery and Data Mining (PAKDD2022), held in Chengdu, China, during May 16–19, 2022. Starting in 1997, PAKDD has long established itself as one of the leading international conferences in data mining and knowledge discovery. PAKDD provides an international forum for researchers and industry practitioners to share their new ideas, original research results, and practical development experiences from all Knowledge Discovery and Data Mining (KDD) related areas. In response to the COVID-19 pandemic and the need for social distancing, PAKDD 2022 was held as a hybrid conference for both online and onsite attendees.

Our gratitude goes first and foremost to the researchers, who submitted their work to PAKDD 2022. We would like to deliver our sincere thanks for their efforts in research, as well as in preparing high-quality presentations. We also thank all the collaborators and sponsors for their trust and cooperation. It is our great honor that three eminent keynote speakers joined the conference: Jian Pei (Simon Fraser University, Canada), Bernhard Schölkopf (Max Planck Institute for Intelligent Systems, Germany) and Ji-Rong Wen (Renmin University, China). They were extremely professional and have high reputations in their respective areas. We enjoyed their participation and talks, which made the conference one of the best academic platforms for knowledge discovery and data mining.

We would like to express our sincere gratitude to the contributions of Steering Committee members, Organizing Committee members, Program Committee members and anonymous reviewers, led by Program Committee Co-chairs: João Gama (University of Porto), Tianrui Li (Southwest Jiaotong University), and Yang Yu (Nanjing University). We are also grateful for the hosting organization Southwest Jiaotong University which is continuously providing institutional and financial support to PAKDD 2022. We feel beholden to the PAKDD Steering Committees for their constant guidance and sponsorship of manuscripts.

Finally, our sincere thanks go to all the participants and volunteers. We hope all of you enjoyed PAKDD 2022.

April 2022

Enhong Chen
Yu Zheng

PC Chairs' Preface

It is our great pleasure to present at the 26th Pacific-Asia Conference on Knowledge Discovery and Data Mining (PAKDD 2022) as the Program Committee Chairs. PAKDD is one of the longest established and leading international conferences in the areas of data mining and knowledge discovery. It provides an international forum for researchers and industry practitioners to share their new ideas, original research results, and practical development experiences from all KDD related areas, including data mining, data warehousing, machine learning, artificial intelligence, databases, statistics, knowledge engineering, big data technologies and foundations.

This year PAKDD received 627 submissions, among which 69 submissions were rejected at a preliminarily stage due to the policy violations. There were 320 Program Committee members and 45 Senior Program Committees members involved in reviewing process. Each submission was reviewed by at least three different reviewers. Over 67% of those submissions were reviewed by four or more reviewers. Eventually, 121 submissions were accepted and recommended to be published, resulting in an acceptance rate of 19.30%. Out of these, 29 submissions were about applications, 4 submissions were related to big data technologies, 46 submissions were on data science and 42 submissions were about foundations. We would like to appreciate all PC members and reviewers, who offered a high-quality program with diligence on PAKDD 2022.

The conference program featured keynote speeches from distinguished researchers in the community, most influential paper talks, cutting-edge workshops and comprehensive tutorials.

We wish to sincerely thank all PC members and reviewers for their invaluable efforts in ensuring a timely, fair, and highly effective PAKDD 2022 program.

April 2022

João Gama
Tianrui Li
Yang Yu

Organization Committee

Honorary Co-chairs

Dan Yang	Southwest Jiaotong University, China
Zhi-Hua Zhou	Nanjing University, China

General Co-chairs

Enhong Chen	University of Science and Technology of China, China
Yu Zheng	JD.com, China

Program Committee Co-chairs

Joao Gama	University of Porto, Portugal
Tianrui Li	Southwest Jiaotong University, China
Yang Yu	Nanjing University, China

Workshop Co-chairs

Gill Dobbie	University of Auckland, New Zealand
Can Wang	Griffith University, Australia

Tutorial Co-chairs

Gang Li	Deakin University, Australia
Tanmoy Chakraborty	Indraprastha Institute of Information Technology Delhi, India

Local Arrangement Co-chairs

Yan Yang	Southwest Jiaotong University, China
Chuan Luo	Sichuan University, China
Xin Yang	Southwestern University of Finance and Economics, China

Sponsor Chair

Xiaobo Zhang	Southwest Jiaotong University, China

Publicity Co-chairs

Xiangnan Ren	Group 42, United Arab Emirates
Hao Wang	Zhejiang Lab, China
Junbo Zhang	JD.com, China
Chongshou Li	Southwest Jiaotong University, China

Proceedings Chair

Fei Teng	Southwest Jiaotong University, China

Web and Content Co-chairs

Xiaole Zhao	Southwest Jiaotong University, China
Zhen Jia	Southwest Jiaotong University, China

Registration Chairs

Hongmei Chen	Southwest Jiaotong University, China
Jie Hu	Southwest Jiaotong University, China
Yanyong Huang	Southwestern University of Finance and Economics, China

Steering Committee

Longbing Cao	University of Technology Sydney, Australia
Ming-Syan Chen	NTU
David Cheung	University of Hong Kong, China
Gill Dobbie	University of Auckland, New Zealand
Joao Gama	University of Porto, Portugal
Zhiguo Gong	University of Macau, China
Tu Bao Ho	Japan Advanced Institute of Science and Technology, Japan
Joshua Z. Huang	Shenzhen Institutes of Advanced Technology, Chinese Academy of Sciences, China
Masaru Kitsuregawa	Tokyo University, Japan
Rao Kotagiri	University of Melbourne, Australia
Jae-Gil Lee	Korea Advanced Institute of Science & Technology, South Korea

Ee-Peng Lim	Singapore Management University, Singapore
Huan Liu	Arizona State University, USA
Hiroshi Motoda	AFOSR/AOARD and Osaka University, Japan
Jian Pei	Simon Fraser University, Canada
Dinh Phung	Monash University, Australia
P. Krishna Reddy	International Institute of Information Technology, Hyderabad, India
Kyuseok Shim	Seoul National University, South Korea
Jaideep Srivastava	University of Minnesota, USA
Thanaruk Theeramunkong	Thammasat University, Thailand
Vincent S. Tseng	NCTU
Takashi Washio	Osaka University, Japan
Geoff Webb	Monash University, Australia
Kyu-Young Whang	Korea Advanced Institute of Science & Technology, South Korea
Graham Williams	Australian National University, Australia
Min-Ling Zhang	Southeast University, China
Chengqi Zhang	University of Technology Sydney, Australia
Ning Zhong	Maebashi Institute of Technology, Japan
Zhi-Hua Zhou	Nanjing University, China

Host Institute

Contents – Part II

Foundations

Text2Chart: A Multi-staged Chart Generator from Natural Language Text

Md. Mahinur Rashid, Hasin Kawsar Jahan, Annysha Huzzat, Riyasaat Ahmed Rahul, Tamim Bin Zakir, Farhana Meem, Md. Saddam Hossain Mukta, and Swakkhar Shatabda(✉)

Department of Computer Science and Engineering, United International University, Dhaka, Bangladesh

{mrashid171045,hjahan171054,ahuzzat171034,rrahul171089,tzakir171032, fmeem171031}@bscse.uiu.ac.bd

{saddam,swakkhar}@cse.uiu.ac.bd

Abstract. Generation of scientific visualization from analytical natural language text is a challenging task. In this paper, we propose Text2Chart, a multi-staged chart generator method. Text2Chart takes natural language text as input and produces visualization as two-dimensional charts. Text2Chart approaches the problem in three stages. Firstly, it identifies the axis elements, known as x and y entities, of a chart from the given text. Next, it finds a mapping of x-entities with its corresponding y-entities. Subsequently, it generates a chart type among bar, line, or pie, which is suitable for the given text. Combination of these three stages is capable of generating visualization from the given statistical text. We have also constructed a dataset for this problem. Experiments show that Text2Chart performs best with BERT based encodings with LSTM models in the first stage to label x and y entities, Random Forest classifier in the mapping stage and fastText embedding with LSTM in the chart type prediction stage. In our experiments, all the stages show satisfactory results and effectiveness considering the formation of charts from analytical text, achieving a commendable overall performance.

Keywords: Chart generation · Natural Language Processing · Information retrieval · Neural network · Automated visualization

1 Introduction

In recent years, advances in Natural Language Processing (NLP) have made huge progress in extracting information from natural language texts. Among them, a few example tasks are: document summarization [1], title or caption generation from texts, generating textual descriptions of charts [2], named entity recognition [3], etc. There have been several attempts to generate graphs or structural elements from natural language texts or free texts [4–6]. Scientific charts (bar, line, pie, etc.) are visualizations that are often used in communication. However,

J. Gama et al. (Eds.): PAKDD 2022, LNAI 13281, pp. 3–16, 2022.
https://doi.org/10.1007/978-3-031-05936-0_1

automated generation of charts from natural language text has always been a challenging task.

There are very few works in the literature addressing the exact problem of scientific chart generation from natural language text [7,8]. In [7], the authors have presented an infographic generation technique from natural language statements. However, their method is limited to single entity generation only. Text2Chart extends it to multiple entity generation and thus can generate more complex charts. Nevertheless, Generative Pre-trained Transformer 3 (GPT-3)[8] has been a recent popular phenomenon in the field of deep learning. OpenAI has designed this third-generation language model that is trained using neural networks. To the best of our knowledge, there has been an attempt to make a simple chart building tool using GPT-3. As its implementation is not accessible yet, the field of information extraction regarding chart creation can still be considered unexplored to some extent. Moreover, the dataset used in GPT-3 is a very large one, and the training is too expensive.

In this paper, we propose Text2Chart, a multi-staged technique that generates charts from analytical natural language text. Text2Chart works in a combination of three stages. In the first stage, it recognizes x-axis and y-axis entities from the input text. In the second stage, it maps x-axis entities with their corresponding y-axis entities, and in the third stage, it predicts the best-suited chart type for the particular text input. Text2Chart is limited to three types of charts: bar charts, line charts and pie charts. Tasks in each stage are formulated as supervised learning problems. We have created our own dataset which is labeled for all three stages of Text2Chart. We have used a wide range of evaluation metrics for all the three stages and different combinations of word embeddings and classifiers. The experimental results shows that the best results in the first stage are obtained using BERT embedding and Bidirectional LSTM, achieving an $F1$-score of 0.83 for x-entity recognition and 0.97 for y-entity recognition in the test set. In the mapping stage, Random Forest achieves the best results of 0.917 of Area under Receiver Operating Characteristic Curve (auROC) in the test set. In the third stage, the model fastText with LSTM layers performs the best to predict the suitable chart type. Here, Text2Chart achieves the best results of auROC 0.64 for pie charts and auROC 0.91 for line charts. The experimental analysis of each stage and in combination shows the overall effective performance of Text2Chart for generating charts from given natural language charts.

2 Related Work

Recent developments in the field of NLP is advancing information extraction in general. One of the first and foremost steps in NLP is the proper vectorization of the input corpora. One of the breakthroughs in this area is word2vec proposed in [9]. Word2Vec maps words with similar meaning to adjacent points in a vector space. The embedding is learnt using a neural network on a continuous bag of words or skip-gram model. A character-level word embedding is proposed in [10]. Recently, Bidirectional Encoder Representations from Transformers (BERT) is proposed in [11]. BERT is trained on a large corpora and enables pre-trained models to be applicable to transfer learning to a vast area of research. BERT has been successfully applied to solve problems like Named Entity Recognition (NER) [3], text summarization [1], etc.

Text based information processing has been a long quest in the field [12]. Kobayashi et al. [12] have presented a NLP based modeling for line charts. A Hidden Markov Model based chart (bar, line, etc.) recognition method is proposed in [13]. Graph neural networks have been employed in [4] to generate logical forms with entities from free text using BERT. In a very recent work [5], Obeid et al. have used transformer based models for text generation from charts. For this work, they have also constructed a large dataset extracting charts from Statista. However, their work focuses on chart summarizing and hence called 'Chart-to-Text'. In an earlier work [14], authors have proposed a method for generating ground truth for chart images. Both of the works are limited to bar charts and line charts only. A Generative Adversarial Network, AttnGAN is proposed in [15] that can generate images from text descriptions. Balaji et al. [2] has proposed an automatic chart description generator. CycleGT has been proposed recently that works in both directions: text to graphs and graphs to text [6]. Kim et al. [16] has proposed a pipeline to generate an automatic question answering system based on charts.

Automated visualization has always been a very fascinating area. A survey of Machine Learning based visualization methods has been presented in [17]. Deep Eye is proposed in [18] to identify best visualizations from pie chart, bar chart, line chart and scatter chart for a given data pattern. 'Text-to-Viz' is proposed in [7] that generates excellent infographics from given text. However, their method is limited to a single entity only. GPT-3 [8] has been a recent phenomenon in the field which has been reported to generate charts from natural language texts. However, GPT-3 implementation is not open yet. Moreover, it is trained on an extremely large corpora and an extremely large transformer based model which requires huge resources. In the light of the review of the existing methods, we believe there is a significant research gap to be addressed in this area.

3 Proposed Method

Text2Chart consists of three stages as shown in Fig. 1. It takes a free text as input containing the analytical information. Then it produces x and y axis entities followed by a mapping generation among these elements. In stage 3, the chart type is predicted. A combination of these three are then passed on to the chart generation module. This section presents the detailed procedure of these stages.

3.1 Stage 1: x-Axis and y-Axis Label Entity Recognition

In the first stage, we identify the potential candidate words for both x-axis and y-axis entities of a two dimensional chart. We have formulated the problem as a supervised machine learning task. Here, input to the problem is a paragraph

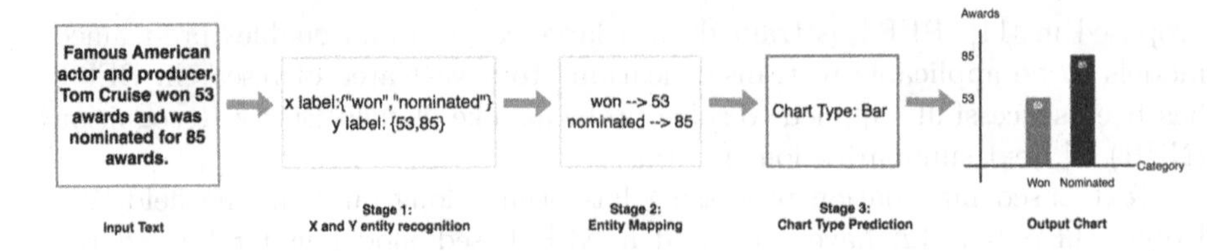

Fig. 1. The overall methodology of Text2Chart.

or natural language text and output is a list of words labeled as x-entity and y-entity.

To identify x-entity and y-entity, we build a neural network with different word embeddings and sequence representations. We have employed and experimented with two different strategies - i) detecting both types of entities at once and ii) using separate models for recognizing x and y entities. Detecting both x and y entities at once shows a drawback as there lies a possibility that a certain type of entity may outperform the loss function of the other types as observed in the experiments (Sect. 4.3).

We have experimented with both of the strategies using word embedding like Word2Vec [9], fastText [10] and the sequence output of the pre-trained model provided by BERT [11]. For each sample text in the dataset, we take the generated embedding and use it as an input to our model. Then we use layers of Bi-directional LSTM networks. On top of that, we use the time-distribution layer and dense layer to classify each word index that falls into a category of a respected entity or not.

3.2 Stage 2: Mapping of x and y Label Entities

After identifying the x and y entities in Stage 1, we map each of the identified x entity with its corresponding y entity. For example, if we have an x entity set for a text as $\{x_1, x_2, \cdots, x_M\}$ and y entity set of that text is $\{y_1, y_2, \cdots, y_N\}$ and their mapping is as follows $\{(x_1, \phi(x_1)), (x_2, \phi(x_2)), \cdots, (x_M, \phi(x_M)\}$. Please note, here x_i, y_j denotes their position in the sequence. Here the mapping function, $\phi(x_i)$ maps an entity x_i to another entity, y_k. However, there is often found that the entity set lengths are not same $M \neq N$ and often the sequential order is not maintained. For two x entities x_i, x_j if they maps to y_k, y_l, then a sequential mapping ϕ guarantees, $i \leq j, k \leq l$ whereas the non-sequential mapping will not guarantee that. However, in our observation, non-sequential mapping is not that frequent. In order to address these issues, we propose that the mapping is dependent on the distances between the corresponding entities. We call it our baseline model for this task. From the training dataset, we learn the probability distribution for positive and negative likelihood for distances between x and y entities which are $P(d(x_i, y_k)|\phi(x_i) = y_k)$ and $P(d(x_i, y_k)|\phi(x_i) \neq y_k)$ respectively. For the missing values in the range, nearest neighbor smoothing is used to estimate the likelihood values and then normalized to convert it to a probability distribution. The baseline model defines the mapping as in the following equation:

$$\phi(x_i) = \underset{k}{\operatorname{argmax}} \frac{P(d(x_i, y_k)|\phi(x_i) = y_k)}{P(d(x_i, y_k)|\phi(x_i) = y_k) + P(d(x_i, y_k)|\phi(x_i) \neq y_k)} \tag{1}$$

For a particular entity x_i and a particular y_k entity, we take the two other entities, one immediately before (x_{i-1}, y_{k-1}) and the next one (x_{i+1}, y_{k+1}) to create the feature vector. For 6 such entity positions, we generate 15 possible pairs and take pairwise distances among them. Note that, for two similar type entities we take unsigned distance and for different entities signed distances are taken to encode their relative positions into the feature vector. With this feature vector, we train two models: SVM and Random Forests, where the latter works slightly better. As this is an argmax based calculation, the probability distribution of the Random Forest classifier was more consistent than that of SVM. The reason for the inconsistency of the distribution with the scores in SVC is that the 'argmax' of the scores may not be the argmax of the probabilities. Therefore we take the auROC as the primary evaluation matrix for this stage. We take the harmonic mean of auROC of both training and validation so that the measure is balanced and they do not outperform each other.

3.3 Stage 3: Chart Type Prediction

Generally, a bar chart is the most commonly accepted chart type for any statistical data. However, for better visualization and understanding, pie charts and line charts are also used. Pie charts are suitable if the entities conform to a collection/composition. Line charts are suitable for the cases where the entities themselves form a continuous domain. For this stage, we have applied fastText word embeddings to build two models with LSTM layers and dense layers. Each model performs binary classification; one is to predict if a pie chart is suited for the text or not, and the other is for the line chart. When neither of these two chart types are fitting, only the bar chart is assigned to the text.

4 Experimental Analysis

Text2Chart is implemented using Tensorflow version 2.3. All the experiments have run using Google Colab and the cloud GPU provided with it. The hardware environment of our work requires a CPU of 2.3 GHz, GPU 12 GB, RAM 12.72 GB and Disk of 107 GB. All the experiments have run at least 5 times with different random seeds and only the average results are reported in this section. Source codes and the dataset of Text2Chart will be made available via a public repository (at the time of publication).

4.1 Dataset Construction

While starting this work, no datasets were available for this particular task of automatic generation of a chart out of a natural language text. Text2Chart requires a specific dataset from which the text samples are suitable for recognizing the chart information. Here chart information refers to the x-axis entities

Table 1. Summary of datasets used in the experiments.

Dataset	Text samples	x, y Entity prediction				Mapping pairs	Chart type	
		x Tokens	y Tokens	x Labels	y Labels		Pie	Line
Training	464	3411	3614	1984	1909	1984	73	58
Validation	116	985	1058	548	529	548	20	11
Test	137	988	1075	574	561	574	20	15

and the corresponding y-axis values respectively. The text samples must contain all these entities to construct the particular chart. We have collected text samples from Wikipedia, other statistical websites and crowdsourcing. We have used crowdsourcing to label the data so that the texts are labeled for all three stages. All the labeled data are then cross checked by a team of volunteers and only the consensus labels are taken. In total, 717 text samples are taken in the final dataset with 30,027 words/tokens. The average length of the text samples is 53 words and the maximum length is 303 words in a single text. This final dataset is then split in the train, validation and test sets each containing 464, 116 and 137 samples respectively. A summary of the dataset is shown in Table 1. Please note that in the first stage the token number is higher than labels since a particular x or y entity/label might consist of two words or tokens. All the texts are labeled to be suitable for bar charts and only the statistics for pie and line charts are shown in the table.

4.2 Performance Evaluation

All the methods are trained using the training set and the performance are validated using the validation set. Only after the final model is selected, the model is tested on the test set. For the axis entity recognition task in the first stage, we adopt the $F1$-score and its variant the harmonic mean of f1-scores. We observe the Receiver Operating Characteristic (ROC) curve and the area under curve (auROC) in order to summarize and compare the performances of the classifiers in the second stage of entity mapping. Finally for chart type prediction, we adopt Matthews Correlation Coefficient (MCC) evaluation metric, as MCC being a more reliable statistical rate than F1-score and accuracy in binary classification evaluation for an imbalanced dataset.

4.3 Axis Label Recognition Task

The first stage of our work is x-axis and y-axis label entity recognition. Here we predict whether a given word from the text input can be an x-axis or y-axis entity. We have experimented with our neural architecture model of bidirectional LSTM combining several embeddings, such as fastText, Word2Vec and BERT in order to recognize these entities. For each of the embeddings, we have used two different approaches. In the first approach, x-entity and y entity prediction is considered as separate prediction tasks. Here we have the two models, one for each of the tasks. In the second approach, they are considered together as a combined prediction task.

Experiments with fastText Embedding. For both of the approaches using fastText (individual and combined), we have used a neural architecture with 4 hidden layers and a dense output layer. The first two hidden layers consist of bidirectional LSTM layers of 512 neurons and 128 neurons followed by a time-distributed dense layer of 64 neurons and a dense hidden layer with 1024 neurons. Epoch and batch size are kept fixed at 8 for all the models considered here. Experimental results of fastText experiments are given in the first four rows of Table 2. Note that we have reported precision, recall and $F1$-score for x and y entity predictions. Also the harmonic mean of $F1$-score is reported. Note that, the individual approach achieves $F1$-score for x and y entities of 0.66 and 0.85 respectively in the validation set which is improved in the combined approach being 0.66 and 0.89. It is clear that the prediction or recognition of x axis entities is a much more difficult task compared to y axis entity recognition. Here, we can conclude that both models perform almost similarly which is also reflected in the harmonic mean of $F1$-score respectively 0.74 and 0.76.

Table 2. Experimental results for the axis label prediction task in the frist stage of Text2Chart.

Model	Dataset	Precision (x)	Recall (x)	Precision (y)	Recall (y)	$F1$-score (x)	$F1$-score (y)	Harmonic $F1$-score
fastText	Training	0.81	0.80	0.93	0.88	0.80	0.90	0.84
Individual	Validation	0.68	0.64	0.89	0.81	0.66	0.85	0.74
fastText	Training	0.81	0.73	0.89	0.97	0.77	0.93	0.84
Combined	Validation	0.73	0.60	0.86	0.93	0.66	0.89	0.76
word2Vec	Training	0.90	0.88	1.00	1.00	0.89	1.00	0.94
Individual	Validation	0.72	0.62	0.79	0.77	0.67	0.78	0.72
word2Vec	Training	0.99	0.99	1.00	1.00	0.99	1.00	0.99
Combined	Validation	0.72	0.64	0.83	0.74	0.68	0.78	0.73
BERT	Training	0.99	0.99	.99	0.99	0.99	0.99	0.99
Individual	Validation	**0.89**	**0.86**	0.95	**0.98**	**0.87**	**0.97**	**0.92**
BERT	Training	0.99	1.00	0.99	1.00	0.99	0.99	0.99
Combined	Validation	0.86	0.78	**0.96**	0.97	0.82	**0.97**	0.89
Best	**Test**	0.85	0.82	0.96	0.98	0.84	0.97	0.89

Experiments with word2vec Embedding. The word2vec embedding represents the word tokens in the corpus by representing the words with common context in a close proximity in the vector space as well. Similar to the experiments of fastText we have two approaches employed here: individual and combined. For word2vec embedding, the network structure is kept the same as in the fastText experiments. However, for training we have used 16 epochs and a batch size of 8. The experimental results are shown in the second four rows of Table 2. From Table 2, we can see that this combined approach is giving $F1$-score of the x and y entity recognition task as 0.68 and 0.78 respectively which is almost similar to the performance of the individual approach (0.67 and 0.78 respectively). The performance only differs in the x entity recognition task which is also observed in the harmonic mean of $F1$-score. Note that the overall performance of word2vec embedding is significantly worse compared to fastText embedding. Also note that the higher level of overfitting of the word2vec model has reflected in the high values of precision, recall and $F1$ score in all the tasks in the training dataset which is not repeated in validation.

Experiments with BERT Embedding. We have also experimented with BERT embeddings on the same architecture proposed in Sect. 3. However, in these experiments the network structure is different with the same number of layers. Here too we have used two approaches: individual and combined. In the individual approach, the first two hidden layers of the neural architecture are bidirectional LSTM with 1024 neurons in each followed by a time-distributed dense layer with 1024 neurons and a dense layer with 256 neurons. In the case of x entity recognition, we have used a batch size of 2 and 80 epochs for training. In the case of y entity recognition, the batch size was 8. In the combined approach, the architecture structure has differed only in the last hidden dense layer. Here the number of neurons is 1024. We have used an online training for this combined approach. The experimental results with BERT embedding is reported in the third four rows of Table 2. From the results shown there, we can notice that for BERT embedding, the performances in the individual approach outperform the combined approach in x entity prediction performance. The results in y entity recognition is almost similar for both of the approaches. Thus the both harmonic mean and $F1$-score of x entity recognition are superior in combined approach which are 0.87 and 0.92 respectively compared to those of 0.82 and 0.89 in the individual approach.

To summarize, we can note that the results in BERT embedding are superior to two other embeddings. The best achieved values are shown in boldfaced fonts in the Table. Thus, we take the BERT embedding individual x and y entity prediction approach with bidirectional LSTM as the best performing model among those used in the experiments. With the best model, we have also tested its performance on the test dataset. The results are shown in the last row of Table 2. Here, it is interesting to note that the learned model is not overfitting and the performances in the validation set and test set are not much different.

4.4 Mapping Task

After recognizing the x and y entities with high precision and recall in stage 1, the second stage sets the target to map them in an ordered way. We have first used a transfer model from the best performing model in the first stage to see if that helps. However, the very low $F1$-score of 0.41 and auROC of 0.64 have discouraged us from proceeding in this way. It is evident that the same architecture is not suitable for the different stages due to differences in the type of the task. Note that this task is highly imbalanced as the number of positive mappings are very small compared to negative mappings. Thus the model often gets biased towards the negative model and might show poor performance in the positive prediction.

Table 3. Experimental results for the mapping task in the second stage.

Model	Dataset	Class	Precision	Recall	$F1$-score	Harmonic $F1$-score	auROC
Baseline	Training	0 (−ve)	0.94	0.94	0.94	0.84	0.908
		1 (+ve)	0.76	0.76	0.76		
	Validation	0 (−ve)	0.95	0.95	0.95	0.82	0.914
		1 (+ve)	0.73	0.73	0.73		
SVM	Training	0 (−ve)	0.93	0.93	0.93	0.81	0.897
		1 (+ve)	0.72	0.72	0.72		
	Validation	0 (−ve)	**0.96**	**0.96**	**0.96**	**0.86**	0.924
		1 (+ve)	**0.78**	**0.78**	**0.78**		
Random Forest	Training	0 (−ve)	0.95	0.95	0.95	0.85	0.913
		1 (+ve)	0.77	0.77	0.77		
	Validation	0 (−ve)	**0.96**	**0.96**	**0.96**	0.84	**0.930**
		1 (+ve)	0.77	0.77	0.77		
Best	**Test**	0 (−ve)	0.94	0.94	0.94	0.85	0.917
		1 (+ve)	0.77	0.78	0.77		

Our baseline model is a simple argmax calculation of the likelihood based on Eq. (1). The results of the baseline model are presented in the first four rows of Table 3. In this table, we have reported precision, recall and $F1$-score for both of the classes and also the auROC. Note that the results of the baseline model is encouraging with a high auROC of 0.908. However, note that the positive class performance is poor compared to the negative class which leaves room for improvement.

Next we have experimented with the supervised learning approach described in Sect. 3 using Support Vector Machine (SVM) and Random Forest classifiers. In Table 3, we notice that the performance in both of the classes are improved using this approach in both of the classes compared to the baseline model. We note that the performance in the negative class is the same. However, the $F1$-score of the Random Forest classifier is slightly lower in the positive case which is not that significant (0.77 vs 0.78). The fact is evident in auROC. There we see significant improvement achieved by the Random Forest classifier compared to SVM. The best values are shown in boldface font in the table. Thus we conclude that Random Forest is the best performing model for stage 2.

Finally, we have tested the performance of the best performing Random Forest model on the test set and the results are shown in the last row of Table 3. We see that the performances in the test set are stable and similar to the validation set.

4.5 Chart Type Prediction Task

At the third stage, the task is to predict the suitable chart type from the given text. Note that for all the texts in the dataset, the bar chart is common and thus we exclude it from classification models. We train two separate models: one for the pie chart and another for the line chart. This model uses fastText embedding

Table 4. Experimental results for chart type prediction task.

Problem	Dataset	Specificity	Sensitivity	MCC	auROC
Pie chart	Training set	0.742	0.944	0.51	0.86
	Validation set	0.6945	0.714	0.32	0.66
	Test set	0.573	0.75	0.22	0.64
Line chart	Training set	0.9634	0.963	0.96	0.96
	Validation set	0.990	0.933	0.92	0.98
	Test set	0.893	0.733	0.51	0.91

with bidirectional LSTM layers. The network architecture and structure is kept the same for both of the classifiers. The neural network has three hidden layers. The first two layers are the LSTM layers with 128 neurons each followed by a dense layer of 512 neurons. The output layer is a simple sigmoid layer. We have used the RMSprop algorithm to train the models.

For pie chart recognition, we set the batch size to 128 and the learning rate to 4e-4. As we have a highly imbalanced dataset, we achieve good enough results in terms of MCC, scoring 0.22 in the test set as shown in Table 4. The obtained auROC for pie charts is 0.64 in the test set. We have achieved a better result in terms of recall or sensitivity of 0.94 in the training set, 0.71 in the validation set and 0.75 in the test set. For line charts, we set the batch size to 256 and the learning rate remains as default to 1e-3. In Table 4, we find outstanding results in terms of auROC score of 0.96 in the training set, 0.98 in the validation set and over 0.91 in the test set. Our obtained MCC in the train, validation and test sets is 0.96, 0.92 and 0.51 which is a better score than the prediction of pie charts.

4.6 Overall Performance

In order to discuss the overall performance of our work, we have created a pipeline same as shown in Fig. 1. Our pipeline merges all the stages of our work and outputs the results we have already discussed and shown in this section. After obtaining the final results, we have checked for all possible errors that occur after completion of each stage. After completing stage 1, if both of the entity sets have a similar number of entities ($N = M$) then we consider 1-to-1 sequential mapping. The cumulative frequency of error count for each of the stages is shown in Fig. 2. This plot shows how each stage cumulatively produces error in the pipeline. However, we notice that although we have a good number of samples without error, there is a room to improve and as shown in the figure, the most error-prone task is task 3 due to the poor performance in pie chart type prediction. We also show one partially correct and one fully correct chart example generated by Text2Chart in Table 5.

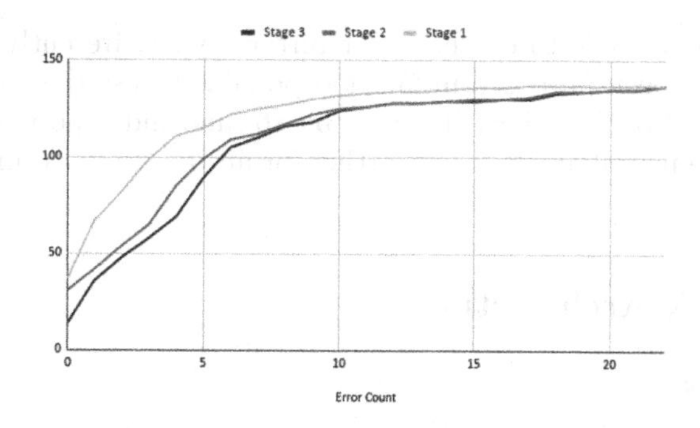

Fig. 2. Cumulative frequency of error of three states put in a pipeline on the test set.

Table 5. Sample input and outputs of Text2Chart.

Input	Sample text	Tzuyu is a gaming expert . She surveyed 200 individuals to judge the popularity of the video games among her all time favorites . After her survey she concluded that 25 people voted for World of Warcraft , 46 voted for Black Ops , 12 voted for Overwatch , 25 for Modern Warfare , 30 for PUBG , 50 for Sims and 40 for Assassin ' s Creed
Output	x Entities	['World of Warcraft', 'Black Ops', 'Overwatch', 'Modern Warfare', 'PUBG', 'Sims', 'Assassin', 's Creed']
	y Entities	['25', '12', '25', '30', '50', '50', '40', '40']
	Chart type	['bar']
Input	Sample text	Mr . Jamal worked in the Meteorological Department for 8 years . He noticed a strange thing in recent times . On certain days of the month , the weather varied strongly . He wrote down the information to make a pattern of the event . The information of the paper is as follows : on the 3rd day of the month the temperature is 36 °C , 7th day is 45 °C , 9th day is 18 °C , 11th day is 21 °C , 17th day is 9 °C , 19th day is 45 °C , 21st day is 36 °C , 27th day is 21 °C and 29th day is 45 °C . He finds a weird pattern in these dates and makes a report and sends it to his senior officer
Output	x Entities	['3rd day', '7th day', '9th day', '11th day', '17th day', '19th day', '21st day', '27th day', '29th day']
	y Entities	['36', '45', '18', '21', '9', '45', '36', '21', '45']
	Chart type	['bar', 'Line']

5 Conclusion

In this paper we have presented Text2Chart, an automatic multi-staged technique that is able to generate charts from human written analytical text. Our technique has been tested on a dataset curated for this task. Despite having a short corpora, Text2Chart provides satisfactory results in every stage regarding automatic chart generation. One of the limitations of our work is the size of the dataset. With a larger dataset, we believe the methodology presented in this paper will provide further improved results. Text2Chart is currently limited to the prediction of only three basic chart types: bar charts, pie charts and line charts. It is possible to extend it for further types. Recently a dataset for chart-to-text has been proposed in [5]. It is possible to use that dataset for the reverse problem also. We believe it is possible to tune and experiment with more types of suitable neural architecture further for all the stages to improve overall accuracy.

A Network Architectures

See Figs. 3 and 4.

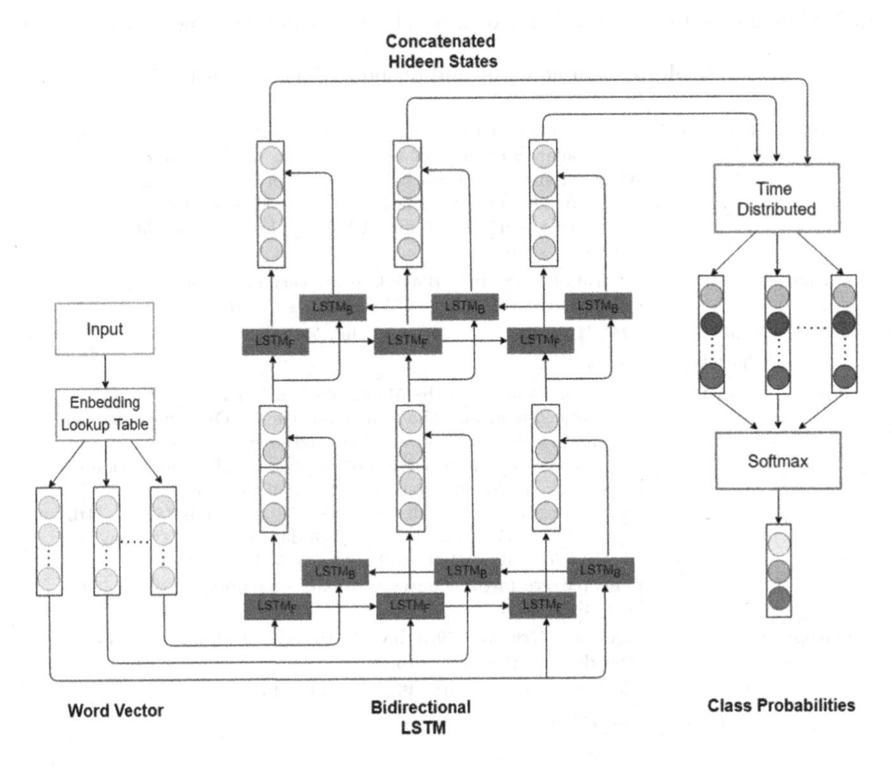

Fig. 3. Proposed neural architecture for recognition of x-axis and y-axis entities.

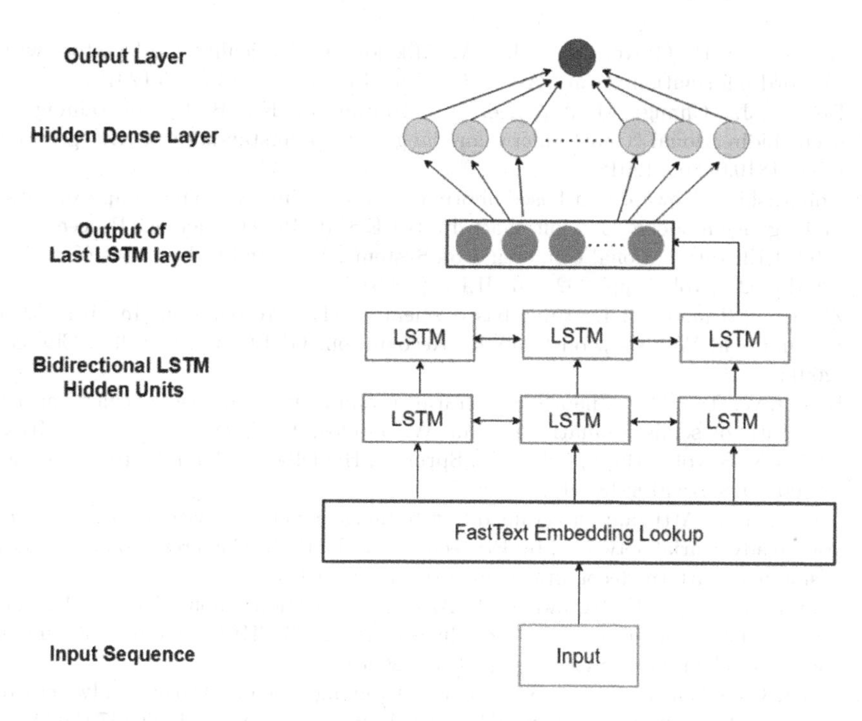

Fig. 4. Proposed neural network architecture for chart type prediction.

References

1. Liu, Y., Lapata, M.: Text summarization with pretrained encoders. arXiv preprint arXiv:1908.08345 (2019)
2. Balaji, A., Ramanathan, T., Sonathi, V.: Chart-text: a fully automated chart image descriptor. arXiv preprint arXiv:1812.10636 (2018)
3. Sang, E.F., De Meulder, F.: Introduction to the conll-2003 shared task: language-independent named entity recognition. arXiv preprint arXiv:cs/0306050 (2003)
4. Shaw, P., Massey, P., Chen, A., Piccinno, F., Altun, Y.: Generating logical forms from graph representations of text and entities. arXiv preprint arXiv:1905.08407 (2019)
5. Obeid, J., Hoque, E.: Chart-to-text: generating natural language descriptions for charts by adapting the transformer model. arXiv preprint arXiv:2010.09142 (2020)
6. Guo, Q., Jin, Z., Qiu, X., Zhang, W., Wipf, D., Zhang, Z.: Cyclegt: unsupervised graph-to-text and text-to-graph generation via cycle training. arXiv preprint arXiv:2006.04702 (2020)
7. Cui, W., et al.: Text-to-viz: automatic generation of infographics from proportion-related natural language statements. IEEE Trans. Visualiz. Comput. Graph. **26**(1), 906–916 (2019)
8. Brown, T.B., et al.: Language models are few-shot learners. arXiv preprint arXiv:2005.14165 (2020)
9. Mikolov, T., Chen, K., Corrado, G., Dean, J.: Efficient estimation of word representations in vector space. arXiv preprint arXiv:1301.3781 (2013)

10. Bojanowski, P., Grave, E., Joulin, A., Mikolov, T.: Enriching word vectors with subword information. Trans. Assoc. Comput. Linguist. **5**, 135–146 (2017)
11. Devlin, J., Chang, M.-W., Lee, K., Toutanova, K.: Bert: pre-training of deep bidirectional transformers for language understanding. arXiv preprint arXiv:1810.04805. (2018)
12. Kobayashi, I.: Toward text based information processing: with an example of natural language modeling of a line chart. In: IEEE SMC 1999 Conference Proceedings. 1999 IEEE International Conference on Systems, Man, and Cybernetics (Cat. No. 99CH37028), vol. 5, pp. 202–207. IEEE (1999)
13. Zhou, Y., Tan, C.L.: Learning-based scientific chart recognition. In: 4th IAPR International Workshop on Graphics Recognition, GREC, pp. 482–492. Citeseer (2001)
14. Huang, W., Tan, C.L., Zhao, J.: Generating ground truthed dataset of chart images: automatic or semi-automatic? In: Liu, W., Lladós, J., Ogier, J.-M. (eds.) GREC 2007. LNCS, vol. 5046, pp. 266–277. Springer, Heidelberg (2008). https://doi.org/10.1007/978-3-540-88188-9_25
15. Xu, T., et al.: Attngan: fine-grained text to image generation with attentional generative adversarial networks. In: Proceedings of the IEEE Conference on Computer Vision and Pattern Recognition, pp. 1316–1324 (2018)
16. Kim, D.H., Hoque, E., Agrawala, M.: Answering questions about charts and generating visual explanations. In: Proceedings of the 2020 CHI Conference on Human Factors in Computing Systems, pp. 1–13 (2020)
17. Wang, Q., Chen, Z., Wang, Y., Qu, H.: Applying machine learning advances to data visualization: a survey on ml4vis. arXiv preprint arXiv:2012.00467 (2020)
18. Luo, Y., Qin, X., Tang, N., Li, G.: Deepeye: towards automatic data visualization. In: 2018 IEEE 34th International Conference on Data Engineering (ICDE), pp. 101–112. IEEE (2018)

ENDASh: Embedding Neighbourhood Dissimilarity with Attribute Shuffling for Graph Anomaly Detection

Qizhou Wang[1(✉)], Mahsa Salehi[1], Jia Shun Low[2], Wray Buntine[1], and Christopher Leckie[2]

[1] Monash University, Clayton 3168, Australia
{qizhou.wang,mahsa.salehi,wray.buntine}@monash.edu
[2] The University of Melbourne, Parkville 3010, Australia
{jason.low,caleckie}@unimelb.edu.au

Abstract. Recent unsupervised GNN based graph anomaly detection (GAD) methods adopt specific mechanisms designed for anomaly detection. This is in contrast to earlier methods that utilise components such as graph autoencoders that were designed for more general use-cases. However, these newer methods only lead to a modest increase in detection accuracy at the cost of complicated optimisation schemes and higher runtimes. To overcome these issues, we propose **E**mbedding **N**eighbourhood **D**issimilarity (END) with **A**ttribute **Sh**uffling (ENDASh), a simple but scalable and effective GAD framework. ENDASh utilises our proposed END measure to quantify the degree of abnormality of nodes using GraphSAGE embeddings that were optimised with Attribute Shuffling, a data augmentation method designed to project anomalies away from inliers in the latent space. Extensive experiments on real-world benchmarks demonstrate the competitive GAD performance of our ENDASh while being computationally efficient and capable of operating in an inductive environment.

Keywords: Anomaly detection · Attributed graph · Graph mining

1 Introduction

Attributed graphs are an expressive means of representing complex real-world interactions, such as publication citations and social media relationships [8,19]. Graph Anomaly Detection (GAD) [1,14] on attributed graphs is the task of identifying anomalous graph components that deviate from the majority inlier components. GAD has a wide range of applications in various domains such as cybersecurity, fraud detection and social network analysis [19].

Neural network based GAD methods have proven to be more effective than non-network based methods [5,13], due to the ability of GNNs to learn more meaningful representations of graphs. Earlier GNN based GAD methods suffer from substantial drawbacks such as design choices that are not explicit for GAD

J. Gama et al. (Eds.): PAKDD 2022, LNAI 13281, pp. 17–29, 2022.
https://doi.org/10.1007/978-3-031-05936-0_2

or scalability issues. We discuss these issues in detail in Sect. 2. Recently, newer GNN based GAD methods have overcome these issues by adopting mechanisms such as contrastive learning [13,22] and adversarial training [2,4] to perform optimizations that explicitly aim to model anomalies. They also improve scalability by processing the graph into smaller segments to avoid resource bottlenecks. However, these modifications can involve complicated graph processing steps and prolonged training and inference time while achieving relatively modest gains in GAD accuracy compared to earlier methods.

To date, there has been limited work on investigating the use of node embeddings for GAD. In this paper, we tackle the GAD problem by obtaining discriminative information from node embeddings. The approach we take improves runtime and space scalability while having increased accuracy compared to current baselines. We also leverage a data augmentation step to generate node embeddings that are better suited to GAD and use them in the optimization of a specific function aimed at GAD. To the best of our knowledge, this is the first paper to demonstrate the utility of data augmentation for GAD.

In this paper, we propose a new framework for GAD that utilizes a novel anomaly measure called **E**mbedding **N**eighbourhood **D**issimilarity (END), and a new data augmentation method we term Attribute Shuffling. Our proposed framework combines END and Attribute Shuffling, so we term it ENDASh. END is a localized measure that assesses the level of inconsistency between a node and its immediate neighbours using embeddings generated by a GraphSAGE model [7]. Using END naively as an anomaly score can already outperform state of the art methods in terms of accuracy on various benchmark datasets while being faster to train and make predictions. However, END based anomaly detection does not perform well on graphs with high degrees and number of attributes. To overcome this problem, we utilize Attribute Shuffling as a way to make the training process more explicit for anomaly detection. We term this approach Attribute Shuffling as it shuffles the attribute matrix associated with an attribute graph. The combined ENDASh framework exhibits higher accuracy for the GAD task and is scalable to large graphs because it operates in mini-batch mode without preprocessing, and has low computational complexity and empirical runtime. ENDASh is also capable of being used in an inductive setting, so it can generalise to unseen nodes during training. We conduct experiments on a wide range of real-world datasets to demonstrate the effectiveness of ENDASh and validate the feasibility of using Attribute Shuffling to improve GAD.

2 Related Work

Early GNN based GAD methods [5,12] utilise reconstruction errors to detect anomalies. However, the reconstruction loss function is not explicitly designed for anomaly detection, resulting in models occasionally being insufficiently discriminative for detection [6,18] (i.e., they can also reconstruct outliers well). Later works utilise more sophisticated mechanisms to make the training more targeted towards GAD. Adversarial training is used by methods such as [2,4] to model anomalies, so a discriminator can be trained to distinguish them from normal data. These adversarial training methods are generative in nature and are not a form of data augmentation. However, they have drawbacks such as non-convergence and slower training from adversarial training. Furthermore, finding the appropriate stopping point for training without a labelled validation set is challenging. Besides adversarial training, Zhou *et al.*[23] proposed pseudo-labelling nodes according to the deviations between node embeddings. However, its optimisation effectively assumes that inliers come from a single distribution (i.e., located in a hypersphere), which may be unrealistic for real-world graphs. Contrastive learning and its variants [9,13,22] are also used for generating pseudo labels to develop discriminators. Nevertheless, these methods rely on computationally expensive operations such as graph traversals for preprocessing and multi-round inferences.

Our approach, ENDASh, leverages the processed augmentation method Attribute Shuffling to effectively model anomalous embeddings while introducing only trivial overhead. To the best of our knowledge, this is the first study on applying data augmentation to assist GAD. Due paper length constraints, we do not provide a detailed review on graph data augmentation. For more information, we refer readers to recent works such as [20,21].

3 Preliminaries

3.1 Problem Formulation

Notation. Let $\mathcal{G} = (\mathbf{X}, \mathbf{A})$ denote an attributed graph with $n = |\mathcal{V}|$ nodes, where $\mathcal{V}$ is $\mathcal{G}$'s node set, $\mathbf{X} \in \mathbb{R}^{n \times d}$ is the feature matrix, and $\mathbf{A} \in \mathbb{R}^{n \times n}$ is the adjacency matrix describing the connectivities between nodes.

Definition 1 (Anomaly detection on attributed graphs). In an attributed graph, a node is anomalaous if it deviates either structurally or contextually from the majority of the nodes, known as structural anomalies and contextual anomalies, respectively. Typically, structural anomalies contain non-trivial numbers of links to semantically unrelated nodes, and contextual anomalies possess incongruous attributes with their neighbours. The task of ***anomaly detection on an attributed graph*** $\mathcal{G}$ is to find a scoring mechanism $S(\mathcal{G}, v)$ for $v \in \mathcal{V}$, such that:

$$S(v_i) < S(v_j) \quad \forall (v_i \in \mathcal{V}_{\text{in}}) \wedge (v_j \in \mathcal{V}_{\text{a}}), \tag{1}$$

where $\mathcal{V}_{\text{in}}$ and $\mathcal{V}_{\text{a}}$ are the respective node sets for normal (inlier) and anomalaous nodes, such that $\mathcal{V}_{\text{in}} \cup \mathcal{V}_{\text{a}} = \mathcal{V} \wedge \mathcal{V}_{\text{in}} \cap \mathcal{V}_{\text{a}} = \emptyset \wedge |\mathcal{V}_{\text{in}}| \gg |\mathcal{V}_{\text{a}}|$. To distinguish outliers from inliers, the scoring mechanism S should ideally satisfy the following criterion for a threshold τ, such that $\forall v \in \mathcal{V}, S(v) > \tau \Leftrightarrow v \in \mathcal{V}_{\text{a}}$.

This paper studies anomaly detection in a fully unsupervised setting on non-heterogeneous attributed graphs; that is, no label information is available for developing the model and all nodes of a graph represent the same type of entities. Anomaly detection on heterogeneous graphs is beyond the scope of this paper. GAD can be further categorised into: i) the transductive setting, where all graph elements (nodes and edges) are present during training, and ii) the inductive setting, where only a subset of graph elements are available during training, requiring the model to generalise to unseen nodes during inference time.

3.2 Representation Learning via GraphSAGE

GraphSAGE learns node representations by iteratively aggregating a node's attributes from their K-hop neighbours using trainable and symmetric functions, known as aggregators (denoted as AGGR_k, $\forall k \in \{1, ..., K\}$), where K is a predefined constant. We choose the mean aggregator [7] due to its efficiency and empirical performance. The mean aggregator can be written as:

$$\mathrm{AGGR}_k(v, \mathcal{N}(v)) = \mathrm{MEAN}(\{\mathbf{h}_v^{k-1}\} \cup \{\mathbf{h}_u^{k-1}, \forall u \in \mathcal{N}(v)\}), \quad (2)$$

where $\mathcal{N}(v)$ is the target node v's neighbour set and $\mathbf{h}_v^k$ is the embedding of v at the k-th layer. When $k = 1$, the inputs are attributes of the relevant nodes.

Each layer k consists of an aggregator AGGR_k followed by a linear transformation with weights $\mathbf{W}^k$ and an activation layer σ for information passing between layers. The transformation at layer k can be written as:

$$\mathbf{h}_v^k = \sigma(\mathbf{W}^k \cdot \mathrm{AGGR}_k(v, \mathcal{N}(v))). \quad (3)$$

For the sake of computational efficiency, only some predefined numbers of samples $N_k, \forall k \in K$ are required for the aggregation at each layer [7].

GraphSAGE can be optimised without node class labels, making it applicable for unsupervised GAD. We aim to preserve original neighbourhoods in the embedding space by making node embeddings closer to the embeddings of their neighbours but far from the embeddings of non-neighbouring nodes via the following objective function:

$$\mathcal{L}(\mathbf{z}_u, \mathbf{z}_v) = -\log(\sigma(\mathbf{z}_u^\top \mathbf{z}_v)) - Q \cdot \mathbb{E}_{v_n \sim P_n(v)} \log(\sigma(-\mathbf{z}_u^\top \mathbf{z}_{v_n})), \quad (4)$$

where v is a node within a certain number of hop from the target node u, and σ is the sigmoid function. P_n is a distribution for generating negative samples $\{v_n \mid v_n \in P_n\}$, and Q defines the number of negative examples for each target node.

4 Proposed Methods

In this section, we present our proposed framework Embedding Neighbourhood Dissimilarity (END) with Attribute Shuffling (ENDASh). We explain in detail, two key elements of ENDASh: END and Attribute Shuffling. ENDASh can be viewed as a hybrid method of detecting anomalies inspired by the idea of network homophily [10], which is explored by our END measure on node embeddings. We also introduce a lightweight but effective argumentation method, Attribute Shuffling, with a two-stage training strategy to make GraphSAGE embeddings more discriminative for GAD, which general-purpose graph embedding methods do not address. The efficient characteristics of its components with high synergy empower ENDASh to be fast and efficient.

4.1 Embedding Neighbourhood Dissimilarity

Embedding Neighbourhood Dissimilarity (END) is a localised measure that describes how anomalous a given node is with respect to the embeddings of its randomly sampled 1-hop neighbours. We choose node embeddings because they contain both topological and attributive properties, encapsulating rich information. END is formulated as:

$$\text{END}(\mathbf{z}_u) = \sum_{i=1}^{m} \left(\frac{\|\mathbf{z}_u\|_2 \cdot \|\mathbf{z}_{v_i}\|_2}{\mathbf{z}_u \cdot \mathbf{z}_{v_i}} \right) \cdot m^{-1}, \tag{5}$$

where $\mathbf{z}_u$ is the embedding of the target node u and $\{\mathbf{z}_{v_i} | i \in [1, ..., m]\}$ are embeddings of m randomly selected nodes $\{v_1, ..., v_m\} \subset \mathcal{N}(u)$ from u's immediate neighbours $\mathcal{N}(u)$. The choice of m is discussed in Sect. 5.3.

We propose to apply END to node embeddings generated by GraphSAGE as a simple but effective anomaly scoring mechanism (SAGE + END in short). The intuition is that the inliers' embeddings should be more congruent with their neighbours' in contrast to outliers, which are expected to have higher END scores.

Algorithm 1: Augmented mini-batch training via Attribute Shuffling

Input: $\mathcal{G} = (\mathbf{A}, \mathbf{X})$: the graph; **idx**: target indexes; M_0: a GraphSAGE model.
Output: M_1: the GraphSAGE model after tuning
$\mathbf{idx}_{\text{pos}} \leftarrow$ random sample one positive example for idx in **idx**;
$\mathbf{Z}_u, \mathbf{Z}_v \leftarrow M_0(\mathcal{G}, \mathbf{idx}), M_0(\mathcal{G}, \mathbf{idx}_{\text{pos}})$;
$\tilde{\mathbf{X}} \leftarrow \text{random\_row\_shuffle}(\mathbf{X})$; $\mathcal{G}' \leftarrow (\mathbf{A}, \tilde{\mathbf{X}})$;
$\mathbf{Z}_{\text{aug}} \leftarrow M_0(\mathcal{G}', \mathbf{idx})$;
$M_1 \leftarrow \text{ADAM}(M_0, \mathcal{L}_{\text{aug}}(\mathbf{Z}_u, \mathbf{Z}_v, \mathbf{Z}_{\text{aug}}))$ // Tuning the model using Eq.6

4.2 Augmenting Training via Attribute Shuffling

Although SAGE + END can be effective on some datasets, it does not work well on complex graphs (i.e., graphs with high degrees and high-dimensional features). Additionally, GAD is not addressed directly by the objective in Eq. 4, potentially compromising the model's GAD effectiveness.

To address these issues, we introduce a simple but effective data augmentation technique, termed Attribute Shuffling, for generating more discriminating GraphSAGE embeddings for GAD. The overall idea is that we first create negative samples to model the anomalous embeddings in the latent space by aggregating misplaced feature vectors (in terms of their position in the original feature matrix) according to the original computational graph. Simply put, we shuffle the rows of the attribute matrix $\mathbf{X}$ but keep the adjacency matrix $\mathbf{A}$ for aggregations. We then encourage GraphSAGE to project inliers to distant regions from such negative embeddings.

Algorithm 1 describes how Attribute Shuffling can be used to tune a GraphSAGE model. At each mini-batch, for each target node u, we generate: i) $\mathbf{z}_u$, which is an embedding of u, ii) $\mathbf{z}_v$, which is an embedding of a randomly selected node v that co-occurs within a random walk of length c from u as its positive embedding and iii) u's negative embedding via Attribute Shuffling $\mathbf{z}_{\text{aug}}$. These are involved in tuning GraphSAGE according to the following objective:

$$\mathcal{L}_{\text{aug}}(\mathbf{z}_u, \mathbf{z}_v, \mathbf{z}_{\text{aug}}) = -\log(\sigma(\mathbf{z}_u^\top \mathbf{z}_v)) - 0.5 \cdot \log(\sigma(-\mathbf{z}_v^\top \mathbf{z}_{\text{aug}})) - 0.5 \cdot \log(\sigma(-\mathbf{z}_u^\top \mathbf{z}_{\text{aug}})). \quad (6)$$

This objective function encourages the model to project anomalies to isolated regions from their corresponding inliers' and their neighbours' embeddings in the latent space, resulting in more discriminating embeddings for END calculation.

4.3 Two Stage Training

We find that performing END calculations on embeddings output by a GraphSAGE model tuned via Attribute Shuffling after being first trained on the objective defined in Eq. 4, generally yields better performance than GraphSAGE models tuned using Attribute Shuffle from scratch. Therefore, we introduce a two-stage training strategy that integrates Attribute Shuffling with SAGE + END as ENDASh. We first tune the model using Eq. 4 for an initial number of epochs and then tune the model via the augmented training objective defined in Eq. 6 for an additional number of epochs.

4.4 Runtime and Scalability

The runtime of ENDASh is dominated by embedding generations via GraphSAGE operations. The complexity of generating a single embedding is $\mathcal{O}(\prod_{k=1}^{K} N_k)$ [7], where N_k is the size of the set of sampled neighbours for aggregation at the k-th layer. Accordingly, the complexity of computing the END of one node is $\mathcal{O}((m+1) \cdot \prod_{k=1}^{K} N_k)$ in the worst case scenario, where none of its neighbours' embedding is available. However, in a more realistic case where predictions are made on a non-trivial proportion of nodes, embeddings may be required by more than one computation, thus can be cached once computed and retrieved later, reducing $m+1$ by some factors depending on the proportion. In the best case, where the predictions on all nodes are needed, $m+1$ can be further reduced to 1. Attribute Shuffling attracts only a trivial runtime overhead because random row shuffling can be performed by simply creating a random index mapping.

ENDASh operates in a mini-batch fashion for both training and inference, therefore is free from any major form of resource bottleneck, allowing great scalability. In addition, unlike some methods, ENDASh does not preprocess the graph, thus requiring no additional overhead for storing processed data.

5 Experiments

In this section, we perform an empirical evaluation on a wide range of datasets to test the anomaly detection performance, inductivity, scalability and runtime of our methods, followed by ablation studies and parameter analysis.

Table 1. A summary of datasets and the numbers of injected anomalies.

	ACM [16]	BlogCatalog [17]	Flickr [17]	ogbn-arxiv [8]	ogbn-products [8]
node num.	16,484	5,196	7,575	169,343	2,449,029
edge num.	71,980	171,743	239,738	1,166,243	61,859,140
anomaly num.	600	300	450	6,000	90,000

Datasets. We use five real-world attributed graph benchmarks for our experiments (Table 1). In particular, we extend our experiments to Open Graph Benchmark (OGB) benchmarks, in addition to the three benchmarks that are used by most baselines, for scalability evaluation because of their substantially larger scales. Experiments in inductive settings are performed on ogbn datasets subject to the availability of graph split information (Table 4).

Anomaly Injection. To the best of our knowledge, there is no known homogeneous attributed graph dataset for anomaly detection with ground truth GAD information, therefore we have to inject synthetic anomalies. We utilize an injection scheme that is common practice in many papers, including those that introduced our baseline methods [2,4,5,13,15]. Specifically, structural anomalies are injected by creating q random fully connected cliques of size p. Contextual anomalies are injected by overwriting a node's attributes with the most different attributes in r random sampled node attributes from the entire graph in terms of the Euclidean distance. We repeat this procedure to inject multiple contextual anomalies. For ACM, Flickr, BlogCatalog and ogbn-arxiv, we use identical injection parameter settings as used in the baseline papers. Since ogbn-products has not been used by any baseline, we use the ogbn-arxiv's setting for it, besides the total number of anomalies, which varies based on the graph (with the ratio of anomaly kept similar). For each dataset, we inject the same number of anomalies ($p \times q$) for each type to make the composition balanced. On all datasets, we set $p = 15$ and $r = 50$. We set q to 10, 15, 20, 200 and 3000 for BlogCatalog, Flickr, ACM, ogbn-arxiv and ogbn-products, respectively.

Baseline Methods. We choose state of the art GNN based GAD methods for the same application setting as baselines, including DOMINANT [5], COLA [13], AEGIS [4], GGAN [2] and SL-GAD [22]. Due to the lack of publically available implementations, we report the results of baseline methods as the reported values in the original papers. Our anomaly injection scheme is identical to the schemes in the baseline papers, therefore we believe it is reasonable to use those values.

Implementation Details. Our method is implemented using Pytorch. ENDASh uses single-layer GraphSAGE models with ELUs ($\alpha = 0.5$) [3], trained with the ADAM [11] optimiser. For all datasets, we set embedding dimensions to 256, END sample set size m to 50, and the length of random walk c to 2. For simplicity, we report the two stage training setting with the following 5-tuple: (stage 1 epoch number, stage 2 epoch number, stage 1 learning rate, stage 2 learning rate. batch size) and summary settings of all datasets as follows: BlogCatalog (20, 30, 3e−4, 3e−4, 512), Flickr (20, 30, 2e−4, 2e−4, 512), ACM (5, 5, 1e−3, 1e−4, 1024), ogbn-arxiv (5, 5, 1e−3, 1e−4, 1024), and products (5, 5, 1e−3, 1e−3, 2048).

Table 2. Performance comparisons between our methods and baseline methods in terms of AUROC ↑. We use "-" to indicate results that are not reported by the original paper where we cannot rerun experiments due to lack of availability of their implementation. "OOM" abbreviates "out-of-memory" in our hardware setting.

Dataset	Baselines					Ours	
	DOMINANT	COLA	AEGIS	GAAN	SL-GAD	SAGE + END	ENDASh
ACM	0.7494	0.8237	0.7600	0.8770	0.8538	0.9441	**0.9527**
Flickr	0.7490	0.7513	0.7740	0.7530	0.7966	0.6988	**0.8527**
BlogCatalog	0.7813	0.7854	0.8170	0.7650	0.8184	0.7871	**0.8252**
ogbn-arxiv	OOM	0.8073	–	–	OOM	0.9257	**0.9273**
ogbn-products	OOM	OOM	–	–	OOM	0.8824	**0.8988**

Table 3. A summary of runtimes. DOMINANT is measured using full-batch setting as it cannot run in a mini-batch setting. Others are measured in mini-batch setting.

	T_{train} (sec) ↓			$T_{\text{inference}}$ (sample/sec) ↓		
	Flickr	BlogCatalog	ACM	Flickr	BlogCatalog	ACM
DOMINANT	138.84	51.54	52.56	1.11×10^{-4}	3.54×10^{-5}	1.34×10^{-4}
COLA	521.77	343.73	1299.41	4.22×10^{-2}	4.10×10^{-2}	4.76×10^{-2}
SL-GAD	1299.07	854.93	2936.82	1.03×10^{-1}	9.84×10^{-2}	1.08×10^{-1}
ENDASh	29.25	25.56	7.53	1.99×10^{-4}	2.69×10^{-4}	1.52×10^{-4}

5.1 Empirical Performance

Transductive Anomaly Detection. Table 2 shows that ENDASh outperforms all baseline methods on all five datasets, with substantial improvements on ACM, Flickr and ogbn-arxiv datasets. It is also worth noting that the performance of SAGE + END can be comparable to ENDASh on some datasets, suggesting that GraphSAGE embeddings are sufficiently informative for GAD when an appropriate similarity function is used, such as END. SAGE + END can be seen as a simplified version of ENDASh without adopting Attribute Shuffling and the two-stage training strategy to enhance the existing data. It can be used as an alternative to ENDASh for graphs on which the embeddings of a GraphSAGE trained without using Attribute Shuffling can result in reasonably discriminative END measures for anomaly detection. More discussion on the effectiveness is provided in Sect. 5.2.

Furthermore, it should be noted that our proposed methods are highly scalable, as they can achieve competitive performance on ogbn datasets, which are considerably larger than the other datasets, particularly with the ogbn-products dataset that contains over a hundred times more nodes than non-ogbn datasets. This high scalability is made possible because the steps in our proposed methods can be done strictly in a mini-batch fashion for graphs of any size without significant resource bottlenecks.

Transductive Runtime. We perform all experiments using the same hardware configuration (single NVIDIA V100 GPU and four cores of Intel Xeon CPU @ 2.60 GHz) for fairness. We only compare to baseline methods that have publicly available code. For training, we use the same settings as described in the original publications. The performance is measured by the total training time. For inference, we set consistent batch size to 300 for all methods and report average inference time per node. Note that the focus of the comparisons is with mini-batch methods, but for completeness, we include DOMINANT despite it only working in a full-batch setting. Table 3 displays training and inference runtimes on selected datasets. Our method can be trained in significantly less time than other GNN based methods and also demonstrates substantially faster inference speed than baselines that operate in mini-batch, making it more applicable for larger scale anomaly detection and real-time deployment. Notably, while DOMINANT has a faster inference time than us on some datasets, it requires the entire graph to be loaded in memory, even if just to make a prediction on one node. This results in us being out of memory when using DOMINANT on the obgn datasets. ENDASh however does not have such constraints as detailed in Sect. 4.4, because it uses a small amount of extra time on mini-batch processing to enable greater scalability and flexibility, which is essential for inductive learning and deployment in evolving environments.

Table 4. Inductive datasets split info.

	Number of nodes			
	Total	Train	Test	Split Info.
ogbn-arxiv	169,343	90,941	48,603	time evol
ogbn-products	2,449,029	196,615	2,213,091	sales rank

Table 5. Inductive performance comparisons.

	Trans.	Induc.
ogbn-arxiv	0.9273	0.9277
ogbn-products	0.8988	0.9002

Inductive Anomaly Detection. Our model naturally operates in an inductive setting, so we extend our investigation to the task of inductive GAD. Unfortunately, not all benchmarks include graph split information, which is essential for splitting a graph in a semantically meaningful way (e.g., according to timestamp information) for inductive graph anomaly detection. Randomly splitting a graph can result in partitions being significantly different from the whole graph structurally and the splits themselves lacking semantic meaning. Therefore, we only perform inductive evaluation on the ogb datasets as they contain semantically sensible node splits (Table 4). Table 5 shows the comparisons of GAD performance between the transductive and the inductive settings. ENDASh achieves almost identical performance for both settings, demonstrating its strong capability for inductive detection. Notably, on the ogbn-products datasets, ENDASh can generalise to unseen nodes after being trained using only a very small percentage of nodes, which is advantageous for efficient and timely training on large scale graphs in practice.

5.2 Ablation Study

This section discusses the effect of Attribute Shuffling and the two-stage training strategy. Specifically, we compare the GAD performances of ENDASh with its variant optimised via Attribute Shuffling only and SAGE + END. Since no ground truth information is available for validation, the performance of SAGE + END at two points are reported: i) after training for the epoch number of the stage 1, and ii) after a further number of stage 2 epochs.

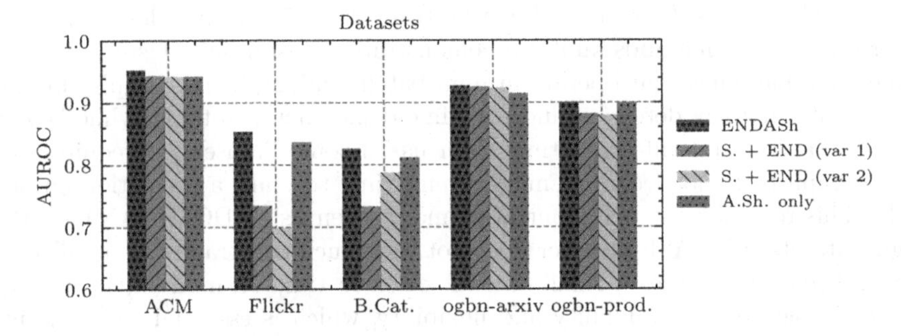

Fig. 1. Performance comparison of ENDASh (the first bar from left) versus SAGE + END's two variants (the second and the third bar, respectively) and ENDASh optimised using Attribute Shuffling only (the fourth bar) on each dataset.

Figure 1 shows the aforementioned comparisons-the two-stage training strategy yields better performance than the other methods on all datasets. The performance gains on the Flickr and the BlogCatalog datasets are substantial, especially compared to the performance of models optimised without augmentation, implying that Attribute Shuffling can provide improved training data for anomaly detection beyond the original graph. By examining the second bar and the third bar for each dataset, it is noticeable that prolonged training without Attribute Shuffling is unlikely to improve the performance on most datasets.

The benefit of Attribute Shuffling is fairly minor on the ogbn-arxiv datasets. We attribute this phenomenon to the limitation of the augmented samples' quality. Note that the proposed augmentation method is relatively straightforward and is done rapidly with a minimal level of processing, so it potentially does not generate augmented samples that can precisely model all true outliers. In rare cases, some of the augmented samples may be more similar to inliers, introducing unwanted noise impacting the performance. Therefore, the usefulness of the augmented training method on a given dataset depends on the performance that can be achieved by SAGE + END, which is consistent with the observation that the augmented training offers a greater performance gain for datasets on which SAGE + END is less effective.

The two-stage training also achieves higher performance than training from scratch using the augmented objective. The stage one training can be regarded as a form of pretraining or initialisation and the second stage as fine tuning.

5.3 Parameter Analysis

This section investigates how essential parameters affect GAD performance of ENDASh. We use the three smaller datasets for illustration purposes.

Number of Neighouring Samples for GraphSAGE Aggregation. Figure 2 (a) shows the anomaly detection performance versus the number of samples of GraphSAGE aggregation. As the number grows, a trend of performance improvement with a decreased margin can be noted, with no further improvement after some number of points, proving that a subset of each node's neighbourhood is adequately representative without examining the entire neighbourhood during aggregation. The improvement is more trivial on the ACM dataset than the other two datasets, attributed to its higher sparsity, as fewer samples are needed to model neighbourhoods well, suggesting that the number of samples can be set following the sparsity of the graph to avoid redundant computation.

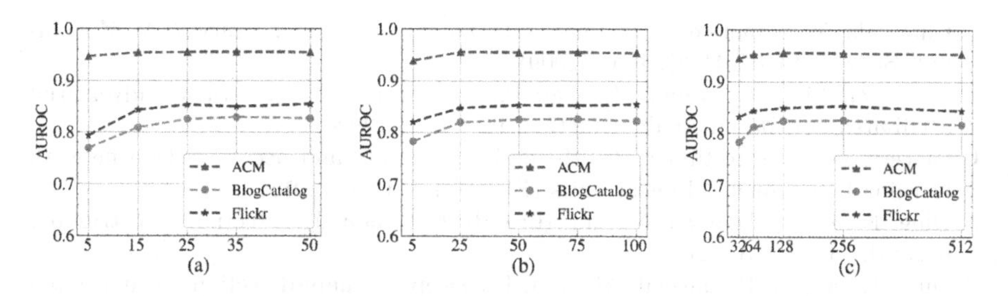

Fig. 2. Anomaly detection performance using different: (a) GraphSAGE sample sizes; (b) END sample sizes; and (c) embeddings dimensions.

Number of Samples for END Calculation. Figure 2 (b) depicts the relationship between the performance of the proposed model and the number of samples for END computation. Similar to the number of samples for GraphSAGE aggregation, the performance increases at a diminishing rate as the sample size increases on all three datasets. We can also conclude that highly connected datasets require a larger number of neighbours to obtain better results.

Dimensionality of Embeddings. Intuitively, the appropriate embedding dimension for a graph positively correlates to its connectivity and complexity of its node attributes, aligning with the empirical results shown in Fig. 2(c). For example, the Flickr and the BlogCatalog datasets have much higher dimensional feature vectors and average degrees than the ACM dataset. Therefore, when the dimensionality of embeddings is small, we can observe that increasing it improves performance more on the two datasets. In practice, the embedding dimension should be set according to the graphs' feature dimension and degree. In general, a number between 128 to 512 is a reasonable range for the initial attempt.

6 Conclusion

This paper introduces a novel framework, ENDASh, that leverages the END measure with Attribute Shuffling for anomaly detection on attributed graphs. On a wide range of datasets, ENDASh outperforms state-of-the-art baselines in terms of AUROC while demonstrating high scalability and competitive training and inference runtime. ENDASh also generalises to unseen nodes, thus making it suitable for both transductive and inductive settings. For future work, it will be interesting to explore more sophisticated augmentation techniques with more advanced embedding methods with higher scalability.

Acknowledgements. This research was supported by The University of Melbourne's Research Computing Services and the Petascale Campus Initiative.

References

1. Chandola, V., Banerjee, A., Kumar, V.: Anomaly detection: a survey. ACM Comput. Surv. (CSUR) **41**(3), 1–58 (2009)
2. Chen, Z., Liu, B., Wang, M., Dai, P., Lv, J., Bo, L.: Generative adversarial attributed network anomaly detection. In: CIKM (2020)
3. Clevert, D.A., Unterthiner, T., Hochreiter, S.: Fast and accurate deep network learning by exponential linear units (ELUs). In: ICLR (2016)
4. Ding, K., Li, J., Agarwal, N., Liu, H.: Inductive anomaly detection on attributed networks. In: IJCAI (2020)
5. Ding, K., Li, J., Bhanushali, R., Liu, H.: Deep anomaly detection on attributed networks. In: SDM (2019)
6. Gong, D., et al.: Memorizing normality to detect anomaly: memory-augmented deep autoencoder for unsupervised anomaly detection. In: CVPR (2019)
7. Hamilton, W.L., Ying, R., Leskovec, J.: Inductive representation learning on large graphs. In: NIPS 2017 (2017)
8. Hu, W., et al.: Open graph benchmark: datasets for machine learning on graphs. Adv. NeurIPS **33**, 22118–22133 (2020)
9. Jin, M., Liu, Y., Zheng, Y., Chi, L., Li, Y.F., Pan, S.: Anemone: graph anomaly detection with multi-scale contrastive learning. In: CIKM (2021)
10. Kim, K., Altmann, J.: Effect of homophily on network formation. Commun. Nonlinear Sci. Numer. Simul. **44**, 482–494 (2017)

11. Kingma, D.P., Ba, J.: Adam: a method for stochastic optimization. In: ICLR (2015)
12. Li, Y., Huang, X., Li, J., Du, M., Zou, N.: SpecAE: spectral autoencoder for anomaly detection in attributed networks. In: CIKM (2019)
13. Liu, Y., Li, Z., Pan, S., Gong, C., Zhou, C., Karypis, G.: Anomaly detection on attributed networks via contrastive self-supervised learning. In: TNNLS, pp. 1–15 (2021)
14. Noble, C.C., Cook, D.J.: Graph-based anomaly detection. In: SIGKDD (2003)
15. Song, X., Wu, M., Jermaine, C., Ranka, S.: Conditional anomaly detection. TKDE **19**(5), 631–645 (2007)
16. Tang, J., Zhang, J., Yao, L., Li, J., Zhang, L., Su, Z.: ArnetMiner: extraction and mining of academic social networks. In: SIGKDD (2008)
17. Tang, L., Liu, H.: Relational learning via latent social dimensions. In: SIGKDD (2009)
18. Wang, Q., Erfani, S.M., Leckie, C., Houle, M.E.: A dimensionality-driven approach for unsupervised out-of-distribution detection, pp. 118–126 (2021)
19. Wu, Z., Pan, S., Chen, F., Long, G., Zhang, C., Yu, P.S.: A comprehensive survey on graph neural networks. TNNLS **32**(1), 4–24 (2021)
20. You, Y., Chen, T., Sui, Y., Chen, T., Wang, Z., Shen, Y.: Graph contrastive learning with augmentations. In: NeurIPS (2020)
21. Zhao, T., Liu, Y., Neves, L., Woodford, O., Jiang, M., Shah, N.: Data augmentation for graph neural networks. In: AAAI (2021)
22. Zheng, Y., Jin, M., Liu, Y., Chi, L., Phan, K.T., Chen, Y.P.P.: Generative and contrastive self-supervised learning for graph anomaly detection. In: TKDE (2021)
23. Zhou, S., Tan, Q., Xu, Z., Huang, X., Chung, F.L.: Subtractive aggregation for attributed network anomaly detection. In: CIKM (2021)

Convergence and Applications of ADMM on the Multi-convex Problems

Junxiang Wang(✉) and Liang Zhao

Emory University, 201 Dowman Drive, Atlanta, GA, USA
{jwan936,lzhao41}@emory.edu

Abstract. In recent years, although the Alternating Direction Method of Multipliers (ADMM) has been empirically applied widely to many multi-convex applications, delivering an impressive performance in areas such as nonnegative matrix factorization and sparse dictionary learning, there remains a dearth of generic work on proposed ADMM with a convergence guarantee under mild conditions. In this paper, we propose a generic ADMM framework with multiple coupled variables in both objective and constraints. Convergence to a Nash point is proven with a sublinear convergence rate $o(1/k)$. Two important applications are discussed as special cases under our proposed ADMM framework. Extensive experiments on ten real-world datasets demonstrate the proposed framework's effectiveness, scalability, and convergence properties. We have released our code at https://github.com/xianggebenben/miADMM.

1 Introduction

Due to the advantages and popularity of non-differentiable regularized and distributive computing for complex optimization problems, the Alternating Direction Method of Multipliers (ADMM) has received a great deal of attention in recent years [5]. The standard ADMM was originally proposed to solve the following separable convex optimization problem:

$$\min_{x,z} f(x) + g(z) \quad s.t. \; Ax + Bz = c.$$

where $f(x)$ and $g(z)$ are closed convex functions, A and B are matrices and c is a vector. There are extensive reports in the literature exploring the theoretical properties for convex optimization problems related to ADMM and its variants, including multi-block ADMM [11], Bregman ADMM [28], fast ADMM [13,17], and stochastic ADMM [22]. ADMM has now been extended to cover a wide range of nonconvex problems, and it has achieved outstanding performance in many practical applications [38].

Unlike convex problems, the convergence theory on the nonconvex ADMM is much more difficult, and considerable progress has been made on this problem, please refer to Sect. 2 for a detailed summary. Recently, however, there has been an increasing number of real-world applications where the objective functions are multi-convex (i.e. nonconvex for all the variables but convex for each when all the others are fixed). For example, nonnegative matrix factorization, which aims to decompose a matrix into a product of

J. Gama et al. (Eds.): PAKDD 2022, LNAI 13281, pp. 30–43, 2022.
https://doi.org/10.1007/978-3-031-05936-0_3

two matrices, has been applied widely in computer vision, machine learning, and various other fields [18]; A bilinear matrix inequality problem has been designed for the analysis of linear and nonlinear uncertain systems [15].

All of the above applications can be considered as special cases of multi-convex optimization problems. However, such problems have not yet been rigorously and systematically investigated by ADMM. Moreover, the convergence properties of the ADMM required to solve such problems remain unknown. In this work, we propose mild conditions to ensure the convergence of ADMM to a Nash point on the multi-convex problems with a sublinear convergence rate $o(1/k)$. We also discuss how our ADMM is applied to two important applications. Extensive experiments show the effectiveness of our proposed ADMM. Our contributions in this paper include:

- We propose an ADMM framework to solve the multi-convex problem, and we investigate the convergence properties of the proposed ADMM. Specifically, we prove that the objective value and the residual are convergent. Moreover, any limit point generated by the proposed ADMM is a Nash point of the original problem. The convergence rate of the proposed ADMM is $o(1/k)$.
- We demonstrate two important and promising applications that are special cases of our proposed ADMM framework and benefit from its theoretical properties. Specifically, we show how these applications can be transformed equivalently to fit into the proposed ADMM framework.
- We conduct extensive experiments to validate our proposed ADMM. Experiments on ten real-world datasets demonstrate its effectiveness, scalability, and convergence properties.

The rest of this paper is summarized as follows: Sect. 2 summarizes previous work related to this paper. Section 3 introduces the ADMM algorithm and its convergence properties. In Sect. 4, the proposed ADMM algorithm is applied to several important applications. Extensive experiments are described in Sect. 5. The paper concludes with a summary of the work in Sect. 6.

2 Related Work

Multi-convex Optimization Problems: Some works studied multi-convex problems. The earliest work required that the objective function was differentiable continuous and strictly convex [36]. Various conditions on separability and regularity on the objective functions have been discussed in [26,27]. In the most recent work, Xu and Yin presented three types of multi-convex algorithms and analyzed convergence based on either Lipschitz differentiability or strong convexity assumption [37]. For a comprehensive survey, please refer to [24].

Convergence Analysis of ADMM: Existing literature on the convergence analysis of ADMM can be categorized into two classes: the convex ADMM and the nonconvex ADMM. The convex ADMM is investigated relatively well compared with the nonconvex ADMM. Existing works either study suitable stepsizes of the convex ADMM or extend ADMM to the stochastic version. For example, Bai et al. proposed a generalized symmetric ADMM to solve the multi-block separable objective by updating the Lagrange multiplier twice with suitable stepsizes [3]; Gu et al. extended contractive Peaceman-Rachford splitting method to ADMM with larger stepsizes [14]; Ouyang et al. proposed a stochastic ADMM with a convergence rate $O(\frac{1}{\sqrt{t}})$. Despite the outstanding performance of the nonconvex ADMM, its convergence theory is not well established due to the complexity of both coupled objectives and various (inequality and equality) constraints. Most existing works discussed the convergence of the nonconvex ADMM on separable objectives: they provided convergence guarantee to the stationary solutions with different assumptions [4,8,9,19]. Some works explored more difficult cases where the objectives are coupled: for example, Wang et al. presented mild convergence conditions of the nonconvex ADMM where the objective can be nonsmooth [35]; Gao et al. explored the convergence condition of ADMM on multi-affine constraints [12]; Wang et al. gave the convergence proofs of ADMM in the nonconvex deep learning problems [29,31,32]; while experiments by Wang and Zhao showed that the ADMM was not necessarily convergent in the nonlinear-constrained problems [33].

3 ADMM on the Multi-convex Problems

In this section, we present an ADMM framework to solve Problem 1.

3.1 Preliminaries

First, the definition of Lipschitz differentiability is shown as follows [7]:

Definition 1 (Lipschitz Differentiability). *Any arbitrary differentiable function $G_1 : \mathbb{R}^m \to \mathbb{R}$ is Lipschitz differentiable if for any $x^{'}, x^{''} \in \mathbb{R}^m$,*

$$\|\nabla G_1(x^{'}) - \nabla G_1(x^{''})\| \leq D\|x^{'} - x^{''}\|.$$

where $D \geq 0$ is constant and $\nabla G_1(x)$ denotes the gradient of $G_1(x)$.

The following defines strong convexity, which is indispensable for the proof of convergence to a Nash point.

Definition 2 (Strong Convexity). *A convex function $G_4(x)$ is strongly convex if there exists $H > 0$ such that for $\forall x^{'}, x^{''} \in dom(G_4)$, the following holds*

$$G_4(x^{''}) \geq G_4(x^{'}) + (v^{'})^T(x^{''} - x^{'}) + (H/2)\|x^{''} - x^{'}\|_2^2.$$

where $\forall v^{'} \in \partial G_4(x^{'})$ is a subdifferential of G_4 at $x^{'}$.

Finally, the Nash point is defined as follows [37]:

Definition 3 (Nash Point). *Given $G_5(a_1, a_2, \cdots, a_m)$, a Nash point $(a_1^*, a_2^*, \cdots, a_m^*)$ satisfies the following property:*

$$G_5(a_1^*, \cdots, a_{i-1}^*, a_i^*, a_{i+1}^*, \cdots, a_m^*) \leq G_5(a_1^*, \cdots, a_{i-1}^*, a_i, a_{i+1}^*, \cdots, a_m^*),$$
$$\forall (a_1^*, \cdots, a_{i-1}^*, a_i, a_{i+1}^*, \cdots, a_m^*) \in dom(G_5),\ (i = 1, \cdots, m).$$

Naturally, when we optimize one variable while fixing others, the Nash point ensures the optimality of this variable [37]. Without loss of generality, we assume that Problem 1 has at least a Nash point, and in the next section, we will prove that any limit point generated by ADMM converges to a Nash point.

3.2 The ADMM Algorithm

The following problem is our focus in this paper:

Problem 1

$$\min_{x_1,\cdots,x_n,z} F(x_1,\cdots,x_n,z) = f(x_1,\cdots,x_n)+h(z) \quad s.t. \sum\nolimits_{i=1}^{n} A_i x_i - z = 0.$$

where $x_i \in \mathbb{R}^{p_i}(i=1,\cdots,n)$, $z \in \mathbb{R}^q$, $f(x_1,\cdots,x_n): \mathbb{R}^p \to \mathbb{R} \cup \{\infty\}(p = \sum_{i=1}^{n} p_i)$ are proper, continuous, multi-convex and possibly nonsmooth functions, $h(z)$ is a proper, differentiable and convex function. $A_i \in \mathbb{R}^{q \times p_i}(i=1,\cdots,n)$ are matrices. Obviously, the domain of F is $dom(F) = \{(x_1,\cdots,x_n,z)| \sum_{i=1}^{n} A_i x_i - z = 0\}$. Without the loss of generality, the objective of Problem 1 is assumed to be bounded from below.

To ensure the convergence of the proposed ADMM, some mild assumptions are imposed, which are shown as follows:

Assumption 1 (Lipschitz Differentiability). *$h(z)$ is Lipschitz differentiable with constant $H \geq 0$.*

Most loss functions such as the cross-entropy loss and the square loss are Lipschitz differentiable [32]. In order to propose the ADMM algorithm, the augmented Lagrangian function can be formulated mathematically as follows:

$$L_\rho(x_1,\cdots,x_n,z,y) = F(x_1,\cdots,x_n,z)+y^T(\sum\nolimits_{i=1}^{n} A_i x_i - z)+(\rho/2)\|\sum\nolimits_{i=1}^{n} A_i x_i - z\|_2^2. \tag{1}$$

where y is a dual variable and $\rho > 0$ is a penalty parameter. The proposed ADMM aims to optimize the following $n+1$ subproblems alternately.

$$\begin{aligned} x_i^{k+1} &\leftarrow \arg\min\nolimits_{x_i} f(\cdots, x_{i-1}^{k+1}, x_i, x_{i+1}^{k}, \cdots) + (y^k)^T A_i x_i \\ &+ (\rho/2)\|\sum\nolimits_{j=1}^{i-1} A_j x_j^{k+1} + A_i x_i + \sum\nolimits_{j=i+1}^{n} A_j x_j^{k} - z^k\|_2^2. \qquad (2) \\ z^{k+1} &\leftarrow \arg\min\nolimits_z L_\rho(\cdots, x_n^{k+1}, z, y^k) \qquad (3) \\ &= \arg\min\nolimits_z h(z) - (y^k)^T z + (\rho/2)\|\sum\nolimits_{i=1}^{n} A_i x_i^{k+1} - z\|_2^2. \end{aligned}$$

Algorithm 1 is presented for Problem 1. Concretely, Lines 3–5 and 6 update $x_i^{k+1}(i = 1, \cdots, n)$ and z^{k+1}, respectively. Line 7 updates the primal residual r^{k+1}, which is defined in accordance with the standard ADMM [5]: it measures how the linear constraint $\sum_{i=1}^{n} A_i x_i - z = 0$ is violated. Line 8 updates the dual variable y^{k+1}, which follows the routine of the standard ADMM. Line 10 uses the norm of the primal residual r as a condition to terminate the algorithm, where $\delta > 0$ is a threshold. Each subproblem is convex and implicitly assumed to be solvable.

Algorithm 1. The Proposed ADMM to Solve Problem 1

Require: $A_i(i = 1, \cdots, n), \delta > 0$.
Ensure: $x_i(i = 1, \cdots, n), z$.
1: Initialize ρ, $k = 0$.
2: **repeat**
3: **for** i=1 to n **do**
4: Update x_i^{k+1} in Eq. (2).
5: **end for**
6: Update z^{k+1} in Eq. (3).
7: $r^{k+1} \leftarrow \sum_{i=1}^{n} A_i x_i^{k+1} - z^{k+1}$. # update primal residual
8: $y^{k+1} \leftarrow y^k + \rho r^{k+1}$.
9: $k \leftarrow k + 1$.
10: **until** $\|r^{k+1}\| \leq \delta$.
11: Output $x_i(i = 1, \cdots, n), z$.

3.3 Convergence Analysis

This section focuses on the convergence of the proposed ADMM algorithm. Specifically, the first lemma states that the augmented Lagrangian L_ρ keeps decreasing, which is stated as follows.

Lemma 1 (Objective Descent). *If $\rho > 2H$ so that $C_1 = \rho/2 - H/2 - H^2/\rho > 0$, then there exists $C_2 = \min(\rho/2, C_1)$ such that*

$$L_\rho(x_1^k, \cdots, x_n^k, z^k, y^k) - L_\rho(x_1^{k+1}, \cdots, x_n^{k+1}, z^{k+1}, y^{k+1})$$
$$\geq C_2(\|z^{k+1} - z^k\|_2^2 + \sum\nolimits_{i=1}^{n} \|A_i(x_i^{k+1} - x_i^k)\|_2^2). \tag{4}$$

Lemma 1 holds under Assumption 1, and its proof can be found in the supplementary materials[1]. The next lemma states that the augmented Lagrangian is bounded from below, as shown below:

Lemma 2 (Objective Bound). *If $\rho > 2H$, then $L_\rho(x_1^k, \cdots, x_n^k, z^k, y^k)$ is lower bounded.*

The proof of Lemma 2 can be found in the supplementary materials (See footnote 1). Now we can prove that the proposed ADMM converges globally in the following theorem.

Theorem 1 (Residual and Objective Convergence). *If $\rho > 2H$, then for the bounded sequence $(x_1^k, \cdots, x_n^k, z^k, y^k)$, then it has the following properties:*
a). Residual convergence. This means that as $k \to \infty$, $r^k \to 0$, where r^k is defined in Algorithm 1.
b). Objective convergence. This means that as $k \to \infty$, $F(x_1^k, \cdots, x_n^k, z^k)$ converges.

[1] The supplementary materials are available at https://github.com/xianggebenben/miADMM/blob/main/multi_convex_ADMM-13-18.pdf.

Theorem 1 guarantees the convergence of the proposed ADMM, whose proof is in the supplementary materials (See footnote 1). However, $x_i^k(i = 1, \cdots, n)$ and z^k are not necessarily shown to be convergent. The next theorem states that any limit point is a feasible Nash Point of Problem 1.

Theorem 2 (Convergence to a Nash Point). *Let $\rho > 2H$, if **either** of two assumptions hold: (a). $A_i(i = 1, \cdots, n)$ have full rank. (b). F is strongly convex with regard to x_i. Then for bounded variables $(x_1^k, \cdots, x_n^k, z^k)$, it has at least a limit point $(x_1^*, \cdots, x_n^*, z^*)$, and any limit point $(x_1^*, \cdots, x_n^*, z^*)$ is a feasible Nash point of F defined in Problem 1. That is*

$$\sum A_i x_i^* - z^* = 0. \text{ (feasibility)}$$
$$F(x_1^*, \cdots, x_n^*, z^*) \leq F(x_1^*, \cdots, x_{i-1}^*, x_i, x_{i+1}^*, \cdots, x_n^*, z^*),$$
$$\forall (x_1^*, \cdots, x_{i-1}^*, x_i, x_{i+1}^*, \cdots, x_n^*, z^*) \in dom(F), (i = 1, \cdots, n).$$
$$F(x_1^*, \cdots, x_n^*, z^*) \leq F(x_1^*, \cdots, x_n^*, z) \forall (x_1^*, \cdots, x_n^*, z) \in dom(F) \text{ (Nash point).}$$

The proof of Theorem 2 is detailed in the supplementary materials (See footnote 1). The third theorem states that our proposed ADMM can achieve a sublinear convergence rate of $o(1/k)$ under Assumption 1, despite the nonconvex and complex nature of Problem 1. Such a rate is state-of-the-art even compared to those methods for simpler convex problems. The theorem is shown as follows:

Theorem 3 (Convergence Rate). *If $\rho > 2H$, for a bounded sequence $(x_1^k, \cdots, x_n^k, z^k, y^k)$, define $u_k = \min_{0 \leq l \leq k}(\|z^{l+1} - z^l\|_2^2 + \sum_{i=1}^n \|A_i(x_i^{l+1} - x_i^l)\|_2^2)$, then the convergence rate of u_k is $o(1/k)$.*

The proof of this theorem is in the supplementary materials (See footnote 1). The $o(1/k)$ convergence rate of the proposed ADMM is consistent with much existing work analyzing the convex ADMM, including [11,16,21]. Our contribution in term of convergence rate is that we extend the guarantee of $o(1/k)$ into the multi-convex Problem 1.

Our proposed ADMM is more general than some influential works in terms of formulation. The relations between our proposed ADMM and previous works are summarized as follows:

1. Generalization of Block Coordinate Descent (BCD) for multi-convex problems. When the linear constraint $\sum_{i=1}^n A_i x_i = z$ is removed in Problem 1, then the proposed ADMM is reduced to the Block Coordinate Descent [37].

2. Generalization of multi-block ADMM. When $f(x_1, \cdots, x_n) = 0$, the proposed ADMM is reduced to the convex multi-block ADMM [25], i.e. the ADMM with no less than three variables.

Apart from general formulations, the convergence guarantees of our proposed ADMM cover more applications than previous literature. For example, [35] requires the coupled objective $f(x_1, \cdots, x_n)$ to be Lipschitz differentiable. However, some important applications such as weakly-constrained multi-task learning (Sect. 4.1) and learning with signed-network constraints (Sect. 4.2) do not satisfy this condition. But they are covered by our convergence guarantees of the multi-convex ADMM to a Nash point.

4 Applications

In this section, we apply our proposed ADMM to two real-world applications, both of which conform to Problem 1 and benefit from the convergence properties of the proposed ADMM.

4.1 Weakly-Constrained Multi-task Learning

In multi-task learning problems, multiple tasks are learned jointly to achieve a better performance compared with learning tasks independently [30]. Most work on multi-task learning has tended to enforce the assumption of similarity among the feature weight values across tasks [2,10,30,34,41] because this makes it possible to use convex regularization terms like $\ell_{2,1}$ norms [34] and Graph Laplacians [41]. However, this assumption is usually too strong and is seldom satisfied by the real-world data. Instead of requiring feature weights to be similar in magnitude, a more conservative but probably more reasonable assumption is that multiple tasks share similar polarities for the same feature, which means that if a feature is positively relevant to the output of a task, then its weight will also be positive for other related tasks. This assumption is appropriate for many applications. For example, the feature 'number of clinic visits' will be positively related to flu outbreaks, while the feature 'popularity of vaccination' will be negatively related to them, even though their feature weights can vary dramatically for different countries (namely tasks here). This is achieved by enforcing the requirement for every pair of tasks with neighboring indices to have the same weight sign. This optimization objective is shown as follows:

$$\begin{aligned} &\min_{w_1,\cdots,w_n} \sum\nolimits_{i=1}^{n} (Loss_i(w_i) + \Omega_i(w_i)) \\ &s.t.\ w_{i,j} w_{i+1,j} \geq 0\ (i = 1, 2, \cdots, n-1, j = 1, 2, \cdots, m). \end{aligned} \tag{5}$$

where n and m denote the number of tasks and features, respectively, $w_{i,j}$ is the weight of the j-th feature in the i-th task, w_i is the weight of the i-th task, and $Loss_i(w_i)$ and $\Omega_i(w_i)$ are the loss function and the regularization term of the i-th task, respectively. The inequality constraint implies that the i-th and the $i+1$-th tasks share the same sign for their weights. Equation (5) is rewritten in the following form to fit in our proposed ADMM framework:

$$\begin{aligned} &\min_{w_1,\cdots,w_n,z} \sum\nolimits_{i=1}^{n} (Loss_i(w_i) + \Omega_i(z_i)) + \lambda_1 \sum\nolimits_{i=1}^{n-1} \sum\nolimits_{j=1}^{m} c_1(w_{i,j} w_{i+1,j}) \\ &s.t.\ z_i = w_i\ (i = 1, 2, \cdots, n). \end{aligned} \tag{6}$$

where $z = [z_1; \cdots; z_n]$ is an auxiliary variable, and $\lambda_1 > 0$ is a tuning parameter. Notice that the inequality constraint $w_{i,j} w_{i+1,y} \geq 0$ is transformed to a quadratic penalty $c_1(x)$ such that $c_1(x) = \begin{cases} x^2 & x < 0 \\ 0 & x \geq 0 \end{cases}$ which makes the formulation consistent with Problem 1. The proposed ADMM algorithm for this case is shown in the supplementary materials (See footnote 1).

4.2 Learning with Signed-Network Constraints

The application of network models for social network analysis has attracted the attention of a large number of researchers [6]. For example, influential societal events often spread across many social networking sites and are expressed in different languages. Such multi-lingual indicators usually transmit similar semantic information through networks and have thus been utilized to facilitate social event forecasting [39]. The problem with network constraints is formulated as follows:

$$\min_{\beta_1,\cdots,\beta_n} Loss(\beta_1,\cdots,\beta_n)+\sum\nolimits_{i=1}^{n}\omega_i(\beta_i)$$
$$s.t.\ \exists(\beta_i,\beta_j)\in E_s, \exists(\beta_p,\beta_q)\in E_d\ (1\leq i,j,p,q\leq n).$$

where β_i is the weight of the i-th node. $Loss(\beta_1,\cdots,\beta_n)$ is a loss function and $\omega_i(\beta_i)$ is a regularization term for the i-th node. $E_s = \{(\beta_i,\beta_j)|\beta_i\beta_j\geq 0\}$ and $E_d = \{(\beta_p,\beta_q)|\beta_p\beta_q\leq 0\}$ are two edge sets to represent two opposite relationships: $(\beta_i,\beta_j)\in E_s$ means that $\beta_i\beta_j\geq 0$, while $(\beta_p,\beta_q)\in E_d$ means that $\beta_p\beta_q\leq 0$. The constraint means that some pair (β_i,β_j) satisfies the edge set E_s, and that some pair (β_p,β_q) satisfies the edge set E_d. For example, in the problem of social event forecasting with French and English, E_s and E_d are edge sets of synonyms and antonyms between French and English, and the weight pair of the French word "bien" and the English word "good" belongs to E_s. The problem with network constraints can be reformulated approximately to the following:

$$\min_{\beta_1,\cdots,\beta_n,z} Loss(\beta_1,\cdots,\beta_n)+\sum\nolimits_{i=1}^{n}\omega_i(z_i)+\lambda_2\Big(\sum\nolimits_{(\beta_i,\beta_j)\in E_s}c_2(\beta_i,\beta_j)$$
$$+\sum\nolimits_{(\beta_p,\beta_q)\in E_d}c_3(\beta_p,\beta_q)\Big) s.t.\ z_i=\beta_i\ (i=1,2,\cdots,n) \quad (7)$$

where $z=[z_1;\cdots;z_n]$ is an auxiliary variable, and $\lambda_2>0$ is a tuning parameter. The constraint $(\beta_i,\beta_j)\in E_s$ and $(\beta_p,\beta_q)\in E_d (1\leq i,j,p,q\leq n)$ are transformed to two quadratic penalties $c_2(\beta_i,\beta_j)$ and $c_3(\beta_p,\beta_q)$ as follows:

$$c_2(\beta_i,\beta_j)=\begin{cases}(\beta_i\beta_j)^2 & (\beta_i,\beta_j)\notin E_s\\ 0 & (\beta_i,\beta_j)\in E_s\end{cases}, c_3(\beta_p,\beta_q)=\begin{cases}(\beta_p\beta_q)^2 & (\beta_p,\beta_q)\notin E_d\\ 0 & (\beta_p,\beta_q)\in E_d\end{cases}.$$

The proposed ADMM for this case is also shown in the supplementary materials .

5 Experiments

In this section, we test our proposed ADMM using ten real-world datasets on two applications detailed in Sect. 4. Scalability, effectiveness, and convergence properties are compared with several existing state-of-the-art methods on ten real datasets. All experiments were conducted on a 64-bit machine with Intel(R) Core(TM) processor (i7-6820HQ CPU@ 2.70 GHZ) and 16.0 GB memory.

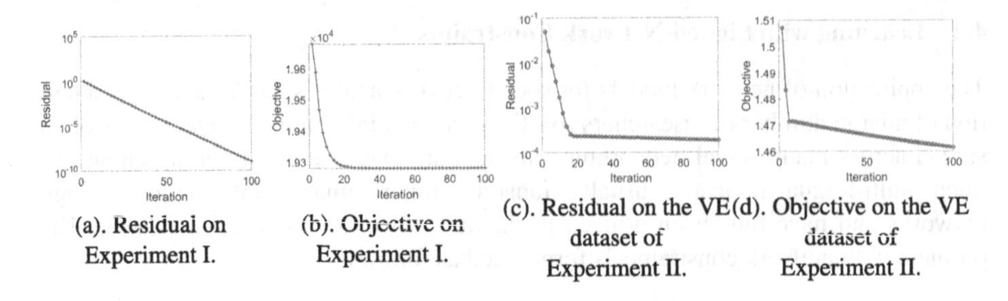

(a). Residual on Experiment I. (b). Objective on Experiment I. (c). Residual on the VE dataset of Experiment II. (d). Objective on the VE dataset of Experiment II.

Fig. 1. Convergence curves on Experiments I and II.

5.1 Experiment I: Weak-Constrained Multi-task Learning

To evaluate the effectiveness of our method on the application of weak-constrained multi-task learning described in Eq. (6), a real-world school dataset is used to evaluate the effectiveness of our proposed ADMM. It consists of the examination scores in three years of 15,362 students from 139 secondary schools, which are treated as tasks for examination scores prediction based on 27 input features such as year of the examination, school-specific features, and student-specific features. The dataset is publicly available and the detailed description can be found in the original paper [20]. ρ was set to 1000. Here we chose two kinds of λ_1: (1) $\lambda_1^k = 10^5$; (2) $\lambda_1^{k+1} = \lambda_1^k + 10$ with $\lambda_1^k = 1$. $\lambda_1(1)$ and $\lambda_1(2)$ are the first and the second choice of λ_1, respectively.

Metrics. In this experiment, five metrics were utilized to evaluate model performance. Mean Squared Error (MSE) measures the average of the squares of the difference between observation and estimation. Different from MSE, Mean Squared Logarithmic Error (MSLE) measures the ratio of observation to estimation. Mean Absolute Error (MAE) is also an error measurement but computed in the absolute value. The less the above three metrics are, the better a regression model is. Explained Variance (EV) computes the ratio of the variance of the error to that of observation. The coefficient of determination or R2 score is the proportion of the variance in the dependent variable that is predictable from the independent variable. The higher score of EV and R2 are, the better a regression model is.

Baselines. To validate the effectiveness of the proposed ADMM, five benchmark multi-task learning models served as comparison methods. Loss functions were set to least square errors. The number of iterations was set to $5,000$. The regularization parameter α was set based on 5-fold cross-validation on the training set.

1. multi-task learning with Joint Feature Selection (JFS) [2,41] . JFS is one of the most commonly used strategies in multi-task learning. It captures the relatedness of multiple tasks by a constraint of a weight matrix to share a common set of features. α was set to 100.

2. Clustered Multi-Task Learning (CMTL) [40,41]. CMTL assumes that multiple tasks are clustered into several groups. Tasks in the same group are similar to each other. α was set to 1.
3. multi-task Lasso (mtLasso) [41]. mtLasso extends the classic Lasso model to the multi-task learning setting. α was set to 10.
4. a convex relaxation of Alternating Structure Optimization (cASO) [1,41]. cASO decomposes each task into two components: task-specific feature mapping and task-shared feature mapping. α was set to 0.01.
5. Block Coordinate Descent (BCD) [37]. BCD is an intuitive method to solve multi-convex problems, which optimizes each variable alternately. α was set to 10.

Performance. As discussed in Sect. 4.1, the convergence of our proposed ADMM is guaranteed based on our theoretical framework. To verify this, Figs. 1(a) and 1(b) illustrate the residual and objective values in different iterations, which demonstrates the convergence of the proposed ADMM on this nonconvex problem. Then the performance of examination score prediction on this dataset is illustrated in Table 1. Table 1 shows the mean and the standard deviation of all methods, which were repeated 10 times by initializing parameters randomly, to make experimental evaluation robust. It shows that $\lambda_1(2)$ outperforms $\lambda_1(1)$ in four out of five metrics for the proposed ADMM. In addition, the proposed ADMM achieves the best performance in all the metrics, compared to all comparison methods. Moreover, the standard deviation of the proposed ADMM is about 30% smaller than any other comparison method. This is because our method only enforces that the sign of the feature weight across different tasks is the same, while comparison methods typically perform too aggressive assumptions on the similarity among tasks. For example, CMTL enforces that the correlated tasks need to have similar feature weights using squared regularization on the difference between feature weights. JFS and mtLasso still tend to enforce similar weights on features in different tasks by $\ell_{2,1}$ norm. Because their enforcement is weaker than CMTL, their performance is better. cASO gets relatively weak performance because it optimizes an approximation of a nonconvex problem, and thus the approximate solution may be distant from the true solution to the original problem. Finally, the BCD performs the worst among all methods, even though it shares the same formulation with our proposed ADMM. This reflects the advantage of our proposed ADMM algorithm: dual information in one iteration can be passed to its following iteration by dual variables, which yields better performance.

Table 1. Performance in Experiment I.

Mean					
Method	MSE	MSLE	MAE	EV	R2
JFS	114.1052	0.4531	8.4349	0.2948	0.2948
CMTL	114.9892	0.4647	8.4756	0.2876	0.2875
mtLasso	115.3143	0.4625	8.4725	0.2873	0.2873
cASO	137.8336	0.5204	9.3450	0.1606	0.1605
BCD	149.2313	0.5577	9.8057	0.1299	0.0777
ADMM($\lambda_1(1)$)	113.6975	**0.4423**	8.4024	0.2950	0.2960
ADMM($\lambda_1(2)$)	**113.2400**	0.4428	**8.3943**	**0.3002**	**0.3002**
Standard deviation					
Method	MSE	MSLE	MAE	EV	R2
JFS	2.02	0.02	0.06	0.02	0.02
CMTL	1.85	0.02	0.05	**0.01**	**0.01**
mtLasso	1.77	0.02	0.05	**0.01**	**0.01**
cASO	7.26	0.01	0.06	**0.01**	**0.01**
BCD	1.41	0.01	0.06	0.15	**0.01**
ADMM($\lambda_1(1)$)	**0.83**	**0.005**	**0.03**	**0.01**	**0.01**
ADMM($\lambda_1(2)$)	0.95	0.01	0.04	0.02	0.02

Scalability. To investigate the scalability of the proposed ADMM compared with all baselines in Experiment I, we measured the training time of them in the school dataset when the number of features varies. The training time was averaged by running 20 times. Figure 2 shows the training time of all methods when the number of features ranges from 10 to 28. The training time of all methods increased linearly concerning the number of features. cASO was the most efficient of all methods, while the proposed ADMM was ranked second. mtLasso and JFS also trained a model within 5 s on average. CMTL was time-consuming for training, which spent more than 10 s.

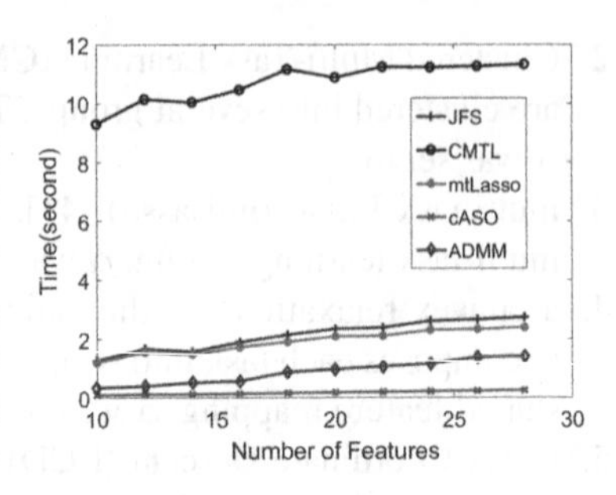

Fig. 2. The training time of all methods in Experiment I.

5.2 Experiment II: Event Forecasting with Multi-lingual Indicators

Datasets. To evaluate the performance of our proposed ADMM on the application in Sect. 4.2, extensive experiments on nine real-world datasets have been performed. The dataset is obtained by randomly sampling 10% (by volume) of the Twitter data from Jan 2013 to Dec 2014. The data in the first and second years are used and training and test set, respectively. For the topic (i.e., social unrest) of interest, 1,806 keywords in the three major languages in Latin America, namely English, Spanish, and Portuguese, were provided by the paper [39]. Their translation relationships have also been labeled as semantic links among them, such as "protest" in English, "protesta" in Spanish, and "protesto" in Portuguese. The event forecasting results were validated against a labeled event set, known as the Gold Standard Report (GSR), which is publicly available [23].

Metric and Baselines. The metric used to evaluate the performance is Area Under the receiver operating characteristic Curve (AUC). Five comparison methods including the state-of-the-art Multi-Task learning (MTL), Multi-Resolution Event Forecasting (MREF), and Distant-supervision of Heterogeneous Multitask Learning (DHML) as well as classic methods Logistic Regression (LogReg) and Lasso. ρ was set to 10. Here we chose two kinds of λ_2: (1) $\lambda_2^k = 10^5$; (2) $\lambda_2^{k+1} = \lambda_2^k + 10$ with $\lambda_2^k = 1$. $\lambda_2(1)$ and $\lambda_2(2)$ are the first and the second choice of λ_2, respectively. All the hyperparameters were tuned by 5-fold cross-validation.

Performance. As shown in Table 2, $\lambda_2(2)$ outperforms $\lambda_2(1)$ marginally in seven out of nine datasets for the proposed ADMM, and they generally perform the best among all the methods, with DHML and BCD the second-best performer.

Table 2. Event forecasting performance in AUC in each of the 9 datasets.

	BR	CL	CO	EC	EL	MX	PY	UY	VE
LogReg	0.686	0.677	0.644	0.599	0.618	0.661	0.616	0.628	0.667
LASSO	0.685	0.677	0.648	0.603	0.636	0.665	0.615	0.666	0.669
MTL	0.722	0.669	0.810	0.617	0.772	0.795	0.600	0.811	0.771
MREF	0.714	0.563	0.515	0.784	0.612	0.693	0.658	0.681	0.588
DHML	0.845	0.683	0.846	0.839	0.780	0.793	0.737	0.835	0.835
BCD	0.847	0.668	0.850	0.830	0.773	0.800	0.736	0.835	0.856
ADMM ($\lambda_2(1)$)	0.864	0.699	0.870	0.848	0.794	0.820	0.746	**0.850**	**0.867**
ADMM ($\lambda_2(2)$)	**0.867**	**0.701**	**0.872**	**0.851**	**0.798**	**0.823**	**0.747**	0.847	0.865

They all outperform the others typically by at least 5%–10%. This is because they leverage the multilingual correlation among the features to boost up the model's generalizability. Thanks to the framework of multi-task learning, MTL and MREF obtained a competitive performance with AUC typically over 0.7, which outperforms simple methods like LogReg and LASSO by 5% on average.

Efficiency. In Experiment II, we also compared the training time of the proposed ADMM in comparison with all baselines on 9 datasets. The training time was averaged by running 5 times. The training time was shown in Table 3. We do not show BCD because its training time is similar to the proposed ADMM. Overall, the proposed ADMM was the most efficient of all methods for all datasets. It consumed no more than 30 s on all datasets. MTL ranked second, but it spent hundreds of seconds on some datasets, like BR and MX. As the most time-consuming baselines, LogReg and MREF trained a model in thousands of seconds or more.

Table 3. Comparison of running time (in seconds) on 9 datasets in Experiment II.

	LogReg	LASSO	MTL	MREF	DHML	ADMM
BR	30193	1535	233	25889	332	**14**
CL	2981	242	35	6521	852	**11**
CO	8060	780	108	14714	87	**31**
EC	312	295	**17**	4332	46	25
EL	551	261	17	4669	33	**3**
MX	17712	2043	853	31349	175	**29**
PY	7297	527	40	9495	242	**5**
UY	748	336	20	5305	82	**3**
VE	5563	1008	49	5769	179	**28**

6 Conclusions

We propose an ADMM framework for multi-convex problems with multiple coupled variables. It not only inherits the merits of general ADMMs but also provides advantageous theoretical properties on convergence conditions and properties under mild conditions. Besides, several machine learning applications of recent interest are discussed as special cases of our proposed ADMM. Extensive experiments have been conducted on ten real-world datasets, and demonstrate the effectiveness, scalability, and convergence properties of our proposed ADMM.

Acknowledgement. This work was supported by the National Science Foundation (NSF) Grant No. 1755850, No. 1841520, No. 2007716, No. 2007976, No. 1942594, No. 1907805, a Jeffress Memorial Trust Award, Amazon Research Award, NVIDIA GPU Grant, and Design Knowledge Company (subcontract No: 10827.002.120.04).

References

1. Ando, R.K., Zhang, T.: A framework for learning predictive structures from multiple tasks and unlabeled data. J. Mach. Learn. Res. **6**, 1817–1853 (2005)
2. Argyriou, A., Evgeniou, T., Pontil, M.: Multi-task feature learning. In: Advances in Neural Information Processing Systems, pp. 41–48 (2007)
3. Bai, J., Li, J., Xu, F., Zhang, H.: Generalized symmetric ADMM for separable convex optimization. Comput. Optim. Appl. **70**(1), 129–170 (2017). https://doi.org/10.1007/s10589-017-9971-0

4. Boţ, R.I., Nguyen, D.-K.: The proximal alternating direction method of multipliers in the nonconvex setting: convergence analysis and rates. Math. Oper. Res. **45**(2), 682–712 (2020)
5. Boyd, S., Parikh, N., Chu, E., Peleato, B., Eckstein, J.: Distributed optimization and statistical learning via the alternating direction method of multipliers. Foundations Trends® Mach. Learn. **3**(1), 1–122 (2011)
6. Carrington, P.J., Scott, J., Wasserman, S.: Models and Methods in Social Network Analysis, vol. 28. Cambridge University Press, Cambridge (2005)
7. Cavalletti, F., Rajala, T.: Tangent lines and Lipschitz differentiability spaces. Anal. Geom. Metr. Spaces **4**(1), 85–103 (2016)
8. Chao, M.T., Zhang, Y., Jian, J.B.: An inertial proximal alternating direction method of multipliers for nonconvex optimization. Int. J. Comput. Math. **98**(6), 1199–1217 (2021)
9. Chartrand, R., Wohlberg, B.: A nonconvex ADMM algorithm for group sparsity with sparse groups. In: 2013 IEEE International Conference on Acoustics, Speech and Signal Processing (ICASSP), pp. 6009–6013 (2013)
10. Chen, J., Zhou, J., Ye, J.: Integrating low-rank and group-sparse structures for robust multi-task learning. In: Proceedings of the 17th ACM SIGKDD International Conference on Knowledge Discovery and Data Mining, pp. 42–50 (2011)
11. Deng, W., Lai, M.-J., Peng, Z., Yin, W.: Parallel multi-block ADMM with o (1/k) convergence. J. Sci. Comput. **71**(2), 712–736 (2017)
12. Gao, W., Goldfarb, D., Curtis, F.E.: ADMM for multiaffine constrained optimization. Optim. Methods Softw. **35**(2), 257–303 (2020)
13. Goldstein, T., O'Donoghue, B., Setzer, S., Baraniuk, R.: Fast alternating direction optimization methods. SIAM J. Imag. Sci. **7**(3), 1588–1623 (2014)
14. Gu, Y., Jiang, B., Han, D.: A semi-proximal-based strictly contractive Peaceman-Rachford splitting method. arXiv preprint arXiv:1506.02221, pp. 1–20 (2015)
15. Hassibi, A., How, J., Boyd, S.: A path-following method for solving BMI problems in control. In: Proceedings of the 1999 American Control Conference, vol. 2, pp. 1385–1389 (1999)
16. He, B., Yuan, X.: On the o(1/n) convergence rate of the Douglas-Rachford alternating direction method. SIAM J. Numer. Anal. **50**(2), 700–709 (2012)
17. Kadkhodaie, M., Christakopoulou, K., Sanjabi, M., Banerjee, A.: Accelerated alternating direction method of multipliers. In: Proceedings of the 21th ACM SIGKDD International Conference on Knowledge Discovery and Data Mining, pp. 497–506 (2015)
18. Lee, D.D., Seung, H.S.: Algorithms for non-negative matrix factorization. In: Advances in Neural Information Processing Systems, pp. 556–562 (2001)
19. Li, G., Pong, T.K.: Global convergence of splitting methods for nonconvex composite optimization. SIAM J. Optim. **25**(4), 2434–2460 (2015)
20. Li, Y., Tian, X., Liu, T., Tao, D.: Multi-task model and feature joint learning. In: International Joint Conference on Artificial Intelligence, pp. 3643–3649 (2015)
21. Lin, T.-Y., Ma, S.-Q., Zhang, S.-Z.: On the sublinear convergence rate of multi-block ADMM. J. Oper. Res. Soc. China **3**(3), 251–274 (2015)
22. Ouyang, H., He, N., Tran, L., Gray, A.: Stochastic alternating direction method of multipliers. In: International Conference on Machine Learning, pp. 80–88 (2013)
23. Reed, T.: Open source indicators project (2017). https://doi.org/10.7910/DVN/EN8FUW
24. Shen, X., Diamond, S., Udell, M., Gu, Y., Boyd, S.: Disciplined multi-convex programming. In: 2017 29th Chinese Control and Decision Conference (CCDC), pp. 895–900 (2017)
25. Tao, M., Yuan, X.: Convergence analysis of the direct extension of ADMM for multiple-block separable convex minimization. Adv. Comput. Math. **44**(3), 773–813 (2017). https://doi.org/10.1007/s10444-017-9560-x
26. Tseng, P.: Dual coordinate ascent methods for non-strictly convex minimization. Math. Program. **59**(1–3), 231–247 (1993)

27. Tseng, P.: Convergence of a block coordinate descent method for nondifferentiable minimization. J. Optim. Theory Appl. **109**(3), 475–494 (2001)
28. Wang, H., Banerjee, A.: Bregman alternating direction method of multipliers. In: Advances in Neural Information Processing Systems, pp. 2816–2824 (2014)
29. Wang, J., Chai, Z., Cheng, Y., Zhao, L.: Toward model parallelism for deep neural network based on gradient-free ADMM framework. In: 2020 IEEE International Conference on Data Mining (ICDM), pp. 591–600 (2020)
30. Wang, J., Gao, Y., Züfle, A., Yang, J., Zhao, L.: Incomplete label uncertainty estimation for petition victory prediction with dynamic features. In: 2018 IEEE International Conference on Data Mining (ICDM), pp. 537–546 (2018)
31. Wang, J., Li, H., Chai, Z., Wang, Y., Cheng, Y., Zhao, L.: Towards quantized model parallelism for graph-augmented MLPS based on gradient-free ADMM framework. arXiv preprint arXiv:2105.09837 (2021)
32. Wang, J., Yu, F., Chen, X., Zhao, L.: ADMM for efficient deep learning with global convergence. In: Proceedings of the 25th ACM SIGKDD International Conference on Knowledge Discovery & Data Mining, pp. 111–119 (2019)
33. Wang, J., Zhao, L.: Nonconvex generalization of alternating direction method of multipliers for nonlinear equality constrained problems. In: Results in Control and Optimization, p. 100009 (2019)
34. Wang, L., Li, Y., Zhou, J., Zhu, D., Ye, J.: Multi-task survival analysis. In: 2017 IEEE International Conference on Data Mining (ICDM), pp. 485–494 (2017)
35. Wang, Yu., Yin, W., Zeng, J.: Global convergence of ADMM in nonconvex nonsmooth optimization. J. Sci. Comput. **78**(1), 29–63 (2019)
36. Warga, J.: Minimizing certain convex functions. J. Soc. Ind. Appl. Math. **11**(3), 588–593 (1963)
37. Yangyang, X., Yin, W.: A block coordinate descent method for regularized multiconvex optimization with applications to nonnegative tensor factorization and completion. SIAM J. Imag. Sci. **6**(3), 1758–1789 (2013)
38. Xu, Z., De, S., Figueiredo, M., Studer, C., Goldstein, T.: An empirical study of ADMM for nonconvex problems. In: NIPS 2016 Workshop on Nonconvex Optimization for Machine Learning: Theory and Practice (2016)
39. Zhao, L., Wang, J., Guo, X.: Distant-supervision of heterogeneous multitask learning for social event forecasting with multilingual indicators. In: Proceedings of the AAAI Conference on Artificial Intelligence, vol. 32(1), April 2018
40. Zhou, J., Chen, J., Ye, J.: Clustered multi-task learning via alternating structure optimization. In: Advances in Neural Information Processing Systems, pp. 702–710 (2011)
41. Zhou, J., Chen, J., Ye, J.: MALSAR: multi-task learning via structural regularization. Arizona State University, vol. 21 (2011)

Prototypical Classifier for Robust Class-Imbalanced Learning

Tong Wei[1], Jiang-Xin Shi[1], Yu-Feng Li[1](✉), and Min-Ling Zhang[2,3]

[1] National Key Laboratory for Novel Software Technology, Nanjing University, Nanjing 210023, China
{weit,shijx,liyf}@lamda.nju.edu.cn

[2] School of Computer Science and Engineering, Southeast University, Nanjing 210096, China
zhangml@seu.edu.cn

[3] Key Laboratory of Computer Network and Information Integration (Southeast University), Ministry of Education, Nanjing, China

Abstract. Deep neural networks have been shown to be very powerful methods for many supervised learning tasks. However, they can also easily overfit to training set biases, i.e., label noise and class imbalance. While both learning with noisy labels and class-imbalanced learning have received tremendous attention, existing works mainly focus on one of these two training set biases. To fill the gap, we propose *Prototypical Classifier*, which does not require fitting additional parameters given the embedding network. Unlike conventional classifiers that are biased towards head classes, Prototypical Classifier produces balanced and comparable predictions for all classes even though the training set is class-imbalanced. By leveraging this appealing property, we can easily detect noisy labels by thresholding the confidence scores predicted by Prototypical Classifier, where the threshold is dynamically adjusted through the iteration. A sample reweighting strategy is then applied to mitigate the influence of noisy labels. We test our method on both benchmark and real-world datasets, observing that Prototypical Classifier obtains substaintial improvements compared with state of the arts.

Keywords: Noisy labels · Class imbalance · Contrastive learning

1 Introduction

Deep neural networks (DNNs) have been widely used for machine learning applications. Despite of their success, it has been shown that the training of DNNs requires large-scale labeled and *unbiased* data. However, in many real-world

T. Wei and J.-X. Shi—Co-first authors. This work was done when Tong Wei was a student at Nanjing University.

Supplementary Information The online version contains supplementary material available at https://doi.org/10.1007/978-3-031-05936-0_4.

J. Gama et al. (Eds.): PAKDD 2022, LNAI 13281, pp. 44–57, 2022.
https://doi.org/10.1007/978-3-031-05936-0_4

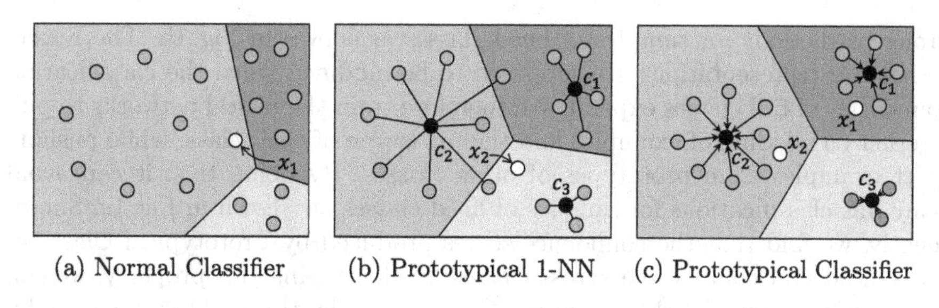

Fig. 1. Illustration of normal classifier and Prototypical Classifier.

applications, training set biases are prevalent [9,21,27,28], which typically have two types: i) class-imbalanced data distribution; and ii) noisy labels. For example, in autonomous driving, the vast majority of the training data is composed of standard vehicles but models also need to recognize rarely seen classes such as emergency vehicles or animals with very high accuracy. This will sometime lead to biased training models that do not perform well in practice. Moreover, large-scale high-quality data annotations are expensive and time-consuming to obtain. Although coarse labels are cheap and of high availability, the presence of noise will hurt the model performance. Therefore, it is desirable to develop machine learning algorithms that can accommodate not only class-imbalanced training set, but also the presence of label noise.

Both learning with noisy labels and class-imbalanced learning (a.k.a. long-tailed learning) have been studied for many years. When dealing with label noise, the most popular approach is sample selection where correctly-labeled examples are identified by capturing the training dynamics of DNNs [11,29]. When dealing with class imbalance, many existing works propose to reweight examples or design unbiased loss functions by taking into account the class distribution of training set [3,8,26]. However, most existing methods focus on only one of these two training set biases.

In this paper, we address both training set biases simultaneously. As shown in Fig. 1a, it is known that the classifier directly learned on class-imbalanced data is biased towards head classes [8,32] which results in poor generalization on tail classes. Moreover, using sample loss/confidence produced by biased classifiers fails to detect label noise, because both clean and noisy samples of tail classes have large loss and low confidence. To solve this problem, we propose to use *Prototypical Classifier* which is demonstrated to produce balanced predictions even through the training set is class-imbalanced. Our basic idea is that there exists an embedding in which examples cluster around a single prototype representation for each class. In order to do this, we learn a non-linear mapping of the input into an embedding space using a neural network and take a class's prototype to be the normalized mean vector of examples in the embedding space. Classification is then performed for an embedded test example by simply finding the nearest class prototype. Notably, Prototypical Classifier does not need additional learnable parameters given embedding of examples. Unfortunately, it is easy to observe that simply using prototypes for classification may lead to many wrong predictions for samples of head classes as shown in Fig. 1b. The reason is that the representations are supposed to be modified when the classification boundaries of tail classes expand. We therefore train the neural networks to pull together embedding of examples and the prototype of their class, while pushing apart examples from prototypes of other classes. By doing this, it can avoid many mis-classifications for samples of head classes, as shown in Fig. 1c. Subsequently, we find that the confidence scores produced by Prototypical Classifier is balanced and comparable across classes. By leveraging this property, we can simply detect noisy labels via thresholding where the threshold is dynamically adjusted, followed by a sample re-weighting strategy.

In summary, our key contributions of this work are:

- We propose to learn from training set with mixed biases, which is practical but has been understudied;
- Our approach, Prototype Classifier, is simple yet powerful. It produces more balanced predictions over all classes than normal classifiers even when the training set is class-imbalanced. This property further benefits the detection of label noise.
- On both simulated datasets and a real-world dataset Webvision with label noise, Prototype Classifier achieves substantial performance improvement.

2 Related Work

Class-Imbalanced Learning. Recently, many approaches have been proposed to handle class-imbalanced training set. Most extant approaches can be categorized into three types by modifying (i) the inputs to a model by re-balancing the training data [16,22,32]; (ii) the outputs of a model, for example by post-hoc adjustment of the classifier [8,17,25]; and (iii) the internals of a model by modifying the loss function [2,6,20,23]. Each of the above methods are intuitive, and have shown strong empirical performance. However, these methods assume the training examples are correctly-labeled, which is often difficult to obtain in real-world applications. Instead, we study a realistic problem to learn from class-imbalanced data with label noise.

Label Noise Detection. Plenty of methods have been proposed to detect noisy labels [4,7,10]. Many works adopt the small-loss trick, which treats samples with small training losses as correctly-labeled. In particular, MentorNet [7] reweights samples with small loss so that noisy samples contribute less to the loss. Co-teaching [4] trains two networks where each network selects small-loss samples in a mini-batch to train the other. DivideMix [10] fits a Gaussian mixture model on per-sample loss distribution to divide the training data into clean set and noisy set. In addition, AUM [19] introduces a margin statistic to identify noisy samples by measuring the average difference between the logit values for a sample's assigned class and its highest non-assigned class. The above methods only consider class-balanced training sets, thus is not directly applicable for class-imbalanced problems. Ref. [12] observes that real-world dataset with label noise also has imbalanced number of samples per-class. Nevertheless, they only inspect a particular setup of class imbalance.

3 Prototypical Classifier with Dynamic Threshold

3.1 Motivation

Consider a binary classification problem with the data generating distribution $\mathbb{P}_{XY}$ being a mixture of two Gaussians. In particular, the label Y is either positive (+1) or negative (−1) with equal probability (i.e., $\frac{1}{2}$). Condition on $Y = +1, \mathbb{P}(X \mid Y = +1) \sim \mathcal{N}(\mu_1, \sigma_1)$ and similarly, $\mathbb{P}(X \mid Y = -1) \sim \mathcal{N}(\mu_2, \sigma_2)$. Without loss of generality, let $\mu_1 > \mu_2$. It is straightforward to verify that the optimal Bayes's classifier is $f(x) = sign(x - \frac{\mu_1+\mu_2}{2})$ [30], i.e., classify x as +1 if $x > \frac{\mu_1+\mu_2}{2}$. This reminds us the nearest neighbor classifier, whose classification boundary is at the middle of two data points (i.e., balanced classification boundary). For general multi-class tasks, this motivates us to measure the distance of samples to class prototypes, which is empirically observed to produce balanced classification boundary even though the training set is class-imbalanced, as shown in Fig. 2.

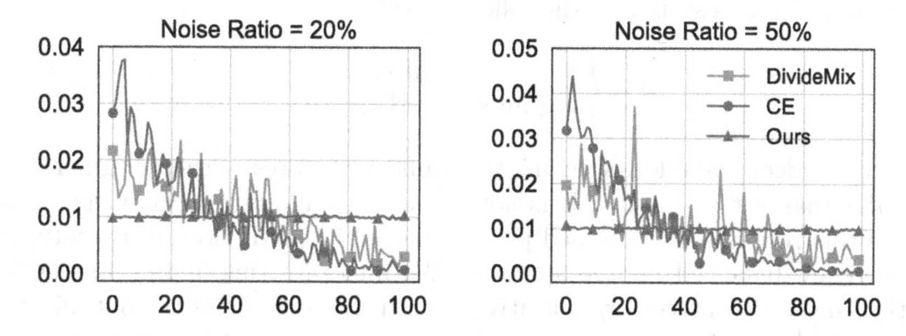

Fig. 2. Experiment on CIFAR-100-LT. x-axis is the class labels with decreasing training samples and y-axis is the marginal likelihood $p(y)$ on the test set.

In order to do this, we learn a non-linear mapping of the input into an embedding space using a neural network f_θ parameterized by θ using training set $\mathcal{D} = \{(\boldsymbol{x}_i, y_i)\}_{i=1}^N$. The class prototype is taken as the normalized mean vector of the embedded examples belonging to its class. For example, the prototype for class $k \in \{1, \ldots, K\}$ is computed as:

$$\boldsymbol{c}_k = \text{Normalize}\left(\frac{1}{|\mathcal{D}_k|}\sum_{i \in \mathcal{D}_k} f_\theta(\boldsymbol{x}_i)\right), \mathcal{D}_k = \{i \mid y_i = k\}. \tag{1}$$

Prototypical Classifier produces a distribution over classes for sample $\boldsymbol{x}$ based on a softmax over distances to the prototypes in the embedding space. In particular, when use cosine similarity as distance measure, we have:

$$\mathbb{P}_\theta(Y = k \mid \boldsymbol{x}) = \frac{\exp\left(f_\theta(\boldsymbol{x})^\top \mathbf{c}_k\right)}{\sum_{k'} \exp\left(f_\theta(\boldsymbol{x})^\top \mathbf{c}_{k'}\right)}. \tag{2}$$

Learning proceeds by minimizing the negative log-probability $J(\theta) = -\log \mathbb{P}_\theta(Y = k \mid \mathbf{x})$ of the true class label k via SGD. Notably, the model in Eq. (2) is equivalent to a linear model with a particular parameterization [18]. To see this, expand the term in the exponent:

$$\mathbf{c}_k^\top f_\theta(\boldsymbol{x}) = \mathbf{w}_k^\top f_\theta(\boldsymbol{x}) + b_k, \text{ where } \mathbf{w}_k = \mathbf{c}_k \text{ and } b_k = 0. \quad (3)$$

Our results indicate that Prototypical Classifier is effective despite the equivalence to a linear model. We hypothesize this is because all of the required non-linearity can be learned within the embedding function [24]. Indeed, this is the approach that modern neural network classification systems currently use.

3.2 Dynamic Thresholding for Label Noise Detection

However, the existence of label noise may hurt the representation learning of the network. To tackle this issue, it is a common practice to correct noisy labels. Let $\hat{\boldsymbol{y}} = [\hat{y}_1, \cdots, \hat{y}_K] = \mathbb{P}_\theta(Y \mid \boldsymbol{x})$ be the prediction of Prototypical Classifier, the labels are refined as stated by the following rule:

$$\tilde{y} = \begin{cases} y_i & \text{if } \hat{y}_{y_i} > \tau_t \\ \arg\max_j \hat{y}_j & \text{otherwise.} \end{cases} \quad (4)$$

In words, we deem samples as clean if the confidence scores on their original labels is greater than a threshold τ_t. It is notably that using normal classifiers cannot achieve this goal due to its biased predictions, while predictions of Prototypical Classifier are balanced and comparable. We illustrate this finding in Fig. 3. We then need to construct τ_t. Intuitively, with the increase of the optimization iteration t, the predictive confidence also increases in general, so that τ_t is also required to increase. Mathematically, we set the dynamic threshold τ_t as an increasing function of t, which is given by:

$$\tau_t = \gamma^t \tau_0. \quad (5)$$

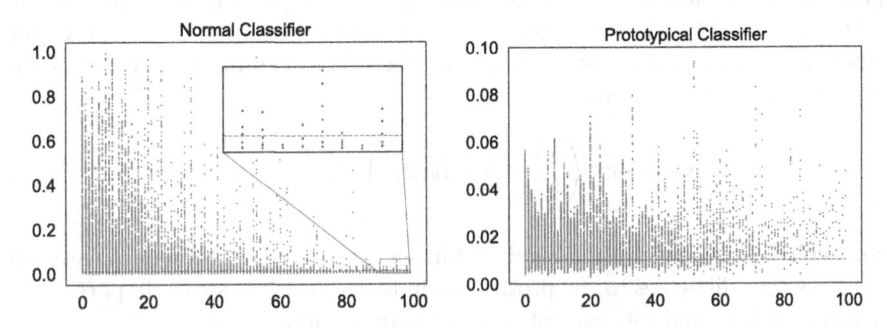

Fig. 3. Experiment on CIFAR-100-LT. x-axis is the class labels with decreasing training samples and y-axis is the confidence scores of classifiers on training set.

Here, τ_0 is the initial threshold and γ is set to 1.005 in our experiments. We provide more analysis about τ_t in supplementary materials. Lemma 1 summarizes the performance bound of the label noise detection method.

Lemma 1. *With probability at least p, the* F_1*-score of detecting noisy labels in* $\mathcal{D}_j$ *by thresholding the predictive scores of Prototypical Classifier is at least* $1 - \frac{e^{-v}\max(N^-,N^+)+\alpha}{N^-}$ *when the noise ratio is known, where* $p = \int_{-1}^{\mu^{true}-\mu^{false}-\Delta} f(t)dt$, $f(t)$ *is the probability density function of the difference of two independent beta-distributed random variables* $\beta_1 - \beta_2$*, where* $\beta_1 \sim \text{Beta}(N^-, 1)$, $\beta_2 \sim \text{Beta}(\alpha+1, N^+ - \alpha)$.

Lemma 1 shows that the performance of noise detection depends on the intraclass concentration of clean samples in the embedding space (denoted by $\frac{\Delta^2}{v}$), which is optimized by the prototypical contrastive loss defined in Eq. (6). We refer the reader to Ref. [33] for the proof of Lemma 1. We further justify the effectiveness of our method in Fig. 4, which produces high F_1-score for both head and tail classes.

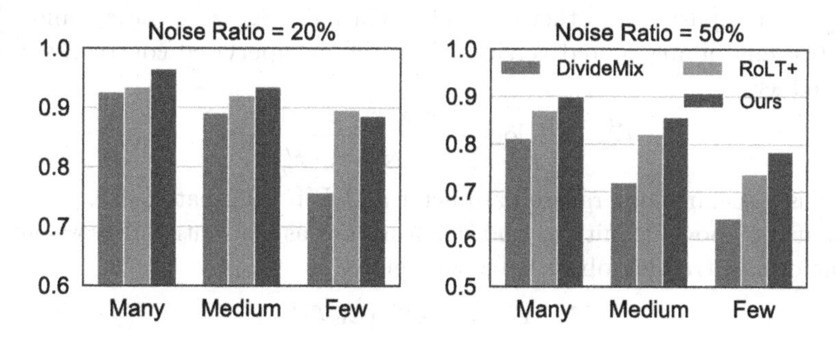

Fig. 4. Experiment on CIFAR-100-LT. We show the F_1-score of clean examples selection module for many, medium and few classes.

3.3 Example Reweighting

In standard training, we aim to minimize the expected loss for the training set, where each input example is weighted equally. Here we aim to learn a reweighting of the inputs to cope with hard mislabeled samples whose labels are not correctly refined, where we minimize a weighted loss:

$$\mathcal{L}_{\text{pc}} = \frac{-1}{\sum_{i=1}^{N} w_i} \sum_{i=1}^{N} w_i \log \frac{\exp\left(f_\theta(\boldsymbol{x}) \cdot \boldsymbol{c}_{y_i}/\tau\right)}{\sum_{k=1}^{K} \exp\left(f_\theta(\boldsymbol{x}) \cdot \boldsymbol{c}_k/\tau\right)}. \tag{6}$$

With a slight abuse of the notation, we re-define w_i to be the weight for the i-th example and τ is a temperature parameter. We expect the weights can reflect the likelihood of examples being correctly-labeled. In that regard, we devise a weighted version for computing prototypes as:

$$c_k = \text{Normalize}\left(\frac{1}{\sum_{i\in\mathcal{D}_k} w_i}\sum_{i\in\mathcal{D}_k} w_i f_\theta(\boldsymbol{x}_i)\right), \mathcal{D}_k = \{i \mid y_i = k\}. \tag{7}$$

Recall that, one appealing property of Prototypical Classifier is balanced predictions across all classes, as opposite to biased normal classifiers. We therefore simply set examples weights as the predicted score of Prototypical Classifier on the training label, i.e., for the i-th example, we set $w_i = \mathbb{P}_\theta(Y = y_i \mid \boldsymbol{x}_i)$ where y_i is the training label of $\boldsymbol{x}_i$. For samples whose labels are rectified, we update their weights by $w' = \frac{\tau_t - w}{2}$ to reflect the uncertainty. The modified example weights are always positive since the label is refined if and only if $w = \mathbb{P}(Y = y_i \mid \boldsymbol{x}_i) \le \tau_t$. The optimization of $\mathcal{L}_{\text{pc}}$ is realized by contrastive learning, which has been demonstrated effective in learning representations [13]. Observing that the presence of label noise may have negative effect on representation learning, we train networks to optimize the unsupervised contrastive loss, which does not use the biased training labels. The basic idea of unsupervised contrastive learning is to pull together two embeddings of the same example, while pushing apart from other examples. Formally, let $\boldsymbol{z}_i = f_\theta(\boldsymbol{x}_i)$ and $\boldsymbol{z}_i'$ be the embedding of augmented version of $\boldsymbol{x}_i$, the unsupervised contrastive loss is computed as:

$$\mathcal{L}_{\text{cc}}^i = -\log\frac{\exp(\boldsymbol{z}_i\cdot\boldsymbol{z}_i'/\tau)}{\sum_{b=0}^{B}\exp(\boldsymbol{z}_i\cdot\boldsymbol{z}_b'/\tau)}, \tag{8}$$

where τ is a scalar temperature parameter and B is mini-batch size.

Given the above definitions and denoting $\mathcal{L}^{\text{ce}}$ as conventional cross-entropy loss, the overall training objective is written as:

$$\mathcal{L} = \mathcal{L}^{\text{ce}} + \lambda_1\mathcal{L}^{\text{cc}} + \lambda_2\mathcal{L}^{\text{pc}}, \tag{9}$$

where hyperparameters λ_1 and λ_2 are trade-off parameters. We adopt DNNs as feature extractor and a linear layer as projector to generate latent feature representation $\boldsymbol{z}_i$. Another linear layer following the feature extractor is used as classifier. When minimizing $\mathcal{L}_{\text{pc}}$, we apply mixup [31] to improve the generalization which has been shown to be effective for learning with noisy labels [29].

4 Experiments

We perform experiments on CIFAR-10 and CIFAR-100 datasets by controlling label noise ratio and imbalance factor of the training set. Additionally, we perform experiments on a commonly used dataset Webvision with real-world label noise.

4.1 Results on Simulated Datasets

Class-Imbalanced Dataset Generation. Formally, for a dataset with K classes and N training examples for each class, by assuming the imbalance factor is ρ, the number of examples for the k-th class is set to $N_k = N/\rho^{\frac{k-1}{K-1}}$.

Label Noise Injection. Let Y denote the variable for the clean label, $\bar{Y}$ the noisy label, and X the instance/feature, the transition matrix $T(X = x)$ is defined as $T_{ij}(X) = \mathbb{P}(\bar{Y} = j \mid Y = i, X = x)$. In this work, we follow the setup in RoLT+ [28] by setting $T(X = x)$ according to the estimated class priors $\mathbb{P}(y)$, e.g., the empirical class frequencies in the training dataset. Formally, given the noise proportion $\gamma \in [0, 1]$, we define:

$$T_{ij}(X) = \mathbb{P}(\bar{Y} = j \mid Y = i, X = x) = \begin{cases} 1 - \gamma & i = j \\ \frac{N_j}{N - N_i}\gamma & \text{otherwise.} \end{cases} \tag{10}$$

Here, N is the size of training set and N_j is frequency of class j.

Table 1. Test accuracy (%) on CIFAR-10. * denotes ensemble models.

Noise ratio		0.2			0.5		
Imbalance factor		10	50	100	10	50	100
(1) CE	Best	77.86	64.38	61.79	60.72	46.50	38.43
	Last	74.00	61.38	55.69	44.29	32.69	27.78
(2) LDAM	Best	83.48	72.01	66.41	63.57	38.92	34.08
	Last	82.91	71.23	66.22	62.13	37.97	32.56
(3) LDAM-DRW	Best	84.98	76.77	73.24	69.53	49.90	42.60
	Last	84.71	75.98	72.46	68.76	47.71	40.47
(4) DivideMix*	Best	88.79	75.34	66.90	87.54	67.92	61.81
	Last	88.10	73.48	63.76	86.88	65.22	59.65
(5) RoLT+*	Best	87.95	77.26	72.31	**88.17**	**75.11**	64.42
	Last	87.54	75.90	69.12	**87.45**	**73.92**	61.15
(6) Prototypical Classifier	Best	**90.92**	**84.12**	**79.54**	84.04	71.44	**66.33**
	Last	**90.81**	**83.71**	**78.34**	83.51	71.44	**64.69**

Result. We train the PreAct ResNet-18 network using SGD optimizer with momentum 0.9 for all methods. We set $\lambda_1 = 1$ and $\lambda_2 = 5$. We use $\tau_0 = 0.1$ for CIFAR-10 and $\tau = 0.01$ for CIFAR-100. Tables 1 and 2 respectively summarize the results for CIFAR-10 and CIFAR-100 datasets. We compare our methods with several commonly used baselines for long-tailed learning (1–3) and learning with noisy labels (4–5). As shown in the results, previous methods dreadfully degrade their performance as the noise ratio and imbalance factor increase, while our methods retain robust performance. In particular, compared with CE, Prototypical Classifier improves the test accuracy by 9% on average. It can be observed that the improvement becomes more significant when the noise ratio is high, benefiting from proposed noise detection method.

As DivideMix [10] and RoLT+ [28] are two strong baselines in this task, (4) and (5) obtain much higher performance than (1–3), particularly when noise ratio is high. Although (4) and (5) use an ensemble of two networks, our method (6) outperforms them in most cases. On CIFAR-100, Prototypical Classifier achieves the best results among all the approaches and outperforms others by a large margin for both head and tail classes in Fig. 5.

Table 2. Test accuracy (%) on CIFAR-100. * denotes ensemble models.

Noise ratio		0.2			0.5		
Imbalance factor		10	50	100	10	50	100
(1) CE	Best	45.97	33.41	29.85	28.70	18.49	16.24
	Last	45.75	33.12	29.58	23.70	16.56	14.19
(2) LDAM	Best	47.30	35.70	32.67	27.86	17.62	15.68
	Last	47.12	35.50	32.60	24.20	17.50	14.73
(3) LDAM-DRW	Best	47.85	36.29	33.38	27.86	17.91	15.68
	Last	47.68	36.01	32.99	24.45	17.81	15.07
(4) DivideMix*	Best	63.79	49.64	43.91	49.35	36.52	31.82
	Last	63.17	48.37	42.59	48.87	35.72	31.05
(5) RoLT+*	Best	64.22	51.01	45.35	53.31	39.78	35.29
	Last	63.31	49.40	43.16	52.44	39.27	34.43
(6) Prototypical Classifier	Best	**65.23**	**51.73**	**47.38**	**57.65**	**42.51**	**38.42**
	Last	**65.14**	**51.46**	**47.12**	**57.65**	**42.51**	**38.36**

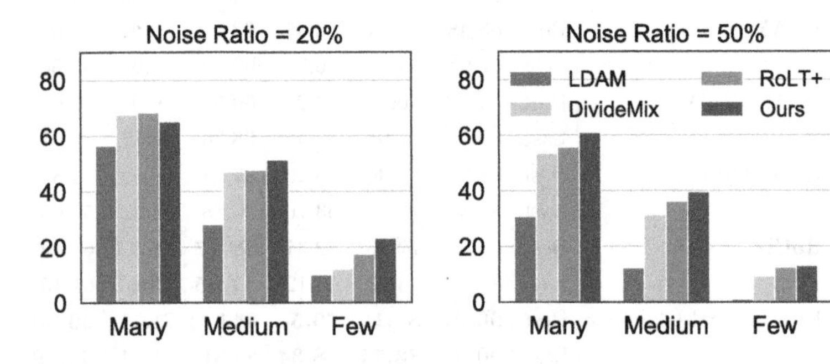

Fig. 5. Experiment on CIFAR-100-LT. We show the accuracy for many (#inst >100), medium (#inst $\in$ [20, 100]) and few (#inst < 20) classes.

4.2 Results on Real-World Dataset

We test the performance of our method on a real-world dataset. WebVision [14] contains 2.4 million images collected from Flickr and Google with real noisy and class-imbalanced data. Following previous literature, we train on a subset, mini WebVision, which contains the first 50 classes. In Table 3, we report results comparing against state-of-the-art approaches, including MentorNet [7], Co-teaching [4], ELR [15], HAR [1], and DivideMix [10]. We use InceptionResNet-v2 for all methods. We set $\tau_0 = 0.05$, $\lambda_1 = 1$ and $\lambda_2 = 2$ in all experiments. From the results, we can see that, by using a single model, the proposed method achieves competitive performance with DivideMix and outperforms other baselines.

4.3 Ablation Studies

We examine the effectiveness of the each module of our method by removing it and comparing its performance with the full framework. The results are reported

Table 3. Accuracy (%) on WebVision and ImageNet. * denotes ensemble models.

		MentorNet	Co-teaching	ELR	HAR	DivideMix*	**Ours**
Webvision	top1	63.00	63.58	76.26	75.5	**77.32**	**77.32**
	top5	81.40	85.20	91.26	90.7	91.64	**92.60**
ImageNet	top1	57.80	61.48	68.71	70.3	**75.20**	75.12
	top5	79.92	84.70	87.84	90.0	90.84	**91.92**

in Table 4. Generally, it is easy to see that removing any part of the method significantly drops the performance or even fails in some cases. The performance of re-weighting and dynamic threshold shows their great effectiveness for dealing with label noise. Though we do not use the normal classifier trained via $\mathcal{L}_{ce}$, it is observed to help improve the representation learning. We have a similar observation for the unsupervised contrastive loss $\mathcal{L}_{ce}$. The strong augmentation method AugMix [5] also provides substaintial improvement.

Additionally, we also test our method on class-balanced training sets with label noise in Table 5. Prototypical Classifier outperforms other methods in most cases, even though both DivideMix and RoLT+ uses an ensemble of two networks, which shows the generality of Prototypical Classifier.

Table 4. Ablation studies. $\rho = 0.5$ and $\gamma = 100$. ▼ (▲) indicate performance loss (gain) compared with Prototypical Classifier.

Method		CIFAR-10	CIFAR-100
w/o re-weighting	Best	61.69 (▼4.64)	–
	Last	58.57 (▼6.12)	–
w/o dynamic threshold	Best	63.85 (▼2.48)	39.04 (▲0.62)
	Last	56.01 (▼8.68)	38.67 (▲0.25)
w/o mixup	Best	52.79 (▼13.54)	33.09 (▼5.33)
	Last	51.43 (▼13.26)	32.57 (▼5.79)
w/o AugMix	Best	62.51 (▼3.82)	36.11 (▼2.31)
	Last	55.21 (▼9.48)	35.68 (▼2.68)
w/o $\mathcal{L}_{cc}$	Best	55.34 (▼9.35)	32.65 (▼5.71)
	Last	53.17 (▼11.52)	32.39(▼5.97)
w/o $\mathcal{L}_{ce}$	Best	57.61 (▼7.08)	35.25 (▼3.11)
	Last	53.24 (▼11.45)	35.02 (▼3.34)

Table 5. Accuracy (%) on class-balanced datasets. * denotes ensemble models.

Noise ratio		CIFAR-10		CIFAR-100	
		0.2	0.5	0.2	0.5
DivideMix*	Best	92.79	**95.03**	77.25	73.84
	Last	92.41	**94.63**	77.03	73.42
RoLT+*	Best	92.46	94.59	78.60	74.11
	Last	92.01	94.41	78.14	73.35
Prototypical Classifier	Best	**95.93**	92.55	**79.41**	**75.50**
	Last	**95.80**	92.40	**79.41**	**75.10**

5 Conclusion

We propose Prototypical Classifier for learning with training set biases. Prototypical Classifier is shown to produce balanced predictions for all classes even when learned on class-imbalanced training set. This appealing property provides a way of detecting label noise by thresholding the predicted scores of examples. Experiments demonstrate the superiority of the proposed method. We believe Prototypical Classifier can motivate solutions to more problems with class-imbalanced training sets, for instance semi-supervised learning and self-supervised learning.

Acknowledgments. The authors wish to thank the anonymous reviewers for their helpful comments and suggestions. This research was supported by the NSFC (62176118).

A Ablations on Dynamic Threshold

Figure 6 shows a comparison of fixed threshold and the dynamic threshold τ_t with $\tau_0 = 0.1$. We consider both exponential scheduler controlled by γ and linear scheduler controlled by the threshold of last iteration τ_T.

We test the performance of different choice of parameters and the results are reported in Table 6. From the results, we have two observations: i) when using fixed threshold or the dynamic threshold grows too slow, performance drops in the last iterations because many noisy labels are incorrectly flagged as clean; and ii) when dynamic threshold grows too fast, the network cannot achieve best performance, because many clean labels are incorrectly flagged as noisy.

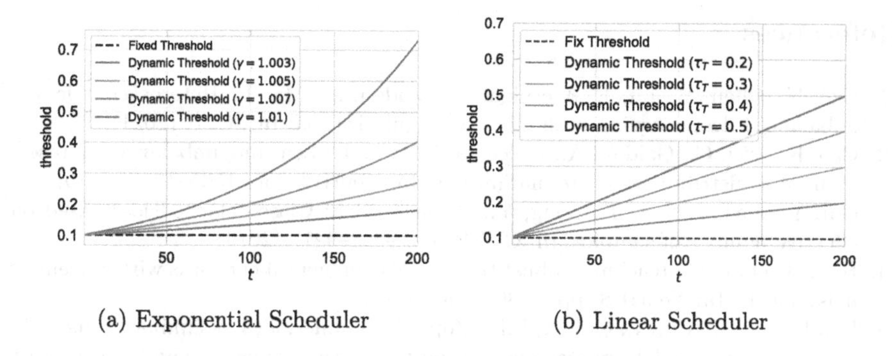

(a) Exponential Scheduler

(b) Linear Scheduler

Fig. 6. Comparison of fixed threshold and dynamic threshold. Fix threshold $\tau = 0.1$, exponential dynamic threshold $\tau_t = 0.1\gamma^t$ and linear dynamic threshold $\tau_t = 0.1 + \frac{\tau_T - 0.1}{T} t$.

Table 6. Test accuracy (%) on CIFAR-10-LT with imbalance factor 100 and noise ratio 50%.

	Ours ($\gamma = 1.005$)	Fix	Exponential			Linear			
			1.003	1.007	1.01	0.2	0.3	0.4	0.5
Best	66.33	66.01	66.27	63.47	56.81	65.18	66.09	61.78	59.41
Last	64.69	61.37	63.57	58.93	35.84	63.40	65.11	57.84	55.12

B Results on Clean Datasets

Although our method is particularly designed learning with noisy labels, it is interesting to study its performance on clean but class-imbalanced datasets. In this experiment, we do not use sample re-weighting and label noise correction. We report the results in Table 7. For fair comparison, we do not apply AugMix in this experiment. Intriguingly, Prototypical Classifier consistently outperforms all baselines by a large margin, showing the superiority of our proposed representation learning method.

Table 7. Test accuracy (%) on clean datasets with different imbalanced factor.

	CIFAR-10			CIFAR-100		
Imbalance factor	10	50	100	10	50	100
CE	88.42	79.56	73.43	60.14	45.79	41.87
LDAM	87.43	80.32	74.50	59.84	47.61	42.59
LDAM-DRW	88.15	83.18	79.43	60.40	48.90	43.63
cRT	88.26	79.22	73.61	60.69	46.67	42.26
NCM	89.45	83.06	79.36	61.46	49.36	45.49
Prototypical Classifier	92.78	86.03	83.11	68.71	56.60	50.94

References

1. Cao, K., Chen, Y., Lu, J., Arechiga, N., Gaidon, A., Ma, T.: Heteroskedastic and imbalanced deep learning with adaptive regularization. In: ICLR (2021)
2. Cao, K., Wei, C., Gaidon, A., Aréchiga, N., Ma, T.: Learning imbalanced datasets with label-distribution-aware margin loss. In: NeurIPS, pp. 1565–1576 (2019)
3. Cui, Y., Jia, M., Lin, T., Song, Y., Belongie, S.J.: Class-balanced loss based on effective number of samples. In: CVPR, pp. 9268–9277 (2019)
4. Han, B., et al.: Co-teaching: robust training of deep neural networks with extremely noisy labels. In: NeurIPS, pp. 8536–8546 (2018)
5. Hendrycks, D., Mu, N., Cubuk, E.D., Zoph, B., Gilmer, J., Lakshminarayanan, B.: AugMix: a simple data processing method to improve robustness and uncertainty. In: ICLR (2020)
6. Jamal, M.A., Brown, M., Yang, M.H., Wang, L., Gong, B.: Rethinking class-balanced methods for long-tailed visual recognition from a domain adaptation perspective. In: CVPR, pp. 7610–7619 (2020)
7. Jiang, L., Zhou, Z., Leung, T., Li, L.J., Fei-Fei, L.: MentorNet: learning data-driven curriculum for very deep neural networks on corrupted labels. In: ICML, pp. 2304–2313 (2018)
8. Kang, B., et al.: Decoupling representation and classifier for long-tailed recognition. In: ICLR (2020)
9. Karthik, S., Revaud, J., Boris, C.: Learning from long-tailed data with noisy labels. CoRR abs/2108.11096 (2021)
10. Li, J., Socher, R., Hi, S.C.: DivideMix: learning with noisy labels as semi-supervised learning. In: ICLR (2020)
11. Li, J., Xiong, C., Hoi, S.C.: Learning from noisy data with robust representation learning. In: ICCV, pp. 9485–9494 (2021)
12. Li, J., Xiong, C., Hoi, S.C.: MOPRO: webly supervised learning with momentum prototypes. In: ICLR (2021)
13. Li, J., Zhou, P., Xiong, C., Hoi, S.C.H.: Prototypical contrastive learning of unsupervised representations. In: ICLR (2021)
14. Li, W., Wang, L., Li, W., Agustsson, E., Gool, L.V.: Webvision database: visual learning and understanding from web data. CoRR abs/1708.02862 (2017)
15. Liu, S., Niles-Weed, J., Razavian, N., Fernandez-Granda, C.: Early-learning regularization prevents memorization of noisy labels. In: NeurIPS, pp. 20331–20342 (2020)
16. Liu, Z., Miao, Z., Zhan, X., Wang, J., Gong, B., Yu, S.X.: Large-scale long-tailed recognition in an open world. In: CVPR, pp. 2537–2546 (2019)
17. Menon, A.K., Jayasumana, S., Rawat, A.S., Jain, H., Veit, A., Kumar, S.: Long-tail learning via logit adjustment. In: ICLR (2021)
18. Mensink, T., Verbeek, J.J., Perronnin, F., Csurka, G.: Distance-based image classification: generalizing to new classes at near-zero cost. IEEE Trans. Pattern Anal. Mach. Intell. **35**(11), 2624–2637 (2013)
19. Pleiss, G., Zhang, T., Elenberg, E.R., Weinberger, K.Q.: Identifying mislabeled data using the area under the margin ranking. In: NeurIPS, pp. 17044–17056 (2020)
20. Ren, J., et al.: Balanced meta-softmax for long-tailed visual recognition. In: NeurIPS, pp. 4175–4186 (2020)
21. Ren, M., Zeng, W., Yang, B., Urtasun, R.: Learning to reweight examples for robust deep learning. In: ICML, pp. 4331–4340 (2018)

22. Shen, L., Lin, Z., Huang, Q.: Relay backpropagation for effective learning of deep convolutional neural networks. In: Leibe, B., Matas, J., Sebe, N., Welling, M. (eds.) ECCV 2016. LNCS, vol. 9911, pp. 467–482. Springer, Cham (2016). https://doi.org/10.1007/978-3-319-46478-7_29
23. Shu, J., et al.: Meta-weight-net: learning an explicit mapping for sample weighting. In: NeurIPS, pp. 1917–1928 (2019)
24. Snell, J., Swersky, K., Zemel, R.S.: Prototypical networks for few-shot learning. In: NeurIPS, pp. 4077–4087 (2017)
25. Tang, K., Huang, J., Zhang, H.: Long-tailed classification by keeping the good and removing the bad momentum causal effect. In: NeurIPS, pp. 1513–1524 (2020)
26. Wang, Y., Ramanan, D., Hebert, M.: Learning to model the tail. In: NeurIPS, pp. 7029–7039 (2017)
27. Wei, T., Li, Y.F.: Does tail label help for large-scale multi-label learning? IEEE Trans. Neural Netw. Learn. Syst. **31**(7), 2315–2324 (2020)
28. Wei, T., Shi, J., Tu, W., Li, Y.: Robust long-tailed learning under label noise. CoRR abs/2108.11569 (2021)
29. Wu, Z.F., Wei, T., Jiang, J., Mao, C., Tang, M., Li, Y.F.: NGC: a unified framework for learning with open-world noisy data. In: ICCV, pp. 62–71 (2021)
30. Yang, Y., Xu, Z.: Rethinking the value of labels for improving class-imbalanced learning. In: NeurIPS, pp. 19290–19301 (2020)
31. Zhang, H., Cisse, M., Dauphin, Y.N., Lopez-Paz, D.: Mixup: beyond empirical risk minimization. In: ICLR (2017)
32. Zhou, B., Cui, Q., Wei, X., Chen, Z.: BBN: bilateral-branch network with cumulative learning for long-tailed visual recognition. In: CVPR, pp. 9716–9725 (2020)
33. Zhu, Z., Dong, Z., Cheng, H., Liu, Y.: A good representation detects noisy labels. arXiv preprint arXiv:2110.06283 (2021)

Quantum Entanglement Inspired Correlation Learning for Classification

Junwei Zhang[1(✉)], Zhao Li[2,4(✉)], Juan Wang[1], Yinghui Wang[1], Shichang Hu[3], Jie Xiao[4], and Zhaolin Li[5]

[1] College of Intelligence and Computing, Tianjin University, Tianjin, China
junwei@tju.edu.cn, wangyinghui@tju.edu.cn
[2] Zhejiang University, Zhejiang, China
zhao_li@zju.edu.cn
[3] Alibaba Group, Zhejiang, China
shichang.hsc@alibaba-inc.com
[4] Hangzhou Yugu Technology Co., Ltd., Hangzhou, China
[5] Tianjin Xiniu Huaan Technology Co., Ltd., Tianjin, China

Abstract. Correlation is an important information resource, which is often used as a fundamental quantity for modeling tasks in machine learning. Since correlation between quantum entangled systems often surpasses that between classical systems, quantum information processing methods show superiority that classical methods do not possess. In this paper, we study the virtue of entangled systems and propose a novel classification algorithm called Quantum Entanglement inspired the Classification Algorithm (QECA). Particularly, we use the joint probability derived from entangled systems to model correlation between features and categories, that is, Quantum Correlation (QC), and leverage it to develop a novel QC-induced Multi-layer Perceptron framework for classification tasks. Experimental results on four datasets from diverse domains show that QECA is significantly better than the baseline methods, which demonstrates that QC revealed by entangled systems can improve the classification performance of traditional algorithms.

Keywords: Quantum correlation · Quantum-inspired algorithms · Classification algorithm

1 Introduction

In machine learning, correlation is considered an important information resource and is often used as a fundamental quantity in the modeling process of learning tasks. Correlation is any statistical association, although it usually refers to the degree to which a pair of variables is linearly related [6].

In recent years, quantum information technology have been developed by leaps and bounds [4,9]. Quantum information processing has the advantages that classical information processing cannot match, and can complete information processing tasks that cannot be achieved by classical methods [16,18,19],

J. Gama et al. (Eds.): PAKDD 2022, LNAI 13281, pp. 58–70, 2022.
https://doi.org/10.1007/978-3-031-05936-0_5

such as quantum teleportation, quantum communication, etc. Quantum Correlation (QC) in quantum composite systems [1,17], namely, non-classical statistical correlation, has become more and more important, because it is the core quantum resource and it is stronger than classical statistical correlation [14]. In fact, the reason why quantum information processing has the superiority that classical information processing does not possess is because there is a correlation between quantum systems that is beyond classical correlation [12].

Since quantum theory is not widely used in classical machine learning tasks, here we give answers to several questions that readers may be concerned about. Although quantum theory is generally regarded as a micro-physical theory, its connotation is about information rather than physics. Since Hardy [7], the informational nature of quantum mechanics has gradually become more and more rigorous. Therefore, the laws of quantum mechanics should not only be regarded as the laws of the micro-physical world, but should be regarded as the general rules of information processing [3,8].

In this paper, we study the virtue of quantum entangled systems in the classification tasks and propose a novel classification algorithm called **Quantum Entanglement inspired the Classification Algorithm (QECA)** to learn the statistical correlation between the features and categories. Particularly, base on the Multi-layer Perceptron (MLP), we develop a novel QC-induced classification framework. The framework uses a fully connected layer to learn the parameters of observations of the subsystems, and then uses the weighted sums to integrate the measured probability values of each entangled state. In short, it can be understood that the hidden layer neurons (nodes) of the MLP are replaced with a measurement process of entangled states. This replacement makes QECA has the ability to learn the QC between features and categories during training process. We validate the effectiveness of proposed QECA on four machine learning datasets, and the experimental results show that QECA not only significantly outperforms the baseline method MLP, but also achieves the best performance than the other comparing methods in most cases.

The contribution of this paper is to apply QC revealed by quantum entanglement into traditional classification tasks of machine learning and leverage it to develop a novel QC-induced classification algorithm. Moreover, this paper theoretically analyzes that the framework used has the ability to violate Bell inequality, which proves that the framework has the ability to reproduce QC. Finally, this paper experimentally verifies that QC learned by the framework is effective for classification tasks and combining QC into traditional classification frameworks can further boost the classification performance.

2 Theoretical Analysis and Verification by Bell Inequality

In quantum theory, when several particles interact, the properties of each particle will be integrated into the properties of the overall system, and the properties of each particle can only describe the properties of the overall system. This phenomenon is called Quantum Entanglement (QE). QE could also be defined as one multi-body quantum system in which tensor decomposition is not possible [13]. First, let us give the basic definition of entanglement for bipartite systems (namely, 2-qubit).

Definition 1. *Let $\mathcal{H}_1$ and $\mathcal{H}_2$ be two Hilbert spaces and $|\psi\rangle \in \mathcal{H}_1 \otimes \mathcal{H}_2$*[1]*. Then $|\psi\rangle$ is said to be disentangled, or separable or a product state if $|\psi\rangle = |\psi_1\rangle \otimes |\psi_2\rangle$, for some $|\psi_1\rangle \in \mathcal{H}_1$ and $|\psi_2\rangle \in \mathcal{H}_2$. Otherwise, $|\psi\rangle$ is said to be entangled.*

We begin with an arbitrary bipartite entangled state in the bases $\sigma_3|\pm\rangle = \pm|\pm\rangle$[2] that

$$|\psi\rangle = \alpha|+-\rangle + \beta|-+\rangle \tag{1}$$

where α and β are the normalization condition with $|\alpha|^2 + |\beta|^2 = 1$ but $\alpha, \beta \neq 0$. Without losing generality, α and β can be parameterized as $\alpha = e^{i\eta}\sin(\xi)$, $\beta = e^{-i\eta}\cos(\xi)$, where i is the imaginary number with $i^2 = -1$ and η, ξ are two real parameters but $\sin(\xi), \cos(\xi) \neq 0$. The density matrix of the entangled state, $\rho = |\psi\rangle\langle\psi|$, can be separated to the local and non-local parts [15], $\rho = \rho_{lc} + \rho_{nlc}$. The local part

$$\rho_{lc} = \sin^2(\xi)|+-\rangle\langle+-| + \cos^2(\xi)|-+\rangle\langle-+|, \tag{2}$$

describes the classic statistical correlation between subsystems (or properties), which belongs to the classical statistics. The non-local part

$$\rho_{nlc} = \sin(\xi)\cos(\xi)\left(e^{i2\eta}|+-\rangle\langle-+| + e^{-i2\eta}|-+\rangle\langle+-|\right) \tag{3}$$

describes the phenomenon of interference between subsystems (or properties), which belongs to the quantum statistics.

2.1 The Measurement on Density Matrix

The observable of the subsystem r of the bipartite entangled system, say a and b, is defined as:

$$M_r = \begin{bmatrix} \cos(\theta_r) & e^{-i\phi_r}\sin(\theta_r) \\ e^{i\phi_r}\sin(\theta_r) & -\cos(\theta_r) \end{bmatrix} \tag{4}$$

where θ and ϕ are two arbitrary real parameters and $r \in \{a, b\}$. The observable has a spectral decomposition, $M_r = \sum_m m P_r^m$, where P_r^m is the projector onto

[1] The widely used Dirac notations are used in this paper, in which a unit vector $\boldsymbol{v}$ and its transpose $\boldsymbol{v}^T$ are denoted as a ket $|v\rangle$ and a bra $\langle v|$, respectively. $\otimes$ denotes the tensor product.

[2] $\{|+\rangle, |-\rangle\}$ denotes an arbitrary orthonormal basis of the 1-qubit Hilbert space $\mathbb{C}^2$. $\sigma_3 = \sigma_z$ denotes Pauli matrix, and Pauli matrix refers to four common matrices, which are 2×2 matrix, each with its own mark, namely $\sigma_x \equiv \sigma_1 \equiv X$, $\sigma_y \equiv \sigma_2 \equiv Y$, $\sigma_z \equiv \sigma_3 \equiv Z$ and $\sigma_0 \equiv I$.

the eigenspace of M_r with eigenvalue m. The possible outcomes of the measurement correspond to the eigenvalues, m, of the observable. Upon measuring the state $|\varphi\rangle$, the probability of getting result m is given by

$$p(m_r) = Tr[P_r^m(|\varphi\rangle\langle\varphi|)] = \langle\varphi|P_r^m|\varphi\rangle \tag{5}$$

where Tr denotes the trace of the matrix.

Projective measurements have many nice properties. In particular, it is very easy to calculate average values for projective measurements. By definition, the average value of the measurement is

$$E(M) = \sum_m mp(m) = \sum_m m\langle\varphi|P_m|\varphi\rangle = \langle\varphi|M|\varphi\rangle. \tag{6}$$

The average value of the observable M is often written $\langle M\rangle \equiv \langle\varphi|M|\varphi\rangle$.

Therefore, the joint probability derived from QE is obtained as:

$$p(+_a, +_b) = Tr[(P_a^+ \otimes P_b^+)\rho]. \tag{7}$$

It can be also divided into the local (classical probability) and non-local (quantum probability) parts

$$p_{all}(+_a, +_b) = Tr[(P_a^+ \otimes P_b^+)(\rho_{lc} + \rho_{nlc})] \tag{8}$$

$$= Tr[(P_a^+ \otimes P_b^+)\rho_{lc}] + Tr[(P_a^+ \otimes P_b^+)\rho_{nlc}] \tag{9}$$

$$= p_{lc}(+_a, +_b) + p_{nlc}(+_a, +_b). \tag{10}$$

Accordingly, the probability of other combinations, i.e., $p_{lc}(\pm_a, \pm_b)$, $p_{lc}(\mp_a, \pm_b)$, $p_{nlc}(\pm_a, \pm_b)$ and $p_{nlc}(\mp_a, \pm_b)$, can also be obtained. Moreover, the average values of a and b in the classical and quantum cases are

$$\langle ab\rangle_{lc} = -\cos(\theta_a)\cos(\theta_b) \tag{11}$$

and

$$\langle ab\rangle_{nlc} = \sin(\theta_a)\sin(\theta_b)\sin(2\xi)\cos(\phi_a - \phi_b + 2\eta), \tag{12}$$

respectively.

2.2 Verification by Bell Inequality

The theoretical tool for verifying QE is the Bell inequality [2]. Violating (or Destroying) Bell inequality is a sufficient condition for the existence of QE. The Bell inequality has many well-known promotion forms, the first and simple of which is the Clauser-Horne-Shimony-Holt (CHSH) inequality [11]. The form of the CHSH inequality is simpler and more symmetrical than many other Bell inequalities that are later proposed. The specific form of the CHSH inequality is

$$|E(Q, S) + E(R, S) + E(R, T) - E(Q, T)| \leq 2 \tag{13}$$

where E denotes the average value and Q, R, S and T denote observable.

The main conclusions and their proofs are given below:

Conclusion 1. *The local part of the joint probability derived from QE satisfies the CHSH inequality, which belongs to the local hidden variables theory.*

Proof. Let $E(a,b) = \langle ab \rangle_{lc}$, i.e. Equation (11), the simple formula transformation and the absolute value inequality can prove that the CHSH inequality holds, i.e.

$$|\langle QS \rangle_{lc} + \langle RS \rangle_{lc} + \langle RT \rangle_{lc} - \langle QT \rangle_{lc}| \leq 2. \tag{14}$$

It indicates that p_{lc} is a classical probability.

Conclusion 2. *The non-local part (quantum interference term) of the joint probability derived from QE does not satisfy the CHSH inequality, which belongs to the quantum mechanics theory.*

Proof. Let $E(a,b) = \langle ab \rangle_{nlc}$, i.e. Equation (12), we use a counterexample to prove that the non-local part can violate the CHSH inequality. For example, when $\theta_Q = \theta_R = \theta_S = \theta_T = \frac{\pi}{2}$, $\phi_Q = \frac{\pi}{3}$, $\phi_R = \phi_S = \frac{\pi}{6}$, $\phi_T = 0$, $\xi = \frac{\pi}{4}$ and $\eta = 0$, then

$$|\langle QS \rangle_{nlc} + \langle RS \rangle_{nlc} + \langle RT \rangle_{nlc} - \langle QT \rangle_{nlc}| \approx 2.232 \nleq 2. \tag{15}$$

It indicates that p_{nlc} is a quantum probability.

Conclusion 3. *The joint probability derived from QE does not satisfy the CHSH inequality, which belongs to the quantum mechanics theory.*

Proof. Let $E(a,b) = \langle ab \rangle_{all} = \langle ab \rangle_{lc} + \langle ab \rangle_{nlc}$, the CHSH inequality can also be violated. For example, when $\theta_Q = 0$, $\theta_R = \frac{\pi}{2}$, $\theta_S = \frac{5\pi}{4}$, $\theta_T = \frac{7\pi}{4}$, $\phi_Q = \phi_R = \phi_S = \phi_T = 0$, $\xi = \frac{\pi}{4}$ and $\eta = 0$, then

$$|\langle QS \rangle_{all} + \langle RS \rangle_{all} + \langle RT \rangle_{all} - \langle QT \rangle_{all}| = 2\sqrt{2} \nleq 2. \tag{16}$$

It indicates that $p_{all} = p_{lc} + p_{nlc}$ is a quantum probability.

2.3 Analysis

Almost all books on quantum mechanics have discussions about the double-slit experiment, that is, electrons passing through two open slits. See also Ref. [13]. Let A_k denote an event of passing through the slit with label k, here $k = 1, 2$. Interpretation of the results of this experiment has led to the following formula for the probability:

$$p(A_1 \cup A_2) = p(A_1) + p(A_2) + 2\sqrt{p(A_1)p(A_2)}\cos(\theta) \tag{17}$$

where p is a symbol of probability and θ is a certain parameter. Generally, $2\sqrt{p(A_1)p(A_2)}\cos(\theta)$ is interpreted as the self-interference inherent to the wave nature of an electron. It will be convenient to give another form to Eq. (17). Set

$C = A_1 \cup A_2$ where $A_1 \cap A_2 = \emptyset$ and rewrite Eq. (17) as a nonclassical (quantum) total probability formula:

$$p(C) = p(C|A_1)p(A_1) + p(C|A_2)p(A_2) \tag{18}$$

$$+ 2\sqrt{p(C|A_1)p(A_1)p(C|A_2)p(A_2)}\cos(\theta). \tag{19}$$

where, as usual, $p(C|A_i) = p(CA_k)/p(A_k)$ and $p(A_k) > 0$, $k = 1, 2$.

Based on the quantum total probability formula, a natural judgment can be drawn that the quantum effect can be described as a quantum interference term for classical probability. In this paper, we decompose the quantum joint probability derived from QE into the classical probability and the quantum interference term, that is, we present the specific form of QC (or called strong statistical correlation) revealed by QE and the way it works, and use the CHSH inequality to verify its correctness. In the following, we will experimentally verify the role of this interference term in classical tasks.

3 Implement Classification Algorithm by the Framework

From the analysis of the previous theoretical section, we can get the following cognition: The essential reason that QC revealed by QE can be stronger than the classical correlation is that the quantum interference term described by the phase information is added. In this section, we will construct a classification algorithm based on the mathematical formalization of QE to verify the validity of QC revealed by QE in classification tasks.

This section is organized as follows: First, we will describe how to calculate the quantum joint probability between features and categories. Second, we describe how to use a fully connected layer to learn the parameters in the subsystems of an entangled system, that is, how to construct QECA. Formally, it can be understood as replacing the output layer of the MLP with the measurement operation of the entangled state. Finally, the learning method of parameters in the model is given.

3.1 Calculate Joint Probability Between Features and Categories

Entanglement arises from the measurement process of entangled systems (states), that is, obtaining the quantum joint probability not only requires entangled systems, but also requires to define the observables of the entangled systems.

We choose the quantum system with the maximum entanglement under two qubits as the entangled system, e.g., Bell states, and its form is

$$|\Psi\rangle = \frac{1}{\sqrt{2}}(|0\rangle \otimes |0\rangle + |1\rangle \otimes |1\rangle) = \frac{1}{\sqrt{2}}(|00\rangle + |11\rangle). \tag{20}$$

The reason for choosing the entangled system of two qubits is that we want to describe one qubit as the attribute (feature) and the other as the label (category). It can be seen that if there are N attributes in each instance (sample), N Bell states are needed. Since there are only two eigenvalues in each set of orthogonal bases of a qubit, it is suitable for binary classification tasks. Of course, one can achieve multi-class classification tasks by adding the number of qubits of the label, but this is not the focus of this paper. Moreover, from the form of the quantum interference term, the probability amplitude and phase information of the entangled system can be fully reflected by the polar and azimuth angles of the measurement operator. In order to reduce the number of parameters of QECA, we choose the maximum entangled state, i.e., Bell states, to represent the entangled system.

We define the observable of the subsystem of the entangled system as Eq. (4), and the spectral decomposition of the observable is

$$M_r(\theta_r, \phi_r) = P_r^+(\theta_r, \phi_r) - P_r^-(\theta_r, \phi_r) = |+_r\rangle\langle +_r| - |-_r\rangle\langle -_r| \tag{21}$$

with

$$|+_r\rangle = \cos\left(\frac{\theta_r}{2}\right)|0\rangle + e^{i\phi_r}\sin\left(\frac{\theta_r}{2}\right)|1\rangle \tag{22}$$

$$|-_r\rangle = \sin\left(\frac{\theta_r}{2}\right)|0\rangle - e^{i\phi_r}\cos\left(\frac{\theta_r}{2}\right)|1\rangle \tag{23}$$

where the polar and azimuth angles, θ_r and ϕ_r, are the arbitrary real parameters. For the measurement operator of the attribute, P_{att}, we use ϕ_r to represent the parameter value of the attribute, and θ_r to represent the degree of freedom of the attribute, e.g., weight. For the measurement operator of the label, P_{lab}, we use the determined measurement operator to represent the label, e.g., $\theta_{lab} = \frac{\pi}{2}$ and $\phi_{lab} = 0$,

$$P_{lab}^+ = \frac{1}{2}(\sigma_1 + \sigma_0), \quad P_{lab}^- = \frac{1}{2}(\sigma_1 - \sigma_0). \tag{24}$$

In fact, any set of eigenstates can be chosen to represent the label, only to satisfy the orthogonality. The reason why we select a set of orthogonal bases to represent the label is that the positive and negative examples (samples) of the two-class classification task are (often) binary opposition.

Now we can formally define the measurement operator of the entangled system, i.e., Eq. (20). Assuming that each instance (sample) has N attributes and one label, the positive and the negative measurement operators for the entangled system consists of the n-th attribute and the label are

$$\mathcal{P}_n^\pm(\theta_n, \phi_n) = P_n^+(\theta_n, \phi_n) \otimes P_{lab}^\pm. \tag{25}$$

P_n^+ can also be replaced by P_n^-, the effect is the same. Applying $\mathcal{P}_n^+$ and $\mathcal{P}_n^-$ separately to each entangled system, i.e., Eq. (20), the probability values of both positive and negative examples will be obtained,

$$p_n^\pm(\theta_n, \phi_n) = Tr[\mathcal{P}_n^\pm(\theta_n, \phi_n)(|\Psi\rangle\langle\Psi|)] = \langle\Psi|\mathcal{P}_n^\pm(\theta_n, \phi_n)|\Psi\rangle. \tag{26}$$

Based on this formal framework, we can calculate the quantum joint probability between the label and any attribute, and then construct QECA.

3.2 Constructing QECA by Quantum Joint Probability

In QECA, we use a fully connected layer to learn the parameters of the observations of the entanglement system. Formally, it can be understood as replacing the output layer of the MLP with a measurement operation of the entangled system. To make it easier for readers to understand the structure of QECA, we use the illustrated method to give the architecture of QECA, see Fig. 1.

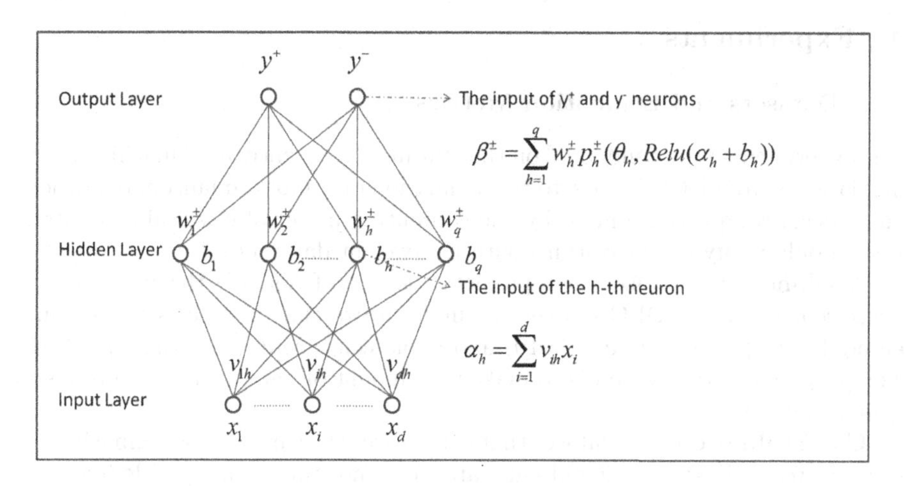

Fig. 1. Schematic diagram of QECA.

We perform weighting summation on the attributes of each instance, $x \in \mathbb{R}^d$, to get the input of the hidden layer neurons, which is $\alpha_h = \sum_{i=1}^{d} v_{ih}x_i$ where $v_h \in \mathbb{R}^d$ represents the weight. α_h plus the bias $b_h \in \mathbb{R}$, and then apply the activation function *ReLU* (Rectified Linear Unit) [10] to get the parameters of the measurement operator of the entangled system, which is

$$\phi_h = ReLU(\alpha_h + b_h). \tag{27}$$

Together with the defined degrees of freedom, $\theta_h \in \mathbb{R}$, the measurement operator of the entangled system can be obtained, which is $\mathcal{P}_h^{\pm}(\theta_h, \phi_h)$, i.e., Eq. (25). By applying this measurement operator to the entangled state, i.e., Eq. (20), the joint probability value of the entangled state, i.e., Eq. (26), can be obtained.

Finally, perform weighting summation on $p_h^{\pm}(\theta_h, \phi_h)$ to get the final output value

$$y^{\pm} = \beta^{\pm} = \sum_{h=1}^{q} w_h^{\pm} p_h^{\pm}(\theta_h, \phi_h) \tag{28}$$

where $w^{\pm} \in \mathbb{R}^q$ represents the weight. $\beta^{\pm}$ represent the input value of the output layer neurons, as shown in Fig. 1.

3.3 Parameter Learning

Machine learning uses the loss function to improve the performance of the model. This process of improvement is called optimization. QECA uses the classical cross-entropy loss function to act on its loss function. Since Adaptive Moment Estimation (Adam) defines a clear range of learning rates per iteration, making the parameters change smoothly, we use Adam as the optimizer for QECA.

4 Experiments

4.1 Datasets and Evaluation Metrics

The experiments were conducted on the four most frequently used machine learning datasets from UCI [5]. Due to the simulation of complex quantum operations on classical computers, limited by the computing power of classical computers, we can only verify our algorithms with lightweight datasets.

Abalone[3] is a dataset that predicts the age of abalone through physical measurements. Since QECA is verified under a two-class task, it is divided into an adult group (covering $age \geq 10$) and adolescent group (covering $age < 10$). The purpose of this division is to make the amount of data in the two groups as close as possible.

Car Evaluation[4] is a dataset that categorizes the car by a few simple indicators. We reclassified the original four categories into two, unacceptable (covering *unacc*) and acceptable (covering *acc*, *good* and *vgood*).

Wine Quality[5] is a dataset that scores on wine quality. We divide the scores less than or equal to 5 into one class, and the others into another.

Breast Cancer[6] is a dataset that is diagnosed by the patient's physiological indicators, which is a two-class dataset.

All experiments use the 5-fold Cross-Validation method to divide the training set and test set. The experimental evaluation metrics, *F1-score*, ACC (Accuracy) and AUC (Area Under Curve), are taken as the average of 5 results.

4.2 Compared with Classical Classification Algorithms

Baselines: QECA is built on the basis of the standard MLP. Compared with the MLP of the same structure and setting, it can truly reflect the superiority of QECA. Both QECA and MLP uses an architecture of a single hidden layer, in order to compare them in a fair manner (or less interference). The number of neurons in the input layer is equal to the number of attributes; the number of neurons in the hidden layer is twice the number of input layers; because it is a binary classification task, the number of neurons in the output layer is two.

[3] http://archive.ics.uci.edu/ml/datasets/Abalone.
[4] http://archive.ics.uci.edu/ml/datasets/Car+Evaluation.
[5] http://archive.ics.uci.edu/ml/datasets/Wine+Quality.
[6] http://archive.ics.uci.edu/ml/datasets/breast+cancer+wisconsin+(original).

Both use the cross-entropy loss function to evaluate the model and the optimizer *Adam* to optimize the parameters.

We also conduct a comprehensive comparison across a wide range of machine learning algorithms, including Logistic Regressive (LR), Decision Tree (DT), Naive Bayesian Model (NBM), K-Nearest Neighbor (KNN), Support Vector Machine (SVM), Linear Discriminant Analysis (LDA), Quadratic Discriminant Analysis (QDA), Gradient Boosting Decision Tree (GBDT) and Ada Boosting Decision Tree (ABDT).

Parameter Settings: QECA has three hyper-parameters, which are *learning rate*, *mini-batch* and *training epoch*, respectively, and uses the same settings on all datasets: the *learning rate* is 0.0001, the *mini-batch* is 1 and the *training epoch* is 500. Their *weights* are initialized to a truncated positive distribution, and the *biases* to 0.01. The permutation and combination method is used to select the hyper-parameters.

The hyper-parameters in the baselines are set to: in LR, *penalty* is $L2$; in DT, *min-samples-split* is 10; in SVM, C is 1.0 and *kernel* is rbf; in KNN, *n-neighbors* is 10; in LDA, *solver* is *svd* and *store-covariance* is $True$; in QDA, *store-covariance* is $True$; in MLP, *activation* is *relu* and *solver* is *adam*; in GBDT, *n-estimators* is 20; in ABDT, *n-estimators* is 20. Other hyper-parameters not listed use the default value of the framework scikit-learn[7].

Experiment Results: Inspired by the quantum double-slit experiment, we also use the quantum interference term to characterize the strong statistical correlation revealed by QE and design an algorithm to verify the role of the quantum interference term in the classification task. Table 1 presents the experiment results under Abalone, Breast Cancer, Wine Quality (Red) and Car Evaluation respectively, where bold values are the best performances out of all algorithms. From the experimental results, the most metrics of QECA on four datasets are significantly better than the majority of machine learning algorithms. The basic conclusion that QECA has excellent classification ability can be drawn. This proves the effectiveness of QECA from a holistic perspective.

Moreover, the comparison with the MLP can explain that QECA is improved on the basis of MLP, and it shows that the quantum interference term plays an important role on QECA, that is, the quantum interference term described by the quantum phase has learned a strong statistical correlation between attributes and labels. Below we will analyze the entire learning process to determine whether the learning (or classification) ability is stable rather than accidental.

4.3 Comparison with the Training Process of Standard MLP

In order to analyze QECA's learning ability in more detail, we compared the training process of QECA with the baseline method MLP. We use the validation

[7] https://scikit-learn.org/stable/index.html.

Table 1. Experiment results: the best-performed values for each dataset are in bold.

Dataset	Abalone			WQ (Red)			Car evaluation			Breast cancer		
Algorithm	F1-score	ACC	AUC	F1-score	ACC	AUC	F1-score	ACC	AUC	F1-score	ACC	AUC
LR	0.7704	0.7708	0.7708	0.7559	0.7410	0.7404	0.7582	0.8611	0.8224	0.9494	0.9648	0.9594
DT	0.7153	0.7170	0.7170	0.7340	0.7191	0.7187	0.9626	0.9774	0.9756	0.8957	0.9298	0.9151
NBM	0.7387	0.7345	0.7345	0.7451	0.7298	0.7292	0.8519	0.9062	0.9043	0.9450	0.9604	0.9628
KNN	0.7791	0.7842	0.7842	0.6682	0.6447	0.6428	0.9422	0.9670	0.9477	0.9493	0.9648	0.9593
SVM	0.7678	0.7541	0.7543	0.7299	0.7110	0.7095	0.9565	0.9733	0.9732	0.9444	0.9589	0.9655
LDA	0.7443	0.7603	0.7601	0.7566	0.7273	0.7224	0.7698	0.8657	0.8323	0.9422	0.9605	0.9513
QDA	0.7390	0.7560	0.7558	0.7532	0.7204	0.7146	0.8500	0.9161	0.8816	0.9344	0.9516	0.9570
MLP	0.7896	0.7864	0.7865	0.7361	0.7373	0.7412	0.9384	0.9629	0.9564	0.8980	0.9280	0.9214
GBDT	0.7880	0.7859	0.7860	0.7609	0.7467	0.7460	0.9316	0.9571	0.9616	0.9279	0.9471	0.9477
ABDT	0.7884	0.7842	0.7843	0.7536	0.7335	0.7312	0.9155	0.9490	0.9409	0.9301	0.9517	0.9445
QECA	**0.8018**	**0.8027**	**0.8027**	**0.7633**	**0.7536**	**0.7544**	**0.9759**	**0.9855**	**0.9841**	**0.9630**	**0.9736**	**0.9758**
over MLP	1.54%↑	2.07%↑	2.05%↑	3.69%↑	2.21%↑	1.78%↑	3.99%↑	2.34%↑	2.89%↑	7.23%↑	4.91%↑	5.90%↑

set divided by the 5-fold Cross-Validation method as the test set to obtain the accuracy curve during the training process. The hyper-parameter settings of both QECA and MLP are exactly the same as those in Experiment Sect. 4.2. We selected three representative datasets for experiments: Wine Quality (Red) and Wine Quality (White) have the same data structure, but the amount of data in Wine Quality (Red) is balanced and Wine Quality (White) is not balanced; Moreover, in order to illustrate the effect of the number of attributes on the training effect, we use Abalone to compare with Wine Quality.

The experiment results are shown in Fig. 2. From the accuracy curve of the training process under the three datasets, compared with the MLP, QECA has significant improvement and its contribution is obvious. The experimental results of this section can prove that the quantum interference term plays an important role in QECA. It also further shows that QC revealed by QE can play an important role in the classic classification task.

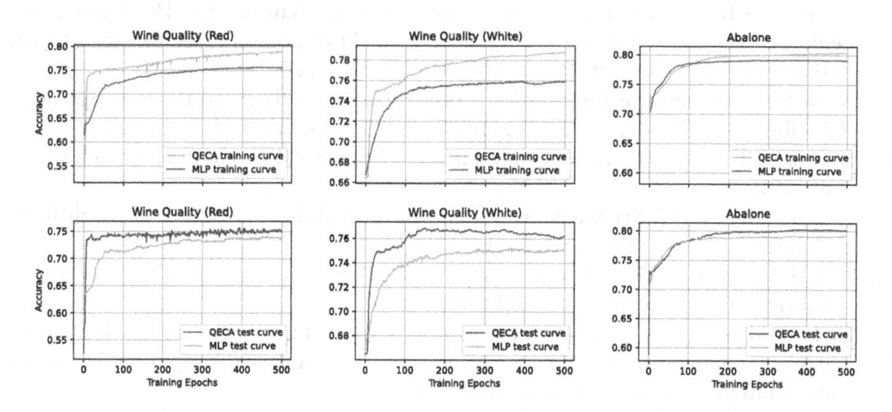

Fig. 2. Experiment results: the left column is the accuracy curve on the training set, and the right column on the test set.

5 Conclusion and Future Work

In this paper, we propose a novel classification framework, called Quantum Entanglement inspired the Classification Algorithm (QECA), to learn a strong statistical correlation (i.e., QC) between features and categories and leverage it to improve the classification performance by integrating QC into MLP. QECA achieved excellent results on the four machine learning datasets compared with the baseline method MLP, which only uses the statistical correlation described by classical theory. More importantly, QECA also outperforms the other competitive methods in most metric. These results prove the effectiveness of QC in classification tasks.

References

1. Bai, Y.K., Yang, D., Wang, Z.: Multipartite quantum correlation and entanglement in four-qubit pure states. Phys. Rev. A **76**(2), 022336 (2007)
2. Bell, J.S.: On the einstein podolsky rosen paradox. Phys. Physique Fizika **1**(3), 195 (1964)
3. Chiribella, G., Spekkens, R.W.: Quantum Theory: Informational Foundations and Foils. Springer, Dordrecht (2016). https://doi.org/10.1007/978-94-017-7303-4
4. Dowling, J.P., Milburn, G.J.: Quantum technology: the second quantum revolution. Philos. Trans. R. Soc. Lond. Ser. A Math. Phys. Eng. Sci. **361**(1809), 1655–1674 (2003)
5. Dua, D., Graff, C.: UCI machine learning repository (2017). http://archive.ics.uci.edu/ml
6. Hall, M.A.: Correlation-based feature selection for machine learning (1999)
7. Hardy, L.: Quantum theory from five reasonable axioms. arXiv preprint quant-ph/0101012 (2001)
8. Kochen, S.B.: A reconstruction of quantum mechanics. In: Bertlmann, R., Zeilinger, A. (eds.) Quantum [Un]Speakables II. TFC, pp. 201–235. Springer, Cham (2017). https://doi.org/10.1007/978-3-319-38987-5_12
9. Hayashi, M.: Quantum Information Theory. Graduate Texts in Physics, Springer, Heidelberg (2017). https://doi.org/10.1007/978-3-662-49725-8
10. Hinton, G.E.: Rectified linear units improve restricted Boltzmann machines Vinod Nair (2010)
11. Khrennikov, A.: CHSH inequality: quantum probabilities as classical conditional probabilities. Found. Phys. **45**(7), 1–15 (2014)
12. Kok, P., Lovett, B.W.: Introduction to Optical Quantum Information Processing. Cambridge University Press, Cambridge (2010)
13. Nielsen, M.A., Chuang, I.: Quantum computation and quantum information (2002)
14. Wittek, P.: Quantum Machine Learning: What Quantum Computing Means to Data Mining. Academic Press, Cambridge (2014)
15. Zhang, H., Wang, J., Song, Z., Liang, J.Q., Wei, L.F.: Spin-parity effect in violation of bell's inequalities for entangled states of parallel polarization. Mod. Phys. Lett. B **31**(04), 1750032 (2017)
16. Zhang, J., et al.: Quantum correlation revealed by bell state for classification tasks. In: 2021 International Joint Conference on Neural Networks (IJCNN), pp. 1–8 (2021)
17. Zhang, J., Hou, Y., Li, Z., Zhang, L., Chen, X.: Strong statistical correlation revealed by quantum entanglement for supervised learning. In: ECAI (2020)
18. Zhang, J., Li, Z.: Quantum contextuality for training neural networks. Chin. J. Electron. **29**, 1178–1184 (2020)
19. Zhang, J., Li, Z., He, R., Zhang, J., Wang, B., Li, Z., Niu, T.: Interactive quantum classifier inspired by quantum open system theory. In: 2021 International Joint Conference on Neural Networks (IJCNN), pp. 1–7 (2021)

Self-paced Safe Co-training for Regression

Fan Min[1,2](✉), Yu Li[1], and Liyan Liu[1]

[1] School of Computer Science, Southwest Petroleum University, Chengdu, China
minfan@swpu.edu.cn, 201921000431@stu.swpu.edu.cn
[2] Institute for Artificial Intelligence, Southwest Petroleum University, Chengdu 610500, China

Abstract. In semi-supervised learning, co-training is successfully in augmenting the training data with predicted pseudo-labels. With two independently trained regressors, a co-trainer iteratively exchanges their selected instances coupled with pseudo-labels. However, some low-quality pseudo-labels may significantly decrease the prediction accuracy. In this paper, we propose a self-paced safe co-training for regression (SPOR) algorithm to enrich the training data with unlabeled instances and their pseudo-labels. First, a safe mechanism is designed to enhance the quality of pseudo-labels without side effects. Second, a self-paced learning technique is designed to select "easy" instances in the current situation. Third, a "qualifier-based" treatment is designed to remove "weak" instances selected in previous rounds. Experiments were undertaken on nine benchmark datasets. The results show that SPOR is superior to both popular co-training regression methods and state-of-the-art semi-supervised regressors.

Keywords: Co-training · Self-paced learning · Semi-supervised regression · Safe learning

1 Introduction

Semi-supervised regression (SSR) [26] aims to use additional unlabeled data to improve learning performance. However, some unlabeled data can help build the model, while others cannot [1,17]. Therefore, the quality of both added unlabeled data and assigned pseudo-labels is critical to learning performance. Co-training [6] is a well-known form of semi-supervised learning. To efficiently select unlabeled data and assign pseudo-labels, co-training trains two different classifiers and changes their pseudo-labels in an iterative manner. Co-training is also especially effective in dealing with regression task. Different from the classifications tasks, the co-training regression method [25] selects suitable unlabeled data based on the improvement of labeled data's quality after adding the unlabeled data. In this way, more hidden insights in unlabeled data can be distinguished through two different regressors.

Despite the advantages of co-training in SSR, popular methods still have major drawback. That is, they assume that unlabeled instances with high-confidence

J. Gama et al. (Eds.): PAKDD 2022, LNAI 13281, pp. 71–82, 2022.
https://doi.org/10.1007/978-3-031-05936-0_6

are credible for the model. For co-training, most methods assume that unlabeled instances with high-confidence are credible for the model. However, this assumption is too subjective to be fully satisfied in applications. In fact, the regressors may not be robust enough in the initial selection process. They may choose weak unlabeled instances and assign pseudo-labels to them. These weak instances and pseudo-labels cannot improve learning performance and will remain in the labeled data throughout the whole process. Consequently, it is highly desirable to reduce the impact of weak instances and labels in the co-training process.

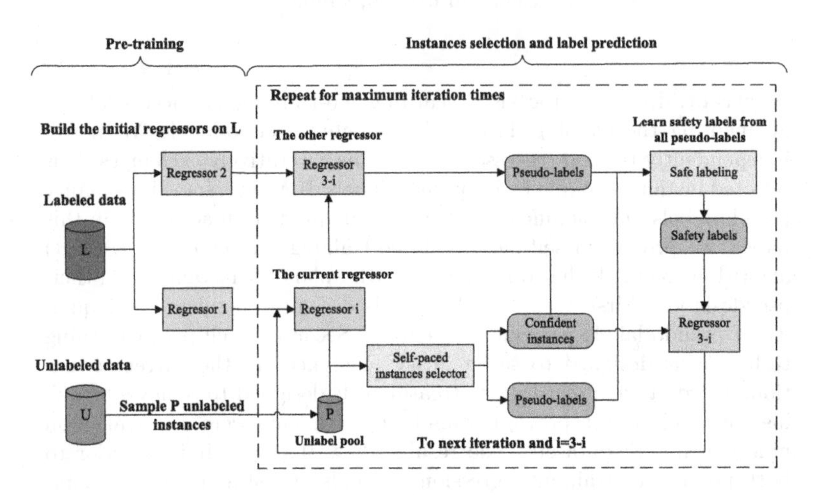

Fig. 1. The algorithm framework. 1) In the pre-training process, SPOR uses labeled data L to construct 2 base regressors. 2) Each time we select a confident instance using Regressor i from the unlabeled pool. This instance is predicted by all regressors. All these predictions are fused using a labeling safe mechanism to obtain the pseudo-label. The selected instance coupled with the pseudo-label augment the training data of the next regressor. Because there are 2 regressors and T iterations, this process repeats $2 \times T$ times.

In this paper, we propose a self-paced safe co-training for regression (SPOR) algorithm to handle the above issue. Figure 1 shows the main framework of SPOR. First, the two base regressors are trained on labeled data only. Then, these regressors will select unlabeled instances for each other among the sampled unlabeled data. In this step, the instances selected in the previous process will be re-evaluated. Once the previously added instances are weaker than the current selection, these instances will be removed from the labeled data. Finally, the pseudo-labels of all regressors will be merged to learn the safety label. This learned safety label will not be worse than the pseudo-label. The safety labels will be added to the labeled data of the next regressor along with the selected instances to help build the model.

Our main contributions can be summarized as follows:

- A safe co-regression mechanism is designed to improve the pseudo-label quality of unlabeled instances. This mechanism provides a label enhancement, which can improve label quality without side effects.
- A self-paced learning paradigm is inserted into the selection process to mine more appropriate unlabeled instances. This paradigm corresponds with the insights of self-paced implementation and selects unlabeled instances with an "easy-to-hard" way.
- An adaptive and "qualifier-based" treatment is adopted to reduce the impact of weak instances. This treatment corresponds to the improvement of co-training, which can remove the weak unlabeled instances and keeping the confident ones in labeled data.

2 Related Work

This section introduces the three related studies of SPOR, including self-paced learning, co-training regression and safe learning.

Self-paced learning (SPL) [13] is a general methodology for problem-solving especially in data mining and machine learning. It is rooted in the curriculum learning model [5], which learns the model by including samples from easy to complex in the training process. Due to its versatility, SPL is widely used in various tasks, e.g., object tracking [20], image classification [12] and multimedia event detection [9,11]. SPL also has been connected with other theories to form new ones, such as self-paced curriculum learning [12], self-paced co-training [18,19].

The Co-training [6] method has been proposed to ameliorate the shortage of self-training. It estimates unlabeled instances through two different base models that are trained from different perspectives of labeled instances. Co-training regression can be roughly spirited as multi-view [4,22,23] and single-view [10, 25] paradigm. Besides, Co-Expectation Maximization Algorithm (Co-EM) [7] extends the co-training regime on the data sets without natural feature split. Balcan et al. [2] proposed "expansion assumption", which reduced the strong assumption requirements of co-training.

The safe mechanism [24] is a special mechanism of semi-supervised learning (SSL) which concerns the problem that the exploitation of unlabeled data might hurt learning performance. Its goal is to improve learning performance without the negative consequences of adding more unlabeled data. The safe scheme is related to two branches of studies, safe classification [15,17] and safe regression [16]. Kwok et al. [14] established a general safe classification framework for different performance metrics such as F1 scores, AUC, and Top-k. In addition, Balsubramani et al. [3] explored a robust prediction with the highest accuracy based on the supervised prediction restricted to a specific candidate set.

3 The Proposed Algorithm

In this section, we introduce the details of the proposed method, self-paced co-training for regression (SPOR). We first present the mathematical form of

SPOR, and then introduce the designed safe regression mechanism. Finally, we will introduce the details of the algorithm.

3.1 The Model

This section presents the mathematical details of SPOR, which extends the self-paced safe co-training framework for regression. Let l denote the number of labeled instances and u the number of unlabeled instances, $\mathbf{V} = \{\mathbf{v}^{(1)}, \mathbf{v}^{(2)}\}$ be the set of selection vectors. The selection vector of j-th regressor is denoted by $\mathbf{v}^{(j)} = (v_{l+1}^{(j)}, \ldots, v_{l+u}^{(j)})$ where $v_{l+i}^{(j)} = 1$ if $\mathbf{x}_{l+i}$ is selected and 0 otherwise. Let $\mathbf{G} = \{g_1, g_2\}$ be the set of regressors, g_j be the j-th learner where $j = \{1, 2\}$, $g_j(\mathbf{x}_i)$ be the prediction of $\mathbf{x}_i$, $\overline{g}_j(\mathbf{x}_i)$ be the safety label of $\mathbf{x}_i$, $\mathcal{L}(y_i, g_j(\mathbf{x}_i))$ be the loss function. Our optimization objective is

$$\min_{\mathbf{V},\mathbf{G}} E = \sum_{j=1}^{2}\sum_{i=1}^{l} \mathcal{L}(y_i, g_j(\mathbf{x}_i)) + \sum_{j=1}^{2}\sum_{i=l+1}^{l+u} v_i^{(j)} \left(\mathcal{L}\left(\overline{g}_j(\mathbf{x}_i), g_j(\mathbf{x}_i)\right) - \lambda^{(j)}\right), \quad (1)$$

where $\lambda^{(j)}$ is the age parameter which controls the size of selected unlabeled instances for j-th regressor in each training iteration.

From the viewpoint of self-paced learning, SPOR is a self-paced regression method with two semi-supervised regressors by committee. The inner hard self-paced regularizer $\lambda^{(j)} v_i^{(j)}$ encodes the robustness of regressor. With the training progress, the age parameters $\lambda^{(j)}$ will decrease, and the robustness of the regressor increases. More unlabeled instances can be learned according to the enhanced regressor. This is in line with the principle of SPL and the SSR method of using unlabeled instances to increase learning performance.

From the viewpoint of co-training, SPOR is a co-training regression method with adaptive confidence criteria. The maximum confidence level of the instances selected by the co-regressor varies in different selection processes. The confidence of the weak instances selected earlier may be lower than the current confidence criterion. By removing the weak instances, better performance can be obtained for labeled data. This finely suits the idea SPOR's idea of recorrecting the weak instances to improve learning performance.

3.2 The Safe Technique

Safe regression [16] aims to learn a safety pseudo-label $\overline{y}$ that performs no worse than its baseline. In SPOR, we use the confidence-based method to help select the appropriate instances. Let $g_j(\mathbf{x})$ denote the semi-supervised prediction of the j-th regressor on $\mathbf{x}$, where $j = 1, 2$, and $g_j^{(0)}(\mathbf{x})$ is the supervised prediction of $\mathbf{x}$, y denote the true label of $\mathbf{x}$. The performance of safety prediction $\overline{g}_j(\mathbf{x})$ can be measured by the difference of mean squared error against y

$$\max_{y \in \mathbb{R}} \left(\left(g_j^{(0)}(\mathbf{x}) - y\right)^2 - \left(\overline{g}_j(\mathbf{x}) - y\right)^2\right). \quad (2)$$

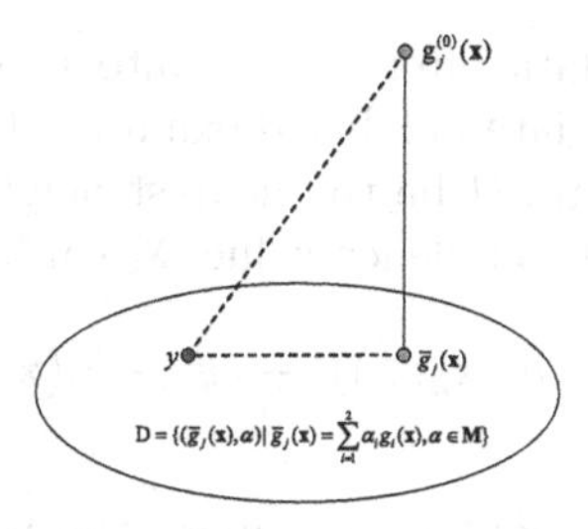

Fig. 2. The safe strategy is described through a geometric projection way. The safe strategy aims to learn a projection of $g_j^{(0)}(\mathbf{x})$ onto a label plane D. The projection label $\overline{g}_j(\mathbf{x})$ is significant closer than $g_j^{(0)}(\mathbf{x})$ to real label y.

However, in reality, the truth label y and the optimal weights α_i are hardly to obtain.

To simplify above problem, α_i can be assumed from a convex liner set $\mathbf{M} = \{(\alpha_1, \alpha_2) \mid \alpha_1 + \alpha_2 = 1, \alpha_i > 0, i = 1, 2\}$ and the worst-case performance gain in co-training process similarity as [3,17]

$$\max_{y\in\mathbb{R}} \min_{\alpha\in\mathrm{M}} \sum_{i=1}^{2} \alpha_i \left(\left(g_j^{(0)}(\mathbf{x}) - g_i(\mathbf{x}) \right)^2 - \left(\overline{g}_j(\mathbf{x}) - g_i(\mathbf{x}) \right)^2 \right). \tag{3}$$

Eq. (3) can gain a closed-form solution by setting the derivative w.r.t y to zero

$$\overline{g}_j(\mathbf{x}) = \sum_{i=1}^{2} \alpha_i g_i(\mathbf{x}). \tag{4}$$

Substituting the Eq. (4) to Eq. (3), the equivalent form only relates to α

$$\min_{\alpha\in\mathrm{M}} \left(\sum_{i=1}^{2} \alpha_i g_i(\mathbf{x}) - g_j^{(0)}(\mathbf{x}) \right)^2, \tag{5}$$

which can be viewed as a geometric projection problem as illustrated in Fig. 2. Expanding Eq. (5) and the problem can be solved as a simple convex quadratic program

$$\min_{\alpha\in\mathrm{M}} \left(\boldsymbol{\alpha}^{\mathrm{T}} \mathbf{F} \boldsymbol{\alpha} - \mathbf{g}^{\mathrm{T}} \boldsymbol{\alpha} \right), \tag{6}$$

where $\mathbf{F} \in \mathbb{R}^{2\times 2}$ is a kernel matrix of y_i, and $F_{ij} = y_i \cdot y_j$ for any $i, j \in \{1, 2\}$, $\mathbf{g} = [2g_1(\mathbf{x})g_j(\mathbf{x}), 2g_2(\mathbf{x})g_j(\mathbf{x})]$. It is effective to obtain the optimal weights α_i^* with the help of optimization solvers. After that, we can obtain the learned safe prediction $\overline{g}_j(\mathbf{x}) = \sum_{i=1}^{2} \alpha_i^* g_i(\mathbf{x})$ according to the Eq. (4).

3.3 The Instances Selection Strategy

In SPOR, our "qualifier-based" treatment combines instance confidence with a self-paced paradigm for instance selection. The instance confidence represents the promotion of labeled data after adding into labeled data. The self-paced paradigm can select appropriate unlabeled examples to add to the labeled data.

Let L be the labeled data, Ω be the nearest neighbor set of $\mathbf{x}_i$ in L, $\mathbf{x}_k$ be the k-th element of Ω_i. The confidence value Δ_i can be calculated according to

$$\Delta_i = \sum_{\mathbf{x}_k \in \Omega_i} \left(y_k - g_j(\mathbf{x}_k; L)\right)^2 - \left(y_k - g_j(\mathbf{x}_k; L \cup \{\mathbf{x}_i\})\right)^2. \tag{7}$$

Based on the calculated confidence values, the unlabeled instances can be selected according to

$$v_i = \begin{cases} 1, & \text{if } \Delta_i > \lambda \Delta_{\max}; \\ 0, & \text{otherwise,} \end{cases} \tag{8}$$

where v_i is the selection vector of $\mathbf{x}_i$, λ is the self-paced regularizer to control the number of selection instances, $\Delta_{\max}$ is the maximum confidence value of unlabeled instances.

The confidence value Δ_i represents the improvement of L after adding $\mathbf{x}_i$. The higher the Δ_i, the greater the value of $\mathbf{x}_i$ to the labeled data. In particular, when $\mathbf{x}_i$ is an outlier, Δ_i is 0. The self-paced regularizer λ indicates the robustness of regressors. As the training progresses, the robustness of the regressor increases, λ decreases, and more instances can be included in the labeled data. The change of λ actually reflects the increase in the robustness of the regressor.

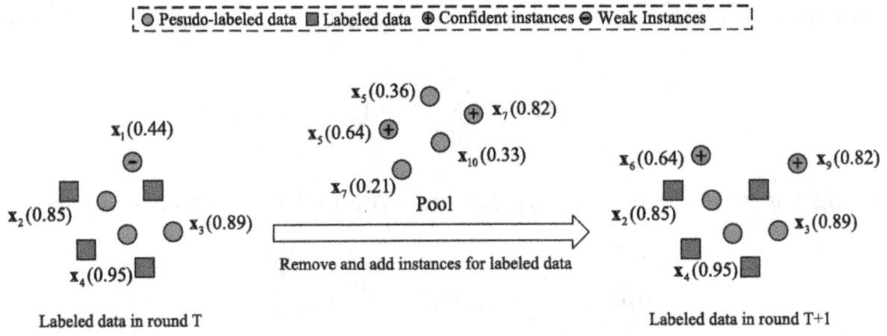

Fig. 3. The example of our "qualifier-based" treatment.

Figure 3 illustrates a simple example of our selection strategy. In round T, the self-paced regularizer is 0.7, and the confidence standard is 0.56. In unlabeled pool, $\mathbf{x}_5$ and $\mathbf{x}_7$ will be treated as confident instances and included in labeled data. In addition, in the label data, $\mathbf{x}_1$ will be considered as a weak instance to be removed, because its confidence is lower than the current standard. The updated labeled data will be used for the $T+1$ round.

3.4 Algorithm Description

Algorithm 1 presents the general framework of SPOR. The first step is to initialize the model parameters. Line 1 sets the $\mathbf{v}^{(1)}$ and $\mathbf{v}^{(2)}$ to 0, indicating that no unlabeled instances are selected. The $\lambda^{(1)}$ and $\lambda^{(2)}$ are both set to 1, indicating that only the highest-confidence unlabeled instance can be selected. Line 2 builds two supervised regressors g_1 and g_2 only on the original labeled data.

Algorithm 1. Self-paced Safe Co-training for Regression

Input: labeled data $\mathbf{x}_1, \ldots, \mathbf{x}_l$, unlabeled data $\mathbf{x}_{l+1}, \ldots, \mathbf{x}_{l+u}$, labels $y_1, \ldots, y_l$, age parameters $\lambda^{(1)}$ and $\lambda^{(2)}$, selection vectors $\mathbf{v}^{(1)}$ and $\mathbf{v}^{(2)}$, pool size P and max iteration number T.
Output: regressors g_1 and g_2.

1: Initialize $\lambda^{(1)}$, $\lambda^{(2)}$, $\mathbf{v}^{(1)}$, $\mathbf{v}^{(2)}$;
2: Train g_1 and g_2 on labeled data;
3: **while** (some instances can be selected || training iterations $< T$) **do**
4:   **for** ($j \leftarrow 1$ to 2) **do**
5:     Sample P unlabeled instances;
6:     Compute the confident value Δ_i by g_{3-j} according to Eq. (7);
7:     Update $\mathbf{v}^{(j)}$: Select confidenct instances $\mathbf{x}_i$ and update the selection vector $\mathbf{v}^{(j)}$ according to Eq. (8);
8:     Construct a linear kernel martrix $\mathbf{F}$ where $F_{mn} = g_m(\mathbf{x}_i) \cdot g_n(\mathbf{x}_i)$ where $m, n \in \{1, 2\}$;
9:     Derive a vector $\mathbf{g} = \left[2g_1(\mathbf{x}_i) \cdot g_j^{(0)}(\mathbf{x}_i), 2g_2(\mathbf{x}_i) \cdot g_j^{(0)}(\mathbf{x}_i)\right]$ where $g_j^{(0)}$ is the initial regressor of g_j;
10:     Solve the convex quadratic optimization Eq. (6) to gain the optimal weights $\boldsymbol{\alpha}_i = \left[\alpha_i^1, \alpha_i^2\right]$;
11:     Calculate $\overline{g}_j(\mathbf{x}_i) = \sum_{j=1}^{2} \alpha_i^j g_j(\mathbf{x}_i)$;
12:     Update g_j: Add the $\mathbf{x}_i$ and $\overline{g}_j(\mathbf{x}_i)$ into labeled data of g_j and train j-th regressor;
13:     Update $\lambda^{(j)}$: Reduce $\lambda^{(j)}$;
14:   **end for**
15: **end while**
16: **return** g_1 and g_2;

The second step is to select the high-confidence instance according to the current regressor. In lines 5–6, the confidence value of sampled unlabeled instances can be calculated according to Eq. (7). Then, in line 7, SPOR selects $\mathbf{x}_i$ and updates the selection vectors $\mathbf{v}^{(3-j)}$ according to the calculated confidence.

The third step is to learn the safety pseudo-label $\overline{g}_j(\mathbf{x}_i)$. Lines 8–9 construct a label plane by semi-supervised prediction $g(\mathbf{x}_i)$, supervised prediction $g^{(0)}(\mathbf{x}_i)$ and regressor weights $\boldsymbol{\alpha}_i$. Then, in line 10, the regressor weights $\boldsymbol{\alpha}_i$ can be solved by a convex solver as a simple convex quadratic program. In line 11, the safety label $\overline{g}_j(\mathbf{x}_i)$ can be derived according to Eq.(4). Finally, in line 12, the age parameters $\lambda^{(j)}$ will de reduced to allow more confident instance can be included into labeled data. In detail, instead of adjusting $\lambda^{(j)}$ directly, we increase the number of confidence instances that should be picked every 10 training epochs.

The last step is to update the regressors g_{j+1} according to Eq. (1). The above process will be repeated until the regressor cannot select any unlabeled instances or the maximum number of selection iterations is reached.

4 Experiments

In this section, we undertake experiments to answer the following questions:

1) Can the safe strategy effectively improve the quality of pseudo-labels during training?
2) How accurate SPOR is in comparison to the state-of-the-art label propagation algorithm?

The implementation of SPOR is available in the website http://github.com/fansmale/SPOR in which all source code is accesssible.

4.1 Experiment Settings

Table 1 lists nine datasets that have been employed for testing in existing SSR literature. They cover diverse domains including physical measurements (abalone), biography (pollen), engineering (folds5x2), etc.

Table 1. Data sets.

Data set	Size	Feature	Training size	Test size	Source
cpusmall	8192	12	2000	6192	Delve
folds5x2_pp	9565	4	2000	7565	UCI
pollen	3848	5	2000	1848	StatLib
puma8NH	8192	8	2000	6192	UCI
wind	6574	14	2000	4574	Statlib
wine_quality	6497	11	2000	4497	UCI
space_ga	3107	6	2000	1107	StatLib
abalone	4177	8	2000	2177	UCI
kin8nm	8192	8	2000	6192	UCI

For each dataset, we randomly sampled 2000 instances as the training set, while the others were used as the testing set. The training set is further partitioned into labeled and unlabeled parts at a certain ratio. To simulate real cases, 2.5%, 5%, 10% and 20% of the training set were served as the labeled data in different experimental settings.

We compared our approach with four state-of-the-art SSR algorithms, COREG [25], CoBCReg [10], SAFER [16], MSRRA [8] and BHD [21]. To ensure the best performance, we adopted the source code provided by the authors and the best settings given in reference. Besides, in SPOR, we employ two basic networks with 3 hidden layers ($32 \times 32 \times 32$ and $32 \times 64 \times 32$) as the base regressors.

4.2 The Effectiveness of the Safe Strategy

To validate our safe mechanism, we calculate the average RMSE of pseudo-labels and safety labels to real labels in selected instances.

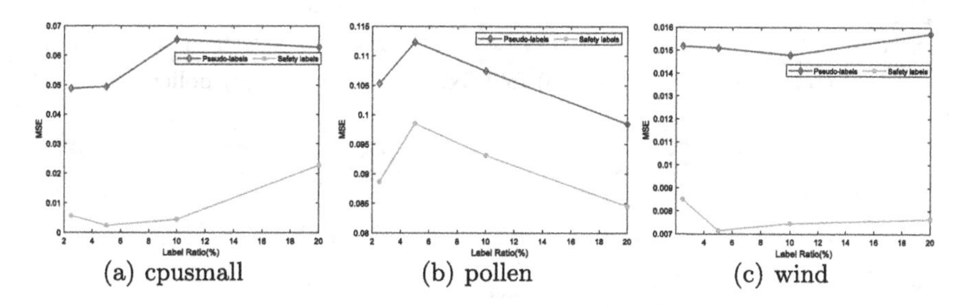

(a) cpusmall (b) pollen (c) wind

Fig. 4. Mean squared errors of pseudo-labels and safety labels to real labels on 2.5%, 5%, 10% and 20% labeled ratio. Safety labels are significant better than pseudo-labels.

Figure 4 shows the test results in dataset cpusmall, pollen and wind. We have the following observations:

- The learned safety labels are significantly better than original pseudo-labels. The safety labels can achieve a 50.41% mean improvement compared to pseudo-labels in all datasets. Even in the pollen, the safety labels still achieve a 13.88% mean improvement than pseudo-labels under different labeled ratios.
- The safe strategy is robust and not greatly affected by the number of labeled instances. The learned safety labels can still achieve a 46.75% mean improvement than pseudo-labels in all datasets under 2.5% labeled ratio. With the labeled ratio increases, the change in performance improvement of safety labels is not significant. This means that our safe strategy is robust and the safe strategy is still effective in improving label quality.

4.3 Comparison with Semi-supervised Methods

To answer our second question of SPOR, we calculate the RMSE and R^2 values of SPOR and comparison methods.

Figure 5 shows the results of RMSE in comparison on to nine real-world datasets. It can be observed that SPOR achieves the best average performance under different labeled ratios, which demonstrates the superiority of our method to comparison methods. We have the following observations:

- Our "qualifier-based" treatment can reduce the influence of weak instances. Under the setting of 2.5% labeled ratio, SPOR achieves an average performance rank of 1.67 (5 best performance out of 9) in all comparison methods.

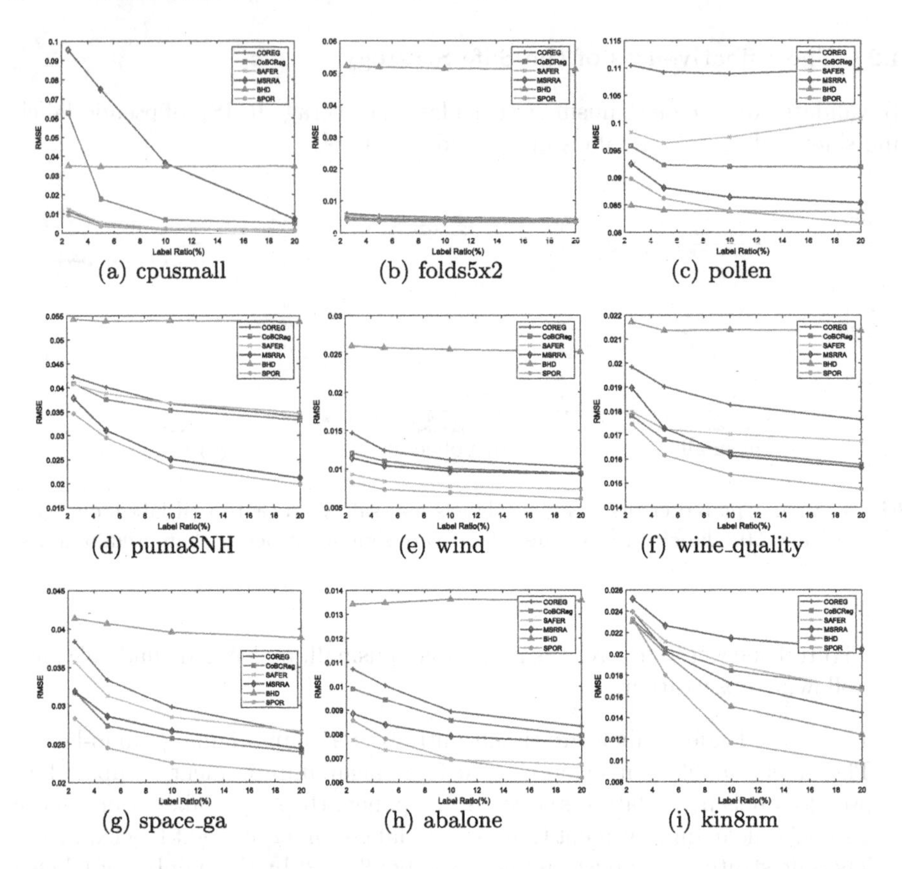

Fig. 5. Root mean squared errors comparison of different models against different labeled instances on the nine datasets.

- The self-paced paradigm is effective in unlabeled instances selection. In the RMSE experiments, the learning performance of all methods increases with the labeled ratio. We can observe that SPOR has the fastest performance improvement of all algorithms. In particular, the SPOR performance increases faster than other methods when the labeled ratio is 20%.

For the R^2 test, different labeled ratios have little effect on the results. Thus, Table 2 only shows the results of the R^2 test under 2.5% labeled ratio. We can observe that SPOR is robust and it outperforms (6 out of 9) other comparison methods in most datasets. On some datasets such as abalone and pollen, SPOR still achieves the second-best performance compared to the other methods. This indicates that SPOR is more robust than the other methods. This makes sense because the "qualifier-based" processing method removes weak instances or outliers from the labeled data. In other words, the R^2 test confirms the idea of SPOR, which reduces the impact of weak instances during the training process.

Table 2. R^2 test for comparison methods and SPOR under 2.5% labeled ratio. The best results for each dataset and the best average performance for all datasets are highlighted with bullets.

Dataset	COREG	CoBCReg	SAFER	MSSRA	BHD	SPOR
cpusmall	0.6778	−0.9331	0.5987	−3.7244	−0.0125	0.6913•
folds5x2_pp	0.8824	0.9071	0.8454	0.9214	-0.0218	0.9245•
pollen	−0.3172	−0.1415	−0.2195	−0.1026•	−0.0181	−0.1362
puma8NH	0.2101	0.2374	0.1914	0.2937	−0.0131	0.3141•
wind	0.5291	0.6365	0.5066	0.6669	−0.0169	0.6819•
wine_quality	0.0636	0.1606	0.1043	0.1305	−0.0255	0.1647•
space_ga	0.0701	0.2295	0.0879	0.2418	−0.0535	0.3132•
abalone	0.2557	0.4168•	0.2867	0.4079	−0.0127	0.3595
kin8nm	0.3065	0.3296•	0.2596	0.2838	−0.0154	0.3066
Mean Rank	4.22	2.78	4.44	2.78	5.22	1.56•

5 Conclusion and Future Work

This paper proposes a self-paced safe co-training regression method, which extends the co-training framework on regression. To enable SPOR to train regressors in a better way, a safe co-regression mechanism is designed to assign better pseudo-labels. In addition, we also analyze the performance of the safe strategy in the training process. Experimental results verify the advantage of SPOR beyond comparison methods.

The research directions for our future work include designing an appropriate self-paced regularizer for regression. Besides, since the valuable of safe strategy in co-training, we can develop a more general safe co-training mechanism to deal with multiple view regression tasks.

Acknowledgment. This work is supported in part by the Central Government Funds of Guiding Local Scientific and Technological Development (No. 2021ZYD0003)

References

1. Balcan, M.F., Blum, A.: A discriminative model for semi-supervised learning. J. ACM **57**(3), 1–46 (2010)
2. Balcan, M.F., Blum, A., Yang, K.: Co-training and expansion: towards bridging theory and practice. In: NIPS, vol. 17, pp. 89–96 (2004)
3. Balsubramani, A., Freund, Y.: Optimally combining classifiers using unlabeled data. In: COLT, vol. 21, pp. 211–225 (2015)
4. Bao, L., Yuan, X.F., Ge, Z.Q.: Co-training partial least squares model for semi-supervised soft sensor development. Chemometr Intell. Lab. Syst. **147**, 75–85 (2015)

5. Bengio, Y., Louradour, J., Collobert, R., Weston, J.: Curriculum learning. In: COLT, pp. 41–48 (2009)
6. Blum, A., Mitchell, T.: Combining labeled and unlabeled data with co-training. In: COLT, pp. 92–100 (1998)
7. Brefeld, U., Scheffer, T.: CO-EM support vector learning. In: ICML, p. 16 (2004)
8. Fazakis, N., Karlos, S., Kotsiantis, S., Sgarbas, K.: A multi-scheme semi-supervised regression approach. Pattern Recogn. Lett. **125**, 758–765 (2019)
9. Gu, Y., Yang, H., Zhou, C.: SelectNet: self-paced learning for high-dimensional partial differential equations. J. Comput. Phys. **441**, 110444 (2021)
10. Hady, M.F.A., Schwenker, F., Palm, G.: Semi-supervised learning for regression with co-training by committee. In: ICANN, pp. 121–130 (2009)
11. Jiang, L., Meng, D.Y., Mitamura, T., Hauptmann, A.G.: Easy samples first: self-paced reranking for zero-example multimedia search. In: ACM MM, pp. 547–556 (2014)
12. Jiang, L., Meng, D.Y., Zhao, Q., Shan, S.G., Hauptmann, A.G.: Self-paced curriculum learning. In: AAAI, vol. 29, pp. 2694–2700 (2015)
13. Kumar, M.P., Packer, B., Koller, D.: Self-paced learning for latent variable models. In: NIPS, vol. 23, pp. 1189–1197 (2010)
14. Li, Y.F., Tsang, I.W., Kwok, J.T., Zhou, Z.H.: Convex and scalable weakly labeled SVMs. J. Mach. Learn. Res. **14**(7), 2151–2188 (2013)
15. Li, Y.F., Wang, S.B., Zhou, Z.H.: Graph quality judgement: a large margin expedition. In: IJCAI, pp. 1725–1731 (2016)
16. Li, Y.F., Zha, H.W., Zhou, Z.H.: Learning safe prediction for semi-supervised regression. In: AAAI, vol. 31, pp. 2217–2223 (2017)
17. Li, Y.F., Zhou, Z.H.: Towards making unlabeled data never hurt. IEEE Trans. Pattern Anal. **37**(1), 175–188 (2014)
18. Ma, F., Meng, D.Y., Xie, Q., Li, Z.N., Dong, X.Y.: Self-paced co-training. In: ICML, vol. 70, pp. 2275–2284 (2017)
19. Ma, F., Meng, D., Dong, X., Yang, Y.: Self-paced multi-view co-training. J. Mach. Learn. Res. **21**(57), 1–38 (2020)
20. Supancic, J.S., Ramanan, D.: Self-paced learning for long-term tracking. In: CVPR, pp. 2379–2386 (2013)
21. Timilsina, M., Figueroa, A., d'Aquin, M., Yang, H.: Semi-supervised regression using diffusion on graphs. Appl. Soft Comput. **104**, 107188 (2021)
22. Wang, W., Zhou, Z.H.: A new analysis of co-training. In: ICML, vol. 12, pp. 1135–1142 (2010)
23. Wang, W., Zhou, Z.H.: Co-training with insufficient views. In: ACML, pp. 467–482 (2013)
24. Yu-Feng, L., Lan-Zhe, G., Zhi-Hua, Z.: Towards safe weakly supervised learning. IEEE Trans. Pattern Anal. **43**(1), 334–346 (2019)
25. Zhou, Z.H., Li, M.: Semi-supervised regression with co-training. In: IJCAI, vol. 5, pp. 908–913 (2005)
26. Zhu, X.J., Goldberg, A.B.: Introduction to semi-supervised learning. Synth. Lect. Artif. Intell. Mach. Learn. **3**(1), 1–130 (2009)

Uniform Evaluation of Properties in Activity Recognition

Seyed M. R. Modaresi[1,2(✉)], Aomar Osmani[2], Mohammadreza Razzazi[1,4], and Abdelghani Chibani[3]

[1] SSRD Lab., Computer Engineering Department, Amirkabir University of Technology, Tehran, Iran
razzazi@aut.ac.ir

[2] LIPN-UMR-CNRS 7030 Lab., Sorbonne University Paris Nord, Paris, France
{modaresi,aomar.osmani}@lipn.univ-paris13.fr

[3] LISSI Lab., Université Paris-Est Créteil, Paris, France
chibani@u-pec.fr

[4] Computer Science, Institute for Research in Fundamental Sciences, Tehran, Iran

Abstract. The main additional problem in activity recognition (AR) systems in contrast to traditional ones is the importance of duration: a predicted concept in AR is durative and can be correct in a period and incorrect in another one. Therefore, it is fundamental to extend the correctness vocabulary and to formalize a new evaluation system including these extensions. Even in similar areas, few empirical attempts are proposed which are confronted with the problems of correctness and completeness. In this paper, we propose the first formal multi-modal evaluation approach for durative concepts. This novel mathematical method evaluates the performance of an AR system from multiple perspectives, including detection, total duration, relative duration, boundary alignment, and uniformity. It extracts the properties considered in the state-of-the-art and redefines the well-known true-positive, false-positive and false-negative terms for durative events. Our proposed method is extensible, interpretable, customizable, open source and improves the expressiveness of the evaluation while its computation complexity remains linear. Comprehensive experimental evaluations are conducted to show the usefulness of our proposed method.

Keywords: Evaluation · Activity recognition · Time series

1 Introduction

Activity Recognition (AR) is expected to be a core component in numerous future Internet of Things applications such as healthcare, smart homes, and security [5,22,23]. Therefore, evaluating the effectiveness of different AR algorithms is essential. Some metrics such as accuracy, observing the recall against precision are common metrics that are easy to understand and interpret even by non-experts. These metrics are well-used for discrete instances and pre-segmented

J. Gama et al. (Eds.): PAKDD 2022, LNAI 13281, pp. 83–95, 2022.
https://doi.org/10.1007/978-3-031-05936-0_7

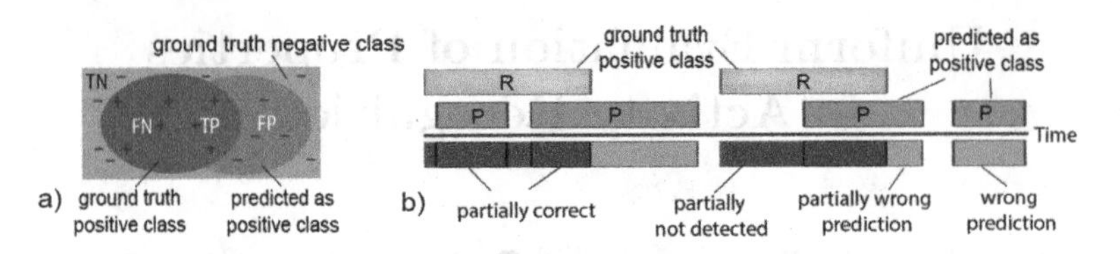

Fig. 1. a) Classical instances b) Durative instances. Durative one may partially correct and partially incorrect while the classical one is either correct or not.

Table 1. The notations used in this paper.

Symbol	Description
TP, TN, FP, FN	True Positive, True Negative, False Positive, False Negative
e=(c, [s:f])	Event e has class c that occurs from time s to time f
GTE, PE	GTE=Ground Truth Event, PE = Predicted Event
E, R, P	E = Event set, R = GTE set, P = PE set
$\|X\|$	The number of element in set X
$\mathcal{T}(e)$, $\mathcal{T}(E)$	$\mathcal{T}(e)$ = Duration of e, $\mathcal{T}(E) = \sum_{e:E} \mathcal{T}(e)$
$e' = e_1 \cap e_2$	$e_1 = (c_1, t_1) \wedge e_2 = (c_2, t_2) \wedge c_1 == c_2 \wedge e' = (c_1, t_1 \cap t_2)$
$e_1 \cap E_2$	$\bigcup_{e_2:E_2} e_1 \cap e_2$
[.]	Iverson bracket. 1 when the enclosed condition is true; otherwise, 0

data sequences [5]; where, a predicted instance is either correct or incorrect. However, concepts in AR are durative; thus, a predicted concept can be correct in one period, incorrect or partially correct in another one [26]. Accordingly, as shown in Fig. 1, previously well-defined terms used in traditional systems such as true positive (TP), false positive (FP), false negative (FN) are not suitable for durative concepts [26].

However, it is often assumed that time-frame, event-based, or classifier performance follows the whole system performance [4,5,22,23]. This assumption neglects practical scenarios and may misleadingly present convincible results (Sect. 2). Despite the importance of evaluating durative concepts, it is not well-developed even in other areas. Still, there is no universally accepted formula for evaluating the effectiveness of systems with durative concepts.

This paper proposes a novel mathematical method for evaluating different properties of AR systems. It redefines TP, FP, and FN to consider various properties such as detection, total duration, relative duration, boundary alignment, and uniformity between ground truth and predicted events. Therefore, confusion matrix based metrics such as recall, precision, and f-score, can be calculated to evaluate and compare different systems. Furthermore, it is simple, time-efficient, extensible, and customizable. It also overcomes the limitations of existing methods. Although, it can select an appropriate algorithm for a new application by prioritizing properties differently. The experiments show that our method can outperform state-of-the-art methods with enhanced generalization capability.

2 Preliminaries and Related Work

Evaluating the performance of AR systems is usually done by comparing predicted events (PEs) with the ground-truth events (GTEs) [16]. It can be viewed as the matching of two time-series. However, it is not easy to determine the time boundaries of ground truth labels perfectly; moreover, the distinction between activities is not always clear [5]. Therefore, some decision functions accommodate offsets using *ambiguous range* [10], *fuzzy event boundaries* [20], time series matching techniques (such as *dynamic time warping*, *longest common sub-sequences* [7]), or *categorical probability distribution* [9]; however, they fail to distinguish different types of errors (e.g., fragmentation) [27]. Common approaches to evaluate AR systems include time-frame, event-based, and classifier performance [12,17,23]. *Time-frame based* methods uses fixed period interval as atomic units and facilitate comparing different AR algorithms since each frame is independent of both the GTEs and PEs [12,17]. Nevertheless, the interpretation of errors is not the same in different applications. Hence, each frame's error is classified to *insertion* (detection of an activity when nothing actually happened), *overfill* (time before and after the occurrence time of an activity that is incorrectly identified as part of the activity), and *merge* (covering multiple GTEs by a single PE) as sources of FP errors and *deletion* (failure to detect an activity), *substitutions* (wrongly detected with another class), *underfill* (not detected duration at the beginning and end of the activity), and *fragmentation* (detecting a GTE by multiple PEs) as sources of FN errors [17]. Moreover, event based methods are also essential to be considered as well as time-frame [27]. Event based errors are categorized as insertion, deletion, fragmentation, merge and fragmented-merge (occurrence of both merge and fragmentation errors) [27]. However, an expert must do a time-consuming analysis of these massive and heterogeneous diagrams, matrices, and information. Therefore, combining them as a scalar metric is complex. Besides, These approaches also consider the total duration of positional errors and do not provide an event-based tunable model for it.

From the behavior analysis perspective, evaluating each activity needs a different evaluation method [1]. e.g., duration sensitive activities need to be evaluated differently from frequency sensitive ones. Timeliness is another metric used for online and realtime prediction [24]. It is defined as the duration continuous correct prediction of an activity without switching to an inaccurate prediction. To compare different AR algorithms in a similar situation, a competition is held and time frame f_1-score, recognition delay, installation complexity, user acceptance, and interoperability are used as the evaluation criteria [8].

In sound event detection (SED) [4], video action detection [3], anomaly detection [26], and video abnormal event detection [11], etc., concepts are also durative. The IEEE Audio and Acoustic Signal Processing challenge [25] highlights the need for an appropriate metric in SED. Still, researchers mainly used collar, segment (time-frame based), and PSDS (polyphonic sound detection score) methods [4,16]. However, they can not show the different sources of errors. Our recent work dedicated to multimodal metrics in SED system [18] provides some evaluation approaches depending on the hypothesis and constraints on SED applications. National Institute of Standards and Technology (NIST) developed a challenge for detecting activities in video (ActEV) [3]. It firstly used false alarms rate (instance based) and missed detections probability (instance based) as evaluation metrics. However, In 2019, it uses time-frame method for calculating false alarm rate [3]. Other metrics in abnormal event detection in video are false rejection rate, equal error rate, decidability index, receiver operating characteristic curves, and area under the Curve [6,11]. However, equal error rate can be misleading in the anomaly detection setting [15]. Numenta anomaly benchmark [14] is designed to evaluate different anomaly detection algorithms. It uses a scaled sigmoidal scoring function for the relative position of each detection; however, it ignores fragmented predictions. To resolve previously mentioned issues, researchers in [26] redefine precision and recall for time-series (particularly on anomaly detection). They need some functions to be explicitly defined for a given application. Those functions are: γ (to consider fragmented events), δ (to consider the positional relation between PE and GTE), overlap (the rate of the correctly detected events (e.g., overlap(x, y,δ()) $= \mathcal{T}(x \cap y)/\mathcal{T}(x)$), and α which is a coefficient. They are formulated in Eq. (1) using notations of Table 1.

$$\mathrm{exist}(e, X) = [e \cap X \neq \emptyset], \qquad \mathrm{score}(e, X) = \gamma(e, X) \times \sum_{x \in X} \mathrm{overlap}(e, e \cap x, \delta()), \tag{1}$$
$$\mathrm{Recall} = \frac{1}{|R|} \sum_{r \in R} \alpha \times \mathrm{exist}(r, P) + (1 - \alpha) \times \mathrm{score}(r, P), \qquad \mathrm{Precision} = \frac{1}{|P|} \sum_{p \in P} \mathrm{score}(p, R)$$

Issues in [26] (Eq. (1)) are analysed deeply in the following:

1. It surprisingly ignores the α (coefficient) in calculating precision. Therefore, it gives inconsistent weights to *overlap* function in calculating recall and precision. Therefore, to prevent misled interpretation, they can not be used as complementary (e.g., in calculating f1 score).
2. Fragmented PEs have significant positive score in precision. e.g., in Fig. 2, the precision of (a) is much higher than (b). Similar situation happens for recall.
3. It normalizes the duration of events to avoid the duration impacts. Briefly, the precision calculation is $\underset{p \in P}{avg}(\frac{TP}{\mathcal{T}(p)})$ and the recall calculation is $\underset{r \in R}{avg}(\frac{TP}{\mathcal{T}(r)})$. This normalization looks well for a single PE and GTE; however, in total, it gives different values for TP in recall and precision. Therefore, they are not calculated in a similar mathematical model and they can not be used as complementary (e.g., for f1-score). Equation (2) presents these calculations for Fig. 2 (d).

$$\mathrm{Precision} = \frac{\frac{TP_1}{P_1} + \frac{TP_2}{P_2}}{1+1} = \frac{\Sigma \text{normalized TPs based on PEs}}{\Sigma \text{normalized PEs}} \tag{2}$$
$$\mathrm{Recall} = \frac{\frac{TP_1}{R_1} + \frac{TP_2}{R_2} + \frac{0}{R_3}}{1+1+1} = \frac{\Sigma \text{normalized TPs based on GTEs}}{\Sigma \text{normalized GTEs}}$$

4. Defining an appropriate cardinality function is complex. Furthermore, it is difficult to adjust and tune this formula since the dependencies between cardinality, position, and overlap are not clear [10]. e.g., in Fig. 2 (c), the first

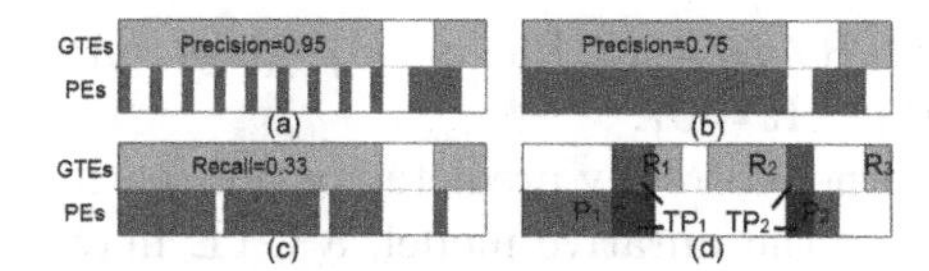

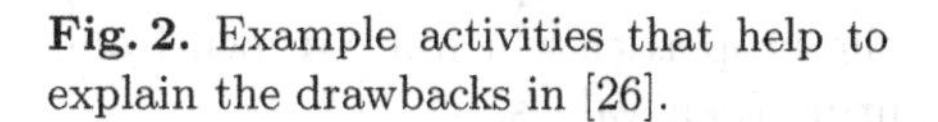

Fig. 2. Example activities that help to explain the drawbacks in [26].

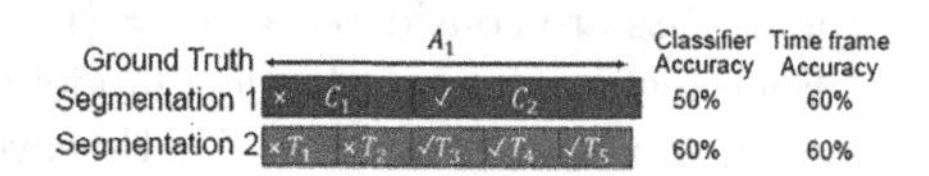

Fig. 3. Evaluation of AR systems that use different segmentation approach.

and second GTEs have the same recall (0.33) (using $\gamma(e, X) = |e \cap X|^{-1}$ as suggested by authors). It is similar for calculating precision for merged PEs.
5. This approach can not be applied to duration-sensitive activities[1].
6. Adding a new property (e.g., total duration) is not straightforward.

Issue in classifier metrics is the inability to compare algorithms in a unified space since AR systems may use various segmentation (windowing) algorithms. Figure 3 is an illustration of two algorithms. Activity A_1 is not detected in segments C_1, T_1 and T_2. Thus, the classifier accuracy in the first approach is 50% while it is 60% in the second one. Clearly, the difference in their performances are due to the effects of the different segmentation procedures. Accordingly, it may misleadingly present convincing results and it can not capture duration specific properties, although it is widely used in several papers [5,7,13,19,23]. Time frame accuracy is more consistent metric [12]; however, it can not displays different property of an AR system such as uniformity, detection of each event or the boundary alignment. Additionally, a long event affect the whole result.

As a result, a new metric is needed to better evaluate AR algorithms while paying attention to the peculiarities of the applications and activities.

3 Proposed Metric

An evaluation method should determine the different properties of AR algorithms. We define a measurement (in terms of recall and precision) for each property, and all together constitute our proposed metrics. A weighted combination of them can produce a scalar value, or they can be used collectively as a multi-objective metric. Because of our approach's modularity, it can be easily extended to include a measurement for a new property. Our metric is based on the following assumptions: 1- R and P are given as input. 2- Times in concepts are durative and specified. 3- The acceptable time shift of PEs to be assumed as detected is within the GTE range. i.e., PEs and GTEs are relevant when they have some overlap. 4- Only a single activity class is exist. For multi-class cases, all classes are evaluated individually as a positive class and the rest as a negative one. This allows using different parameters for each activity class which is an necessary feature for AR [1]. 5- One instance of an activity class occur at a time.

We use ground truths as references in the normalization process because they are independent of predictions of different algorithms. Therefore, we cluster

GTEs and PEs in such a way $C = \{(r, ps) | r \in R \wedge ps = \{p \in P | r \cap p \neq \emptyset\} \wedge ps \neq \emptyset\})$. Orphan PEs are considered as $\overline{C} = \{p \in P | p \cap R = \emptyset\}$.

Each instance in the classical model is either correctly predicted or not (each TP, FP or FN is either 0 or 1). However, in the durative model, a GTE may be partially covered by positive PEs. Therefore, we allow partial value for TP, FP, and FN. In the following, we present the properties which are drawn from state-of-the-art and our formulas for measuring their values.

Detection (D) Property calculates the detection of a GTE even by a small (at least θ [10]) PE (It checks for the existence of overlaps between PEs and GTEs). A GTE is TP if it is detected at least once and is FN if it is not. PEs that don't have any intersection with any GTEs are considered as FP. This property is useful in applications like alarm systems [26].

$$TP^D = \sum_{(r,ps):C} \left[\sum_{p:ps} \frac{\mathcal{T}(r \cap p)}{\mathcal{T}(r)} > \theta_{tp} \right], \quad FP^D = \sum_{(r,ps):C} \left[\sum_{p:ps} \frac{\mathcal{T}(p) - \mathcal{T}(r \cap p)}{\mathcal{T}(r)} > \theta_{fp} \right] + |\overline{C}| \tag{3}$$

$$FN^D = |R| - TP^D,$$

Therefore, a GTE is considered as TP when at least θ_{tp} fraction of it is correctly identified; otherwise, it will be considered as FN. FP counts not detected PEs ($|\overline{C}|$) plus the PEs which the rate of its wrong prediction part is higher than θ_{fp}.

Uniformity (U) Property considers the detection of GTE by a single PE instead of multiple fragmented ones. e.g., in a taking medicine event, detecting two taking medicine events instead of one shows a disorder; therefore, the duration is not as important as the number of occurrences. Researchers in [26,27] consider uniformity as an essential property; however, they do not formulate it. Event analysis [27] leads us to consider a GTE as a TP if it is identified by only one PE. In this case, all other PEs are considered as FP or FN.

$$TP^U = \sum_{(r,ps):C} [|ps \cap R| = 1], \quad FN^U = \sum_{(r,ps):C} [|ps \cap R| > 1], \quad FP^U = |P| - |\overline{C}| - TP^U \tag{4}$$

Thus, the recognized GTEs are considered as TP if each is detected by one PE and that PE does not identify any other GTEs; otherwise, they are considered as FN. Similarly, a PE, that is neither TP nor orphan, is considered as FP.

Total Duration (T) Property is well-known and is similar to time-frame-based methods. It divides the PEs and GTEs by their boundaries; therefore, each frame is either TP, FP, FN, or TN [12].

$$TP^T = \sum_{(r,ps):C} \mathcal{T}(r \cap ps), \qquad FN^T = \mathcal{T}(R) - TP^T, \qquad FP^T = \mathcal{T}(P) - TP^T \tag{5}$$

Relative Duration (R) Property normalizes the duration of each event individually to lessen the effect of varying durations of events.

$$\mathrm{TP^R} = \sum_{(\mathrm{r,ps}):\mathrm{C}} \frac{\mathcal{T}(\mathrm{r} \cap \mathrm{ps})}{\mathcal{T}(\mathrm{r})}, \qquad \mathrm{FP^R} = \sum_{(\mathrm{r,ps}):\mathrm{C}} \min(1, \sum_{\mathrm{p:ps}} \frac{\mathcal{T}(\mathrm{p}) - \mathcal{T}(\mathrm{r} \cap \mathrm{p})}{\mathcal{T}(\mathrm{r})}),$$
$$\mathrm{FN^R} = |\mathrm{C}| - \mathrm{TP^R} \tag{6}$$

Consequently, TP (FN) is the sum of normalized durations of correctly detected (incorrectly undetected) parts of GTEs. The FP calculation is similar; however, FP of each cluster can not exceed 1.

Boundary Alignment (B_t) Property rewards TP when PEs GTE's boundaries precisely match the boundaries of its related PEs; otherwise, it loses some score by FN (underfill error[1]), or FP (overfill error (see footnote 1)) [27]. This property concentrates only on the alignment error and is related to the needs considered in [26,27]. The parameter t specifies the kind of alignment (start ($\mathrm{B_s}$) or end ($\mathrm{B_e}$)).

$$\begin{aligned}
\forall \mathrm{t}: \{\mathrm{start}, \mathrm{end}\}: \quad \mathrm{fn_1(r,ps)} &= \textbf{if } \mathrm{ps} \neq \emptyset \textbf{ then } 1 - e^{-\beta_t \frac{\mathrm{underfill_{t(r,ps)}}}{\mathcal{T}(\mathrm{r})}} \textbf{ else } 0 \\
\mathrm{fp_1(r,ps)} &= \textbf{if } \mathrm{ps} \neq \emptyset \textbf{ then } 1 - e^{-\beta_t \frac{\mathrm{overfill_{t(r,ps)}}}{\mathcal{T}(\mathrm{r})}} \textbf{ else } 0 \\
\mathrm{TP^{B_t}} &= \sum_{(\mathrm{r,ps}):\mathrm{C}} \max(0, 1 - \mathrm{fp_1(r,ps)} - \mathrm{fn_1(r,ps)}) \\
\mathrm{FN^{B_t}} &= \sum_{(\mathrm{r,ps}):\mathrm{C}} \mathrm{fn_1(r,ps)}, \qquad \mathrm{FP^{B_t}} = \sum_{(\mathrm{r,ps}):\mathrm{C}} \mathrm{fp_1(r,ps)}
\end{aligned} \tag{7}$$

Accordingly, TP of each cluster is justified by the alignment error between predictions and ground truths. In addition, errors increase exponentially (adjustable with β_t) by the distance between the boundaries of PEs and GTEs. Increasing parameter β_t gives more penalties to longer positional errors.

Precision, Recall, and F-Score are calculated using the following known formula using TPs, FPs, and FNs that were defined earlier for each AR properties.

$$\forall \mathrm{f} \in \{\mathrm{D}, \mathrm{T}, R, B_s, B_e, U\}: \text{ //Abbreviation of properties} \tag{8}$$
$$\mathrm{Recall^f} = \frac{\mathrm{TP^f}}{\mathrm{TP^f} + \mathrm{FN^f}}, \quad \mathrm{Precision^f} = \frac{\mathrm{TP^f}}{\mathrm{TP^f{+}FP^f}}, \quad \mathrm{F_1^f} = 2\frac{\mathrm{Precision^f.Recall^f}}{\mathrm{Precision^f} + \mathrm{Recall^f}}$$

Computation Complexity of the presented formulas is $O(|R| \times |P|)$ because elements of both sets of P and R are iterated. Since each element of R needs only related P; the interval tree helps us to optimize it to $O(|R|log|R| + |P|log|P|)$. In the case that P and R are sorted by time, this complexity can be reduced to $O(|R| + |P|)$ by considering the time relationships of P and R.

[1] $\mathrm{overfill_{start}(r,ps)} = \max(0, \mathrm{start(r)} - \mathrm{start(ps)})$ $\mathrm{underfill_{start}(r,ps)} = \max(0, \mathrm{start(ps)} - \mathrm{start(r)})$
$\mathrm{overfill_{end}(r,ps)} = \max(0, \mathrm{end(ps)} - \mathrm{end(r)})$ $\mathrm{underfill_{end}(r,ps)} = \max(0, \mathrm{end(r)} - \mathrm{end(ps)})$.

4 Experimental Results

This section presents an experimental study of our metric. The first experiment is done on small visualizable data. The second one compares two algorithms in a real-world dataset. The parameters of each property of our metric are as follows. The $\theta_{\text{tp}}, \theta_{fp}$ are needed to have an appropriate detection property. In this experiment, if a PE has any overlap with GTE ($\theta_{\text{tp}} = 0$), we consider it as TP; additionally, if an incorrect part of a PE is longer than the related GTE's duration ($\theta_{\text{fp}} = 1$), we consider it as FP. We also use ($\beta_t = 2$) to consider near linear boundary error. The codes and datasets are existed in our repository at https://github.com/modaresimr/AR-MME-EVAL.

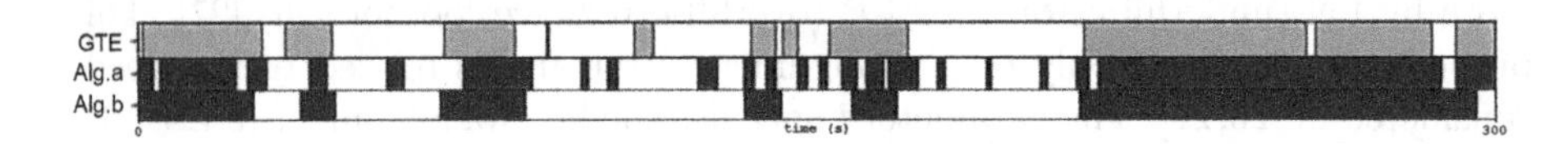

Fig. 4. Ground truths and output of two algorithms used in [27].

Table 2. Details of our metric for algorithms of Fig. 4. The spider chart (right image) shows the f1-score on each property for those algorithms.

Algorithm	Alg.a			Alg.b		
Property	recall	precision	f1	recall	precision	f1
detection	0.73	0.50	0.59	0.73	1.00	0.84
uniformity	0.75	0.43	0.55	0.62	0.83	0.71
total duration	0.78	0.77	0.77	0.84	0.90	0.87
relative duration	0.73	0.81	0.77	0.83	0.85	0.84
boundary start	0.81	0.93	0.86	0.87	0.84	0.85
boundary end	0.99	0.78	0.87	0.85	0.87	0.86

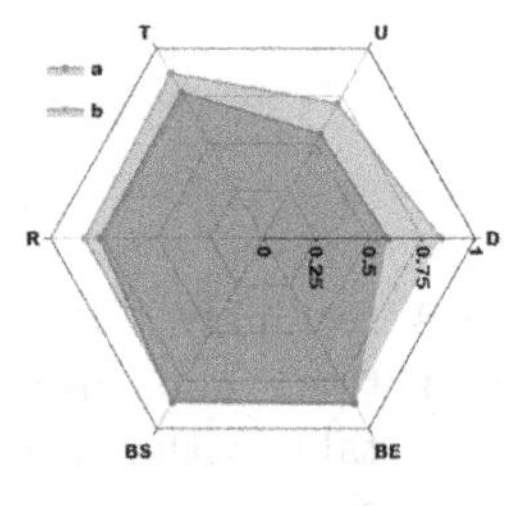

Our Proposed Metric on Small Data is explored in this experiment for simplicity in visualization. This data contains a subset of 13 relations between two intervals in Allen's interval algebra [21]. This data and our metrics' outputs are illustrated on Fig. 4 and Table 2. Clearly, more PEs of Alg.a are incorrectly predicted than Alg.b in Fig. 4, while the number of undetected GTEs is the same. The precision and recall in *detection* measurement confirm this observation. The *uniformity* of Alg.b is higher than Alg.a since most of the GTEs detected with a single PE in Alg.b instead of multiple fragmented PEs. For the *total duration* measurement, we can see that the correctly predicted time frames (TP) in Alg.b are more than Alg.a, while it is inverse for the incorrect ones. The *relative duration* normalizes events independently and applies the total duration measurement. It shows Alg.b predict more part each recognized concept than Alg.a. Since the concepts' duration are similar, the *total duration* shows similar result. In the *boundary* measurement, we can observe that almost all predictions of Alg.a cover the end boundary of GTEs. Therefore, the end part of all GTEs are well-detected (recall = 0.99); however, there are some part of predictions after end of the GTE's boundary that are incorrectly predicted (prediction = 0.78).

Our Proposed Metric on a Public Dataset is explored in this experiment. We compare non-overlapping sliding time window of 30 s (SW)[2] with Hierarchical Hidden Markov model (H-HMM) [2] to show how our metric works. WSU CASAS Home1 dataset [13] that contains 32 sensors, 400,000 events and about 3000 durative concepts (activities) is used in this experiment. We use its first 20% for test and the remaining for training.[3] Then we evaluate the effectiveness of *take medicine* activity and the macro average of all classes.[4] We compare [26] and [27] metrics with ours. The classifier metric issues is discussed in Sect. 2.

Table 5 (b) shows that 50% of times, HHMM algorithm do not detect the concepts and 29% of times it can not detect the start boundary while almost none of its prediction is incorrect. For SW algorithm, it shows great performance except around 16% of times the prediction is fragmented. However, our metric (Table 3) shows this observation is not complete. Analysing the data shows that the duration of 5% of concepts is equal to the others. Therefore, they dominate the system's quality when using the time frame metrics (e.g., Ward's time metrics) and classifier metrics[5]. Table 5(a) helps to understand more about the predictions with event analysis perspective. It displays that 28% and 40% of predictions in SW and HHMM algorithms are incorrectly predicted (in contrast to the observation from Table 5(b)). However almost all of the concepts are recognized by SW algorithm and nearly half of them are not recognized at all in the HHMM algorithm. It also shows that the predicted concepts in both HHMM and SW algorithm are mostly uniform (have few fragmented or merged predictions). These observation is clearly shown in our detection and uniformity property in Table 3. Our proposed metric also correctly shows the quality of detecting the boundaries of concepts while Table 5 (b) display these information totally. Since the duration of this class is much less than the total duration of this dataset while this class constitutes 13% of concepts in this dataset, the last four errors in Table 5 (b) are close to zero. Relative duration properties in Table 3 shows SW either recognize a whole ground truth concept (recall = 0.92) or does not recognize the concept at all; however, its prediction exceed the boundaries (precision < 0.6).

Table 4 shows the metric proposed in [26] with the different parameters. We can observe that γ function, which considers fragmented and merged predictions, has a small affect on the recall and precision. As it is observable from our uni-

[2] We use feature extraction in [13] and three layers perceptron for classifier step.

[3] The internal steps are not important since the concentration is on the metrics.

[4] For saving the space, the analysis of other classes are existed in our repository.

[5] If the used segmentation algorithm generates more segments for longer events which is the case with the well-used sliding window method.

Table 3. Our metric and the spider chart of f1 over two algorithms for one class.

Algorithm	HHMM			SW		
Property	recall	precision	f1	recall	precision	f1
detection	0.53	0.51	0.52	0.97	0.49	0.65
uniformity	0.95	0.97	0.96	0.86	0.89	0.88
total duration	0.19	0.32	0.24	0.80	0.41	0.55
relative duration	0.39	0.58	0.47	0.92	0.54	0.68
boundary start	0.70	0.63	0.66	1.00	0.48	0.65
boundary end	0.86	0.54	0.66	0.92	0.34	0.49

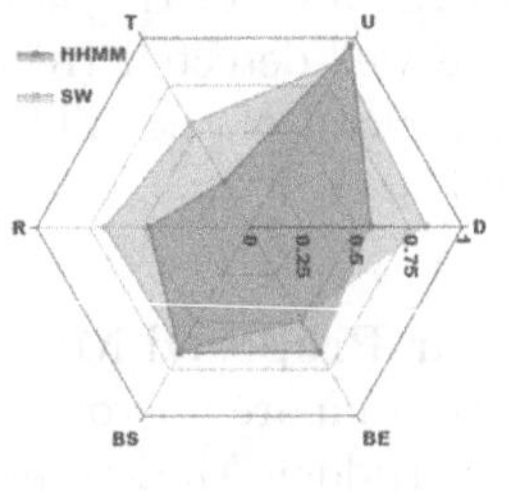

Table 4. Tatbul metric [26] with several parameters and its f1 chart for one class.

Algorithm	HHMM			SW		
Parameter	recall	precis.	f1	recall	precis.	f1
α=0, γ=1, δ=back	0.42	0.29	0.34	0.93	0.27	0.42
α=0, γ=1, δ=middle	0.39	0.37	0.38	0.92	0.36	0.52
α=0, γ=1, δ=front	0.37	0.37	0.37	0.92	0.34	0.50
α=0, γ=1, δ=flat	0.39	0.33	0.36	0.92	0.31	0.46
α=1, γ=1, δ=flat	0.53	0.33	0.41	0.97	0.31	0.47
α=0, γ=reci, δ=flat	0.39	0.33	0.36	0.92	0.30	0.45

Table 5. Ward's proposed metrics for evaluating two algorithms for one class

(a) Event metrics	HHMM	SW
Deletions / $\|R\|$	0.47	0.03
Merged / $\|R\|$	0.03	0.13
Fragmented / $\|R\|$	0	0.04
Frag. and merged / $\|R\|$	0	0
Correct / $\|R\|$	0.51	0.80
Insertions / $\|P\|$	0.40	0.28
Merging / $\|P\|$	0.02	0.05
Fragmenting / $\|P\|$	0	0.06
Frag. and merging / $\|P\|$	0	0
Correct / $\|P\|$	0.58	0.61

(b) Time metrics	HHMM	SW
True positive rate	0.19	0.80
Deletion rate	0.50	0
Fragmenting rate	0	0.16
Start underfill rate	0.29	0
End underfill rate	0.02	0.03
1-false positive rate	1.00	1.00
Insertion rate	0	0
Merge rate	0	0
Start overfill rate	0	0
End overfill rate	0	0

formity property in Table 3, we can see the predictions of both algorithms are uniform but HHMM works better. This observation, can not be captured from Tatbul's metric. As analysed at the end of Sect. 2, the main issue of Tatbul's metric is that recall and precision are not calculated in similar model and can not be used as complementary (e.g., changing α parameter has effect only on recall.). Lastly, δ parameter in Table 4 is proposed by them to consider the boundary alignment errors; however, changing that does not provide significant changes in

Table 6. Macro average of all classes by our metric over two algorithms.

Algorithm	HHMM (macro avg)			SW (macro avg)		
Property	recall	precision	f1(m)	recall	precision	f1(m)
detection	0.44	0.42	0.41	0.86	0.34	0.51
uniformity	0.98	0.92	0.95	0.97	0.85	0.9
total duration	0.31	0.46	0.34	0.58	0.4	0.48
relative duration	0.37	0.87	0.47	0.67	0.78	0.73
boundary start	0.8	0.92	0.82	0.92	0.83	0.85
boundary end	0.94	0.89	0.9	0.89	0.79	0.81

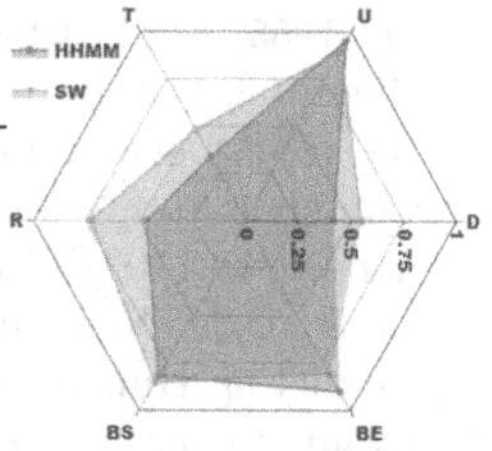

recall and precision while our boundary properties in (Table 3) clearly provide the situation of predictions. This experiment ends with Table 6 that compares the macro average of our metric across all classes of this dataset.

5 Conclusions

In general, activity events are durative in AR. Choosing an appropriate evaluating metric is an essential step to compare AR systems. However, due to the absence of an appropriate one, researchers often use time-frame, event-based, or classifier performance, which can misleadingly present convincible performance for an AR system. This paper proposes a new mathematical model to evaluate AR algorithms which is expressive (by capturing several properties of AR algorithm such as detection, total duration, relative duration, boundary alignment, and uniformity), customizable (the adjustable parameters can support a wide range of applications and can give more weights to some properties of AR algorithms), extensible (adding a new property is straightforward and independent of others). Although our method can give more meaningful information about AR algorithms, its computation complexity remains linear on the size of predictions and ground truths. Our metric has been tested on several datasets, and its ability to measure different AR algorithm properties has been shown. One exciting outcome of this formulation is the possibility to generate a profile (in terms of properties) for each algorithm. Therefore, it can be used as a heuristic for faster algorithm selection which will be explored more in future researches. We are also interested in including fuzziness in our properties.

References

1. Alemdar, H., Tunca, C., Ersoy, C.: Daily life behaviour monitoring for health assessment using machine learning: bridging the gap between domains. Pers. Ubiquit. Comput. **19**(2), 303–315 (2014). https://doi.org/10.1007/s00779-014-0823-y
2. Asghari, P., Soleimani, E., Nazerfard, E.: Online human activity recognition employing hierarchical hidden Markov models. J. Ambient. Intell. Humaniz. Comput. **11**(3), 1141–1152 (2020). https://doi.org/10.1007/s12652-019-01380-5

3. Awad, G., et al.: TRECVID 2020: a comprehensive campaign for evaluating video retrieval tasks across multiple application domains. In: Proceedings of TRECVID, pp. 1–55. NIST, USA (2021)
4. Bilen, C., Ferroni, G., Tuveri, F., Azcarreta, J., Krstulovic, S.: A framework for the robust evaluation of sound event detection. In: ICASSP, IEEE International Conference on Acoustics, Speech and Signal Processing, pp. 61–65 (2020). https://doi.org/10.1109/ICASSP40776.2020.9052995
5. Cook, D.J., Narayanan, C.K.: Activity Learning: Discovering, Recognizing, and Predicting Human Behavior from Sensor Data. Wiley Series on Parallel and Distributed Computing, 1st edn. Wiley (2015)
6. Dutta, J., Banerjee, B.: Online detection of abnormal events using incremental coding length. In: AAAI Conference on Artificial Intelligence (2015). https://ojs.aaai.org/index.php/AAAI/article/view/9799
7. Fu, T.C.: A review on time series data mining. Eng. Appl. Artif. Intell. **24**(1), 164–181 (2011). https://doi.org/10.1016/j.engappai.2010.09.007
8. Gjoreski, H., et al.: Competitive live evaluations of activity-recognition systems. IEEE Pervasive Comput. **14**(1), 70–77 (2015). https://doi.org/10.1109/MPRV.2015.3
9. Hein, A., Kirste, T.: Generic performance metrics for continuous activity recognition. In: Bach, J., Edelkamp, S. (eds.) KI 2011. LNCS (LNAI), vol. 7006, pp. 139–143. Springer, Heidelberg (2011). https://doi.org/10.1007/978-3-642-24455-1_13
10. Hwang, W.S., Yun, J.H., Kim, J., Kim, H.C.: Time-series aware precision and recall for anomaly detection. In: Proceedings of the 28th ACM International Conference on Information and Knowledge Management, pp. 2241–2244. ACM, New York (2019). https://doi.org/10.1145/3357384.3358118
11. Ionescu, R.T., Khan, F.S., Georgescu, M.I., Shao, L.: Object-centric auto-encoders and dummy anomalies for abnormal event detection in video. In: IEEE/CVF Conference on Computer Vision and Pattern Recognition (CVPR), vol. 2019-June, pp. 7834–7843. IEEE (2019). https://doi.org/10.1109/CVPR.2019.00803, https://ieeexplore.ieee.org/document/8954309/
12. Kasteren, T.V., Alemdar, H., Ersoy, C.: Effective performance metrics for evaluating activity recognition methods. In: ARCS (2011)
13. Krishnan, N.C., Cook, D.J.: Activity recognition on streaming sensor data. Pervasive Mobile Comput. **10**(PART B), 138–154 (2014). https://doi.org/10.1016/j.pmcj.2012.07.003
14. Lavin, A., Ahmad, S.: Evaluating real-time anomaly detection algorithms - the Numenta Anomaly Benchmark. In: IEEE 14th International Conference on Machine Learning and Applications (ICMLA), pp. 38–44. IEEE (2015). https://doi.org/10.1109/ICMLA.2015.141, http://ieeexplore.ieee.org/document/7424283/
15. Lu, Y., Kumar, K.M., Nabavi, S.S., Wang, Y.: Future frame prediction using convolutional VRNN for anomaly detection. In: IEEE International Conference on Advanced Video and Signal Based Surveillance (AVSS). IEEE (2019). https://doi.org/10.1109/AVSS.2019.8909850
16. Mesaros, A., Heittola, T., Virtanen, T.: Metrics for polyphonic sound event detection. Appl. Sci. (Switzerland) **6**(6) (2016). https://doi.org/10.3390/app6060162
17. Minnen, D., Westeyn, T.L., Starner, T., Ward, J.A., Lukowicz, P.: Performance metrics and evaluation issues for continuous activity recognition. In: Performance Metrics for Intelligent Systems, pp. 141–148. NIST, Gaithersburg (2006)

18. Modaresi, S., Osmani, A., Razzazi, M., Chibani, A.: Multimodal evaluation method for sound event detection. In: IEEE International Conference on Acoustics, Speech and Signal Processing, (ICASSP). IEEE (2022)
19. Ni, Q., García Hernando, A., de la Cruz, I.: The elderly's independent living in smart homes: a characterization of activities and sensing infrastructure survey to facilitate services development. Sensors **15**(5), 11312–11362 (2015). https://doi.org/10.3390/s150511312
20. NIST: TRECVID 2004 Evaluation (2004). https://www-nlpir.nist.gov/projects/tv2004/index.html
21. Osmani, A.: STCSP: a representation model for sequential patterns. Foundations and Applications of Spatio-Temporal Reasoning (FASTR) (2003). https://www.aaai.org/Library/Symposia/Spring/2003/ss03-03-010.php
22. Perera, C., Zaslavsky, A., Christen, P., Georgakopoulos, D.: Context aware computing for the Internet of Things: a survey. IEEE Commun. Surv. Tutor. **16**(1), 414–454 (2014). https://doi.org/10.1109/SURV.2013.042313.00197, http://ieeexplore.ieee.org/document/6512846/
23. Qian, H., Pan, S.J., Miao, C.: Latent independent excitation for generalizable sensor-based cross-person activity recognition. In: Proceedings of the AAAI Conference on Artificial Intelligence, vol. 35, no. 13, pp. 11921–11929 (2021). https://ojs.aaai.org/index.php/AAAI/article/view/17416
24. Ross, R.J., Kelleher, J.: Accuracy and timeliness in ML based activity recognition. In: Proceedings of the 13th AAAI Conference on Plan, Activity, and Intent Recognition, AAAIWS'13-13, vol. WS-13-13, pp. 39–46. AAAI Press (2013). https://doi.org/10.5555/2908241.2908247
25. Stowell, D., Giannoulis, D., Benetos, E., Lagrange, M., Plumbley, M.D.: Detection and classification of acoustic scenes and events. IEEE Trans. Multimedia **17**(10), 1733–1746 (2015). https://doi.org/10.1109/TMM.2015.2428998
26. Tatbul, N., Lee, T.J., Zdonik, S., Alam, M., Gottschlich, J.: Precision and recall for time series. In: Neural Information Processing Systems (NIPS) (2018). https://papers.nips.cc/paper/7462-precision-and-recall-for-time-series
27. Ward, J.A., Lukowicz, P., Gellersen, H.W.: Ward: performance metrics for activity recognition. ACM Trans. Intell. Syst. Technol. (2011). https://doi.org/10.1145/1889681.1889687

Effect of Different Encodings and Distance Functions on Quantum Instance-Based Classifiers

Alessandro Berti(✉), Anna Bernasconi, Gianna M. Del Corso, and Riccardo Guidotti

University of Pisa, Pisa, Italy
alessandro.berti@phd.unipi.it,
{anna.bernasconi,gianna.delcorso,riccardo.guidotti}@unipi.it

Abstract. In the last years, we have witnessed the increasing usage of machine learning technologies. In parallel, we have observed the raise of quantum computing, a paradigm for computing making use of quantum theory. Quantum computing can empower machine learning with theoretical properties allowing to overcome the limitations of classical computing. The translation of classical algorithms into their quantum counter-part is not trivial and hides many difficulties. We illustrate and implement alternatives for the quantum nearest neighbor classifier focusing on the challenges related to data preparation and their effect on the performance. We show that, with certain data preparation strategies, quantum algorithms are comparable with the classic version, yet allowing for a theoretical reduction of the complexity for distances calculation.

Keywords: Quantum KNN · Encodings · Quantum Machine Learning

1 Introduction

Machine Learning (ML) gained a lot of attention in the latest years due to its effective usage in many applications [18]. These applications typically require high performance in terms of accuracy and low computational time. Another research field for which we have observed significant advancements is Quantum Computing (QC) [11]. QC is the exploitation of properties of quantum states, such as superposition and entanglement, to perform computation. The idea is that with QC it is possible to solve certain problems substantially faster than with classical computing. Nowadays, both ML and QC are playing a crucial role in how society deals with information and data. Therefore, their combination seems something natural. Indeed, Quantum Machine Learning (QML) summarizes approaches that use synergies between machine learning and quantum computing [9].

Among existing QML approaches, we focus on those using QC to process classical datasets [15]. They require a classical-quantum "interface" that is typically realized through ad-hoc data transformation procedures. These procedures

J. Gama et al. (Eds.): PAKDD 2022, LNAI 13281, pp. 96–108, 2022.
https://doi.org/10.1007/978-3-031-05936-0_8

are designed according to the task that we want to solve and they can be different depending on the structure of the circuits of the QML algorithm employed.

We focus here on the Quantum K-Nearest Neighbor (QKNN) algorithm. The theoretical advantage of QKNN with respect to KNN is that QKNN can calculate the distances between the test instance and all the records in the training set at the same time. In the literature we can find various versions of QKNN [1,3,8,13,14,16,20] implementing different distance functions and requiring different data encoding. However, none of them allows to easily reproduce the results to understand which are the pros and cons of the different quantum circuits. In this paper, we clearly describe the quantum circuits implementing two versions of QKNN that adopt different distance functions. We design the circuits responsible for the calculus of the distances and those responsible for storing the data. We analyze all these circuits in detail by examining also their complexity, and we empirically experiment with the different QKNN methods. Our analysis illustrates the theoretical and empirical differences on various datasets. In particular, we want to assess if the accuracy scores of QKNN solutions are promising, and, thus, employable for real world tasks. The results show that, with appropriate data encoding and training strategies, QKNN is comparable or even better than the classic KNN. However, current technological limitations do not allow to empirically reach the better theoretical complexity of QKNN. Indeed, the experiments highlight that the challenges in the usage of QKNN lie in the data preparation and its encoding.

2 Related Works

In the following, we review representative works on QKNN. We refer the reader to [15] for a comprehensive overview on key concepts, ideas and QML algorithms.

In [16] is designed one of the first proposal of QKNN that relies on a binary representation of the data combined with Hamming distance. The data is encoded in the qubits according to the so called *basis encoding*. The overall algorithm runs in polynomial time $O(Tmn)$ where n is the number of features, m is the number of training examples and T is the accuracy threshold. Another QKNN based on Hamming distance and basis encoding is discussed in [13]. Here, the features of the training and test instances are extracted, stored as bit vectors and mapped to quantum ground states [11]. The differences between test sample and vectors of the training set are computed in quantum parallelism using CNOT gates. Hamming distances are then computed, exploiting the addition circuit proposed in [6]. The outcome is identified modifying the corresponding ancilla qubit according to a distance threshold. The time cost is $O(n^3)$, with n number of features. This evaluation does not take into account the cost of the initial state preparation. Another QKNN based on Hamming distance has been proposed in [8]. Hamming distances are computed as in [13]. Instead, the selection of the nearest neighbor is performed through a quantum sub-algorithm for searching the minimum of an unsorted integer sequence [5]. The overall time complexity for a constant value of k and $n \ll m$, is $O(\sqrt{m} \log m)$, showing a quadratic speedup over its classical counterpart.

A QKNN encoding data using *amplitude encoding* is proposed in [14]. The training and test data are stored in superposition using a number of qubits logarithmic in nm. Then, an Hadamard gate is applied to an ancilla qubit resulting in an interference between each of the training data with the test data. Finally, a conditional measurement allows to select the entries of the training set closest to the test data. Another approach using amplitude encoding is presented in [20] for s-sparse datasets, i.e., only s features are non-zero, while, an application of QKNN based on amplitude encoding to image classification has been presented in [3]. The QKNN in [1] employs a controlled swap and an Hadamard on an ancilla qubit to statistically estimate the Fidelity, a distance measure corresponding to cosine similarity. In contrast with other approaches, the class label is not encoded in qubits. Thus, the test data is labeled after a classic majority voting. From the state-of-the-art it is clear that QKNN has valuable properties that make it more efficient than classic KNN, at least from a theoretical perspective. However, most of the aforementioned work does not provide many details about the quantum circuits nor about the data preprocessing needed to run the QKNN algorithms. Our work, in addition to providing a deep comparison of the performance of various QKNNs, clarifies various implementation aspects that make different QKNN procedures reproducible and comparable.

3 Setting the Stage

A classification dataset $\mathcal{D} = \langle X, Y \rangle$ consists of a set X of m instances described by n features, and a set Y of m labels $y_i \in \mathbb{N}$ each assigned to an instance $x^{(i)} \in X$. Each label (or class) y_i is chosen among l available labels in V, i.e., $l = |V|$. In ML, given a dataset $\mathcal{D}$, the objective is to learn a function f that assigns to an unseen instance x a label y, i.e., $y = f(x)$ such that $y = \hat{y}$ where $\hat{y}$ is the real class of x. Our objective is to show how this problem can be solved with QML procedures modeling the well known K-Nearest Neighbor (KNN) classifier [18]. We keep our paper self-contained by summarizing here the key concepts necessary to comprehend our analytical proposal.

Classical KNN. KNN is a supervised ML algorithm implementing function f. KNN takes as input a set of training examples $\mathcal{D}$ and the number of nearest neighbors k. Then, it works as follows [18]. For each test example x, given a distance function d, it computes the distance between x and all the instances in $\mathcal{D}$. Then it selects from $\mathcal{D}$ the k instances $\mathcal{D}_x \subseteq \mathcal{D}$ having the smallest distance from x, i.e., $\mathcal{D}_x$ are the nearest neighbors of x. Finally, it assigns to x a label based on the majority class of the nearest neighbors. Despite being simple, KNN can be characterized with many variants [18]. First, the parameter k controls KNN sensitivity and can affect the classification outcome. In many applications $k = 1$ is effectively adopted [16]. For 1NN the test instance is assigned the label of the closest record in the training set. Second, the distance function d must be selected. The most commonly used distances are the Euclidean distance for continuous features, and the Hamming distance for categorical ones. Third, the selection of the majority class can be weighted. KNN has a main weakness. Given a test instance x, KNN must calculate the distance between x and every record in the training set. Thus, its computational complexity is $O(nm)$. To reduce the number of training records we can use *centroids* instead of real data. The test instance x is compared with a small set of records representative for the classes of the problem typically obtained by averaging the n dimensions. As an alternative, KNN can refer to a sample $\mathcal{D}' \subset \mathcal{D}$ of instances randomly selected, with $m' < m$ and $m' = |\mathcal{D}'|$. In any case, even though the number of training instances can be reduced, KNN still requires to calculate m' distances on n features. QKNN can overcome this limitation calculating all the distances simultaneously.

Amplitude Encoding and Basis Encoding. We can encode information into qubits in different ways [15]. We report here the two encoding for QKNN. *Amplitude encoding* associates classical data with the amplitudes of a quantum state. Given the normalized vector $x = (x_0, \ldots, x_{n-1})$, we prepare the quantum state $|\psi\rangle = \sum_j x_j |j\rangle$. If n is a power of two, it can be encoded in a quantum state on exactly $\hat{n} = \log_2 n$ qubits while shorter vectors must be padded with zeros to reach a dimension which is a power of two. Amplitude encoding only requires $\log(nm)$ qubits to encode a dataset with m records and n features. *Basis encoding* associates a classical ℓ-bit-string with a computational basis state of an ℓ-qubit system. For instance, the string 0011 is encoded as the basis state $|0011\rangle$. In this way, if we use s bits to encode a feature, each n-feature record is represented with $\ell = n\,s$ bits. Thus, given a *binary* dataset X where each record $x^{(i)} \in X$ is a binary string of the form $x^{(i)} = (x_0^{(i)}, \ldots, x_{\ell-1}^{(i)})$, we can prepare a superposition of basis states $|x^{(i)}\rangle$ where each qubit corresponds to a bit of the binary input $|X\rangle = \frac{1}{\sqrt{m}} \sum_{i=0}^{m-1} |x^{(i)}\rangle$. The amplitude vector of $|X\rangle$ has values $1/\sqrt{m}$ for basis states associated with a record of X, and zero otherwise. Since the number of amplitudes 2^ℓ is larger than the number of nonzero amplitudes m, basis encoding datasets generate sparse amplitude vectors.

Initial State Preparation. This step is required by both encodings to load the data into the states $|\psi\rangle$ and $|X\rangle$, respectively. To preserve the advantages of quantum algorithms, quantum state preparation should be performed efficiently. A standard approach is proposed in [17]. A different one is based on the use of *Quantum Random Access Memory* (QRAM) [10,12,19]. A QRAM has the same three basic components as the RAM: a memory array, an address register, and an output register. Address and output registers are composed of qubits instead of bits, while the memory array can be either quantum or classical. The *Flip-Flop* QRAM (FF-QRAM) proposed in [12] can read unsorted classical data stored in memory cells, and superpose them in the computational basis states with non-uniform probability amplitudes to create a specific input state for a quantum algorithm. FF-QRAM allows to encode discrete or continuous classical information as quantum bits or as probability amplitudes of a quantum state. For amplitude encoding, the final state is obtained after a post-selection step. For registering or updating classical data consisting of m entries, each with n features, FF-QRAM requires $O(m\,n)$ quantum operations.

4 Quantum KNN Algorithms

Generally, all the QKNN procedures work on a single test instance. Indeed the computation of the quantum circuits destroys the initial quantum states that have to be re-prepared to classify a new test instance [14]. On the other hand, there are two fundamental aspects for which QKNN algorithms can differ: *(i)* the *distance function* used, and *(ii)* the *data encoding*. The choice of a distance function highly impacts the data encoding.

Another aspect to address is that several QKNN approaches use $k = 1$, while some other do not have a well-defined notion of k. This choice is mainly due to the fact that *(i)* using $k > 1$ can require more computational resources, *(ii)* it is not trivial to find the k closest instances and run a majority voting using quantum circuits, *(iii)* distances are calculated in parallel among different instances. Among the various QKNN implementations briefly discussed in Sect. 2, only [3] and [1] allow to specifically set $k > 1$. However, in [1] the quantum circuit is only used to calculate the distances. Therefore, the computation needed to identify the k nearest neighbors and the majority voting is performed with classic calculus. In [3] the distances are moved from the amplitudes to the basis using amplitude estimation [2], but the paper does not present enough details to reproduce the quantum circuit and the code used is not available. Hence, since our goal is to observe the performance of a complete QML procedure, and to have a completely reproducible research, we focus on $k = 1$ for the basis encoding and to the distance with the closest class for amplitude encoding, and leave a detailed analysis of QKNN with $k > 1$ for future studies. We highlight that, independently from the distance function adopted and from the data encoding, because of the limited computational resources available, it is not typically possible to exploit all the instances and features of a given dataset. Thus, in the rest of paper, we denote by $m' \leq m$ the number of training instances encoded in the quantum circuit of QKNN, and with $n' \leq n$ the number of features used.

QKNN with Amplitude Encoding. We describe here an implementation of QKNN inspired by [14] that uses the *Euclidean distance*. We use aQKNN as a short name for amplitude encoding-based QKNN.

Quantum Euclidean Distance. To compute the Euclidean distance $d(\delta, \phi)$ of two quantum states $|\delta\rangle, |\phi\rangle$ stored in a register D, we need to use an additional ancilla qubit A that will be entangled with the two states $|\delta\rangle$ and $|\phi\rangle$: the state $|0\rangle_A$ will be entangled with $|\delta\rangle$, while $|1\rangle_A$ with $|\phi\rangle$. This can be accomplished by first applying an Hadamard gate on the ancilla, and then by loading in the register D the two states conditioned on the ancilla: in the branch where the ancilla is $|0\rangle_A$ we load $|\delta\rangle$, and in the other branch we load $|\phi\rangle$, as shown in Fig. 1. After this step, the initial state $|0\rangle_A|00\cdots0\rangle_D$ becomes $\frac{1}{\sqrt{2}}(|0\rangle_A|\delta\rangle_D+|1\rangle_A|\phi\rangle_D)$. Finally, an Hadamard applied to the ancilla generates the state $\frac{1}{2}\big(|0\rangle_A(|\delta\rangle_D + |\phi\rangle_D) + |1\rangle_A(|\delta\rangle_D - |\phi\rangle_D)\big)$. The probability of measuring the ancilla in the state $|0\rangle_A$ is given by $p_A = \frac{1}{4}\|\delta+\phi\|_2^2$ that in turns corresponds to $p_A = 1-\frac{1}{4}\|\delta-\phi\|_2^2 = 1 - \frac{1}{4}d(\delta, \phi)$, since δ and ϕ are unit vectors.

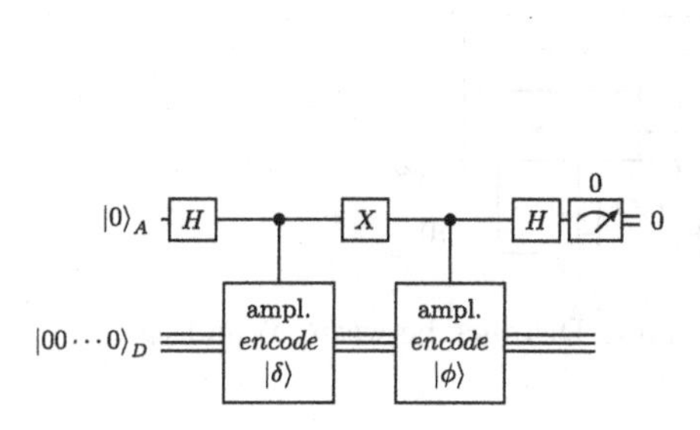

Fig. 1. Circuit for the Euclidean distance.

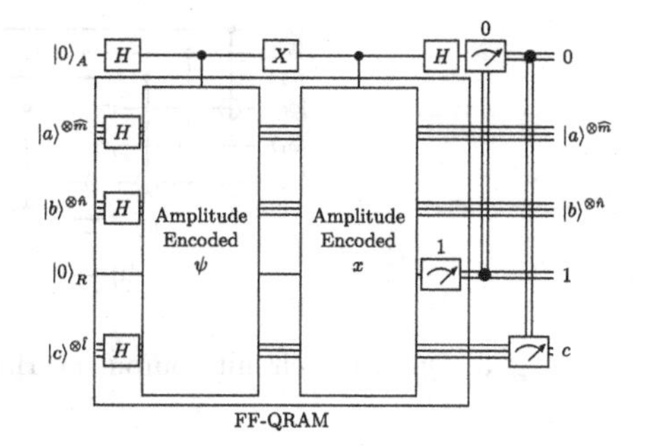

Fig. 2. Classif. with Euclidean distance.

Quantum Data Encoding. The quantum data encoding adopted to use the Euclidean distance is done through Amplitude Encoding. In particular, we adopt the FF-QRAM [12] to store the amplitude encoded values in the circuit. Given $\mathcal{D}$ with m instances and a single feature, we need $\hat{m} = \log_2(m)$ qubits denoted by $|a^{(i)}\rangle = |a_0^{(i)} a_1^{(i)} \dots a_{\hat{m}-1}^{(i)}\rangle$ to address a specific instance i, where $0 \leq i < m$, and an additional qubit $|.\rangle_R$ storing the feature value $x^{(i)}$ of instance i via amplitude encoding. To avoid overloading the notation, we denote by $|x^{(i)}\rangle_R$ the qubit which amplitude-encodes the feature $x^{(i)}$. We model the FF-QRAM as:

$$QRAM(\mathcal{D}) = \sum_{i=0}^{m-1} |a^{(i)}\rangle |x^{(i)}\rangle_R, \tag{1}$$

where $|a\rangle$ acts as a *register* index that identifies a given memory address.

To store the feature value $x^{(i)}$ in the amplitude of register R we first normalize it in the range $[-1, 1]$ and then perform a rotation of an angle $\theta_i = \arcsin x^{(i)}$ around the Bloch Sphere $\overrightarrow{y}$ axis. Once applied the rotation $R_{\overrightarrow{y}}(\theta_i)$ along $\overrightarrow{y}$ we retrieve the amplitude $x^{(i)}$ by post-selection of $|x^{(i)}\rangle_R = \cos\theta_i|0\rangle_R + \sin\theta_i|1\rangle_R$ on qubit $|1\rangle_R$. We highlight that the FF-QRAM rotates (i.e., stores the feature value $x^{(i)}$) the same qubit $|.\rangle_R$, but it does not overwrite the previous stored value $x^{(i-1)}$ since the register index switched from $|a^{(i-1)}\rangle$ to $|a^{(i)}\rangle$. In some sense, even if we are rotating the same qubit in register R, every time that the register index switches, we are targeting a brand-new qubit.

To extend the FF-QRAM of Eq. 1 to handle more than one feature, we add a "second level addressing" using additional qubits. If n is the number of features, then $\hat{n} = \log_2(n)$ qubits are needed to locate a given feature j, where $0 \leq j < n$, of a given instance i. We denote these qubits as $|b^{(i)}\rangle = |b_0^{(i)} b_1^{(i)} \dots b_{\hat{n}-1}^{(i)}\rangle$. Also, since we are dealing with training instances, we also need to load their class. Thus, $\hat{l} = \log_2(l)$ qubits are needed and $|c^{(i)}\rangle = |c_0^{(i)} c_1^{(i)} \dots c_{\hat{l}-1}^{(i)}\rangle$ is the class index of the instance $x^{(i)}$, for $0 \leq i < m$. Thus, we have

$$QRAM(\mathcal{D}) = \sum_{i=0}^{m-1} \sum_{j=0}^{n-1} |a^{(i)}\rangle |c^{(i)}\rangle |b_j^{(i)}\rangle |x_j^{(i)}\rangle_R. \tag{2}$$

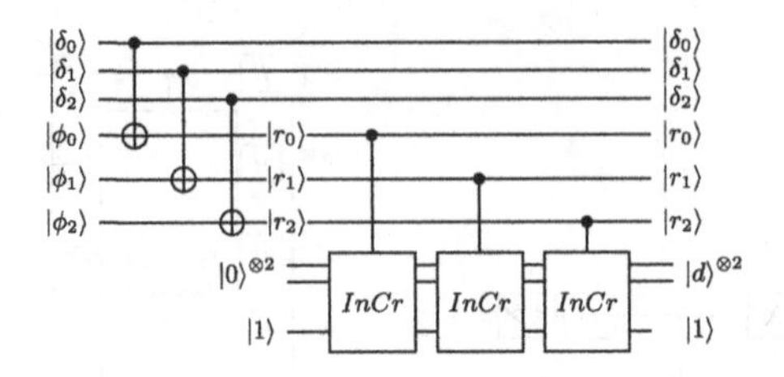

Fig. 3. Quantum circuit computing Hamming Distance between $|\delta\rangle$ and $|\phi\rangle$.

As already observed, from a computational resources perspective it is typically difficult to simulate a training process of a QML algorithm over an entire dataset $\mathcal{D}$ exploiting each feature. This is the reason why we decided to select a subset $m' \leq m$ training instances and $n' \leq n$ feature of a given dataset $\mathcal{D}$. Hence, to be precise, we need to change the indexes of m and n in Eq. 2 to m' and n', respectively. Then, the dataset $\mathcal{D}$ can be encoded into a quantum state which has an amplitude vector of dimensions $n'm'$ that is constructed by concatenating all the training inputs. It follows that the amplitude vector to represent training and test inputs requires $O(\log_2(n') + \log_2(m'))$ qubits.

Quantum Classification with Euclidean Distance. Given $\mathcal{D}$, we want to classify a test instance x exploiting quantum parallelism by means of the quantum Euclidean distance. We denote as $|a\rangle^{\otimes \hat{m}}$, $|b\rangle^{\otimes \hat{n}}$ and $|c\rangle^{\otimes \hat{l}}$ the registers exploited to amplitude encode the test instance and each of the training instances. We use amplitude-encoding to load the test x and $\mathcal{D}$. $|x\rangle$ and $|\psi\rangle$ are the encode of x and of the superposition of $\mathcal{D}$, respectively. Note that, for the test instance x we encode a superposition of classes instead of a specific one. To implement the Quantum Euclidean distance, we then need an additional ancilla qubit with ground state $|0\rangle_A$ entangled with $|x\rangle$, and an excited state $|1\rangle_A$ entangled with the superposition $|\psi\rangle$ of the training set. Finally, we apply the H-gate to the ancilla, thus computing all Euclidean distances among x and $\mathcal{D}$ in just one single shot as shown in Fig. 2. The post-selection of states with $R = 1$ and $A = 0$ and the measure of register C allows to assign the class to x.

Indeed, measuring the ancilla in the state $|0\rangle_A$ produces a state with amplitudes depending on the distances between the test vector and all the training ones, so that a measure on C gives as most probable outcome the class of the training instances closest to x. This method relates to KNN when setting $k = m$ and weighing the training instances by the distance measure [14]. The computational cost of aQKNN is $O(m'n')$ and is the same of the classic KNN. The cost is dominated by the quantum encoding of the classical data. However, once the data have been encoded, the classification is very fast as it consists in just two Hadamard gates and conditional measurements.

QKNN with Basis Encoding. We use bQKNN to refer to QKNN using *Hamming distance* and the *basis encoding* described in this section.

Quantum Hamming Distance. The Hamming distance between two binary strings δ and ϕ of equal length ℓ is given by the number of positions at which the two strings differ and can be evaluated by first computing the bitwise XOR of the two strings and then counting the number of 1s in the obtained sequence. The Hamming distance circuit thus consists in the following steps:

1. Base encode the two binary vectors δ and ϕ in the two registers $|\delta\rangle$ and $|\phi\rangle$.
2. Perform the bitwise XOR between $|\delta\rangle$ and $|\phi\rangle$ using CNOT gates. In fact, $\text{CNOT}(|\delta_j\rangle, |\phi_j\rangle) = |\delta_j\rangle|\delta_j \oplus \phi_j\rangle$, for $0 \leq j \leq \ell - 1$. Set $|\delta_j \oplus \phi_j\rangle$ as $|r_j\rangle$.
3. Counting the 1s in $|r\rangle = |r_0\rangle, \dots, |r_{\ell-1}\rangle$ returns the Hamming distance. The sum of the 1s in $|r\rangle$ is computed with the circuit discussed in [6] that takes in input the register $|r\rangle$ and a register $|d\rangle$ of $h = \lceil \log_2 \ell \rceil$ qubits needed to represent the sum of the 1s in the string r.

The steps above are summarized in Fig. 3 assuming $\ell = 3$ and $h = 2$ qubits.

With basis encoding the number m of records in the dataset $\mathcal{D}$ does not impact in the encoding as all records are in superposition, while the number n of features impacts on the representation. For this reason, it becomes crucial to use less features than those available if we can use at most ℓ qubits. Indeed, if we reserve s bits for each feature, we need to reduce the number of features from n to n' s.t. $n' s \leq \ell$. Thus, before running bQKNN we have to compress the information captured by the n features into ℓ bits. We perform this task following two different strategies. The first one consists in the discretization of the numerical attributes through the *Recursive Minimal Entropy Partitioning* (RMEP) method [4], if necessary preceded by a PCA run. The second strategy consists in using a hash function preserving Hamming distance. In particular, we adopt *Locality-Sensitive Hashing* (LSH) [7]. The problem with both RMEP and LSH is that more instances can be associated to the same binary representation.

We overcome this limitation by removing binarized duplicate instances from the training set, and adopting a majority voting to determine the class label.

Once the data have been encoded as binary strings, we can initialize the quantum register via basis encoding following the strategy of [19]. In order to memorize the m binary instances of length ℓ, the quantum circuit requires $2\ell+1$ qubits, logically organized as follows: an ℓ-qubit register $|x^{(i)}\rangle$ for the m binary instances; a register $|c^{(i)}\rangle$ of two control qubits used to determine which states are affected by a particular operator; and an additional $\ell - 1$ garbage qubit register used to implement Multi-Control Toffoli (MCT) gate with ℓ control qubits through a chain of $\ell - 1$ classical Toffoli gates. Observe that in the first register there is one qubit for each bit in the instance to be stored, and therefore any possible binary pattern can be represented appropriately flipping the qubits.

Quantum Classification with Hamming Distance. Given $\mathcal{D}$, we want to classify a test instance x by means of the Hamming distance [8]. First, we need to prepare and load the dataset, eventually reducing the number of features to n'. Once computed the superposition $|X\rangle$ of the m training instances as in Eq. (1) in an ℓ-qubit register, with $\ell = n' s$, we base encode x in the register $|x\rangle = |x_0 x_1 \dots x_{\ell-1}\rangle$. We then compute the Hamming distances in one shot, exploiting quantum parallelism. After this step, the register $|d\rangle$ contains the superposition of all the Hamming distances between the test instance x and each training instance $x^{(i)}$. The last step is to find the training instance $x^{(i)}$ minimizing the Hamming distance with x. This last problem can be depicted as "searching the minimum of an unordered integer sequence" and it is detailed in [8]. The class of $x^{(i)}$, is the bQKNN prediction.

The cost of the bQKNN is of order $O(m'\,\ell + \ell\,\log^2 \ell + \sqrt{m'}\,\log \ell\,\log m')$, where the terms account for the initial state preparation, the Hamming distance computation, and the computation of the instance with the minimum Hamming distance from the test vector. The cost of the input preparation is linear in the number of training instances and in the number of qubits. Again the most expensive step is data preparation, and the algorithm will become competitive only when the encoding of classical data will be done more efficiently.

5 Experiments

Experiments for QKNN are run both using a quantum simulator and (when possible) using the resources offered by IBM Quantum Experience[1]. We ran experiments on three well-known open source datasets: `iris`, `cancer`, and `mnist`[2]. We used 70% of the datasets for training and the remaining 30% for evaluation.

Due to limited computational resources, we cannot train QKNN algorithms on the whole training set. Thus, we set as $n' \leq n$ the number of features used by aQKNN and as $\ell \leq n\,s$ the number of bits used by bQKNN. When possible, we experiment with $n' = n$. However, since this is typically not feasible, we perform PCA [18] to work with $n' < n$ features. Moreover, still due to limited computational resources, we set the size of the training set as $m' \leq m$. For each test instance x, we train the models with $m' = l \cdot i$ for $i \in [1, \ldots, 32]$, where $l = |V|$ is the number of class labels. We select the $m' \leq m$ training instances employing two different strategies named *sampling* and *prototypes*, respectively. Inspired by [1,16], for the *sampling* strategy we sample uniformly at random m' training instances. Since the selection is random, for each test record we repeat the classification at least 50 times using different samples of m' training instances. We report the results by averaging the measures observed over the various experiments. For the *prototypes* strategy if $i = 1$ and $m' = l$, then, like in [20], we use a single prototypical instance for each class obtained as the average value for each feature. If $i > 1$, then we use i prototypes for each class. We derive these prototypes as the centroids returned by centroid-based clustering algorithms [18] applied separately for each subset of instances belonging to the same class. We exploit K-Means [18] for preparing the prototypes for the amplitude encoding and K-Modes [18] for the binary encoding. To have a fair comparison we adopt the same strategy for the experiments also with the classic KNN.

---

[1] Python code available at: https://github.com/Brotherhood94/quantum_knn. We implemented QKNN using the `qiskit` library: https://qiskit.org/.

[2] https://scikit-learn.org/stable. For `mnist` we focus on the task of classification between "0" and "8".

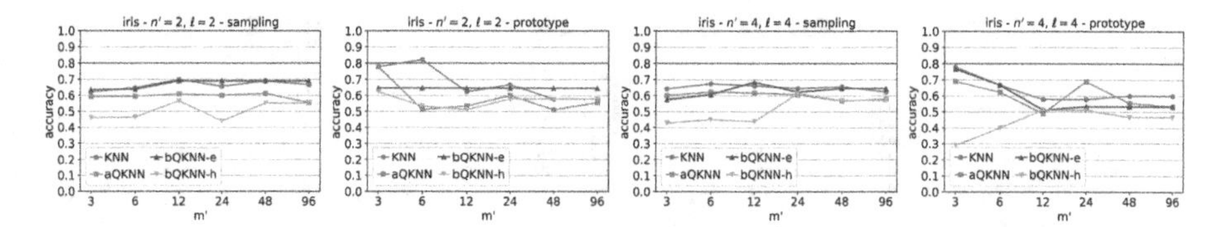

Fig. 4. Performance comparison in terms of accuracy on the `iris` dataset.

Results. We report the performance in terms of *accuracy* [18] for QKNN in the amplitude (aQKNN) and basis (bQKNN) versions, compared with the classic KNN for the Aer QASM quantum simulator of `qiskit` with 8192 shots. For bQKNN we consider the two different data preparation approaches presented: RMEP and LSH identified with *-e* and *-h*, respectively. All the methods are evaluated with the same experimental setting varying the number of features n' and the number of training instances m'.

Iris. In Fig. 4 we observe the performance for the `iris` dataset. We note that, when considering less instances, the performance is markedly lower w.r.t. classic KNN using the whole training set as illustrated with the red continuous line. However, in the prototype setting with PCA and $n' = 2$, KNN achieves accuracy comparable with the accuracy on the whole training set for $m' \leq 6$. Also, we observe that for `iris` the RMEP encoding has always better performance than the LSH one. In all the plots is that aQKNN is on average half point under KNN, but shows promising behavior. Indeed, for $n' = n = 4$, i.e., when we are exploiting all the available features, and thus no PCA is applied, aQKNN is even better than KNN with $m' = 24$ in the prototype setting. This is a signal that the quantum circuit and the necessary data manipulation do not necessarily affect too much the performance with respect to a classic setting. Also, when $n' = 2$ using sampling we notice that bQKNN-e is comparable to KNN. This is probably due also to the binary encoding that makes different instances collapsing into the same representation. On the other hand, when $n' = 2$ using prototypes bQKNN methods have markedly lower performance than KNN and aQKNN. Finally, we notice an improvement in the performance of bQKNN-h when $m' > 12$. We can conclude that, on some setting, depending on the data preparation, the performance of QKNN and KNN can be comparable.

On the `iris` dataset we were also able to run experiments on a real quantum machine. We run experiments on *ibmq_16_melbourne*, the one with the highest number of qubits among those at our disposal, i.e. 15 qubits. Unfortunately, we were only able to perform the experiment with $n' = 2$, $m' = 2$ in the prototype setting. The performance are not extraordinary as we reached an accuracy of less than 0.5 versus a score of about 0.6 obtained in the simulator. This is probably due to the fact that the `qiskit` circuits run on the real machine, that does not optimize for the quantum noise that can interfere during the calculus [16]. Probably, a re-ordering of the quantum wires would reduce real interference bringing the performance close to those obtained with the simulator.

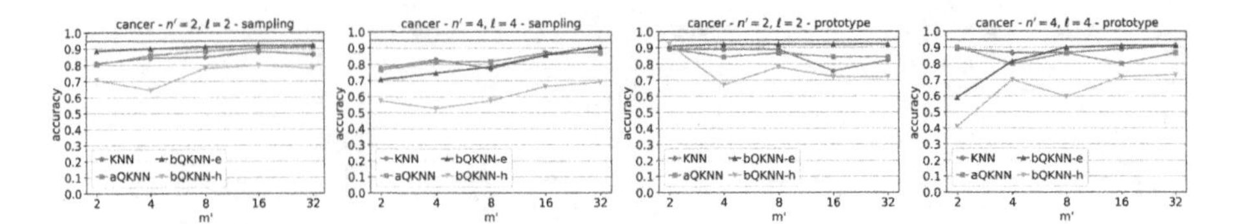

Fig. 5. Performance comparison in terms of accuracy on the `cancer` dataset.

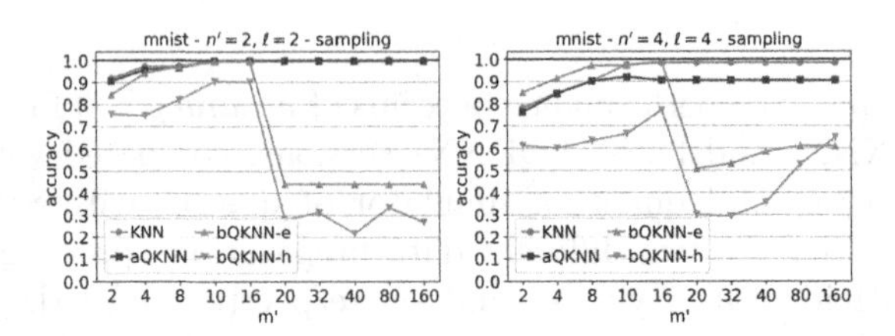

Fig. 6. Performance comparison in terms of accuracy on the `mnist` dataset.

Breast Cancer. In Fig. 5 we show the accuracy for `cancer` dataset. We notice that the performance of quantum approaches are in line with those of the classic one (red line). This is also probably due to the fact that for `cancer` we are facing a binary classification problem. QKNN approaches are comparable and even better than the classic KNN trained with comparable settings. Moreover, it is clear from the plots using sampling strategy that when the number of training instances m' increases, then the accuracy of both aQKNN and bQKNN increases. The bQKNN version seems better than the aQKNN one when using prototypes. However, we were not able to run experiments for bQKNN with $n' > 4$, while the results for aQKNN are not reported due to lack of space. Differently from the previous results, in these cases KNN is slightly better than aQKNN, but we have an overall drops in the accuracy. Finally, as for `iris`, the performance of bQKNN-e are markedly better than those of bQKNN-h.

Mnist. In Fig. 6 we show the accuracy for the `mnist` dataset. We report results only for the sampling strategy as the prototype one, like for the datasets illustrated above, leads to worse accuracy scores. The first aspect that we notice is that, when $m' > 16$ there is a consistent drop in the performance of bQKNN. As before, the LSH version of bQKNN has worse performance than the RMEP one. When $m' < 16$ the performance of the various algorithms are comparable and both aQKNN and bQKNN reach the same level of KNN when $m' = 16$ and $n' = 2$. Thus, the dimensionality reduction adopted, i.e., the PCA, probably helps in identifying discriminating attributes. Results for $n' = 2$ are slightly better than $n' = 4$ for aQKNN. Hence, probably the usage of more information insert distortions into the amplitudes of the data used by aQKNN.

6 Conclusion

We have presented in detail the similarities and differences between alternatives for QKNN algorithms and we have experimented with them on different datasets. Results are promising as they show that, exploiting quantum-parallelism, the QKNN can theoretically overcome current state-of-the-art ML approaches. Also, from an empirical perspective, the QKNN has comparable scores with respect to classical KNN. In addition, due to limited computational resources, we had to employ PCA on the features of the three datasets. In general, PCA seems to have a positive impact on the classification tasks. However, it is important to remark two aspects. First, QKNN has a lower complexity than KNN in calculating the distances but requires a higher cost at data preparation time. Several future research directions are possible. We would like to study the impact of QKNN with different values of k. Also, we would like to analyze the complexity of the data preparation. Finally, we would like to perform a similar analysis on other QML algorithms such as SVM, or Neural Networks.

Acknowledgment. This work is supported by Università di Pisa under the "PRA - Progetti di Ricerca di Ateneo" (Institutional Research Grants) - Pr. no. PRA_2020-2021_92, "Quantum Computing, Technologies and Applications", and the work of Gianna Del Corso was supported also by GNCS-INdAM.

References

1. Afham, A., et al.: Quantum K-nearest neighbor machine learning algorithm. arXiv:2003.09187 (2020)
2. Brassard, G., et al.: Quantum amplitude amplification and estimation. Quantum Comput. Inf. **305**, 53–74 (2002)
3. Dang, Y., Jiang, N., Hu, H., Ji, Z., Zhang, W.: Image classification based on quantum K-nearest-neighbor algorithm. Quantum Inf. Proc. **17**(9), 239 (2018). https://doi.org/10.1007/s11128-018-2004-9
4. Dougherty, J., Kohavi, R., Sahami, M.: Supervised and unsupervised discretization of continuous features. In: ICML, pp. 194–202 (1995)
5. Durr, C., Hoyer, P.: A quantum algorithm for finding the minimum. arXiv preprint arXiv:quant-ph/9607014v2 (1999)
6. Kaye, P.: Reversible addition circuit using one ancillary bit with application to quantum computing. arXiv, p. 0408173, August 2004
7. Leskovec, J., et al.: Mining of Massive Datasets. Cambridge University Press, Cambridge (2014)
8. Li, J., Lin, S., Kai, Y., Guo, G.: Quantum K-nearest neighbor classification algorithm based on hamming distance. arXiv preprint arXiv:2103.04253 (2021)
9. Lloyd, S., Mohseni, M., Rebentrost, P.: Quantum algorithms for supervised and unsupervised machine learning. arXiv preprint arXiv:1307.0411 (2013)
10. Möttönen, M., et al.: Transformation of quantum states using uniformly controlled rotations. Quantum Inf. Com. **5**, 467–473 (2005)
11. Nielsen, M., et al.: Quantum Computation and Quantum Information. CUP, Cambridge (2016)

12. Park, D.K., Petruccione, F., Rhee, J.-K.K.: Circuit-based quantum random access memory for classical data. Sci. Rep. **9**(1), 1–8 (2019)
13. Ruan, Y., et al.: Quantum algorithm for K-nearest neighbors classification based on the metric of hamming distance. IJTP **56**(11), 3496–3507 (2017)
14. Schuld, M., et al.: Implementing a distance-based classifier with a quantum interference circuit. EPL (Europhys. Lett.) **119**(6), 60002 (2017)
15. Schuld, M., et al.: Supervised Learning with Quantum Computers. Springer, Cham (2018). https://doi.org/10.1007/978-3-319-96424-9
16. Schuld, M., Sinayskiy, I., Petruccione, F.: Quantum computing for pattern classification. In: Pham, D.-N., Park, S.-B. (eds.) PRICAI 2014. LNCS (LNAI), vol. 8862, pp. 208–220. Springer, Cham (2014). https://doi.org/10.1007/978-3-319-13560-1_17
17. Shende, V.V., Bullock, S.S., Markov, I.L.: Synthesis of quantum-logic circuits. IEEE Trans. Comput. Aided Des. Integr. Circ. Syst. **25**(6), 1000–1010 (2006)
18. Tan, P., et al.: Introduction to Data Mining. Addison-Wesley, Boston (2005)
19. Ventura, D., et al.: Quantum associative memory. Inf. Sci. **124**(1–4), 273–296 (2000)
20. Wiebe, N., Kapoor, A., Svore, K.M.: Quantum nearest-neighbor algorithms for machine learning. Quantum Inf. Comput. **15**, 318–358 (2018)

Attention-to-Embedding Framework for Multi-instance Learning

Mei Yang[1], Yu-Xuan Zhang[1], Mao Ye[3], and Fan Min[1,2](✉)

[1] School of Computer Science, Southwest Petroleum University, Chengdu, China
{yangmei,minfan}@swpu.edu.cn

[2] Institute for Artificial Intelligence, Southwest Petroleum University, Chengdu, China

[3] School of Computer Science and Engineering, University of Electronic Science and Technology, Chengdu, China

Abstract. We present an attention-to-embedding framework that explicitly addresses the challenge posed by multi-instance learning (MIL) classification tasks, where learning objects are bags containing various numbers of instances. Two key issues of this work are to extract relevant information by determining the relationship between the bag and its instances, and to embed the bag into a new feature space. To respond to these problems, a network with the popular attention mechanism is designed that assigns a new representation and a class probability vector to a given instance in the bag. In addition, compared with the traditional MIL methods, we offer a new embedding function according to the assigned results of instances to process the bag embedding that is unrelated to the distance metric. As a result, MIL challenges will be reduced to single-instance learning (SIL) problems that can be solved using basic machine learning algorithms such as SVM. Extensive experiments on thirty-four data sets demonstrate that our proposed method has the best overall performance over other state-of-the-art MIL methods. This strategy, in particular, has a substantial advantage on web data sets and better stability. *Source codes are available at* https://github.com/InkiInki/AEMI.

Keywords: Attention · Embedding · Multi-instance learning · Network

1 Introduction

Multi-instance learning (MIL) was originally designed for drug activity prediction [4]. In contrast to traditional single-instance learning (SIL), each object in MIL is a bag containing various numbers of instances. A label is assigned to the bag, but not to the individual instances. To date, MIL has also been frequently utilized in a variety of applications, such as image classification [15], text categorization [14], sentiment analysis [1], and web index recommendation [10].

J. Gama et al. (Eds.): PAKDD 2022, LNAI 13281, pp. 109–121, 2022.
https://doi.org/10.1007/978-3-031-05936-0_9

Over the years, many excellent algorithms for MIL classification tasks have been proposed. Traditional MIL covers but is not limited to the following solutions: a) Instance-based approaches calculate the bag label by predicting the instance label and combining MIL assumptions [6]; b) Bag-based approaches treat each bag as an atom and train a classifier based on bag-level metrics, such as graph kernel [19] and isolation set-kernel [14]; and c) Embedding-based approaches transform bags into a new feature space and establish the learning process with SIL methods [13]. Neural network-based MIL can be categorized into two types [11]: a) mi-Net uses an instance-level classifier to obtain the instance probabilities. As a result, the bag label is derived using instance probabilities and the convex max operator (or max operator); and b) MI-net builds a fixed-length vector as the new representation of the bag and learns a bag-level classifier directly to obtain the bag label.

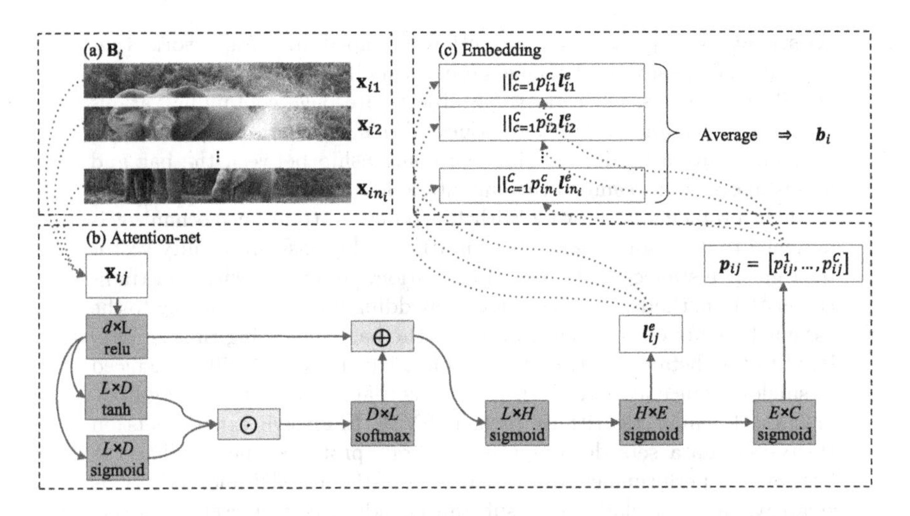

Fig. 1. The overall framework of AEMI: a) The original bag $\mathbf{B}_i$ with a series of unlabeled instances $\mathbf{x}_{ij}$; b) The attention-net for extracting instance information; and c) The bag-level embedding is used to transform each bag into a new feature space. In addition, d is the dimension; L, D, H and E are the number of nodes; C is the number of classes; $\odot$ and $\oplus$ are element-wise multiplication and addition, respectively; $\boldsymbol{l}_{ij}^{e}$ and $\boldsymbol{p}_{ij}$ are the new representation and class probability vector of $\mathbf{x}_{ij}$, respectively.

In this paper, we propose a new attention-to-embedding framework (AEMI) to handle multi-instance learning classification tasks. Figure 1 shows the AEMI's overall framework, which innovatively combines the attention mechanism derived from neural networks and the MIL embedding method. The first part is a sample image that can be regarded as a bag, with each sub-area corresponding to one of the bag's instances. This part reflects the challenge that the neural network must face when applied to MIL: The label of the image is known, but the label of the instance contained within it is unknown. In the previous MIL neural network-based approaches, the relationship between instances and the bag is commonly determined via pooling functions or attention coefficients. While in our configuration, the embedding function needs to ensure that the embedded bags can be distinguished.

Therefore, we provide an attention network whose input is the instance and outputs are its new representation and class probability vector. By designing an embedding function and controlling the size of the representation, each instance can be embedded as a vector, and each bag can be transformed into the same space by the arithmetic average of all embedded vectors, as shown in the third part. As a result, any traditional SIL classifier can be employed to train a model.

The contributions of this work are summarized as follows:

- We convert MIL tasks into SIL ones by designing a framework that connects MIL neural networks with the embedding method. This has the advantage of alleviating the issue of neural network classification instability caused by random parameter initialization.
- We design a network that is not reliant on the MIL pooling function and an embedding function without distance metrics. Specifically, the class probability vector of each instance in the bag is introduced into the embedding process to improve the distinguishability of embedding bags.

Experiments are undertaken on thirty-four MIL classification data sets to quantify the performance of AEMI. These data sets come from a variety of fields, including drug activity prediction, text classification, image classification tasks and web index recommendation tasks. In most cases, the experimental results show that AEMI outperforms state-of-the-art algorithms and has demonstrated significant benefits on web data sets.

2 Related Work

In this section, we will briefly introduce the related work, including MIL attention neural networks and embedding methods.

2.1 MIL Attention Neural Networks

The attention mechanism is commonly employed in deep learning for text analysis [2] or image recognition [5]. However, few studies have focused on the attention mechanism of MIL. Attention-net [8] incorporates interpretability into the MIL method and increases its flexibility. Loss-attention [9] connects the attention mechanism with the loss function. Unlike prior techniques, we exclusively employ neural networks to obtain the new representations and class probability vector for each instance in the bag. Therefore, the network we designed does not depend on the MIL pooling function, and its input is the instance assigned as the corresponding bag label.

2.2 MIL Embedding Methods

Multi-instance embedding methods' core idea is to transform the bag into a new feature space and train a model with traditional machine learning methods. miFV [12] extracts information from the instance space using the Gaussian mixture model and derives the embedded vector using the Fisher vector, while the time complexity of this technique increases as the dimensionality of the data set grows. MILDM [13] designs an instance evaluation function to select instances with the most discriminativeness and builds a mapping pool to embed bags. StableMIL [17] builds upon identifying a novel connection between MIL and the potential outcome framework in causal effect estimation. The majority of these methods rely on distance metrics, such as the bag-level average Hausdorff distance [16] and the bag-instance minimum distance [13]. However, the adaptability of different distance measures to different types of data sets may be completely different. Therefore, we design an embedding function by combining the instance's new representation and probability vector derived from the attention-net.

3 Methodology

In this section, we will first describe the preliminaries of the proposed AEMI algorithm. Then, the attention-net and bag-level embedding are introduced as part of AEMI. Finally, we have some discussions about this method.

Let $\mathcal{B} = \{\mathbf{B}_i\}_{i=1}^N$ be a MIL data set with N bags, where $\mathbf{B}_i = \{\mathbf{x}_{ij}\}_{j=1}^{n_i} \in \mathcal{B}$ is a bag with n_i instances, $\mathbf{x}_{ij} \in \mathbf{B}_i$, $n_i = |\mathbf{B}_i|$ and d is the dimension. Let $\mathbf{Y} = [y_1, \ldots, y_N]$ be the label vector corresponding $\mathcal{B}$, where $y_i \in \{1, \ldots, C\}$ is the label of $\mathbf{B}_i$ and C is the number of classes. With the basic MIL assumption [4], label y_i is supposed that: a) $\mathbf{B}_i$ is labeled as c-th class iff it contains at least one c-th class instance; and b) $\mathbf{B}_i$ contains an instance belonging to two or more classes is impossible.

Our goal is to transform the MIL tasks into SIL ones by connecting the attention-net with the bag-level embedding.

3.1 The Attention-Net

The core of MIL attention network [8] is to calculate an attention coefficient for each instance $\mathbf{x}_{ij}$:

$$\alpha_{ij} = \boldsymbol{w}^{\mathrm{T}}(\tanh(\boldsymbol{V}\boldsymbol{p}_{ij}^{\mathrm{T}}) \odot \mathrm{sigmoid}(\boldsymbol{U}\boldsymbol{p}_{ij}^{\mathrm{T}})), \tag{1}$$

where $\boldsymbol{w} \in \mathbb{R}^{L\times 1}$ and $\boldsymbol{V}, \boldsymbol{U} \in \mathbb{R}^{L\times C}$ are parameters of neural network, L is the number of fully connected layer's nodes, C is the number of classes, $\boldsymbol{p}_{ij}$ is the class probability vector of the instance $\boldsymbol{x}_{ij}$ and $\odot$ is element-wise multiplication.

To extract the information of $\boldsymbol{x}_{ij}$ and construct the embedding function, we modify this mechanism to generate the attentional representation of $\mathbf{x}_{ij} \in \mathbf{B}_i$

$$\boldsymbol{h}_{ij}^{a} = softmax((\tanh(\boldsymbol{h}_{ij}^{r}\boldsymbol{W}^{t}) \odot \mathrm{sigmoid}(\boldsymbol{h}_{ij}^{r}\boldsymbol{W}^{s}))\boldsymbol{W}^{o}), \tag{2}$$

where

$$\boldsymbol{h}_{ij}^{r} = \text{relu}(\mathbf{x}_{ij}\boldsymbol{W}^{r}), \tag{3}$$

and $\boldsymbol{W}^t, \boldsymbol{W}^s \in \mathbb{R}^{L\times D}$, $\boldsymbol{W}^o \in \mathbb{R}^{L\times D}$ and $\boldsymbol{W}^r \in \mathbb{R}^{d\times L}$ are parameters of the designed attention-net $f_\psi(\cdot)$, D is the number of nodes and d is the dimension. In addition, $\boldsymbol{h}_{ij}^{a}$ and $\boldsymbol{h}_{ij}^{r}$ are merged into

$$\boldsymbol{h}_{ij}^{m} = \boldsymbol{h}_{ij}^{a} \oplus \boldsymbol{h}_{ij}^{r}, \tag{4}$$

where $\oplus$ is element-wise addition. To improve the information extraction capabilities of the network, and get the new representation and class probability vector of instance, we add the following fully connected layers:

$$\begin{aligned} \boldsymbol{l}_{ij}^{h} &= \text{sigmoid}(\boldsymbol{h}_{ij}^{m}\boldsymbol{W}^{h}), \\ \boldsymbol{l}_{ij}^{e} &= \text{sigmoid}(\boldsymbol{l}_{ij}^{h}\boldsymbol{W}^{e}), \\ \boldsymbol{p}_{ij} &= \text{sigmoid}(\boldsymbol{l}_{ij}^{e}\boldsymbol{W}^{c}), \end{aligned} \tag{5}$$

where $\boldsymbol{W}^h \in \mathbb{R}^{L\times H}$, $\boldsymbol{W}^e \in \mathbb{R}^{H\times E}$, $\boldsymbol{W}^c \in \mathbb{R}^{H\times C}$, H and E are the number of nodes.

Network $f_\psi(\cdot)$ will participate in the construction of the embedding function, and the most basic requirement it needs to meet is to determine the class of $\mathbf{x}_{ij}$. Therefore, we set the input of $f_\psi(\cdot)$ to $\mathbf{x}_{ij}$, and its label will be assigned as y_i. The benefits include the following: a) Instance label and bag label to ensure uniformity; b) The training process is not affected by the bag structure; and c) With appropriate modifications, most existing MIL neural networks can be applied. Finally, we define the loss function as

$$\ell = -\sum_{i=1}^{N}\sum_{j=1}^{n_i} \log \frac{\exp(p_{ij}^{y_i})}{\sum_c \exp(p_{ij}^{c})}, \tag{6}$$

where $p_{ij}^{c} \in \boldsymbol{p}_{ij}$.

3.2 The Bag-Level Embedding

The embedding function is used to transform a bag into a new feature space, and its general definition is as follows:

$$\mathcal{F}^B(\mathbf{B}_i) \mapsto \mathbf{B}_i = [d(\mathbf{B}_1, \boldsymbol{K}_1), \ldots, d(\mathbf{B}_{|\mathcal{K}|}, \boldsymbol{K}_{|\mathcal{K}|})], \tag{7}$$

where $\mathcal{K}$ is a key sample set derived from the bag space $\mathcal{B}$, K_i is the i-th sample of $\mathcal{K}$, and $d(\cdot,\cdot)$ is the distance between the bag and the key sample. By specifying the size of $\mathcal{K}$, the bag $\mathbf{B}_i$ can be embedded as a vector $\mathbf{b}_i$ in the new feature space.

One disadvantage of Eq. (7) is that the employed $d(\cdot,\cdot)$ has a significant impact on embedding results. Therefore, by considering the probability distribution of instances in the bag, we design a new bag-level embedding without distance metrics as

$$\mathcal{F}^N(\mathbf{B}_i) \mapsto \mathbf{b}_i = \frac{1}{n_i}\sum_{j=1}^{n_i} \mathcal{F}^I(\mathbf{x}_{ij}), \tag{8}$$

where

$$\mathcal{F}^I(\mathbf{x}_{ij}) = \|_{c=1}^{C} p_{ij}^c \boldsymbol{l}_{ij}^e, \tag{9}$$

where $\|$ denotes concatenation. For example, for an input instance $\mathbf{x}_{ij}$, we assume that the corresponding outputs are $\boldsymbol{p}_{ij} = [0.8, 0.2]$ and $\boldsymbol{l}_{ij}^e = [0.4, 0.5]$. As a result, $\mathcal{F}^I(\mathbf{x}_{ij}) = [0.32, 0.4]\|[0.08, 0.1] = [0.32, 0.4, 0.08, 0.1]$. The advantage of this strategy is that the difference between the embedding results of the two instances will be positively correlated with their class probability vectors.

Algorithm 1 presents the pseudo code of the AEMI algorithm. Line 1 generates the instance space $\mathcal{X}$ by collecting all instances of the data set $\mathcal{B}$, and uses it as the input to the attention-net $f_\psi(\cdot)$. Line 2 assigns the label of instance $\mathbf{x}_{ij}$ as $y_i \in \mathbf{Y}$ with the goal of allowing $f_\psi(\cdot)$ to distinguish as accurately as possible the instances in different classes of bags. Line 3 generates a single-instance label vector $\mathcal{L}$ and uses it to participate in the loss calculation. Line 4 trains $f_\psi(\cdot)$ with these generations. Lines 5–9 embed each bag $\mathbf{B}_i$ into $\boldsymbol{b}_i$ according to the designed embedding function, and merge it into $\mathbf{X}$.

Based on the set of embedding vector $\mathbf{X}$ and its corresponding label vector $\mathbf{Y}$, we can train a classification model $\mathcal{M}$ with a traditional single-instance classifier.

Algorithm 1. The AEMI algorithm

Input:

Data set $\mathcal{B}$;

Label vector $\mathbf{Y}$;

Output:

Single-instance classifier $\mathcal{M}$;

The trained neural network $f_\psi(\cdot)$;

1: $\mathcal{X} = \{\mathbf{x}_{ij} | i \in [1..N], j \in [1..n_i]\}$, where $\mathbf{x}_{ij} \in \mathbf{B}_i \in \mathcal{B}$, $N = |\mathcal{B}|$ and $n_i = |\mathbf{B}_i|$;
2: Assign the label of instance $\mathbf{x}_{ij}$ as y_i, where $y_i \in Y$ is the label of $\mathbf{B}_i$;
3: Generate the single-instance label vector $\mathcal{L}$ by collecting all the assigned instance labels;
4: Train a neural network $f_\psi(\cdot)$ with $\mathcal{X}$ and $\mathcal{L}$;
5: $\mathbf{X} = \emptyset$;
6: **for** $(i \in [1..N])$ **do**
7: Embed bag $\mathbf{B}_i$ into $\mathbf{b}_i$ according to Eq. (8) with $f_\psi(\cdot)$;
8: $\mathbf{X} \leftarrow \mathbf{X} \cup \{\mathbf{b}_i\}$;
9: **end for**
10: Train a single-instance classifier $\mathcal{M}$ with $\mathbf{X}$ and $\mathbf{Y}$;
11: Output $\mathcal{M}$ and $f_\psi(\cdot)$;

Proposition 1. *The time complexity of Algorithm 1 is $O(\varepsilon dn)$, where ε, d, and n are the number of epochs, dimensions, and total instances in all bags, respectively.*

Proof. Let the number of bags be N. The instance space $\mathcal{X}$ and its corresponding label vector $\mathcal{L}$ are generated in Lines 1–3. Their time complexity is $O(n)$. In Line 4, the training of neural network costs $O(\varepsilon dn)$. Lines 5–9 embed each bag into a new feature space, which costs $O(dn)$. Line 10 trains a single-instance classifier, which costs $O(EN)$. Generally, we have $E < d$ and $N \ll n$. Therefore, the total time complexity of AEMI is $O(\varepsilon dn)$.

3.3 Scheme Analysis

The following are two characteristics of the designed attention-net: a) Adaptability: The structure of this attention-net is adaptively adjusted according to the size of the data set, i.e., any dimensional instance can be represented in a vector. To put it another way, this network can function normally with the default parameter settings; and b) Interpretability: Ideally, $\forall \mathbf{x}_{ij} \in \mathbf{B}_i, y_i = c, p_{ij}^c \geq p_{ij}^k$, where $k \in [1..C]$. Therefore, our goal is to construct an embedding function with higher distinguishability by making the training results of the designed network fit this state as much as possible.

By combining embedding-based approaches with neural networks, the AEMI algorithm is designed to transform MIL tasks to the SIL ones. With this algorithm, each bag $\mathbf{B}_i$ can be embedded as a vector $\mathbf{b}_i \in \mathbb{R}^{CE}$ in the new feature space. According to embedding results, we may encounter such a dilemma. When $CE \geq d$, where d is the dimension, the increased dimensionality of the embedding vector may cause some noise.

4 Experiments

In this section, we will firstly describe the used data sets and the comparison algorithms. Then, the AEMI algorithm is put to the test in comparison against seven state-of-the-art approaches in a series of experiments.

4.1 Data Sets

We conducted experiments on four types of MIL data sets: Drug activity prediction, text classification, image classification data sets, and web index recommendation data sets. All of these data sets can be found at https://blog.csdn.net/weixin_44575152/article/details/104769348.

Drug Activity Prediction. The benchmark data sets musk1 and musk2 are commonly used in drug activity prediction tasks [4]. Its goal is to predict whether a new molecule can be used to make a drug. In the MIL domain, a musk molecule is represented as a bag with a variable number of 166-dimensional instances. According to the basic MIL assumption, a molecule is positive iff it possesses at least one instance that can be used to make a drug; otherwise negative.

Text Categorization. To conduct experiments, we employed ten text data sets derived from the Newsgroups corpus. Each data set contains 50 positive and 50 negative bags. Each positive bag contains 3% of posts from the specified positive class and the rest from other classes, whereas instances of negative bags are randomly drawn from the non-main class. Each instance is also represented by a 200-dimensional TFIDF feature.

Image Classification. Corel with 100 categories is a famous database for the image classification task [3]. Each category contains 100 images in JPG format with a shape of 187×126 or 126×187. Elephant and tiger are from the Corel database, and all of them have been preprocessed by the Blobworld bag generator. To consider a more challenging scenario, we built ten mnist-bag data sets with the mnist classification data set. Take mnist0 as an example. The generation details are as follows: a) Set the number of positive and other class bags to 100; b) Set the minimum and maximum size of bags to 10 and 50, respectively; c) Set the minimum and maximum number of the positive instances in the positive bag to 2 and 8, respectively; d) Each positive instance is an image randomly selected from the 0-th class of the mnist data set, while the other instance is from the other classes; and e) The selected image will be stretched as a 786-dimensional instance. The random seed of the generating algorithm will be fixed for experimental fairness.

Web Index Recommendation. The purpose of web index recommendation is to recommend interesting web page indexes to particular users. Each of the nine sub data sets in the web data set corresponds to a user's evaluation of a web page [18]. Each web page serves as a bag and links on the page serves as instances. Since web page processing is connected to word frequency, web data sets have high-dimensionality and sparsity.

4.2 Comparative Algorithms

As a comparison, we employed seven start-of-the-art MIL classification algorithms listed below: a) BAMIC [16] and miVLAD [12] use the clustered centers of bag- and instance-level kMeans as key samples, respectively; b) miFV [12] uses the Gaussian mixture model to extract information of the data set; c) MILDM [13] selects the key samples with the discriminative instance evaluation criterion; d) MILFM [7] treats all instances of positive bags and the clustered centers of other bags as key samples; and e) Attention-net [8] and loss-attention [9] are two popular MIL networks. Table 1 shows the parameter settings for AEMI and the above algorithms.

Table 1. Parameter settings.

Algorithm	Parameter	Setting
AEMI	Epoch for musk, elephant and tiger	100
	Epoch for others	5
	Learning rate	0.001
	Number of nodes E	Number of bags N
BAMIC	Number of clustering centers	N
	Distance metric	Average Hausdorff with Euclidean distance [16]
miFV	Components of Gaussian mixture model	1
miVLAD	Size of Code book	1
MILDM	Distance metric	Bag-instance maximum distance [13] with gamma 1
	Instance selection mode	Global
	Number of discriminative instances	N
MILFM	Number of clustering centers	50
	Distance metric	Same as MILDM
Attention-net	Epoch and Learning rate	Same as AEMI
Loss-attention	Epoch	Same as AEMI
	Learning rate	0.0001

4.3 Experimental Results

Tables 2 shows the experimental results of the AEMI and seven rival algorithms based on three SIL classifier kNN, SVM and J48. The best accuracy value for each data set is highlighted with "•". Average ($d < 1000/d \geq 1000$) denotes the average classification performance across data sets in the specified dimension range. The results demonstrate that the AEMI algorithm has a significant advantage on web recommendation and the mnist data sets. Specifically, the accuracy of AEMI is about 10% greater than that of competing methods on some data sets, such as mnist9 and web4, and the average ones are 3.6% than in second place and 30.4% than in penultimate place when $d < 1000$. The following reasons may apply: a) The proposed attention-net can effectively extract information from web data sets and generate the new representation and class probability vector for instances; and b) The embedding mechanism converts the bag into the new feature space while preserving as much information as possible.

Furthermore, some results necessitate further care. a) On the text categorization data sets, AEMI achieves a moderate outcome, while BAMIC, miFV, and miVLAD get relatively large advantages. For example, miFV has a considerable edge on the news.mf data set. The reason for this could be that the Gaussian mixture model of miFV can effectively mine the information of this type of data sets. While the news.mf's embedding results of AEMI may contain some noise; and b) MILDM and MILFM have inadequate impacts on text and web data sets. All three methods find key instances in the specified instance space. Take the "flower"/"other" images as an example, the number of "flower"-instances is usually less than the number of "other"-instances. As a result, these selected key instances may not be "key".

Table 2. Performance comparison between AEMI and rival algorithms. Experiments were run 5 times 10CV and an average of the classification accuracy (± the standard deviation) is reported.

Data set	(d)	BAMIC	miFV	miVLAD	MILDM	MILFM	Attention-net	Loss-attention	**AEMI**
Musk1	(166)	0.891 ± 0.011	0.920 ± 0.008•	0.847 ± 0.011	0.824 ± 0.025	0.871 ± 0.005	0.884 ± 0.022	0.890 ± 0.020	0.867 ± 0.019
Musk2	(166)	0.860 ± 0.011•	0.848 ± 0.015	0.780 ± 0.054	0.826 ± 0.016	0.822 ± 0.035	0.822 ± 0.047	0.848 ± 0.019	0.804 ± 0.010
News.aa	(200)	0.852 ± 0.010	0.834 ± 0.016	0.836 ± 0.027	0.510 ± 0.050	0.510 ± 0.000	0.862 ± 0.019	0.874 ± 0.016•	0.808 ± 0.023
News.cg	(200)	0.812 ± 0.004•	0.802 ± 0.008	0.790 ± 0.014	0.526 ± 0.052	0.504 ± 0.010	0.610 ± 0.017	0.644 ± 0.033	0.782 ± 0.016
News.mf	(200)	0.696 ± 0.019	0.736 ± 0.016•	0.716 ± 0.029	0.488 ± 0.037	0.510 ± 0.006	0.666 ± 0.022	0.716 ± 0.032	0.676 ± 0.037
News.rm	(200)	0.808 ± 0.016	0.877 ± 0.020•	0.812 ± 0.016	0.546 ± 0.041	0.530 ± 0.026	0.854 ± 0.021	0.871 ± 0.026	0.818 ± 0.013
News.rsh	(200)	0.828 ± 0.010	0.884 ± 0.010	0.894 ± 0.010	0.442 ± 0.021	0.500 ± 0.000	0.872 ± 0.005	0.914 ± 0.012•	0.884 ± 0.022
News.sc	(200)	0.774 ± 0.010	0.750 ± 0.018	0.818 ± 0.023•	0.518 ± 0.042	0.512 ± 0.004	0.780 ± 0.014	0.802 ± 0.036	0.800 ± 0.026
News.se	(200)	0.938 ± 0.004•	0.926 ± 0.005	0.918 ± 0.008	0.574 ± 0.061	0.530 ± 0.000	0.554 ± 0.010	0.572 ± 0.036	0.864 ± 0.014
News.tpmd	(200)	0.830 ± 0.000	0.799 ± 0.016	0.832 ± 0.015	0.554 ± 0.019	0.554 ± 0.048	0.836 ± 0.016	0.844 ± 0.012•	0.788 ± 0.033
News.tpmi	(200)	0.690 ± 0.011	0.752 ± 0.015	0.766 ± 0.015•	0.482 ± 0.037	0.506 ± 0.005	0.720 ± 0.013	0.482 ± 0.022	0.710 ± 0.011
News.trm	(200)	0.728 ± 0.008	0.740 ± 0.014	0.786 ± 0.022•	0.466 ± 0.048	0.510 ± 0.011	0.606 ± 0.060	0.514 ± 0.064	0.720 ± 0.026
Elephant	(230)	0.762 ± 0.012	0.852 ± 0.013	0.853 ± 0.010	0.765 ± 0.022	0.817 ± 0.023	0.848 ± 0.014	0.872 ± 0.005	0.875 ± 0.010•
Tiger	(230)	0.704 ± 0.011	0.789 ± 0.006	0.843 ± 0.008•	0.692 ± 0.008	0.754 ± 0.006	0.810 ± 0.031	0.819 ± 0.011	0.814 ± 0.009
Mnist0	(786)	0.913 ± 0.018	0.820 ± 0.009	0.873 ± 0.002	0.484 ± 0.015	0.507 ± 0.002	0.979 ± 0.005	0.995 ± 0.003•	0.985 ± 0.010
Mnist1	(786)	0.978 ± 0.004	0.724 ± 0.013	0.845 ± 0.013	0.803 ± 0.012	0.975 ± 0.006	0.873 ± 0.146	0.992 ± 0.004•	0.980 ± 0.003
Mnist2	(786)	0.773 ± 0.021	0.858 ± 0.009	0.910 ± 0.008	0.462 ± 0.008	0.496 ± 0.028	0.959 ± 0.019	0.966 ± 0.005	0.973 ± 0.005•
Mnist3	(786)	0.865 ± 0.015	0.787 ± 0.005	0.863 ± 0.004	0.556 ± 0.009	0.580 ± 0.024	0.940 ± 0.006	0.942 ± 0.019	0.956 ± 0.006•
Mnist4	(786)	0.855 ± 0.006	0.757 ± 0.011	0.810 ± 0.017	0.451 ± 0.012	0.520 ± 0.034	0.931 ± 0.014	0.896 ± 0.024	0.937 ± 0.011•
Mnist5	(786)	0.759 ± 0.023	0.759 ± 0.016	0.831 ± 0.009	0.487 ± 0.028	0.496 ± 0.008	0.922 ± 0.020	0.838 ± 0.039	0.964 ± 0.006•
Mnist6	(786)	0.914 ± 0.006	0.837 ± 0.017	0.852 ± 0.007	0.466 ± 0.026	0.460 ± 0.034	0.927 ± 0.037	0.963 ± 0.007•	0.959 ± 0.006
Mnist7	(786)	0.908 ± 0.012	0.859 ± 0.013	0.855 ± 0.006	0.530 ± 0.027	0.629 ± 0.037	0.986 ± 0.004•	0.974 ± 0.004	0.975 ± 0.008
Mnist8	(786)	0.786 ± 0.035	0.731 ± 0.020	0.808 ± 0.005	0.494 ± 0.021	0.507 ± 0.002	0.879 ± 0.035	0.749 ± 0.062	0.926 ± 0.014•
Mnist9	(786)	0.837 ± 0.017	0.742 ± 0.017	0.797 ± 0.005	0.583 ± 0.023	0.516 ± 0.017	0.845 ± 0.010	0.771 ± 0.048	0.958 ± 0.005•
Web1	(5863)	0.844 ± 0.016•	0.838 ± 0.007	0.813 ± 0.018	0.838 ± 0.007	0.824 ± 0.012	0.811 ± 0.015	0.811 ± 0.140	0.809 ± 0.013
Web2	(6519)	0.806 ± 0.024	0.826 ± 0.007	0.818 ± 0.013	0.833 ± 0.009	0.820 ± 0.023	0.807 ± 0.019	0.820 ± 0.006	0.838 ± 0.018•
Web3	(6306)	0.815 ± 0.024	0.826 ± 0.009	0.827 ± 0.012•	0.826 ± 0.007	0.815 ± 0.020	0.813 ± 0.009	0.813 ± 0.007	0.813 ± 0.023
Web4	(6059)	0.765 ± 0.004	0.807 ± 0.012	0.844 ± 0.015	0.806 ± 0.015	0.804 ± 0.020	0.844 ± 0.027	0.785 ± 0.009	0.916 ± 0.011•
Web5	(6407)	0.789 ± 0.004	0.782 ± 0.061	0.822 ± 0.014	0.787 ± 0.021	0.781 ± 0.011	0.822 ± 0.015	0.776 ± 0.011	0.895 ± 0.009•
Web6	(6417)	0.809 ± 0.019	0.778 ± 0.005	0.847 ± 0.016	0.846 ± 0.021	0.816 ± 0.022	0.811 ± 0.020	0.782 ± 0.005	0.920 ± 0.012•
Web7	(6450)	0.558 ± 0.016	0.687 ± 0.030	0.742 ± 0.012	0.602 ± 0.037	0.566 ± 0.018	0.713 ± 0.021	0.485 ± 0.031	0.786 ± 0.019•
Web8	(5999)	0.504 ± 0.032	0.706 ± 0.021	0.727 ± 0.021	0.544 ± 0.016	0.576 ± 0.032	0.713 ± 0.012	0.466 ± 0.050	0.806 ± 0.024•
Web9	(6279)	0.500 ± 0.015	0.753 ± 0.022	0.758 ± 0.021	0.551 ± 0.021	0.591 ± 0.021	0.724 ± 0.039	0.503 ± 0.021	0.809 ± 0.026•
Average ($d < 1000$)		0.823 ± 0.012	0.808 ± 0.013	0.831 ± 0.014	0.564 ± 0.028	0.588 ± 0.015	0.832 ± 0.025	0.823 ± 0.023	0.868 ± 0.014•
Average ($d \geq 1000$)		0.710 ± 0.017	0.778 ± 0.019	0.800 ± 0.016	0.737 ± 0.017	0.733 ± 0.020	0.784 ± 0.020	0.693 ± 0.031	0.844 ± 0.017•

Table 3 shows the comparison results of the maximum and minimum classification accuracy of the AEMI algorithm and an attention network method. The terms "net" represents the gate-attention network [8] used for comparison and "our" denotes specifically to the comparison of AEMI's SVM classification results. The symbol $\diamond/\star$ means that the difference between the maximum value minus the minimum value is greater than or equal to 0.05/0.1. The results show that AEMI can alleviate the instability of the neural network caused by parameter initialization without reducing the classification performance. In the mnist2 data set, for example, the net method's accuracy varies by 37.5%, while ours varies by only 1%.

Table 3. Performance comparison between AEMI and gate-attention network. Experiments were run 5 times 10CV and minimum/maximum of the classification accuracy (± the standard deviation) is reported.

Data set	Net (Min)	Net (Max)	Our (Min)	Our (Max)
Mnist0	⋄0.915	⋄0.970	0.970	0.995
Mnist1	⋆0.585	⋆0.960	0.975	0.985
Mnist2	0.940	0.950	0.965	0.980
Mnist3	0.935	0.980	0.955	0.965
Mnist4	⋄0.895	⋄0.945	0.920	0.950
Mnist5	⋆0.860	⋆0.960	0.955	0.970
Mnist6	⋆0.775	⋆0.960	0.950	0.965
Mnist7	0.975	0.990	0.965	0.985
Mnist8	0.875	0.915	0.910	0.945
Mnist9	⋄0.820	⋄0.870	0.950	0.960
Web1	0.800	0.846	0.791	0.827
Web2	0.773	0.818	⋄0.809	⋄0.864
Web3	0.800	0.836	⋄0.791	⋄0.855
Web4	⋄0.818	⋄0.873	0.900	0.927
Web5	⋄0.773	⋄0.855	0.882	0.900
Web6	⋄0.791	⋄0.855	0.900	0.927
Web7	0.700	0.746	0.764	0.809
Web8	0.709	0.745	⋄0.764	⋄0.836
Web9	⋄0.682	⋄0.736	⋄0.773	⋄0.846

5 Conclusion and Further Work

We propose the AEMI algorithm to train an attention-net based on the relationship between the bag and its instances, and use an embedding function to transform MIL tasks into SIL ones. The experimental results of studies prove that AEMI is superior to state-of-the-art MIL classification methods, has significant advantages, especially on web data sets, and has relatively stable classification performance. In addition, the majority of the rival MIL methods perform poorly on MIL web recommendation and mnist, and the neural network-based methods' outcomes of successive experiments may be substantially different due to the random setting of the parameter initialization.

The following topics deserve further investigation:

- More flexible embedding functions. Web data sets with thousands of features can be effectively reduced in dimensionality using the proposed embedding function. However, this may increase the dimensionality of these relatively low-dimensional data sets after embedding. As a result, these may be a factor in their moderate performance on some data sets, such as musk1 and tiger.

– More efficient neural networks. On most data sets, the designed attention-net only requires 5 epochs of training to achieve good results, but on few data sets like musk1, it requires more epochs. Some details are shown in Table 1.

Acknowledgements. This work was supported in part by the National Key R&D Program of China (2018YFE0203900), National Natural Science Foundation of China (61773093), Sichuan Science and Technology Program (2020YFG0476), Important Science and Technology Innovation Projects in Chengdu (2018-YF08-00039-GX), and Central Government Funds of Guiding Local Scientific and Technological Development (2021ZYD0003).

References

1. Angelidis, S., Lapata, M.: Multiple instance learning networks for fine-grained sentiment analysis. Trans. Assoc. Comput. Linguist. **6**, 17–31 (2018)
2. Bahdanau, D., Cho, K., Bengio, Y.: Neural machine translation by jointly learning to align and translate (2014)
3. Chen, Y.X., Bi, J.B., Wang, J.Z.: MILES: multiple-instance learning via embedded instance selection. IEEE Trans. Pattern Anal. Mach. Intell. **28**(12), 1931–1947 (2006)
4. Dietterich, T.G., Lathrop, R.H., Lozano-Pérez, T.: Solving the multiple instance problem with axis-parallel rectangles. Artif. Intell. **89**(1–2), 31–71 (1997)
5. Fu, J.L., Zheng, H.L., Mei, T.: Look closer to see better: recurrent attention convolutional neural network for fine-grained image recognition. In: Computer Vision and Pattern Recognition, pp. 4438–4446 (2017)
6. He, C.K., Shao, J., Zhang, J.S., Zhou, X.M.: Clustering-based multiple instance learning with multi-view feature. Expert Syst. Appl. **162**, 113027 (2020)
7. Hong, R.C., Wang, M., Gao, Y., Tao, D.C., Li, X.L., Wu, X.D.: Image annotation by multiple-instance learning with discriminative feature mapping and selection. IEEE Trans. Cybern. **44**(5), 669–680 (2014)
8. Ilse, M., Tomczak, J., Welling, M.: Attention-based deep multiple instance learning. In: International Conference on Machine Learning, pp. 2127–2136 (2018)
9. Shi, X.S., Xing, F.Y., Xie, Y.P., Zhang, Z.Z., Cui, L., Yang, L.: Loss-based attention for deep multiple instance learning. In: Association for the Advancement of Artificial Intelligence, pp. 5742–5749 (2020)
10. Tarragó, D.S., Cornelis, C., Bello, R., Herrera, F.: A multi-instance learning wrapper based on the Rocchio classifier for web index recommendation. Knowl.-Based Syst. **59**, 173–181 (2014)
11. Wang, X.G., Yan, Y.L., Tang, P., Bai, X., Liu, W.Y.: Revisiting multiple instance neural networks. Pattern Recogn. **74**, 15–24 (2016)
12. Wei, X.S., Wu, J.X., Zhou, Z.H.: Scalable algorithms for multi-instance learning. IEEE Trans. Neural Netw. Learn. Syst. **28**(4), 975–987 (2017)
13. Wu, J., Pan, S., Zhu, X., Zhang, C., Wu, X.: Multi-instance learning with discriminative bag mapping. IEEE Trans. Knowl. Data Eng. **30**(6), 1065–1080 (2018)
14. Xu, B.C., Ting, K.M., Zhou, Z.H.: Isolation set-kernel and its application to multi-instance learning. In: Special Interest Group on Knowledge Discovery and Data Mining, pp. 941–949 (2019)
15. Yang, M., Zhang, Y.X., Wang, X.Z., Min, F.: Multi-instance ensemble learning with discriminative bags. IEEE Trans. Syst.Man Cybern. Syst., 1–12 (2021)

16. Zhang, M.L., Zhou, Z.H.: Multi-instance clustering with applications to multi-instance prediction. Appl. Intell. **31**(1), 47–68 (2009)
17. Zhang, W.J., Liu, L., Li, J.Y.: Robust multi-instance learning with stable instances, pp. 1682–1689. arXiv:1902.05066 (2020)
18. Zhou, Z.H., Jiang, K., Li, M.: Multi-instance learning based web mining. Appl. Intell. **22**, 135–147 (2005)
19. Zhou, Z.H., Sun, Y.Y., Li, Y.F.: Multi-instance learning by treating instances as non-I.I.D. samples. In: International Conference on Machine Learning, pp. 1249–1256 (2009)

Multi-instance Embedding Learning Through High-level Instance Selection

Mei Yang[1], Wen-Xi Zeng[1], and Fan Min[1,2(✉)]

[1] School of Computer Science, Southwest Petroleum University, Chengdu, China
yangmei@swpu.edu.cn

[2] Institute for Artificial Intelligence, Southwest Petroleum University, Chengdu 610500, China
minfan@swpu.edu.cn

Abstract. Multi-instance learning (MIL) handles complex structured data represented by bags and their instances. MIL embedded algorithms based on representative instance selection transform bags into a single-instance space. However, they may select weak representative instances due to the ignorance of the internal bag structure. In this paper, we propose the multi-instance embedding learning through high-level instance selection (MIHI) algorithm with two techniques. The fast bag-inside instance selection technique obtains instance prototypes of each bag. It fully utilizes the bag information using our new density and affinity metrics. Based on the instance prototypes, the high-level instance selection technique chooses instances using the peak density metric. It obtains high-level instances with higher representative power than the instance prototypes. Experiments were conducted on six learning tasks and nine comparison algorithms. The results confirmed that MIHI achieved better performance in terms of efficiency and classification accuracy. This method, in particular, has a substantial advantage in image retrieval and web data sets.

Keywords: Embedding · High-level instance · Instance selection · Multi-instance learning

1 Introduction

Compared with traditional single-instance learning (SIL), multi-instance learning (MIL) is the study of bags containing multiple instances. Taking the drug activity prediction as an example, molecules and their isomers are viewed as bags and instances, respectively. The task is to predict whether the new molecule is suitable for making drugs. A molecule is positive if at least one of its isomers can be used to make drugs, otherwise it is negative. Furthermore, multi-instance problems are common in real-world application scenarios, such as image retrieval [2], text classification [21], and web index recommendation [14].

In recent years, many embedded MIL algorithms based on instance prototypes have been widely proposed. Their common strategies tend to perform clustering in the entire instance space to select instance prototypes [10,15]. MILFM

J. Gama et al. (Eds.): PAKDD 2022, LNAI 13281, pp. 122–133, 2022.
https://doi.org/10.1007/978-3-031-05936-0_10

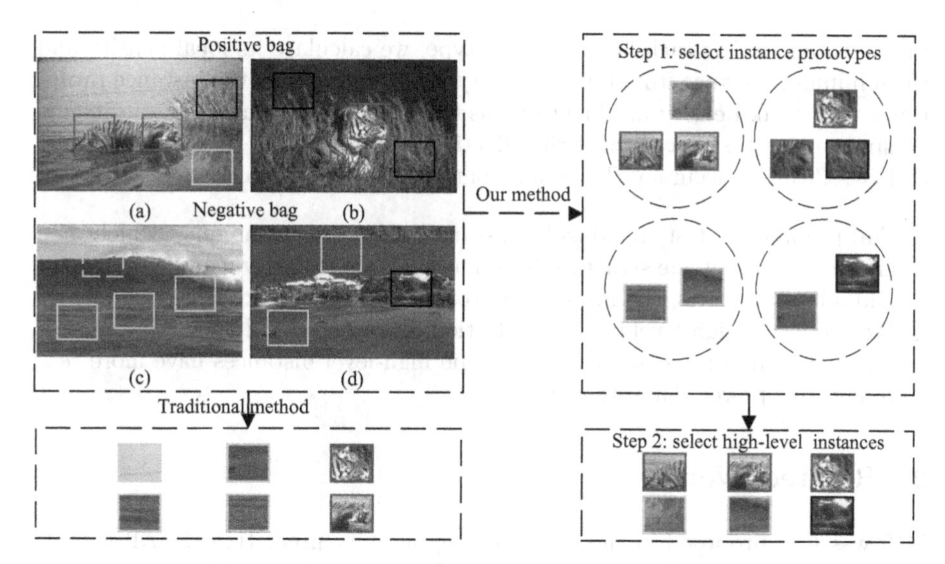

Fig. 1. The main framework of MIHI is compared with traditional methods. Traditional methods usually use clustering algorithms to select instance prototypes in the entire data space. Our method first selects the instance prototypes from the bag, and then selects the high-level instances from the instance prototypes.

[10] first selects instance prototypes in the entire instance space, and selects cluster instances from the negative bags. CMIL [9] only divides the instances of the positive bag into multiple clusters, and selects the instances with the high score in each cluster as the instance prototypes. However, two dilemmas will be encountered: 1) The cardinality of the instance space is much larger than that of the bag space; and 2) The number of negative instances is far greater than that of positive instances. As a result, the classification effectiveness may be reduced. Figure 1 shows an example of tiger image classification task. In subgraphs (a) and (b), there are tigers, grass and water. In subgraphs (c) and (d), there are only grass and water. Obviously, grass and water occupy a large proportion of the entire feature space. The instances prototypes chosen by traditional methods are more likely to be grass and water than tigers. However, the selected instances have weak representativeness due to ignoring the internal structure of the bag. Therefore, the selection of highly representative instance prototypes is the key to the embedded MIL algorithms.

In this paper, we propose the multi-instance embedding learning through high-level instance selection (MIHI) algorithm to handle these issues with two techniques. Figure 1 shows the main framework of MIHI. The goal is to select high-level instances with strong representativeness. In Step 1, the fast bag-inside instance selection technique is designed to select instance prototypes from each bag. This technique takes into account the density and affinity of instances in the bag. The instance prototypes highlight the bag's internal structure information. Accordingly, the high-level instance selection technique chooses global representative instances. For each instance prototype, we calculate its local density and the minimum distance from higher-density prototypes. Then the instance prototypes with peak density are identified as the high-level instances. Experiments on six learning tasks confirmed the effectiveness of MIHI in terms of efficiency and classification accuracy. The main contributions of our work are:

- We propose a fast bag-inside instance selection technique, which can effectively exploit the structure information of the bag. By using new density and affinity metrics, the instance prototypes of the bag are found.
- We propose a high-level instance selection technique based on instance prototypes. Through peak density metric, the high-level instances have more representative power than other prototypes.

2 Related Work

MIL was first proposed in the study of drug activity prediction [7]. After that, many MIL algorithms have been proposed. They are mainly divided into two categories: 1) Basic methods predict the bag label based on the structural characteristics of bag [21] or instance [8] spaces; and 2) Embedding methods transform MIL into SIL based on representative samples [3,17].

The basic methods mainly handle MIL problems by designing a bag-level kernel. mi-SVM and MI-SVM [2] treat bags as samples and use support vector machines to handle problems. mi-SVM tries to identify the maximum edge hyperplane for the instances. Its constraint is that at least one instance of each positive bag is located in the positive half space. MI-SVM treats the edge of the most positive instance as the edge of the bag. The purpose is to identify the maximum edge hyperplane of the bag. miGraph [21] proposes an effective bag-level kernel through the affinity matrix. However, it only focuses on the relationship between bags and fails to extract instance-level information.

The bag embedding methods deal with MIL problems by transforming the space. DD-SVM [4] learns a set of instance prototypes by using Diverse Density. Then the bags are embedded into the new feature space based on the instance prototypes. MILES [3] uses a joint strategy based on all instances to implement bag embedding. Bamic [22] selects the representative bags through unsupervised learning. MIKI [19] first trains a weighted multi-class model to select instance prototypes with high positiveness. Then the bag is converted into a vector with instance prototype information. To narrow the gap between the training and testing distribution, the weights of the instance prototypes are combined into the converted bag vector. However, these algorithms directly select instance prototypes in the entire feature space, ignoring the internal structure of the bag. As a result, they may choose weakly representative instances and affect classification performance. MIHI provides a solution for selecting high-level instances through two techniques.

3 The Proposed Algorithm

In this section, we first give the basic symbol definition of MIL. Then we describe the proposed MIHI algorithm process. Furthermore, two key techniques of MIHI are described in detail.

3.1 Algorithm Description

Algorithm 1 reports the detailed process of the proposed MIHI. Let $\mathcal{T} = \{\boldsymbol{B}_i\}_{i=1}^N$ be the MIL data set with N bags, where $\boldsymbol{B}_i = \{\boldsymbol{x}_{ij}\}_{j=1}^{n_i}$ is a bag containing n_i instances. Let $\mathcal{Y} = \{y_i\}_{i=1}^N$ be the label vector corresponding to $\mathcal{T}$, where $y_i \in \{-1, +1\}$ is the label of $\boldsymbol{B}_i$. Lines 2–11 use two techniques to obtain high-level instance set $\boldsymbol{H}$. By analyzing the internal structure of each bag, at least one instance can be selected to construct the instance prototype set $\boldsymbol{C}$. Specifically, Lines 4–5 calculate the representativeness of the instances in each bag $\boldsymbol{B}_i \in \mathcal{T}$. Lines 6–7 select the top-ranked instances as the instance prototypes (IP). Next, our goal is to generate the high-level instance set $\boldsymbol{H}$ by identifying $\boldsymbol{C}$. Lines 9–11 select instances with peak density from $\boldsymbol{C}$ to construct $\boldsymbol{H}$. We design an embedding function to transform each bag into a single instance in the new feature space. Lines 13–17 embed each bag $\boldsymbol{B}_i$ into a new feature vector $\boldsymbol{V}_i$ through $\boldsymbol{H}$. Finally, Line 18 trains the SIL classifier $\mathcal{F}(\cdot)$ through the new data set $\{(\boldsymbol{V}_i, y_i)\}_{i=1}^N$.

3.2 The Fast Bag-Inside Instance Selection Technique

The common method for instance prototype (IP) generation is to select cluster centers [15] or causal instances [18] in the entire feature space. However, these methods have the following two problems: a) High time complexity; and b) The selected instances have no bag structure information. The fast bag-inside instance selection technique chooses instance prototypes of each bag through using its internal structure. The density ρ_{ij} and affinity l_{ij} metric of the instance $\boldsymbol{x}_{ij}$ are computed as follows.

The Density of Instance. For each instance $\boldsymbol{x}_{ij} \in \boldsymbol{B}_i$, the density ρ_{ij} is defined as

$$\rho_{ij} = \sum_{k \neq j}^{n_i} \exp -(\frac{d_{jk}}{d_c})^2, \tag{1}$$

where d_c is a cutoff distance and d_{jk} is the distance between $\boldsymbol{x}_{ij}$ and $\boldsymbol{x}_{ik}$. High-density instances mean that there are more adjacent instances within a given neighborhood radius. Therefore, high-density instances can reflect the local feature distribution of the bag.

In addition, the instances in the bag are not completely independent and distributed [21]. It is not enough to determine the representativeness only based on the density of the instance. Therefore, we use cosine similarity to represent the

Algorithm 1. Multi-instance embedding learning through high-level instance selection.

Input:
The data set $\mathcal{T}$;
The label vector $\mathcal{Y} = \{y_i\}_{i=1}^{N}$;
The proportion of instance prototypes r_c;
The number of high-level instances n_h;
Output:
The SIL classifier $\mathcal{F}(\cdot)$;
The high-level instance set $\boldsymbol{H}$;
1: // Step 1. Select the high-level instances.
2: $\boldsymbol{C} = \emptyset$; // Initialize instance prototype set.
3: **for** $(\boldsymbol{B}_i \in \mathcal{T})$ **do**
4: $k = \lceil r_c \times n_i \rceil$; // The number of instance prototypes of each bag.
5: Compute the score s_{ij} of $\boldsymbol{x}_{ij} \in \boldsymbol{B}_i$ according to Eq. (3);
6: $\boldsymbol{C}'$ = the top-k score instances;
7: $\boldsymbol{C} = \boldsymbol{C} \cup \boldsymbol{C}'$;
8: **end for**
9: $\boldsymbol{H} = \emptyset$; // Initialize high-level instance set.
10: Compute the score λ_i for each prototype $\boldsymbol{c}_i \in \boldsymbol{C}$ according to Eq. (5);
11: $\boldsymbol{H}$ = the set of top-n_h score prototypes;
12: // Step 2. Bag embedding.
13: **for** $(\boldsymbol{B}_i \in \mathcal{T})$ **do**
14: Compute the embedding vector $\boldsymbol{V}_i$ according to Eq. (7) or (8) with $\boldsymbol{B}_i$;
15: $V_{il} \leftarrow sign(V_{il})\sqrt{|V_{il}|}$, where V_{il} represents the l-th attribute of $\boldsymbol{V}_i$;
16: $\boldsymbol{V}_i \leftarrow \boldsymbol{V}_i / \parallel \boldsymbol{V}_i \parallel_2$;
17: **end for**
18: Train the classifier $\mathcal{F}(\cdot)$ with the new data set $\{(\boldsymbol{V}_i, y_i)\}_{i=1}^{N}$;
19: Output $\mathcal{F}(\cdot)$ and $\boldsymbol{H}$;

affinity between instances. The closer the cosine similarity of the two instances is to 1, the more similar they are.

The Affinity of Instance. For each instance $\boldsymbol{x}_{ij} \in \boldsymbol{B}_i$, the affinity l_{ij} is defined as

$$l_{ij} = \sum_{1 \le k \le n_i} \frac{\boldsymbol{x}_{ij} \cdot \boldsymbol{x}_{ik}}{\|\boldsymbol{x}_{ij}\| \|\boldsymbol{x}_{ik}\|}, \tag{2}$$

where $j, k \in [1..n_i]$.

After obtaining the density and affinity of each instance in the bag, the representativeness score s_{ij} of the instance can be computed as

$$s_{ij} = \rho_{ij} \times l_{ij}. \tag{3}$$

According to the MIL assumption, the proportion of positive and negative instances in each bag is different (e.g., tiger, grass and water in Fig. 1). Therefore, we can chose the low/high score instances from the positive/negative bag as the IP. Finally, we can obtain the instance prototype set $\boldsymbol{C} = \{\boldsymbol{c}_1, \cdots, \boldsymbol{c}_{n_c}\}$, where n_c is the cardinality of $\boldsymbol{C}$.

By considering the solution interval of the optimization objective, we design three types of instance prototypes selection modes as follows:

- **Global (G)** selects $\lceil r_c \times n_i \rceil$ instance prototypes from each bag.
- **Positive (P)** only selects from all positive bags.
- **Negative (N)** only selects from all negative bags.

The time complexity of the fast bag-inside instance selection technique is $O(dn)$, where d is the dimension and n is the number of all instances. The time complexity of instance selection based on the entire space is $O(dn^2)$. In contrast, our complexity is only linearly related to the number of instances rather than square related.

3.3 High-level Instance Selection Technique

In order to explore the characteristics of the instance space, high-level instance selection technique is proposed. Based on $\boldsymbol{C} = \{\boldsymbol{c}_1, \cdots, \boldsymbol{c}_{n_c}\}$, we can obtain high-level instances (HI).

For each $\boldsymbol{c}_i$, we calculate two quantities: its local density δ_i and its minimum distance β_i from the higher density prototypes. The local density δ_i is computed by Eq. (1). The difference is that the calculation interval is migrated from each bag to $\boldsymbol{C}$. The distance β_i is measured by computing the minimum distance between the $\boldsymbol{c}_i$ and any other IP with higher density:

$$\beta_i = \min_{j:\delta_j > \delta_i} (d_{ij}). \tag{4}$$

Particularly, for the IP with highest density, its distance is $\beta_i = \max_j(d_{ij})$. Finally the score λ_i of IP is calculated as

$$\lambda_i = \delta_i \times \beta_i. \tag{5}$$

With the scores of all IP calculated by Eq. (5), we select the top-n_h IP as the HI. Finally, we can obtain $\boldsymbol{H} = \{\boldsymbol{h}_1, \cdots, \boldsymbol{h}_{n_h}\}$, where n_h the cardinality of $\boldsymbol{H}$.

3.4 Embedding Technique via HI

After getting $\boldsymbol{H}$, we design the following method to embed the bags into a new feature space. Firstly, each instance $\boldsymbol{x}_{ij} \in \boldsymbol{B}_i$ is assigned to its nearest $\boldsymbol{h}_k$, denoted by $NH(\boldsymbol{x}_{ij}) = \boldsymbol{h}_k$. Then, each bag $\boldsymbol{B}_i$ can be expressed by n_h local vectors $\boldsymbol{v}_{ik}$:

$$\boldsymbol{v}_{ik} = \sum_{\boldsymbol{x}_{ij} \in \Omega} \boldsymbol{x}_{ij} - \boldsymbol{h}_k, \tag{6}$$

where $\Omega = \{\boldsymbol{x}_{ij} | NH(\boldsymbol{x}_{ij}) = \boldsymbol{h}_k\}$. Finally, the embedding vector $\boldsymbol{V}_i$ of bag $\boldsymbol{B}_i$ is a D-dimensional vector composed of concatenated local vectors [15]:

$$\boldsymbol{V}_i = \Big\|_{k=1}^{n_h} \boldsymbol{v}_{ik}, \tag{7}$$

where $D = n_h \times d$ and d is the dimension of instance $\boldsymbol{x}_{ij}$. However, the above embedding method will embed each bag into a high-dimensional space. Therefore, we design the second embedding method, which superimposes all the local vectors to get the embedding vector:

$$\boldsymbol{V}_i = \sum_{k=1}^{n_h} \boldsymbol{v}_{ik}. \tag{8}$$

Furthermore, each element of $\boldsymbol{V}_i$ is processed by $V_{il} \leftarrow sign(V_{il})\sqrt{|V_{il}|}$, and then the embedding vector is normalized by $\boldsymbol{V}_i \leftarrow \boldsymbol{V}_i / \parallel \boldsymbol{V}_i \parallel_2$ [11]. After getting the $\boldsymbol{V}_i$ for each $\boldsymbol{B}_i$, we can predict the bag label by processing $\boldsymbol{V}_i$ with any single-instance classifier $\mathcal{F}(\cdot)$ (e.g., SVM).

4 Experiments

In this section, we conducted experiments on MIHI and 9 comparison algorithms for six learning tasks. To ensure the validity of the experiment, we used 10 times 10-fold cross-validation to calculate the average accuracy. The averaged results (mean) and standard deviation (std) of each algorithm is reported.

4.1 Comparison Algorithms

We compared MIHI with 9 state-of-the-art algorithms: a) MILES [3] embeds bags based on the bag-instance similarity measure and all instances; b) BAMIC [22] embeds bags by employing bag-level k-means, with the parameters including average Hausdorff distance and the number of clustering centers ($r \times \min\{N, 100\}$, where r is enumerated in $\{0.1, \cdots, 1.0\}$); c) MILFM [10] uses AdaBoost to select the bag features embedded by instance prototypes, with the parameters including the number of cluster centers (40); d) Simple-MI [1] uses the arithmetic mean of instances in the bag as the representation of the bag itself. e) miFV [15] extracts the instance information with the Gaussian mixture model (GMM), with the parameters including the number of components for GMM (enumerate in $\{1, 2, 3\}$); f) miVLAD [15] embeds bags based on the instance-level k-means, with the parameters including the number of clustering centers (enumerate in $\{1, 2\}$); g) MILDM [16] selects the discriminative instances via instance evaluation criteria, with the parameters including the size of discriminative instance pool (the number of bags); h) StabelMIL [18] embeds bags based on causal instances, with the parameters including the scale variable (0.25); and i) ELDB [17] selects more representative bags with the discriminative analysis and reinforcement technique, and finally obtains more distinguishable single vectors.

4.2 Experimental Data Sets

Six fields of learning tasks across 26 data sets are used to validate MIHI. We briefly introduce the domain knowledge of these data sets: 1) **Image retrieval:**

Content-based image retrieval problems include identifying the expected target object in the image [2]. In our experiments, elephant, fox, and tiger data sets are used; 2) **Mutagenicity prediction:** Mutagenesis is a drug activity prediction problem. There are two versions, easy (1) and hard (2), of the data set [13]; 3) **Medical image:** Messidor is a medical classification problem data set, which consists of 1,200 fundus images from 546 healthy and 654 diabetic patients [5]; 4) **Newsgroups:** The newsgroups is a text categorization data set [21]. Posts from different newsgroups form a bag. Each category has 50 positive bags and 50 negative bags; 5) **Web recommendation:** The question is whether to classify web pages as interesting web pages [20]. There are a total of 9 users who rate the web page this way, so there are 9 different data sets; A web page is a bag, and the links on the web page are instances; and 6) **Biocreative:** Biocreative is a large-scale text classification data set [12]. The task is to decide whether some genetic ontology (GO) code should be used to annotate a given pair.

4.3 Performance Comparison

Table 1. Accuracy ($\%, mean_{\pm std}$) with standard deviations on 26 MIL data sets. The highest average accuracy is marked with •.

Datasets	(d)	MILES	BAMIC	MILFM	Simple-MI	miFV	miVLAD	MILDM	StableMIL	ELDB	**MIHI**
Elephant♣	(230)	$81.2_{\pm 2.36}$	$75.8_{\pm 0.12}$	$81.7_{\pm 0.23}$	$80.2_{\pm 0.08}$	$84.2_{\pm 0.09}$	$84.1_{\pm 0.13}$	$76.1_{\pm 0.29}$	$84.2_{\pm 4.23}$	$75.8_{\pm 3.21}$	$90.4_{\pm 0.93}$•
Fox♣	(230)	$58.5_{\pm 3.63}$	$51.9_{\pm 0.33}$	$45.4_{\pm 0.24}$	$62.6_{\pm 0.13}$	$61.9_{\pm 0.12}$	$62.3_{\pm 0.16}$	$58.8_{\pm 0.41}$	$55.3_{\pm 2.83}$	$60.7_{\pm 2.02}$	$65.5_{\pm 2.46}$•
Tiger ♣	(230)	$77.1_{\pm 2.65}$	$69.2_{\pm 0.14}$	$73.3_{\pm 3.09}$	$79.3_{\pm 0.07}$	$77.2_{\pm 0.06}$	$84.8_{\pm 0.12}$	$64.3_{\pm 0.14}$	$60.7_{\pm 2.24}$	$72.2_{\pm 2.00}$	$85.7_{\pm 1.10}$•
Mutagenesis1◊	(7)	$88.3_{\pm 2.11}$	$75.8_{\pm 0.09}$	$84.8_{\pm 0.21}$	$66.6_{\pm 0.06}$	$79.2_{\pm 0.08}$	$77.6_{\pm 0.13}$	$81.1_{\pm 0.28}$	$88.3_{\pm 2.11}$•	$84.9_{\pm 1.71}$	$70.0_{\pm 1.91}$
Mutagenesis2◊	(7)	$84.2_{\pm 1.60}$	$82.8_{\pm 0.17}$	$83.5_{\pm 0.12}$	$68.8_{\pm 0.17}$	$79.0_{\pm 0.17}$	$78.8_{\pm 0.32}$	$81.7_{\pm 0.34}$	$85.3_{\pm 0.02}$•	$82.8_{\pm 0.83}$	$82.0_{\pm 1.50}$
Messidor♦	(687)	$50.3_{\pm 3.33}$	$62.0_{\pm 0.05}$	$54.5_{\pm 0.00}$	$55.9_{\pm 0.03}$	$71.5_{\pm 0.05}$•	$67.5_{\pm 0.05}$	$54.5_{\pm 0.24}$	$54.5_{\pm 0.01}$	$63.8_{\pm 0.45}$	$68.6_{\pm 0.39}$
alt.atheism▲	(200)	$50.9_{\pm 0.30}$	$84.9_{\pm 0.05}$	$52.9_{\pm 0.07}$	$83.4_{\pm 0.11}$	$82.4_{\pm 0.17}$	$85.6_{\pm 0.18}$	$53.9_{\pm 0.50}$	$52.5_{\pm 5.37}$	$85.6_{\pm 2.01}$	$88.5_{\pm 1.22}$•
comp.graphics▲	(200)	$49.4_{\pm 1.28}$	$80.7_{\pm 0.10}$	$52.7_{\pm 0.15}$	$77.3_{\pm 0.05}$	$80.1_{\pm 0.11}$	$78.8_{\pm 0.12}$	$52.0_{\pm 0.49}$	$51.4_{\pm 2.97}$	$81.1_{\pm 1.02}$	$83.5_{\pm 1.76}$•
comp.os.ms▲	(200)	$51.9_{\pm 1.64}$	$72.1_{\pm 0.16}$	$46.6_{\pm 0.26}$	$53.2_{\pm 0.29}$	$72.5_{\pm 0.12}$	$68.8_{\pm 0.26}$	$47.7_{\pm 0.29}$	$47.8_{\pm 3.34}$	$73.7_{\pm 1.33}$•	$73.0_{\pm 4.24}$
comp.sys.mac▲	(200)	$51.0_{\pm 4.45}$	$80.0_{\pm 0.13}$	$52.3_{\pm 0.46}$	$77.6_{\pm 0.09}$	$77.3_{\pm 0.11}$	$78.2_{\pm 0.15}$	$51.5_{\pm 0.43}$	$51.2_{\pm 3.79}$	$81.1_{\pm 1.71}$•	$80.5_{\pm 1.73}$
comp.window.x▲	(200)	$64.3_{\pm 4.12}$	$77.9_{\pm 0.08}$	$53.0_{\pm 0.10}$	$66.0_{\pm 0.11}$	$85.4_{\pm 0.11}$•	$82.1_{\pm 0.14}$	$58.2_{\pm 0.55}$	$53.4_{\pm 2.91}$	$79.7_{\pm 1.41}$	$83.9_{\pm 0.94}$
misc.forsale▲	(200)	$50.3_{\pm 1.49}$	$67.3_{\pm 0.11}$	$51.2_{\pm 0.19}$	$56.2_{\pm 0.36}$	$72.5_{\pm 0.25}$•	$71.8_{\pm 0.23}$	$45.5_{\pm 0.53}$	$49.3_{\pm 5.51}$	$70.2_{\pm 0.63}$	$68.5_{\pm 1.86}$
rec.motorcycles▲	(200)	$50.7_{\pm 0.46}$	$78.4_{\pm 0.10}$	$52.5_{\pm 0.45}$	$45.6_{\pm 0.22}$	$86.7_{\pm 0.13}$•	$81.2_{\pm 0.12}$	$53.8_{\pm 0.41}$	$55.4_{\pm 3.99}$	$79.7_{\pm 2.40}$	$83.3_{\pm 1.49}$
rec.sport▲	(200)	$52.9_{\pm 4.09}$	$83.1_{\pm 0.05}$	$50.0_{\pm 0.00}$	$74.8_{\pm 0.12}$	$85.1_{\pm 0.10}$	$82.9_{\pm 0.16}$	$48.5_{\pm 0.50}$	$49.5_{\pm 3.93}$	$82.2_{\pm 1.01}$	$90.2_{\pm 1.17}$•
sci.crypt▲	(200)	$51.4_{\pm 0.66}$	$76.8_{\pm 0.07}$	$51.1_{\pm 0.10}$	$73.4_{\pm 0.08}$	$75.6_{\pm 0.14}$	$81.1_{\pm 0.16}$	$47.7_{\pm 0.35}$	$50.7_{\pm 5.24}$	$77.1_{\pm 1.02}$	$89.6_{\pm 1.36}$•
sci.med▲	(200)	$53.7_{\pm 3.82}$	$82.5_{\pm 0.05}$	$55.0_{\pm 0.57}$	$71.1_{\pm 0.09}$	$83.1_{\pm 0.08}$	$82.2_{\pm 0.15}$	$50.9_{\pm 0.36}$	$50.4_{\pm 3.85}$	$82.7_{\pm 0.83}$	$89.9_{\pm 0.83}$•
web1▼	(5,863)	$82.1_{\pm 2.71}$	$81.2_{\pm 0.06}$	$81.5_{\pm 0.04}$	$79.0_{\pm 0.11}$	$81.5_{\pm 0.06}$	$79.9_{\pm 0.09}$	$82.5_{\pm 0.09}$•	$82.4_{\pm 1.15}$	$82.5_{\pm 2.04}$	$81.2_{\pm 1.22}$
web2▼	(6,519)	$81.5_{\pm 0.58}$	$81.4_{\pm 0.06}$	$82.4_{\pm 0.15}$	$79.4_{\pm 0.12}$	$81.5_{\pm 0.06}$	$80.2_{\pm 0.07}$	$83.1_{\pm 0.08}$•	$80.5_{\pm 2.16}$	$82.9_{\pm 2.27}$	$81.5_{\pm 0.60}$
web3▼	(6,306)	$82.1_{\pm 2.19}$	$81.2_{\pm 0.04}$	$83.2_{\pm 0.15}$•	$79.5_{\pm 0.17}$	$81.6_{\pm 0.08}$	$81.2_{\pm 0.08}$	$82.9_{\pm 0.04}$	$81.2_{\pm 0.82}$	$81.4_{\pm 0.68}$	$81.7_{\pm 1.43}$
web4▼	(6,059)	$78.9_{\pm 2.75}$	$77.7_{\pm 0.07}$	$79.5_{\pm 0.17}$	$78.1_{\pm 0.08}$	$77.7_{\pm 0.06}$	$81.7_{\pm 0.14}$	$79.3_{\pm 0.16}$	$77.6_{\pm 0.45}$	$79.8_{\pm 1.31}$	$83.9_{\pm 0.91}$•
web5▼	(6,407)	$78.8_{\pm 0.71}$	$79.3_{\pm 0.05}$	$78.8_{\pm 0.12}$	$77.2_{\pm 0.09}$	$77.1_{\pm 0.08}$	$82.1_{\pm 0.11}$	$78.6_{\pm 0.27}$	$78.1_{\pm 0.61}$	$78.1_{\pm 1.22}$	$82.5_{\pm 0.68}$•
web6▼	(6,417)	$81.7_{\pm 2.71}$	$77.3_{\pm 0.15}$	$81.8_{\pm 0.23}$	$79.6_{\pm 0.07}$	$77.7_{\pm 0.06}$	$82.5_{\pm 0.14}$	$83.6_{\pm 0.20}$	$73.3_{\pm 0.34}$	$80.9_{\pm 2.33}$	$84.1_{\pm 1.30}$•
web7▼	(6,450)	$56.4_{\pm 1.55}$	$42.9_{\pm 0.31}$	$61.5_{\pm 0.16}$	$58.4_{\pm 0.47}$	$68.5_{\pm 0.23}$	$73.5_{\pm 0.26}$	$63.6_{\pm 0.34}$	$62.0_{\pm 2.75}$	$52.8_{\pm 4.56}$	$75.7_{\pm 2.03}$•
web8▼	(5,999)	$56.4_{\pm 2.86}$	$48.5_{\pm 0.44}$	$61.5_{\pm 0.37}$	$58.0_{\pm 0.52}$	$71.0_{\pm 0.33}$	$73.8_{\pm 0.28}$	$57.0_{\pm 0.37}$	$59.0_{\pm 3.19}$	$50.5_{\pm 3.57}$	$78.4_{\pm 1.56}$•
web9▼	(6,279)	$59.5_{\pm 2.61}$	$41.5_{\pm 0.41}$	$59.8_{\pm 0.28}$	$58.1_{\pm 0.28}$	$71.5_{\pm 0.37}$	$76.1_{\pm 0.11}$	$56.5_{\pm 0.32}$	$54.9_{\pm 3.73}$	$49.3_{\pm 4.42}$	$78.5_{\pm 1.99}$•
component★	(200)	N/A	$92.2_{\pm 0.01}$	N/A	$69.6_{\pm 0.04}$	$91.5_{\pm 0.01}$	$92.9_{\pm 0.01}$	N/A	N/A	N/A	$93.4_{\pm 0.04}$•
function★	(200)	N/A	$95.6_{\pm 0.01}$	N/A	$71.7_{\pm 0.04}$	$94.9_{\pm 0.01}$	$95.8_{\pm 0.01}$	N/A	N/A	N/A	$96.6_{\pm 0.05}$•
process★	(200)	N/A	$96.0_{\pm 0.00}$	N/A	$66.7_{\pm 0.04}$	$95.8_{\pm 0.00}$	$96.8_{\pm 0.00}$	N/A	N/A	N/A	$97.1_{\pm 0.01}$•
Mean rank		6.43	6.57	5.89	7.21	4.79	4.43	6.21	6.64	4.11	2.71•

♣image retrieval, ◊mutagenicity prediction, ♦medical image, ▲newsgroups, ▼web recommendation, ★biocreative.

Table 1 shows the experimental results of MIHI and comparison algorithms. The best performance value for each data set is highlighted with a small black bullet.

Mean rank represents the ranking of the average performance of the current algorithm on each data set [6]. The symbol "N/A" means that the algorithm cannot get experimental results.

The experimental results show that the MIHI algorithm has achieved the best experimental results on more than 70% of the data sets. And its mean rank is 2.71, which is superior to 9 traditional algorithms. Specifically, the accuracy of MIHI on some data sets is about 10% higher than other algorithms, such as elephant, rec.sport.hockey and web4. The reason may be that the our instance selection techniques can effectively select the instance with the largest amount of information from each bag. On image retrieval data, MIHI performed well on the three image data sets. However, MIHI performed poorly on the two mutagenicity prediction data sets, which may be caused by the low dimensionality of mutagenicity. StableMIL performs very well on mutagenicity. The reason may be that StableMIL can obtain the most informative causal instance from the super low-dimensional positive bag. From the performance of newsgroups, web recommendation and large-scale data sets, MIHI can get better results whether it is low-dimensional or high-dimensional data. We only compare MIHI with the four algorithms on large-scale data sets, because the time complexity of MILES, MILFM, MILDM, StableMIL and ELDB is relatively high.

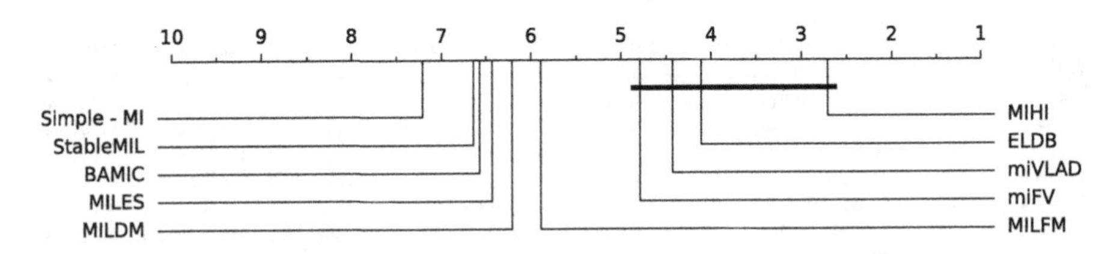

Fig. 2. Comparison of MIHI with 9 comparison algorithms with Bonferroni-Dunn test. Algorithms not connected to MIHI in the CD plot were considered to have significant performance of the control algorithm (CD = 2.24, significance level 0.05).

We also applied the post hoc Bonferroni-Dunn test [6] to test whether MIHI achieves competitive performance among the 9 compared algorithms. Figure 2 reports the critical difference (CD) plot at the 0.05 significance level. The mean accuracy ranks for each algorithm are marked along the axis (lower grades on the left). In addition, algorithms with an mean ranking within one CD of MIHI are connected by thick lines. Otherwise, any MIHI-independent algorithm is considered significantly different.

4.4 Parameter Analysis

Figure 3 shows the experimental results of parameter analysis on elephant data set. The symbols "S" and "C" respectively represent the two modes of bag embedding: superimpose and concatenate; "G", "P" and "N" respectively represent three instance selection modes. For all these subgraphs, the abscissa indicates the mode selected by the instance prototypes, and the ordinate indicates

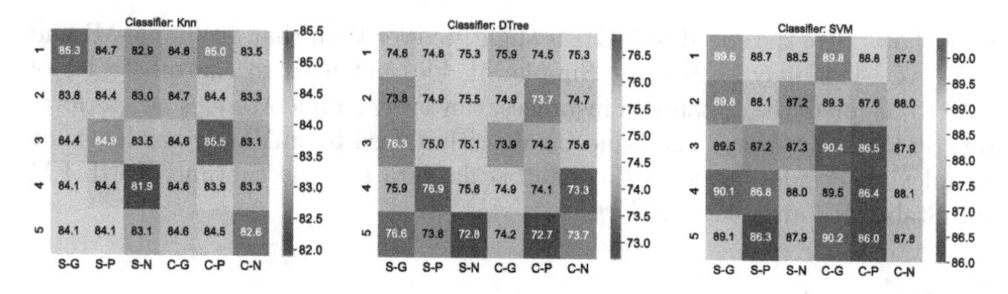

Fig. 3. Parameter analysis of MIHI with the number of instance prototype, three instance selection modes, two bag embedding modes and three classifiers for elephant data set. The best parameter settings of elephant are: 3 instance prototypes, instance selection mode "G" and bag embedding mode "C".

the number of instance prototypes. The three subgraphs show the classification accuracy on the classifier Knn, Decision Tree (DTree) and SVM respectively. The darkest colored table of the heat map indicates the highest accuracy. The following summarizes the impact of parameters on MIHI:

- **Bag embedding modes:** The classification performance of the two bag embedding modes is equivalent. However, since mode "C" embeds each bag in a high-dimensional space, we choose mode "S" for bag embedding in subsequent experiments.
- **Instance selection modes:** The results of the elephant in the classifier SVM show that the classification performance of mode "G" is better than the other two modes. However, in the other two classifiers, it is the best in mode "P".
- **The number of instance prototypes:** MIHI can achieve the best performance in most cases when the number of instance prototypes is 3–5.
- **Classifier:** SVM is more suitable for these data sets than DTree and Knn.

4.5 Efficiency Comparison

Table 2. The CPU runtime (in seconds) of one 10CV of the comparison algorithm on the 4 MIL classification data set.

Data sets	$(d/n/N)$	MILES	BAMIC	MILFM	Simple-MI	miFV	miVLAD	MILDM	StableMIL	ELDB	**MIHI**
Time complexity		$O(dn^2)$	$O(dN^2)$	$O(dn^2)$	$O(dN)$	$O(dn)$	$O(dn)$	$O(dn^2)$	$O(dn^2)$	$O(dn^2)$	$O(dn)$
Fox	$(230/1,320/200)$	2.378	1.512	8.514	0.151	4.202	1.284	5.425	13.959	3.712	1.034
alt.atheism	$(200/5,443/100)$	24.870	20.200	46.422	0.124	4.694	1.827	30.625	48.483	39.465	15.417
comp.graphics	$(200/3,094/100)$	8.939	6.651	25.217	0.104	3.520	1.533	11.635	43.796	13.543	5.526
web4	$(6,059/3,423/113)$	27.216	25.228	163.253	1.205	406.751	20.005	39.843	610.798	52.981	8.134
Mean rank		5.50	4.50	8.75	1.00	5.50	2.50	7.00	10.00	7.25	3.00

Table 2 shows the time complexity and runtime of MIHI compared with 9 competing algorithms. For MIHI, the construction of the high-level instances cost $O(dn)$, where d is the dimension and n is the cardinality of instance space. Table 2 compares the CPU running time of these methods on four data sets. The mean rank shows that the speed of MIHI is slightly lower than that of Simple-MI and miVLAD. This may be because Simple-MI does not need to consume a lot of time to calculate the distance of instances. However, Simple-MI does not perform well on these data sets. The k-means algorithm used by miVLAD has low time complexity. Besides, even on the small scale data set, the runtime of MILFM and StableMIL are relatively large.

5 Conclusion

In this paper, we proposed the MIHI algorithm to select high-level instances. MIHI fully utilizes the structure information of the bag-inside and effectively explores the characteristics of the instance space. The experiments were conducted on 26 MIL data sets. According to Table 1, the MIHI algorithm has achieved the best accuracy on more than 70% of the data sets. Its mean rank is 2.71, which is superior to 9 traditional algorithms. In addition, MIHI has linear time complexity, and its efficiency is slightly lower than that of Simple-MI and miVLAD.

Acknowledgements. This work was supported by the National Natural Science Foundation of China (62006200), Natural Science Foundation of Sichuan Province (2019YJ0314), Sichuan Province Youth Science and Technology Innovation Team (2019JDTD0017), and Central Government Funds of Guiding Local Scientific and Technological Development (2021ZYD0003).

References

1. Amores, J.: Multiple instance classification: review, taxonomy and comparative study. Artif. Intell. **201**(4), 81–105 (2013)
2. Andrews, S., Tsochantaridis, I., Hofmann, T.: Support vector machines for multiple-instance learning. In: NIPS, pp. 561–568 (2002)
3. Chen, Y.X., Bi, J.B., Wang, J.Z.: MILES: multiple-instance learning via embedded instance selection. IEEE Trans. Pattern Anal. Mach. Intell. **28**(12), 1931–1947 (2006)
4. Chen, Y.X., Wang, J.Z.: Image categorization by learning and reasoning with regions. J. Mach. Learn. Res. **5**, 913–939 (2004)
5. Decencière, E., et al.: Feedback on a publicly distributed image database: the messidor database. Image Anal. Stereol. **33**(3), 231–234 (2014)
6. Demšar, J.: Statistical comparisons of classifiers over multiple data sets. J. Mach. Learn. Res. **7**, 1–30 (2006)
7. Dietterich, T.G., Lathrop, R.H., Lozano-Pérez, T.: Solving the multiple instance problem with axis-parallel rectangles. Artif. Intell. **89**(1–2), 31–71 (1997)
8. Faria, A.W., et al.: MILKDE: a new approach for multiple instance learning based on positive instance selection and kernel density estimation. Eng. Appl. Artif. Intell. **59**, 196–204 (2017)
9. He, C.K., Shao, J., Zhang, J.S., Zhou, X.M.: Clustering-based multiple instance learning with multi-view feature. Expert Syst. Appl. **162**, 113027 (2020)

10. Hong, R.C., et al.: Image annotation by multiple-instance learning with discriminative feature mapping and selection. IEEE Trans. Cybern. **44**(5), 669–680 (2014)
11. Jorge, S., Florent, P., Thomas, M., Jakob, V.: Image classification with the fisher vector: theory and practice. Int. J. Comput. Vis. **105**(3), 222–245 (2013)
12. Ray, S., Craven, M.: Learning statistical models for annotating proteins with function information using biomedical text. BMC Bioinform. **6**(1), 1–9 (2005)
13. Srinivasan, A., Muggleton, S., King, R.: Comparing the use of background knowledge by inductive logic programming systems. In: ILP, pp. 199–230 (1995)
14. Tarragó, D.S., Cornelis, C., Bello, R., Herrera, F.: A multi-instance learning wrapper based on the Rocchio classifier for web index recommendation. Knowl. Based Syst. **59**(0950–7051), 173–181 (2014)
15. Wei, X.S., Wu, J.X., Zhou, Z.H.: Scalable algorithms for multi-instance learning. IEEE Trans. Neural Netw. Learn. Syst. **28**(4), 975–987 (2017)
16. Wu, J., Pan, S.R., Zhu, X.Q., Zhang, C.Q., Wu, X.D.: Multi-instance learning with discriminative bag mapping. IEEE Trans. Knowl. Data Eng. **30**(6), 1065–1080 (2018)
17. Yang, M., Zhang, Y.X., Wang, X.Z., Min, F.: Multi-instance ensemble learning with discriminative bags. IEEE Trans. Syst. Man Cybern. Syst., 1–12 (2021)
18. Zhang, W., Li, J., Liu, L.: Robust multi-instance learning with stable instances (2019)
19. Zhang, Y.L., Zhou, Z.H.: Multi-instance learning with key instance shift. In: IJCAI, pp. 3441–3447 (2017)
20. Zhou, Z.H., Jiang, K., Li, M.: Multi-instance learning based web mining. Appl. Intell. **22**(2), 135–147 (2005)
21. Zhou, Z.H., Sun, Y.Y., Li, Y.F.: Multi-instance learning by treating instances as non-I.I.D. samples. In: ICML, pp. 1249–1256 (2009)
22. Zhou, Z.H., Zhang, M.L.: Multi-instance clustering with applications to multi-instance prediction. Appl. Intell. **31**(1), 47–68 (2009)

High Average-Utility Itemset Sampling Under Length Constraints

Lamine Diop(✉)

University of Tours, 3 Place Jean Jaurès, 41029 Blois, France
lamine.diop@univ-tours.fr

Abstract. High Utility Itemset extraction algorithms are methods for discovering knowledge in a database where the items are weighted. Their usefulness has been widely demonstrated in many real world applications. The traditional algorithms return the set of all patterns with a utility above a minimum utility threshold which is difficult to fix, while top-k algorithms tend to lack of diversity in the produced patterns. We propose an algorithm named HAISampler to sample itemsets where each itemset is drawn with a probability proportional to its average-utility in the database and under length constraints to avoid the long and rare itemsets with low weighted items. The originality of our method stems from the fact that it combines length constraints with qualitative and quantitative utilities. Experiments show that HAISampler extracts thousands of high average-utility patterns in a few seconds from different databases.

1 Introduction

High Utility Itemset mining (HUIM) [21] is an extension of the frequent pattern mining [1] which takes into account the quality and the quantity of an item in a transaction (the price for instance). Its usefulness has been widely demonstrated in many real life applications like user behavior analysis [17], marketing analysis [14], mobile commerce [16], stream web clicks [6] and interactive pattern mining [2]. Interactive pattern mining is a process that requires a short loop with rapid interaction between the system and the user [13]. Indeed, the instant discovery imposes a constraint on the response time of a few seconds to extract a representative set of patterns. Complete methods do not provide the relevant patterns in such a short time. Methods based on a condensed representation [20] or on top-k patterns [18] are also used to find the best patterns. Therefore, they often focus on the same part which contains slightly different patterns and then leads to a lack of diversity. The latter is crucial to present to the user a set of varied patterns at each iteration in order to improve his/her view and help the system to know his/her interest.

To solve this problem, we propose to benefit from the pattern sampling techniques [3,4]. It consists in providing a representative sample of patterns according to a distribution proportional to an interestingness measure chosen by the user while ensuring good diversity between the sampled patterns. Weighted utilities

J. Gama et al. (Eds.): PAKDD 2022, LNAI 13281, pp. 134–148, 2022.
https://doi.org/10.1007/978-3-031-05936-0_11

are proposed in [5,9], but independently of any transaction. In other words, these methods do not work in our case where the utility of an occurrence of a pattern depends on the transaction in which it appears and its length. To the best of our knowledge, this paper is the first to address the problem of high average-utility itemset sampling while integrating length constraints on the sampled patterns.

The rest of this paper is organized as follows: Sect. 2 presents a state-of-the-art on HUIM and pattern sampling methods. Section 3 gives basic notions and formalizes the problem. Section 4 presents HAISampler (High Average-utility Itemset Sampler) and Sect. 5 evaluates its accuracy and complexity. Finally, we present some experiments in Sect. 6 and conclude in Sect. 7.

2 Related Works

HUIM is one of the most difficult tasks to extract useful patterns in pattern mining area. Two of its main challenges are the control of the returned candidate patterns and the time cost of computing the utility of each pattern of the database. To solve these problems, many efficient methods are proposed [16,18,21]. Unfortunately, the efficiency of the exhaustive HUIM often depends on the size of the database on which they are applied. Nowadays, the used databases are very large and contain information as rich as their variety. The diversity of the information that a database contains is proportional to the cardinality of its pattern language. But, the more the number of pattern the more it is difficult to explore the corresponding database. Another problem encountered by HUIM methods is the long tail where patterns containing low weighted items have high utilities thanks to their length. In [15], the authors propose an average-based utility measure to avoid the long tail problem. However, this one favors itemsets of length 1, since they are not affected by the division, and therefore the returned patterns become obvious to the user. An alternative consists in setting a minimum frequency threshold in order to avoid the long patterns. However, it is very difficult for the user to set a minimum frequency threshold.

Pattern sampling [3] is a non-exhaustive method for discovering relevant patterns while offering strong statistical guarantees thanks to its randomness. Its usefulness has been widely demonstrated in many applications such as classification [4,7], anomaly detection [9,11] and instant discovery [10,13]. It has also been applied to many types of structured data like graphs [3], itemsets [4], numerical data [12] and sequences [8]. In [7] the authors weight each pattern with a norm-based utility (regardless of any sequence) to avoid the long tail problem. However, the output sampling is much more difficult in the case where the draw of a pattern is not proportional to its frequency in the database, and even more when length constraints are integrated.

In this paper, we propose an original method of output pattern sampling to address the high average-utility itemsets under length constraints. Contrary to the methods which are based on heuristic algorithms [19] to find the top-k high utility itemsets, the sampling method that we propose is exact. It draws an itemset proportionally to its average-utilities from the set of all patterns of the database that respect the length constraints.

3 Preliminaries and Problem Formulation

This section begins by presenting some basic notions and notations as well as the necessary definitions for the understanding of the subject. It ends with a formalization of the problem that we want to solve in this paper.

Let $\mathcal{I} = \{e_1, \cdots, e_N\}$ be a finite set of literals called items with an arbitrary total order $>_\mathcal{I}$ between items : $e_1 >_\mathcal{I} \cdots >_\mathcal{I} e_N$. An itemset or pattern, denoted by $\varphi = e_{i_1} \cdots e_{i_n}$, with $n \leq N$, is a none empty subset of $\mathcal{I}$, $\varphi \subseteq \mathcal{I}$. The pattern language corresponds to $\mathcal{L} = 2^{|\mathcal{I}|} \backslash \emptyset$ and the length of a pattern $\varphi \in \mathcal{L}$ denoted by $|\varphi|$, is the number of items it contains (its cardinality). A transactional database $\mathcal{D}$ corresponds to a set of couple (j, t) where $j \in \mathbb{N}$ is the unique identifier of a transaction and $t = e_1 \cdots e_n$ is an itemset of length $|t| = n$ defined in $\mathcal{I}$. We denote by $\mathcal{L}(\mathcal{D})$ the set of all patterns that appear in $\mathcal{D}$. In the rest of this paper, a transaction identified by j is denoted by t_j. In addition, for a transaction $t_j = e_{j_1} \cdots e_{j_n}$, we denote by $t_j^i = e_{j_{i+1}} \cdots e_{j_n}$ an itemset formed by discarding the i first items of t. So we have $|t_j^i| = |t_j| - i$. The ith item of the transaction t_j is $t_j[i] = e_{j_i}$. In this paper, each item e_{j_i} of a transaction t_j has a weight, a strict positive real, which depends on this transaction, called its utility. For instance, in the case of a transaction which represents the set of all items purchased by a customer, the utility of an item e_i in the transaction t can be the product of its quantity $q(e_i, t)$ and its unit price $p(e_i)$. To be simpler on the rest of this paper, we associate each item e_i with its quantity in the transaction t that it appears, $e_i : q(e_i, t)$. Table 1 shows a database $\mathcal{D}$ with 5 transactions t_1, t_2, t_3, t_4 and t_5 defined on the set of items $\mathcal{I} = \{A, B, C, D, E, F\}$. We suppose that $A >_\mathcal{I} B >_\mathcal{I} C >_\mathcal{I} D >_\mathcal{I} E >_\mathcal{I} F$. In the database $\mathcal{D}$, the unit prices are: $p(A) = 25$, $p(B) = 30$, $p(C) = 10$, $p(D) = 5$, $p(E) = 15$ and $p(F) = 10$. With the transaction t_1, we have the following quantities $q(A, t_1) = 2$, $q(B, t_1) = 3$ and $q(C, t_1) = 2$.

Table 1. Example of database $\mathcal{D}$ with utilities on items

	$\mathcal{D}$					
t_1	A:2	B:3	C:2			
t_2		B:2		D:4		
t_3	A:1		C:1	D:1		
t_4	A:3	B:1				
t_5	A:1	B:2		D:1	E:1	F:1

Items	Price
A	25
B	30
C	10
D	5
E	15
F	10

This toy database will be used in the rest of this paper to give illustrations. Since items in the transaction t_1 have not the same weight, then the occurrences of patterns in t_1 may not have the same utility in t_1.

Definition 1 (Occurrence of a pattern). *Let φ be a pattern defined on a language $\mathcal{L}$ of a database $\mathcal{D}$. If it exists a transaction t_j of $\mathcal{D}$ such that $\varphi \subseteq t_j$, then φ_j is an occurrence of the pattern φ in the transaction t_j. The utility of the pattern φ in the transaction t_j, denoted by $\texttt{uOcc}(\varphi, t_j)$, is equal to 0 if $\varphi \not\subseteq t_j$ or $\varphi = \emptyset$, else $\texttt{uOcc}(\varphi, t_j) = \sum_{e \in \varphi} (q(e, t_j) \times p(e))$.*

There are also utilities that are independent of any database such as length-based utilities [9]. In the following, we consider the length-based utility defined by $\texttt{uLen}_{[m..M]}(\varphi) = 1/|\varphi|$ if $|\varphi| \in [m..M]$ and 0 otherwise, with m and M two positive integers. Thus, a pattern whose length is larger than M or smaller than m will be deemed useless.

Definition 2 (Average-Utility of a pattern under length constraints). *Let $\mathcal{D}$ be a database, $\mathcal{L}$ its language, m and M two integers such that $m \leq M$. The average-utility of the pattern $\varphi \in \mathcal{L}$ in $\mathcal{D}$ under minimum m and maximum M length constraints, denoted by $u^{avg}_{[m..M]}(\varphi, \mathcal{D})$, is the product of the sum of utilities of its occurrences and its length-based utility. Formally, $u^{avg}_{[m..M]}(\varphi, \mathcal{D}) = (\sum_{(j,t) \in \mathcal{D} \wedge \varphi \subseteq t} \texttt{uOcc}(\varphi, t)) \times \texttt{uLen}_{[m..M]}(\varphi)$.*

It is important to note that $u^{avg}_{[m..M]}$ is not a length-based utility.

Example 1. Let's consider the database $\mathcal{D}$ in Table 1 and the length constraints $m = 1$ and $M = 2$. We note that the pattern AC belongs in t_1 and t_3 only. So, AC has only two occurrences in $\mathcal{D}$: AC_1 et AC_3. We also have $\texttt{uLen}_{[m..M]}(AC) = 1$ because $|AC| = 2 \in [1..2]$. So, $u^{avg}_{[1..2]}(AC, \mathcal{D}) = (\texttt{uOcc}(AC, t_1) + \texttt{uOcc}(AC, t_3))/2 = ((2 \times 25 + 2 \times 10) + (1 \times 25 + 1 \times 10)) \times 1/2 = (70 + 35)/2 = 52.5$. By the same way, we have $u^{avg}_{[1..2]}(ABEF, \mathcal{D}) = 110 \times 0 = 0$. Indeed, $|ABEF| = 4 \notin [1..2]$.

In this paper, we want to solve the problem formulated as follows:
Given $\mathcal{D}$ a transactional database with weighted items (quantity and/or quality), two positive integers m and M such that $m \leq M$, our main goal is to draw a pattern φ from the language $\mathcal{L}$ with a probability exactly equal to:

$$\mathbb{P}(\varphi, \mathcal{D}) = \frac{u^{avg}_{[m..M]}(\varphi, \mathcal{D})}{\sum_{\varphi' \in \mathcal{L}(\mathcal{D})} u^{avg}_{[m..M]}(\varphi', \mathcal{D})}.$$

The notations of this paper are summarized in Table 2.

4 Two-Phase Sampling of High Average-Utility Itemsets

In this section, we will first present the basics of our method (detailing the weighting phase and the drawing phase of a pattern) before presenting the HAISampler algorithm that we propose to sample high average-utility itemsets under length constraints.

4.1 Basics of Our Sampling Method

The high average-utility itemset sampling method that we propose in this paper uses a position-based length weighting system to weight each transaction. It is a system which consists in weighting each item of a given transaction according to the position it occupies there. The item weights are then used to draw a pattern using a conditional probability.[1]

Table 2. Notations

Symbol	Definition
$\mathtt{uOcc}(\varphi, t)$	Utility of the pattern φ in the transaction t
$\mathtt{uLen}_{[m..M]}(\varphi)$	Length-based utility of φ equal to $1/\lvert\varphi\rvert$ if $\lvert\varphi\rvert \in [m..M]$ and 0 otherwise
$u^{avg}_{[m..M]}(\varphi, \mathcal{D})$	Average-utility of the pattern φ in $\mathcal{D}$. It is equal to 0 if $\lvert\varphi\rvert \notin [m..M]$
t^i	Itemset formed by discarding the i first items of t, $t^i = t[i+1]\cdots t[n]$
$\omega^+_\ell(t[i], t)$	Sum of occurrences' utilities of length ℓ in t^{i-1} with item $t[i]$
$\omega^-_\ell(t[i], t)$	Sum of occurrences' utilities of length ℓ in t^i (without item $t[i]$)
$\omega^{avgU}_{[m..M]}(t)$	Sum of average-utilities of all occurrences in t
$\mathbb{P}^t_\ell(t[i] \mid \varphi, \ell')$	Probability to draw item $t[i]$ in the transaction t after drawing $\ell - \ell'$ items and storing them in φ

Transaction Weighting: Let t be a transaction of length n defined on a set of items $\mathcal{I}$ endowed with a total order relation $>_{\mathcal{I}}$, m and M maximum and minimum length constraints respectively. The ith item of the transaction t, $t[i]$, is associated with two lists of values $\omega^+_\ell(t[i], t)$ and $\omega^-_\ell(t[i], t)$, for $\ell \in [m..M]$.

Definition 3. *The weight $\omega^+_\ell(t[i], t)$ is the sum of utilities of the occurrences of length $\ell - 1$ in the transaction $t^i = t[i+1]\cdots t[n]$ to which we add the item $t[i]$, and the weight $\omega^-_\ell(t[i], t)$ is that of occurrences of length ℓ in t^i.*

$$\omega^+_\ell(t[i], t) = \sum_{\varphi \subseteq t^i \wedge |\varphi| = \ell - 1} \mathtt{uOcc}(\{t[i]\} \cup \varphi, t) \quad \text{and} \quad \omega^-_\ell(t[i], t) = \sum_{\varphi \subseteq t^i \wedge |\varphi| = \ell} \mathtt{uOcc}(\varphi, t).$$

Property 1 gives a formalization of the weights based on Definition 3.

Property 1 (Item weights $\omega^\bullet_\ell(t[i], t)$). The weights $\omega^+_\ell(t[i], t)$ and $\omega^-_\ell(t[i], t)$ of the item $t[i]$, for all $\ell \in [m..M]$, may be formally written as follows:[2] $\omega^+_\ell(t[i], t) = \omega^+_1(t[i], t) \times \binom{|t^i|}{\ell-1} + \sum_{\star \in \{+,-\}} \omega^\star_{\ell-1}(t[i+1], t)$ and $\omega^-_\ell(t[i], t) = \sum_{\star \in \{+,-\}} \omega^\star_\ell(t[i+1], t)$, with $\omega^+_1(t[i], t) = \mathtt{uOcc}(t[i], t)$ for all $i \in [1..|t|]$ and $\omega^\star_\ell(t[i], t) = 0$ for all $i > |t|$.

Indeed, the weights of an item $t[i]$ are deduced from those of $t[i+1]$. Using Property 1, we can easily compute the weight of a transaction under length constraints. By definition, the average-utility of an occurrence $\varphi \subseteq t$ is $\mathtt{uOcc}(\varphi, t)/|\varphi|$.

[1] Proof of theoretical results are available in Sect. A.

[2] By convention $\binom{n}{k} = 0$ if k>n and 1 if k=0.

Property 2 (Transaction weight). The weight of a transaction t under minimum m and maximum M length constraints, denoted by $\omega^{avgU}_{[m..M]}(t)$, is the sum of average-utilities of the occurrences that it contains. Formally,

$$\omega^{avgU}_{[m..M]}(t) = \sum_{\ell=m}^{M} \left(\frac{1}{\ell} \sum_{i=1}^{|t|} \omega^{+}_{\ell}(t[i], t) \right) = \sum_{\ell=m}^{M} \frac{1}{\ell} \left(\omega^{+}_{\ell}(t[1], t) + \omega^{-}_{\ell}(t[1], t) \right).$$

Example 2. Let's consider transaction $t_1 = \{A{:}2, B{:}3, C{:}2\}$. The prices of its items are $p(A) = 25$, $p(B) = 30$ and $p(C) = 10$. Following the total order relation $>_{\mathcal{I}}$, the occurrences that start with A are : $\{A, AB, AC, ABC\}$, those who start with B are : $\{B, BC\}$, and finally only a pattern begins with C : $\{C\}$. The sum of the utilities of the occurrences of length $\ell \in [1..3]$ that start with $t_1[i]$, $i \in [1..3]$, in the transaction t_1 is: $\omega^{+}_{1}(A, t_1) = \texttt{uOcc}(A, t_1) = 2 \times 25 = 50$. Using Property 2 we have: $\omega^{+}_{2}(A, t_1) = \texttt{uOcc}(AB, t_1) + \texttt{uOcc}(AC, t_1) = (50 + 90) + (50 + 20) = 210$. In an identical way, we have the following weights for the transaction t_1:

t_1	A			B			C		
ω^{+}_{ℓ}	50	210	160	90	110	0	20	0	0
ω^{-}_{ℓ}	110	110	0	20	0	0	0	0	0

From this weighting, we deduce the weight of the transaction t_1 under the minimum $m = 1$ and maximum $M = 3$ length constraints which is equal to: $\omega^{avgU}_{[1..3]}(t_1) = (50 + 110)/1 + (210 + 110)/2 + (160 + 0)/3 = 373.33$.

We are going to show how to draw an occurrence from our weighting system.

Drawing an Itemset from a Transaction: Drawing a pattern from a position-based weighted transaction can be done using conditional probability. Lemma 1 gives an idea on the computation of the probability of drawing a given item knowing that we have already drawn (or not) higher items according to the order relation $>_{\mathcal{I}}$ introduced in Sect. 3.

Lemma 1. *Let ℓ be the length of the itemset to output and $\mathbb{P}^t_\ell(t[i]|\varphi, \ell')$ the probability to draw the item $t[i]$ in the transaction t after drawing $\ell - \ell'$ items and storing them in φ, with $e >_{\mathcal{I}} t[i]$ for all $e \in \varphi$. The probability to draw $t[i]$ knowing φ and ℓ' can be formulated as follows:*

$$\mathbb{P}^t_\ell(t[i]|\varphi, \ell') = \frac{\sum_{\varphi' \subseteq t^i \wedge |\varphi'| = \ell' - 1} \mathit{uOcc}(\varphi \cup \{t[i]\} \cup \varphi', t)}{\sum_{\varphi' \subseteq t^{i-1} \wedge |\varphi'| = \ell'} \mathit{uOcc}(\varphi \cup \varphi', t)}.$$

Property 3. The probability to draw the item $t[i]$ in the transaction t knowing the itemset φ and the length ℓ', with $|\varphi| = \ell - \ell'$, denoted by $\mathbb{P}^t_\ell(t[i]|\varphi, \ell')$, is given by the following formula:

$$\mathbb{P}^t_\ell(t[i]|\varphi, \ell') = \frac{\left(\sum_{k<i \wedge t[k] \in \varphi} \omega_1(t[k], t)\right) \times \binom{|t^i|}{\ell' - 1} + \omega^{+}_{\ell'}(t[i], t)}{\left(\sum_{k<i \wedge t[k] \in \varphi} \omega_1(t[k], t)\right) \times \binom{|t^{i-1}|}{\ell'} + \left(\sum_{\star \in \{+,-\}} \omega^{\star}_{\ell'}(t[i], t)\right)}.$$

The probability that the item $t[i]$ is not drawn knowing φ and ℓ' is $1 - \mathbb{P}^t_\ell(t[i]|\varphi, \ell')$.

The proofs of these two formulas follow from the fact that the probability of drawing $t[i]$ depends on the utilities of the items already drawn and those of the items which follow it to form a pattern of length ℓ.

Example 3. Suppose we need to draw a pattern of length $\ell = 2$ from the transaction t_1. The probability to draw $\varphi = AC$ is computed as follow: $\mathbb{P}^{t_1}(AC|\ell) = \mathbb{P}^{t_1}_{\ell}(A|\varphi = \emptyset, \ell' = 2) \times (1 - \mathbb{P}^{t_1}_{\ell}(B|\varphi = A, \ell' = 1)) \times \mathbb{P}^{t_1}_{\ell}(C|\varphi = A, \ell' = 1)$. Which gives us $\mathbb{P}^{t_1}(AC|\ell) = \frac{0+210}{0+320} \times (1 - \frac{50\times\binom{1}{1-1}+90}{50\times\binom{2}{1}+90+20}) \times \frac{50\times\binom{0}{1-1}+20}{50\times\binom{1}{1}+20} = \frac{210}{320} \times (1 - \frac{140}{210}) \times \frac{70}{70} = \frac{210}{320} \times \frac{70}{210} \times \frac{70}{70} = \frac{70}{320}$.

We are now going to formalize and present our two-phase approach for drawing patterns from a transactional database whose items are weighted.

4.2 HAISampler: High Average-utility Itemset Sampler Algorithm

As we described it in Sect. 4.1, our approach is done in two phases: preprocessing and drawing. The phase of drawing an itemset is divided into several steps: drawing a transaction t, drawing a length ℓ and finally, drawing an itemset of length ℓ based on the conditional probability. This phase is repeated K times to draw K patterns.

Algorithm 1. HAISampler (<u>High</u> <u>A</u>verage-utility <u>I</u>temset <u>Sampler</u>)

Input: A transactional database $\mathcal{D}$ having weighted items with a total order relation $>_{\mathcal{I}}$ and minimum m and maximum M length constraints
Output: φ a pattern drawn proportionally to its average-utility: $\varphi \sim u^{avg}_{[m..M]}(\mathcal{L}, \mathcal{D})$
//Phase 1: Preprocessing
1: Compute the weight of each transaction t in $\mathcal{D}$: $\omega^{avgU}_{[m..M]}(t)$
//Phase 2: Drawing
2: Draw a transaction proportionally to its weight: $t \sim \omega^{avgU}_{[m..M]}(\mathcal{D})$
3: Draw a length ℓ according to its weight $\sum_{\star\in\{+,-\}} \omega^{\star}_{[\ell..\ell]}(t[1], t)$: $\ell \sim \omega^{avgU}_{[m..M]}(t)$
4: $\varphi \leftarrow \emptyset$ ▷ Empty initialization of the pattern to return
5: $y \leftarrow 0$
6: $i \leftarrow 1$
7: **while** $\ell > 0$ **do**
8: $z \leftarrow y \times \binom{|t^i|}{\ell} + \omega^{+}_{\ell}(t[i], t) + \omega^{-}_{\ell}(t[i], t)$
9: $x \leftarrow random() \times z$ ▷ Randomly draw a real number between 0 and z
10: **if** $x \leq y \times \binom{|t^i|}{\ell-1} + \omega^{+}_{\ell}(t[i], t)$ **then**
11: $\varphi \leftarrow \varphi \cup \{t[i]\}$
12: $y \leftarrow y + \omega^{+}_{1}(t[i], t)$
13: $\ell \leftarrow \ell - 1$
14: $i \leftarrow i + 1$
15: **return** φ ▷ A pattern drawn proportionally to its average-utility in $\mathcal{D}$

Algorithm 1 takes as input a transactional database defined over a set of items with a total order relation $>_{\mathcal{I}}$ and minimum m and maximum M length constraints. We start with a preprocessing phase which computes the weight of each transaction (line 1) using Property 1 and Property 2. To draw a pattern, we first draw a transaction t proportionally to its weight $\omega^{avgU}_{[m..M]}(t)$ (line 2). Second, we draw a length ℓ proportionally to the sum of average-utilities of the occurrences of length ℓ that appear in the transaction t previously drawn (line 3). Lines 4 to 14 allow us to randomly draw an occurrence of length ℓ with a probability proportional to its utility in t. In line 8, we compute the total sum, z, of the utilities of the itemsets that start with $\varphi \cup \{t[i]\}$ following the order relation $>_{\mathcal{I}}$ in the transaction t. Then, we randomly draw a real number between 0 and z (line 9). If the drawn number is smaller than the sum of utilities of the ordered items, which starting with φ also contain the item $t[i]$ of transaction t, then we add $t[i]$ in the set of items to output (lines 10 and 11). In that case, the sum of utilities of the drawn items is updated in the variable y (line 12) and the number of remaining items decrements (line 13). When $\ell = 0$, we return on line 15 an itemset φ drawn proportionally to its average-utility in the database $\mathcal{D}$.

5 Theoretical Analysis of the Method

This section shows in Property 4 that HAISampler performs an exact draw of a pattern and gives finally its time complexity.

Property 4 (Soundness). Let $\mathcal{D}$ be a transactional database having utilities on items with a total order relation $>_{\mathcal{I}}$, and m and M two integers such that $m \leq M$. HAISampler randomly draws a pattern φ from the language $\mathcal{L}(\mathcal{D})$ with a probability equal to $u^{avg}_{[m..M]}(\varphi, \mathcal{D})/Z$ where $Z = \sum_{\varphi' \in \mathcal{L}(\mathcal{D})} u^{avg}_{[m..M]}(\varphi', \mathcal{D})$.

The complexity of our method, can be split into two parts: the complexity of preprocessing and that of drawing a pattern. It is important to note that the combination values are computed incrementally and stored in memory.

Preprocessing: To weight a transaction our method, HAISampler, spends a time of $O(|\mathcal{I}| \times (M - m) \times 2)$. Consequently, it weights all the transactions of the database in a complexity of $O(|\mathcal{D}| \times |\mathcal{I}| \times (M - m))$.

Drawing a Pattern: HAISampler starts by drawing a transaction with a complexity in $O(\log(|\mathcal{D}|))$. After, it draws a pattern proportionally to its utility in $O(|\mathcal{I}|)$. So, the complexity of drawing a pattern is $O(\log(|\mathcal{D}|) + |\mathcal{I}|)$. The draw of K patterns is then done in $O(K \times (\log(|\mathcal{D}|) + |\mathcal{I}|))$, hence equal to that of [4].

6 Experiments

In this experimental section, we study the efficiency of our method and present the dispersion of the average-utilities of the sampled patterns according to their length. Finally, we give some memory storage cost of HAISampler. The experiments[3] were carried out on 6 datasets including 3 from the UCI: `Adult`, `Chess`

[3] HAISampler (Python 3.8) https://github.com/HAISampler/haisampler-src.

and Mushroom with preprocessed versions[4], and 3 real datasets from SPMF[5]: BMS, Foodmart and Retail. Table 3 details the characteristics of the benchmarks. The value of the minimum length constraint is fixed at $m = 1$ throughout the experiments. All the experiments were performed on a 2.11 GHz 2 Core CPU PC with 32 GB memory.

Table 3. Characteristics of the benchmarks: the number of transactions, the number of distinct items, the minimum, the maximum and the average length of the transactions, the minimum, maximum and average weight of the items

$\mathcal{D}$	$\lvert\mathcal{D}\rvert$	$\lvert\mathcal{I}\rvert$	$\lvert t\rvert_{\min}$	$\lvert t\rvert_{\max}$	$\lvert t\rvert_{avg}$	$p(e,t)_{\min}$	$p(e,t)_{\max}$	$p(e,t)_{avg}$
Adult	48,842	97	12	15	14.87	1.0	99.0	50.04
BMS	59,602	497	1	267	2.51	7.0	9,000.0	724.79
Chess	28,056	58	7	7	7.00	1.0	99.0	50.05
Foodmart	4,141	1,559	1	14	4.42	50.0	2,166.0	655.66
Mushroom	8,124	90	22	23	22.69	1.0	99.0	50.02
Retail	88,162	16470	1	76	10.31	1.0	140.0	16.41

Speed of the Method. The average execution times that we are going to present were obtained by repeating the program 100 times for each case. The standard deviations obtained are very low (mostly equal to 0 in the drawing phase), that is why we have omitted them.

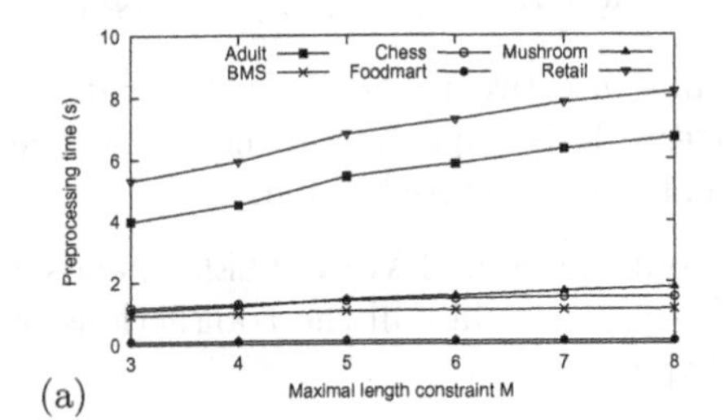

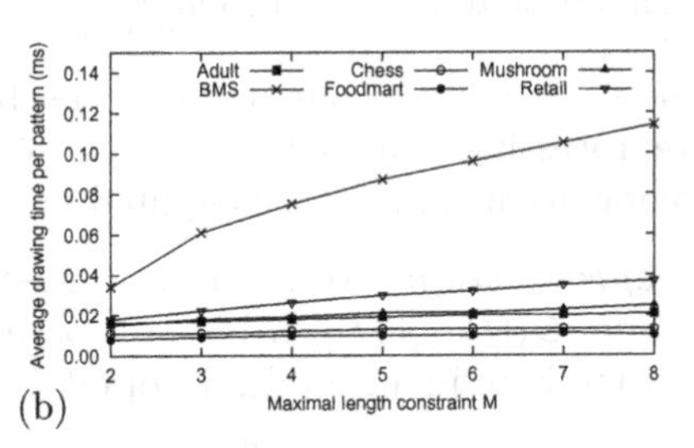

Fig. 1. Preprocessing time (a) and the drawing time of a pattern (b) according to M

Preprocessing Time: Figure 1-(a) shows the preprocessing time according to the maximal length constraint $M \in [3..8]$ of the 6 datasets in Table 3. First, it shows that the preprocessing time varies according to the maximum length constraint. However, it remains less than 9 s in all our datasets with a maximum length constraint $M = 8$, which is already too high if we want to avoid the long tail phenomenon. It is also important to note that the time to preprocess the datasets by HAISampler increases with the size of the database and the average length

[4] Each item was associated with a utility taken randomly between 1 and 100.
[5] http://www.philippe-fournier-viger.com/spmf.

of transactions. Finally, we can say that the method HAISampler consumes low time for preprocessing datasets.

Drawing Time per Pattern: Figure 1-(b) shows the evolution of the drawing time per pattern according to the maximum length constraint $M \in [2..8]$. First, we note that the drawing times change slightly depending on the maximum length constraint. Then, the curves show that the drawing time depends on the size of the database and the maximum length of transactions. Indeed, longest transactions consume a lot of time especially when the maximum length constraint is high. Finally, the time to draw a pattern remains less than 0.15 ms on all datasets we use here. It is less than 0.04 ms when $M \leq 8$ except in `BMS`. This means that HAISampler manages to draw thousands of patterns in a few seconds.

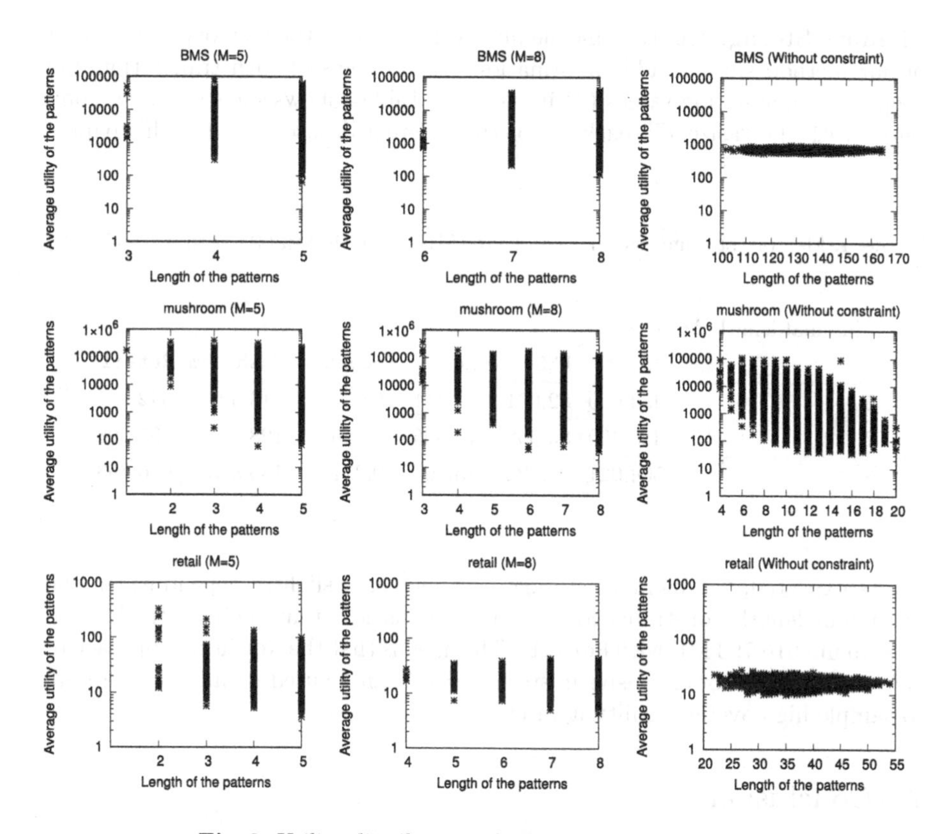

Fig. 2. Utility distribution of 10,000 sampled patterns

Impact of Length Constraints on the Sampled Patterns. We will show how the maximum length constraint can impact the utility of the sampled patterns from databases reaching the long tail curse. To do this, we have chosen two real datasets `BMS` and `Retail`, and one synthetic dataset `Mushroom`. The maximum length constraint M is tested with 5, 8 and ∞ (without constrained).

Figure 2 shows the distribution of the average-utilities of the patterns according to their length for the three chosen datasets (each dot represents a sampled pattern). `BMS` and `Retail` clearly show that if the maximum length constraint is very high or even unused, the drawn patterns are very long and formed by low weighted items (this is the long tail phenomenon). Besides, none of the returned patterns by the unconstrained method has a length smaller than 100 for `BMS`, 20 for `Retail`, and 4 for `Mushroom`. The latter is less impacted by the fact that all of its transactions have almost the same length. So we can say that it is very interesting to use length constraints to sample high utility patterns from a database suffering from the long tail curse, which is often the case with real data.

Memory Storage Cost. Someone may wonder about the memory storage cost of our method since it adds information on the items of each transaction and keeps the combination values $\binom{n}{k}$ in memory. Table 4 shows some statistics computed with the "asizeof"[6] package for the different datasets used in this paper.

Table 4. Memory storage cost in Mega Byte (MB) of HAISampler with $M \in \{5, 7, 8\}$

Maximal length M	$\mathcal{D}$					
	`Adult`	`BMS`	`Chess`	`Foodmart`	`Mushroom`	`Retail`
5	434.629	82.681	109.411	10.051	113.593	542.965
7	483.499	85.658	113.906	10.256	128.513	595.098
8	504.021	86.822	113.906	10.271	135.323	616.793

As expected, the memory storage cost increases slightly depending on the maximum length constraint M, and it remains less than 1 GB with $M = 8$ (maximum 616.793 MB with `Retail`). This means that the weighting approach of HAISampler is not expensive in storage. So it can be used with larger datasets to sample high average-utility itemsets.

7 Conclusion

This paper presents the first method for sampling high average-utility itemsets under length constraints. We have shown that HAISampler is exact and efficient at drawing thousands of patterns in a few seconds on real and synthetic datasets with reasonable preprocessing time. The experiments carried out show the value of length constraints for sampling patterns that have good utilities.

[6] https://code.activestate.com/recipes/546530-size-of-python-objects-revised/.

Moreover, we can easily adapt our approach to situations where the utility of an occurrence of a pattern in a transaction is the product of the utilities of its items. In perspective, we would like to extend our approach to other complex data structures such as sequences [7] and graphs [3]. In the short term, we intend to show how our position-based length weighting system can be extended for sampling high average-utility itemsets on data streams [6].

A Appendix (Proof of Theoretical Results)

Proof (**Property** 1). Let's start by showing that $\omega_\ell^-(t[i],t) = \sum_{\star\in\{+,-\}} \omega_\ell^\star(t[i+1],t)$. By definition, $\omega_\ell^-(t[i],t)$ is the sum of the utilities of the set of patterns of length ℓ in t^i, $\omega_\ell^-(t[i],t) = \sum_{\varphi\subseteq t^i \wedge |\varphi|=\ell} \mathtt{uOcc}(\varphi,t)$. This set can be split into two parts: the one that contains the patterns starting with the item $t[i+1]$ whose sum of their utilities is equal to $\omega_\ell^+(t[i+1],t)$ by definition, and the one that contains the patterns not starting with $t[i+1]$ and whose sum of their utilities is equal to $\omega_\ell^-(t[i+1],t)$. It implies that

$$\sum_{\varphi\subseteq t^i \wedge |\varphi|=\ell} \mathtt{uOcc}(\varphi,t) = \omega_\ell^+(t[i+1],t) + \omega_\ell^-(t[i+1],t) = \sum_{\star\in\{+,-\}} \omega_\ell^\star(t[i+1],t). \quad (1)$$

Let's show that $\omega_\ell^+(t[i],t) = \omega_1^+(t[i],t) \times \binom{|t^i|}{\ell-1} + \sum_{\star\in\{+,-\}} \omega_{\ell-1}^\star(t[i+1],t)$. We know by definition that $\omega_\ell^+(t[i],t)$ is the sum of utilities of itemsets of length ℓ in t^{i-1} which start with $t[i]$ following the total order relation $>_\mathcal{I}$. Formally, we have: $\omega_\ell^+(t[i],t) = \sum_{\varphi\subseteq t^i \wedge |\varphi|=\ell-1} \mathtt{uOcc}(\{t[i]\} \cup \varphi,t)$. But $\mathtt{uOcc}(\{t[i]\} \cup \varphi,t) = \mathtt{uOcc}(\{t[i]\},t) + \mathtt{uOcc}(\varphi,t)$ by definition. Then, $\omega_\ell^+(t[i],t) = \sum_{\varphi\subseteq t^i \wedge |\varphi|=\ell-1} (\mathtt{uOcc}(\{t[i]\},t) + \mathtt{uOcc}(\varphi,t))$. This implies: $\omega_\ell^+(t[i],t) = \sum_{\varphi\subseteq t^i \wedge |\varphi|=\ell-1} \mathtt{uOcc}(\{t[i]\},t) + \sum_{\varphi\subseteq t^i \wedge |\varphi|=\ell-1} \mathtt{uOcc}(\varphi,t)$. However, we know on the one hand that $\sum_{\varphi\subseteq t^i \wedge |\varphi|=\ell-1} \mathtt{uOcc}(\{t[i]\},t) = \mathtt{uOcc}(\{t[i]\},t) \times \binom{|t^i|}{\ell-1}$ and $\mathtt{uOcc}(\{t[i]\},t) = \omega_1^+(t[i],t)$ by definition, so $\sum_{\varphi\subseteq t^i \wedge |\varphi|=\ell-1} \mathtt{uOcc}(\{t[i]\},t) = \omega_1^+(t[i],t) \times \binom{|t^i|}{\ell-1}$. On the other hand, $\sum_{\varphi\subseteq t^i \wedge |\varphi|=\ell-1} \mathtt{uOcc}(\varphi,t)$ is the sum of utilities of the set of patterns of length $\ell-1$ in the transaction t^i. From (1), we can also say that $\sum_{\varphi\subseteq t^i \wedge |\varphi|=\ell-1} \mathtt{uOcc}(\varphi,t) = \sum_{\star\in\{+,-\}} \omega_{\ell-1}^\star(t[i+1],t)$. Then we have: $\omega_\ell^+(t[i],t) = \omega_1(t[i],t) \times \binom{|t^i|}{\ell-1} + \sum_{\star\in\{+,-\}} \omega_{\ell-1}^\star(t[i+1],t)$. Hence the result. □

Proof (**Property** 2). By definition, the weight of the transaction t is the sum of the average-utilities of the pattern occurrences it contains. According to Property 1, the weight of the transaction t under the minimum m and maximum M length constraints is nothing more than the sum of the average-utilities of pattern occurrences that start with the item $t[1]$ and respect the imposed length constraints, $\sum_{\ell=m}^{M}(\frac{1}{\ell} \times \omega_\ell^+(t[1],t))$, and that of the patterns that do not start with the item $t[1]$ but respect the length constraints, $\sum_{\ell=m}^{M}(\frac{1}{\ell} \times \omega_\ell^-(t[1],t))$. However, we know that $\sum_{\ell=m}^{M}(\frac{1}{\ell} \times \omega_\ell^+(t[1],t)) + \sum_{\ell=m}^{M}(\frac{1}{\ell} \times \omega_\ell^-(t[1],t)) = \sum_{\ell=m}^{M} \frac{1}{\ell} \times \left(\omega_\ell^+(t[1],t) + \omega_\ell^-(t[1],t)\right)$. Hence the result. □

Proof (**Lemma** 1). By definition, the probability to draw the item $t[i]$ of the transaction t after having drawing from it $\ell-\ell'$ items and store them in φ is nothing more than the probability of drawing a pattern that begins with $\varphi \cup \{t[i]\}$, according to the order relation $>_{\mathcal{I}}$, among the set of patterns that start with φ. On the one hand, we know that the set of patterns of length ℓ that start with $\varphi \cup t[i]$ is defined by $\{\varphi'' \subseteq t : (\varphi'' = \varphi \cup \{t[i]\} \cup \varphi')(\varphi' \subseteq t^i)(|\varphi'| = \ell' - 1)\}$. The sum of utilities of the patterns of this set is equal to $\sum_{\varphi' \subseteq t^i \wedge |\varphi'|=\ell'-1} \mathtt{uOcc}(\varphi \cup \{t[i]\} \cup \varphi', t)$. On the other hand, we know that the set of patterns of length ℓ that start with φ is defined by $\{\varphi'' \subseteq t : (\varphi'' = \varphi \cup \varphi')(\varphi' \subseteq t^{i-1})(|\varphi'| = \ell')\}$. The sum of utilities of the patterns of this set is equal to $\sum_{\varphi' \subseteq t^{i-1} \wedge |\varphi'|=\ell'} \mathtt{uOcc}(\varphi \cup \varphi', t)$. So $\mathbb{P}^t_\ell(t[i]|\varphi, \ell') = \frac{\sum_{\varphi' \subseteq t^i \wedge |\varphi'|=\ell'-1} \mathtt{uOcc}(\varphi \cup \{t[i]\} \cup \varphi', t)}{\sum_{\varphi' \subseteq t^{i-1} \wedge |\varphi'|=\ell'} \mathtt{uOcc}(\varphi \cup \varphi', t)}$. Hence the result. □

Proof (**Property** 3). From Lemma 1, we have: $\mathbb{P}^t_\ell(t[i]|\varphi, \ell') = \frac{\sum_{\varphi' \subseteq t^i \wedge |\varphi'|=\ell'-1} \mathtt{uOcc}(\varphi \cup \{t[i]\} \cup \varphi', t)}{\sum_{\varphi' \subseteq t^{i-1} \wedge |\varphi'|=\ell'} \mathtt{uOcc}(\varphi \cup \varphi', t)}$. First, by definition we have $\mathtt{uOcc}(\varphi \cup \{t[i]\} \cup \varphi', t) = \mathtt{uOcc}(\varphi, t) + \mathtt{uOcc}(\{t[i]\} \cup \varphi', t)$. Let $z_i = \sum_{\varphi' \subseteq t^i \wedge |\varphi'|=\ell'-1} \mathtt{uOcc}(\varphi \cup \{t[i]\} \cup \varphi', t)$. It implies that $z_i = \sum_{\varphi' \subseteq t^i \wedge |\varphi'|=\ell'-1} (\mathtt{uOcc}(\varphi, t) + \mathtt{uOcc}(\{t[i]\} \cup \varphi', t))$. Then we have: $z_i = \sum_{\varphi' \subseteq t^i \wedge |\varphi'|=\ell'-1} \mathtt{uOcc}(\varphi, t) + \sum_{\varphi' \subseteq t^i \wedge |\varphi'|=\ell'-1} \mathtt{uOcc}(\{t[i]\} \cup \varphi', t)$. But $\sum_{\varphi' \subseteq t^i \wedge |\varphi'|=\ell'-1} \mathtt{uOcc}(\varphi, t) = \mathtt{uOcc}(\varphi, t) \times \binom{|t^i|}{\ell'-1}$ and $\sum_{\varphi' \subseteq t^i \wedge |\varphi'|=\ell'-1} \mathtt{uOcc}(\{t[i]\} \cup \varphi', t) = \omega^+_{\ell'}(t[i], t)$ by definition. Then $z_i = \mathtt{uOcc}(\varphi, t) \times \binom{|t^i|}{\ell'-1} + \omega^+_{\ell'}(t[i], t)$. We also know that $\mathtt{uOcc}(\varphi, t) = \sum_{k<i \wedge t[k] \in \varphi} \omega^+_1(t[k], t)$. So, $z_i = \left(\sum_{k<i \wedge t[k] \in \varphi} \omega^+_1(t[k], t)\right) \times \binom{|t^i|}{\ell'-1} + \omega^+_{\ell'}(t[i], t)$. Second, we have $\mathtt{uOcc}(\varphi \cup \varphi', t) = \mathtt{uOcc}(\varphi, t) + \mathtt{uOcc}(\varphi', t)$. By setting $Z_i = \sum_{\varphi' \subseteq t^{i-1} \wedge |\varphi'|=\ell'} \mathtt{uOcc}(\varphi \cup \varphi', t)$, we get then $Z_i = \sum_{\varphi' \subseteq t^{i-1} \wedge |\varphi'|=\ell'} \mathtt{uOcc}(\varphi, t) + \sum_{\varphi' \subseteq t^{i-1} \wedge |\varphi'|=\ell'} \mathtt{uOcc}(\varphi', t)$. But $\sum_{\varphi' \subseteq t^{i-1} \wedge |\varphi'|=\ell'} \mathtt{uOcc}(\varphi, t) = \mathtt{uOcc}(\varphi, t) \times \binom{|t^{i-1}|}{\ell'} = \left(\sum_{k<i \wedge t[k] \in \varphi} \omega^+_1(t[k], t)\right) \times \binom{|t^{i-1}|}{\ell'}$ et $\sum_{\varphi' \subseteq t^{i-1} \wedge |\varphi'|=\ell'} \mathtt{uOcc}(\varphi', t) = \sum_{\star \in \{+,-\}} \omega^\star_{\ell'}(t[i], t)$, so $Z_i = \left(\sum_{k<i \wedge t[k] \in \varphi} \omega^+_1(t[k], t)\right) \times \binom{|t^{i-1}|}{\ell'} + \sum_{\star \in \{+,-\}} \omega^\star_{\ell'}(t[i], t)$. Finally, $\mathbb{P}^t_\ell(t[i]|\varphi, \ell') = \frac{z_i}{Z_i} = \frac{\left(\sum_{k<i \wedge t[k] \in \varphi} \omega^+_1(t[k], t)\right) \times \binom{|t^i|}{\ell'-1} + \omega^+_{\ell'}(t[i], t)}{\left(\sum_{k<i \wedge t[k] \in \varphi} \omega^+_1(t[k], t)\right) \times \binom{|t^{i-1}|}{\ell'} + \sum_{\star \in \{+,-\}} \omega^\star_{\ell'}(t[i], t)}$. □

Proof (**Property** 4). Let m be the minimum and M the maximum length constraints, the probability of drawing the pattern φ of length ℓ in the database $\mathcal{D}$ denoted by $\mathbb{P}_{[m..M]}(\varphi, \mathcal{D})$, and Z a normalization constant defined by $Z = \sum_{\varphi' \in \mathcal{L}(\mathcal{D})} u^{avg}_{[m..M]}(\varphi', \mathcal{D})$. We know that : $\mathbb{P}_{[m..M]}(\varphi, \mathcal{D}) = \sum_{(j,t) \in \mathcal{D}} \left(\mathbb{P}_{[m..M]}(t_j, \mathcal{D}) \times \mathbb{P}_{[m..M]}(\varphi, t_j)\right)$. But $\mathbb{P}_{[m..M]}(t_j, \mathcal{D}) = \frac{\omega^{avgU}_{[m..M]}(t_j)}{Z}$, then

$$\mathbb{P}_{[m..M]}(\varphi, \mathcal{D}) = \sum_{(j,t) \in \mathcal{D}} \left(\frac{\omega^{avgU}_{[m..M]}(t_j)}{Z} \times \mathbb{P}_{[m..M]}(\varphi, t_j)\right). \tag{2}$$

We also know that:

$$\mathbb{P}_{[m..M]}(\varphi, t_j) = \mathbb{P}_{[m..M]}(\ell|t_j) \times \mathbb{P}^{t_j}_{[m..M]}(\varphi|\ell). \tag{3}$$

$\mathbb{P}_{[m..M]}(\ell|t_j) = \frac{\omega^{avgU}_{[\ell..\ell]}(t_j)}{\omega^{avgU}_{[m..M]}(t_j)}$ and $\mathbb{P}^{t_j}_{[m..M]}(\varphi|\ell) = \frac{\texttt{uOcc}(\varphi,t_j)}{\omega^{avgU}_{[\ell..\ell]}(t_j)\times\ell}$ then by substituting the two terms in (3), $\mathbb{P}_{[m..M]}(\varphi, t_j) = \frac{\omega^{avgU}_{[\ell..\ell]}(t_j)}{\omega^{avgU}_{[m..M]}(t_j)} \times \frac{\texttt{uOcc}(\varphi,t_j)}{\omega^{avgU}_{[\ell..\ell]}(t_j)\times\ell} = \frac{\texttt{uOcc}(\varphi,t_j)}{\omega^{avgU}_{[m..M]}(t_j)\times\ell}$. Now, if we replace $\mathbb{P}_{[m..M]}(\varphi, t_j)$ in (2) by its last expression, we get: $\mathbb{P}_{[m..M]}(\varphi, \mathcal{D}) = \sum_{(j,t)\in\mathcal{D}} \left(\frac{\omega^{avgU}_{[m..M]}(t_j)}{Z} \times \frac{\texttt{uOcc}(\varphi,t_j)}{\omega^{avgU}_{[m..M]}(t_j)\times\ell}\right) = \frac{1}{Z} \times \frac{\sum_{(j,t)\in\mathcal{D}}\texttt{uOcc}(\varphi,t_j)}{\ell}$. But by definition, we have $\frac{\sum_{(j,t)\in\mathcal{D}}\texttt{uOcc}(\varphi,t_j)}{\ell} = u^{avg}_{[m..M]}(\varphi, \mathcal{D})$, so $\mathbb{P}_{[m..M]}(\varphi, \mathcal{D}) = \frac{u^{avg}_{[m..M]}(\varphi,\mathcal{D})}{Z}$. Hence the result. □

References

1. Agrawal, R., Srikant, R.: Fast algorithms for mining association rules in large databases. In: VLDB'94, pp. 487–499. Morgan Kaufmann Publishers Inc. (1994)
2. Ahmed, C.F., Tanbeer, S.K., Jeong, B., Lee, Y.: Efficient tree structures for high utility pattern mining in incremental databases. IEEE Trans. Knowl. Data Eng. **21**(12), 1708–1721 (2009)
3. Al Hasan, M., Zaki, M.J.: Output space sampling for graph patterns. Proc. VLDB Endow. **2**(1), 730–741 (2009)
4. Boley, M., Lucchese, C., Paurat, D., Gärtner, T.: Direct local pattern sampling by efficient two-step random procedures. In: Proceedings of the 17th ACM SIGKDD, pp. 582–590 (2011)
5. Boley, M., Moens, S., Gärtner, T.: Linear space direct pattern sampling using coupling from the past. In: Proceedings of the 18th ACM SIGKDD International Conference on Knowledge Discovery and Data Mining, pp. 69–77. ACM (2012)
6. Chu, C.J., Tseng, V.S., Liang, T.: An efficient algorithm for mining temporal high utility itemsets from data streams. J. Syst. Softw. **81**(7), 1105–1117 (2008)
7. Diop, L., Diop, C.T., Giacometti, A., Li, D., Soulet, A.: Sequential pattern sampling with norm-based utility. Knowl. Inf. Syst. **62**(5), 2029–2065 (2019). https://doi.org/10.1007/s10115-019-01417-3
8. Diop, L., Diop, C.T., Giacometti, A., Li Haoyuan, D., Soulet, A.: Sequential pattern sampling with norm constraints. In: IEEE International Conference on Data Mining (ICDM), Singapore, November 2018
9. Diop, L., Diop, C.T., Giacometti, A., Soulet, A.: Pattern on demand in transactional distributed databases. Inf. Syst. **104**, 101908 (2022)
10. Dzyuba, V., Leeuwen, M.V., Nijssen, S., De Raedt, L.: Interactive learning of pattern rankings. Int. J. Artif. Intell. Tools **23**(06), 1460026 (2014)
11. Giacometti, A., Soulet, A.: Anytime algorithm for frequent pattern outlier detection. Int. J. Data Sci. Anal. **5**, 119–130 (2016). https://doi.org/10.1007/s41060-016-0019-9
12. Giacometti, A., Soulet, A.: Dense neighborhood pattern sampling in numerical data. In: Proceedings of SDM 2018, pp. 756–764 (2018)
13. Leeuwen, M.: Interactive data exploration using pattern mining. In: Holzinger, A., Jurisica, I. (eds.) Interactive Knowledge Discovery and Data Mining in Biomedical Informatics. LNCS, vol. 8401, pp. 169–182. Springer, Heidelberg (2014). https://doi.org/10.1007/978-3-662-43968-5_9
14. Li, H., Huang, H., Chen, Y., Liu, Y., Lee, S.: Fast and memory efficient mining of high utility itemsets in data streams. In: 2008 Eighth IEEE International Conference on Data Mining, pp. 881–886, December 2008

15. Lin, J.C.W., Li, T., Fournier-Viger, P., Hong, T.P., Zhan, J., Voznak, M.: An efficient algorithm to mine high average-utility itemsets. Adv. Eng. Inform. **30**(2), 233–243 (2016)
16. Shie, B.-E., Hsiao, H.-F., Tseng, V.S., Yu, P.S.: Mining high utility mobile sequential patterns in mobile commerce environments. In: Yu, J.X., Kim, M.H., Unland, R. (eds.) DASFAA 2011. LNCS, vol. 6587, pp. 224–238. Springer, Heidelberg (2011). https://doi.org/10.1007/978-3-642-20149-3_18
17. Shie, B.E., Yu, P.S., Tseng, V.S.: Mining interesting user behavior patterns in mobile commerce environments. Appl. Intell. **38**(3), 418–435 (2013)
18. Singh, K., Singh, S.S., Kumar, A., Biswas, B.: TKEH: an efficient algorithm for mining top-k high utility itemsets. Appl. Intell. **49**(3), 1078–1097 (2019). https://doi.org/10.1007/s10489-018-1316-x
19. Song, W., Zheng, C., Huang, C., Liu, L.: Heuristically mining the top-k high-utility itemsets with cross-entropy optimization. Appl. Intell., 1–16 (2021). https://doi.org/10.1007/s10489-021-02576-z
20. Tseng, V.S., Wu, C., Fournier-Viger, P., Yu, P.S.: Efficient algorithms for mining the concise and lossless representation of high utility itemsets. IEEE Trans. Knowl. Data Eng. **27**(3), 726–739 (2015)
21. Yao, H., Hamilton, H.J., Butz, C.J.: A foundational approach to mining itemset utilities from databases. In: Proceedings of the Third SIAM International Conference on Data Mining, pp. 482–486 (2004)

Divide and Imitate: Multi-cluster Identification and Mitigation of Selection Bias

Katharina Dost[1(✉)], Hamish Duncanson[1], Ioannis Ziogas[2], Patricia Riddle[1], and Jörg Wicker[1]

[1] University of Auckland, Auckland, New Zealand
{kdos481,hdun603}@aucklanduni.ac.nz, {p.riddle,j.wicker}@auckland.ac.nz
[2] University of Mississippi, Oxford, USA
ziogas@olemiss.edu

Abstract. Machine Learning can help overcome human biases in decision making by focussing on purely logical conclusions based on the training data. If the training data is biased, however, that bias will be transferred to the model and remains undetected as the performance is validated on a test set drawn from the same biased distribution. Existing strategies for selection bias identification and mitigation generally rely on some sort of knowledge of the bias or the ground-truth. An exception is the Imitate algorithm that assumes no knowledge but comes with a strong limitation: It can only model datasets with one normally distributed cluster per class. In this paper, we introduce a novel algorithm, MIMIC, which uses Imitate as a building block but relaxes this limitation. By allowing mixtures of multivariate Gaussians, our technique is able to model multi-cluster datasets and provide solutions for a substantially wider set of problems. Experiments confirm that MIMIC not only identifies potential biases in multi-cluster datasets which can be corrected early on but also improves classifier performance.

1 Introduction

Throughout the years, Machine Learning and Data Mining have gained influence into a wide variety of applications, typically under the assumption that they ideally overcome conscious and unconscious human biases, prejudices, and emotions in decision making. To overcome limitations of our own knowledge and experience, Machine Learning learns concepts from – hopefully unbiased – data and thereby discovers latent knowledge. As such, it has been applied to domains with large amounts of data that are no longer humanly processable and require us to rely, up to a certain degree, on the models trained in automated settings, e.g., credit scoring [9], medical diagnoses [15], or crime risk assessment [7].

In reality, although these models improve in accuracy, the data is often flawed and induces biases in the models that are largely overlooked since the performance is evaluated against equally biased test data. Existing bias mitigation strategies not only require the user to be aware of the bias but also to have a

J. Gama et al. (Eds.): PAKDD 2022, LNAI 13281, pp. 149–160, 2022.
https://doi.org/10.1007/978-3-031-05936-0_12

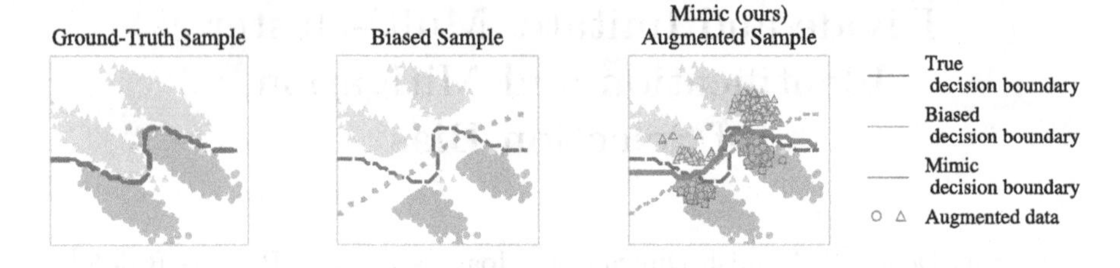

Fig. 1. Decision Boundaries of Support Vector Machines trained on three different datasets: a sample representative for the ground-truth (left), a biased subset (center), and the biased subset augmented with our algorithm, MIMIC (right).

certain knowledge of the ground-truth. But what if the user does not suspect any bias? In this case they will use the data and train a biased model delivering poor performance when applied to previously unseen or underrepresented cases trusting in the quality of its predictions.

Biases are easily induced during the data gathering phase, for example in clinical trials [15] where the data is collected from local volunteers that might not represent the entire population. However, the resulting model will be used to predict the reactions to treatments or drugs for the entire population. Knowledge of the bias early in the development process cannot only help improve the data quality, but can also mitigate its effect on the learned model.

In order to identify and mitigate selection biases where no additional information is available, Dost *et al.* [5] proposed Imitate, a technique that, given a biased dataset, aims to estimate the ground-truth distribution and generate data points to augment the dataset accordingly. While the authors demonstrate Imitate's ability to improve model performance through pre-augmentation on several examples, it is limited by a major assumption: the underlying ground-truth is expected to be normally distributed. In practice, this strongly limits the applicability of Imitate as it is neither flexible enough to model non-Gaussian distributions nor can it capture datasets consisting of several clusters.

In this paper, we introduce MIMIC *(Multi-IMItate Bias Correction)*, a multi-cluster solution for identification and mitigation of selection biases that exploits Imitate as a building block. Modeling data as a mixture of possibly biased and overlapping multivariate Gaussians, MIMIC overcomes Imitate's limitations and greatly increases its applicability. The parameters of these Gaussians bridge between the estimated and the present distribution and can indicate under-represented regions in the data that are likely to correspond to a selection bias. Generating points in these regions helps mitigate the effect of the bias and pushes the decision boundary towards the ground-truth (see Fig. 1). Our contributions are as follows:

- We propose MIMIC, a novel selection bias identification and mitigation strategy that does not require any knowledge of the bias or the ground-truth.
- In contrast to existing approaches, MIMIC is able to function in a multi-cluster setting and hence drastically increases the range of datasets and distributions that can be modeled.

- In a set of experiments, we demonstrate the shortcomings of existing techniques and highlight the potential of Mimic in these scenarios. We made our Python+sklearn [16] implementation publicly available[1].

The remainder of this paper is organized as follows: Sections 2 and 3 review the problem statement including the notation and the related research fields, respectively. We introduce our proposed method in Sect. 4 before evaluating it in Sect. 5. Section 6 concludes the paper.

2 Problem Statement

Assuming that we are facing a biased dataset, we aim to generate additional data points that are able to mitigate the bias. Following the notation in [5], this key idea can be formalized as the following problem statement:

Reconstruction Problem. *Let $D \subset \mathbb{R}^n$ be an n-dimensional dataset (potentially with class labels) that is representative of an underlying distribution which we consider to be the ground-truth. Given only a biased subset $B \subset D$, the task is to approximate $I := D \setminus B$ with a generated dataset $\hat{I}$ such that a model trained on the augmented dataset $B \cup \hat{I}$ is minimally different from one trained on D.*

The problem was first introduced by Dost *et al.* [5] where D is required to be normally distributed (when split into classes). This assumption is well motivated due to two factors: First, Bareinboim *et al.* [2] prove theoretically that the true class label distribution cannot be recovered from the biased dataset alone without utilising additional data or assumptions, so some assumption is necessary. Second, following the Central Limit Theorem[2], numerical real-world observations frequently are approximately Gaussian which makes normal distributions very common [13]. In this paper, however, we relax the requirement of normal distributions and assume each class of D consists of a mixture of Gaussians. In other words, we assume that each class of the dataset can be represented by a set of possibly overlapping Gaussian clusters.

3 Related Work

Apart from Imitate, to the best of our knowledge, there does not exist any research attempting to solve the problem defined in the previous section. However, methods have been proposed that solve the problem under additional assumptions. This section provides an overview of related research areas.

Bias Mitigation Using Additional Information. If only a subset of the variables is affected by a selection bias, *Missing Value Imputation* techniques [17] can impute these values. For a dataset X with labels Y, however, they

[1] Implementation and Supplementary Material: https://github.com/KatDost/Mimic.

[2] The Central Limit Theorem states that a sequence of independent and identically distributed (i.i.d.) random variables converges almost surely to a Gaussian [10]. Since we can typically assume that real-world measurements are not perfectly i.i.d. but rather combinations of different effects, we will often observe this effect.

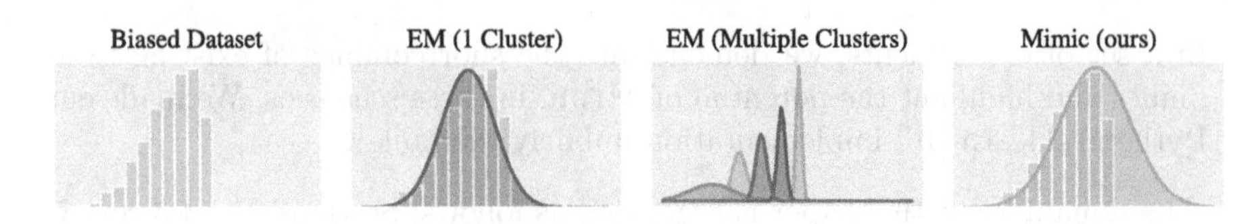

Fig. 2. When facing a biased sample (1st plot from left), the EM algorithm will fit one (2nd) or multiple (3rd; here controlled by BIC) Gaussians to minimize the error on the presented data. Imitate and Mimic (4th) instead use the histogram bin heights as weights for the fitting procedure and capture the underlying ground-truth more closely.

operate under the assumption that $\mathbb{P}[X|Y]$ and $\mathbb{P}[Y|X]$ are unchanged between the training and the test set. The *Selection Bias* literature [14] widely assumes that all data points in D are known (or at least their distribution), but only a biased subset of the labels is available. More general is the field of *Covariate Shift Correction* [18] where $\mathbb{P}[Y|X]$ is assumed to be shared whereas $\mathbb{P}[X|Y]$ can differ between the training and the test set and will be "shifted". Methods in both fields typically operate model-free and require an unbiased sample to estimate the bias and assign more weight to data points in underrepresented regions during training [20,21]. In the field of *Fairness in Machine Learning*, different techniques to test for biases in models have been proposed, e.g., using the AI Fairness 360 toolkit [3]. These methods require the user to decide which attributes in the dataset might be critical and need to be protected, e.g., gender, and the detected biases can be validated using additional data if possible [19].

If a researcher does not suspect a concrete bias or deals with a numerical tabular dataset without ground-truth information, none of the above mentioned approaches are feasible. Dataset visualization [12] can be considered here, but it is either limited to simple biases or requires inherent bias detection mechanisms to decide upon the kind of visualization, and it detects biases rather than mitigates them. Hence, in the situation of the Reconstruction Problem (see Sect. 2), the Imitate algorithm is, to the best of our knowledge, the only option if neither the ground-truth nor the bias are known.

Imitate. When facing a biased dataset B, Imitate [5] splits it into classes $c \in C$ and treats each resulting subset B_c separately. The dataset B_c is transformed using *Independent Component Analysis (ICA)* [11] to obtain statistically independent components that reveal non-Gaussian densities and allow individual analysis. For each of these components d, the data is represented as a histogram h_d, or using kernel density estimators, and the bin heights are exploited as weights for a least squares optimizer fitting a Gaussian g_d to the histogram. Note that this design puts more emphasis on the existing data points than potentially missing ones and therefore yields fundamentally different results than typical Expectation-Maximization fitting if a selection bias is present (see Fig. 2 for an example). Once all components have been processed, additional data points are generated such that the gaps between g_d and h_d are filled and the distributions g_d are preserved. Then the new points are back-transformed into the original data space. These data points not only indicate a potential selection bias if focussed on certain areas in the space, but can also be used to augment B_c and mitigate the effect the bias has on subsequent modeling tasks (see [5] for details). Due to the particular design of Imitate that uses ICA for component-wise fitting of one Gaussian, the algorithm is restricted to one normally distributed cluster per class only. In this paper, we relax this restriction.

4 Proposed Method

Aiming to provide a bias mitigation strategy for a wide range of problems, in this paper, we assume that ground-truth data consists of a mixture of multivariate Gaussians. Although this is still a limiting assumption, it substantially widens the range of datasets that can be modeled when compared to existing techniques, i.e., the Imitate algorithm (see Sect. 2 for a discussion of that assumption). Before analyzing each Gaussian for potential biases, we need to find a suitable mixture model for the ground-truth based solely on the biased dataset.

If no bias is present in the dataset, *Gaussian Mixture Models (GMMs)* can fulfill the task as they are able to identify the optimal Gaussians to describe a presented dataset given suitable initial cluster centers. These centers (and the number of clusters) could be found using, for example, the *Bayesian information criterion (BIC)* [8]. In the case of a selection bias, however, one biased cluster might be split into several Gaussian clusters as that mixture fits the presented dataset better, as shown in Fig. 2. Assume a clinical study testing the impact of a new drug on test and control groups. While GMM breaks the group of participants into many small clusters as it models the presented datasets, we need to find clusters that give an indication of where some data might be missing and thereby indicating that, e.g., women below a certain age did not participate due to safety concerns. Therefore, we need to develop a novel strategy to cluster biased datasets into separate potentially overlapping Gaussians that capture the ground-truth rather than the biased presented data.

The central idea for MIMICis simple: We start with a large number of clusters and let Imitate indicate where data might be missing. In contrast to Agglomerative Clustering, we operate on a point-basis rather than by subsequently merging clusters. If data is available in another cluster to fill in the gap, we let the cluster grow by assigning these data points until it is approximately normally distributed or no suitable data points can be found. In this case, we found a potential selection bias and generate data points to mitigate it. Once all initial clusters have been fully grown, a merging procedure purges duplicates and combines suitable clusters to overcome locally optimal solutions. This process is carried out for every class of the initial dataset (if any) separately, but we describe it for only one class in the following in order to simplify. See Algorithm 1 for an overview and the following for a detailed discussion of the components.

Initialization [Algorithm 1; Lines 1–2]**.** Starting with only the biased dataset B, the Initialization step divides it into a large number of initial clusters that MIMIC uses to search each of them for non-normality. It then uses this information to "steal" data points from other clusters into this one and grow it. If the initial

Algorithm 1. MIMIC

Input: biased dataset B
Output: parameters $\theta_i = (\mu_i, \Sigma_i)$ for each cluster i; a set P of generated points to mitigate the bias

▷ remove outliers using LOF
1: $B' \leftarrow$ `removeOutliers`(B)
▷ initialize clustering using KMeans with large K
2: $l \leftarrow$ `initializeClustering`(B')
3: $\theta \leftarrow \emptyset$
4: $L \leftarrow$ `largestValidCluster`(l)
▷ Grow every valid cluster. A cluster is valid if it is large and dense enough and has neither been processed before nor subsumed by a previous iteration
5: **while** L exists **do**
6: $l, \theta_L \leftarrow$ `growCluster`(L, B', l)
7: $\theta \leftarrow \theta \cup \theta_L$
▷ select the largest valid cluster based on the updated labels l (if possible)
8: $L \leftarrow$ `largestValidCluster`(l)
▷ merge clusters if it improves normality
9: $\theta \leftarrow$ `merge`(θ, B')
▷ generate data to mitigate the bias
10: $P \leftarrow$ `augment`(θ, B)
11: **return** θ, P

Algorithm 2. `growCluster`

Input: label L to be grown, outlier-free dataset B' with labels l
Output: updated labels l, parameters θ_L for cluster L

1: **repeat**
2: $B'_L \leftarrow B'\big|_{l=L}$ ▷ cluster L
▷ run Imitate on L to obtain G_L (grid representing where data might be missing), n_L (number of missing points), θ_L (parameters of the fitted Gaussian)
3: $G_L, n_L, \theta_L \leftarrow$ `Imitate`(B'_L)
▷ score all remaining data points based on if they are likely to help improve the fit of the Gaussian
4: $s \leftarrow$ `score`$(B' \setminus B'_L, G_L, \theta_L)$
▷ identify n_L suitable candidates in batches b_i; sample based on s
5: **for** batches b_i with $\sum_i b_i = n_L$ **do**
6: $C_i \leftarrow$ `sample`$(B' \setminus B'_L, b_i, s)$
▷ assign a batch of candidates to the cluster if it improves the likelihood of the model fitting the data
7: **if** $\mathbb{P}[\theta_L \mid B'_L \cup C_i] > \mathbb{P}[\theta_L \mid B'_L]$ **then**
8: $l(C_i) \leftarrow L$ ▷ update l for accepted C_i
9: **until** l did not change
10: **return** l, θ_L

clusters are already sufficiently normal, no direction for growth can be identified. Therefore, after pre-processing the data with *Local Outlier Factor (LOF)* [4] for higher cluster quality, MIMIC starts off with non-Gaussian initial clusters like those obtained from KMeans. A high number of initial clusters increases the probability that for each true cluster, a less overlapping part is captured in an initial cluster that can later be grown, even if overlaps exist. In order to use a sufficient number of initial clusters, we use twice the number that maximizes the Silhouette score [8], and split further if we detect two density peaks in a histogram instead of one. From here on, the outlier-free dataset is denoted as B' and is passed on to the next step together with the initial labels l.

Identifying Valid Clusters [Algorithm 1; Lines 4, 8]**.** Once a large number of initial clusters has been found, MIMIC grows them into Gaussian clusters where possible using points from B. Aiming to secure reliable performance during the subsequent fitting of a multivariate normal distribution, we filter out all clusters that are either (i) too small (fewer than 10 data points in our implementation) or (ii) too widespread with low density (that is, if the cluster's LOF lies below the 3σ-interval of the average cluster LOF). Note that the latter is a necessary measure as we can expect to obtain unreliable results when fitting a normal distribution to a set of singletons. Additionally, we reduce the computational burden by ensuring that no cluster is grown more than once and no cluster that has been fully subsumed in previous iterations is processed. Thereby, we reduce the number of duplicate clusters we obtain and focus on the most promising ones. Each iteration selects the largest valid cluster and grows it as described below, until no valid clusters remain.

Adapting Imitate to Our Needs [Algorithm 2; Line 3]. Given a cluster L, Imitate estimates a multivariate Gaussian (see Sect. 3) and indicates based on a grid where (and how many) points need to be generated in order to smooth out the cluster's density and have it resemble the fitted Gaussian. Note that the Imitate algorithm as described in the original paper continues to operate on the grid representation which would result in a high complexity given our repeated Imitate calls and does not allow for precise probability assignments, hence we adjust: Assume we fitted one Gaussian (μ_i, σ_i^2) per component i' in the ICA-transformed space. Since the components are independent, this results in a multivariate Gaussian with mean $\mu = (\mu_1, \ldots, \mu_d)$ for d dimensions and the covariance matrix $\Sigma \in \mathbb{R}^{d \times d}$ with diagonal $(\sigma_1^2, \ldots, \sigma_d^2)$ and 0 elsewhere. Let $I \in \mathbb{R}^{d \times d}$ be the ICA transformation matrix. The multivariate Gaussian (μ, Σ) can then be back-transformed into the original space and yields the Gaussian $(I^{-1}\mu, I^{-1}\Sigma(I^{-1})^T)$. We refer to Suppl. A for the proofs of both claims.

Additionally, we adjusted Imitate's method of selecting the grid granularity: Instead of repeating the entire modeling and augmentation process and using the results with the highest confidence score (see the original paper), we use the *corrected Akaike Information Criterion (AICc)* [8] (see Suppl. C for additional experiments justifying this choice) to select, for each dimension, the grid over which a histogram represents the data best. This adjustment is necessary since MIMIC uses repeated calls of the Imitate fitting procedure and the inflicted computational expense of the confidence-based strategy would be infeasible.

Growing Clusters [Algorithm 2]. For a cluster L, Imitate provides us with a multivariate Gaussian θ_L, a grid G_L indicating where and how much (n_L) data might be missing. As outlined in Algorithm 2, both are passed on to a scoring function that estimates for each point p outside L how well it contributes to filling in the gap between the present (h) and fitted (f) density (first term), and how likely it belongs to that distribution (second term):

$$s(p) = d \log[\max\{f(p) - h(p), 0\} + 1] + \log[f(p) + 1]$$

where d denotes the number of features and puts more emphasis on filling the gap for higher dimensions. Using the score, MIMIC then searches for n_L fitting candidates in batches b_i to overcome locally optimal solutions. A batch of candidates C_i is drawn randomly with probabilities based on the score values s and added to the cluster if it fulfills $\mathbb{P}\left[\theta_L \mid B'|_{l=L} \cup C_i\right] > \mathbb{P}\left[\theta_L \mid B'|_{l=L}\right]$, that is, if adding the candidates to the cluster improves the likelihood of the fitted Gaussian given the assigned data points (see Suppl. for the calculations). In our implementation, we restart the sampling (with replacement) of a rejected batch twice in order to avoid "unlucky" choices. If points have been added, MIMIC fits another multivariate Gaussian and repeats the process until no further points are added. The parameters of the last fitted Gaussian represent this cluster.

Merging [Algorithm 1; Line 9]. Once the parameters for all clusters have been obtained, we make sure not to have duplicate clusters or those that are locally optimally normal but can be combined into a better fit. Additionally, MIMIC risks overgrowing clusters if the initial clustering was particularly poor, e.g., if it captures the overlapping area of two clusters. Here, the point density is higher and the Imitate procedure will demand to grow the cluster in all directions simultaneously such that it never reaches a Gaussian-like shape and continues to grow, absorbing more and more data. Such a cluster L is typically characterized by a very wide probability distribution reaching low density values for all points, such that the points p with $L = \arg\max_i \mathbb{P}[p \mid \theta_i]$ exhibit a substantially larger distance to each other than average and can be detected as such. We identify and remove these overgrown clusters as a first step of the merging procedure.

The overlap of two clusters can be quantified by counting the points in the dataset for which the cluster membership is not entirely clear and weighting them using their probabilities. MIMIC calculates the overlap between each combination of two clusters and merges greedily until no further merge improves the fitting of the Gaussians (see Suppl. A for details).

Data Augmentation [Algorithm 1; Line 10]**.** After receiving the final cluster parameter sets from the merging step, MIMIC probabilistically assigns the data points to the clusters and generates points for each cluster separately to "fill in the gap" between the found and the fitted distribution as in Imitate.

Assumptions and Expectations. Selection Biases cannot be reconstructed without making some kind of assumption regarding the ground-truth and/or the nature of the bias. Hence, MIMIC assumes a ground-truth that can be modeled by a mixture of (possibly overlapping) multivariate Gaussians which, in contrast to existing techniques, requires neither a ground-truth sample nor knowledge of the bias. This freedom, however, comes at a cost and forces some implicit requirements: (i) The data cannot contain categorical, binary, or discrete features with a very small number of values as fitting a Gaussian would not be meaningful, (ii) B itself cannot consist only of Gaussian clusters or MIMIC will not be able to identify growth directions, (iii) several strongly overlapping biased clusters might not be disentangled correctly, and (iv) the bias in each cluster is expected to have a convex shape as our component-wise analysis fails otherwise. Lastly, biases can be misleading pointing towards a different Gaussian than the true one and causing MIMIC to introduce new biases into the data. We aim to suppress that behavior by refusing to take action if the Gaussians do not fit reasonably well (see the Imitate paper for details). This, however, causes conservative results with bias reconstructions pointing towards the right locations rather than correcting entirely which is the reason for only small improvements in classification accuracy (as can be seen in the experimental results). In practice, however, this is enough to point a practitioner towards potential problems in the data that can be corrected upon confirmation.

5 Experiments and Discussion

In order to investigate MIMIC's ability to improve classifier performance, we set up all experiments similarly: we train three classifiers on a biased training set B, the augmented biased training set $B \cup \hat{I}$, and an unbiased training set D.

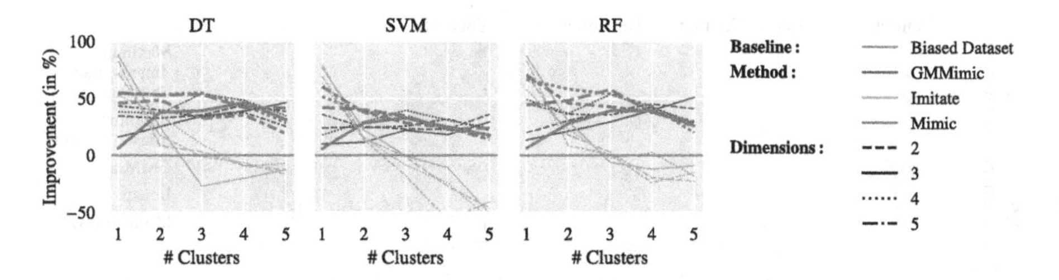

Fig. 3. For each classification method, we compare the impact of the dataset dimensionality and the number of clusters on the performance.

The accuracy acc_B, $acc_{B \cup \hat{I}}$, and acc_D of all three classifiers, respectively, is then evaluated on an unbiased test set with the hope that $acc_{B \cup \hat{I}} > acc_B$. After providing details on the experimental setup, we assess the impact of different characteristics of datasets on the performance.

Experimental Setup. In our experiments, we compare MIMIC not only to the biased accuracy as a baseline, but also for augmented biased datasets $B \cup \hat{I}$ where $\hat{I}$ is obtained using (i) augmentation with Imitate, (ii) clustering and augmentation with MIMIC, and (iii) clustering with GMM and augmentation with MIMIC which we denote as "GMMimic". GMM selects the number of clusters (from 1 to 20) that achieve the best BIC and initializes using KMeans. As classifiers, we use Decision Trees (DT), Support Vector Machines with RBF-kernel (SVM), and Random Forests (RF) with 100 trees. All parameters are kept at sklearn's default values. We use synthetic datasets since they allow us a high level of control, and real-world datasets to demonstrate that MIMIC is indeed applicable in practice. Real-world datasets are taken from the UCI Machine Learning Repository [1,6,22]. Semi-artificial biases are created as in [5] by splitting into B and I using a decision stump (the larger subset is taken for B). This way, the impact on the classification accuracy is guaranteed (see Suppl. B for details). All synthetic experiments are repeated 30 times to compensate for the randomness in the dataset generation, and we report the median results. Experiments on real-world datasets are repeated 10 times as there is no dataset generation step involved. Here, we report the mean together with 90% confidence intervals. We measure the performance as the *improvement over the biased accuracy* and normalize using the unbiased accuracy, i.e., $(acc_{B \cup \hat{I}} - acc_B)/(acc_D - acc_B)$.

Unbiased Datasets. Being able to mitigate a selection bias is important, however, if MIMIC is presented with an unbiased dataset, it should not "correct" it. Experiments (Suppl. C) show that substantially fewer data points (none after purging the noise) are generated for the unbiased datasets.

Dimensionality. The dimensionality of synthetic datasets is closely related to their difficulty as higher dimensions naturally increase the distance between clusters even while under the same cluster-to-center distances. Figure 3 demonstrates this, as lower dimensionalities typically exhibit poorer performance than higher ones, but this effect vanishes with larger numbers of clusters. GMMimic and

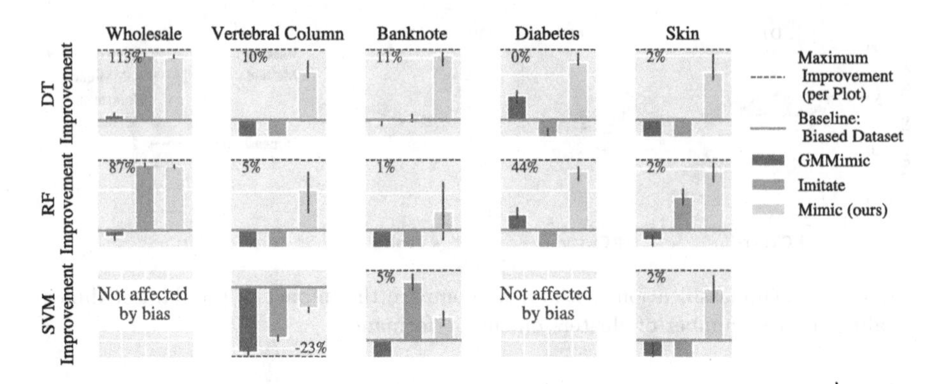

Fig. 4. We compare the degree to which the classifier accuracy can improve when different augmentation techniques are used. The baseline (red line) represents the accuracy when the classifiers are trained on the biased dataset alone. 100% corresponds to training on a ground-truth sample. Note that we omit the y-axes labels and replace them with the dashed line indicating the maximum improvement (maximum y-value) for each plot. The bottom of the plots is cut off unless MIMIC's performance is displayed there for an easier comparison. The black lines are 90% confidence intervals and indicate significant differences from the baseline if they do not touch it. (Color figure online)

MIMIC show similar performances for a larger number of clusters while MIMIC clearly dominates when only a small number of clusters is present, regardless of the dimensionality. Imitate shows strong performance in this case too, but decreases rapidly since it operates with only one cluster.

Cluster Overlap. The center-to-cluster distances directly affect the difficulty of the clustering task as they control the overlap. In experiments (Suppl. C) GMMimic and MIMIC both show improvements even for a large number of clusters and high overlaps. MIMIC demonstrates its strength particularly for better isolated clusters where it improves the classification accuracy by up to 50%.

Real-Life Datasets. Figure 4 summarizes the results on five real-world datasets. For most datasets, we can see MIMIC's potential to improve the classifier accuracy substantially, in most cases more than its competitors. A few observations are noteworthy: On the Wholesale dataset, Imitate performs well since it consists of only one cluster per class. The Vertebral Column dataset seems particularly hard for all methods as the semi-synthetic bias removes 70% of the majority class points (which therefore cannot be reconstructed by any method), leaving an almost balanced classification problem with full overlap and an imbalanced test set. Here, the tree-based methods essentially select the majority class, and MIMIC is able to tip the scales favorably, but cannot help the SVM. Overall, although GMMimic demonstrates solid performance on the synthetic dataset, it does not seem to generalize well to the real-world datasets.

Discussion. Overall, the experiments show that application of an augmentation technique can provide a meaningful improvement on a biased dataset. While Imitate is designed for datasets with only one cluster per class, GMMimic and MIMIC can improve upon its performance when dealing with multi-cluster datasets. The experiments on synthetic datasets with artificial biases point towards a similar performance of GMM- and MIMIC-based data augmentation. On the real-world datasets, however, we do not see this confirmed: MIMIC can further improve the classification performance. Further research will investigate where which method tends to be superior and particularly if a symbiosis of both can be beneficial, e.g., with GMM as an initial model and a MIMIC-inspired merging strategy and augmentation.

MIMIC relaxes Imitate's assumption that the ground-truth dataset consists of only one Gaussian per class. Instead, it can model multiple Gaussian clusters or even approximate non-Gaussian clusters with mixture models. This makes MIMIC applicable to a substantially wider range of datasets. However, not all distributions can be approximated well as a mixture of Gaussians. Future extensions should include an automated test of applicability as well as approaches applicable to a wider range of distributions.

6 Conclusion

Machine Learning models inherit selection biases from datasets causing them to predict inaccurately if the biases remain undetected. Existing bias mitigation strategies require certain kinds of knowledge of the bias or the ground-truth. In real-world scenarios, however, this requirement often cannot be met. A first attempt to detect and mitigate selection biases in a "blind" setting has been made with the Imitate algorithm, although it is limited to datasets with only one Gaussian cluster per class.

In this paper, we introduced MIMIC, a technique that uses Imitate as a building block but overcomes these limitations and can model a wider range of datasets exploiting mixtures of Gaussians. As such, multi-cluster modeling of many non-normally distributed datasets is now possible.

Although limitations still exist as discussed in Sect. 5, we believe that MIMIC is a major step forward towards automated bias identification and mitigation in the case that no knowledge of the bias or the ground-truth exists.

References

1. Abreu, N.: Análise do perfil do cliente Recheio e desenvolvimento de um sistema promocional. Mestrado em marketing, ISCTE-IUL, Lisbon (2011)
2. Bareinboim, E., Tian, J., Pearl, J.: Recovering from selection bias in causal and statistical inference. In: Proceedings of the 28th AAAI Conference on Artificial Intelligence, June 2014 (2014)
3. Bellamy, R.K.E., et al.: AI fairness 360: an extensible toolkit for detecting and mitigating algorithmic bias. IBM J. Res. Develop. **63**(4/5), 4:1–4:15 (2019). https://doi.org/10.1147/JRD.2019.2942287
4. Breunig, M.M., Kriegel, H.P., Ng, R.T., Sander, J.: LOF: identifying density-based local outliers. In: Proceedings of the 2000 ACM SIGMOD International Conference on Management of Data, pp. 93–104 (2000). https://doi.org/10.1145/342009.335388
5. Dost, K., Taskova, K., Riddle, P., Wicker, J.: Your best guess when you know nothing: identification and mitigation of selection bias. In: 2020 IEEE International Conference on Data Mining (ICDM), pp. 996–1001. IEEE (2020). https://doi.org/10.1109/ICDM50108.2020.00115

6. Dua, D., Graff, C.: UCI ML repository (2017). http://archive.ics.uci.edu/ml
7. Goel, N., Yaghini, M., Faltings, B.: Non-discriminatory machine learning through convex fairness criteria. In: Proceedings of the 32nd AAAI Conference on Artificial Intelligence, April 2018 (2018)
8. Granichin, O., Volkovich, Z.V., Toledano-Kitai, D.: Cluster validation. In: Randomized Algorithms in Automatic Control and Data Mining. ISRL, vol. 67, pp. 163–228. Springer, Heidelberg (2015). https://doi.org/10.1007/978-3-642-54786-7_7
9. Hassani, B.K.: Societal bias reinforcement through machine learning: a credit scoring perspective. AI Ethics **1**(3), 239–247 (2020). https://doi.org/10.1007/s43681-020-00026-z
10. Hoeffding, W., Robbins, H.: The central limit theorem for dependent random variables. Duke Math. J. **15**(3), 773–780 (1948). https://doi.org/10.1215/S0012-7094-48-01568-3
11. Hyvärinen, A., Oja, E.: Independent component analysis: algorithms and applications. Neural Netw. **13**(4), 411–430 (2000). https://doi.org/10.1016/S0893-6080(00)00026-5
12. Lavalle, A., Maté, A., Trujillo, J.: An approach to automatically detect and visualize bias in data analytics. In: CEUR Workshop Proceedings of the 22nd International Workshop on Design, Optimization, Languages and Analytical Processing of Big Data, vol. 2572. CEUR (2020)
13. Lyon, A.: Why are normal distributions normal? Br. J. Philos. Sci. **65**(3), 621–649 (2014). https://doi.org/10.1093/bjps/axs046
14. Mehrabi, N., Morstatter, F., Saxena, N., Lerman, K., Galstyan, A.: A survey on bias and fairness in machine learning. ACM Comput. Surv. **54**(6), 1–35 (2021). https://doi.org/10.1145/3457607
15. Panch, T., Mattie, H., Atun, R.: Artificial intelligence and algorithmic bias: implications for health systems. J. Glob. Health **9**(2), 010318 (2019). https://doi.org/10.7189/jogh.09.020318
16. Pedregosa, F., Varoquaux, G., Gramfort, A., Michel, V., Thirion, B., et al.: Scikit-learn: machine learning in Python. J. Mach. Learn. Res. **12**, 2825–2830 (2011)
17. Poulos, J., Valle, R.: Missing data imputation for supervised learning. Appl. Artif. Intell. **32**(2), 186–196 (2018). https://doi.org/10.1080/08839514.2018.1448143
18. Rabanser, S., Günnemann, S., Lipton, Z.: Failing loudly: an empirical study of methods for detecting dataset shift. Adv. Neural Info. Process. Syst. **32**, 1396–1408 (2019)
19. Rezaei, A., Liu, A., Memarrast, O., Ziebart, B.D.: Robust fairness under covariate shift. In: Proceedings of the AAAI Conference on Artificial Intelligence, vol. 35, pp. 9419–9427 (2021)
20. Smith, A.T., Elkan, C.: Making generative classifiers robust to selection bias. In: Proceedings of the 13th ACM SIGKDD International Conference on Knowledge Discovery and Data Mining, pp. 657–666 (2007). https://doi.org/10.1145/1281192.1281263
21. Stojanov, P., Gong, M., Carbonell, J., Zhang, K.: Low-dimensional density ratio estimation for covariate shift correction. Proc. Mach. Learn. Res. **89**, 3449–3458 (2019)
22. Strack, B., Deshazo, J., Gennings, C., Olmo Ortiz, J.L., Ventura, S., et al.: Impact of HbA1c measurement on hospital readmission rates: analysis of 70,000 clinical database patient records. BioMed Res. Int. **2014**, 781670 (2014). https://doi.org/10.1155/2014/781670

Hypersphere Neighborhood Rough Set for Rapid Attribute Reduction

Yu Fang[1,2], Xue-Mei Cao[1], Xin Wang[1], and Fan Min[1,3](✉)

[1] School of Computer Science, Southwest Petroleum University, Chengdu 610500, China
{fangyu,minfan}@swpu.edu.cn

[2] School of Computing, Faculty of Engineering, Universiti Teknologi Malaysia (UTM), 81310 Johor Bahru, Johor, Malaysia

[3] Institute for Artificial Intelligence, Southwest Petroleum University, Chengdu 610500, China

Abstract. Neighborhood rough set (NRS) has been successfully applied to attribute reduction for numeric data. Most existing algorithms have a time complexity of at least $O(MN^2)$. In this paper, we propose a hypersphere neighborhood rough set (HNRS) algorithm with a time complexity of $O(MN)$. HNRS adaptively generates the neighborhood radius without manual setting. First, a set of hyperspheres is built to accurately describe the decision boundary on the original data. Second, the hypersphere radius serves as the neighborhood radius to obtain the positive region. Therefore, we avoid the time-consuming grid searching of the NRS algorithm for radius optimization. Third, according to the change of objects within the positive region, the redundant attributes can be reduced efficiently. Experimental results show that HNRS outperforms state-of-the-art attribute reduction methods in terms of both efficiency and classification accuracy.

Keywords: Neighborhood rough sets · Hypersphere · Support vector data description · Attribute reduction · Decision boundary

1 Introduction

With the rapid increase of data dimension and volume, traditional data mining algorithms are facing challenges from both the perspective of data storage and computation. The attributes of these data are often correlated, redundant, or even noisy, which can lead to adverse effects such as high computational complexity and poor performance [3]. Attribute reduction is one of the most effective data preprocessing strategies to deal with this issue [1]. It directly removes redundant attributes from the original feature space for a compact and accurate representation. Additionally, attribute reduction helps build simpler learning models, improve learning performance and data quality [18].

J. Gama et al. (Eds.): PAKDD 2022, LNAI 13281, pp. 161–173, 2022.
https://doi.org/10.1007/978-3-031-05936-0_13

Rough set theory was originally presented by Pawlak as an effective mathematical tool to deal with uncertain information [10]. In recent years, this theory has been widely applied to attribute reduction in the field of artificial intelligence [9,13,15]. Rough set derives from the indiscernibility relation of the attribute subset of the universe. The upper and lower approximations are defined according to this relation. For a classification problem, the upper and lower approximation sets contain all elements that may and exactly belong to a class, respectively. According to different definitions of the indiscernibility relation, various rough set models have been proposed, such as the classical rough set [10], fuzzy rough set [2], and other rough set models [7,19].

Neighborhood rough set (NRS) [4,16] is one of the most important rough set models. From the extensions of the classical rough set, NRS uses neighborhood relations to generate a family of neighborhood granules from the universe, and then these neighborhood granules can be used to approximate decision classes [5]. The neighborhood relation is characterized by the size of the neighborhood. However, the size of the neighborhood is influenced by the distribution of the dataset [6]. This will lead to changes in the results of the NRS. Therefore, the neighborhood's size is a parameter that is typically optimized by grid searching. Moreover, the process of parameter optimization is very time-consuming, and this is unacceptable when training on large-scale datasets.

To alleviate this issue, Liu et al. [17] proposed the FHARA algorithm with the concept of hash bucket, which divides the data into different regions and only calculates the neighborhood relations of objects in the same region. However, this algorithm lacks the ability to optimize the neighborhood radius. In fact, optimizing the neighborhood radius refers to automatically generating the ideal radius for each dataset. This is very effective for the performance of attribute reduction, especially when the scale of the provided dataset is large. Although the GBNRS algorithm proposed by Xia et al. [14] attempts to solve this optimization problem, the algorithm exhibits instability and requires multiple experiments to obtain the best results. Thus, it is of vitally important for NRS to take an effective strategy to generate the neighborhood radius adaptively and stably.

Motivated by the above observations, we propose to perform neighborhood partition for each instance in an adaptive fashion. To be specific, we generate the optimal neighborhood radius parameter adaptively rather than setting a fixed value. An illustration of the proposed adaptive neighborhood partition is shown in Fig. 1. In essence, we study (1) how to rapidly generate the optimal neighborhood radius based on the different distribution of the dataset. (2) how to perform attribute reduction stably and rapidly under the generated neighborhood relation. To address these two research issues, we propose a new NRS model for rapid attribute reduction in classification, called hypersphere neighborhood rough set (HNRS). The main contributions are summarized as follows:

- We establish a novel model of the hypersphere neighborhood rough set by introducing hypersphere computing into NRS, and the time complexity of HNRS is only $O(MN)$.

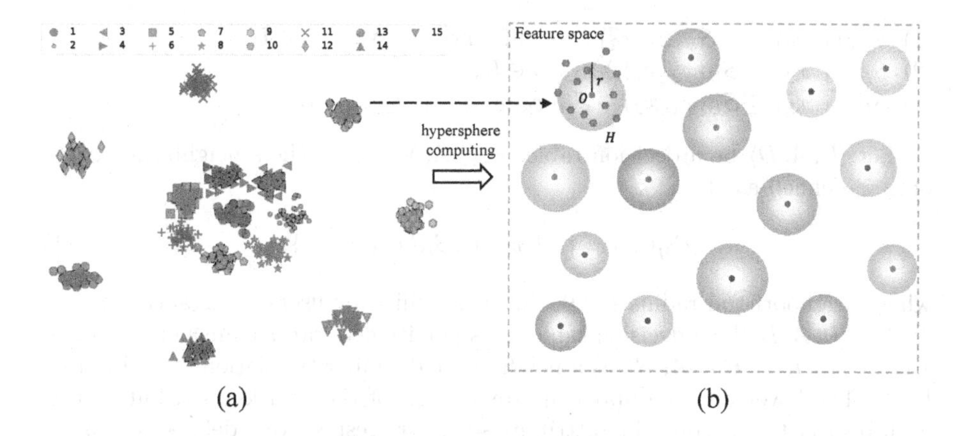

Fig. 1. Illustration of the hypersphere neighborhood rough set. (a) is the original distribution of a dataset with 15 classes. (b) is a set of hyperspheres built for the data in (a) by hypersphere computing. The hypersphere contains the majority of equivalence class instances while excluding outliers. The boundary points of the hypersphere are support vectors. The radius r and center O of each hypersphere are calculated based on support vectors.

- HNRS is a parameter-free model which adaptively generates the neighborhood radius, making it more flexible and generalizable. Furthermore, HNRS is quite stable and does not contain any randomness.
- We validated the effectiveness of the HNRS model on public benchmark datasets in the domain.

2 Preliminaries

In this section, we simply review the mathematical foundations of the two theories used, hypersphere computing and neighborhood rough sets.

2.1 Neighborhood Rough Set

In practical applications, a dataset is always given by a data decision table and denoted as $\langle U, A, D\rangle$, where $U = \{x_1, x_2, \dots, x_N\}$ is a nonempty finite set of objects, $A = \{a_1, a_2, \dots, a_m\}$ is a condition attribute set in which the attributes are real-valued and D is a decision attribute whose values are nominal. There are many ways to define the neighborhood of an object. However, regardless of the type of neighborhood, a metric is required to calculate the distance between objects. The following is a description of the distance metric.

Let $\langle U, A, D\rangle$ be a decision table, let $U = \{x_1, x_2, \dots, x_N\}$ be a non-empty finite set of real space, $B \subseteq A$ is an attribute subset, $\mathcal{S}_B : U \times U \rightarrow R$ is a binary function. $\mathcal{S}_B$ is known as a metric, if it satisfies following conditions:

(1) $\mathcal{S}_B(x_1, x_2) \geq 0, \mathcal{S}_B(x_1, x_2) = 0 \quad \text{iff} \quad x_1 = x_2, \forall x_1, x_2 \in U$;
(2) $\mathcal{S}_B(x_1, x_2) = \mathcal{S}_B(x_2, x_1), \forall x_1, x_2 \in U$;
(3) $\mathcal{S}_B(x_1, x_3) \leq \mathcal{S}_B(x_1, x_2) + \mathcal{S}_B(x_2, x_3), \forall x_1, x_2, x_3 \in U$.

Let $\langle U, A, D\rangle$ be a decision table, $B \subseteq A$, $\forall x_i \in U$, the ε-neighborhood R_B^{ε} of x_i is defined as [4]:

$$R_B^{\varepsilon}(x_i) = \{x \mid x \in U, \mathcal{S}_B(x, x_i) \leq \varepsilon\}, \tag{1}$$

where neighborhood radius $\varepsilon > 0$, that is specified by users in advance.

Let $\langle U, A, D\rangle$ be a decision table, U is partitioned into r equivalence classes: $E_1, E_2, \ldots, E_r$. $\forall B \subseteq A$, R_B^{ε} is a neighborhood similarity relation on U induced by B. The lower and the upper approximation of the decision attribute set D with respect to the condition attribute set B are respectively defined as [13]:

$$\underline{R}_B^{\varepsilon}(D) = \bigcup_{i=1}^{r} \underline{R}_B^{\varepsilon}(E_k), \tag{2}$$

$$\overline{R}_B^{\varepsilon}(D) = \bigcup_{i=1}^{r} \overline{R}_B^{\varepsilon}(E_k), \tag{3}$$

where $\underline{R}_B^{\varepsilon}(E_k) = \{x_k \mid R_B^{\varepsilon}(x_k) \subseteq E_k, x_k \in U\}$, $\overline{R}_B^{\varepsilon}(E_k) = \{x_k \mid R_B^{\varepsilon}(x_k) \bigcap E_k \neq \emptyset, x_k \in U\}$, $POS_B^{\varepsilon}(D) = \underline{R}_B^{\varepsilon}(D)$ is the sample domain of consistent decisions in all neighborhoods. The dependency function of D associated with B is formulated as $\gamma_B^{\varepsilon}(D) = |POS_B^{\varepsilon}(D)|/|U|$, where $|\cdot|$ indicates the cardinality of a set.

Classical NRS calculates the neighborhood similarity relation between any two objects within a given radius on a condition attribute set. Concretely, NRS searches for neighbors in the given neighborhood of each object on a condition attribute set. The object is added into the positive region if the neighbors has the same label with the queried object. Therefore, the time complexity of NRS is $O(MN^2)$. Unfortunately, the neighborhood radius must be provided by the user, and the user does not know the optimal radius for the dataset in advance. Generally, the optimal neighborhood radius for each dataset is found by grid searching. However, parameter optimization is a very time-consuming process that significantly affects the performance of model. To alleviate this issue, we introduce hypersphere computing into NRS, which is a stable and adaptive method of generating radius.

2.2 Hypersphere

The theoretical foundation of hypersphere computing is support vector data description (SVDD) [12]. SVDD is a machine learning technique that is widely used for single-class classification and outlier detection. Its purpose is to use a set of support vectors to find the hypersphere with the minimum volume that contains the most target objects. We suppose that $\{x_1, x_2, \ldots, x_n\}$ is a set of target objects. Therefore, the mathematical formulation of the problem is to find a nonnegative vector $\boldsymbol{\alpha}$ that contains Lagrange multipliers for all data points to maximize the optimization problem:

$$L = \sum_i \alpha_i (x_i \cdot x_i) - \sum_{i,j} \alpha_i \alpha_j (x_i \cdot x_j). \tag{4}$$

The x_i's for which $0 < \alpha < C$ for a preselected $0 < C < 1$ lie on the boundary are called support vectors.

Let $\langle U, A, D\rangle$ be a decision table, $U/D = \{E_1, E_2, \ldots, E_k\}$, $E_k = \{x_1, x_2, \ldots, x_n\}$, $\boldsymbol{\alpha}$ is the Lagrange multipliers obtained by the quadratic planning, $\|\boldsymbol{\alpha}\|_1 = 1$. We generate a hypersphere H with O as its center and r as its radius on E_k as follows [12]:

$$O = \sum_i \alpha_i x_i, \tag{5}$$

$$r = \sqrt{(x_k \cdot x_k) - 2\sum_i \alpha_i (x_k \cdot x_i) + \sum_{i,j} \alpha_i \alpha_j (x_i \cdot x_j)}, \tag{6}$$

where x_k is any support vector, $(x_i \cdot x_j)$ is the inner product of x_i and x_j. Generally, the inner product $(x_i \cdot x_j)$ is replaced by the Gaussian kernel, $\kappa(x_i, x_j) = \exp(-\frac{\|x_i - x_j\|_2^2}{2\sigma^2})$. To be specific, it is to map the original data space to a high-dimensional feature space by the Gaussian kernel.

In fact, most datasets have multiple equivalence classes, and we serve the instances of each equivalence class as the target objects to generate its hypersphere. Therefore, the dataset's number of hyperspheres equals the number of equivalence classes. Although each hypersphere covers many instances, it consists of only two properties, the center and the radius. Consequently, the dataset's representation becomes quite simple. In addition, the decision boundary can be accurately described by the hyperspheres generated by the support vectors [12]. This not only ensures the hypersphere's purity, but also efficiently eliminates contentious instances. This can be seen directly in the hypersphere H in Fig. 1(b). Fortunately, Jiang et al. [8] presented a fast method for obtaining support vectors (FISVDD). This allows us to swiftly construct hyperspheres from large-scale datasets.

3 Hypersphere Neighborhood Rough Set

In this section, we will describe the details of the proposed hypersphere neighborhood rough set.

3.1 Theory and Mathematical Models

We have shown the process of hypersphere neighborhood rough set in Fig. 1. The multiple hyperspheres with different radii are adaptively generated by hypersphere computing. Furthermore, we can ensure that all instances within a hypersphere have the same label by generating a hypersphere for each equivalence class. In other words, the purity of the hypersphere is guaranteed to be 1. We consider the centers of all hyperspheres as positive region for the following three reasons:

1) The center of the hypersphere is far away from the boundary and completely within the class. Therefore, it has no impact on the dataset's decision boundary;
2) There is no randomness in the process of generating the centers, which helps to generate a stable positive region;
3) The instances inside a range at the center all have the same label with it, which is identical to the definition of the positive region in NRS.

Since the hypersphere is the smallest hypersphere that contains the target objects, and non-target objects or abnormal points are strictly excluded, the positive region has no risk of being contaminated. The following is the mathematical definition of our hypersphere neighborhood rough set.

Definition 1 (Neighborhood). *Let $\langle U, A, D \rangle$ be a decision table, $U/D = \{E_1, E_2, \dots, E_k\}$, $B \subseteq A$, and let the generated hyperspheres cover the entire U. The hypersphere generated by the E_k is H_k, it's center and radius are O_k and r_k, respectively. For $x_i \in H_k$, we define the neighborhood of x_i as:*

$$R_B^{r_k}(x_i) = \{x \mid \forall x \in H_k, \mathcal{S}_B(x, O_k) \leq r_k\}, \tag{7}$$

where $\mathcal{S}_B(x, O_k)$ is the distance from x to O_k under the condition attribute set B, and

$$\mathcal{S}_B(x, O_k)^2 = \kappa(x \cdot x) - 2\sum_i \alpha_i \kappa(x \cdot x_i) + \sum_{ij} \alpha_i \alpha_j \kappa(x_i \cdot x_j). \tag{8}$$

Definition 2 (Lower approximation). *Let $\langle U, A, D \rangle$ be a decision table, $U/D = \{E_1, E_2, \dots, E_k\}$, $B \subseteq A$, $R_B^{r_k}(x_i)$ is the neighborhood of x_i on U induced by B, and let the generated hyperspheres cover the entire U. The k-th hypersphere under the condition attribute set is $H_k(B)$. The lower approximation set of E_k with respect to a attribute set B is defined as:*

$$\underline{R}_B^{r_k}(E_k) = \{x = \sum_i \alpha_i x_i \mid x_i \in H_k(B), R_B^{r_k}(x) \subseteq E_k\}. \tag{9}$$

Definition 3 (Positive region). *Let $\langle U, A, D \rangle$ be a decision table, $B \subseteq A$, $U/D = \{E_1, E_2, \dots, E_k\}$, and let the generated hyperspheres cover the entire U. The positive region is defined as:*

$$HPos_B(D) = \bigcup_{i=1}^{k} \underline{R}_B^{r_k}(E_k). \tag{10}$$

Definition 4 (Relative redundancy attribute). *Let $\langle U, A, D\rangle$ be a decision table. For a condition attribute set $B \subseteq A$, $a \in B, B \neq \emptyset$. a is a relative redundancy attribute of C if a satisfies:*

$$HPos_{B-\{a\}}(D) = HPos_B(D) \tag{11}$$

Due to the dependency function in NRS only considers the lower approximation, we introduce the lower approximation and the positive region of HNRS in Definition 2 and in Definition 3, respectively. As described in Definition 2, the lower approximation of each equivalence class is composed of its corresponding hypersphere center. Therefore, the generated positive region is composed of the center of each hypersphere. In particular, the center of the hypersphere is a virtual point, which is generated rather than selected. In Definition 4, redundant attributes are accurately removed by positive region changes.

3.2 Algorithm Design

The proposed HNRS algorithm is mainly divided into two stages. First, the positive region on the entire dataset is generated. Second, the relative redundancy attributes are deleted by the positive region.

In the first stage, we adopt FISVDD to rapidly obtain the support vectors of each equivalence class to generate hyperspheres. The center and radius of each hypersphere are generated adaptively. Additionally, the hypersphere radius is served as the neighborhood radius to generate the neighborhood of the instance, and then the lower approximation and the positive region are obtained. Concretely, the positive region consists of the centers of all hyperspheres.

In the second stage, we calculate the distance between each instance and each hypersphere's center after removing an attribute, and partition these instances into their nearest hypersphere. In particular, the center belongs to the generated positive region and the distance is calculated according to Eq. (8). This partition will generate new hyperspheres. If the purity of these new hyperspheres is all 1, the removed attribute is a relative redundancy attribute that can be deleted; otherwise, it should be retained. Additionally, the hyperspheres need to be reconstructed after deleting an attribute. As a reminder, the purity of a hypersphere is the percentage of the majority instance in the hypersphere. All conditional attributes are checked in turn, and finally a reduct set is generated.

Algorithm 1. With the above description, the pseudo code of the proposed HNRS framework is summarized in Algorithm 1. From steps 2 to 4, the initial positive region on the original data is generated. From steps 5 to 6, the instances are repartitioned into the nearest hypersphere after removing an attribute. From steps 7 to 17, we determine whether the removed attribute is redundant by whether the positive region has changed. In particular, the change in the positive region is assessed by the purity of the hyperspheres. When an attribute is removed, the support vector describing the decision boundary is affected. As a result, the hyperspheres will need to be rebuilt.

Time Complexity. The support vectors describing the decision boundary can be found rapidly with FISVDD, and the time complexity is $O(N)$. Specifically,

Algorithm 1: Hypershpere Neighborhood Rough Set (HNRS)

input : A decision table $\langle U, A, D \rangle$, $A = \{a_1, a_2, \ldots, a_m\}$;
output: A reduced attribute set B;
1 Initialization: $B = A$;
2 Generate a hyperspheres for each equivalence class on B by FISVDD [8];
3 Generate the center and the radius of each hypersphere on B by Eq. (5) and Eq. (6), respectively;
4 Generate the positive region of U on B by Eq. (10);
5 Remove a condition attribute a_i in B;
6 Partition each instance into the nearest hypersphere on B by Eq. (8) based on the positive region in step 4;
7 Calculate the purity of the newly generated hyperspheres;
8 **if** *the purity of each hypersphere is 1* **then**
9   Go to step 2; // a_i is a relative redundancy attribute of B
10 **else**
11   Add the condition attribute a_i to B; // a_i should be retained
12   **if** *all attributes in B have been checked* **then**
13   Return B;
14   **else**
15   Remove a new condition attribute in B and go to step 6;
16   **end**
17 **end**

the time complexity of a hypersphere with a center and a radius constructed by the support vectors is $O(N)$. Therefore, HNRS is substantially more efficient than NRS at generating positive region. This can be intuitively understood as: NRS needs to search the neighborhood of each object when generating a positive region. In contrast, HNRS merely needs to traverse the center of each hypersphere. Although the number of hyperspheres is small, they are generated by adaptive dataset distribution and can accurately describe the decision boundary. Due to we need to check all conditional attributes, the time complexity of HNRS is $O(MN)$. As a result, HNRS performs better, as we have demonstrated in experiments.

4 Experiments

In this section, we conduct a series of experiments to evaluate the performance of the algorithm of HNRS. The first experiment aims to validate the effectiveness of HNRS in removing redundant attributes. Then, six representative datasts are used to verify the efficiency of HNRS. All the code for the experiment is available at https://github.com/diadai/HNRS.

4.1 Experimental Setup

To illustrate the effectiveness and superiority of HNRS, it is compared with classical NRS [16] and the current state-of-the-art NRS algorithms, including fast hash attribute reduct algorithm (FHARA) [17], and granular ball neighborhood rough sets (GBNRS) [14].

For a fair comparison, the neighborhood radius of NRS and FHARA is tuned from $\{0.025, 0.05, \ldots, 1\}$ by grid searching, GBNRS runs ten times to avoid randomness, and Gaussian kernel parameter $\sigma \in [0.6, 1, 4]$ for HNRS. Specifically, we set σ in three cases based on experience: a) when $N \leq 1000$, $\sigma = 0.6$; b) when $1000 < N \leq 50000$, $\sigma = 1$; c) when $N > 50000$, $\sigma = 4$, where N is the number of instances in the dataset. For each comparing algorithm, we report their best classification accuracies on the reduct set.

Table 1. Properties of datasets.

Datasets	Number of attributes	Number of instances	Number of classes	Datasets	Number of attributes	Number of instances	Number of classes
Hepatiti	19	155	2	Ring	20	7,400	2
Wine	13	178	3	Pendigits	16	10,992	10
Ionoshpere	35	351	2	Online	17	12,330	2
Derm	34	366	6	Dry-Bean	17	13,611	7
Vote	16	436	2	Letter	15	20,000	16
Wdbc	30	569	2	Bank	17	45,211	2
Australian	14	690	2	Adult	14	45,222	2
Crx	15	690	2	Sensorless	48	58,509	11
Vehicle	18	846	4	Miniboone	50	130,064	2
Segmentation	18	2,310	7	Har	18	165,632	5

4.2 Effectiveness

The first experiment aims to validate the effectiveness of HNRS in removing redundant attributes. We use 20 datasets from UCI (http://archive.ics.uci.edu/ml/datasets.php) to conduct experiments, and these datasets are summarized in Table 1. For each dataset, the following things were reported: a) The average accuracy of ten times 10-fold cross validation (10CV); b) The standard deviation of 10CV (the value with $\pm$). Additionally, the mean rank was obtained by applying the Friedman test, which is the most well-known non-parametric test [11]. The Friedman test analyzes whether there are significant differences among the algorithms.

Table 2 shows the accuracy comparison of HNRS and rival attribute reduction algorithms. The black dots highlight the best results, and the original denotes the classification result of the original data on kNN. From Table 2, we observe that our algorithm outperforms other algorithms on 12 out of 20 datasets and also achieves very competitive accuracy on the remaining datasets in comparison with the state-of-the-art algorithms. Furthermore, HNRS can efficiently improve the overall stability of the classifier. On the other hand, these comparison algorithms

Table 2. Comparison of experimental accuracy on kNN

Datasets	Original	HNRS	NRS	FHARA	GBNRS
Hepatiti	$56.67_{\pm 13.74}$	$64.00_{\pm 7.42}$•	$60.67_{\pm 5.54}$	$63.33_{\pm 8.56}$	$60.67_{\pm 10.52}$
Wine	$95.88_{\pm 5.91}$	$97.65_{\pm 3.90}$•	$97.65_{\pm 3.95}$•	$97.65_{\pm 3.90}$•	$94.12_{\pm 5.26}$
Ionoshpere	$86.29_{\pm 5.08}$	$92.29_{\pm 4.44}$•	$91.43_{\pm 4.04}$	$90.57_{\pm 4.44}$	$88.29_{\pm 6.32}$
Derm	$97.50_{\pm 2.31}$	$97.50_{\pm 2.62}$•	$97.22_{\pm 2.78}$	$93.33_{\pm 4.84}$	$95.00_{\pm 2.72}$
Vote	$91.86_{\pm 2.80}$	$95.35_{\pm 1.80}$•	$92.09_{\pm 3.15}$	*	$93.95_{\pm 3.32}$
Wdbc	$96.96_{\pm 1.96}$	$97.32_{\pm 1.65}$•	$97.32_{\pm 1.83}$•	$96.96_{\pm 2.53}$	$94.64_{\pm 2.88}$
Australian	$82.75_{\pm 5.00}$	$84.64_{\pm 4.06}$	$84.20_{\pm 4.65}$	$84.20_{\pm 4.65}$	$85.36_{\pm 3.07}$•
Crx	$85.65_{\pm 5.59}$	$85.51_{\pm 4.20}$	$85.36_{\pm 4.87}$	$85.36_{\pm 4.87}$	$85.07_{\pm 4.81}$
Vehicle	$70.48_{\pm 5.32}$	$73.33_{\pm 5.89}$•	$70.83_{\pm 4.94}$	$70.71_{\pm 5.79}$	$68.57_{\pm 5.14}$
Segmentation	$96.15_{\pm 1.09}$	$96.15_{\pm 1.09}$•	$96.15_{\pm 0.68}$•	$96.10_{\pm 1.02}$	$96.06_{\pm 1.10}$
Ring	$71.89_{\pm 2.41}$	$77.39_{\pm 2.30}$	$84.69_{\pm 1.54}$•	$84.47_{\pm 1.21}$	$71.89_{\pm 2.41}$
Pendigits	$99.34_{\pm 0.24}$	$99.33_{\pm 0.21}$	$99.31_{\pm 0.15}$	$97.82_{\pm 0.42}$	$99.34_{\pm 0.24}$•
Online	$84.36_{\pm 1.20}$	$85.95_{\pm 0.94}$•	$85.95_{\pm 1.19}$•	$84.62_{\pm 1.46}$	$84.51_{\pm 1.15}$
Dry-Bean	$91.54_{\pm 0.63}$	$91.83_{\pm 0.58}$	$91.93_{\pm 0.46}$	$91.95_{\pm 0.56}$•	$91.64_{\pm 0.48}$
Letter	$65.73_{\pm 0.86}$	$67.52_{\pm 0.43}$•	*	$65.73_{\pm 0.86}$	$65.73_{\pm 0.86}$
Bank	$88.58_{\pm 0.31}$	$88.34_{\pm 0.42}$	*	$88.58_{\pm 0.30}$•	$88.58_{\pm 0.30}$•
Adult	$81.42_{\pm 0.49}$	$81.95_{\pm 0.55}$•	*	$81.01_{\pm 0.47}$	$81.42_{\pm 0.49}$
Sensorless	$99.02_{\pm 0.09}$	$99.02_{\pm 0.09}$	*	$99.30_{\pm 0.10}$•	$99.01_{\pm 0.06}$
Miniboone	$87.90_{\pm 0.15}$	$87.42_{\pm 0.14}$	*	*	*
Har	$99.47_{\pm 0.04}$	$99.47_{\pm 0.04}$•	*	*	*
Meanrank	2.6	1.55•	2.55	2.55	2.85
Win/Tie/Lost	4/4/12	–	2/4/14	4/1/15	3/0/17

* indicates that the experiment results cannot be obtained.

are unable to get classification results for large-scale data sets such as miniboone in a reasonable amount of time and memory. Consequently, HNRS is an efficient attribute reduction method that can be applied to many domains, especially to large-scale datasets.

4.3 Efficiency

To evaluate the efficiency of the proposed Hypersphere neighborhood rough set algorithm, 6 large-scale datasets in Table 2 are used, including Online, Dry_Bean, Letter, Adult, Sensorless and Har. In these experiments, we gradually increased the number of instances from 10% to 100% of a dataset. Because NRS and FHARA are very slow for large-scale datasets, the neighborhood radius in this experiment is fixed at 0.01 and no grid searching is performed. The attribute reduction time consumption variation curves are illustrated in Fig. 2. As shown in Fig. 2, comparison algorithms trend to achieve a significant increase in time consumption with the increase of instances in data, and HNRS can consistently outperform the state-of-the-art.

Furthermore, the partially incomplete curve in Fig. 2 indicates that the corresponding algorithm was unable to obtain the reduct set due to memory constraints. On the six large-scale datasets, HNRS not only runs normally, but also has the best performance. However, the rival algorithms cannot be executed when the number of instances reaches a certain level. For example, in the largest experimental dataset, Har in Fig. 2(f), NRS can not run normally at all. Additionally, on 80% of Har's data instances, FHARA and GBNRS take more than 10, 000 and 1, 000 s, respectively; In contrast, HNRS only takes 34 s, a more than 95% improvement. The results in Fig. 2 demonstrate the significant superiority of HNRS in terms of processing large-scale data.

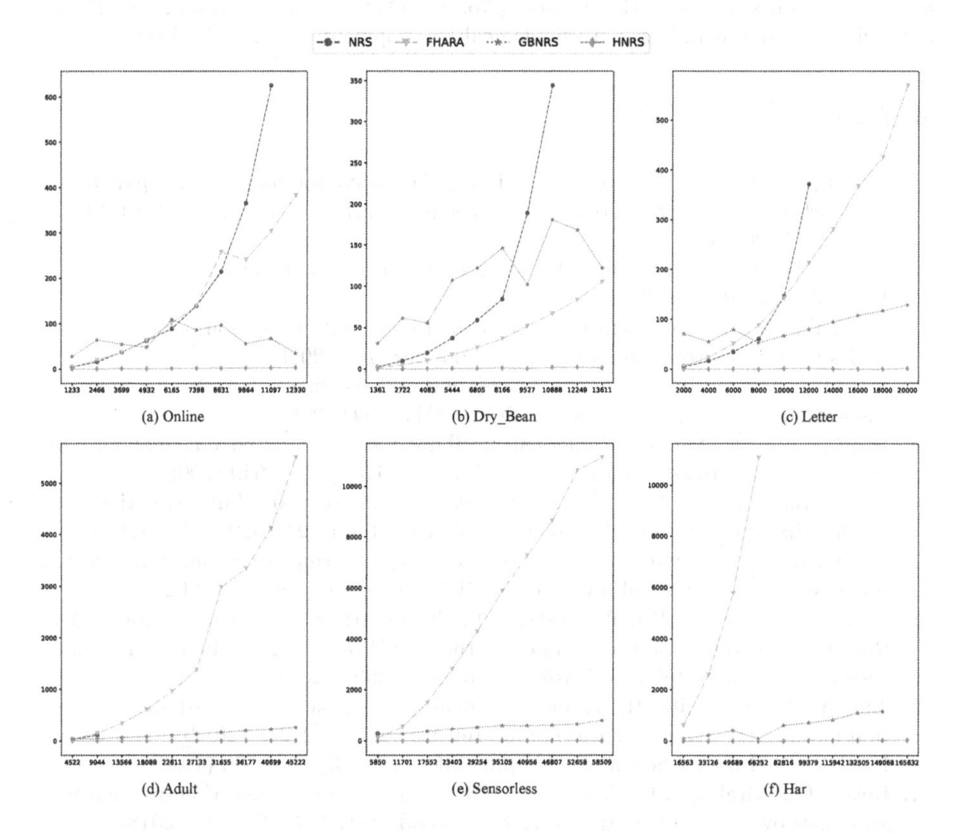

Fig. 2. Efficiency comparison. The y-axis is the computational time (s), and the x-axis is the number of instances.

5 Conclusion

In this paper, we proposed a novel rough set algorithm with $O(MN)$ time complexity, called HNRS. HNRS is a parameter-free algorithm that introduces hypersphere computing into NRS. Furthermore, HNRS adaptively generates neighborhood radius, which solves the problem of optimizing radius for grid searching in NRS. We conduct experiments on different datasets, and compare HNRS with the state-of-the-art attribute reduction algorithms. The results demonstrate the superiority of HNRS over others in terms of effectiveness and efficiency.

Acknowledgement. This work was supported by the National Natural Science Foundation of China (62006200); the Sichuan Province Youth Science and Technology Innovation Team, China (2019JDTD0017); the Southwest Petroleum University Postgraduate English Course Construction Project (No. 2020QY04); Central Government Funds of Guiding Local Scientific and Technological Development (No. 2021ZYD0003).

References

1. Dai, J., Hu, H., Wu, W., Qian, Y., Huang, D.: Maximal-discernibility-pair-based approach to attribute reduction in fuzzy rough sets. IEEE Trans. Fuzzy Syst. **26**(4), 2174–2187 (2018)
2. Dubois, D., Prade, H.: Rough fuzzy sets and fuzzy rough sets. Int. J. Gen Syst **17**(2–3), 191–209 (1990)
3. Fang, Y., Min, F.: Cost-sensitive approximate attribute reduction with three-way decisions. Int. J. Approximate Reason. **104**, 148–165 (2019)
4. Hu, Q., Yu, D., Liu, J., Wu, C.: Neighborhood rough set based heterogeneous feature subset selection. Inf. Sci. **178**(18), 3577–3594 (2008)
5. Hu, Q., Yu, D., Xie, Z.: Numerical attribute reduction based on neighborhood granulation and rough approximation. J. Softw. **19**(3), 640–649 (2008)
6. Hu, Q., Zhao, H., Yu, D.: Efficient symbolic and numerical attribute reduction with neighborhood rough sets. Pattern Recogn. Artif. Intell. **21**(6), 732–738 (2008)
7. Huang, Z., Li, J., Qian, Y.: Noise-tolerant fuzzy covering based multigranulation rough sets and feature subset selection. IEEE Trans. Fuzzy Syst. (2021)
8. Jiang, H., Wang, H., Hu, W., Kakde, D., Chaudhuri, A.: Fast incremental SVDD learning algorithm with the Gaussian kernel. In: Proceedings of the AAAI Conference on Artificial Intelligence, vol. 33, pp. 3991–3998 (2019)
9. Jing, Y., Li, T., Fujita, H., Wang, B., Cheng, N.: An incremental attribute reduction method for dynamic data mining. Inf. Sci. **465**, 202–218 (2018)
10. Pawlak, Z.: Rough sets. Int. J. Comput. Inf. Sci. **11**(5), 341–356 (1982)
11. Reyes, O., Altalhi, A.H., Ventura, S.: Statistical comparisons of active learning strategies over multiple datasets. Knowl. Based Syst. **145**, 274–288 (2018)
12. Tax, D.M., Duin, R.P.: Support vector data description. Mach. Learn. **54**(1), 45–66 (2004)
13. Wang, C., Huang, Y., Shao, M., Hu, Q., Chen, D.: Feature selection based on neighborhood self-information. IEEE Trans. Cybern. **50**(9), 4031–4042 (2019)
14. Xia, S., Zhang, Z., Li, W., Wang, G., Giem, E., Chen, Z.: GBNRS: a novel rough set algorithm for fast adaptive attribute reduction in classification. IEEE Trans. Knowl. Data Eng. **34**, 1231–1242 (2020)

15. Xue, B., Zhang, M., Browne, W.N.: Particle swarm optimization for feature selection in classification: a multi-objective approach. IEEE Trans. Cybern. **43**(6), 1656–1671 (2012)
16. Yao, Y.: Relational interpretations of neighborhood operators and rough set approximation operators. Inf. Sci. **111**(1–4), 239–259 (1998)
17. Yong, L., Wenliang, H., Yünliang, J., Zhiyong, Z.: Quick attribute reduct algorithm for neighborhood rough set model. Inf. Sci. **271**, 65–81 (2014)
18. Zhao, S., Chen, H., Li, C., Zhai, M., Du, X.: RFRR: robust fuzzy rough reduction. IEEE Trans. Fuzzy Syst. **21**(5), 825–841 (2013)
19. Zhu, W.: Generalized rough sets based on relations. Inf. Sci. **177**(22), 4997–5011 (2007)

A Novel Clustering Algorithm with Dynamic Boundary Extraction Strategy Based on Local Gravitation

Jiangmei Luo[1], Qingsheng Zhu[1(✉)], Junnan Li[1], Dongdong Cheng[2], and Mingqiang Zhou[1]

[1] College of Computer Science, Chongqing University, Chongqing, China
{jmluo,qszhu,zmqmail}@cqu.edu.cn

[2] College of Big Data and Intelligent Engineering, Yangtze Normal University, Chongqing, China
cdd@cqu.edu.cn

Abstract. Clustering has been widely used in visual analysis, pattern recognition, privacy protection and other fields. In recent years, numerous clustering methods have received increasing attention. However, discovering arbitrarily shaped clusters, determining the location and number of clustering cores and dealing with fuzzy boundaries is tough for most algorithms. We propose a novel clustering algorithm with dynamic boundary extraction strategy based on local gravitation (DBELG) which extracts boundary in a natural way, rather than mechanically defining a few core points. In order to identify fuzzy boundaries, a novel gravity model that makes use of three significant information about the data objects is proposed. The structure of the reserved core groups is clear and easy to cluster. On this basis, the core group clustering (CGC) is further proposed to cluster the core points. The experimental results show that DBELG achieves better performance than existing methods in handling datasets with fuzzy boundaries and complex structures.

Keywords: Clustering · Natural neighbor · Boundary extraction · Fuzzy boundaries · Local gravitation

1 Introduction

As one of the essential techniques for mining and describing the intrinsic properties of data [1], cluster analysis has been widely used in document clustering [2], image segmentation [3], medical services [4,5], etc.

K-means [6] and K-medoids [7] are typical center-based algorithms. This type of algorithm works by randomly selecting or finding the densest points as cluster cores and assigning the remaining data objects according to heuristic rules. Most

Supported by grants from the Graduate Scientific Research and Innovation Foundation of Chongqing, China (No. CYB20063), and the National Natural Science Foundation of China (No. 62006029).

J. Gama et al. (Eds.): PAKDD 2022, LNAI 13281, pp. 174–186, 2022.
https://doi.org/10.1007/978-3-031-05936-0_14

of the above methods can cluster the data with Gaussian distribution effectively, but they are not suitable for non-spherical clusters. In addition, predetermining the number of clusters is also an unavoidable problem. Some methods attempt to solve the above problems.

DBSCAN [8] is one of representative examples due to several advantages: (a) it is not necessary to specify the number of clusters, (b) it is less susceptible to interference from outliers, (c) clusters of arbitrary shapes can be found. However, the shortcomings of difficult parameter tuning and sensitivity to fuzzy boundaries cannot be ignored. HDBSCAN [9] expands DBSCAN by converting it into a hierarchical clustering algorithm so that the algorithm is no longer sensitive to radius *Eps*. However, the issue of fuzzy border sensitivity remains.

DPC [10] is an algorithm that has been proposed in recent years and has attracted extensive research interest. DPC not only automatically determines the number of clusters, but also detects outliers. However, it requires setting a fixed cutoff distance dc, which is predefined by the user. In addition, DPC does not have the ability to distinguish clusters of arbitrary shapes. SNN-DPC [11] addresses the effect of the cutoff distance dc, but SNN-DPC also requires a decision graph to select the clustering centers. In addition, the number of shared nearest neighbors k must be set manually.

The above algorithms are not applicable to complex patterns of manifold structures. DCore [12] is a new algorithm that can detect complex patterns. DCore is based on the assumption that each cluster has a density core that does not include outliers, boundaries and edges. Nevertheless, Dcore requires setting five global parameters. It is time-consuming to find five suitable values at the same time. In addition, the core of a cluster is often not one or several points but some sets of points without a clear shape and structure. Therefore, it is unnatural to simply define several cluster cores.

In recent years, some innovative clustering algorithms have been proposed to solve these problems, one of which is Border-Peeling Clustering (BP for short) [13]. BP not only correctly identifies the true structure of most clusters, but also automatically detects outliers. Although BP exhibits strong modeling capabilities, it also has some shortcomings. The input parameter k, i.e., the number of neighbors, is difficult to determine, especially when the shape of the clusters is complex.

Unlike SNN-DPC and BP, FSNN [14] and HCLCS [15] introduce the state-of-the-art concept of neighbors, natural neighbors (NaN) [16]. Due to the introduction of NaN, which is parameter-free and effective, the clustering strategies proposed in [14] and [15] are able to identify both spherical and manifold clusters. Nevertheless, both HCLCS and FSNN struggle to maintain satisfactory performance on clusters with outliers and fuzzy boundaries. Obviously, the clear border between clusters is used as the implied condition in numerous algorithms. However, more or less data are often distributed among different clusters in the actual situation, if some sparse data points are exactly between two clusters, they may serve as bridges between clusters and bind them together (leading to redundant merging of clusters).

In summary, determining the location and number of clustering cores, discovering arbitrarily shaped clusters and dealing with fuzzy boundaries is tough formost algorithms. To address the issues discussed above, we propose a boundary extraction strategy. Unlike previous work, our method does not directly define the cluster core but continuously peels the border points and explores the cluster center naturally, thereby avoiding the assumption of the location and number of cluster cores. Secondly, the border of after-peeled cluster is clear, and the clusters formed by the remaining points are easily distinguished and aggregated. The contributions of this research are as follows:

- A novel gravity model is proposed which makes use of three significant information about the data objects in dense and sparse regions to estimate the local density of data objects. Local density estimation has advantages in avoiding over-segmentation of clusters.
- Unlike algorithms based on *knn*-neighbourhoods, and unlike algorithms with ϵ-neighbourhoods, our algorithm uses state-of-the-art natural neighbor and therefore does not require the specification of neighbor parameters.
- A new clustering framework is proposed. There are no strong assumptions about the structure and density distribution of the data points, such as a single density peak. We focus on extracting boundary in a natural way. Unlike previous work, such as DPC [10], Dcore [12], this is a reverse and natural strategy. In addition, our core points are not defined globally, but through an iterative process of sensing the gravity of the data.

The rest of this paper is organized as follows. In the second section, the concept of natural neighbor is introduced. The third section describes the boundary extraction clustering algorithm based on data gravity (DBELG for short). Experimental results are reported in the next section, followed by conclusions and future work in the last section.

2 Natural Neighbor

The Natural Neighbor (NaN) [16] is a widely used [17,18] and effective neighbor concept. The formal definition of the natural neighbor can be given as follows.

Natural Neighbor: Given a set of n data points $X = \{x_1, x_2, \cdots, x_n\}$ in $\mathbb{R}$ and $x_i, x_j \in X$. If x_i belongs to the *supk*-th nearest neighbors of x_j and x_j belongs to the *supk*-th nearest neighbors of point x_i, then x_i and x_j are considered to be Natural Nearest Neighbor of each other.

$$x_j \in NaN(x_i) \Leftrightarrow (x_i \in NN_{supk}(x_j)) \wedge (x_j \in NN_{supk}(x_i)) \tag{1}$$

where $NN_{supk}(x_i)$ is the set of *supk* nearest neighbors of x_i. The objective of natural neighbor searching is to achieve a natural stable state, where each data object has at least one mutual neighbor. The number of reverse nearest neighbors $nb(x_i)$ of each point x_i is calculated in each iteration. Before the searching state is stable, the natural neighbor of each data point is searched by continuously expanding the neighbor searching round r, that is, the k value in KNN. The more detail of NaN-Searching is described in Algorithm 1 of [16]. Specially, we use *kd*-tree to speed up the NaN searching process.

Table 1. Description of notations.

Notation	Description
X	The set of all data
$X_p^{(t)}$	The set of peeled data at the t-th iteration
$X_{up}^{(t)}$	The core groups at the t-th iteration
$\hat{F_{ij}}$	Local gravitation between x_i and x_j
m_i	The mass of data object x_i
$\hat{F}_i^k$	The local resultant force of data object x_i
$NaNLGF_i^{(t)}$	The local gravitation fluctuation of x_i
$\nabla NaNLGFmax^{(t)}$	The fluctuation truncation value
$B_i^{(t)}$	The border classification value

3 Methodology

The main process of DBELG is visually shown in Fig. 1. First, as shown in Fig. 1(a), the border points (red points in the figure) are peeled from the data set by the proposed NaNLGF, thereby the interior points (gray points in the figure) are preserved. At the same time, the relationship between border points and interior points is recorded. Next, as shown in Fig. 1(b), the interior points are used to construct the initial cluster. Finally, according to the connection relationship between border points and interior points, the previously peeled points are assigned to the cluster where the associated points are located. Detailed steps of the proposed DBELG are described in Algorithm 1. For ease of understanding, we list the important notation used in this paper and their corresponding descriptions are listed in Table 1.

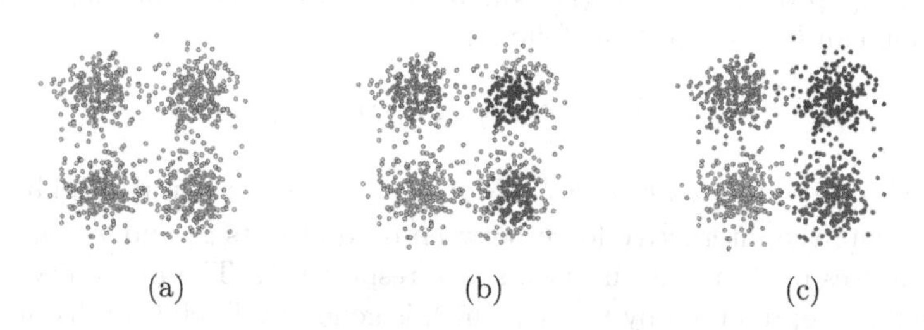

Fig. 1. Illustration of the main processes of DBELG. (Color figure online)

3.1 Dynamic Boundary Extraction

In order to extract the border points, we define a novel density influence value called local gravitation fluctuation based on natural neighbor (NaNLGF). This notion is originate from three significant differences between the neighbors of dense region and sparse region.

Algorithm 1. DBELG

Require: X: A set of points $X = \{x_1, x_2, \cdots, x_n\} \in \mathbb{R}$.
Ensure: Cluster indices $C = \{C_1, C_2, \cdots, C_m\}$.
1: Initializing $X_{up}^{(1)} \leftarrow X$.
2: **for** peeling iteration $1 \leq t \leq T$ **do**
3: **for** each point $x_i \in X_{up}^{(t)}$ **do**
4: λ, nb, NaN=NaN-searching($X_{up}^{(t)}$);
5: $NaNLGF_i^{(t)} = \sum_{k=1}^{\max(nb)} \left|\left|\hat{F}_i^k\right| - \left|\hat{F}_i^{k+1}\right|\right|$;
6: **end for**
7: Compute $\nabla NaNLGF\max^{(t)}$;
8: $X_p^{(t)} \leftarrow \left\{x_i : B_i^{(t)} = 1 \wedge x_i \in X_{up}^{(t-1)}\right\}$; //dynamic boundary extraction, see Sect. 3.1
9: $X_{up}^{(t)} \leftarrow X_{up}^{(t-1)} \backslash X_p^{(t)}$;
10: **for** each peeled point $x_i \in X_p^{(t)}$ **do**
11: $\gamma_i \leftarrow Association\left(x_i, X_{up}^{(t)}, NaN\right)$; // association strategy, see Sect. 3.2
12: **end for**
13: **end for**
14: $\tilde{C} \leftarrow CGC\left(X_{up}^{(t)}\right)$; // core group clustering, see Sect. 3.3
15: $C \leftarrow ComputeFinalResult\left(X_p, \tilde{C}, \gamma\right)$. // linking border points to interior points

- The sum of distances between a point in sparse region (outliers and border points) and its nearest neighbors is usually larger than that in dense region.
- The number of natural neighbors of the sparse area is usually less than the point of the dense area.
- The distribution of neighbors in intensive region is more uniform than points in sparse area.

Like Newton's law of universal gravitation, in our model, each object is regarded as point with mass in the data space. Two different objects are attracted to each other, and the magnitude of the attraction is proportional to the mass and inversely proportional to the square of the distance. Therefore, the data gravitation can be computed as follows:

$$\hat{F}_{ij} = m_i m_j h(\xi(i,j))^2 \hat{u}_{ij} \tag{2}$$

where $\xi(i,j) > 0$ and $x_i, x_j \in X_{up}^{(t)}$. Moreover, x_j is a natural neighbor of x_i. $\hat{F}_{ij}$ represents the interactive force between data objects x_i and x_j. m_i and m_j are the masses of x_i and x_j in data space respectively. The proximity between data points is represented by $\xi(i,j)$, which is generally Euclidean distance. $h(\cdot)$ denotes an inverse proportion function and $\hat{u}_{ij}$ is the unit vector from x_i to x_j. In our method, the mass is calculated as follow:

$$m_i = h\left(\sum_{j=1}^{k} \xi(i,j)\right) nb(x_i) \tag{3}$$

According to formula 3, the value of mass is large for points in dense region and small in sparse region. As previously mentioned, in a dense area, the distance between data objects and its neighbor is small, which means that the sum of the distances between point x_i and its k nearest neighbors (expressed by $\sum_{j=1}^{k} \xi(i,j)$) is small. Therefore, the value of $h(\sum_{j=1}^{k} \xi(i,j))$ is large. The data in sparse areas is the opposite. In addition, $nb(x_i)$ is used as the weight factor to further enlarge the difference of mass between the border points and the data objects in the core area. $nb(x_i)$, the number of reverse nearest neighbors of x_i in the stable state of natural neighbor search, is obtained by NaN-searching algorithm described in Sect. 2. nb in sparse areas is significantly less than that in dense areas. This is consistent with the scale-free characteristics of natural neighbors [16].

Unlike Newton's law of universal gravitation, we assume that data objects are attracted only by neighbors in a local region. In general, the distance between a point and its neighbors does not change significantly. Thus, the local gravitation of point x_i with its natural neighbors can be computed as:

$$\hat{F}_i^k = \sum_{j=1}^{k} \hat{F}_{ij} = h(m_i) \sum_{j=1}^{k} \hat{u}_{ij} \tag{4}$$

$\hat{F}_i^k$ reveals the resultant force of k natural neighbors of x_i and encapsulates the three significant differences mentioned earlier. Based on the above analysis, NaNLGF defined to quantify the local density. It is noteworthy that our method does not require parameters, which is substantially different from the density influence value defined by BP [13]. Mathematically, the local gravitation fluctuation of point x_i can be expressed as:

$$NaNLGF_i^{(t)} = \sum_{k=1}^{\max(nb)} \left| \left| \hat{F}_i^k \right| - \left| \hat{F}_i^{k+1} \right| \right| \tag{5}$$

For each data point $x_i \in X_{up}^{(t)}$, $B_i^{(t)}$ is used to represent the border classification value of x_i. If x_i is a border point, the value is 1, otherwise 0.

$$B_i^{(t)} = \begin{cases} 1, if NaNLGF_i^{(t)} > \nabla NaNLGFmax^{(t)} \\ 0, \qquad\qquad otherwise. \end{cases} \tag{6}$$

The cutoff values $\nabla NaNLGFmax^{(t)}$ can be specified manually, or as we describe below. If the maximum difference of adjacent numbers after descending sorting of $NaNLGF_i^{(t)}$ is recorded as $NaNLGFmax$, and the difference comes from point p and point q, then $\nabla NaNLGFmax^{(t)}$ can be set as $NaNLGF_q^{(t)}$.

As for the number of the iteration of peeling, denoted as T, can be specified manually, or following the strategy: in each iteration t, we tracks the value set of border points to be peeled: $\left\{NaNLGF_i^{(t)} \middle| x_i \in X_p^{(t)}\right\}$, and measure the mean value of that set, denoted by $NaNLGF_p^{(t)}$. The termination condition is $\frac{NaNLGF_p^{(t)}}{NaNLGF_p^{(t-1)}} - \frac{NaNLGF_p^{(t-1)}}{NaNLGF_p^{(t-2)}} > 0.15$.

3.2 Association Strategy

While recognizing the border points at the t-th iteration, we associate each identified point with a non-border point according to the following rules.

$$\gamma_i = \begin{cases} x_j , & x_j \in NaN(x_i) \; and \; x_j \in X_{up}^{(t)} \\ \gamma_m , & otherwise \end{cases} \tag{7}$$

where $x_i \in X_p^{(t)}$ and γ_i represents the combination point of x_i. If x_j is the natural neighbor to x_i, and x_j in the set of $X_{up}^{(t)}$, then $\gamma_i = x_j$. In the other case, there are no natural neighbors of x_i in the set of $X_{up}^{(t)}$, that means, all points in $NaN(x_i)$ are identified as border points, then x_i is combination with the association point of the closest point x_m ($\gamma_m = x_j$), so $\gamma_i = \gamma_m = x_j$.

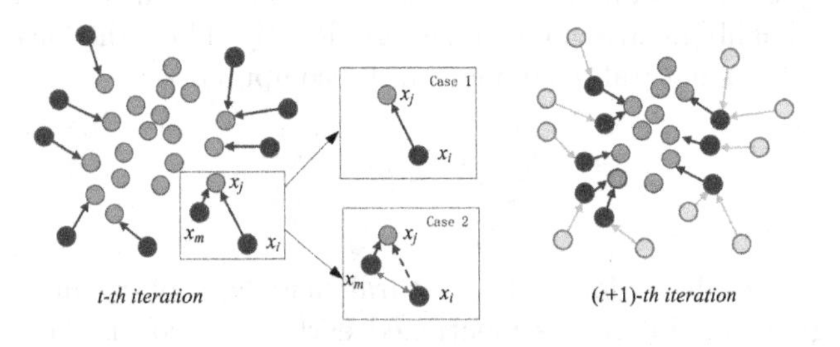

Fig. 2. The combination process of border points. (Color figure online)

Obviously, Fig. 2 visually illustrates this process. Such an association forms a transfer relationship. The figure on the left shows the correlation between border points (indicated in red) and non-border points (indicated in green) identified in the t-th iteration. The two subgraphs illustrate two cases of γ_i. The figure on the right shows the relationship formed in the next iteration. The peeled points are ignored in the next peeling. Finally, each of the peeled points has the same label as the point they are associated with. In particular, to identify outliers, we extract outliers from the $X_p^{(1)}$. The NaNLRF of outliers is in top β percent.

3.3 Core Group Clustering

Aiming at clustering the points of core groups automatically, we designed a novel method Core group clustering, denoted as CGC, see Algorithm 2 for pseudo code. To make it easier to understand, we give the formal definition as follows:

Extended Natural Neighbor: For each point $x_i \in X_{up}^{(T)}$, the extended natural neighborhood of x_i is composed of the natural neighbors of x_i and the reverse neighbors, denoted as $ENN(x_i)$.

$$ENN(x_i) = NaN(x_i) \cup RNN(x_i) \tag{8}$$

Reachable Extended Natural Neighbor: If there is a path $\langle x_0, x_1, x_2, \cdots x_k \rangle$ from the interior core x_i to the point x_j , where $x_0 = x_i, x_k = x_j$, and

Algorithm 2. CGC

```
Require: X: A set of points X = {x_1, x_2, ··· , x_n} ∈ ℝ.
Ensure: Interior cluster indices C = {C_1, C_2, ··· , C_m}.
 1: Initializing X' ← X, C ← C_r = ∅, label(x_i) ← unvisited, x_i ∈ X.
 2: while X' ≠ ∅ do
 3:    for each point x_i ∈ X' do
 4:       if label(x_i) = unvisited and nb(x_i) = max_{j∈X'}(nb(x_j)) then
 5:          Inp(r) = x_i
 6:          C_r ← {Inp(r)}
 7:          label(x_i) = visited
 8:          for each reachable extended natural neighbors of Inp(r) x_re do
 9:             if label(x_re) = unvisited then
10:                C_r ← C_r ∪ {x_re}
11:                label(x_re) = visited
12:             end if
13:          end for
14:       end if
15:    end for
16:    C ← C ∪ C_r
17:    X' ← X' − C_r
18:    r = r + 1
19: end while
```

$i = 0, 1, \cdots k-1, x_{i+1} \in ENN(x_i)$, then x_j is the reachable extended natural neighbor of the interior core x_i.

Interior Core: If the number of neighbors in the reachable extended neighbor of point x_i is the largest, $x_i \in X_{up}^{(T)}$, then x_i is an interior core, abbreviated as $Inp(i)$.

The CGC is an inside-out diffusion method. Moreover, CGC explore each data point of the reachable extended natural neighbor of interior core. First, an interior core $Inp(1)$ is selected as the start point, where $Inp(1) \in X_{up}^{(T)}$. Next, visit all the neighbors $x_2, x_3, ..., x_r$ from the $ENN(Inp(1))$ in turn, and then visit all the reachable extended natural neighbors of $Inp(1)$ and $x_2, x_3, ..., x_r$ that have not been visited before. Repeat this process until all reachable extended neighbors of $Inp(1)$ have been visited, then the construction of an initial cluster is completed. Next, select the next interior core $Inp(2)$ from the set of remaining unvisited points, and repeat the above process until all interior points are visited. Finally, the number of interior cores is the number of clusters, and an interior core and its reachable extended neighbors are an initial cluster.

3.4 The Complexity Analysis

DBELG mainly includes the following parts: (a) border points extraction; (b) clustering of core groups; and (c) the distribution of border points. The calculation of NaNLGF is the key of border points recognition, which mainly depends on NaN-searching. Since we introduced *kd*-tree, the time complexity is $O(n*log(n))$. Suppose that the number of non-border points is m, the complexity of core group clustering is $O(m)$. The relation between each border point and an interior point will be recorded in the identification stage, so the complexity of the distribution of border points is $O(n-m)$. Therefore, the complexity of DBELG is $O(n*log(n))$.

4 Experimental Evaluation

In order to evaluate the performance of DBELG, we compare DBELG with six benchmark algorithms including one classic algorithms (K-means) and five excellent methods (HDBSCAN, DPC, Dcore, SNN-DPC, BP) proposed recently. Besides, we use two popular criteria: Accuracy (ACC) and Adjusted Rand Index (ARI). They are well known external criteria, a larger value represents a better clustering result. All experiments are run on a PC with an AMD R7 37000X, 24 G memory, 3.60 GHz CPU, Windows 10, and Python 3.8.

Table 2. Data characteristics of 8 synthetic datasets.

Datasets	Instances	Dimensions	Clusters	Source
Dataset 1	622	2	4	[19]
Dataset 2	1064	2	2	[19]
Dataset 3	1427	2	4	[19]
Dataset 4	1916	2	6	[19]
Dataset 5	8000	2	6	[19]
Dataset 6	8000	2	6	[19]
Dataset 7	8533	2	7	[19]
Dataset 8	10000	2	9	[19]

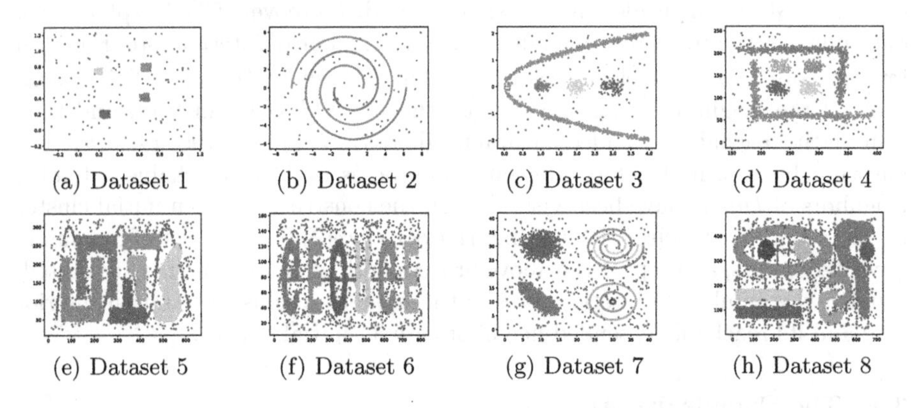

Fig. 3. Clustering results of DBELG on 8 synthetic datasets. The values of β are (20%, 8%, 10%, 13%, 17%, 20%, 6%, 27%) respectively

4.1 Experiment on Synthetic Datasets

In this section, we discuss the experimental results of 6 benchmark algorithms and DBELG on 8 synthetic datasets. All datasets have different shapes, densities, and overlapping degrees of fuzzy boundaries. Figure 3 shows the clustering results of DBELG on the eight datasets. One color represents one cluster, and the

Table 3. Performances of 7 algorithms on 8 synthetic datasets.

Datasets	Metric	K-means	HDBSCAN	DPC	DCore	SNN-DPC	BP	DBELG
Dataset 1	ACC	0.720	0.906	0.906	0.720	0.907	0.915	**0.978**
	ARI	0.649	0.910	0.786	0.650	0.789	0.911	**0.983**
Dataset 2	ACC	0.005	0.823	0.767	0.008	0.830	0.529	**0.904**
	ARI	0.007	0.887	0.792	0.011	0.800	0.514	**0.948**
Dataset 3	ACC	0.414	0.926	0.215	0.860	0.330	0.489	**0.980**
	ARI	0.178	0.965	0.212	0.917	0.100	0.109	**0.997**
Dataset 4	ACC	0.476	**0.957**	0.672	0.836	0.726	0.663	0.946
	ARI	0.285	**0.979**	0.546	0.836	0.539	0.328	0.969
Dataset 5	ACC	0.611	0.927	0.746	0.868	0.892	0.739	**0.956**
	ARI	0.521	0.952	0.588	0.865	0.820	0.625	**0.970**
Dataset 6	ACC	0.821	0.916	0.697	0.835	0.881	0.789	**0.941**
	ARI	0.727	0.910	0.516	0.832	0.745	0.744	**0.946**
Dataset 7	ACC	0.653	0.952	0.730	0.847	0.936	0.785	**0.990**
	ARI	0.555	0.965	0.629	0.712	0.895	0.643	**0.993**
Dataset 8	ACC	0.595	0.926	0.719	0.865	0.879	0.631	**0.938**
	ARI	0.381	0.913	0.449	0.857	0.837	0.484	**0.947**

number of different colors corresponds to the number of clusters. In particular, outlier points are indicated in gray. The detailed characteristics of the data are listed in the Table 2.

As for Kmeans, we provide all correct number of clusters. For BP, we tune the major parameter k (i.e., the number of neighbors), and the value of k varies from 15 to 25. HDBSCAN has two major parameters, k (i.e., the number of neighbors) and Nc (i.e., smallest size to be considered a cluster). We select the parameter Nc from 5 to 20. In DPC and SNN-DPC, density peaks are manually selected from the decision graph. DPC has one parameter dc (i.e., a cutoff distance), the dc is 2%. SNN-DPC must set the number of iterations k, we set the parameter of SNN-DPC varies from 5 to 30 in 1 increments. The results of DCore are affected by five parameters, we attempt to use different parameter settings to ensure better results.

The experimental results of DBELG on 8 synthetic datasets are shown in Fig. 3. The scores of evaluation metric are listed in Table 3. As shown in Table 3 and Fig. 3, DBELG performs well on datasets with fuzzy boundaries. Besides, It can also handle both spherical and non-spherical datasets with outliers. DBELG performs better than other algorithms on seven datasets. It is slightly inferior to HDBSCAN on Dataset 4, but all outperform the other five algorithms. In summary, DBELG has advantages over other advanced algorithms.

4.2 Experiment on Real-World Datasets

To further test the effectiveness of DBELG, we design experiments on 8 real-world datasets. The comparison algorithms are the same as in the previous part.

The evaluation scores of each algorithm on different datasets are listed in Table 4. DBELG performs well on six real datasets (Iris, Seeds, Glass, Ecoli, Inosphere, Dermatology) in terms of two criteria, and its score on two datasets (Wine, Segmentation) is slightly lower than that of other algorithms. For Seeds dataset, the scores of SNN-DPC and DBELG are the same. In terms of Ecoli dataset, the performance of BP and DBELG achieve the same score, which is higher than other algorithms. Due to the completely different calculation methods of ACC and ARI, they do not always consistent. On the Wine dataset, DBELG works best in ACC, SNN-DPC obtains the highest score in terms of ARI. However, DBELG scores higher than other algorithms. As for Segmentation, the score of ARI, DBELG is higher than other algorithms, but SNN-DPC is the best in ACC.

The running time of each algorithm on the 8 datasets is listed in Table 5. The running time of DBELG is not the fastest, but it is not the slowest either. Overall, DBELG is competitive with the other 6 algorithms. In summary, it can be concluded that DBELG is more effective than the one classic algorithms (K-means), more competitive than the four novel algorithms (HDBSCAN, DPC, Dcore, SNN-DPC), and better than the original border peeling algorithm (BP).

Table 4. Performances of 7 algorithms on 8 real-world datasets.

Datasets	Metric	K-means	HDBSCAN	DPC	Rcore	SNN-DPC	BP	DBELG
Iris	ACC	0.758	0.734	0.653	0.462	0.900	0.723	**0.907**
	ARI	0.730	0.568	0.453	0.298	0.904	0.556	**0.913**
Wine	ACC	0.430	0.384	0.565	0.356	0.893	0.412	**0.944**
	ARI	0.371	0.242	0.505	0.403	**0.915**	0.375	0.837
Seeds	ACC	0.695	0.443	0.719	0.417	**0.768**	0.671	**0.768**
	ARI	0.717	0.280	0.745	0.160	**0.811**	0.688	**0.811**
Glass	ACC	0.322	0.380	0.313	0.312	0.313	0.321	**0.724**
	ARI	0.166	0.252	0.212	0.177	0.161	0.235	**0.651**
Segmentation	ACC	0.571	0.541	0.668	0.465	**0.691**	0.355	0.653
	ARI	0.399	0.268	0.537	0.035	0.535	0.101	**0.687**
Ecoli	ACC	0.616	0.426	0.587	0.636	0.671	**0.712**	**0.712**
	ARI	0.426	0.413	0.437	0.497	0.732	**0.753**	**0.753**
Iononsphere	ACC	0.135	0.272	0.134	0.256	0.405	0.376	**0.897**
	ARI	0.178	0.135	0.213	0.253	0.520	0.367	**0.832**
Dermatology	ACC	0.103	0.522	0.659	0.438	0.761	0.157	**0.855**
	ARI	0.027	0.322	0.602	0.166	0.540	0.049	**0.836**

Table 5. Running time of 7 algorithms on 8 real-world datasets.

Datasets	K-means	HDBSCAN	DPC	DCore	SNN-DPC	BP	DBELG
Iris	0.023	0.007	0.016	0.152	0.031	0.099	0.018
Wine	0.024	0.006	0.006	0.060	0.033	0.055	0.027
Seeds	0.025	0.010	0.017	0.021	0.030	0.109	0.023
Glass	0.027	0.105	0.003	0.026	0.084	0.079	0.044
Segmentation	0.034	0.045	0.005	0.080	0.031	0.089	0.033
Ecoli	0.044	0.018	0.008	0.325	0.061	0.171	0.084
Ionosphere	0.028	0.022	0.015	0.381	0.065	0.131	0.188
Dermatology	0.040	0.018	0.010	0.023	0.107	0.176	0.115

5 Conclusions

In this paper, a new clustering algorithm DBELG has been proposed. The key of DBELG is to iteratively identify and peel the border points according to proposed NaNLGF to eliminate the border points. At the same time, the core of the cluster is revealed. And then, we cluster the points of core groups by proposed CGC and assign the border points to the clusters. The preponderance of DBELG is that DBELG does not require the assumptions about the location and the number of cluster centers and not affected by fuzzy boundaries because of the peeling of border points. In addition, due to the introduction of natural neighbors, DBELG does not need to preset the neighbor parameter. Numerous experiments on synthetic and real-world datasets demonstrate that DBELG can not only recognize spherical clusters and manifold clusters effectively, but also is not easily disturbed by fuzzy boundaries. Our future work focus on apply DBELG into some practical applications.

References

1. Wang, R., Fung, B.C., Zhu, Y.: Heterogeneous data release for cluster analysis with differential privacy. Knowl. Based Syst. **201**, 106047 (2020)
2. AlMahmoud, R.H., Hammo, B., Faris, H.: A modified bond energy algorithm with fuzzy merging and its application to Arabic text document clustering. Exp. Syst. Appl. **159**, 113598 (2020)
3. Hu, F., Chen, H., Wang, X.: An intuitionistic kernel-based fuzzy c-means clustering algorithm with local information for power equipment image segmentation. IEEE Access **8**, 4500–4514 (2020)
4. Bose, A., Mali, K.: Type-reduced vague possibilistic fuzzy clustering for medical images. Pattern Recogn. **112**, 107784 (2021)
5. Bu, F., Hu, C., Zhang, Q., Bai, C., Yang, L.T., Baker, T.: A cloud-edge-aided incremental high-order possibilistic c-means algorithm for medical data clustering. IEEE Trans. Fuzzy Syst. **29**(1), 148–155 (2020)

6. Kanungo, T., Mount, D.M., Netanyahu, N.S., Piatko, C.D., Silverman, R., Wu, A.Y.: An efficient k-means clustering algorithm: analysis and implementation. IEEE Trans. Pattern Anal. Mach. Intell. **24**(7), 881–892 (2002)
7. Sheng, W., Liu, X.: A genetic k-medoids clustering algorithm. J. Heuristics **12**(6), 447–466 (2006)
8. Ester, M., Kriegel, H.P., Sander, J., Xu, X., et al.: A density-based algorithm for discovering clusters in large spatial databases with noise. In: KDD, vol. 96, pp. 226–231 (1996)
9. Campello, R.J.G.B., Moulavi, D., Sander, J.: Density-based clustering based on hierarchical density estimates. In: Pei, J., Tseng, V.S., Cao, L., Motoda, H., Xu, G. (eds.) PAKDD 2013. LNCS (LNAI), vol. 7819, pp. 160–172. Springer, Heidelberg (2013). https://doi.org/10.1007/978-3-642-37456-2_14
10. Rodriguez, A., Laio, A.: Clustering by fast search and find of density peaks. Science **344**(6191), 1492–1496 (2014)
11. Liu, R., Wang, H., Yu, X.: Shared-nearest-neighbor-based clustering by fast search and find of density peaks. Inf. Sci. **450**, 200–226 (2018)
12. Chen, Y.: Decentralized clustering by finding loose and distributed density cores. Inf. Sci. **433**, 510–526 (2018)
13. Averbuch-Elor, H., Bar, N., Cohen-Or, D.: Border-peeling clustering. IEEE Trans. Pattern Anal. Mach. Intell. **42**(7), 1791–1797 (2019)
14. Yuan, M., Zhu, Q.: Spectral clustering algorithm based on fast search of natural neighbors. IEEE Access **8**, 67277–67288 (2020)
15. Shi, J., Zhu, Q., Li, J., Liu, J., Cheng, D.: Hierarchical clustering based on local cores and sharing concept. In: 2021 IEEE 45th Annual Computers, Software, and Applications Conference (COMPSAC), pp. 284–289. IEEE (2021)
16. Zhu, Q., Feng, J., Huang, J.: Natural neighbor: a self-adaptive neighborhood method without parameter k. Pattern Recogn. Lett. **80**, 30–36 (2016)
17. Srinilta, C., Kanharattanachai, S.: Application of natural neighbor-based algorithm on oversampling smote algorithms. In: 2021 7th International Conference on Engineering, Applied Sciences and Technology (ICEAST), pp. 217–220. IEEE (2021)
18. Wu, Z., Zeng, Y., Li, D., Liu, J., Feng, L.: High-volume point cloud data simplification based on decomposed graph filtering. Autom. Constr. **129**, 103815 (2021)
19. Dua, D., Graff, C., et al.: UCI machine learning repository (2017)

Modelling Zeros in Blockmodelling

Laurence A. F. Park[1(✉)], Mohadeseh Ganji[2], Emir Demirovic[3], Jeffrey Chan[4], Peter Stuckey[5], James Bailey[6], Christopher Leckie[6], and Rao Kotagiri[6]

[1] Centre for Research in Mathematics and Data Science, Western Sydney University, Sydney, Australia
lapark@westernsydney.edu.au

[2] ANZ, Melbourne, Australia

[3] TU Delft, Delft, The Netherlands

[4] School of Computing Technologies, RMIT University, Melbourne, Australia

[5] Department of Data Science and AI, Monash University, Clayton, Australia

[6] School of Computing and Information Systems, The University of Melbourne, Parkville, Australia

Abstract. Blockmodelling is the process of determining community structure in a graph. Real graphs contain noise and so it is up to the blockmodelling method to allow for this noise and reconstruct the most likely role memberships and role relationships. Relationships are encoded in a graph using the absence and presence of edges. Two objects are considered similar if they each have edges to a third object. However, the information provided by missing edges is ambiguous and therefore can be measured in different ways. In this article, we examine the effect of the choice of block metric on blockmodelling accuracy and find that data relationships can be position based or set based. We hypothesise that this is due to the data containing either Hamming noise or Jaccard noise. Experiments performed on simulated data show that when no noise is present, the accuracy is independent of the choice of metric. But when noise is introduced, high accuracy results are obtained when the choice of metric matches the type of noise.

1 Introduction

Relationships between objects can be represented as a graph, where the graph vertices represent the objects and the edges represent the relationships between the objects. Many algorithms have been proposed for clustering/partitioning graph vertices based on their relationships (e.g. Spectral Clustering [9]). These algorithms allow us to identify clusters of objects that are closely related and are useful for tasks such as identifying a group of employees who work in the same department, a group of people who attended the same school, or a set of video games that are made by the same company.

Graphs also contain a deeper level of information that allows us to identify the roles of the objects. Roles are not identified by the similarity of objects, but they are identified by the relationships that the objects share with others. For example a set of employees within a department might have the role of Manager. Each manager is not likely to be connected to each other, but their relationships to others in the department are likely to be similar (each is likely to be acting as gateway between senior management and the other employees within the department).

J. Gama et al. (Eds.): PAKDD 2022, LNAI 13281, pp. 187–198, 2022.
https://doi.org/10.1007/978-3-031-05936-0_15

Blockmodelling allows us to discover clusters of objects that have the same or similar role in the graph. The name comes from its process of revealing blocks within the graph adjacency matrix, where a block is a set of objects that share the same links to other objects in the graph. It is not clear, however, if the similarity between two objects should be stronger if they have missing edges in common.

In this article, we examine the effect of the chosen block metric on blockmodelling accuracy. Experiments show that high accuracy is obtained using position based block metrics on some data and set based block metrics on other data. We hypothesise that these results are due to the noise within the data being either Hamming or Jaccard noise and run simulations examine this hypothesis.

The contributions of this article are:

- A presentation of a seriation based approach for blockbodelling, allowing block metric selection (Sect. 2).
- An analysis of the effect of Hamming and Jaccard noise when using each metric on specific block structures, and their interaction with the number of observations, number of roles and noise level (Sect. 3).

We also identify that block metrics from the same category behave similarly, and therefore, we conjecture that relational data can be either position based or set based, and that the block metric should be chosen to match the data noise type.

The article is organised as follows: Sect. 2 describes the initial investigation in to the effect of block metric on blockmodel accuracy. Section 3 continues the investigation by examining if the effect of the metrics are due to the type of noise. Section 4 examines the results.

2 Blockmodelling with a Chosen Block Metric

A graph G with vertices V and edges E, can be represented by its adjacency matrix A, where each element $a_{ij} \in A$ depicts the weight of the edge directed from vertex j to i. If many vertices have the same in or out edges, they form a block in A, where the block represents a role (a set of objects that have similar relationships to the remainder of the graph). If the rows and columns of A are ordered correctly, we are able to visualise the block, unfortunately identifying the correct permutation is difficult and so the existence of a particular block may not be obvious.

Both Stochastic [5] and Spectral [8] forms of blockmodels exist, where the stochastic form allows us to identify the underlying sampling distribution, while the spectral form provides a hard or soft clustering of objects into roles. We focus on the spectral form. The adjacency matrix $A \in \{0,1\}^{n \times n}$ of a blockmodel with k roles, by definition, can be decomposed into $A = CMC'$ where $M \in \{0,1\}^{k \times k}$ contains the blockmodel structure and $C \in \{0,1\}^{n \times k}$ contains the membership of each of n objects to one of the k roles (such that the rows contain one 1 and the rest 0). Many methods of approximating this decomposition have been derived as gradient based optimisation problems that compute C and M by minimising a function of the error [10]. But the optimisation is difficult due to the binary nature of the problem [2]. Current methods in Spectral Blockmodelling [1] encourage sparsity by separately weighting errors on absent and present edges.

2.1 Blockmodelling Metric

Blockmodelling can be thought of as clustering based on secondary relationships; two objects are found in the same block if they have the same relationship to a third object. If the relationships between objects are binary, then the associated graph provides an edge for a weight of 1 and no edge for a weight of 0. If we find that objects x_1 and x_2 both have edges to x_3, then the similarity between x_1 and x_2 should increase. If both x_1 and x_2 don't have edges to x_3, it is not clear if that information should increase or decrease their similarity. For example if the edge to x_3 represents if an object likes x_3, then no edge from both x_1 and x_2 show that they both don't like x_3, which increases their similarity. On the other hand, if an edge to x_3 represents if an object knows x_3, then it is unclear how both x_1 and x_2 not knowing the person (having no edge to x_3) influences their similarity.

To examine this problem, we will investigate the effect of block metric choice on blockmodelling accuracy. This requires us to easily change the block metric without effecting the blockmodelling algorithm. Therefore we will perform blockmodelling using Seriation, taking inspiration from the cluster visualisation family xVAT [6], which permute the rows and columns of a relational matrix to visualise clustering and identify the number of clusters.

The structural matrix M shows association between roles. The membership matrix C simply replicates and permutes the structure in M to form the graph adjacency matrix A. If the rows of the membership matrix were ordered such that objects with the same role membership are placed together, we would see the shape of the structural matrix in A. Unfortunately, C is not likely to be ordered, and so the structure is difficult to observe in A. But this implies that if we apply the correct permutation π to the rows of C to obtain πC, or equivalently, the rows and columns of A to obtain $\pi A \pi^t$, then we can easily recover M and πC from the visible block structure and hence C. Discovery of this permutation π is a *seriation* problem [3].

Seriation is the process of computing a permutation for a set of objects, such that the similarity between each object and its neighbours is maximised. Given an appropriate measure of similarity, we are able to reveal the block structure and expose roles using seriation. Unfortunately, seriation only provides the permutation of the objects; further processing of the adjacency matrix is required to cluster the objects, but we know that the clustering can be performed by partitioning the ordered set of objects.

A common method for seriation is to use hierarchical clustering with optimal leaf ordering [4]. This is a two stage process, where 1) hierarchical clustering is applied to the dissimilarity matrix (based on a given metric), then 2) the permutation provided by the dendrogram is optimised by maximising the similarity between each pair of neighbouring objects. This reordering process is performed by swapping the children at nodes of the dendrogram, ensuring that all objects remain in the clusters they were assigned to. For example, the hierarchical clustering $\{\{\{1,2\},\{3\}\},\{4,5\}\}$ can be permuted to $\{\{4,5\},\{\{2,1\},\{3\}\}\}$, where the clustering has not changed, but the similarity between each point and its neighbour has changed. To obtain the set of block clusters, the process is:

1. Create a dissimilarity matrix of the objects, where the dissimilarity is measured in terms of the object connectivity,

2. Apply hierarchical clustering to the dissimilarity matrix to obtain a dendrogram,
3. Reorder the dendrogram leaves using optimal leaf ordering,
4. Partition the dendrogram leaves into clusters.

We must decide upon a metric for our data; candidates are presented in the following section.

2.2 Candidate Block Metrics

To perform blockmodelling, we cluster objects that have similar roles, implying similar in and out links in the graph. Therefore, a blockmodelling metric must compare the in and out links of a pair of objects. We define the blockmodelling distance between vertex v_x and v_y as

$$\Delta(v_x, v_y) = d(\vec{e}_{\cdot,x}, \vec{e}_{\cdot,y}) + d(\vec{e}_{x,\cdot}, \vec{e}_{y,\cdot})$$

where $\vec{e}_{\cdot,x} = [e_{1,x}\ e_{2,x}\ \dots\ e_{N,x}]$ (the xth column of the adjacency matrix) is the vector of edge weights $e_{i,x}$ directed from vertex v_x to vertex v_i, and $\vec{e}_{x,\cdot} = [e_{x,1}\ e_{x,2}\ \dots\ e_{x,N}]$ (the xth row of the adjacency matrix) is the vector of edge weights $e_{x,i}$ directed from vertex v_i to vertex v_x (if no edge exists, the weight is zero).

The graphs we will be examining are unweighted, therefore $\vec{e}_{\cdot,x}$ and $\vec{e}_{x,\cdot}$ will be binary vectors, or set membership vectors. We will examine the position based metrics Hamming and Euclidean, and the set based metrics Cosine, Jaccard and Dice, as candidates for $d(\cdot, \cdot)$.

Position based

$$d_{\text{Ham}}(\vec{x}, \vec{y}) = \frac{1}{N}\|\vec{x} - \vec{y}\|_1$$

$$d_{\text{Euc}}(\vec{x}, \vec{y}) = \|\vec{x} - \vec{y}\|_2$$

Set based

$$d_{\text{Jac}}(\vec{x}, \vec{y}) = \frac{\|\vec{x} - \vec{y}\|_1}{N - (\vec{1} - \vec{x})'(\vec{1} - \vec{y})}$$

$$d_{\text{Cos}}(\vec{x}, \vec{y}) = 1 - \frac{\vec{x}'\vec{y}}{\|\vec{x}\|_2\|\vec{y}\|_2}$$

$$d_{\text{Dic}}(\vec{x}, \vec{y}) = \frac{\|\vec{x} - \vec{y}\|_1}{N - (\vec{1} - \vec{x})'(\vec{1} - \vec{y}) + \vec{x}'\vec{y}}$$

where $\vec{x}$ and $\vec{y} \in \{0, 1\}^N$ are binary vectors (containing either 0 or 1), N is the vector length, $\|\vec{x}\|_1$ is the l_1 norm of $\vec{x}$, and $\|\vec{x}\|_2$ is the l_2 norm of $\vec{x}$. Note that both Hamming and Euclidean metrics treat vectors as positions, where Cosine, Jaccard and Dice treat the vectors as representing sets. The major difference between these two categories is how they treat zeros.

2.3 Effect of Blockmodelling Metric on Real Data

We begin our investigation by examining how the choice of block metric effects the blockmodelling accuracy on real data commonly used in assessing community structure algorithms. The data used in this experiment (Sampson: $n = 18$, $k = 4$; Polbooks: $n = 105$, $k = 2$; Polblogs: $n = 1490$, $k = 2$; Karate: $n = 34$, $k = 2$; Football: $n = 115$,

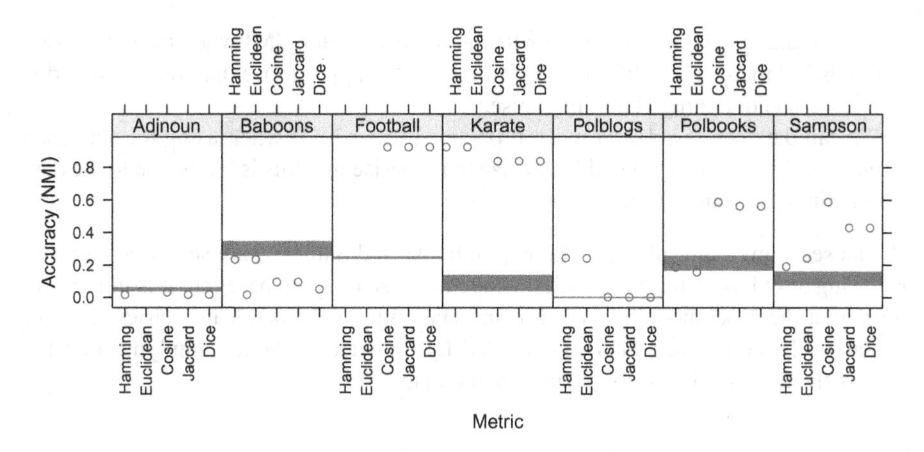

Fig. 1. NMI accuracy for blockmodel clustering using seriation with the given metric (x axis) and data set. The grey region is the 95% confidence interval for mean Coord NMI.

$k = 12$; Baboons: $n = 14$, $k = 2$; Adjnoun: $n = 112$, $k = 2$) are found on Mark Newman's homepage and the Pajek repository.[1] Initial experiments were performed to identify the effect of the hierarchical clustering merging method, and we found that Weighted merging consistently provided high accuracy results, so we focus on that merging method in this paper to reduce the number of variables in each experiment.

Baseline results were computed using Projected Gradient Descent (Grad) and Coordinate Descent (Coord) [1]. Note that the baseline methods are dependent on their initialisation, so we repeated the baseline clustering 10 times for each graph, using random initialisation. The clustering accuracy is presented in Fig. 1 containing the NMI when blockmodelling using each metric on each data set. The greyed out region in each plot shows the 95% confidence interval for the Coord mean. The 95% confidence interval for the Grad mean was also computed but it was lower than the interval for Coord, and so left off the plot.

The results in Fig. 1 show that the five metrics can be placed in two groups; the position based metrics (Hamming and Euclidean metrics) provide similar NMI for each data set, and the set based metrics (Jaccard, Dice and Cosine metrics) provide similar results for each data set. The data where set based metrics are preferred, show significant improvement over the state-of-the-art. The results for the position based metrics are generally equivalent in accuracy to the state-of-the-art. It is known that the Football data contains little noise, while the Adjnoun data contains large amounts of noise, and so we see that the accuracy of each are independent of the metric. It is likely that the difference in results is due to the different noise distributions in each network. This leads us to the definitions:

[1] www-personal.umich.edu/~mejn/ vlado.fmf.uni-lj.si/pub/networks/pajek/.

- We call data *position based data* when it obtains greater NMI when using the position based metrics (Euclidean and Hamming). We hypothesise that this is likely due to the data containing Hamming noise.
- We call data *set based data* when it obtains greater NMI when using the set based metrics (Cosine, Jaccard and Dice). We hypothesise that this is likely due to the data containing Jaccard noise.

We can see from the results that Karate, Polbooks and Sampson are set based data and that using a set based metric provides a huge increase over the baseline. It is interesting to see that the blockmodelling when using Hamming and Euclidean metrics are very similar to the confidence interval provided by the state-of-the-art, implying that the state-of-the-art is designed for position based data.

3 Simulated Data with Hamming and Jaccard Noise

The previous experiment revealed that there were two types of network data and stated that the difference is likely due to the noise distribution being different. In this section we will examine the validity of this assumption by simulating the noise and examining the effect of a set of parameters on the blockmodelling results. By simulating data, we are able to control the data parameters and hence examine the effect of seriation on the accuracy, given the block structure and noise type. To begin, we first describe the basic block structures, and then present an analysis using simulated data with Hamming and Jaccard noise.

3.1 Simulated Block Structures

When simulating data for this experiment, we use ring, star and tree block structures for the structure matrix M. A ring structure arranges the roles so that each is connected to exactly two other roles, forming a ring. A star structure assigns one role as a hub to which all remaining roles are connected. Finally, the tree structure requires that each role has a parent role, and at most two children roles, where one role (the root), has no parent role.

Simulated data was generated using the following parameters: *type* was chosen from *Ring*, *Star* or *Tree*; *the number of objects* was 50, 100, 200, 500, or 1000; and *the number of roles* (clusters) was 2, 4, 8, 16, or 32. We also generated three replicates of graphs using each parameter combination, providing 1,125 random graphs. An initial baseline experiment was run to examine the use of seriation on data with no noise. Experiments were then run to examine the effect of increasing Hamming and Jaccard noise in the data.

3.2 Generating Noise

Hamming noise is simple to generate, and so more likely to be used in blockmodel simulations. For a given binary vector of length n we can generate Hamming noise with expected Hamming distance of np by flipping each 0 or 1 value to a 1 or 0 value

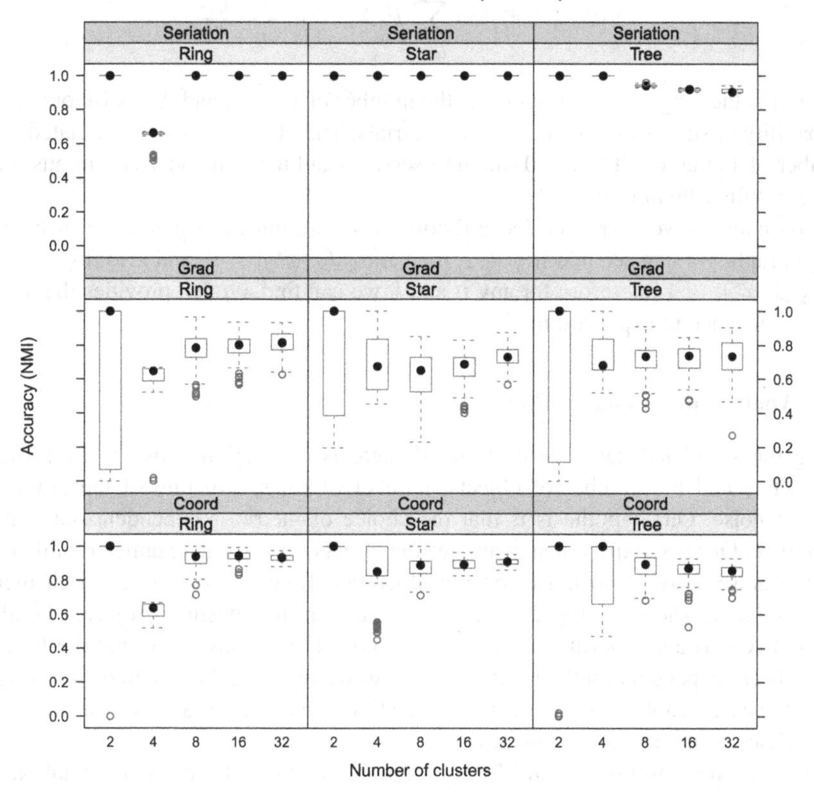

Fig. 2. Clustering accuracy versus cluster count (zero noise) for Ring, Star and Tree graphs. The results on the first row cover seriation using each metric, implying that the choice of metric has no effect on accuracy when no noise is present. The second and third rows contain baseline results.

with probability p. The probability of k flips is a Binomial distribution with n trials and probability of success p. Therefore, we can obtain the expected number and standard deviation number of flips from the Binomial distribution. For a graph, n is the number of vertices and so the noise level is controlled by the proportion p.

On the other hand, Jaccard noise applied to a vector of length n is dependent on the number of 1s in the vector. The Jaccard coefficient is the size of the intersection divided by the size of the union. When applying noise, the size of the intersection can only reduce (when a 1 is flipped to a 0). The size of the union can only increase (when a 0 is flipped to a 1). If we let q be the probability of a flip, we find that the change in intersection and union are independent of each other [7]. If we were to flip edges of $\vec{x}$ with probability q to obtain $\vec{x}^{\star}$, the expected Jaccard coefficient between binary vectors $\vec{x}$ and $\vec{x}^{\star}$ is:

$$\mathbb{E}[d_{\text{Jac}}(\vec{x}, \vec{x}^{\star})] = \sum_{b=0}^{n-l} P(X_u = b) \frac{(1-q)l}{n-b}$$

where n is the length of the vector, l is the number of 1s in $\vec{x}$, and X_u is Binomial with probability of success $1 - q$ and with $n - l$ trials. The $(1 - q)l$ term is associated to the number of 1s that don't flip to 0 (the intersection) and the remainder is n minus the 0s that don't flip (the union).

To obtain a given expected Jaccard noise level, we must compute q for each vertex in the graph. We can see that if $q = 1$, then $\mathbb{E}[d_{\text{Jac}}(\vec{x}, \vec{x}^{\star})] = 0$, and also if $q = 0$ then $\mathbb{E}[d_{\text{Jac}}(\vec{x}, \vec{x}^{\star})] = 1$, therefore for any n and l, we can find a q that provides the desired Jaccard distance in expectation.

3.3 Analysis of Simulated Data

Using the simulated data, we examine if there is interaction between the seriation parameters and the number of objects, number of roles, structure of the graph and level of noise. Our hypothesis is that the choice of metric is dependent only on the noise type. Our first experiment examines the effect of each data parameter while holding the noise at zero (no noise) on the blockmodelling accuracy using each metric. The results are shown in Fig. 2 with a comparison to the existing Projected Gradient Descent (Grad) and Coordinate Descent (Coord) [1] methods. Note that the baseline methods are dependent on their initialisation, so we repeated the baseline clustering 10 times for each graph, using random initialisation. Cluster accuracy is measured using Normalised Mutual Information (NMI).

The box plots in Fig. 2 show the variation due to each of the experimental parameters, while holding the noise at 0. Individual results for each seriation metric are not shown because the variance between each method was minimal or zero. It is surprising to see that each of the seriation methods provides perfect results for each graph containing identifiable clusters, independent of the metric, number of objects and roles. The unidentifiable graphs lead to lower accuracy at 4 clusters for the Ring data, and 8, 16, and 32 clusters for the tree data. The baselines show higher variation and lower mean accuracy. This shows that there is no interaction between the seriation metric and the data parameters when there is no data noise, except for the slight interaction with the number of roles in the tree data due to the leaf unidentifiable roles.

The small variance in the seriation method, with respect to each non-noise parameter, is ideal for examining the effects of noise. Therefore, we focus on the seriation blockmodelling method for the remainder of the article.

Our second experiment examines the robustness of blockmodelling using each metric, to *Hamming noise* (using all simulated 1,125 graphs). The results for seriation are shown in Fig. 3. To make the plots more visually appealing, we limited the data to graphs containing 16 clusters; results for the other cluster sizes have a similar trend. As expected, we find that increasing the Hamming noise reduces the accuracy of each blockmodelling method. It can be seen that results can be grouped in terms of position based and set based metrics; for each block structure, set based metrics are less tolerant of Hamming noise.

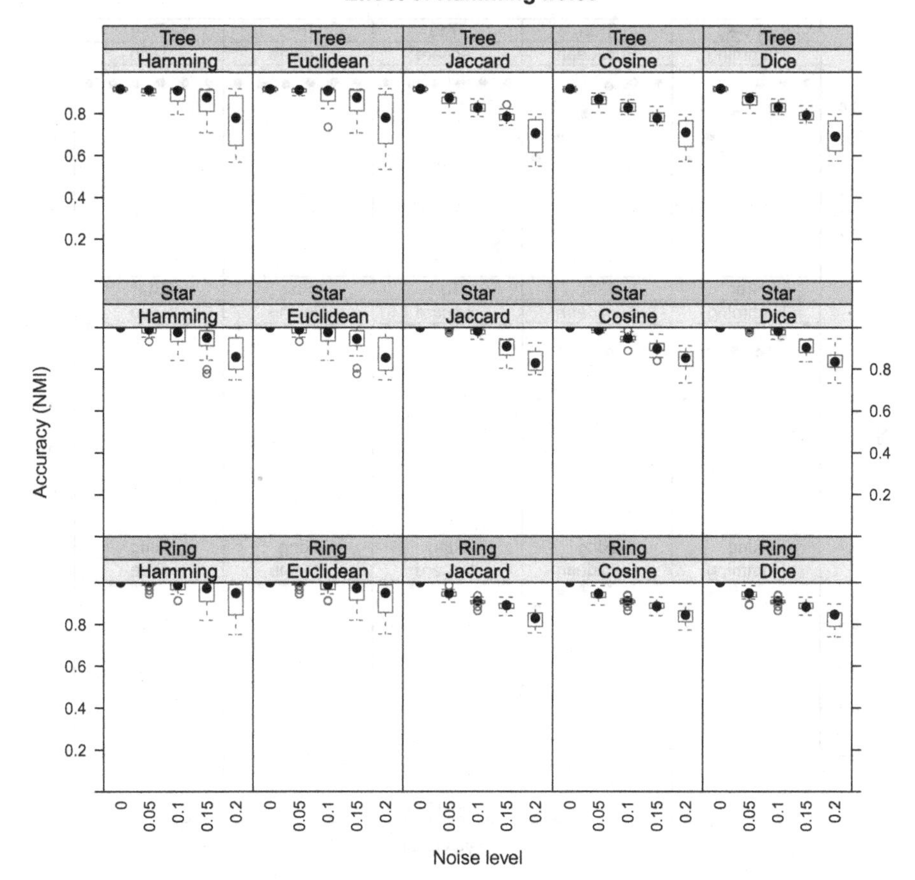

Fig. 3. Clustering accuracy versus Hamming noise level for seriation using each metric on tree, star and ring block structured data. Each data set contained 16 clusters.

Our third experiment examines the robustness of each blockmodelling method to *Jaccard noise* (using all simulated 1,125 graphs). The results for the seriation model are shown in Fig. 4. We find that the set based metrics provide a partitioning that is very robust to Jaccard noise (even when the expected Jaccard noise is 0.5), while the accuracy when using position based metrics drops as the noise level increases.

To identify the effect of Hamming and Jaccard noise on blockmodelling with each metric, we have computed the NMI decay rate (the expected drop in NMI when the data noise increases by 0.1). An NMI decay rate of 0 implies that noise has no effect on NMI, while a large NMI decay rate means that an increase in noise causes a large drop in NMI. The set of NMI decay rates are provided in Table 1. We find that the NMI noise decay rate is lowest for the position based metrics when the data contains Hamming noise, and for the set based metrics when the data contains Jaccard noise.

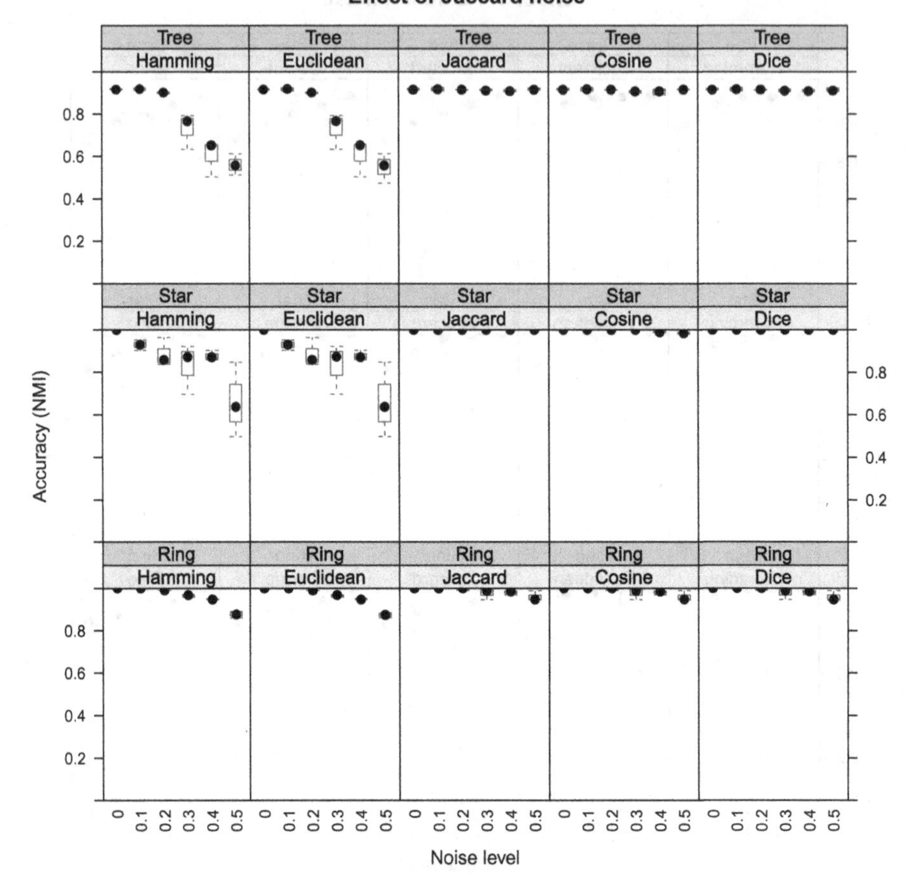

Fig. 4. Clustering accuracy versus Jaccard noise level for seriation using each metric on tree, star and ring block structured data. Each data set contained 16 clusters.

A paired difference permutation significance test was performed to compare the NMI noise decay rate between set based and position based metrics for each type of data. The results showed a significant difference in each case.

4 Discussion

The major difference between position and set based metrics, and the associated Hamming and Jaccard noise, are their treatment of True Negatives (missing edges remaining missing edges). Position based metrics use true negatives as evidence of similarity between the two items; if two vertices are both not connected to a third vertex, it means that the two are similar since they both have a poor relationship with the third. While set based metrics ignore true negatives. This distinction is important when missing edges have different meanings. For social networks, a missing edge might represent

Table 1. The NMI noise decay rate (expected drop in NMI when the noise increases by 0.1) using data with Hamming and Jaccard noise for blockmodelling with each metric. A value of zero implies that the noise level has no effect on the accuracy. The bolded set provide a statistically significant difference to the unbolded set for each network type.

	Euclid	Hamming	Jaccard	Cosine	Dice
Hamming noise					
Tree	**0.0534**	**0.0532**	0.0913	0.1007	0.0902
Ring	**0.0298**	**0.0295**	0.0616	0.0561	0.0624
Star	**0.0339**	**0.0342**	0.0589	0.0570	0.0564
Jaccard noise					
Tree	0.8427	0.8235	**0.0057**	**0.0129**	**0.0111**
Ring	0.2321	0.2271	**0.0777**	**0.0782**	**0.0786**
Star	0.5436	0.5437	**0.0000**	**0.0292**	**0.0000**

a poor relationship between the two items, or it might mean that the relationship was not measured (missing information). For these cases, the missing edges should be treated differently.

This leads us to investigate the meaning of the edges in the network data where set based metrics provided high accuracy. We found that of the seven network data sets from Fig. 1, three provided greater NMI when using set based metrics. Further investigation of these three data sets showed: **Sampson**: only the top three and bottom three relationships between monks were provided, so a missing edge represents an unknown relationship. **Polbooks**: an edge represents if the associated books were bought together. A missing edge does not mean that the books are not related. **Karate**: an edge represents interaction of the members outside of the club. A missing edge does not imply a poor relationship. So for each of these network data sets, a missing edge represents unknown information, not a poor relationship, hence set based metrics are ideal for these particular network data.

Simulations using Hamming noise showed that the position based metrics were more robust to the noise, while simulations using Jaccard noise showed that set based metrics were more robust to the noise. These results reinforced our belief that the network data contained either Hamming or Jaccard noise. We also found that the number of objects, number of roles, and basic structure of the data has little effect on NMI when using seriation for blockmodelling. Therefore there is a direct link to the robustness for a given metric and the noise type.

Based on our results and observations, we conjecture that if network data missing edges represent a poor relationship, then it is likely that it contains Hamming noise, and so position based metrics should be used. If missing edges represent missing information, then the noise is likely to be Jaccard and so set based metrics should be used.

Finally, the experiments showed that when there is little to no noise in the data (regardless of the type), that all metrics performed equally well when using seriation. When a sufficient noise level was reached, there was a difference in accuracy when using the different metrics. The Football data from Fig. 1 exhibits that same behaviour (high accuracy, with no difference in metrics). An edge in that data represents that a game was held between two teams and since most games in the season are held within a conference, there are only few inter conference labels, and hence little noise. This supports our simulation result.

5 Conclusion

Blockmodelling is the process of clustering similar roles within a graph, which are visualised as blocks in the graph adjacency matrix. Edges in the graph increase the strength of block relationships, but it is unclear if missing edges in common should increase or decrease the strength of a relationship. In this article, we examined the effect of the choice of block metric on blockmodelling accuracy. We found that block metrics can be categorised into position based and set based metrics. Experiments on simulated data showed that the blockmodelling accuracy is independent of the block metric when no noise is present, but when noise was introduced, high accuracy results are obtained when the choice of block metric matched the noise type.

References

1. Chan, J., Liu, W., Kan, A., Leckie, C., Bailey, J., Kotagiri, R.: Discovering latent block-models in sparse and noisy graphs using non-negative matrix factorisation. In: CIKM, pp. 811–816. ACM (2013)
2. Fiala, J., Paulusma, D.: The computational complexity of the role assignment problem. In: Baeten, J.C.M., Lenstra, J.K., Parrow, J., Woeginger, G.J. (eds.) ICALP 2003. LNCS, vol. 2719, pp. 817–828. Springer, Heidelberg (2003). https://doi.org/10.1007/3-540-45061-0_64
3. Hahsler, M.: An experimental comparison of seriation methods for one-mode two-way data. Eur. J. Oper. Res. **257**(1), 133–143 (2017)
4. Hurley, C.B.: Clustering visualizations of multidimensional data. J. Comput. Graph. Stat. **13**(4), 788–806 (2004)
5. Karrer, B., Newman, M.E.: Stochastic blockmodels and community structure in networks. Phys. Rev. E **83**(1), 016107 (2011)
6. Park, L.A.F., Bezdek, J.C., Leckie, C., Kotagiri, R., Bailey, J., Palaniswami, M.: Visual assessment of clustering tendency for incomplete data. IEEE TKDE **28**(12), 3409–3422 (2016)
7. Park, L.A.F., Read, J.: A blended metric for multi-label optimisation and evaluation. In: Berlingerio, M., Bonchi, F., Gärtner, T., Hurley, N., Ifrim, G. (eds.) ECML PKDD 2018. LNCS (LNAI), vol. 11051, pp. 719–734. Springer, Cham (2019). https://doi.org/10.1007/978-3-030-10925-7_44
8. Reichardt, J., White, D.R.: Role models for complex networks. The Eur. Phys. J. B **60**(2), 217–224 (2007)
9. Von Luxburg, U.: A tutorial on spectral clustering. Stat. Comput. **17**(4), 395–416 (2007)
10. Zhang, Y., Yeung, D.Y.: Overlapping community detection via bounded nonnegative matrix tri-factorization. In: Proceedings of the 18th ACM SIGKDD, pp. 606–614. ACM (2012)

Towards Better Generalization for Neural Network-Based SAT Solvers

Chenhao Zhang[1], Yanjun Zhang[1], Jeff Mao[1], Weitong Chen[1], Lin Yue[1], Guangdong Bai[1](✉), and Miao Xu[1,2]

[1] University of Queensland, Brisbane, QLD 4072, Australia
chenhao.zhang@uq.net.au,
{yanjun.zhang,w.chen9,l.yue,g.bai,miao.xu}@uq.edu.au
[2] RIKEN, Tokyo 103-0027, Japan

Abstract. Neural network (NN) has demonstrated its astonishing power in many data mining tasks. Recently, NN is adapted to the boolean satisfiability (SAT) problem as a solver, which is trained on a dataset containing the satisfiable annotation of a series of logical expressions. In SAT problem, each expression is composed of the conjunction of logical variables. The effectiveness of NN as a solver to SAT has been verified empirically when the training and test data contain the same group of logical variables. However, when the test set contains more logical variables, test performance significantly degenerates; that is, the generalization performance on the test set containing more logical variables is far below expected. In this paper, we conjecture that the degeneration may be due to that a non-trivial way is requested to calibrate the continuous output by NN. Based on the conjecture, we design a generalized framework that is expected to improve the NN solver's performance when the test data include much more variables. Specifically, a Temperature-Scaled Neural Network SAT solver (TenSAT) adds a special calibration component to the message-passing NN. Experiments demonstrate the correctness of the conjecture, i.e., TenSAT can stop the test performance from degrading when the test set contains new variables as much as ten folds of the training ones.

Keywords: Neural network · SAT · Generalization · TenSAT

1 Introduction

Neural Networks (NN) [10] offers great promises in data mining tasks, and it is well-recognized for learning a strong model from training data to predict unseen test data. Recently, NN has also shown to be powerful on the NP-complete Boolean satisfiability problem (SAT) which is to determine whether a Boolean formula composed of logic variables in conjunctive normal form (CNF) is satisfiable [16]. NN solvers for the SAT problem has been exploited in data analytic tasks including identifying deterministic finite automata [8,19] and circuit design evaluation [17,20].

J. Gama et al. (Eds.): PAKDD 2022, LNAI 13281, pp. 199–210, 2022.
https://doi.org/10.1007/978-3-031-05936-0_16

Traditional SAT solvers are typically based on backtracking algorithms [4,5], which have difficulty on processing more complex problems efficiently and cannot benefit from advanced machine learning models. The pioneer work [16] proposed the first NN-based SAT solver, namely NeuroSAT, which constructs a bipartite message-passing graph between variables and clauses to consider the permutation invariance and negation invariance required by SAT solver [6,12]. NeuroSAT trains on a training set composed of annotated propositional logical expression. After it has been trained by a recurrent neural network (RNN), test results are given by running the trained model for several iterations over the input test data. Along with the prediction, the model also outputs a continuous probability value for each variable which could then be clipped to discrete value and give the assignment leading to a true expression. Based on NeuroSAT, a series of NN-based SAT solvers are then proposed to further enhance the performance [1,23].

NN-based SAT solvers are efficient and accurate in finding solutions when logical variables in training and testing are identical. However, in real-world applications, such as context-free grammar (CFG) analyzing [2,3], deterministic finite automata (DFA) identifying [8,19], and circuit design evaluation [17,20], unseen logical variables are frequently encountered in test phase. For example, in circuit design, an evaluation model needs to learn from a basic circuit structure which contains less logical gates than a complex circuit. Note that current NN-based SAT solvers' performance significantly degenerates when unseen logical variables appear frequently in the testing phase. That is to say, the generalization performance on the test set containing more logical variables is far below expected. According to our empirical studies, the NeuroSAT degenerate significantly when 500% unseen logical variables are added in the testing phase. The problem may be attributed to its clipping of continuous values to discrete ones, when predicting satisfiability on a continuous spectrum with a cut-off point that distinguishes only two classifications.

To address the aforementioned generalization issue, in this paper, we proposed **Te**mperature scaling **n**eural network SAT solver (TenSAT), a general framework for any NN-based **SAT** solver. The purpose of temperature scaling is to mitigate the neural network calibration, i.e., the predicted probability by neural networks may not represent the true correctness likelihood [7]. To tackle the poor calibration, we develop a Temperature-Scaled styled Neural Network SAT solver (TenSAT) which adds a special calibration component to the message-passing graph. It endows NN-based SAT solvers with the ability to learn and set the optimum threshold for satisfiability on this continuous scale and improve the generalization. Empirical studies on both manually-constructed and real-world data validate the effectiveness of TenSAT when integrated with a series of NN-based SAT solvers. We summarize our contributions by the following.

- We propose a general framework, namely TenSAT, which can be adopted with any existing NN-based SAT solver.
- A temperature scaling component is designed to improve the generalisation of NN-based SAT solvers to tackle unseen logical variables in testing phase.

- The performance of TenSAT is verified on a real-world data set, and we will release the data to facilitate future research in this area upon acceptance.

2 Background and Related Works

2.1 SAT Problem

Satisfiability problems are boolean formulas, and Eq. 1 shows an example.

$$f = (x_1 \vee x_2) \wedge (\neg x_1 \vee x_2) \tag{1}$$

The formula contains *logic variables* x_i or their negations, which are also called *literals*. The operations permitted are limited to conjunctions (logical AND) and disjunctions (logical OR). Here, f exploits the convenient property that all SAT problems have an equivalent representation called conjunctive normal form (CNF) where disjunctions are grouped into *clauses* and each clause is separated by conjunctions. Like all SAT problems, f is said to be satisfiable if and only if there exists an assignment of values to the literals such that f is evaluated to be true. Conversely, f is said to be unsatisfiable if no assignment could enable the formula to be true. In the example Eq. (1), it is easy to see that f is satisfiable if and only if both clauses of f can be satisfied. One possible assignment to satisfy f is $x_1 = F,\ x_2 = T$.

2.2 Neural Network Based SAT Solvers

Existing SAT solvers such as Z3 [12] are mostly deterministic and typically subtle variants of the Conflict-Drive-Clause-Learning (CDCL [11]) backtracking algorithms. They follow the philosophy of traversing a decision tree and thus suffer from an exponential time complexity.

Given the power of deep learning in solving various traditionally complex tasks, bridging deep learning and logic reasoning has become increasingly popular. A landmark study in 2019 proposes a message passing neural network called NeuroSAT [16] that learns to solve SAT problems. In the NeuroSAT, using bipartite graphs to represent problems instead of sequences, the model can tolerate the permutation invariance of SAT problems. The study also shows that neural networks excel in receiving SAT problem instances where only a single literal is negated and the truth evaluation of the entire instance changes. Therefore, by randomly generating SAT problems in this fashion, NeuroSAT is able to learn general "knowledge" from the training dataset as well as learn the subtle difference between a satisfiable and an unsatisfiable problem.

NeuroSAT's success has sparked a considerable increase in attention towards developing various neural network architectures and analyzing their performance on problems across separate domains such as the decision variant of the travelling salesman problem [14]. Instead of requiring all clauses to be satisfiable as NeuroSAT, MAXSAT [21] focuses on maximizing the number of satisfied clauses. In MAXSAT, logical reasoning and deep learning architecture are integrated into an end-to-end network architecture to efficiently compute the forward and backward passes in training, and also be applied to learn the logical structure and rules of a 9×9 Sudoku puzzle.

Not only the groundbreaking creation of NeuroSAT but also its adaptation to the traveling salesman problem (TSP) (also NP-compete) [14] and pseudo-boolean problem [9] are all examples of applying NN to complex combinatorial problems. Classically, deterministic solvers are used to solve SAT problems; yet their inefficiency significantly prevents their ability to solve SAT problems with a large number of variables. Neural networks have shown to be a semi-reliable solver for small NP-complete problem such as the SAT problem. However, if applying an NN based SAT solver trained on small number of variables to solve problems with large number of variables, their generalization ability is not satisfactory, as we will show in Sect. 4.

2.3 SAT Applications

The SAT problem is of central importance in computer science, especially in some fundamental areas of computer science. Compilers of a programming languages and HTTP messages handlers need to parse programs and messages from plain text files. Some parser aims to parse restricted context-free language which is a kind of language generated by a context-free grammar (CFG) [3], which can be converted to CNF and analyzed with SAT solver [2]. In addition, if we treat programs' compiling and HTTP message analyzing progress as deterministic finite automata (DFA), such DFAs can also be analyzed by SAT solvers. Heule et al. [8] and Ulyantsev et al. [19] tried to translate DFA to CNF and then identify DFA by using SAT solver. Besides the above mentioned applications, in circuit designing, logic gates can naturally be expressed in CNF. Tseytin transformation [18] is a method to translate combinatorial logic circuits to CNF. Based on this transformation, more analysis of circuit problems [17,20] can be performed by SAT methods. Roy et al. [15] even tried to reconstruct circuit from CNF to verify if structure of circuit is lost when translate a circuit to CNF. Note that in all these applications, new logical variables may exist in the latter phase of testing the SAT solver. Considering such applications, we may require the generalization of the SAT solver to more logical variables.

3 Method

Consider a SAT problem P with n variables x_i ($i \in \{1, ..., n\}$), and m clauses c_j ($j \in \{1, ..., m\}$). We define $\mathcal{T}$ to denote whether a P is satisfiable. $\mathcal{T}(P)$ is true if and only if P is satisfiable. That is, there exists a valuation of $x_i = v_i$ where $v_i \in \{T, F\}$ ($i \in \{1, ..., n\}$), such that

$$\bigvee(\{v_k | x_k \in \mathcal{S}_j^+\} \cup \{\neg v_k | x_k \in \mathcal{S}_j^-\}) = T \text{ } for \text{ } any \text{ } c_j, \tag{2}$$

where $\mathcal{S}_j^+$ and $\mathcal{S}_j^-$ denote the set of variables and negated variables respectively in the jth clause c_j.

For any problem $P \in \mathcal{P}$, the relationship between literals and clauses are encoded in a matrix $A \in \{0,1\}^{2n \times m}$. $A_{ij} = 1$ means that the ith literal is in the jth clause if $i \leq n$, or otherwise the negation of the $(i-n)$th literal is in the jth clause. Based on the problem-specific matrix A, encoding of all literals and clauses are learned and used to determine whether a given P is satisfiable.

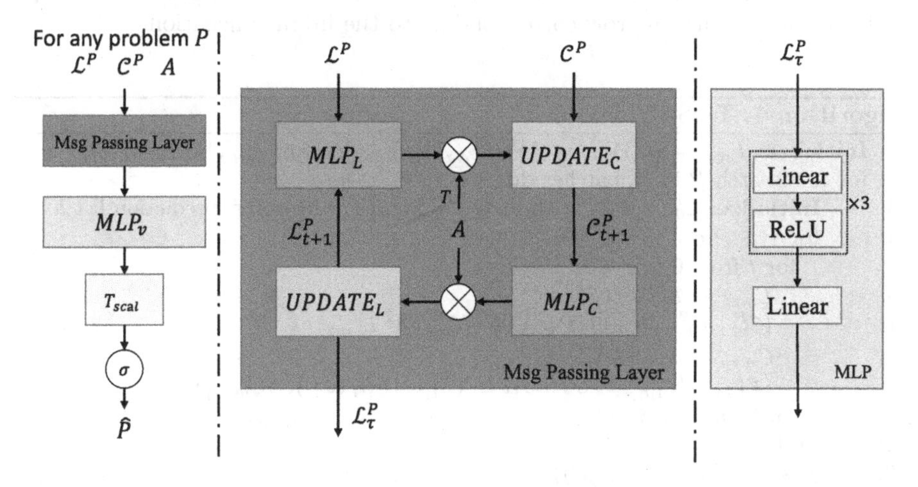

Fig. 1. TenSAT's Architecture. Left: the whole framework of the algorithm; middle: structure of the Message Passing Layer; right: structure of the MLP.

The whole framework is described in Fig. 1 left. To solve the SAT problem, two components are required in existing works: Message Passing Layer and MLP Vote Layer. In the Message Passing Layer (Fig. 1 middle), both $\mathcal{L}^P$ and $\mathcal{C}^P$ are initialized to be all ones. The output of MLP_L is then multiplied with the associate matrix A^T to form the input to the component $UPDATE_C$ which is a neuron of a recurrent neural network. For $UPDATE_C$, the hidden layer output is $\mathcal{C}^P_{t+1}$, which is further encoded by MLP_C, multiplied by A, and used as input to another recurrent neural network neuron $UPDATE_L$ whose hidden layer input is $\mathcal{L}^P_t$ and will output $\mathcal{L}^P_{t+1}$. The output $\mathcal{L}^P_{t+1}$ will be further inputted into MLP_L for another round of embedding. After the iteration continues for τ times, the final output is $\mathcal{L}_\tau$ which will be further inputted into the MLP Vote Layer. The MLP Vote Layer (Fig. 1 right) is used to compute a scalar for each literal which is then averaged to encode a prediction of satisfiability.

Besides these two components existed in classical works such as NeuroSAT [16], another component called Temperature scaling is performed as uniform division on these literal scalars for each problem in each batch during training process. On each problem, the Temperature Scaling is used to calculate the prediction with whom we could further calculate the loss function as

$$Loss = \sum_{P \in \mathcal{P}} BCE\left(\frac{h(P, \mathcal{L}, \mathcal{C}, \mathcal{W})}{T_{scal}}, \mathcal{T}(P)\right). \tag{3}$$

where *Loss* is the loss function, indicating the difference between the predicted satisfiability $h(P, \mathcal{L}, \mathcal{C}, \mathcal{W})$ and the true satisfiability $\mathcal{T}(P)$ for each instance. In optimization, the parameters $\mathcal{W}$ of the neural networks and T_{scal} are optimized alternatively. Specially, we first fix the neural network and optimize T_{scal} and then fix T_{scal} and optimize the neural networks. The whole process of the algorithm is shown in Algorithm 1, in which *Flip* denotes the function that swaps each row of $\mathcal{L}^P$ with the row corresponding to the literal's negation.

Algorithm 1. TenSAT

1: **Initialize:** $T_{scal} = 1$, $(\mathcal{W}_{MLP_L}, \mathcal{W}_{MLP_C}, \mathcal{W}_{MLP_v}, \mathcal{W}_{UPDATE_L}, \mathcal{W}_{UPDATE_C})$
2: **for** mini-batch $\mathcal{P}$ in all batches **do**
3:     **Initialize:** $\mathcal{L}_0^P$ and $\mathcal{C}_0^P$ with ones, $\mathcal{L}_h^P$ and $\mathcal{C}_h^P$ with zeros, prediction list h
4:     **for** $P \in \mathcal{P}$ **do**
5:         **for** t from 0 to τ **do**
6:             $L_{msg} \leftarrow MLP_L(\mathcal{L}_t^P)$
7:             $(\mathcal{C}_{t+1}^P, \mathcal{C}_{t+1,h}^P) \leftarrow UPDATE_C(\mathcal{C}_{t,h}^P, A^T L_{msg})$
8:             $C_{msg} \leftarrow MLP_C(\mathcal{C}_{t+1}^P)$
9:             $(\mathcal{L}_{t+1}^P, \mathcal{L}_{t+1,h}^P) \leftarrow UPDATE_L(\mathcal{L}_{t,h}^P, [Flip(\mathcal{L}_t^P), AC_{msg}])$
10:         **end for**
11:     **end for**
12:     $Loss = \sum BCE(h, \mathcal{T}(P))$
13:     $T_{scal} \leftarrow ADAM(Loss)$
14:     $(\mathcal{W}_{MLP_L}, \mathcal{W}_{MLP_C}, \mathcal{W}_{MLP_v}, \mathcal{W}_{UPDATE_L}, \mathcal{W}_{UPDATE_C}) \leftarrow ADAM(Loss)$
15: **end for**

4 Experiments

4.1 Setting

Data. We create a problem sets with an arbitrary number of problems for any number of variables with the same setting suggested by the NeuroSAT [16]. We use $\boldsymbol{SR}(n)$ to denote the distribution of pairs of random SAT problems on n variables, and $\boldsymbol{SR}(\boldsymbol{U}(n_1, n_2))$ to denote the distribution of problems on n_1 to n_2 variables. Given the number of variables n, NeuroSAT's problem generator samples $k \leq n$ uniformly distributed literals from the entire $2n$ selection of literals into a clause.

Experiment-Setting. To evaluate the performance of the TenSAT, we are comparing the proposed method with two SOTA SAT solvers, including NeuroSAT [16] and NLocalSAT [23], on predicting satisfiability and satisfying assignments problems. In addition, to further analyse the contribution of temperature scaling factor T_{scal} in TenSAT, we investigate the performance of TenSAT under different T_{scal} setting during the test stage.

Training. In this work, we use the terms "round" or "iteration" to represent the round number of message passing, and use the term "epoch" to represent training epoch. We train TenSAT with various recurrent components, including RNN, GRU, and LSTM. We use $TenSAT_{(N,rn)}$ and $TenSAT_{(G,rn)}$ to denote the application of T_{scal} to NeuroSAT and GGCN respectively, where $rn \in \{RNN, GRU, LSTM\}$. We train each model using 10,000 pairs of problems on $\boldsymbol{SR}(\boldsymbol{U}(5,10))$ with 24 rounds (iteration) of message passing and 25 epochs. We perform our experiments on Intel Xeon Platinum 8163 CPU and Tesla V100 GPU (16Gi). All methods are constructed using Pytorch and are publicly available on GitHub[1].

4.2 Results

Predicting Satisfiability. We first conduct a coarse interval granularity experiment. Comparing $NeuroSAT_{(LSTM)}$ (baseline) and $TenSAT_{(N,LSTM)}$ trained with $\boldsymbol{SR}(\boldsymbol{U}(5,10))$, we found that the performances between these two models are similar on $\boldsymbol{SR}(10)$, $\boldsymbol{SR}(20)$ and $\boldsymbol{SR}(160)$, while obvious difference can be seen on $\boldsymbol{SR}(40)$ and $\boldsymbol{SR}(80)$, as shown in Fig. 2. This is because for small-scale problems (e.g., $\boldsymbol{SR}(10)$ and $\boldsymbol{SR}(20)$), both $NeuroSAT_{(LSTM)}$ and $TenSAT_{(N,LSTM)}$ can handle them very well with no difference, while too big problems (e.g., $\boldsymbol{SR}(160)$) are complex under this training condition. We then zoom into $\boldsymbol{SR}(n_f)$ where $n_f \in \{30, 40, 50, 60, 70, 80\}$. We observe that the accuracy of predicting satisfiability stops increasing when the number of iterations are larger than 64. We list the details in Table 1. In light of this, we restrict the iteration time to $t_f \in \{10, 20, 30, 40, 50, 60\}$ in the rest of our experiments.

Table 1. Accuracy for compared methods in iterations $\{60, 70, 80, 90\}$ with SR in $\{30, 40, 50, 60\}$, where *NS* represents NeuroSAT and *TS* represents TenSAT.

Iters	SR30		SR40		SR50		SR60	
	NS	*TS*	*NS*	*TS*	*NS*	*TS*	*NS*	*TS*
60	0.742	**0.792**	0.656	**0.709**	0.602	**0.652**	0.567	**0.608**
70	0.741	**0.793**	0.657	**0.707**	0.602	**0.653**	0.569	**0.615**
80	0.733	**0.799**	0.654	**0.710**	0.604	**0.654**	0.572	**0.615**
90	0.733	**0.801**	0.654	**0.715**	0.603	**0.656**	0.576	**0.616**

After narrowing down the $\boldsymbol{SR}(n)$ and iteration time, fine interval granularity experiments for all 12 models are performed on $\boldsymbol{SR}(n_f)$ and with $t_f \in \{10, 20, 30, 40, 50, 60\}$. Results are shown in Figs. 3 and 4. In general, TenSAT outperforms NeuroSAT and GGCN, especially when the LSTM is used as their recurrent network component (Figs. 3a and 4a). In addition, due to LSTM's capacity in remembering longer sequence, TenSAT with LSTM (as the recurrent component) also outperforms that with RNN and GRU.

[1] https://github.com/ChildEden/TenSAT.

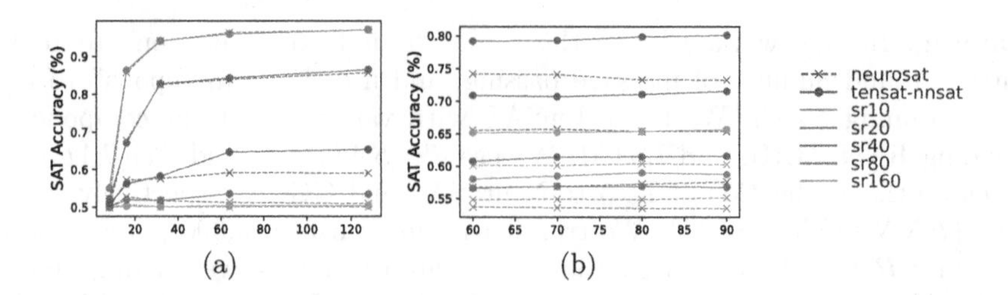

(a) (b)

Fig. 2. Results on the coarse interval granularity problem sets, where x axis is the number of iterations. (a) Neuro-SAT vs TenSAT on LSTM; (b) Stable accuracy when iteration time larger than 60.

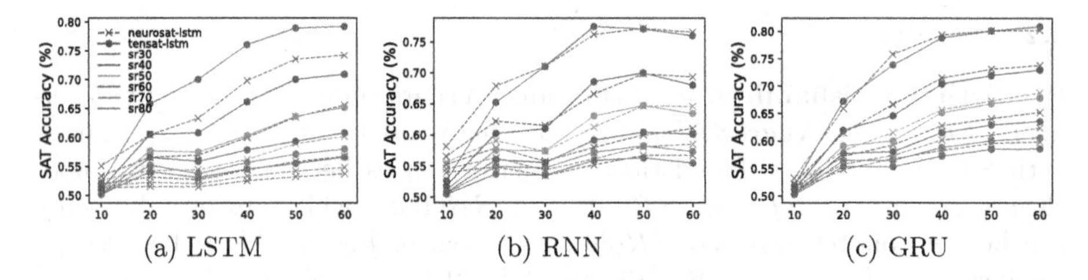

(a) LSTM (b) RNN (c) GRU

Fig. 3. Predicting satisfiability results (NeuroSAT v.s. TenSAT) on the fine interval granularity problem sets.

For results of GGCN with RNN and GRU as its recurrent network component, the best performances appear when iteration time is 20 because this is the nearest from what they are trained with (24 iteration). Since the GGCN uses the symmetrical normalized adjacency matrix and each element in this matrix is in the range from 0 to 1, some gradient issues may happen and weaken the performance when iteration time is larger than 20. Nonetheless, we can still observe the improvement brought by temperature scaling factor T_{scal} when iteration time is 20. When iteration time is 20, GGCN performs better than TenSAT on $\boldsymbol{SR}(30)$ and $\boldsymbol{SR}(40)$, and as the problems become more complex, TenSAT outperforms the GGCN (Fig. 4c).

Since models with LSTM perform stable and better than others, we summarize TenSAT's improvement by calculating the average accuracy difference between TenSAT and baselines in the condition of taking the LSTM as their recurrent component. The improvement brought by TenSAT on NeuroSAT is 3.9%, and 2.1% on GGCN.

Satisfying Assignments. According to Selsam et al. [16], literal votes, which are outputs of the network, can be decoded and get assignment solutions of SAT problems by doing clustering on these votes. In this experiment, we try to decode satisfying assignments for satisfiable problems, and compare such performances of trained models $NeuroSAT_{(rn)}$ and $TenSAT_{(N,rn)}$ where $rn \in \{RNN, GRU, LSTM\}$. It is clear to see that the model trained with temperature scaling factor performs better than original model in Fig. 5a, and their

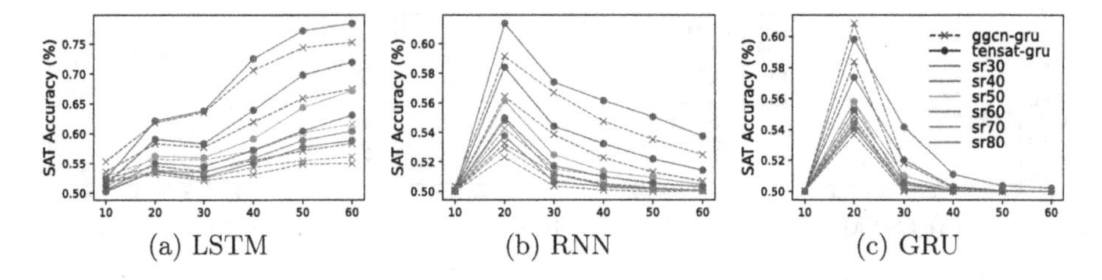

Fig. 4. Predicting satisfiability results (GGCN v.s. TenSAT) on the fine interval granularity problem sets.

difference becomes large as iteration time increases. When the recurrent component is replaced by RNN, as shown in Fig. 5b, the NeuroSAT performs better than TenSAT. This may be because the lack of gated units in the RNN makes the gradient unstable during training and the under-training T_{scal} may amplify such instability. But their performance difference becomes small as iteration time increases. Same as in the predicting satisfiability experiment, we summarize TenSAT's improvement by calculating the average accuracy difference between TenSAT and NeuroSAT in the condition of taking the LSTM as their recurrent component. The improvement brought by TenSAT is 3.5%.

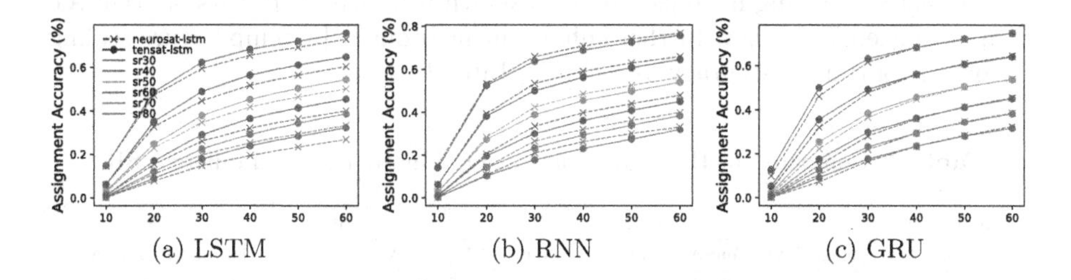

Fig. 5. Satisfying assignments results where x axis is the number of iterations.

4.3 Influence of Temperature Scaling Factor

By implementing T_{scal} as a simple learned scalar in training, we can achieve consistent results for larger problem sets. However, this invites the question whether this can be further improved with the optimization of the scaling factor so that a model trained on n variables can achieve relatively low decrease of accuracy for problems with variables more than n. We train models with T_{scal} start from 1, and as training progresses, the temperature scaling factor decreases, as shown in Table 2.

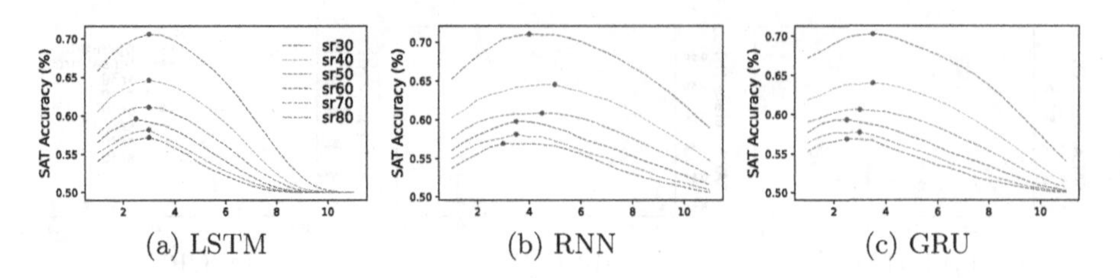

(a) LSTM (b) RNN (c) GRU

Fig. 6. Accuracy vs. T_{scal}. Each figure shows the relation between accuracy and T_{scal} in test stage at iteration 20, where the x axis is T_{scal}.

Table 2. T_{scal} in different training epochs, where *TS* represents TenSAT.

Epoch	*TS*-LSTM	*TS*-RNN	*TS*-GRU	Epoch	*TS*-LSTM	*TS*-RNN	*TS*-GRU
1	1.005	1.002	0.996	30	0.782	0.860	0.797
10	0.851	0.879	0.832	40	0.776	0.857	0.795
20	0.800	0.863	0.806	50	0.770	0.854	0.793

To find out influence of the temperature scaling factor on performance of model, we fix the message passing times at 20 and solve for the optimum scaling factor for each size dataset by computing the maximum accuracy for each scaling factor iterating in steps of 0.5, as shown in Fig. 6. And tests on TenSAT with optimized T_{scal} show further enhancement brought by temperature scaling factor, the optimum accuracy as listed in Table 3.

Table 3. Compared methods' accuracy at Optimized T_{scal} and different SR.

SR	LSTM			RNN			GRU		
	Opt T_{scal}	NeuroSAT	TenSAT	Opt T_{scal}	NeuroSAT	TenSAT	Opt T_{scal}	NeuroSAT	TenSAT
30	3.0	0.605	**0.701**	4.0	0.679	**0.707**	3.5	0.659	**0.698**
40	3.0	0.566	**0.643**	4.5	0.622	**0.647**	3.0	0.614	**0.638**
50	3.0	0.543	**0.612**	3.0	0.591	**0.615**	2.5	0.582	**0.609**
60	3.0	0.532	**0.590**	5.0	0.579	**0.592**	3.5	0.572	**0.586**
70	3.0	0.523	**0.582**	3.5	0.563	**0.577**	3.0	0.557	**0.576**
80	2.5	0.516	**0.575**	3.5	0.549	**0.568**	2.5	0.548	**0.570**

5 Case Study-Circuit Design Evaluation

Experiment Setup. We also evaluate the performance of TenSAT against circuits design evaluation applications. To do this, we use circuit designs from EvoApproxLib [13] and convert these circuit from Verilog files to SAT problems in CNF by Yosys [22], then such SAT problems can be used as input data.

In this experiment, NeuroSAT and TenSAT are still trained with $\boldsymbol{SR}(\boldsymbol{U}(5,10))$, the same as that in previous simulation experiment, and test is performed on distribution $\boldsymbol{S}(n_f)$, which is converted from circuit design data, where $n_f \in \{30, 40, 50, 60, 70, 80\}$. And in this study case, only the LSTM recurrent component is applied.

Results. The experimental results are listed in Table 4. TenSAT outperforms NeuroSAT generally. Since limitation of the number of use cases (less than 10 case included in each $\boldsymbol{SR}(n)$), the accuracy is not stable. Nevertheless, TenSAT still outperforms NeuroSAT in same conditions.

Table 4. Results of experiment on circuit design evaluation, where *NS* represents NeuroSAT and *TS* represents TenSAT.

Iter	SR30		SR40		SR50		SR60		SR70		SR80	
	NS	*TS*	*NS*	*TS*	*NS*	*TS*	*NS*	*TS*	*NS*	*TS*	*NS*	*TS*
10	0.556	**0.556**	0.500	**0.500**	0.500	**0.500**	0.500	**0.500**	0.500	**0.500**	0.500	**0.500**
20	0.778	**0.833**	0.400	**0.400**	0.357	**0.429**	0.250	**0.500**	0.375	**0.375**	0.071	**0.143**
30	0.722	**0.833**	0.500	**0.600**	0.500	**0.714**	**0.875**	0.750	0.500	**0.500**	0.429	**0.571**
40	0.889	**0.889**	0.400	**0.700**	0.643	**0.786**	0.625	**0.625**	0.500	**0.750**	0.571	**0.571**
50	0.833	**0.833**	0.700	**0.900**	**0.714**	0.643	0.625	**0.625**	0.625	**0.750**	0.429	**0.571**
60	0.778	**0.833**	0.700	**0.900**	0.643	**0.643**	**0.750**	0.625	0.625	**0.625**	0.429	**0.500**

6 Conclusion and Discussion

In this work, we propose to use temperature scaling to enhance the generalization of neural networks in solving an typical NP-Complete problem, the SAT problem. Experiments show that our framework with the LSTM as its recurrent component performs more stable and better than that with RNN and GRU. TenSAT outperforms baselines, especially when solving larger SAT problems.

References

1. Amizadeh, S., Matusevych, S., Weimer, M.: PDP: a general neural framework for learning constraint satisfaction solvers. arXiv:1903.01969 (2019)
2. Axelsson, R., Heljanko, K., Lange, M.: Analyzing context-free grammars using an incremental sat solver. In: International Colloquium on Automata, Languages, and Programming (2008)
3. Charniak, E.: Statistical parsing with a context-free grammar and word statistics. In: AAAI (1997)
4. Davis, M., Logemann, G., Loveland, D.: A machine program for theorem-proving. Commun. ACM **5**, 394–397 (1962)
5. Davis, M., Putnam, H.: A computing procedure for quantification theory. J. ACM **7**, 201–215 (1960)
6. Eén, N., Sörensson, N.: An extensible sat-solver. In: Proceedings of the 6th International Conference on Theory and Applications of Satisfiability Testing (SAT) (2003)
7. Guo, C., Pleiss, G., Sun, Y., Weinberger, K.Q.: On calibration of modern neural networks. In: ICML (2017)

8. Heule, M.J., Verwer, S.: Exact DFA identification using sat solvers. In: International Colloquium on Grammatical Inference (2010)
9. Liu, M., Zhang, F., Huang, P., Niu, S., Ma, F., Zhang, J.: Learning the satisfiability of pseudo-Boolean problem with graph neural networks. In: Proceedings of the Principles and Practice of Constraint Programming - 26th International Conference, CP 2020, Louvain-la-Neuve, Belgium, 7–11 September 2020 (2020)
10. Liu, W., Wang, Z., Liu, X., Zeng, N., Liu, Y., Alsaadi, F.: A survey of deep neural network architectures and their applications. Neurocomputing **234**, 11–26 (2017)
11. Marques-Silva, J.P., Sakallah, K.A.: Grasp: a new search algorithm for satisfiability. In: Proceedings of the IEEE/ACM International Conference On Computer Aided Design (1996)
12. de Moura, L., Bjørner, N.: Z3: an efficient SMT solver. In: Proceedings of Tools and Algorithms for the Construction and Analysis of Systems (2008)
13. Mrazek, V., Hrbacek, R., Vasícek, Z., Sekanina, L.: Evoapprox8b: library of approximate adders and multipliers for circuit design and benchmarking of approximation methods. In: Proceedings of Design, Automation & Test in Europe Conference & Exhibition (2017)
14. Prates, M., Avelar, P.H., Lemos, H., Lamb, L.C., Vardi, M.Y.: Learning to solve np-complete problems: a graph neural network for decision TSP. In: AAAI (2019)
15. Roy, J.A., Markov, I.L., Bertacco, V.: Restoring circuit structure from sat instances. In: Proceedings of the International Workshop on Logic and Synthesis (2004)
16. Selsam, D., Lamm, M., Bünz, B., Liang, P., de Moura, L., Dill, D.L.: Learning a SAT solver from single-bit supervision. In: ICLR (2019)
17. Shen, Y., Li, Y., Rezaei, A., Kong, S., Dlott, D., Zhou, H.: Besat: behavioral sat-based attack on cyclic logic encryption. In: Proceedings of the 24th Asia and South Pacific Design Automation Conference (2019)
18. Tseitin, G.S.: On the complexity of derivation in propositional calculus. In: Automation of reasoning, pp. 466–483. Springer, Heidelberg (1983). https://doi.org/10.1007/978-3-642-81955-1_28
19. Ulyantsev, V., Zakirzyanov, I., Shalyto, A.: Symmetry breaking predicates for sat-based DFA identification. arXiv:1602.05028 (2016)
20. Velev, M.N.: Comparison of schemes for encoding unobservability in translation to sat. In: Proceedings of the 2005 Asia and South Pacific Design Automation Conference (2005)
21. Wang, P.W., Donti, P., Wilder, B., Kolter, Z.: Satnet: Bridging deep learning and logical reasoning using a differentiable satisfiability solver. In: ICML (2019)
22. Wolf, C.: Yosys open synthesis suite. http://www.clifford.at/yosys/
23. Zhang, W., et al.: Nlocalsat: boosting local search with solution prediction. In: IJCAI (2020)

Robust and Provable Guarantees for Sparse Random Embeddings

Maciej Skorski, Alessandro Temperoni(✉), and Martin Theobald

Department of Computer Science, University of Luxembourg,
4365 Esch-sur-Alzette, Luxembourg
{maciej.skorski,alessandro.temperoni,martin.theobald}@uni.lu

Abstract. In this work, we improve upon the guarantees for sparse random embeddings, as they were recently provided and analyzed by Freksen at al. (NIPS'18) and Jagadeesan (NIPS'19). Specifically, we show that (a) our bounds are *explicit* as opposed to the asymptotic guarantees provided previously, and (b) our bounds are guaranteed to be *sharper* by practically significant constants across a wide range of parameters, including the *dimensionality*, *sparsity* and *dispersion* of the data. Moreover, we empirically demonstrate that our bounds significantly outperform prior works on a wide range of real-world datasets, such as collections of images, text documents represented as bags-of-words, and text sequences vectorized by neural embeddings. Behind our numerical improvements are techniques of broader interest, which improve upon key steps of previous analyses in terms of (c) tighter estimates for certain types of *quadratic chaos*, (d) establishing extreme properties of *sparse linear forms*, and (e) improvements on bounds for the estimation of *sums of independent random variables.*

Keywords: Sparse random embeddings · Johnson-Lindenstrauss lemma

1 Introduction

1.1 Background: Random Embeddings

The seminal result of Johnson and Lindenstrauss [14] states that *random linear mappings* have nearly isometric properties, and hence are well-suited for embeddings: they *nearly preserve distances* when projecting high-dimensional data into a *lower-dimensional space.* Formally, for an error parameter $\epsilon > 0$, an $m \times n$ matrix A appropriately sampled (e.g., using appropriately scaled Gaussian entries), and any input vector $x \in \mathbb{R}^n$, it holds that

$$1 - \epsilon \leqslant \|A x\|_2 / \|x\|_2 \leqslant 1 + \epsilon \quad \text{with probability } 1 - \delta \tag{1}$$

if the embedding dimension is $m = \Theta\left(\frac{1}{\epsilon^2} \log \frac{1}{\delta}\right)$.

J. Gama et al. (Eds.): PAKDD 2022, LNAI 13281, pp. 211–223, 2022.
https://doi.org/10.1007/978-3-031-05936-0_17

This bound on the dimension m has been shown to be *asymptotically optimal* [13], while the assumptions made on the Gaussian distribution of matrix A can be further replaced by the Rademacher distribution, or relaxed even further by only requiring the sub-Gaussian condition to hold for the construction of the projection matrix A.

The result is a typical *dimension-distortion tradeoff*: one aims to minimize $m \ll n$, while keeping ϵ and δ possibly small. Smaller dimensions m allow for efficient processing of large, high-dimensional datasets, while a small distortion guarantees that analytical tasks can be performed with a similar effect over the embedded data as it is the case for the original data.

Over the past years, variants of the aforementioned *Johnson-Lindenstrauss Lemma* have found important applications to text mining and image processing [2], approximate nearest-neighbor search [1,11], approximation algorithms for clustering high-dimensional data [26], and many others. The focus of this paper is on *linear sparse random embeddings*, where A in Eq. (1) has at most s non-zero entries in each column, which allows for faster computation of the embedded vectors. This setup has been covered by a substantial line of recent research [1,15,24], which established that, for the optimal dimension m, one may set $s = \Theta(m\,\epsilon)$, thereby gaining a factor of ϵ in matrix sparsity[1]. This can be further improved by exploiting *structural properties* of the input data: as shown in recent works [9,12,15], with $v \triangleq \|x\|_\infty/\|x\|_2$, one may set the sparsity to

$$s = \Theta\left(\frac{v^2}{\epsilon}\max\left(\log\frac{1}{\delta}, \frac{\log^2\frac{1}{\delta}}{\log\frac{1}{\epsilon})}\right)\right) \tag{2}$$

while keeping the optimal choice of dimension $m = \Theta\left(\frac{1}{\epsilon^2}\log\frac{1}{\delta}\right)$. This shows that a better sparsity s is feasible when the data-dependent parameter v is small. The parameter v should thus be understood as the *dispersion* of the input vector x, i.e., v is small when the components of x are of comparable magnitude, and it is larger when there are dominating components. Empirically, random embeddings work much better than predicted by their theoretical bounds. The main goal of this work thus is to bridge this frequently observed gap between theory and practice and thereby develop both *robust* and *provable guarantees* for sparse random embeddings.

Recent state-of-the-art analyses [9,12] provide only asymptotic bounds which tend to disguise dependencies on rather large constants [9]. In practice, they often yield trivial results which however limits their usability. Moreover, real-data evaluations from prior works are mostly of qualitative nature: they analyze trends in parameter tradeoffs [12] rather than provable guarantees. Regarding the dispersion measure $v = \|x\|_\infty/\|x\|_2$, which is the key ingredient of recent improvements, no study has evaluated its typical behavior on real-world data to our knowledge so far. The typical value of the dispersion v may also depend on the type of the data (text, images, etc.), which in turn makes the findings harder to generalize.

[1] One may in fact further reduce the sparsity s by a factor of $B > 1$, however at the cost of increasing the dimension m by a factor of $2^{\Theta(B)}$ (i.e., exponentially).

To date, the literature offers no satisfactory treatment of this prevalent gap between provable and practically meaningful guarantees. Some authors [9,33] suggested that very good empirical performance may be an evidence for small constants, but it may well be the case that sparse random embeddings work better than predicted by the underlying theory due to other data properties, not present in any of the analyses. Indeed, while one can expect a low data dispersion to help increasing sparsity, the proposed dispersion measure v is very crude and does not capture this aspect well in a quantitative sense.

1.2 Contributions

We summarize the novel contributions of our work as follows.

Explicit and Efficient Analysis. We re-analyze sparse random embeddings, following the setup of recent state-of-the-art works [9,12], which provide guarantees depending on the data dispersion $v = \|x\|_{\infty}/\|x\|_2$. Our novel bound is a combinatorial expression that is also *computationally fast to evaluate*. More precisely, our expression on the error term ϵ can be evaluated in nearly constant time of $O(\log^4(1/\delta)\log(m/\delta))$ operations.

Robust and Provable Guarantees. We demonstrate that our bounds are very *robust and accurate* over a large variety of practical use-cases as well as over a wide range of dispersion values v and error bounds ϵ. In particular, we give an exhaustive evaluation over both a synthetic benchmark and no less than 10 real-world datasets concerning text in different representations, images of various sizes, and sparse matrices which arise in typical scientific computations. We see improvements by a factor of *more than one order of magnitude* in the projected dimension m and sparsity s of A, and even more in the confidence $1-\delta$.

Improved Estimation of Quadratic Chaos. Virtually all analyses of random embeddings need to estimate quadratic forms in symmetric random variables, which arise due to considering the Euclidean distance of the projected vector. To solve this problem, we develop a *novel bound for the quadratic form* in terms of its linear analogue, with a very good numerical constant. This improves upon direct estimation from prior work [12], as well as (in this context) general-purpose tools such as variants of the Hanson-Wright Lemma [10] and decoupling inequalities.

Extremal Properties of Sparse Linear Chaos. The reduction from a quadratic form leaves us with the task of understanding stochastic properties of certain random sums, namely the inner product of the given weight vector (our input x) and a random vector with entries -1, 1 or 0 (i.e., one row of the matrix A). This problem is related to, but more general than the well-known Khintchine Inequality [17]. In our context (providing bounds based on data dispersion), we explicitly find the *worst-performing set of weights*, as opposed to prior work where only overestimates were obtained [12].

Estimation of Sums of I.I.D. Random Variables. To derive accurate bounds, we rely on a *precise estimation of sums of independent random variables* which goes beyond what is offered by classical Chernoff-Heoffding bounds.

Remarkably, we are able to numerically improve the state-of-the-art bounds due to Latala [22], which further adds to the success of our approach on real-world data.

1.3 Related Work

Due to space constraints, we here only provide a very brief related-work discussion. We refer the reader to [31] for a detailed listed of references.

Theory of Sparse Random Embeddings. Our work improves directly upon [9] (case $s = 1$) and [12] (general s). These works determine the relation between sparsity and data dispersion, thereby building on a long line of earlier works on variants of the Johnson-Lindenstrauss Lemma [1,15,24]. The provided bounds, albeit proven to be asymptotically optimal, suffer from a *lack of explicit constants* which cannot be easily extracted from the previous estimates.

Empirical Evaluation of Random Embeddings. The good empirical performance of random linear embeddings, including sparse variants, has been confirmed many times (cf. [2,33]). These works point out the gap between provable and observed performance, which we are addressing in this work.

Estimation of Quadratic Chaos. Technically speaking, the analysis of errors in random projections can be reduced to the more general problem of estimating quadratic forms of random variables, also called *quadratic chaos*. The literature offers a variety of tools, from variants of the well-known Hanson-Wright Lemma [10,34] to more specialized bounds [3]. However, they would produce worse constants than our direct approach.

Embeddings of Large Collections/Subspaces. Orthogonal to obtaining bounds for a single vector x is the question of how to extend such bounds to hold simultaneously for all x from a finite collection or an entire subspace of input data. This can be done by a black-box reduction using ϵ-net arguments and is solved by a reduction to the single vector case by means of ϵ-net arguments. Such bounds can be also obtained in our case.

2 Robust Guarantees for Sparse Random Projections

We next briefly discuss a number of preliminary concepts, mainly to fixate the notation before we move on to present the main results of our work.

2.1 Preliminaries and Notation

The d-th *norm* of a vector x and a random variable X, respectively, are defined as $\|x\|_d = (\sum_i |x_i|^d)^{\frac{1}{d}}$ and $\|X\|_d = \left(\mathbb{E}\left[|X|^d\right]\right)^{\frac{1}{d}}$; we also define $\|x\|_\infty = \max_i |x_i|$ as usual. $\mathrm{Bern}(p)$ denotes the *Bernoulli distribution* with success probability p, while $\mathrm{Binom}(n,p)$ denotes the *binomial distribution* with n trials and success probability p. The *Rademacher distribution* takes values 1 and -1 with equal probabilities. Moreover, a random variable X is called *symmetric* when it has the same distribution as $-X$. For two vectors $x, y \in \mathbb{R}^n$, we say that x *majorizes* y, denoted by $x \succ y$, when $\sum_{i=1}^k x_i^{\downarrow} \geqslant \sum_{i=1}^k y_i^{\downarrow}$, for $k = 1 \ldots n$. Finally, *Schur-concave* functions f are those that satisfy $f(x) \leqslant f(y)$ whenever $x \succ y$.

2.2 Construction of the Embeddings

Let A be an $m \times n$ matrix which is sampled as follows:

(1) Fix a positive integer $s \leqslant m$, the *column sparsity* of A.
(2) For each column, select s row positions at random (without replacement), place ± 1 uniform-randomly at these positions and 0 at the remaining positions.
(3) Finally, scale all entries of A by $\frac{1}{\sqrt{s}}$.

Remark 1 (Alternative Constructions). The above construction of A is as in [9, 12], but our analysis works also when sampling is done with replacement.

To analyze the error obtained from the respective projection of x by A, we define as in [12]

$$E(x) \triangleq \|Ax\|_2^2 - \|x\|_2^2 = \sum_{r=1}^{m} \sum_{1 \leqslant i \neq j \leqslant n} A_{r,i} A_{r,j} x_i x_j \tag{3}$$

which is then analyzed by looking into individual "row" contributions, namely $E(x) = \frac{1}{s} \sum_{r=1}^{m} E_r(x)$ with

$$E_r(x) \triangleq \; s \sum_{1 \leqslant i \neq j \leqslant n} A_{r,i} A_{r,j} x_i x_j \,. \tag{4}$$

The goal is to identify conditions, such that $\Pr_A[|E(x)| > \epsilon \|x\|_2^2] \leqslant \delta$, as this implies Eq. (1). By scaling, we can assume $\|x\|_2 = 1$. Throughout the paper, we denote $p = \frac{s}{m}$.

2.3 Key Techniques for the Analysis

For the following steps, we leverage two techniques which were not used in prior work, namely (a) *careful use of symmetry properties* and (b) *majorization.*

Quadratic Chaos Estimation. Studying the error $E(x)$, due to pairwise terms, requires the estimation of quadratic forms $\sum_{i \neq j} Z_i Z_j$, with $Z_i = A_{r,i} x_i$. To this end, we develop a useful general inequality, which reduces the problem to (simpler) linear forms.

Lemma 1. *For symmetric and independent random variables Z_i and any positive even d, we have:*

$$\|\sum_{i \neq j} Z_i Z_j\|_d \leqslant 4 \|\sum_i Z_i\|_d^2 \tag{5}$$

Remark 2. The proof (see [31]) establishes more, namely that for a positive integer d (odd or even), we have $\| \sum_{i\neq j} Z_i Z_j \|_d \leqslant 4 \| \sum_{i\neq j} Z_i Z_j' \|_d$ where Z_i' are independent copies of Z_i.

The constant $C = 4$ in Lemma 1 can be further improved. For example, it is easily seen that for $d = 2$ one may choose $C = \sqrt{2}$. For a general d, the use of hypercontractive inequalities may give furthers refinements.

Extremal Properties of Linear Chaos. We now move on to deriving bounds for linear forms of symmetric random variables, which bound quadratic forms. The following lemma gives a geometric insight into their behavior with respect to the input weights.

Lemma 2. *For $x \in \mathbb{R}^n$, define $S(x) = \sum_i x_i Y_i$ where $Y_i \sim^{\text{i.i.d.}} Y$ with $Y \in \{-1, 0, 1\}$ taking values ± 1 each with probability $p/2$ and 0 with probability $1-p$. Then, for every pair of vectors x, x', such that $(x_i^2)_i \succ (x_i'^2)_i$, and positive even integer d, the following inequality holds:*

$$\|S(x)\|_d \leqslant \|S(x')\|_d \tag{6}$$

The lemma yields the following corollary.

Corollary 1. *Let Y_i be as in Lemma 2. For $v \in (0, 1)$, consider all vectors $x \in \mathbb{R}^n$, such that $\|x\|_2 = 1$ and $\|x\|_\infty = v$. Then, $\| \sum_i x_i Y_i \|_d$ for an even $d > 0$ is maximized at $x = x^*$ where:*

$$x_i^* = \begin{cases} v & i = 1 \\ \sqrt{\frac{1-v^2}{n-1}} & i = 2 \dots n \end{cases} \tag{7}$$

Estimation of I.I.D. Sums. The techniques outlined above allow us to bound the row-wise error contributions $E_r(x)$. In order to assemble them into a bound on the overall error $E(x)$, we prove the following lemma.

Lemma 3. *Let $Z_1, \dots, Z_m \sim^{i.i.d.} Z$, where Z is symmetric, and let d be positive and even. Then:*

$$\| \sum_{i=1}^{m} Z_i \|_d \leqslant \min \left\{ t > 0 : \mathbb{E}(1 + Z/t)^d \leqslant \mathrm{e}^{\frac{d}{2m}} \right\} \tag{8}$$

This improves the constant provided in the seminal result of [22] by a factor of $\mathrm{e}^{1/2}$.

2.4 Bounds Based on Error Moments

We first bound the row-wise error contributions $E_r(x)$, defined in Eq. (4), as follows.

Lemma 4. *Suppose that* $\|x\|_2 = 1$ *and* $\|x\|_\infty = v$*, then we have* $\|E_r(x)\|_d \leqslant T_{n,p,d}(v)$ *for any positive and even* d*, where we define*

$$T_{n,p,d}(v) \triangleq 4\left(\sum_{k=0}^{\frac{d}{2}} \binom{d}{2k} p^{\mathbb{I}(k>0)} v^{2k}(1-v^2)^{\frac{d-2k}{2}} \cdot \mathbb{E}(B'-B'')^{d-2k}\right)^{\frac{2}{d}} \tag{9}$$

and $B', B'' \sim^{i.i.d.} \frac{1}{\sqrt{n-1}} \cdot \mathrm{Binom}(n-1, \frac{1-\sqrt{1-2p}}{2})$.

To show this result, we combine Lemma 1 and Lemma 2. When explicitly evaluating $\|\sum_i x_i^* Y_i\|_d$, we thereby arrive at the expression given by Eq. (9).
The following theorem is the main result of our work.

Theorem 1 (Error Moments). *If* $\|x\|_2 = 1$ *and* $\|x\|_\infty = v$*, then for any positive even* d*, we have that*

$$\|E(x)\|_d \leqslant s^{-1} \cdot Q_{n,p,d}(v),$$

where $Q = Q_{n,p,d}(v)$ *solves the equation*

$$\sum_{k=0}^{\frac{d}{2}} \binom{d}{2k} (T_{n,p,2k}(v)/Q)^{2k} = \mathrm{e}^{\frac{d}{2m}} \tag{10}$$

and $T_{n,p,2k}$ *is as in Lemma 4 (with* d *replaced by* $2k$*).*

The detailed proof (see [31]) starts with $E(x) = \frac{1}{s}\sum_{r=1}^{m} E_r(x)$, applies Lemma 3 with $Z_r = E_r(x)$, and finally uses Lemma 4 (similarly to [12]). The subtle points of the proof are summarized below.

- *Correlation of* $E_r(x)$ *for different* r*:* fortunately (due to sampling without replacement), this is a *negative dependency.* Thus, the same moment bounds as for independent random variables can be applied also here [30].
- *Non-symmetric distribution of* $E_r(x)$ *:* we compare the moments of $E_r(x)$ with the moments of a random variable which is symmetric; this allows for applying moment bounds for the sums of symmetric random variables. We remark that this argument also fills a gap in [12].

Corollary 2 (Error Confidence). *For the error*

$$\epsilon = \mathrm{e}\, s^{-1} \cdot Q_{n,p,\lceil \log(1/\delta) \rceil}(v),$$

we have $\Pr[|E(x)| > \epsilon] \leqslant 1-\delta$ *and* (1) *holds.*

The corollary is a direct application of Markov's inequality $\Pr[|E(x)| > \epsilon] \leqslant (s^{-1}Q_{n,p,d}(v)/\epsilon)^d$.

2.5 Discussion

Remark 3 (Computational Efficiency). The time of evaluating the distortion ϵ in Corollary 2 is in

$$\mathrm{TIME} = O(\log^4(1/\delta)\log(m\log(1/\delta))). \quad (11)$$

This is because $T_{n,p,d}(v)$ can be evaluated with $O(d^3)$ operations, utilizing the combinatorial formulas for binomial moments [19]. In turn, $Q_{n,p,d}(v)$ inverts a monotone function which can be computed by bisection in $O(\log(m\,d))$ steps.

Remark 4 (Comparison to State-of-the-Art). The approach of [12] follows the same roadmap, but the critical steps in that work are estimated in a weaker way than in our approach, namely:

1. a weaker analogue of our Lemma 1 is used,
2. in place of our sharp Corollary 1, an overestimation of $\|\sum_i x_i Y_i\|_d$ is obtained,
3. bounds on $E_r(x)$ are assembled to bound $E(x)$ via a weaker variant of [22], which is further weaker than our Lemma 3.

Thus, our bounds are guaranteed to be tighter for all parameter regimes.

Remark 5 (Dependency on n). Although our dependency on n is only asymptotically bounded, we find that—interestingly—it indeed helps improving the bounds on real-world datasets and use-cases, as shown in the next section.

3 Empirical Evaluation

In this section, the present the detailed results of our experimental evaluation. We implemented the bound provided by Theorem 1 in Python 3.6 and tested it in the Google Colab environment using an Intel(R) Xeon(R) CPU @ 2.20 GHz and the default RAM configuration of 13 GB.

Best Bounds in Prior Works. To give a clear and fair comparison, we analyze the best constants in the previous asymptotic analysis [12]. The in-depth analysis gives the value of "optimistic" constants necessary to avoid breaking down the proof (while the actual constants are likely worse).

Remark 6 (Optimistic Constants in Prior Works). The bound of [12] uses the better of the following two lemmas (Lemmas D.1 and D.2, respectively):

1. $\|E_r(x)\|_d \leqslant 2\,C_1 \cdot \left(\sup_{1\leqslant t\leqslant \frac{d}{2}}\left[\frac{dv}{t}\left(\frac{p}{dv^2}\right)^{\frac{1}{2t}}\right]\right)^2$
2. $\|E_r(x)\|_d \leqslant 2\,C_2 \cdot \frac{d}{\log(1/p)}$,

where d is assumed positive and even.

Here, the extra factor of 2 appears as the effect of symmetrization (the random variable $E_r(x)$ must be dominated by a symmetric random variable to conclude the bound on $E(x)$). The best constants satisfy $C_1 \geqslant 4\mathrm{e}$ and $C_2 \geqslant 8$, as it is implied by the analysis of their proof technique.

3.1 Synthetic Benchmark

Setup. The key ingredient of our improvements is the sharper bound on the row-wise error contributions $E_r(x)$ from Lemma 4. In this experiment, we compare this bound (referred to as T_{new}) with its analogue from [12] with the "optimistic" constants as discussed in Remark 6 (referred to as T_{old}). Figures 1 and 2 illustrate the respective ratios of T_{new} and T_{old} with respect to the error contributions $E_r(x)$ for $n = 10^4$ and various ranges of d, v and $p = \frac{s}{m}$. Points with non-even d are interpolated.

Results. Our bounds are better by up to an order of magnitude across a wide range of parameters. Therefore, we should expect similar improvements for our bounds on the overall error $E(x)$ (recall that $E_r(x)$ are aggregated into $E(x)$ using Lemma 3).

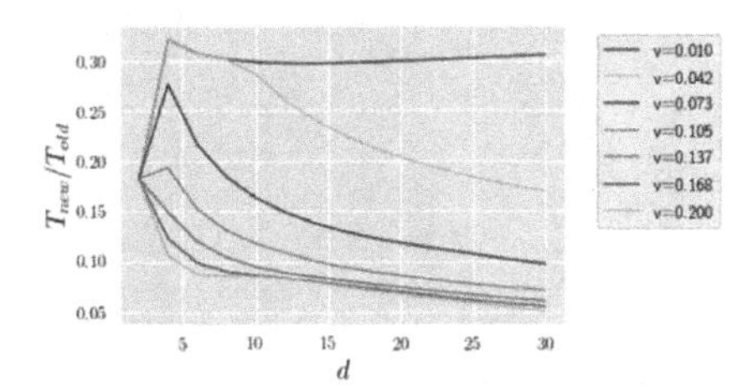

Fig. 1. T_{new}/T_{old} for $n = 10^4$, $p = 10^{-3}$

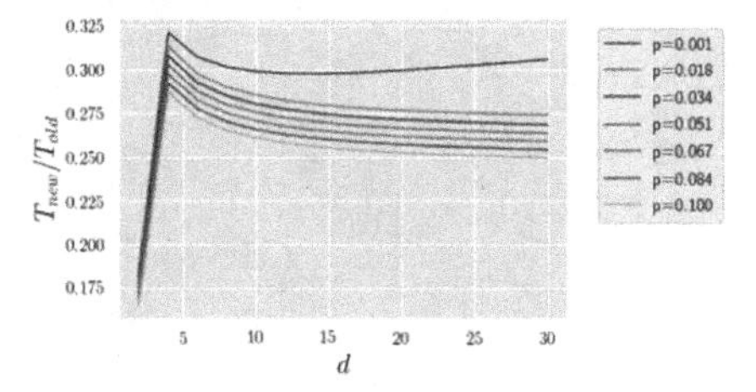

Fig. 2. T_{new}/T_{old} for $n = 10^4$, $v = 10^{-2}$

3.2 Real-World Datasets

Setup. We next consider various real-world datasets of different content types, sizes and numbers of features—as summarized in Table 1.

Table 1. Summary of real-world datasets used in our experiments

Dataset	Content	Comments
NIPS	Text	13,000 words
Word2Vec/Wiki	Text	5M lines/48M words of English Wikipedia articles
News20	Text	20,000 documents/34,000 words of English news [16]
MNIST	Images	60,000 images with 28 × 28 pixels [23]
CIFAR100	Images	60,000 images with 32 × 32 pixels [21]
SVHN	Images	600,000 images with 32 × 32 pixels [27]
Caltech101	Images	9,000 images with 300 × 200 pixels [7]
Cars	Images	16,000 images with 500 × 500 pixels [5]
Goodwin040	Fluid dynamics	18,000 columns/18,000 rows [4]
Mycieliskian17	Undir. graph	98,000 columns/98,000 rows [4]

Dispersion. Since sparsity s depends on the data-dependent dispersion v, results obtained in prior work may be of limited applicability in practice when v is not small. To understand the behavior of v, we evaluate its distribution on our datasets. We conclude that, indeed, the value of v may be quite large, even when n is big; in such cases, using a very small sparsity s is not theoretically justified. Density plots on Figs. 3, 4, 5, and 6 illustrate the distribution of the dispersion $v = \|x\|_\infty / \|x\|_2$ for vectors $x = x_1 - x_2$ over all pairs $x_1, x_2 \in \mathcal{X}$ from a subsample $\mathcal{X}$ of the dataset. Evaluating the dispersion on pairwise differences corresponds to the intended usage of random projections: preserving pairwise distances within a dataset. We used $|\mathcal{X}| = 250$, such that v is estimated based on $\approx 5 \cdot 10^4$ samples. We generally find that, for each dataset, v is sharply concentrated around a "typical" value, whose magnitude is data-dependent.

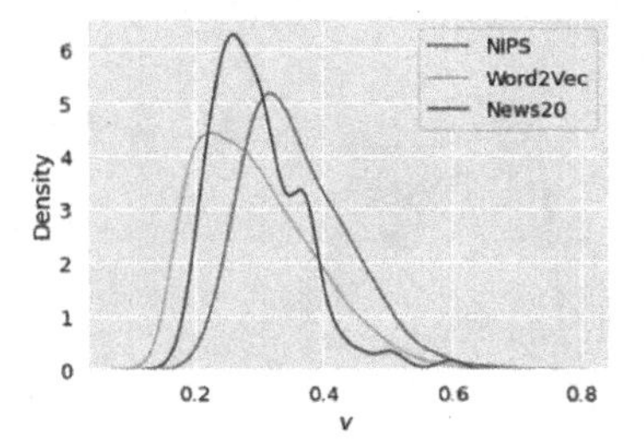

Fig. 3. Dispersion v on text data

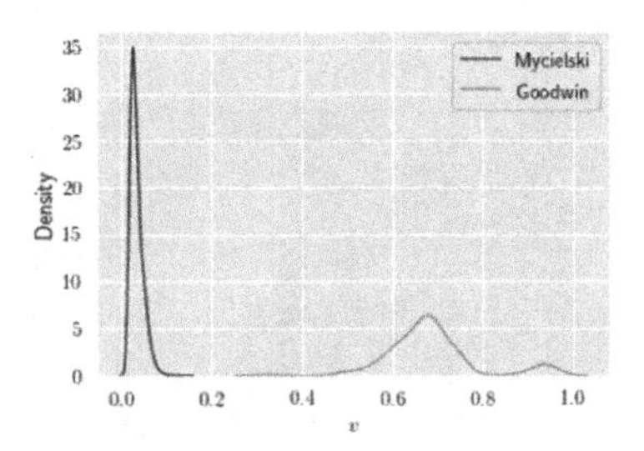

Fig. 4. Dispersion v on sparse-matrix

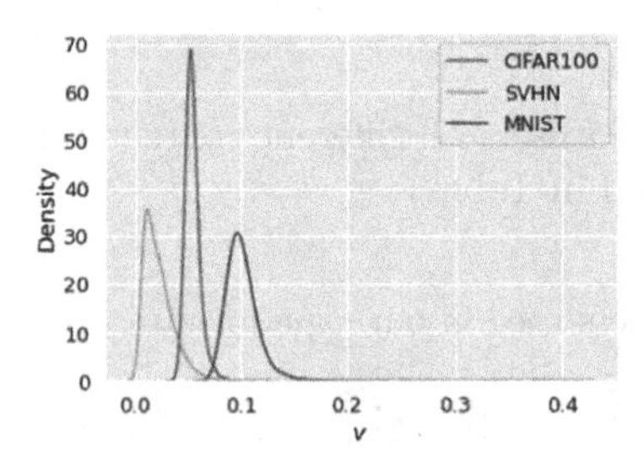

Fig. 5. Dispersion v on small images

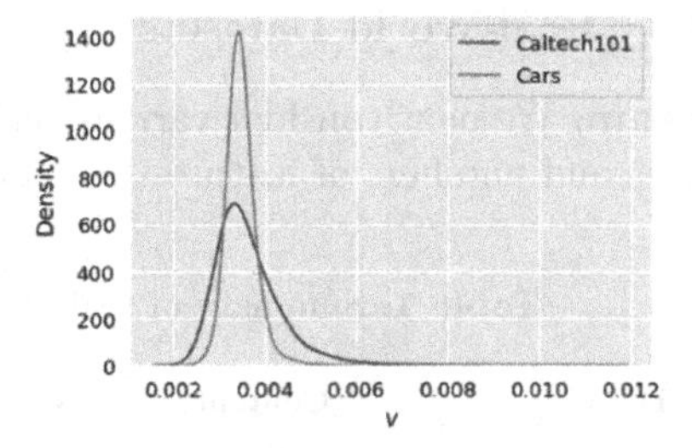

Fig. 6. Dispersion v on large images

Distortion. The next experiment analyzes the confidence $1 - \delta$ as a function of the distortion ϵ of our and previous bounds. We assume $\frac{m}{n} = 0.1$, $\frac{s}{m} = 0.01$. The dispersion v is chosen at the typical most likely value for each dataset. The confidence follows from Theorem 1 by Markov's inequality. The results are illustrated on Fig. 7. Our bounds produce very good results for all datasets with large n, thus outperforming the previous approach by several orders of magnitude in terms of confidence. Remarkably, we also obtain non-trivial bounds when n is small as opposed to the bounds from previous works.

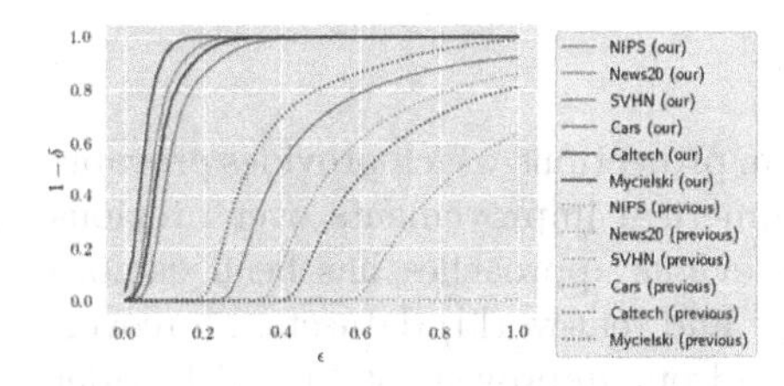

Fig. 7. Confidence $1 - \delta$ vs. distortion ϵ

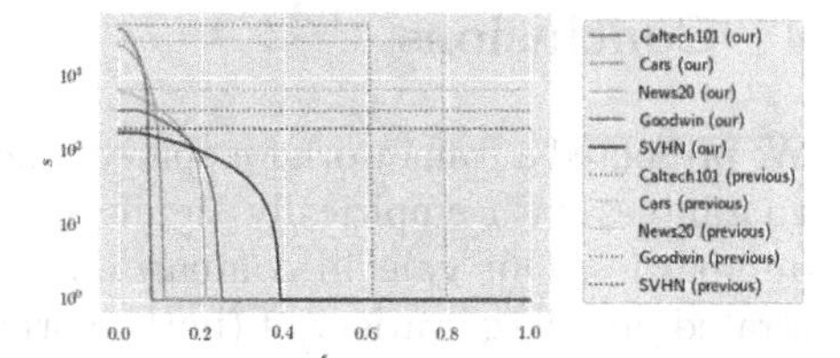

Fig. 8. Sparsity s vs. distortion ϵ

Sparsity. In this experiment, we evaluate the critical value of distortion ϵ, which allows for using non-trivial sparsity $s < m$ such that the confidence $1 - \delta$ is at least $\frac{3}{4}$. For each dataset, we choose as before its typical value v and fix the dimension reduction factor $\frac{m}{n} = 0.1$. The results are summarized in Fig. 8. Note that, for smaller values of ϵ, no $s < m$ can work, which produces flat segments $s = m$. Our bounds offer a non-trivial sparsity s for much smaller distortions, and quickly achieve $s = 1$.

Dimensionality. In the last experiment, we evaluate the minimal non-trivial dimension m. We again consider a fixed sparsity of $\frac{s}{m} = 0.1$ and choose the typical dispersion v for each dataset. Then, for various values of ϵ, we compute the smallest m which still yields a confidence of $1-\delta$ of $\frac{3}{4}$. The results, illustrated in Fig. 9, show that our bounds are better by 10 times or more.

Multiple Data Points. So far the experiments covered the performance on one input vector at a time only; the case of multiple data points reduces to the former one by scaling the confidence accordingly (union bound), where we again compute the smallest m which still yields a confidence $1 - \delta$ of $\frac{3}{4}$ over all points. The result shows the expected logarithmic dependency of the dimensionality m with respect to the data size, as shown in Fig. 10.

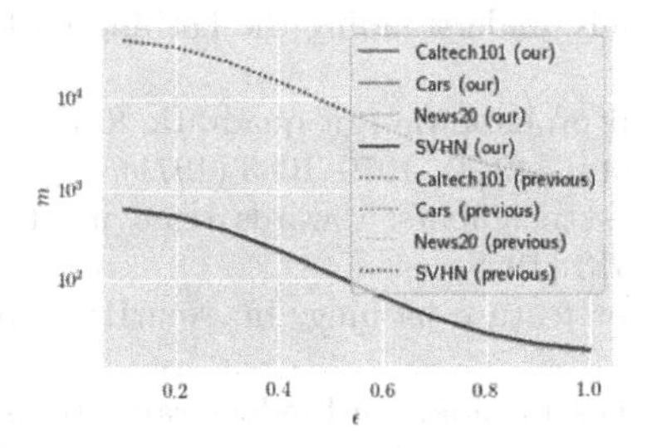

Fig. 9. Dimensionality m vs. distortion

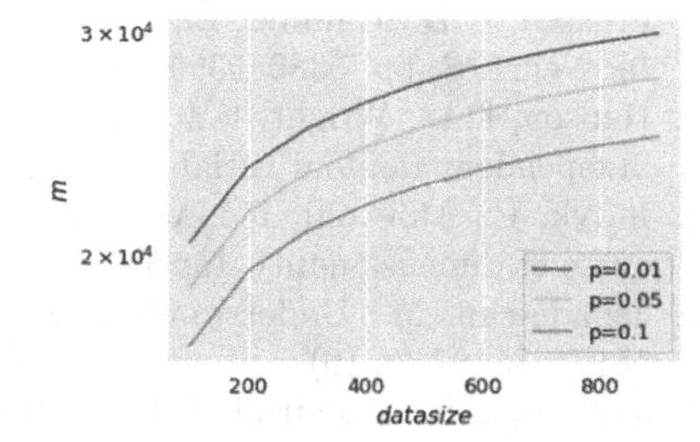

Fig. 10. Dimensionality m vs. data size

4 Conclusions

We presented a framework for sparse random projections which provides provable guarantees with empirically significant numerical improvements over previous approaches. Our gain in comparison to previous approaches has been demonstrated on a large variety of (both synthetic and real-world) datasets. Moreover, we believe that the novel inequalities behind our improvements are of broader interest for a variety of statistical-inference applications.

Acknowledgements. We thank the NVIDIA AI Technology Center (NVAITC) for the fruitful discussion.

References

1. Ailon, N., Chazelle, B.: Approximate nearest neighbors and the fast Johnson-Lindenstrauss transform. In: STOC, pp. 557–563 (2006)
2. Bingham, E., Mannila, H.: Random projection in dimensionality reduction: applications to image and text data. In: SIGKDD, pp. 245–250 (2001)
3. Boucheron, S., Bousquet, O., Lugosi, G., Massart, P., et al.: Moment inequalities for functions of independent random variables. Ann. Probab. **33**(2), 514–560 (2005)
4. Davis, T.A., Hu, Y.: The University of Florida sparse matrix collection. ACM Trans. Math. Softw. **38**(1), 1–25 (2011)
5. Deng, J., Krause, J., Fei-Fei, L.: Fine-Grained Crowdsourcing for Fine-Grained Recognition. In: CVPR, pp. 580–587 (2013)
6. Eaton, M.L.: A note on symmetric Bernoulli random variables. Ann. Math. Stat. **41**(4), 1223–1226 (1970)
7. Fei-Fei, L. Fergus, R., Perona, P.: Learning generative visual models from few training examples: an incremental Bayesian approach tested on 101 object categories. In: CVPR, pp. 178–178 (2004)
8. Fortuin, C.M., Kasteleyn, P.W., Ginibre, J.: Correlation inequalities on some partially ordered sets. Commun. Math. Phys. **22**(2), 89–103 (1971)
9. Freksen, C.B., Kamma, L., Larsen, K.G.: Fully understanding the hashing trick. In: NeurIPS, pp. 5389–5399 (2018)
10. Hanson, D.L., Wright, F.T.: A bound on tail probabilities for quadratic forms in independent random variables. Ann. Math. Stat. **42**(3), 1079–1083 (1971)
11. Indyk, P., Motwani, R.: Approximate nearest neighbors: towards removing the curse of dimensionality. In: STOC, pp. 604–613 (1998)
12. Jagadeesan, M.: Understanding sparse JL for feature hashing. In: NeurIPS, pp. 15177–15187 (2019)
13. Jayram, T.S., Woodruff, D.P.: Optimal bounds for Johnson-Lindenstrauss transforms and streaming problems with subconstant error. Trans. Algor. **9**(3), 1–17 (2013)
14. Johnson, W.B., Lindenstrauss, J.: Extensions of Lipschitz mappings into a Hilbert space. Contemp. Math. **26**(1), 189–206 (1984)
15. Kane, D.M., Nelson, J.: Sparser Johnson-Lindenstrauss transforms. J. ACM **61**(1), 1–23 (2014)
16. Ken, L.: Newsweeder: learning to filter netnews. In: ICML, pp. 331–339 (1995)
17. Khintchine, A.: Über dyadische Brüche. Math. Zeitsch. **18**(1), 109–116 (1923)

18. Klartag, B., Mendelson, S.: Empirical processes and random projections. J. Funct. Anal. **225**(1), 229–245 (2005)
19. Knoblauch, A.: Closed-form expressions for the moments of the binomial probability distribution. SIAM J. Appl. Math. **69**(1), 197–204 (2008)
20. Kolesko, K., Latała, R.: Moment estimates for Chaoses generated by symmetric random variables with logarithmically convex tails. Statist. Prob. Lett. **107**, 210–214 (2015)
21. Krizhevsky, A.: Learning multiple layers of features from tiny images. University of Toronto (Technical Report) (2012)
22. Latała, R., et al.: Estimation of moments of sums of independent real random variables. Ann. Prob. **25**(3), 1502–1513 (1997)
23. LeCun, Y., Cortes, C.: MNIST handwritten digit database (2010)
24. Li, P., Hastie, T.J., Church, K.W.: Very sparse random projections. In: SIGKDD, pp. 287–296 (2006)
25. Lugosi, G., Mendelson, S.: Mean estimation and regression under heavy-tailed distributions: a survey. FoCM **19**(5), 1145–1190 (2019)
26. Makarychev, K., Makarychev, Y., Razenshteyn, I.: Performance of Johnson-Lindenstrauss transform for k-means and k-medians clustering. In: STOC, pp. 1027–1038 (2019)
27. Netzer, Y., Wang, T., Coates, A., Bissacco, A., Wu, B., Ng, A.Y.: Reading Digits in Natural Images with Unsupervised Feature Learning (2011)
28. de la Peña, V.H., Montgomery-Smith, S.J.: Decoupling inequalities for the tail probabilities of multivariate U-statistics. Ann. Prob. 806–816 (1995)
29. Roventa, I.: A note on Schur-concave functions. J. Inequal. Appl. **2012**, 159 (2012)
30. Shao, Q.: A comparison theorem on moment inequalities between negatively associated and independent random variables. J. Theor. Probab. **13**(2), 343–356 (2000)
31. Skorski, M., Temperoni, A., Theobald, M.: Robust and provable guarantees for sparse random embeddings. arXiv:2202.10815v1 (2022)
32. Stanica, P.: Good lower and upper bounds on binomial coefficients. JIPAM **2**(3), 30 (2001)
33. Suresh, V., Qiushi, W.: The Johnson-Lindenstrauss transform: an empirical study. In: ALENEX, pp. 164–173 (2011)
34. Zhou, S.: Sparse Hanson-Wright inequalities for subgaussian quadratic forms. Bernoulli **25**(3), 1603–1639 (2019)

Transferable Interpolated Adversarial Attack with Random-Layer Mixup

Size Ma, Keji Han, Xianzhong Long, and Yun Li(✉)

Jiangsu Key Laboratory of Big Data Security and Intelligent Processing, Nanjing University of Posts and Telecommunications, Nanjing, China
liyun@njupt.edu.cn

Abstract. Deep neural networks have shown their vulnerabilities to adversarial examples crafted by adding imperceptible perturbations to original examples. Despite showing powerful attack strength under the white-box setting, most existing adversarial attack methods can only mislead the black-box model with low attack success rates. In response, a class of image transformation-based attacks has been proposed. Its main idea is to apply transformations to adversarial examples during attack iterations and improve the transferability on the black-box model. However, a major limitation of these transformation-based attacks is that they only apply transformations to input images, while ignoring transformations' usages in hidden representations. Based on our observation that mixup in hidden space can help attack methods achieve higher transferability than in input space, we propose the Random-Layer Mixup Attack Method (RLMAM). Our method interpolates the adversarial examples with clean examples in both input space and hidden space. The interpolated adversarial representations induced by our random-layer mixup can improve representations' diversity in both two spaces and alleviate adversarial examples' overfitting phenomenon on the white-box model. Furthermore, we incorporate RLMAM with our enhanced momentum method. Experimental results on ImageNet and CIFAR-10 datasets demonstrate that our RLMAM outperforms other state-of-the-art black-box attacks.

Keywords: Black-box adversarial attack · Mixup · Manifold mixup

1 Introduction

Deep neural networks (DNNs) have achieved great performance in computer vision [4]. However, recent research [13] has found that DNNs are vulnerable to adversarial examples which are crafted by adding human-imperceptible perturbations to original examples. It raised concern about the security of deep learning, thus it's imperative to investigate how adversarial examples are generated. Besides, the research of adversarial attacks not only helps to evaluate the robustness of DNNs [10], but also makes deep learning more interpretable [6].

Several adversarial attack methods have been proposed to generate adversarial examples such as FGSM [3], DeepFool [11] and C&W Attack [1]. Most of these attack methods are **white-box attacks** which aim to mislead the white-box model with full

J. Gama et al. (Eds.): PAKDD 2022, LNAI 13281, pp. 224–235, 2022.
https://doi.org/10.1007/978-3-031-05936-0_18

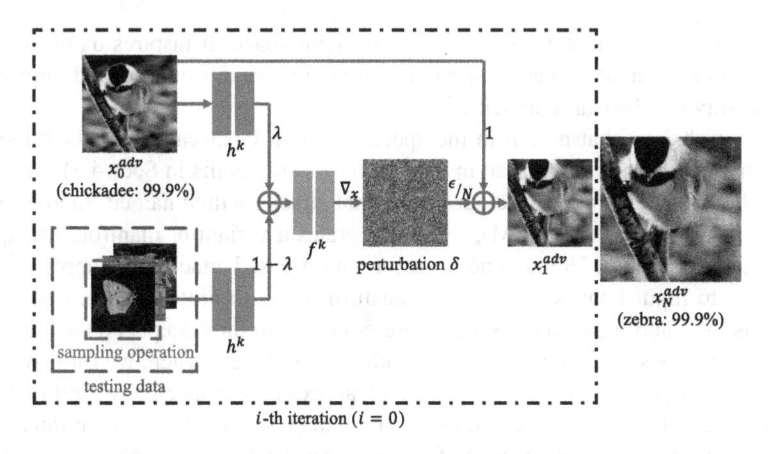

Fig. 1. Given the classifier $f(x) = f^k(h^k(x))$, the general framework of our proposed interpolated adversarial attack.

knowledge of the model's structure and parameters. However, the model's information might be invisible to adversaries in real-world applications [5]. It aroused interest in the research of **black-box attacks** [8] about how to fool the black-box model with limited knowledge. Black-box attacks mainly focus on the transferability, namely whether adversarial examples generated from the white-box model can fool the black-box model. Transferability of adversarial examples doesn't mean the powerful attack strength under the white-box setting, and it rather emphasizes the generalization ability over different models. For example, iterative adversarial attacks such as PGD [10] and I-FGSM [7] can mislead the white-box model with high confidence, but they exhibit low attack success rates under the black-box setting due to the fact that the adversarial examples generated by iterative attacks tend to overfit the white-box model. In contrast, single-step attacks have relatively higher transferability, but they are demonstrated to be less powerful methods than iterative attacks.

To improve the transferability of adversarial attacks, [2] proposed momentum iterative gradient-based methods (MI-FGSM). Unlike traditional iterative attacks using the gradient of current iteration as the update direction, MI-FGSM introduces the momentum term to accumulate the gradients of previous iterations. It can provide stable update directions and escape from local maxima. However, the adversarial examples generated by MI-FGSM still suffer from the overfitting phenomenon, since the gradients of adversarial examples are highly similar during iterations. To achieve better transferability, [9,17,18] proposed a class of transformation based attack methods which combine MI-FGSM with image transformation operations (resize and pad, scale, mixup variant). Applying random transformations to the inputs before propagation at each iteration can alleviate overfitting, as it feeds the white-box model with the randomly transformed inputs and thus increases the input space complexity.

Overall, the transferability of gradient-based attacks has a close connection with the overfitting phenomenon on the white-box model, and such a phenomenon is largely due to the absence of diverse gradient information. Currently, transformation-based attack methods focus on the transform operations in input space. It inspires us to explore if transformations in hidden space can provide more diverse information and improve the transferability of adversarial attacks.

We first observe that mixup in the specific hidden layer can help attack methods achieve higher transferability than in input layer (more details in Sect. 4.3). Motivated by this phenomenon, we propose our adversarial attack method named Random-Layer Mixup Attack Method (RLMAM), which integrates a variant of manifold mixup [16] into adversarial attacks. Unlike other transformation-based attacks only applying transformations to input images, we extend transformations to hidden space. Our method constructs the interpolated adversarial representations in three steps. At each iteration, we first randomly select a layer index (including input layer). Then, we forward propagate the adversarial example and a random clean example respectively until that layer. Last, we interpolate such two examples' intermediate outputs to form the interpolated representation. From the perspective of diversity, our RLMAM is superior to traditional iterative attacks, since our interpolated representations can directly improve the global diversity in both input space and hidden space.

In addition, considering that single-step attacks exhibit higher transfer rates than iterative attacks, we prove that updating perturbations in a granular manner can improve the performance under the white-box setting, but it does harm to the transferability. Thus, we propose Inner-outer Iterations. Our Inner-outer Iterations accumulate gradients at inner iterations and update perturbations at outer iterations, which avoid updating perturbation at each iteration.

Our contributions can be concluded as:

- We propose the random-layer mixup attack method, which integrates a variant of manifold mixup into adversarial attacks. Our method can improve the data diversity in input and hidden spaces, and thus provide more diverse gradient information.
- To further improve the transferability, we propose Inner-outer Iterations to replace the single-loop iterations applied in traditional momentum methods.
- Experimental results in Sect. 4 show that the attack success rate of our method outperforms other methods. It demonstrates that our method can significantly alleviate adversarial examples' overfitting phenomenon on the white-box model.

2 Related Work

In this section, we first introduce the notations related to our work, then we briefly review the input/manifold mixup training principles and several adversarial attack methods. Let x and y^{true} denote an original example and its ground-truth label. Given a classifier $f(x) : x \in \mathcal{X} \rightarrow y \in \mathcal{Y}$, adversarial attacks can be divided into two classes: non-targeted attacks and targeted attacks. The goal of non-targeted attacks is to generate the adversarial example $x^{adv} = x + \delta$ by maximizing the loss function $L(x^{adv}, y^{true}; f)$ such that the prediction of the classifier $f(x^{adv}) \neq y^{true}$. Here $L(\cdot)$ is the classification loss such as cross entropy. Meanwhile, targeted attacks aim to mislead the classifier as $f(x^{adv}) = y'$, where y' is the target class, and $y' \neq y^{true}$. In our work, the perturbation δ is under the ℓ_∞ norm bounded constraint, i.e., $\|\delta\|_\infty \leq \epsilon$, to ensure the perturbation is human-imperceptible.

2.1 Mixup

Empirical risk minimization (ERM) [15] is the fundamental principle in machine learning, which aims to minimize the empirical error on the training dataset. However, research [19] found that the model trained with ERM shows weak generalization when testing distributions slightly differ from the training data. As an alternative, [20] proposed the mixup training principle. For two clean examples (x, y^{true}) and (x^*, y^*) randomly sampled from the training set, mixup constructs virtual training examples as $(\text{Mix}_\lambda(x, x^*), \text{Mix}_\lambda(y^{true}, y^*))$, where $\text{Mix}_\lambda(a, b) = \lambda \cdot a + (1 - \lambda) \cdot b$. Here $\lambda \sim \text{Beta}(\alpha, \alpha)$, and α is the hyperparameter.The loss function of the mixup training method can be defined as

$$L^{\text{Mix}}(x, y^{true}, x^*, y^*; f) = L(\text{Mix}_\lambda(x, x^*), \text{Mix}_\lambda(y^{true}, y^*); f). \quad (1)$$

Mixup can be seen as a data augmentation method, since the virtual examples induce the global linear behavior in-between data manifolds and increase the complexity of training set.

Based on the above input mixup principle, [16] proposed the manifold mixup training principle. For the classifier $f(x) = f^k(h^k(x))$, h^k denotes the prefix part of classifier mapping the input data to the hidden representation at layer k, and f^k denotes the rest part mapping the hidden representation to the output $f(x)$. Specifically, manifold mixup will degrade to input mixup if $k = 0$. The loss function of manifold mixup can be defined as

$$L^{\text{MMix}}(x, y^{true}, x^*, y^*; f) = L(f^k(\text{Mix}_\lambda(h^k(x), h^k(x^*))), \text{Mix}_\lambda(y^{true}, y^*)), \quad (2)$$

since it interpolates two examples in either input space or hidden space during training iterations.

2.2 Adversarial Attack Methods

In this subsection, we provide a brief introduction of the family of fast gradient sign methods.

Fast Gradient Sign Method (FGSM). [3] proposed FGSM as an single-step attack method, which generates an adversarial example x^{adv} along the direction of the loss gradient $\nabla_x L(x, y^{true}; f)$ as

$$x^{adv} = x + \epsilon \cdot \text{sign}(\nabla_x L(x, y^{true}; f)). \quad (3)$$

Iterative Fast Gradient Sign Method (I-FGSM). [7] extended FGSM to an iterative version:

$$\begin{aligned} x_0^{adv} &= x, \\ x_{i+1}^{adv} &= \text{Clip}_x^\epsilon\{x_i^{adv} + \alpha \cdot \text{sign}(\nabla_x L(x_i^{adv}, y^{true}; f))\}, \end{aligned} \quad (4)$$

where Clip_x^ϵ represents that the adversarial example x_{i+1}^{adv} is clipped within the ϵ-ball of the original example x at i-th iteration.

Momentum Iterative Fast Gradient Sign Method (MI-FGSM). MI-FGSM [2] introduces a momentum variable g to accumulate the gradients of previous iterations with a decay factor μ. MI-FGSM improves the transferability under the black-box setting since the gradients' accumulation helps to stabilize update directions and escape from poor local maxima. The updating process can be formulated as

$$\begin{aligned} x_0^{adv} &= x,\ g_0 = 0, \\ g_{i+1} &= \mu \cdot g_i + \frac{\nabla_x L(x_i^{adv}, y^{true}; f)}{\|\nabla_x L(x_i^{adv}, y^{true}; f)\|_1}, \\ x_{i+1}^{adv} &= \mathrm{Clip}_x^{\epsilon}\{x_i^{adv} + \alpha \cdot \mathrm{sign}(g_{i+1})\}. \end{aligned} \tag{5}$$

Momentum Diverse Inputs Iterative Fast Gradient Sign Method (DIM). To further improve the transferability of MI-FGSM, [18] proposed DIM by applying random resize and pad operations to x^{adv} with a given probability p.

Scale-Invariant Nesterov Iterative Fast Gradient Sign Method (SIM). SIM [9] is the combination of two enhanced methods: scale-invariant property and Nesterov accelerated gradient method. Scale-invariance property constructs the scale copies of the input image during iterations, while Nesterov accelerated gradient method substitutes the traditional momentum method.

Admix Attack Method (AAM). AAM [17] proposed to integrate a variant of input mixup with MI-FGSM. For each attack iteration, AAM first randomly sample a clean example x^* and construct the admixed adversarial example $\tilde{x} = \gamma \cdot x^{adv} + \eta' \cdot x^*$, where $\gamma \in [0, 1]$ and $\eta' \in [0, \gamma)$. Then, AAM calculates the gradient on the admixed adversarial example as follows:

$$g_{i+1} = \mu \cdot g_i + \frac{\nabla_x L(\gamma_i \cdot x_i^{adv} + \eta' \cdot x_i^*, y^{true}; f)}{\|\nabla_x L(\gamma_i \cdot x_i^{adv} + \eta' \cdot x_i^*, y^{true}; f)\|_1}. \tag{6}$$

3 Methodology

In this section, we present an illustration of our proposed RLMAM in detail. We illustrate our method in two steps: first we only introduce our random-layer mixup method without momentum, then we show how to combine our method with momentum to further improve the transferability. We follow some notations of manifold mixup in Sect. 2.1.

3.1 Random-Layer Mixup Attack Method Without Momentum

The key idea of our method is about constructing interpolated adversarial representations by taking random-layer interpolation of adversarial examples x^{adv} and clean examples x^* during iterations. Thus, we consider sampling clean examples from all classes except the ground-truth label. In other words, clean examples (x^*, y^*) are randomly sampled from the distribution $p_{\backslash y^{true}}(x, y)$, where $p_{\backslash y^{true}}(x, y)$ is a uniform distribution on the set $\{(x, y) \in (X, Y) | y \neq y^{true}\}$. Here, (X, Y) denote the set of all testing data.

As shown in Fig. 1, the implementation of our proposed RLMAM includes five steps. At each iteration, we first randomly select a layer index k from an eligible set S (more details in Sect. 4.3) and sample a clean example x^*. Then, we feed the adversarial example x^{adv} and the clean example x^* respectively into h^k to get the two examples' intermediate outputs: $h^k(x^{adv})$ and $h^k(x^*)$. Third, we take the interpolation of such two intermediate outputs to form the interpolated adversarial representation $\text{Mix}_\lambda(h^k(x^{adv}), h^k(x^*))$. Unlike the manifold mixup, here λ is a fixed hyperparameter (more details in Sect. 4.3). Fourth, we continue to feed $\text{Mix}_\lambda(h^k(x^{adv}), h^k(x^*)))$ into the rest part of white-box model f^k, and obtain the corresponding gradient. Fifth, we update the perturbation with the gradient. The loss function of our proposed random-layer mixup adversarial attack can be defined as

$$L^{\text{RLM}}(x^{adv}, y^{true}, x^*; f) = L(f^k(\text{Mix}_\lambda(h^k(x^{adv}), h^k(x^*))), y^{true}). \tag{7}$$

Compared with the loss function of two mixup training methods in Eqs. 1 and 2, here we don't require the label y^* of clean examples, since we just aim to minimize the probability of y^{true} for the objective of non-targeted attacks.

The general procedure of RLMAM without momentum is similar to I-FGSM in Eq. 4. It can be reformulated as

$$\begin{gathered} x_0^{adv} = x, \\ x_{i+1}^{adv} = \text{Clip}_x^\epsilon\{x_i^{adv} + \alpha \cdot \text{sign}(\nabla_x L^{\text{RLM}}(x_i^{adv}, y^{true}, x_i^*; f))\}. \end{gathered} \tag{8}$$

3.2 Random-Layer Mixup Attack Method

As discussed above, we have explored how to integrate random-layer mixup into the framework of adversarial attacks. Besides, we can incorporate momentum methods with our proposed random-layer mixup attack to further improve the transferability. We propose two momentum interpolated attack methods: RLMAM with **Single-loop Iterations** (RLMAM_{SI}) and RLMAM with **Inner-outer Iterations** (RLMAM_{II}).

RLMAM_{SI}. RLMAM_{SI} is the simple combination of our proposed random-layer mixup attack and traditional momentum methods with the iteration number K.

RLMAM_{II}. Unlike traditional momentum methods, such as MI-FGSM accumulating the gradients and updating the adversarial examples simultaneously in single-loop iterations, RLMAM_{II} divides the single-loop iterations into inner iterations and outer iterations. We use M, N to denote the number of inner and outer iterations ($K = M \times N$).

Algorithm 1. RLMAM$_{\text{II}}$

Input: f: the white-box model; S: the index set; x: the input; y^{true}: the ground-truth label of x.
Output: x^{adv}: the adversarial example.

1: $x_0^{adv} = x$;
2: $g_0 = 0$; $\tau_0 = 0$;
3: **for** $i = 0$ to $N - 1$ **do**
4: $g_0 = 0$;
5: **for** $j = 0$ to $M - 1$ **do**
6: Randomly select a layer index k from the set S;
7: Randomly sample a clean example x_j^* from the distribution $p_{\setminus y^{true}}(x, y)$;
8: Feed x_i^{adv}, x_j^* into h^k to construct representation $\text{Mix}_\lambda(h^k(x_i^{adv}), h^k(x_j^*))$;
9: Feed the representation into f^k and obtain the gradient $\nabla_x L^{\text{RLM}}$;
10: Update g_{j+1} with $\nabla_x L^{\text{RLM}}$ as Eq. 9;
11: **end for**
12: Add the accumulated g_M to τ_{i+1} as Eq. 10;
13: Update x_{i+1}^{adv} with τ_{i+1} as Eq. 11;
14: **end for**
15: **return** $x^{adv} = x_N^{adv}$.

On one hand, the Inner-outer Iterations method accumulates previous gradients at the j-th inner iteration:

$$g_{j+1} = \mu \cdot g_j + \frac{\nabla_x L^{\text{RLM}}(x_i^{adv}, y^{true}, x_j^*; f)}{\|\nabla_x L^{\text{RLM}}(x_i^{adv}, y^{true}, x_j^*; f)\|_1}, \tag{9}$$

where g is the momentum term proposed in MI-FGSM and μ is the decay factor of inner iterations. On the other hand, Inner-outer Iterations introduces an extra momentum term τ to accumulate the accumulated gradients. After inner iterations finished, we add the accumulated g_M to τ and update the perturbation at the i-th outer iteration:

$$\tau_{i+1} = \eta \cdot \tau_i + g_M, \tag{10}$$

$$x_{i+1}^{adv} = \text{Clip}_x^\epsilon\{x_i^{adv} + \alpha \cdot \text{sign}(\tau_{i+1})\}, \tag{11}$$

where η is the decay factor of outer iterations. The adversarial example x_i^{adv} is perturbed along the direction of the sign of τ_{i+1}. Note that if $M = 1$, RLMAM$_{\text{II}}$ degrades to RLMAM$_{\text{SI}}$.

In summary, we incorporate our random-layer mixup method RLMAM with two momentum methods (II and SI). We report RLMAM$_{\text{II}}$ in Algorithm 1 as an example. Compared with input/manifold mixup, our method sets λ as a fixed hyperparameter and modifies the loss function. Compared with other transformation based methods like AAM, we additionally consider applying transformations in hidden space.

Table 1. The attack success rates (%) of adversarial attacks against three different models on ILSVRC. The adversarial examples are generated from Res-50, VGG-16 and Google. * indicates white-box attacks.

Model	Attack	Res-50	VGG-16	Google
Res-50	I-FGSM	99.1*	24.2	12.6
	MI-FGSM	99.6*	53.3	36.5
	DIM	99.5*	54.0	37.5
	SIM	99.6*	62.1	52.3
	AAM	**100.0***	71.9	55.5
	RLMAM$_{SI}$ **(Ours)**	**100.0***	**77.2**	**62.1**
VGG-16	I-FGSM	10.9	98.2*	5.9
	MI-FGSM	29.4	99.4*	18.9
	DIM	29.0	99.5*	18.7
	SIM	43.7	**100.0***	32.6
	AAM	43.8	**100.0***	30.5
	RLMAM$_{SI}$ **(Ours)**	**45.0**	**100.0***	**33.4**
Google	I-FGSM	8.8	9.2	98.8*
	MI-FGSM	26.5	27.1	99.6*
	DIM	29.4	30.3	99.7*
	SIM	44.7	43.1	**100.0***
	AAM	49.0	50.1	**100.0***
	RLMAM$_{SI}$ **(Ours)**	**54.4**	**56.3**	**100.0***

4 Experiments

4.1 Experiment Setup

Datasets. We evaluate our method on two datasets: ILSVRC 2012 and CIFAR-10. For ILSCRC, we randomly select 5000 images from it as our testing set. For CIFAR-10, we use the whole dataset as our testing set. We conduct our experiments on ILSVRC by default.

Models. For ILSVRC, we consider three normally trained models: ResNet-50 (Res-50) [4], VGG-16 [12] and GoogleNet (Google) [14]. For CIFAR-10, we additionally consider their adversarially trained versions: Res-50 Adv, VGG-16 Adv and Google Adv.

Implementation Details. In our work, we set the maximum perturbation $\epsilon = 8/255$, number of iterations $K = 16$. For all momentum based methods, we follow the default setting in [2] with the decay factor $\mu = 1$. For our RLMAM, the mixup ratio λ is set to 0.7 according to the results in Fig. 2. For our RLMAM$_{II}$, the inner-outer iterations (M, N) are set to (4, 4) to make sure $K = M \times N$, and the decay factor η is set to 0.8 according to the results in Figs. 4 and 5.

Table 2. The attack success rates (%) of adversarial attacks against three normally trained models and their adversarially trained models on CIFAR-10. The adversarial examples are generated from Res-50, VGG-16 and Google. * indicates white-box attacks.

Model	Attack	Res-50	Res-50 Adv	VGG-16	VGG-16 Adv	Google	Google Adv
Res-50	I-FGSM	98.7*	16.9	51.0	22.5	62.7	17.7
	MI-FGSM	99.2*	17.8	70.7	23.3	73.6	19.0
	DIM	99.5*	17.8	58.5	23.5	63.2	19.0
	SIM	99.6*	18.2	72.4	23.7	76.0	19.4
	AAM	99.9*	18.8	82.1	24.2	85.2	20.3
	RLMAM_{SI}(**Ours**)	**100.0***	**20.1**	**87.9**	**25.0**	**89.2**	**21.1**
VGG-16	I-FGSM	64.1	17.2	98.7*	22.6	70.0	18.0
	MI-FGSM	77.8	18.1	99.0*	23.5	78.9	19.2
	DIM	63.6	18.0	98.8*	23.5	65.8	19.2
	SIM	77.6	18.2	99.4*	23.8	80.0	19.7
	AAM	84.8	19.1	99.5*	24.6	87.9	20.5
	RLMAM_{SI}(**Ours**)	**87.2**	**20.6**	**99.7***	**26.4**	**91.2**	**21.3**
Google	I-FGSM	44.7	16.7	40.4	22.0	98.9*	17.3
	MI-FGSM	62.8	17.2	61.0	22.8	99.1*	18.3
	DIM	53.4	17.3	52.4	22.8	99.2*	18.3
	SIM	71.0	17.9	70.1	23.4	99.5*	19.1
	AAM	77.7	18.0	78.2	23.6	99.7*	19.3
	RLMAM_{SI}(**Ours**)	**82.4**	**19.8**	**83.5**	**24.3**	**100.0***	**19.9**

4.2 Attacking Performance

We compare our RLMAM_{SI} with other methods including I-FGSM, MI-FGSM, DIM, SIM and AAM on both ILSVRC and CIFAR-10 dataset. We choose RLMAM_{SI} instead of RLMAM_{II} for fair comparison and highlighting our random-layer mixup method. We also conduct experiments to compare RLMAM_{SI} and RLMAM_{II} in Sect. 4.4.

Table 1 shows the attack success rates of different attack methods against normally trained models on ILSVRC. The success rates are the rates of adversarial examples generated from the white-box model successfully mislead the black-box model. We can observe that our method outperforms other methods with a large margin. For example, adversarial examples generated from Res-50 by our method achieve the success rates of 77.2% against VGG-16 and 62.1% against Google. In contrast, AAM achieves the corresponding success rates of 71.9% and 55.5%. Besides, we find an interesting phenomenon that in black-box attacks, adversarial examples generated from Res-50 achieve higher success rates than generated from other two models.

We also report the attack success rates of different attack methods on CIFAR-10 in Table 2. It can be observed that our method achieves higher success rates than other methods against both normally and adversarially trained models. For example, adversarial examples generated from Res-50 by our method achieve the success rates of 87.9% against VGG-16 and 25.0% against VGG-16 Adv. In contrast, AAM achieves the corresponding success rates of 82.1% and 24.2%.

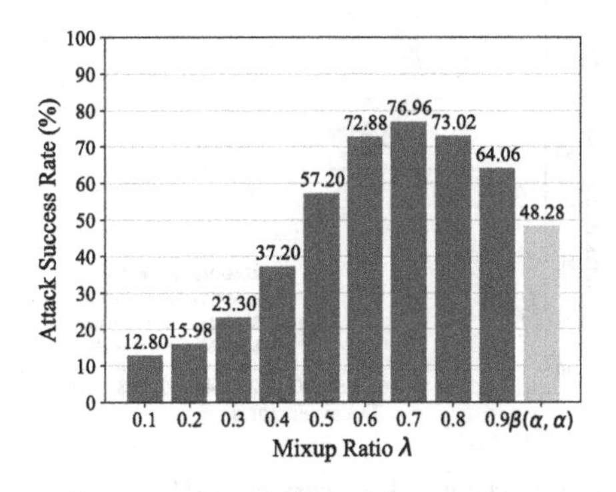

Fig. 2. The attack success rates (%) of our $\mathrm{RLMAM_{II}}$ against VGG-16. The adversarial examples are generated from Res-50 with λ ranging from 0.1 to 0.9 and $\lambda \sim \mathrm{Beta}(\alpha, \alpha)$.

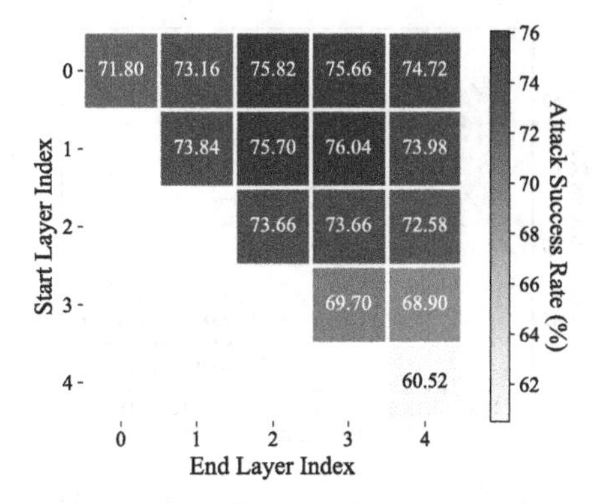

Fig. 3. The attack success rates (%) of our $\mathrm{RLMAM_{II}}$ against VGG-16. The adversarial examples are generated from Res-50 with index k selecting from different intervals.

4.3 Analysis of Random-Layer Mixup

We analyze our random-layer mixup from two aspects: the reason why we set mixup ratio $\lambda = 0.7$ as a fixed hyperparameter and our investigation of random layer selection.

For the first problem, recall that in input/manifold mixup training principles, λ is sampled from a beta distribution, since two clean examples (x, y^{true}) and (x^*, y^*) are equivalent. However, in our attack method, we only consider generating transferable adversarial examples x^{adv}, and we use x^* as a data augmentation to increase the input/hidden space complexity. It means that the two examples x^{adv} and x^* are inequivalent in our method. As shown in Fig. 2, we can see that with λ ranging from 0.1 to 0.9, $\mathrm{RLMAM_{II}}$ achieves the best attack success rate of 76% when $\lambda = 0.7$, and it only achieves 48% when $\lambda \sim \beta(\alpha, \alpha)$.

For the second problem, recall that our method selects a random layer index k from an eligible set S at each iteration. We further assume that S is a continuous interval. In Fig. 3, we report the attack success rates of our $\mathrm{RLMAM_{II}}$ selecting indexes from different intervals. The y and x-axises denote the start and end index of interval. There are five indexes including input layer, since we regard each bottleneck of Res-50 as a layer. The main diagonal shows the results of selecting indexes from the specific layer only. We can first observe that mixup in specific hidden layer can achieve higher transferability than in input layer. For example, $\mathrm{RLMAM_{II}}$ achieves the attack success rate of 74% from layer 1, while it only achieves 71% from input layer (layer 0). Besides, $\mathrm{RLMAM_{II}}$ achieves the best attack success rate of 76% when selecting indexes from the layer interval [0, 3].

4.4 Single Iterations vs. Inner-outer Iterations

As shown in Fig. 4, we compare the performance of Single-loop (SI) Iterations and Inner-outer Iterations (II). The dashed line indicates the baseline performance of

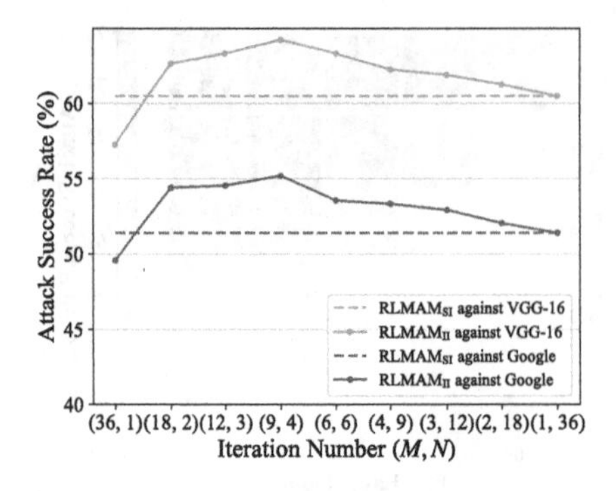

Fig. 4. The attack success rates (%) of our RLMAM_{II} and RLMAM_{SI}. The adversarial examples are generated from Res-50. For II, we divide iteration number $K = 36$ into pairs of inner and outer iteration numbers (M, N).

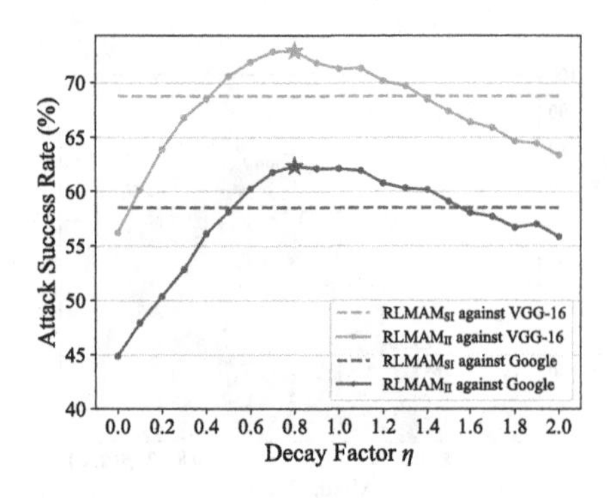

Fig. 5. The attack success rates (%) of our RLMAM_{II} and RLMAM_{SI}. The adversarial examples are generated from Res-50 with η ranging from 0.0 to 2.0.

RLMAM_{SI} with the iteration number $K = 36$. The solid line shows the attack success rate of our enhanced RLMAM_{II} with pairs of inner and outer iteration numbers (M, N), ensuring that $K = M \times N$. Note that we set $\lambda = 0.5$ to underscore the effect of Inner-outer Iterations method. It can be observed that (9, 4) outperforms other pairs of $(M,\ N)$. We can get the empirical conclusion that the appropriately divided Inner-outer Iterations can achieve better transferability than Single-loop Iterations method.

In addition, we further investigate the optimal value of the decay factor η in Inner-outer Iterations. As shown in Fig. 5, we evaluate the attack success rate of our enhanced Inner-outer Iterations method with the decay factor ranging from 0.0 to 2.0. $(M,\ N)$ is set to (4, 4). The success rate increases steadily from $\eta = 0.0$ to 0.8, and then the rate declines gradually. The optimal value of η is achieved when $\eta = 0.8$.

5 Conclusion

In this paper, we propose the RLMAM, which integrates a variant of manifold mixup into adversarial attacks. Our RLMAM leverages the interpolated adversarial representations induced by random-layer mixup to improve the complexity in both input and hidden spaces. Hence, we can obtain diverse gradient information to alleviate the over-fitting phenomenon on the white-box model. Experimental results show that our method can improve the transferability of non-targeted attacks.

Acknowledgement. This work is supported by the National Natural Science Foundation of China Grant (No.61772284, 61906098).

References

1. Nicholas, C., David, A.W.: Towards evaluating the robustness of neural networks. In: IEEE Symposium on Security and Privacy, pp. 39–57. IEEE Computer Society (2017)

2. Yinpeng, D., et al.: Boosting adversarial attacks with momentum. In: IEEE Conference on Computer Vision and Pattern Recognition, pp. 39–57. IEEE Computer Society (2018)
3. Ian, J.G., Jonathon, S., Christian, S.: Explaining and harnessing adversarial examples. In: 3rd International Conference on Learning Representations (2015)
4. Kaiming, H., Xiangyu, Z., Shaoqing, R., Jian, S.: Deep residual learning for image recognition. In: IEEE Conference on Computer Vision and Pattern Recognition, pp. 770–778. IEEE Computer Society (2016)
5. Andrew, I., Logan, E., Anish, A., Jessy, L.: Black-box adversarial attacks with limited queries and information. In: Proceedings of the 35th International Conference on Machine Learning, pp. 2142–2151 (2018)
6. Andrew, I., Shibani, S., Dimitris, T., Logan, E., Brandon, T., Aleksander, M.: Adversarial examples are not bugs, they are features. In: 33th Annual Conference on Neural Information Processing Systems, pp. 125–136 (2019)
7. Alexey, K., Ian, J.G., Samy, B.: Adversarial machine learning at scale. In: 5th International Conference on Learning Representations (2017)
8. Yanpei, L., Xinyun, C., Chang, L., Dawn, S.: Delving into transferable adversarial examples and black-box attacks. In: 5th International Conference on Learning Representations (2017)
9. Jiadong, L., Chuanbiao, S., Kun, H., Liwei, W., John, E.H.: Nesterov accelerated gradient and scale invariance for adversarial attacks. In: 8th International Conference on Learning Representations (2020)
10. Aleksandar, M., Ludwig, S., Dimitris, T., Adrian, V.: Towards deep learning models resistant to adversarial attacks. In: 6th International Conference on Learning Representations (2018)
11. Seyed, M., Moosavi, D., Alhussein, F., Pascal, F.: Deepfool: a simple and accurate method to fool deep neural networks. In: IEEE Conference on Computer Vision and Pattern Recognition, pp. 2574–2582. IEEE Computer Society (2016)
12. Karen, S., Andrew, Z.: Very deep convolutional networks for large-scale image recognition. In: 3rd International Conference on Learning Representations (2015)
13. Christian, S., et al.: Intriguing properties of neural networks. In: 2nd International Conference on Learning Representations (2014)
14. Christian, S., et al.: Going deeper with convolutions. In: IEEE Conference on Computer Vision and Pattern Recognition, pp. 1–9. IEEE Computer Society (2015)
15. Vladimir, V.: The Nature of Statistical Learning Theory. Statistics for Engineering and Information Science. Springer, New York (2000). https://doi.org/10.1007/978-1-4757-3264-1
16. Vikas, V., et al.: Manifold mixup: better representations by interpolating hidden states. In: 36th International Conference on Machine Learning (2019)
17. Xiaosen, W., Xuanran, H., Jingdong, W., Kun, H.: Admix: enhancing the transferability of adversarial attacks. arXiv preprint arxiv:2102.00436 (2021)
18. Cihang, X., et al.: Improving transferability of adversarial examples with input diversity. In: IEEE Conference on Computer Vision and Pattern Recognition, pp. 2730–2739 (2019)
19. Chiyuan, Z., Samy, B., Moritz, H., Benjamin, R., Oriol, V.: Understanding deep learning requires rethinking generalization. In: 5th International Conference on Learning Representations (2017)
20. Hongyi, Z., Moustapha, C., Yann, N.D., David, L.: Mixup: Beyond empirical risk minimization. In: 6th International Conference on Learning Representations (2018)

Deep Depression Prediction on Longitudinal Data via Joint Anomaly Ranking and Classification

Guansong Pang[1(✉)], Ngoc Thien Anh Pham[2], Emma Baker[2], Rebecca Bentley[3], and Anton van den Hengel[2]

[1] Singapore Management University, Singapore, Singapore
gspang@smu.edu.sg
[2] University of Adelaide, Adelaide, Australia
{ngoc.t.pham,emma.baker,anton.vandenhengel}@adelaide.edu.au
[3] University of Melbourne, Parkville, Australia
brj@unimelb.edu.au

Abstract. A wide variety of methods have been developed for identifying depression, but they focus primarily on measuring the degree to which individuals are suffering from depression currently. In this work we explore the possibility of predicting future depression using machine learning applied to longitudinal socio-demographic data. In doing so we show that data such as housing status, and the details of the family environment, can provide cues for predicting future psychiatric disorders. To this end, we introduce a novel deep multi-task recurrent neural network to learn time-dependent depression cues. The depression prediction task is jointly optimized with two auxiliary anomaly ranking tasks, including contrastive one-class feature ranking and deviation ranking. The auxiliary tasks address two key challenges of the problem: 1) *the high within class variance of depression samples*: they enable the learning of representations that are robust to highly variant in-class distribution of the depression samples; and 2) *the small labeled data volume*: they significantly enhance the sample efficiency of the prediction model, which reduces the reliance on large depression-labeled datasets that are difficult to collect in practice. Extensive empirical results on large-scale child depression data show that our model is sample-efficient and can accurately predict depression 2–4 years before the illness occurs, substantially outperforming eight representative comparators.

Keywords: Depression prediction · Anomaly detection · One-class classification · Deep learning

1 Introduction

Major Depressive Disorder (MDD), widely known as depression, is a mental disorder characterized by a severe and persistent feeling of sadness, loss of interest

J. Gama et al. (Eds.): PAKDD 2022, LNAI 13281, pp. 236–248, 2022.
https://doi.org/10.1007/978-3-031-05936-0_19

in activities, or a sense of despair, causing significant impairment in daily life [9]. Globally over 300 million people of all ages are estimated to suffer from depression, and it is a major contributor to nearly 800 thousands suicide deaths per year [23]. There have been many studies [2,13,19,22] demonstrating effective automated diagnosis of depression using machine learning techniques. These studies focus on the detection of ongoing, current depression, using data that relates information about the current state of an individual. We focus here on the more challenging task of predicting depression in advance. This is achieved using longitudinal socio-demographic data. The motivation for this approach is that prediction far enough in advance of upcoming depression might enable early intervention before the condition arises.

The longitudinal socio-demographic data used contains a variety of non-medical information such as education level, socio-economic background, family environment, measured at intervals overtime. There are three major challenges in exploiting these cues for depression prediction. First, this is noisy high-dimensional temporal data that contains thousands of numeric and categorical features, among which only very selective set are relevant to depression prediction. Second, there is no single set of socio-demographic factors that cause depression, leading to high in-class variance for the depression samples. For example, for childhood depression, the root cause may be related to the mental health status of parents, dwelling conditions, or a medical condition. As a result, the depression cases are highly dissimilar and exhibit no single underlying mechanism, or common characteristics. A model that seeks a single common explanation for all cases cannot succeed. Lastly, in practice, only a small amount of labeled data is available, as it is difficult, if not impossible, to collect large depression samples. The available data thus often fails to cover the diverse types of depression cases.

In this work we introduce a novel multi-task learning approach to tackle these challenges, in which a depression classification task is jointly optimized with two auxiliary anomaly ranking tasks, contrastive one-class feature ranking and deviation ranking. Depression samples are treated as anomalous samples in our auxiliary tasks, because the cause-varying depression samples can be widely distributed, and as a result, are difficult to model as a single concrete class. Directly modeling these depression samples with classification models can easily overfit the given depression cases. The two anomaly ranking tasks are devised to enforce compact low-dimensional representations of normal samples and allow variations in the representations of depression samples, presenting effective inductive biases to regularize the classification models. This significantly improves the model's generalization beyond the individual cases presented in the small volume of labeled data available.

In summary, this work makes two key contributions:

- We introduce a novel multi-task learning framework, which harnesses auxiliary anomaly detection tasks to empower the greater classification task. To the best of our knowledge, this is the first multi-task learning approach that

jointly optimizes classification and anomaly ranking tasks, which is an important tool for application problems that are similar to depression prediction.
- We further instantiate the framework into a multi-task recurrent neural network model, termed MTNet, which optimizes the depression model with one-class constraints on its feature space and deviation constraints on its output layer. Extensive empirical results on large-scale child depression data show that MTNet can accurately predict depression two to four years before the illness occurs (*e.g.*, achieving a recall of 0.8), substantially outperforming eight competing methods. Additionally, MTNet can also outperform these competing methods even with largely reduced (50% less) training data.

2 Related Work

Longitudinal Studies. There have been many longitudinal studies of depression [6,8], but they focus on association discovery that identifies factors or predictors associated with depression using traditional bivariate/multivariate statistic models. By contrast, our study is on learning prediction models to predict the occurrence/risk of future depression.

Automated Depression Diagnosis. Current studies focus on the detection of depression using classification models on vocal/visual data taken during clinical interviews, with vocal features like prosodic and cepstral features [2,13,24] and visual features like facial expression, gaze direction, and eye movement [13,22,26]. Depression detection based on social media data using linguistic and network features is also extensively studied [11,19]. However, all these studies focus on the detection of ongoing depression, while we aim at predicting upcoming depression.

An exploratory task is introduced at the eRisk workshop [10] to facilitate the development and evaluation of models for early detection of signs of depression using social media data, resulting in a number of early depression detection models [1,12,18]. Like the aforementioned studies, they are also focused upon detecting *expression of depression cues/symptoms*, while we learn the *hidden root cause factors in socio-demographic features* which are more difficult to learn.

Multi-task Learning. Multi-task learning [16] has been successful in a range of applications. There have been many approaches introduced to jointly learn multiple related tasks (e.g., classification, regression, and clustering) to improve the performance on small labeled data, such as feature learning, task relation learning, task clustering, low-rank and decomposition approaches [25]. Our work uses multi-task feature learning methods with hard parameter sharing as in [4,20,21], but our approach is the first work that uses anomaly ranking at the feature and output layers to regularize the classification.

Anomaly Detection. Anomaly detection techniques have been successfully applied to detect abnormal events/behaviors in many applications [14], but they are rarely used to regularize supervised learning models in multi-task learning as the two paradigms have rather different learning objectives. We show in this work that recent advanced anomaly detection models [15,17] can be adapted to effectively regularize classification models and improve their sample efficiency.

3 Multi-task Recurrent Neural Networks

3.1 The Proposed Framework

Upcoming depression prediction aims to learn a binary depression classification mapping function $\phi : \mathcal{X} \rightarrow \mathcal{Y}$, where $\mathcal{X} = \{\mathbf{X}_1, \mathbf{X}_2, \cdots, \mathbf{X}_N\}$ is a set of longitudinal data of N samples; each sample $\mathbf{X} \in \mathbb{R}^{w \times D}$ is a matrix input of an individual subject, which contains the socio-demographic features of the subject in the recent w *waves* (or time steps) of questionnaire data, *i.e.*, $\mathbf{X} = \{\mathbf{x}_1, \mathbf{x}_2, \cdots, \mathbf{x}_w\}$, where $\mathbf{x}_t \in \mathbb{R}^D$ is a feature vector derived from the t-th wave of questionnaire data; $\mathcal{Y} = \{0, 1\}$ is the output space, with '1' indicating the subject being normal within all the recent w waves of questionnaire but having depression in the next waves of questionnaire, and with '0' indicating the subject being normal in the recent and future waves of questionnaire.

In this work we introduce a novel multi-task learning framework to tackle the problem. An overview of the approach is presented in Fig. 1. Depression classification is our primary task and is jointly optimized with two auxiliary tasks, including deviation score ranking and contrastive one-class feature learning. The auxiliary tasks treat depression samples as anomalies and enforce compact feature representations of normal samples and allow some variations in the representations of depression samples, serving as a regularizer of the classification model. This results in better generalized classification models than that in the single primary task. Formally, let $\tau : \mathcal{X} \rightarrow \mathbb{R}$ be an anomaly ranking function that assigns an anomaly score to each subject; $\psi : \mathcal{X} \rightarrow \mathcal{Q}$ be the one-class feature learning function, where $\mathcal{Q} \in \mathbb{R}^M$ with $M \ll D$ is a new feature space, then our overall objective function can be given as follows.

$$\arg\min_{\Theta_e, \Theta_a, \Theta_o} \sum_{i=1}^{N} \Big[\ell_e\big(\phi(\mathbf{X}_i; \Theta_e), y_i\big) + \alpha \ell_a\big(\tau(\mathbf{X}_i; \Theta_a), y_i\big) + \beta \ell_o\big(\psi(\mathbf{X}_i; \Theta_o), y_i\big) \Big], \quad (1)$$

where ℓ_e, ℓ_a and ℓ_o are respective loss functions for depression classification, deviation ranking and contrastive one-class metric learning, y_i is the class label of $\mathbf{X}_i$, $\Theta = \{\Theta_e, \Theta_a, \Theta_o\}$ is the set of parameters to be learned, α and β are hyperparameters to control the importance of the two auxiliary tasks.

We instantiate the framework into a model called MTNet that leverages a shared LSTM neural network [7] to learn critical temporal changes in longitudinal data for depression classification. Supervised anomaly deviation and one-class support vector data description loss functions are defined to improve the model's generalization. A simple data augmentation method is also introduced to further enhance the generalizability. The modules of MTNet are presented as follows.

3.2 Primary Task: Depression Classification

Our classification model leverages an LSTM neural network layer to learn important temporal dependency in the longitudinal data. An LSTM layer consists of

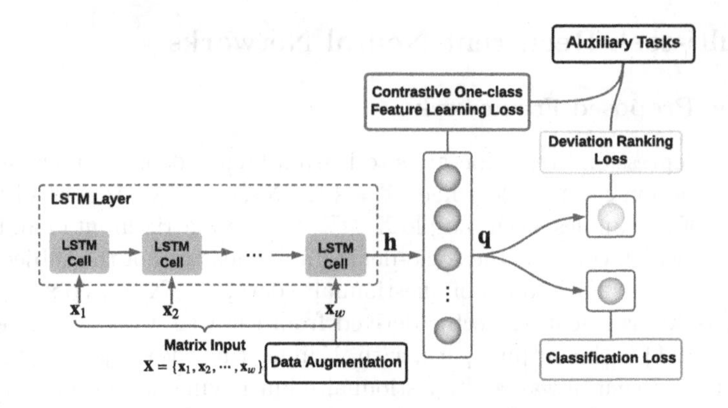

Fig. 1. The proposed multi-task learning framework.

multiple LSTM cells, with each LSTM cell learning temporal-dependent representations of the input data at a specific wave. The full LSTM layer uses the recurrent LSTM cells to encode important temporal changes across all different questionnaire waves into the output vector $\mathbf{h} \in \mathbb{R}^L$ in the last LSTM cell, *i.e.*, the LSTM layer is a mapping function η that performs $\mathbf{h} = \eta(\mathbf{X}; \Theta_l)$, where Θ_l contains all weight parameters in a standard LSTM (see *Supplementary Material*[1] for the full details of LSTM). To learn more expressive representations, a FC layer is used to project $\mathbf{h}$ onto a lower-dimensional feature representation space:

$$\mathbf{q} = \psi(\mathbf{X}; \Theta_l, \mathbf{W}_s, \mathbf{b}_s) = g_s(\mathbf{W}_s \eta(\mathbf{X}; \Theta_l) + \mathbf{b}_s), \tag{2}$$

where $\mathbf{W}_s \in \mathbb{R}^{M \times L}$ and $\mathbf{b}_s \in \mathbb{R}^M$ are the learnable parameters, g_s is an activation function, and $\mathbf{q} \in \mathcal{Q}$ is the final feature representation of $\mathbf{X}$. We then train a classifier on the $\mathcal{Q}$ representation space with a standard binary cross-entropy loss function:

$$\ell_e(\mathbf{X}, y) = -\big(y \log(p) + (1 - y)\log(1 - p)\big), \tag{3}$$

where y is the class label of $\mathbf{X}$ and $p = \phi(\mathbf{X}; \Theta_e) = g_e(\mathbf{W}_e \mathbf{q} + b_e)$, where g_e is a sigmoid activation, $\mathbf{W}_e \in \mathbb{R}^{1 \times M}$ and $b_e \in \mathbb{R}$. $\Theta_e = \{\Theta_l, \mathbf{W}_s, b_s, \mathbf{W}_e, b_e\}$ contains all the network parameters that can be learned in an end-to-end manner.

3.3 Auxiliary Tasks

Two auxiliary tasks, contrastive one-class-based feature ranking and deviation ranking, are incorporated to introduce an inductive bias (*i.e.*, to learn compact normal representations and allow large variations in abnormal representations) to learn more generalized representations of depression samples.

Deviation Ranking. A partially-supervised anomaly ranking task is introduced to enforce the model to assign significantly larger anomaly scores for depression

[1] Supplementary material is available at https://tinyurl.com/MTNetPAKDD22.

samples than that of non-depression samples. Inspired by [15], a prior-driven anomaly ranking loss function, called deviation loss, is used to fulfill this goal. Particularly, a Gaussian prior $\mathcal{N}(\mu, \sigma^2)$ is imposed on the anomaly scores of all samples, which posits that the anomaly scores of non-depression samples are centered around a Gaussian mean value μ while the anomaly scores of depression samples have at least $a * \sigma$ deviations from μ. Formally, we add another network output head with one linear unit to learn an anomaly score for each sample:

$$\tau(\mathbf{X}; \Theta_a) = \mathbf{W}_a \psi(\mathbf{X}; \Theta_r) + b_a, \tag{4}$$

where $\Theta_a = \{\Theta_r, \mathbf{W}_a, b_a\}$ are the parameters to be learned. We then define the deviation using the well-known Z-Score:

$$dev(\mathbf{X}) = \frac{\tau(\mathbf{X}; \Theta_a) - \mu}{\sigma}, \tag{5}$$

The deviation function is then used to define our anomaly ranking loss function:

$$\ell_a(\mathbf{X}, y) = (1 - y)|dev(\mathbf{X})| + y \max\big(0, a - dev(\mathbf{X})\big). \tag{6}$$

By minimizing ℓ_a, our model pushes the anomaly scores of normal samples as close as possible to μ while enforcing at least $a * \sigma$ between μ and the anomaly scores of depression samples in the upper tail of the Gaussian distribution. Following [15], the prior $\mathcal{N}(0, 1)$ is used with $a = 5$ to guarantee significant deviations of depression samples from normal samples.

Contrastive One-class Feature Learning. Unlike the anomaly ranking task that introduces the inductive bias using an output layer independent from the classification output, the one-class feature learning complements the deviation score learning and exerts directly on the feature layer. Particularly, we introduce a contrastive one-class feature learning method, in which we devise a supervised variant of support vector data description (SVDD) [17] by contrasting the one-class center of the normal samples and the depression samples.

$$\ell_o(\mathbf{X}, y) = (1 - y)||\psi(\mathbf{X}; \Theta_r) - \mathbf{n}||_2 + y \max\big(0, m - ||\psi(\mathbf{X}; \Theta_r) - \mathbf{n}||_2\big), \tag{7}$$

where $\Theta_o = \{\Theta_r\}$, $\mathbf{n} \in \mathbb{R}^M$ is the one-class center vector of normal samples and m is a hyperparameter to control the contrast margin. The first term is the original SVDD objective. The second term is added to enforce a large margin between non-depression and depression samples in the ψ-induced representation space, while minimizing the $\mathbf{n}$-centered hypersphere's volume. We found empirically that MTNet can perform well with varying settings of $\mathbf{n}$, *e.g.*, $\mathbf{n} \sim \mathcal{N}(0, 1)$ or $\mathbf{n} \sim \mathcal{U}(0, 1)$. We use $\mathbf{n} \sim \mathcal{N}(0, 1)$ by default, *i.e.*, generating $\mathbf{n}$ by randomly drawing a vector from a standard Gaussian distribution. $m = 1$ is used to enforce a sufficiently large distance margin in the feature representation space.

3.4 Data Augmentation

A simple data augmentation method is introduced to augment depression samples and further enhance the model's generalizability. Specifically, a pair of depression samples are randomly selected, and then a small percentage of randomly selected values in the last wave data of one sample are replaced with the corresponding values in another sample to create a new depression sample. The augmented sample can well retain the original depression-relevant information while at the same time enriching the depression samples. By using this method, we increase the number of depression samples in the training data by a factor of 10. In our experiment, we randomly replaced 5% of the feature values by default.

3.5 The Algorithm of MTNet

The algorithmic procedure of our model MTNet is presented in Algorithm 1. After random initialization of the network parameters in Step 1, stochastic gradient descent is used to optimize the model in Steps 2–8. In Step 4, as the number of depression samples is typically far smaller than that of non-depression samples, we generate sample batches with balanced class distribution to achieve more effective optimization. This shares the same spirit as oversampling in imbalanced learning [5]. Step 5 calculates the batch-wise loss for the three tasks. Step 6 performs gradient descent steps to learn the parameters Θ. Note that Θ_r are shared parameters in $\{\Theta_e, \Theta_a, \Theta_o\}$, and thus, the feature representations in MTNet are jointly optimized by all three tasks. At the inference stage, only the classification function ϕ is used to produce the class label.

Algorithm 1. MTNet

Input: $\mathcal{X} \in \mathbb{R}^{w \times D}$ - training samples, and binary class labels $\mathbf{y}$
Output: $\phi : \mathcal{X} \to \mathcal{Y}$ - a depression classification network
1: Randomly initialize $\Theta = \{\Theta_e, \Theta_a, \Theta_o\}$
2: **for** $j = 1$ to *#epochs* **do**
3: **for** $k = 1$ to *#batches* **do**
4: $\mathcal{B} \leftarrow$ Randomly sample the same number of depression and non-depression samples
5: Calculate the loss using $\frac{1}{|\mathcal{B}|} \sum_{\mathbf{X}_i \in \mathcal{B}} \Big[\ell_e\big(\phi(\mathbf{X}_i; \Theta_e), y_i\big) + \alpha \ell_a\big(\tau(\mathbf{X}_i; \Theta_a), y_i\big) + \beta \ell_o\big(\psi(\mathbf{X}_i; \Theta_o), y_i\big) \Big]$
6: Perform a gradient descent step w.r.t. the parameters in Θ
7: **end for**
8: **end for**
9: **return** ϕ

4 Experiments

4.1 Datasets

Our model is evaluated on a large child depression data dataset based on the Longitudinal Study of Australian Children (LSAC) data [3]. LSAC consists of multiple bi-annual waves of questionnaire-based interview data of 10,090 children across Australia. Initially, children aged from infant to 5 years and their families are interviewed between August 2003 and February 2004. This routine is repeated every two years afterwards. At the time of writing seven waves of data are available. LSAC provides a dataset of 4,983 children aged 4 to 5 years in the first wave of questionnaire. These children are all healthy until 287 children are confirmed to have depression at the 6/7-th wave of interview in the years 2013 to 2015. Depression is measured using parental self-report data. Particularly, in the waves 6 and 7, the primary caregiver is asked "does study child have any of these ongoing (depression) conditions[2]?"; the child is confirmed to have depression if the answer is 'yes'. After a simple feature screening to remove uninformative features (*e.g.*, features with very large percentage of missing values), 210 social-demographic features related to individual growth and development (*e.g.*, age, gender, living location, schooling performance), and family environment (*e.g.*, social, educational, economic, employment, household income, housing conditions) are used (see *Supplementary Material* for the list of the interview questions and a sample of the questionnaire). In the selected features, missing values are filled with the mean/mode value in each feature; categorical features are then converted into numeric features by using one-hot encoding. The resulting dataset contains 762 features in each wave of data. Thus, the dataset used has 4,983 samples, with each sample represented by a 7×762 matrix. We further perform a stratified random split of the dataset into three subsets, including 60% data as a training set, and respective 20% data for validation and testing sets.

4.2 Experimental Setup

We evaluate the performance of predicting the possible occurrence of depression in the near future. To this end, the model is trained and tested using only the first five waves of data when all children are reported to be mentally healthy (*i.e.*, absence of depressive symptoms). The task is to predict whether a child will have depression at the upcoming wave 6 or 7. MTNet is compared with eight temporal and non-temporal methods.

- **Non-temporal Methods**. Three popular classification methods, including logistic regression (LR), support vector machines (SVM) and multi-layer perceptron (MLP) neural networks, are used as baseline methods that are not designed to capture temporal dependence. They are two main ways to apply these methods to the longitudinal data. One way is to build the classification model using the most recent single (*i.e.*, 5-th) wave data only. The second way is to use the data from all the five waves, in which for each subject we concatenate the feature vectors derived from all the waves into one lengthy unified feature vector; the classifiers are then built upon this concatenated data. This way helps capture some temporal-dependent changes. All three

[2] 'Ongoing conditions' means that the conditions "*exist for some period of time (weeks, months or years) or re-occur regularly. They do not have to be diagnosed by a doctor*".

methods are evaluated in both ways, with LR/SVM/MLP-s denoting the classifier using the single wave data and LR/SVM/MLP-m denoting the use of the concatenated multi-wave data.
- **Temporal Methods.** The LSTM-based deep classifier, is used as a competing temporal method. The standard binary cross-entropy loss function is used to train the model. We also compare MTNet with the state-of-the-art anomaly detection model DevNet [15] that is adapted to temporal data with LSTM network.

4.3 Implementation Details

MTNet is implemented with one LSTM layer with 200 units, followed by a fully-connected (FC) layer with 20 units and a classification output layer. The sigmoid function is used in g_r in the LSTM layer by default; the widely-used ReLU activation function is used in g_s in the FC layer. A dropout layer with a dropout rate of 0.5 is applied to the LSTM and FC layers. The competing methods LSTM and DevNet use exactly the same network architecture as MTNet. MLP uses a similar network structure with two hidden layers of respectively 200 and 20 units, with each layer having a dropout rate of 0.5. MTNet, DevNet, LSTM, and MLP are implemented using Keras[3] and optimized using RMSprop with a batch size of 256 and 20 batches per epoch. They are trained with 30 epochs as their performance can converge early. $\alpha = 0.5$ and $\beta = 2.0$ are used in MTNet by default. LR and SVM are taken from the open-source scikit-learn package[4]. Due to a large percentage of irrelevant features presented in the data, our extensive results showed that applying the l_1-norm regularizer to MLP, LR and SVM obtains significantly better performance than the l_2-norm regularizer. Thus, the l_1-norm regularizer is applied to these three classifiers to bring sparsity to the model. The regularization hyperparameter is probed with $\{0.001, 0.01, 0.1, 1\}$, with the best performance reported. The oversampling method in MTNet is used in all competing methods to alleviate the class-imbalanced problem.

4.4 Performance Evaluation Measures

Three widely-used evaluation measures are used, including the Area Under Receiver Operating Characteristic Curve (AUC-ROC), Area Under Precision-Recall Curve (AUC-PR), and F_1-score (F-score for brevity). AUC-ROC summarizes the ROC curve of true positives against false positives, while AUC-PR summarizes the curve of precision against recall. AUC-ROC is popular due to its good interpretability. AUC-PR is more indicative than AUC-ROC in evaluating performance on imbalanced data. F-score is the harmonic mean of precision and recall. We also report the precision and recall results to gain more insights into the performance. The reported results are averaged over five independent runs.

[3] https://keras.io/.
[4] https://scikit-learn.org/.

4.5 Empirical Results

Effectiveness on Real-World Data. The results of depression prediction are shown in Table 1. Our model MTNet is the best performer in AUC-ROC, AUC-PR and F-score. MTNet substantially outperforms all of its competing methods by 2%–38% in AUC-PR and 3%–20% in F-score. Impressively, MTNet obtains a recall of 0.8, achieving at least 7.8% improvement over its contenders. Given a sufficiently high precision of 0.7, the high recall rate in MTNet would enable accurate intervention and pre-treatment of up to 80% upcoming depression cases at a very early stage (2–4 years before the depression occurs), effectively preventing and reducing the depression cases.

Table 1. Performance results (mean±std) of depression prediction.

	AUC-ROC	AUC-PR	F-score	Precision	Recall
LR-s	0.648 ± 0.020	0.595 ± 0.018	0.632 ± 0.021	0.659 ± 0.032	0.611 ± 0.037
LR-m	0.648 ± 0.019	0.596 ± 0.016	0.620 ± 0.023	0.670 ± 0.027	0.579 ± 0.038
SVM-s	0.666 ± 0.012	0.610 ± 0.012	0.646 ± 0.005	0.682 ± 0.023	0.614 ± 0.011
SVM-m	0.684 ± 0.013	0.631 ± 0.010	0.646 ± 0.026	0.729 ± 0.013	0.582 ± 0.045
MLP-s	0.771 ± 0.007	0.780 ± 0.010	0.706 ± 0.023	0.679 ± 0.032	0.742 ± 0.071
MLP-m	0.814 ± 0.009	0.808 ± 0.019	0.718 ± 0.022	0.730 ± 0.049	0.718 ± 0.082
LSTM	0.779 ± 0.012	0.775 ± 0.010	0.662 ± 0.034	$\mathbf{0.751} \pm 0.031$	0.593 ± 0.039
DevNet	0.785 ± 0.013	0.786 ± 0.017	0.688 ± 0.035	0.704 ± 0.044	0.684 ± 0.094
MTNet	$\mathbf{0.818} \pm 0.009$	$\mathbf{0.823} \pm 0.008$	$\mathbf{0.743} \pm 0.037$	0.697 ± 0.006	$\mathbf{0.800} \pm 0.080$

Sample Efficiency. This section examines the model's generalizability from the sample efficiency aspect, *i.e.*, how is the performance if less labeled training data is available? Specifically, each model is trained on a new training set that is a random subset of the original training data, and then it is evaluated using the same test data as that used in Table 1. We focus on comparing to the better contenders LR/SVM/MLP-m and omit the less effective ones – LR/SVM/MLP-s. The results are shown in Fig. 2. Remarkably, MTNet is substantially more sample-efficient than its competing methods; it performs much better than, or comparably well to, the best performance of its competing methods even when it uses 50% less training data. This superiority of MTNet benefits from the integrated anomaly ranking tasks in its multi-task objective function, which enable better generalization to diverse depression cases. This is manifested by the large performance gap between MTNet and LSTM, since the only difference between them is the two auxiliary tasks integrated into MTNet.

Parameter Sensitivity. This section evaluates the sensitivity of MTNet w.r.t. its two hyperparameters, α and β, which respectively adjust the importance of the anomaly ranking loss and one-class metric loss. The test results are presented in Fig. 3. The results show that MTNet generally performs stably with both α

and β in a wide range of setting choices. Relatively small α and large β are needed for MTNet to achieve the best performance. This indicates that MTNet is dependent more on the one-class feature learning than the deviation ranking.

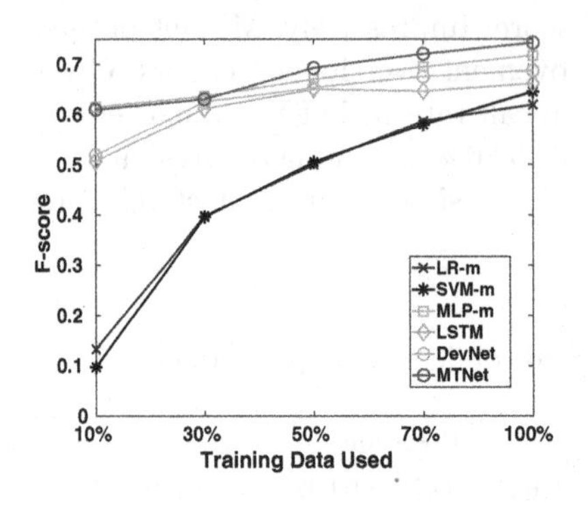

Fig. 2. Sample efficiency

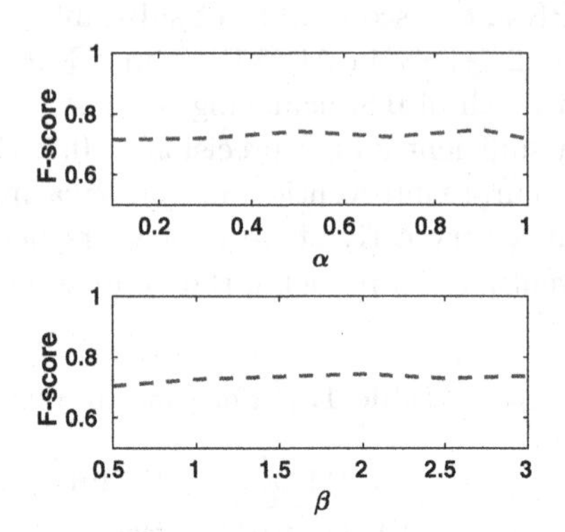

Fig. 3. Parameter sensitivity

Ablation Study. This section evaluates the importance of each module in MTNet. LSTM is used as a baseline to evaluate the effect of incorporating one or more of the following modules, including the deviation ranking loss ℓ_a, the one-class metric learning loss ℓ_o, and the data augmentation (DA). The results are reported in Table 2. It is clear that i) the multi-task learning performs substantially better than the individual tasks, ii) all three modules in MTNet make important contribution to its overall performance, and iii) the deviation ranking loss ℓ_a and the one-class metric learning loss ℓ_o are complementary to each other.

Table 2. Ablation study results. DA is short for data augmentation.

Method	AUC-ROC	AUC-PR	F-score	Precision	Recall
LSTM	0.779	0.775	0.662	**0.751**	0.593
LSTM+ℓ_a	0.785	0.786	0.688	0.704	0.684
LSTM+ℓ_o	0.804	0.821	0.702	0.673	0.737
LSTM+ℓ_a+ℓ_o	0.817	**0.834**	0.721	0.709	0.737
LSTM+ℓ_a+ℓ_o+DA	**0.818**	0.823	**0.743**	0.697	**0.800**

5 Conclusions and Future Work

In this work we propose a novel multi-task learning framework and its instantiation MTNet for depression prediction. MTNet effectively leverages the two auxiliary anomaly ranking tasks to improve the depression prediction model's representations and generalizability. Remarkably, our empirical results show that MTNet is able to accurately predict 80% depression cases 2–4 years before the depression actually occurs in children. The improved generalizability of MTNet is supported by the sample efficiency experiment, in which MTNet requires significantly less labeled depression samples to perform comparably well to, or substantially better than, the competing methods. It should be noted the model has around 70% precision only, indicating 30% false positive predictions. Thus, caution must be taken when using the the model in practice. In future work, to improve the model's accountability and its collaboration with clinical psychologists, we plan to incorporate an interpretation module into our model to provide insightful explanation for each of its depression prediction result.

References

1. Burdisso, S.G., Errecalde, M., Montes-y Gómez, M.: τ-ss3: a text classifier with dynamic n-grams for early risk detection over text streams. Pattern Recogn. Lett. **138**, 130–137 (2020)
2. Gong, Y., Poellabauer, C.: Topic modeling based multi-modal depression detection. In: AVEC, pp. 69–76 (2017)
3. Gray, M., Sanson, A., et al.: Growing up in Australia: the longitudinal study of australian children. Family Matters **72**, 4 (2005)
4. Hassani, K., Haley, M.: Unsupervised multi-task feature learning on point clouds. In: ICCV, pp. 8160–8171 (2019)
5. He, H., Garcia, E.A.: Learning from imbalanced data. IEEE Trans. Knowl. Data Eng. **21**(9), 1263–1284 (2009)
6. Henry, M., et al.: A screening algorithm for early detection of major depressive disorder in head and neck cancer patients post-treatment: longitudinal study. Psycho-oncology **27**(6), 1622–1628 (2018)
7. Hochreiter, S., Schmidhuber, J.: Long short-term memory. Neural Comput. **9**(8), 1735–1780 (1997)
8. Korsten, L.H., et al.: Factors associated with depression over time in head and neck cancer patients: a systematic review. Psycho-oncology **28**(6), 1159–1183 (2019)
9. Lamers, F., Milaneschi, Y., Smit, J.H., Schoevers, R.A., Wittenberg, G., Penninx, B.W.: Longitudinal association between depression and inflammatory markers: results from the Netherlands study of depression and anxiety. Biol. Psychiatry **85**(10), 829–837 (2019)
10. Losada, D.E., Crestani, F., Parapar, J.: Overview of eRisk: early risk prediction on the internet. In: Bellot, P., et al.: (eds.) CLEF 2018. LNCS, vol. 11018, pp. 343–361. Springer, Cham (2018). https://doi.org/10.1007/978-3-319-98932-7_30
11. Mann, P., Paes, A., Matsushima, E.H.: See and read: detecting depression symptoms in higher education students using multimodal social media data. In: ICWSM, vol. 14, pp. 440–451 (2020)
12. Masood, R.: Adapting models for the case of early risk prediction on the internet. In: Azzopardi, L., Stein, B., Fuhr, N., Mayr, P., Hauff, C., Hiemstra, D. (eds.) ECIR 2019. LNCS, vol. 11438, pp. 353–358. Springer, Cham (2019). https://doi.org/10.1007/978-3-030-15719-7_48

13. Nasir, M., Jati, A., Shivakumar, P.G., Nallan Chakravarthula, S., Georgiou, P.: Multimodal and multiresolution depression detection from speech and facial landmark features. In: AVEC, pp. 43–50 (2016)
14. Pang, G., Shen, C., Cao, L., Hengel, A.V.D.: Deep learning for anomaly detection: a review. ACM Comput. Surv. **54**(2), 1–38 (2021)
15. Pang, G., Shen, C., van den Hengel, A.: Deep anomaly detection with deviation networks. In: KDD, pp. 353–362 (2019)
16. Ruder, S.: An overview of multi-task learning in deep neural networks. arXiv preprint arXiv:1706.05098 (2017)
17. Ruff, L., et al.: Deep one-class classification. In: International Conference on Machine Learning, pp. 4393–4402. PMLR (2018)
18. Sadeque, F., Xu, D., Bethard, S.: Measuring the latency of depression detection in social media. In: WSDM, pp. 495–503 (2018)
19. Shen, T., et al.: Cross-domain depression detection via harvesting social media. In: IJCAI, pp. 1611–1617 (2018)
20. Standley, T., Zamir, A., Chen, D., Guibas, L., Malik, J., Savarese, S.: Which tasks should be learned together in multi-task learning? In: ICML, pp. 9120–9132. PMLR (2020)
21. Strezoski, G., Noord, N.V., Worring, M.: Many task learning with task routing. In: ICCV, pp. 1375–1384 (2019)
22. Uddin, M.A., Joolee, J.B., Lee, Y.K.: Depression level prediction using deep spatiotemporal features and multilayer bi-lTSM. IEEE Trans. Affect. Comput. (2020)
23. WHO. Depression and Other Common Mental Disorders: Global Health Estimates. Technical Report, World Health Organization (2017)
24. Yang, Y., Fairbairn, C., Cohn, J.F.: Detecting depression severity from vocal prosody. IEEE Trans. Affect. Comput. **4**(2), 142–150 (2012)
25. Zhang, Y., Yang, Q.: A survey on multi-task learning. arXiv preprint arXiv:1707.08114 (2017)
26. Zhou, X., Jin, K., Shang, Y., Guo, G.: Visually interpretable representation learning for depression recognition from facial images. IEEE Trans. Affect. Comput. (2018)

DeepPAMM: Deep Piecewise Exponential Additive Mixed Models for Complex Hazard Structures in Survival Analysis

Philipp Kopper(✉), Simon Wiegrebe, Bernd Bischl, Andreas Bender, and David Rügamer

Department of Statistics, LMU Munich, Ludwigstr. 33, 80539 Munich, Germany
philipp.kopper@stat.uni-muenchen.de

Abstract. Survival analysis (SA) is an active field of research that is concerned with time-to-event outcomes and is prevalent in many domains, particularly biomedical applications. Despite its importance, SA remains challenging due to small-scale data sets and complex outcome distributions, concealed by truncation and censoring processes. The piecewise exponential additive mixed model (PAMM) is a model class addressing many of these challenges, yet PAMMs are not applicable in high-dimensional feature settings or in the case of unstructured or multimodal data. We unify existing approaches by proposing DeepPAMM, a versatile deep learning framework that is well-founded from a statistical point of view, yet with enough flexibility for modeling complex hazard structures. We illustrate that DeepPAMM is competitive with other machine learning approaches with respect to predictive performance while maintaining interpretability through benchmark experiments and an extended case study.

Keywords: Deep learning · Time-to-event data · Survival analysis · Interpretability · Random effects · Mixed models

1 Introduction

Deep learning (DL) excels in many different areas of application through flexible and versatile network architectures. This has also been demonstrated in survival analysis (SA) [27,33], where it is often not straightforward to apply off-the-shelf machine learning models. Apart from medical applications such as the prediction of time-to-death or the time to disease onset, time-to-event models are also applied in a variety of other domains. Among other fields, SA is successfully employed for predictive maintenance, credit scoring, and customer churn prediction. In practice, time-to-event outcomes are not necessarily observed fully but might be censored, truncated or stem from a competing risks, or a multi-state process. While these aspects relate to the nature of the observation of event times, SA is also challenging due to the typically small amount of observations as well as complex feature effects and dependencies between observations. Medical survival data for instance potentially includes patient data of certain cohorts

A. Bender and D. Rügamer—Contributed equally.

J. Gama et al. (Eds.): PAKDD 2022, LNAI 13281, pp. 249–261, 2022.
https://doi.org/10.1007/978-3-031-05936-0_20

(such as patients from different hospitals with varying levels of patient care), longitudinal data with recurrent events or includes time-varying features such as a patient's vital status. Additionally, data can be multimodal (e.g., tabular patient information paired with medical images).

Our Contribution. In this paper, we introduce a novel method called *DeepPAMM* for continuous time-to-event data that enables the hazard-based learning of survival models via neural networks and supports 1) many common survival tasks, including right-censored, left-truncated, competing risks, or multi-state data as well as recurrent events; 2) the estimation of inherently interpretable feature effects; 3) learning from multiple data sources (e.g., tabular and imaging data); 4) time-varying effects and time-varying features; 5) the modeling of repeated or correlated data using random effects.

2 Related Literature

Various models have been brought forward in SA. We will distinguish between models developed from a statistical point of view (Sect. 2.1), machine learning approaches (Sect. 2.2) and recently proposed deep learning frameworks (Sect. 2.3).

2.1 Piecewise Exponential Additive Models and Cox Proportional Hazard Models

The Cox proportional hazard model (CPH) [11] is the most widely used survival model. Under certain assumptions [42] the Cox PH model is equivalent to the piecewise exponential model (PEM). The original formulation of the PEM, a parametric, linear effects, PH model, goes back to [14]. The general idea is to partition the follow-up time into J intervals and to assume piecewise constant hazards in each interval. The originally proposed PEM requires a careful choice of the number and placement of interval cut-points. The piecewise exponential additive model (PAM) [2,3,9] is an extension of the PEM. PAMs estimate the baseline hazard and other time-dependent effects as smooth functions over time via penalized splines. This leads to more plausible and robust hazard estimates and (indirectly) lower computational cost. PAMs can be further generalized to piecewise exponential additive mixed models (PAMMs) by adding frailty terms (random effects). While PEMs and PAMMs can deal with many types of survival data (see, e.g., [4,6]), they are limited w.r.t. the complexity of feature effects that they can estimate, especially in the case of high-dimensional features and interactions and cannot handle unstructured data.

2.2 Machine Learning Approaches

In recent years a large number of machine learning methods for SA have been put forward. Random forest (RF) based methods include the random survival forest (RSF) [20] and more recently the oblique random survival forests (ORSF) [21]. In contrast to conventional RFs [8], these adaptions make the models applicable to survival data by adjusting the splitting criterion. Next to trees and forests, several boosting methods exist, such as XGBoost [10] or component-wise boosting for accelerated-failure time models [36] and non-parametric hazard boosting [28]. More recently and closest to our work, [4] have proposed a general machine learning approach for various survival tasks based on PEMs and demonstrated its application using the standard XGBoost implementation.

2.3 Deep Learning Approaches

Various deep learning approaches have been proposed for SA, with the first approaches dating back to the mid-1990s (see, e.g., [12]). More recent approaches include both discrete-time methods like DeepHit [27] or Nnet-survival [15] and continuous-time methods such as DeepSurv [22] or CoxTime [24]. DeepHit parametrizes the probability mass function by a neural network and specifically targets competing risks, but is only able to predict survival probabilities for a given set of discrete follow-up time points due to its time-discretization approach. Nnet-survival, by contrast, models discrete hazards and provides flexibility in terms of architecture choice, but it also relies on discretization of event times. DeepSurv is a Cox PH model with the linear predictor replaced by a deep feed-forward neural network. CoxTime further improves upon DeepSurv by allowing for time-varying effects, thereby overcoming the proportional hazards assumption. A deep Gaussian process to predict competing risks is proposed in [1]. While all previous methods focus on tabular data, a few multimodal networks such as [17,23,30,40] have also been proposed as well as survival tasks combined with a generative appraoch [41]. The first combination of PEMs with a NN was proposed by [29]. [7] discussed the estimation of PEM by representing generalized linear models via feed-forward NNs, and [13] proposed the estimation of the shape of the hazard rate with NNs. [25] also discussed the parametrization of the PEM via NNs with application to tabular data. As for PEMs, the choice of cut-points in their framework is crucial for performance and computational complexity. Our framework eliminates this problem.

3 Piecewise Exponential Additive Models

Survival analysis aims to estimate the survival function $S(t) = P(T > t)$. Instead of directly estimating $S(t)$, the hazard function

$$h(t) := \lim_{\Delta t \to 0^+} \frac{P(t < T < t + \Delta t | T \geq t)}{\Delta t} \tag{1}$$

is modeled. The survival function can be derived from $h(t)$ via $S(t) = \exp(-\int_0^t h(s)\,ds)$. A hazard for time point $t \in \mathcal{T}$, conditional on a potentially time-varying feature vector $\boldsymbol{x}(t) \in \mathbb{R}^P$, can be defined by

$$h(t|\boldsymbol{x}(t), k) = \exp\left(\rho(\boldsymbol{x}(t), t, k)\right), k = 1, \ldots, K. \tag{2}$$

The function $\rho(\cdot)$ represents the effect of (time-dependent) features $\boldsymbol{x}(t)$ on the hazard and can itself be potentially time- and transition-specific. k indicates a transition, e.g., from status 0 to status k in competing risks or the transition between two states in the multi-state setting. In the following, we will set K to 1 for better readability and only address the single risk application if not stated otherwise. Further omitting the dependence on t, (2) reduces to the familiar PH form known from the Cox model.

3.1 Data Transformation

PEMs and PAMs approximate (2) via piecewise constant hazards, which requires a specific data transformation, creating one row in the data set for each interval a subject was at risk. Assume observations (subjects) $i = 1, \ldots, n$, for which the tuple $(t_i, \delta_i, \boldsymbol{x}_i)$ with event time t_i, event indicator $\delta_i \in \{0, 1\}$ (1=event, 0=censoring) and feature vector $\boldsymbol{x}_i$ is observed. PAMs partition the follow up into J intervals $(\kappa_{j-1}, \kappa_j]$, $j = 1, \ldots, J$. This implies a new status variable $\delta_{ij} = 1$ if $t_i \in (\kappa_{j-1}, \kappa_j] \wedge \delta_i = 1$, and 0 otherwise, indicating the status of subject i in interval j. Further, we create a variable t_{ij}, the time subject i was at risk in interval j, which will enter the analysis as an offset. Lastly, the variable t_j, (e.g., $t_j := \kappa_j$) is a representation of time in interval j and the feature based on which the model estimates the baseline hazard and time-varying effects. In order to transform the data to the piecewise exponential data format (PED), time-constant features $\boldsymbol{x}_i$ are repeated for each of J_i rows, where J_i, denotes the number of intervals in which subject i was at risk. This data augmentation step transforms a survival task into a standard Poisson regression task. Depending on the setting, e.g., right-censoring, recurrent events, left truncation, etc., the specifics of the data transformation vary, but the general principles remain the same. For more details we refer to [4,5,32].

3.2 Model Estimation

Given the transformed data, PAMs approximate (2) by $h(t|\boldsymbol{x}_i(t)) = \exp(\rho(\boldsymbol{x}_{ij}, t_j)) := h_{ij}, \forall t \in (\kappa_{j-1}, \kappa_j]$, where $\boldsymbol{x}_{ij}$ is the feature vector of subject i in interval j. Assuming $\delta_{ij} \sim \text{Poisson}(\mu_{ij} = h_{ij}t_{ij})$, the log-likelihood contribution of subject i is given by $\ell_i = \sum_{j=1}^{J_i}(\delta_{ij}\log(h_{ij}) - h_{ij}t_{ij})$, where

$$\log(h_{ij}) = \beta_0 + f_0(t_j) + \sum_{p=1}^{P} x_{ij,p}\beta_p + \sum_{l=1}^{L} f_l(x_{ij,l}),$$

with log-baseline hazard $\beta_0 + f_0(t_j)$, linear feature effects β_p of features $x_{ij,p} \subseteq \boldsymbol{x}_{ij}$ and univariate, non-linear feature effects $f_l(x_{ij,l})$ of features $x_{ij,l} \subseteq \boldsymbol{x}_{ij}$. Both f_0 and f_l are defined via a basis representation, i.e., $f_l(x_{ij,l}) = \sum_{m=1}^{M_l}\theta_{l,m}B_{l,m}(x_{ij,l})$ with basis functions $B_{\cdot,m}(\cdot)$ (such as B-spline bases) and basis coefficients $\theta_{\cdot,m}$. To avoid underfitting, the basis dimensions M_0 (for f_0) and M_l (for f_l) are set relatively high. To avoid overfitting, the basis coefficients are estimated by optimizing an objective function that penalizes differences between neighboring coefficients. Let $\boldsymbol{\beta} = (\beta_0, \ldots, \beta_P)^\top$ and $\boldsymbol{\theta}_l = (\theta_{l,1}, \ldots, \theta_{l,M_l})^\top$, $l = 0, \ldots, L$. The objective function minimized to estimate PAMs is the penalized negative log-likelihood given by $-\log\mathcal{L}(\boldsymbol{\beta}, \boldsymbol{\theta}_0, \ldots, \boldsymbol{\theta}_L) + \sum_{l=0}^{L}\psi_l\Psi(\boldsymbol{\theta}_l)$, where the first term is the standard negative logarithmic Poisson likelihood, comprised of likelihood contributions ℓ_i, and the second term $\Psi(\boldsymbol{\theta}_l)$ is a quadratic penalty with smoothing parameter $\psi_l \geq 0$ for the respective spline f_l. Larger ψ_l lead to smoother f_l estimates (see [6,43] for details).

4 Deep Piecewise Exponential Additive Mixed Models

DeepPAMMs extend PAM(M)s with hazard as defined in (2) by allowing for deep neural networks (NN) in the additive predictor. Instead of combining PAMMs with (deep)

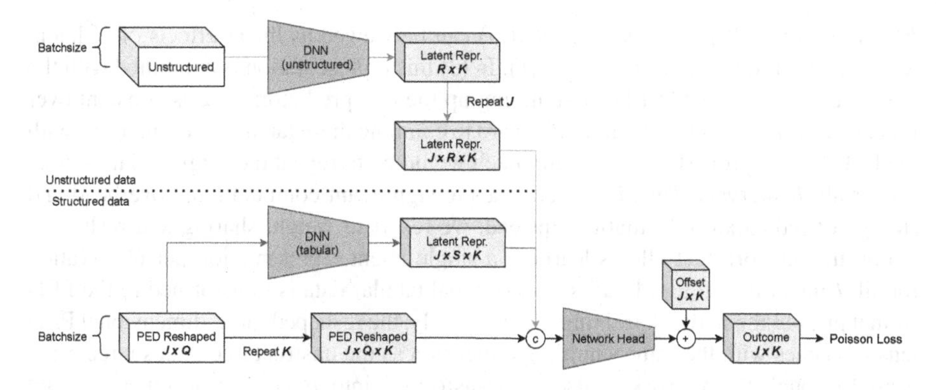

Fig. 1. Exemplary architecture of a DeepPAMM. A DeepPAMM comprises a PAMM (black path) and additionally either a deep neural network (DNN) for unstructured data (yellow path), a DNN for tabular features (blue path), or both. The unstructured data, e.g., images, are summarized to latent representations of size R, repeated J times, and concatenated (c) with the tabular data's latent representation of size S, as well as raw tabular data of size Q. Finally, the offset is added to the output and the network is trained using the Poisson loss for each of the K competing risks.

NNs in a two-stage approach, we embed the PAMM into the NN similar to [35] and train the network based on the (penalized) likelihood in an end-to-end manner.

Network Definition. While PAMMs restrict ρ to structured additive effects, the hypothesis space of DeepPAMMs can also be modeled using a deep NN. Assume that the NN $d(\cdot)$ is used to process a potentially time-varying (unstructured) data source $\boldsymbol{z}(t)$. We first assume a time-constant effect of $\boldsymbol{z}(t)$ and extend the PAMM's definition to

$$h(t|\boldsymbol{x}(t), \boldsymbol{z}(t)) = \exp\{\rho(\boldsymbol{x}(t), t) + d(\boldsymbol{z}(t))\}, \tag{3}$$

by adding one (or several) NN predictor(s) to the structured predictor.

The predictor $d(\boldsymbol{z}(t))$ can be modeled using an arbitrary NN. For example, a DeepPAMM can combine a PAM with an additional NN to explore non-linearities and interactions in tabular features (beyond the ones specified in the structured part). Alternatively, a DeepPAMM can combine different data modalities, e.g., tabular patient data and corresponding medical scans using a convolutional NN for d. By (3), DeepPAMM learns a piecewise constant hazard rate

$$h_{ij} = \exp\{\boldsymbol{B}_{ij}\boldsymbol{w} + \sum_{u=1}^{U} \zeta_{ij,u}\gamma_u\}, \tag{4}$$

for each observation i and each discrete interval j, where $\boldsymbol{B}_{ij}$ subsumes all Q structured features (linear and basis evaluated features) with weights $\boldsymbol{w}$. $\zeta_{ij,1}, \dots, \zeta_{ij,U}$ are $U = R + S$ latent representations learned from the deep network part that processes tabular data (into S latent features) and unstructured data (into R latent features). The network then combines these U latent representations to learn the effect $\gamma_1, \dots, \gamma_U$ of each of these feature effects. Due to the additive structure in predictor (4), the structured terms with linear effects $\boldsymbol{w}$ preserve their interpretability inherited from PAMMs.

PED and Latent Representations. $d(\boldsymbol{z}(t))$ can be viewed as linear effects of U latent representations derived from inputs $\boldsymbol{z}(t)$. In (3) this representation is combined with the structured features in a last layer summing up the two predictors. If $\boldsymbol{z}$ is constant over time, i.e., $\boldsymbol{z}(t) \equiv \boldsymbol{z}$, it is not straightforward to combine these latent representations with the PED format properly. A naive approach would be to repeat the original data source $\boldsymbol{z}$ over all J intervals. This, however, leads to significant computational overhead and storage of redundant information. Instead, we resort to weight-sharing and reshaping within the network that allows learning a single latent representation per observation for all J intervals (cf. Fig. 1). First, the original tabular data is transformed to the PED format prior to the network training. Subsequently, the reshaped three-dimensional PED tensor batches with the same sampling dimension as the unstructured data source $\boldsymbol{z}$ are passed through the network. $\boldsymbol{z}$ itself is transformed into R latent representations and then repeated J times for each interval. This avoids repeating the original unstructured data source multiple times. Finally, we combine these representations with the original tabular data and the S non-linear representations of the structured data part into a joint set of features. While we here focus on time-constant unstructured data, our framework can be extended to allow for time-varying unstructured features by simply also supplying the time t to the deep NN d explicitly, i.e., extending $d(\boldsymbol{z}(t))$ in (3) to $d(\boldsymbol{z}(t), t)$.

Learning Non-proportional Hazards. PAMMs allow for non-proportional hazards via an interaction of features $\boldsymbol{x}$ with a feature that represents time in each of the J intervals. In practice, however, the accompanying computational complexity and manual definition of these interactions are often infeasible. In DeepPAMM, such interactions can be modeled using an appropriate multilayer NN architecture. In particular, interactions between features $\boldsymbol{z}(t)$ and the follow-up t can be expressed by $h(t|\boldsymbol{x}(t), \boldsymbol{z}(t)) = \exp\{\rho\,(\boldsymbol{x}(t), d(\boldsymbol{z}(t)), t)\}$, where ρ now also depends on the the specified NN to model a non-proportional hazard in $\boldsymbol{z}(t)$. As the PH assumption is a helpful inductive bias for applications with small sample sizes, we recommend this extension for larger data sets or in applications where the PH assumption is clearly violated.

Learning Competing Risks Hazards. When modeling competing risks data with K different risks that determine the time-to-event, one is interested in retrieving the cumulative incidence functions of each risk (CIFs). Our architecture allows for a holistic way of modeling the hazard of subject i in interval j and cause k in a joint NN: $h_{ijk} = \exp\{\boldsymbol{B}_{ijk}\boldsymbol{w}_k + \sum_{u=1}^{U} \zeta_{ijk,u}\gamma_{k,u}\}$, where $\boldsymbol{B}_{ijk}$ is equivalent to the input $\boldsymbol{B}_{ij}$, i.e., we repeat $\boldsymbol{B}_{ij}$ K times so that cause-specific weights $\boldsymbol{w}_k$ share the same inputs. Similarly, the latent representations $\zeta_{ijk,u}$ now also depend on the risk $k = 1, \ldots, K$ to yield cause-specific effects $\gamma_{k,u}$ for each latent feature. Figure 1 illustrates the CR case for an exemplary network architecture. Training the network is based on a joint loss summing up all K loss contributions for each CR and weighted by binary interval weights if the observation is still at risk in the jth interval and 0 if not.

Learning Mixed Effects and Recurrent Events. In many SA settings, data comes in clusters. For example, the survival of patients has been observed at different locations. This is typically the case for multi-center studies for which survival may substantially vary between clusters while being more homogeneous within each cluster. A random effect (RE), i.e., a linear effect for each cluster with a normal prior, can account for this within-cluster correlation. REs can also be used to account for repeated measurements and recurrent events. Optimization of NNs with random or mixed effects can be done using an EM-type optimization routine (see, e.g., [45]), by training a Bayesian NN (see, e.g., [19]), or by tuning the prior variance based on the equivalence of a random normal prior and a ridge-penalized effect (see, e.g., [43]). While learning the RE prior variance explicitly is desirable, a carefully chosen ridge penalization should yield similar results (due to their mathematical equivalence) while being more straightforward to incorporate in most NNs.

5 Numerical Experiments

We first explore DeepPAMM by investigating some of the proposed model properties in a simulation study. Additionally, we compare DeepPAMM with state-of-the-art algorithms on various benchmark data sets including real-world medical applications. We examine model performance via the integrated Brier score (IBS) [16], which measures both, discrimination and calibration of predicted survival probabilities. Instead of integrating over the whole time domain, we evaluate the IBS at the first three quartiles (Q25, Q50, Q75) of the observed event times in the test set, in order to assess the performance at different time points. While DL-based approaches usually require large data sets for training, DeepPAMM also works well in small data set regimes. In the worst case, if there is not enough data to train the deep part of our network, the structured network part will dominate the predictions. DeepPAMM will then effectively fall back to estimate a PAMM, which in turn is well suited for small data sets. This property is especially important in SA where most data sets are relatively small.

5.1 Simulation and Ablation Study

The goal of our simulation study is to investigate the performance of DeepPAMM under various controlled settings with a focus on 1) mixed effects, 2) competing risks, 3) multimodal data. For all simulations, the data generating process incorporates both, linear effects and non-linear interactions. For every setting, we repeat this procedure 25 times to account for variance in data generation and model fitting. In the spirit of an ablation study, we compare DeepPAMM with its corresponding PAM(M) to investigate the attribution of performance gains as well as the relation to an ideal model (Optimal).

For **competing risks**, we simulate two competing risks based on two different hazards structures. While cause 1 is based on 5 features and multiple non-linear interaction effects, cause 2 relates to 3 features and a more moderate level of interactions as well as non-linearities.

For **mixed effects**, we simulate repeated measurements by defining 60 clusters and drawing a random effect for each cluster unit from a normal distribution with zero mean and a standard deviation of 1.5. Before training DeepPAMM, we pre-train the random effects of the DeepPAMM with the corresponding PAMM and use the associated ridge penalty as a warm start for tuning.

Table 1. Comparison of the average IBS (with standard deviation in brackets) across the three quartiles Q25, Q50, Q75 (rows) for different methods (columns) in different study settings. The †-symbol indicates methods that can only take tabular data information into account.

	CR (cause 1)			CR (cause 2)			Mixed effects			Multimodal		
	Q25	Q50	Q75	Q25	Q50	Q75	Q25	Q50	Q75	Q25	Q50	Q75
KM	5.1 (0.43)	10.8 (0.49)	16.2 (0.57)	4.3 (0.31))	9.1 (0.52)	13.6 (0.60)	7.0 (0.47)	13.5 (0.57)	18.2 (0.68)	4.1† (0.44)	8.0† (0.49)	12.1† (0.66)
PAMM	3.3 (0.32)	6.0 (0.52)	8.9 (0.52)	2.5 (0.21)	4.5 (0.38)	7.0 (0.61)	3.9 (0.49)	6.9 (0.71)	9.1 (0.89)	3.7† (0.43)	6.3† (0.63)	8.6† (0.71)
Ours	**2.9** (0.41)	**5.4** (0.40)	**8.1** (0.43)	**2.4** (0.38)	**4.4** (0.46)	**6.8** (0.68)	**3.2** (0.41)	**5.7** (0.61)	**7.5** (0.69)	**3.6** (0.43)	**6.1** (0.52)	**8.4** (0.65)
Optimal	2.9 (0.72)	5.4 (0.80)	8.0 (0.79)	2.1 (0.22)	4.1 (0.39)	6.5 (0.63)	2.9 (0.36)	5.2 (0.57)	6.8 (0.68)	3.6 (0.42)	6.1 (0.58)	8.3 (0.68)

For the **multimodal data** scenario, we simulate log-hazards based on linear latent effects from point clouds (PC) based on the data set from ModelNet10 [44]. Each of the PC labels is associated with a different latent coefficient ranging from -0.5 to 0.75. The hazard is defined to depend on these latent coefficients as well as on tabular features. A reduced PointNet [31] is used to model the PCs. This set up has been adapted from [23].

Results. Model comparisons are provided in Table 1. In summary, our proposed model is the best performing method across all three settings and in most cases yields performance values close to the optimal error in terms of the IBS. While performance gains in absolute terms seem small, the decrease in IBS relative to the optimal error is especially noteworthy for CR (cause 1) and the mixed effects setting. Results confirm that DeepPAMM works well in various of the proposed data situations. The ablation study further justifies the deep part of DeepPAMM by its improved performance in comparison to PAMM.

5.2 Benchmark Analysis

We compare our approach with various state-of-the-art methods (Table 2). Comparisons include a tree-based method (ORSF; [21]), a boosting approach (PEMXGB; [4]), as well as (DeepHit; [27]), a well-established deep NN for SA. As baseline models we use a Kaplan-Meier estimator (KM; [11]) and a Cox PH model (CPH; [11]). We restrict our comparison to directly and publicly available SA data sets that have been used in the benchmarks of methods listed above, namely *tumor* [5], *gbsg2* [37], *metabric* (cf. [27]), *breast* [39], *mgus2* [26], and *icu* (cf. [18]). For each method, we perform a random search with 50 configurations and compare the aggregated (mean and std. deviation) test set performances on 25 distinct train-test-splits. The data sets impose different challenges, including CR (icu, mgus2), high-dimensional data (breast), and mixed effects (icu). For these, DeepPAMM is consistently among the best-performing survival models. The main point here is that DeepPAMM is competitive compared to other state-of-the-art methods while maintaining interpretability as illustrated in Sect. 5.3.

Table 2. Performance comparison based on the IBS (↓) at the three quartiles (Q25, Q50, Q75) across different data sets (rows) and models (columns) with best models per row highlighted in bold. Missing entries are due to missing support for CRs.

Data set		KM	Cox PH	ORSF	PEMXGB	DeepHit	DeepPAMM
tumor	Q25	6.6 (0.59)	6.0 (0.58)	**5.5** (0.56)	5.7 (0.63)	5.6 (0.55)	5.7 (0.59)
	Q50	12.3 (0.86)	11.2 (0.82)	**10.8** (0.91)	10.9 (1.05)	11.0 (0.96)	10.9 (0.86)
	Q75	17.6 (0.79)	16.3 (0.77)	16.3 (0.85)	**16.2** (0.92)	16.4 (0.95)	**16.2** (0.81)
gbsg2	Q25	3.1 (0.49)	3.1 (0.45)	**3.0** (0.45)	**3.0** (0.46)	3.1 (0.49)	3.1 (0.41)
	Q50	6.8 (0.80)	6.5 (0.72)	**6.2** (0.70)	6.3 (0.68)	6.6 (0.8)	6.5 (0.69)
	Q75	12.5 (1.04)	11.4 (0.94)	**11.1** (0.95)	11.3 (1.01)	11.9 (0.99)	11.5 (0.95)
metabric	Q25	4.0 (0.22)	4.0 (0.26)	4.1 (0.25)	**3.8** (0.22)	4.0 (0.22)	3.9 (0.27)
	Q50	8.6 (0.51)	8.2 (0.54)	8.9 (0.46)	**7.8** (0.49)	8.4 (0.45)	7.9 (0.45)
	Q75	14.0 (0.38)	12.9 (0.47)	14.7 (1.19)	**12.3** (0.51)	13.5 (0.40)	12.6 (0.44)
breast	Q25	**1.9** (0.61)	–	2.0 (0.59)	2.0 (0.57)	2.1 (0.54)	**1.9** (0.60)
	Q50	4.1 (0.89)	–	**4.0** (0.80)	**4.0** (0.83)	4.2 (0.81)	**4.0** (0.90)
	Q75	7.1 (1.13)	–	**6.7** (0.96)	**6.7** (1.10)	7.1 (1.02)	6.8 (1.33)
mgus2 (cause 1)	Q25	**1.1** (0.21)	–	–	1.9 (0.34)	**1.1** (0.21)	**1.1** (0.21)
	Q50	**2.2** (0.34)	–	–	4.1 (0.55)	**2.2** (0.34)	**2.2** (0.34)
	Q75	**3.4** (0.48)	–	–	6.9 (0.69)	3.5 (0.51)	**3.4** (0.49)
mgus2 (cause 2)	Q25	8.7 (0.52)	–	–	8.6 (0.65)	**8.1** (0.55)	8.3 (0.49)
	Q50	14.4 (0.61)	–	–	13.9 (0.84)	**12.9** (0.66)	13.1 (0.65)
	Q75	18.4 (0.60)	–	–	17.9 (1.04)	**15.8** (0.67)	16.0 (0.67)
icu (cause 1)	Q25	**1.3** (0.06)	–	–	1.4 (0.66)	**1.3** (0.06)	**1.3** (0.06)
	Q50	3.6 (0.14)	–	–	3.6 (0.13)	**3.5** (0.13)	**3.5** (0.13)
	Q75	6.7 (0.19)	–	–	6.7 (0.19)	6.5 (0.20)	**6.4** (0.19)
icu (cause 2)	Q25	3.5 (0.15)	–	–	3.5 (0.14)	**3.4** (0.14)	**3.4** (0.14)
	Q50	7.6 (0.17)	–	–	7.6 (0.17)	**7.3** (0.17)	**7.3** (0.16)
	Q75	12.0 (0.15)	–	–	12.1 (0.17)	11.5 (0.20)	**11.3** (0.17)

5.3 Extended Case Study

In this extended case study, we show how DeepPAMM can be used to obtain interpretable feature effects and at the same time incorporate potentially high-dimensional interactions. To illustrate this, we apply DeepPAMM to spatio-temporal data where the outcome is response times (time-to-arrival) of the London fire brigade to fire-related emergency calls [38]. Additionally, the data includes geographic coordinates of the site of the fire as well as information about the ward from which the truck was deployed and the time of day of the incident. We expect a non-linear effect of the time of day that varies with day and night times as well as traffic hours and a bivariate spatial effect of the location with different hazards in different regions of the city. Therefore, we model the hazard for arrival at time t given time of day t_d, spatial coordinates (c_1 and c_2) and ward $v = 1, \ldots, V$ as

Table 3. Performance comparison based on the IBS (↓) at the three quartiles (Q25, Q50, Q75) across different models (columns) for the data set of [38] with best models per row highlighted in bold. The performance has been assessed using 25 train-test splits.

Quantile	KM	PAMM	DeepPAMM
Q25	12.8 (0.28)	12.3 (0.30)	**12.2** (0.32)
Q50	18.1 (0.18)	16.9 (0.21)	**16.7** (0.23)
Q75	19.9 (0.11)	18.4 (0.14)	**18.2** (0.14)

$$\log(h(t|t_d, c_1, c_2, v)) = \underbrace{\beta_0 + f_0(t) + f_1(t_d) + f_2(c_1, c_2) + b_v}_{\text{structured}} + \underbrace{d(t, t_d, c_1, c_2, v)}_{\text{unstructured}}$$

where $f_1(t_d)$ is estimated as a cyclic spline that enforces equal values of the function at 0 and 24 h, $f_2(c_1, c_2)$ is a bivariate tensor product spline and b_v are random effects for the individual wards. In the unstructured part, we additionally allow for high-dimensional interactions between the features from the structured part. This way, we can investigate whether the predictive performance can be improved beyond the structured part. Structured effects are given in Fig. 2. For interpretation, note that higher hazards imply shorter response times, thus response times are on average longer during night hours and between 12 and 18 p.m. as well as in the periphery of the city. The results w.r.t. the predictive performance are shown in Table 3, where we compare our model with a KM baseline and the respective PAMM. In addition to the PAMM specification, our model includes a NN with three layers (64, 32, 8 neurons) to model feature interactions. The results indicate that on average the performance improves slightly when the unstructured part is added. Given the resulting standard deviations, we conclude that the structured part is sufficient. Further, DeepPAMM's structured effects are in line with results presented in [38]. This shows the strength of DeepPAMM: maintaining interpretability of covariate effects as illustrated in Fig. 2, while also allowing the investigation of additional effects in the unstructured part.

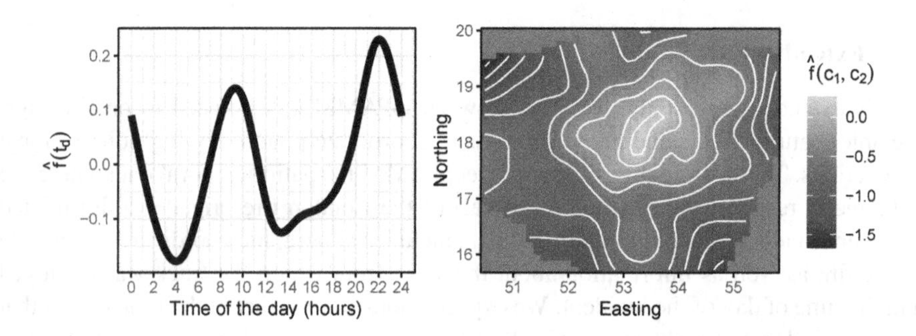

Fig. 2. Smooth cyclic (left) and spatial (right) effect of a DeepPAMM. Effects are from a single of the 25 runs.

6 Concluding Remarks

We present DeepPAMM, a novel semi-structured deep learning approach to survival analysis. Our experiments demonstrate that our model has high predictive capacity and is capable of modeling diverse complex data associations. DeepPAMM allows to include non-linear and feature interaction effects in the model, can be used to model non-proportional hazards, time-varying effects and competing risks, while also accounting for correlation in the data using mixed effects. The deep part of the model further makes estimation in high-dimensional settings possible and can be used to include unstructured data into the survival analysis. The additive predictor in our approach allows for straightforward interpretability and to recover the PAM(M) when no additional deep predictors are necessary. Our method can be fit using existing software solutions (e.g., deepregression [34]).

Acknowledgements. This work has been partly funded by the German Federal Ministry of Education and Research (BMBF) under Grant No. 01IS18036A. The authors of this work take full responsibility for its content.

References

1. Alaa, A.M., van der Schaar, M.: Deep multi-task gaussian processes for survival analysis with competing risks. In: Proceedings of the 31st International Conference on Neural Information Processing Systems, pp. 2326–2334. Curran Associates Inc. (2017)
2. Argyropoulos, C., Unruh, M.L.: Analysis of time to event outcomes in randomized controlled trials by generalized additive models. PLoS ONE **10**(4), e0123784 (2015)
3. Bender, A., Groll, A., Scheipl, F.: A generalized additive model approach to time-to-event analysis. Statist. Model. **18**(3–4), 299–321 (2018)
4. Bender, A., Rügamer, D., Scheipl, F., Bischl, B.: A general machine learning framework for survival analysis. In: Hutter, F., Kersting, K., Lijffijt, J., Valera, I. (eds.) ECML PKDD 2020. LNCS (LNAI), vol. 12459, pp. 158–173. Springer, Cham (2021). https://doi.org/10.1007/978-3-030-67664-3_10
5. Bender, A., Scheipl, F.: pammtools: piece-wise exponential additive mixed modeling tools. arXiv preprint arXiv:1806.01042 (2018)
6. Bender, A., Scheipl, F., Hartl, W., Day, A.G., Küchenhoff, H.: Penalized estimation of complex, non-linear exposure-lag-response associations. Biostatistics **20**(2), 315–331 (2019)
7. Biganzoli, E., Boracchi, P., Marubini, E.: A general framework for neural network models on censored survival data. Neural Netw. **15**(2), 209–218 (2002)
8. Breiman, L.: Random forests. Mach. Learn. **45**(1), 5–32 (2001)
9. Cai, T., Hyndman, R.J., Wand, M.P.: Mixed model-based hazard estimation. J. Comput. Graph. Statist. **11**(4), 784–798 (2002)
10. Chen, T., Guestrin, C.: XGBoost: a scalable tree boosting system. In: Proceedings of the 22nd ACM SIGKDD International Conference on Knowledge Discovery and Data Mining - KDD 2016, pp. 785–794 (2016)
11. Cox, D.R.: Regression models and life-tables. J. Roy. Statist. Soc. Ser. B (Methodological) **34**(2), 187–220 (1972)
12. Faraggi, D., Simon, R.: A neural network model for survival data. Statist. Med. **14**(1), 73–82 (1995)

13. Fornili, M., Ambrogi, F., Boracchi, P., Biganzoli, E.: Piecewise exponential artificial neural networks (PEANN) for modeling Hazard function with right censored data. In: Formenti, E., Tagliaferri, R., Wit, E. (eds.) CIBB 2013 2013. LNCS, vol. 8452, pp. 125–136. Springer, Cham (2014). https://doi.org/10.1007/978-3-319-09042-9_9
14. Friedman, M.: Piecewise exponential models for survival data with covariates. Ann. Statist. **10**(1), 101–113 (1982)
15. Gensheimer, M.F., Narasimhan, B.: A scalable discrete-time survival model for neural networks. PeerJ **7**, e6257 (2019)
16. Graf, E., Schmoor, C., Sauerbrei, W., Schumacher, M.: Assessment and comparison of prognostic classification schemes for survival data. Statist. Med. **18**(17–18), 2529–2545 (1999)
17. Haarburger, C., Weitz, P., Rippel, O., Merhof, D.: Image-based survival prediction for lung cancer patients using CNNS. In: 2019 IEEE 16th International Symposium on Biomedical Imaging (ISBI 2019), pp. 1197–1201. IEEE (2019)
18. Hartl, W.H., Bender, A., Scheipl, F., Kuppinger, D., Day, A.G., Küchenhoff, H.: Calorie intake and short-term survival of critically ill patients. Clin. Nutr. **38**(2), 660–667 (2019)
19. Hernández-Lobato, J.M., Adams, R.: Probabilistic backpropagation for scalable learning of Bayesian neural networks. In: International Conference on Machine Learning, pp. 1861–1869. PMLR (2015)
20. Ishwaran, H., Kogalur, U.B., Blackstone, E.H., Lauer, M.S.: Random survival forests. Ann. Appl. Statist. **2**(3), 841–860 (2008)
21. Jaeger, B.C., et al.: Oblique random survival forests. Ann. Appl. Statist. **13**(3), 1847–1883 (2019)
22. Katzman, J.L., Shaham, U., Cloninger, A., Bates, J., Jiang, T., Kluger, Y.: Deepsurv: personalized treatment recommender system using a cox proportional hazards deep neural network. BMC Med. Res. Methodol. **18**(1), 1–12 (2018)
23. Kopper, P., Pölsterl, S., Wachinger, C., Bischl, B., Bender, A., Rügamer, D.: Semi-structured deep piecewise exponential models. In: Survival Prediction-Algorithms, Challenges and Applications, pp. 40–53. PMLR (2021)
24. Kvamme, H., Borgan, Ø., Scheel, I.: Time-to-event prediction with neural networks and cox regression. arXiv preprint arXiv:1907.00825 (2019)
25. Kvamme, H., Borgan, Ø.: Continuous and discrete-time survival prediction with neural networks. arXiv preprint arXiv:1910.06724 (2019)
26. Kyle, R.A., et al.: A long-term study of prognosis in monoclonal gammopathy of undetermined significance. New Engl. J. Med. **346**(8), 564–569 (2002)
27. Lee, C., Zame, W.R., Yoon, J., van der Schaar, M.: DeepHit: a deep learning approach to survival analysis with competing risks. In: Thirty-Second AAAI Conference on Artificial Intelligence (2018)
28. Lee, D., et al.: Theory and software for boosted nonparametric hazard estimation. In: Survival Prediction - Algorithms, Challenges and Applications, pp. 149–158. PMLR (2021)
29. Liestøl, K., Andersen, P.K., Andersen, U.: Survival analysis and neural nets. Statist. Med. **13**(12), 1189–1200 (1994)
30. Pölsterl, S., Sarasua, I., Gutiérrez-Becker, B., Wachinger, C.: A wide and deep neural network for survival analysis from anatomical shape and tabular clinical data. In: Cellier, P., Driessens, K. (eds.) ECML PKDD 2019. CCIS, vol. 1167, pp. 453–464. Springer, Cham (2020). https://doi.org/10.1007/978-3-030-43823-4_37
31. Qi, C.R., Su, H., Mo, K., Guibas, L.J.: Pointnet: deep learning on point sets for 3d classification and segmentation. In: Proceedings of the IEEE Conference on Computer Vision and Pattern Recognition, pp. 652–660 (2017)
32. Ramjith, J., Bender, A., Roes, K.C., Jonker, M.A.: Recurrent Events Analysis with Piecewise exponential Additive Mixed Models. Preprint at Research Square (2021)

33. Ranganath, R., Perotte, A., Elhadad, N., Blei, D.: Deep survival analysis. In: Proceedings of the 1st Machine Learning for Healthcare Conference, vol. 59, pp. 101–114 (2016)
34. Rügamer, D., et al.: Deepregression: a flexible neural network framework for semi-structured deep distributional regression. arXiv preprint arXiv:2104.02705 (2021)
35. Rügamer, D., Kolb, C., Klein, N.: Semi-structured deep distributional regression: combining structured additive models and deep learning. arXiv preprint arXiv:2002.05777 (2021)
36. Schmid, M., Hothorn, T.: Flexible boosting of accelerated failure time models. BMC Bioinformatics **9**(1), 1–13 (2008)
37. Schumacher, M., et al.: Randomized 2 x 2 trial evaluating hormonal treatment and the duration of chemotherapy in node-positive breast cancer patients. German breast cancer study group. J. Clin. Oncol. **12**(10), 2086–2093 (1994)
38. Taylor, B.M.: Spatial modelling of emergency service response times. J. Roy. Statist. Soc. Ser. A (Statist. Soc.) **180**(2), 433–453 (2017)
39. Ternès, N., Rotolo, F., Heinze, G., Michiels, S.: Identification of biomarker-by-treatment interactions in randomized clinical trials with survival outcomes and high-dimensional spaces. Biometric. J. **59**(4), 685–701 (2017)
40. Vale-Silva, L.A., Rohr, K.: Long-term cancer survival prediction using multimodal deep learning. Sci. Rep. **11**(1), 1–12 (2021)
41. Weber, T., Ingrisch, M., Bischl, B., Rügamer, D.: Towards modelling hazard factors in unstructured data spaces using gradient-based latent interpolation. In: NeurIPS 2021 Workshop on Deep Generative Models and Downstream Applications (2021)
42. Whitehead, J.: Fitting cox's regression model to survival data using glim. J. Roy. Statist. Soc. Ser. C (Appl. Statist.) **29**(3), 268–275 (1980)
43. Wood, S.N.: Generalized Additive Models: An Introduction with R, 2 rev edn. Chapman & Hall/CRC Texts in Statistical Science, Boca Raton (2017)
44. Wu, Z., et al.: 3d shapenets: a deep representation for volumetric shapes. In: Proceedings of the IEEE Conference on Computer Vision and Pattern Recognition, pp. 1912–1920 (2015)
45. Xiong, Y., Kim, H.J., Singh, V.: Mixed effects neural networks (Menets) with applications to gaze estimation. In: Proceedings of the IEEE/CVF Conference on Computer Vision and Pattern Recognition, pp. 7743–7752 (2019)

Assessing Classifier Fairness with Collider Bias

Zhenlong Xu[1], Ziqi Xu[1], Jixue Liu[1], Debo Cheng[1], Jiuyong Li[1](✉), Lin Liu[1], and Ke Wang[2]

[1] University of South Australia, Adelaide, Australia
{Zhenlong.Xu,Ziqi.Xu,Debo.Cheng}@mymail.unisa.edu.au,
{Jixue.Liu,Jiuyong.Li,Lin.Liu}@unisa.edu.au
[2] Simon Fraser University, Burnaby, Canada
wangk@sfu.ca

Abstract. The increasing application of machine learning techniques in everyday decision-making processes has brought concerns about the fairness of algorithmic decision-making. This paper concerns the problem of collider bias which produces spurious associations in fairness assessment and develops theorems to guide fairness assessment avoiding the collider bias. We consider a real-world application of auditing a trained classifier by an audit agency. We propose an unbiased assessment algorithm by utilising the developed theorems to reduce collider biases in the assessment. Experiments and simulations show the proposed algorithm reduces collider biases significantly in the assessment and is promising in auditing trained classifiers.

Keywords: Fairness · Collider bias · Causal inference

1 Introduction

There are increasing concerns over the fairness of decision making algorithms with the wide use of machine learning in various applications, such as job hiring, credit scoring and home loan since discrimination can be inadvertently introduced into machine learning models. To prevent unfairness in a model from spreading in society, audit techniques are needed for the independent authority to audit machine learning models. Figure 1 shows an audit process. An audit agency accesses a model of a company and has its own audit cases for assessing the fairness of the model. The audit agency does

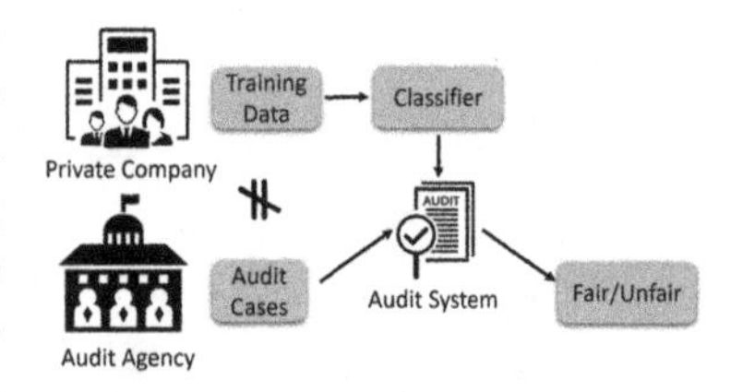

Fig. 1. The process of audit.

Supported by Australian Research Council (DP200101210), Natural Sciences and Engineering Research Council of Canada and Postgraduate Research Scholarship of University of South Australia.

Z. Xu and Z. Xu—contributed equally to this paper.

J. Gama et al. (Eds.): PAKDD 2022, LNAI 13281, pp. 262–276, 2022.
https://doi.org/10.1007/978-3-031-05936-0_21

not have access to the training data set but has the regulatory policy. In this paper, we use a causal graph to represent the regulatory policy. The company may use additional variables that are not specified in the regulatory policy to build its models to improve prediction accuracy.

Situation test has been used in the U.S. to detect discrimination in recruitment [2], which is a controlled experiment approach for analysing employers' decisions on job applicants' characteristics, as illustrated with the following examples. Pairs of research assistants are sent to apply for the same job, and each pair of the pretended applicants have the same qualifications and experience related to the job but have different values for their protected variable, such as male/female or young/old. Discrimination is detected if the favourable decisions are unequal between groups with different protected values.

The above described situation test can be simulated in an audit process, and we call it Naive Situation Test (NST) in this paper. We feed two inputs representing two individuals whose variable values are identical except their protected values to a machine learning model. If the model provides different decisions, NST will detect the model as discriminatory.

NST may produce an incorrect detection. We use the following example to show this. Consider a classifier $f()$ used by a company to determine employees' salaries as $salary = f(race, education, suburb)$. Some predicted outcomes by the model are shown in Table 1. Based on NST, the black people are not discriminated against since with the same education and suburb, both white and black people are predicted to have the same salary. However, Suburb is an irrelevant variable for determining the Salary. Without considering the Suburb, with the same level of Education, 2/3 white people receive a salary higher than 50K while only 1/3 black people receive a salary of 50K or higher. Hence, black people are discriminated against by the model.

Table 1. An example of incorrect detection by NST on a classifier.

Race	Edu	Sub	Predicted.Sal
white	high	A	>50k
black	high	A	>50k
white	high	A	>50k
black	high	B	≤50k
black	high	B	≤50k
white	high	B	≤50k

NST ⇒ "fair"

f(**white**, high, A)=f(**black**, high, A)
f(**white**, high, B)=f(**black**, high, B)

The incorrect detection by NST is caused by collider bias. We use a causal graph, formally defined in Sect. 2, to explain the collider bias. Causal relationships of variables in the above example are shown using the causal graph in Fig. 2 where a directed edge represents a causal relationship. The suburb is a collider since two edges "collide" at it. Conditioning on a collider, an association is formed between the two variables but it is spurious [5]. In the example, the spurious association cancels the association due to the causal relationship between Race and Salary and hides the true discrimination. Collider bias is related to the selection bias [10]. In a classifier, conditioning on a variable is equivalent to selecting sub-populations using the values of the variable. If the

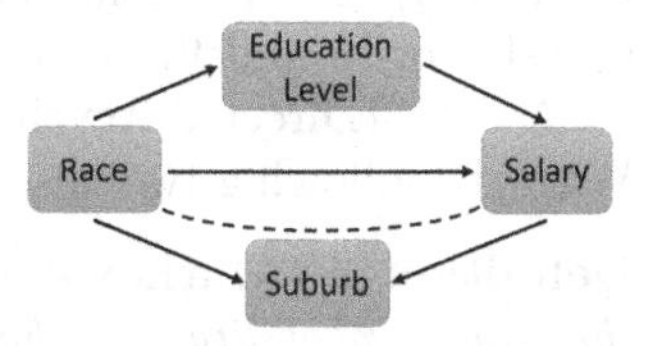

Fig. 2. The causal graph for the above example.

variable is a collider, a selection bias in each sub-population is resulted. We call this bias collider bias in this paper.

There is no existing alternative method to NST to audit classifiers. Most methods, to be reviewed in the related work, need to access the training data set and suffer from collider bias. A causal-based situation test (CST) [25] does not suffer from collider bias, but needs to access the training data set too. The audit cases used by an audit agency are only a small number of individual cases which do not represent the population. Collecting a representative sample of the population needs a significant resource. Therefore, a data-based audit method is not applicable. We make the following contributions in this paper.

- We study collider bias in fairness assessment and present theorems to avoid collider bias. Our theoretical results give a principled guidance on which variables can be used for fairness assessment and also for building fair classifiers.
- We investigate the problem of auditing machine learning models and propose an Unbiased Situation Test (UST) algorithm for auditing without accessing training data or an unbiased sample of the population. Experiments show that UST can effectively reduce collider bias.

2 Background

We present the necessary background of causal inference. We use upper case letters to represent variables and bold-faced upper case letters to denote sets of variables. The values of variables are represented using lower case letters.

Let $\mathcal{G} = (\mathbf{V}, \mathbf{E})$ be a graph, where $\mathbf{V} = \{V_1, \ldots, V_p\}$ is the set of nodes and $\mathbf{E}$ is the set of edges between the nodes, i.e. $\mathbf{E} \subseteq \mathbf{V} \times \mathbf{V}$. A path π is a sequence of distinct nodes such that every pair of successive nodes are adjacent in $\mathcal{G}$. A path π is a directed path if all edges along the path are directed edges. A path between (V_i, V_j) is a backdoor path with respect to V_i if it has an arrow into V_i. Given a path π, V_k is a collider node on π if there are two edges incident like $V_i \rightarrow V_k \leftarrow V_j$. In $\mathcal{G}$, if there exists $V_i \rightarrow V_j$, V_i is a parent of V_j and we use $Pa(V_j)$ to denote the set of all parents of V_j. In a directed path π, V_i is an ancestor of V_j and V_j is a descendant of V_i if all arrows point to V_j.

A DAG (Directed Acyclic Graph) is a directed graph without directed cycles. With the following two assumptions, a DAG links to a distribution.

Definition 1 (Markov condition [17]). *Given a DAG $\mathcal{G} = (\mathbf{V}, \mathbf{E})$ and $P(\mathbf{V})$, the joint probability distribution of $\mathbf{V}$, $\mathcal{G}$ satisfies the Markov condition if for $\forall V_i \in \mathbf{V}$, V_i is probabilistically independent of all non-descendants of V_i, given the parents of V_i.*

When the Markov condition holds, $P(\mathbf{V})$ can be factorised into: $P(\mathbf{V}) = \prod_i P(V_i \mid Pa(V_i))$.

Definition 2 (Faithfulness [20]). *A DAG $\mathcal{G} = (\mathbf{V}, \mathbf{E})$ is faithful to $P(\mathbf{V})$ iff every independence presenting in $P(\mathbf{V})$ is entailed by $\mathcal{G}$ which fulfills the Markov condition. A distribution $P(\mathbf{V})$ is faithful to a DAG $\mathcal{G}$ iff there exists DAG $\mathcal{G}$ which is faithful to $P(\mathbf{V})$.*

With the above two assumptions, we can read the independencies between variables in $P(V)$ from a DAG using the Definition 8 in Appendix A. To conduct causal inference with DAGs, we make the following assumptions.

Definition 3 (Causal sufficiency [20]**).** *A data set satisfies causal sufficiency if for every pair of variables* (V_i, V_j) *in* $\mathbf{V}$*, all their common causes are also in* $\mathbf{V}$.

With a DAG, if we interpret a node's parent as its direct cause, the DAG is known as a causal DAG. We can learn a causal DAG from data when the assumptions of causal sufficiency, faithfulness and Markov condition are satisfied.

An intervention, which forces a variable to take a value, can be represented by a *do* operator. For example, $do(X = 1)$ means X is intervened to take value 1. $P(y \mid do(X = 1))$ is an interventional probability. Let us understand *do* in an ideal experiment.

Definition 4 (Direct effect [17]**).** *The direct effect of* X *on* Y *is* $P(y \mid do(X = x), do(\mathbf{V}_{\setminus XY} = \mathbf{v}))$ *where* $\mathbf{V}_{\setminus XY}$ *means all other variables except* X *and* Y.

In order to study the relationship between X on Y, all other variable are controlled in the ideal experiment. To infer interventional probabilities (by reducing them to normal conditional probabilities) with a causal DAG, the rules of *do*-calculus [17] are necessary. Detailed description of these rules are available in Appendix A, and we used these rules to proof our theorems.

3 Problem Definition

A classifier (prediction model) has been built by a company/organisation from a training data set which contains a binary protected variable A, a binary decision outcome Y, and a set of relevant variables of Y, $\mathbf{X}$, since variables independent of Y are not used for predicting Y. An agency wants to audit the model using some cases. We make the following assumptions about the audit.

Assumption 1 *1. The regulatory policy has specified the causal relationships among the factors and* Y*, and uses a causal DAG to indicate. The factors are ancestral variables of* Y *including all direct causes of* Y.

2. The audit agency has no access to the model training data or an unbiased sample of the population. The agency however has access to the distributional statistics from some sources, such as government census data.

3. The company or organisation has used all the legitimate factors to comply with the regulatory policy. However, some other variables are also used by the model to enhance the prediction performance.

In the theorem development, we assume that there is a DAG that is consistent with the regulatory policy. In the algorithm, we do not need the complete DAG, but ancestral variables of Y and colliders in the descendant nodes of Y.

We first define the criterion for auditing. We use Controlled Direct Effect (CDE) [18] to measure fairness. CDE is extended Definition 4 to simulate an ideal experiment. The alternative definitions are path specific causal effect [4,21] and counterfactual fairness [12], we will discuss why the alternatives have not been used after Definition 5.

The protected variables in this paper include redline variables, which are the descendants of protected variables. The redline variables are recognised as a proxy of protected variables and may cause some discrimination [11]. Some companies or organisations build the models under the concept of fairness through awareness [7], which means the classifier functions may not use the protected variables as input. In this case, the redline variables will be considered as the protected variables.

Definition 5. (Fairness score). *Given a causal DAG $\mathcal{G}$ representing the regulatory policy, A, $\mathbf{X}$, and Y as described above. The fairness score is of an individual (or a subgroup) $\mathbf{X} = \mathbf{x_i}$ is defined by Controlled Direct Effect, $CDE(\mathbf{x_i}) = P(y \mid do(A = 1), do(\mathbf{X} = \mathbf{x}_i)) - P(y \mid do(A = 0), do(\mathbf{X} = \mathbf{x}_i))$, where y denotes $Y = 1$.*

The rationale of the above definition is that we conduct a controlled experiment by intervening the protected variable, and controlling all other variables to $\mathbf{x}_i$. The decision for $\mathbf{x}_i$ is fair if the intervention does not change the outcome.

Unlike previous works [7–9,15], our definition of fairness score is based on the CDE which uses intervention. Thus the spurious association between A and Y caused by conditioning on colliders will be avoided. We do not use counterfactual fairness [12] in our fairness definition since it needs stronger assumptions and poses a practical challenge. To estimate counterfactual outcomes, there is a need for knowing the full causal model and latent background knowledge. Both are not available in our problem setting. Some other definitions [4,21] make use of path specific causal effect. Their solutions also need counterfactual reasoning and they do not fit our problem setting.

Definition 6. (Problem definition). *Given $\mathcal{G}$, A and $\mathbf{X}$ as described above, and classier $\hat{Y} = f(A, \mathbf{X})$. The audit is to determine if a prediction on an individual ($\mathbf{X} = \mathbf{x_i}$) is fair, i.e. $|CDE(\mathbf{x_i})| < \tau$ where τ is a threshold determined by the regulatory policy and Y in $CDE(\mathbf{x_i})$ is replaced by $\hat{Y}$.*

4 Estimating CDE

For the sake of fairness audit, the protected variable A is assumed to be a parent node of Y so we can use CDE for the audit. The results in this section are true in general, not just for auditing classifiers. Due to page limitations, all the proofs of theorems will be presented in Appendix B.

Theorem 1. *DAG $\mathcal{G}$ contains variables A and Y, and variable set $\mathbf{X}$ where $(A \cup Y) \cap \mathbf{X} = \emptyset$. The causal sufficiency is satisfied. $P(y \mid do(A = a), do(\mathbf{X} = \mathbf{x})) = P(y \mid A = a, Pa'(Y) = \mathbf{pa})$ where $Pa'(Y)$ is the set of all parents of Y in $\mathcal{G}$ excluding A.*

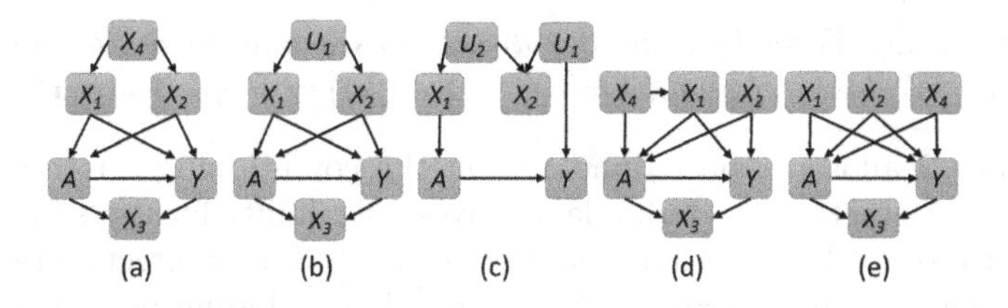

Fig. 3. DAGs for the examples of Theorem/Corollary. X_1 to X_6 are observed variables, and U_1 and U_2 are unobserved variables.

Theorem 1 removes the descendant nodes of Y from the conditioning set in the conditional probabilities for CDE estimation, and this removes possible collider bias. Furthermore, it gives a succinct set of variables for estimating CDE.

For example, in Fig. 3(a), $P(y \mid do(a), do(x_1, x_2, x_3, x_4)) = P(y \mid a, x_1, x_2)$ based on Theorem 1, where we use x_i for $X_i = x_i$. The CDE is determined by conditional probabilities on A, X_1 and X_2. Since X_3 is not used in the conditioning set, there will be no collider bias. Theorem 1 is based on the causal sufficiency assumption, which assumes that there are no unobserved common causes in the data set. In real-world applications, unobserved variables are unavoidable. When there are unobserved variables, how do we estimate CDE? The following corollary will show that they do not invalidate the result of Theorem 1.

Corollary 1. *Let $Ca(Y)$ include all the direct causes and only direct causes of Y except A. $P(y \mid do(A = a), do(\mathbf{X} = \mathbf{x})) = P(y \mid A = a, Ca(Y) = \mathbf{ca})$.*

Corollary 1 indicates that discrimination detection is sound when the audit agency knows all the direct causes of Y and uses them as the conditioning set when calculating CDE. For example, in Fig. 3(b), $P(y \mid do(a), do(x_1, x_2, x_3, U_1)) = P(y \mid a, x_1, x_2)$ based on Corollary 1. Unobserved ancestral variables of Y are blocked off from Y by X_1 and X_2, and they do not affect the probability of Y. The unobservable variables can be in the descendant nodes of Y too, but they do not affect the CDE estimation since they will not be used anyway.

We will further explain why direct causes are necessary for Corollary 1. Let Fig. 3(c) be a true DAG with two unobserved variables U_1 and U_2. X_2 is not a direct cause of Y. Since U_1 and U_2 are unobserved, X_2 is perceived as a parent of Y in the observed data. If X_2 is used to estimate CDE, the estimation will be biased since the back door path (Y, U_1, U_2, X_1, A) is opened when X_2 is conditioned on. In this case, X_1 is necessary to block the path. When both X_1 and X_2 are included, the CDE estimation is unbiased. Sometimes, we need redundancy to prevent such a biased estimation.

Both Theorem 1 and Corollary 1 give a succinct conditioning set for CDE estimation. In fact, a superset of the direct causes works as long as the superset does not contain descendant nodes of Y. In a DAG, ancestral nodes represent the direct causes and indirect causes of Y.

Corollary 2. *Let* $\mathbf{B}$ *include all the direct causes and some (or all) indirect causes of* Y. *We have* $P(y \mid do(A=a), do(\mathbf{X}=\mathbf{x})) = P(y \mid A=a, \mathbf{B}=\mathbf{b})$.

Corollary 2 allows some redundancy in the conditioning set comparing to Corollary 1. In practice, the redundancy gives a flexibility for users to determine the direct causes of Y. Sometimes, a direct cause and an indirect cause are difficult to distinguish, and Corollary 2 indicates that including both does not bias the CDE estimation. For example, in Fig. 3(d), $P(y \mid do(a), do(x_1, x_2, x_3, x_4)) = P(y \mid a, x_1, x_2) = P(y \mid a, x_1, x_2, x_4)$ based on Corollary 2 if X_1 and X_2 are all direct causes of Y. Let us assume that Fig. 3(d) is the true DAG, but a government agency has a DAG as Fig. 3(e) since they do not know which one of X_1 and X_4 is the direct cause of Y. A CDE estimation based on the imprecise DAG in Fig. 3(e), i.e. $P(y \mid do(a), do(x_1, x_2, x_3, x_4)) = P(y \mid a, x_1, x_2, x_4)$ is also unbiased.

5 Implementing Unbiased Situation Test

We summarise the discussion and propose the following unbiased situation test.

Definition 7 (Unbiased Situation Test (UST)). *UST exams whether a classifier* $\hat{Y} = f()$ *is fair for a given case* $\mathbf{x_i}$ *by calculating* $CDE(\mathbf{x_i}) = P(\hat{Y} = 1 \mid A = 1, \mathbf{B} = \mathbf{b_i}) - P(\hat{Y} = 1 \mid A = 0, \mathbf{B} = \mathbf{b_i})$, *where* $\mathbf{B}$ *is the set of direct causes and some (or all) indirect causes of* Y. *The test case* $\mathbf{x}_i$ *is discriminated if* $|CDE(\mathbf{x_i})| \geq \tau$, *where* τ *is a threshold specified by the regulatory policy.*

All variables in the problem except (A, Y) can be categorized into two types: $\mathbf{B}$ and $\mathbf{C}$. $\mathbf{B}$ is the set of ancestral nodes of Y which can be identified by the regulatory policy, and $\mathbf{C}$ includes others. Note that irrelevant variables which are independent of Y are not in $\mathbf{X}$.

To conduct UST as in Definition 7, one problem is that an audit agency cannot obtain the conditional probability $P(\hat{Y} = 1 \mid A = a, \mathbf{B} = \mathbf{b}_i)$ directly since it does not access the training data set or a unbiased sample of the population.

Algorithm 1. Unbiased Situation Test (UST)

Input: Classifier $f()$, $\mathbf{X} = \mathbf{B} \cup \mathbf{C}$ as defined in the text. $P(\mathbf{C} = \mathbf{c_i})$. Test cases D_{Test}. The threshold τ.

Output: L, a list of discriminated cases in D_{Test}.

1: **for** each $r_i \in D_{Test}$ **do**
2: Let r_i' be the record by flipping the value of A in r_i
3: Let $P(\hat{Y} = 1 \mid r_i) = f(r_i)$ and $P(\hat{Y} = 1 \mid r_i') = f(r_i')$
4: Obtain $P(\hat{Y} = 1 \mid A = A(r_i), \mathbf{B} = B(r_i))$ and $P(\hat{Y} = 1 \mid A = A(r_i'), \mathbf{B} = B(r_i'))$ by Equation 1 where $A()$ and $B()$ return values of A and $\mathbf{B}$ in the records respectively
5: Conduct situation test by Definition 7 and update L
6: **end for**
7: Return L

Instead, it can have $P(\hat{Y} = 1 \mid A = a, \mathbf{B} = \mathbf{b_i}, \mathbf{C} = \mathbf{c_i})$ from the classifier $f()$. Therefore, the following marginalisation is used:

$$P(\hat{y} \mid A = a, \mathbf{B} = \mathbf{b_i}) = \sum_{\mathbf{c_i} \in \mathbf{C}} P(\hat{y} \mid a, \mathbf{b_i}, \mathbf{C} = \mathbf{c_i}) P(\mathbf{c_i}) \tag{1}$$

where $\hat{y}$ denotes $\hat{Y} = 1$, and probability $P(\mathbf{c_i})$ can be obtained from some sources, such as government census data. The algorithm for UST is presented in Algorithm 1. The complexity for UST algorithm is $O(n)$, where n is the size of D_{Test}, i.e. linear to the number of test records.

6 Experiments

In this section, we first demonstrate UST algorithm can correct the spurious associations generated by collider. Then, we simulate the audit process by using real-world data set. We only compare UST with NST in population-level sampling since other situation test methods, such as, CST [25], k-NN based situation test [15] need to access training data set which is unavailable in our problem. We also demonstrate that data-based audit method fails in unrepresentative sampling but UST works. Finally, we apply the UST algorithm to compare fairness for different models and guide the audit agency to choose the model. The experimental settings and details can be found in the full version [22].

6.1 Correcting Collider Biases

We construct synthetic data sets including a collider as discussed in the full version [22]. UST has significantly reduced biases in CDE estimation. Bias is used to measure the error between an estimated CDE and the true CDE. Biases of including and not including a collider are shown in Table 2. The later is the UST method which corrects biases of a collider in data by directly using Corollary 2.

Table 2. UST has significantly reduced biases caused by a collider.

Trials	Bias (with collider)	Bias (UST)
1	0.143 ± 0.011	0.072 ± 0.004
2	0.154 ± 0.013	0.074 ± 0.004
3	0.149 ± 0.012	0.066 ± 0.004
4	0.149 ± 0.014	0.067 ± 0.004
5	0.152 ± 0.012	0.069 ± 0.004

Table 3. Suburb variable (collier) improves the accuracy of classification models.

	Acc. w/ Sub	Acc. w/o Sub
DT	89.66%	81.01%
SVM	89.60%	80.93%
RF	89.81%	80.86%
NN	89.64%	80.85%

6.2 Simulating an Audit Process Using Adult Data Set

The Adult data set from UCI Machine Learning Repository [1] is used to simulate audit process as shown in Fig. 4. We use the Adult data set as the population for generating the ground truths. A company has a sample (50%) as the private data to build a model. The red dashed line represents the information that the audit agency has access to. The ground truths are generated from the population and all causes of Y.

Race is the protected variable and Salary is the outcome. Other variables are Education_level, Marriage_statues, Work_hour, Work_class, and they all determine the salary. We simulate a Suburb variable as a collider. The accuracy of a classifier is significantly higher when the Suburb is used than not as shown in Table 3. The accuracy improvement is due to the spurious associations.

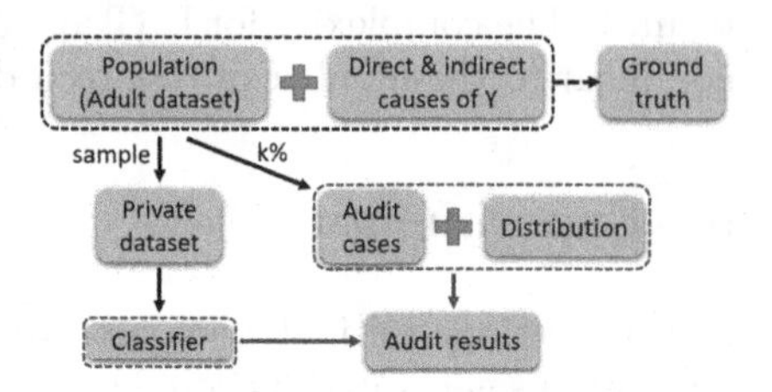

Fig. 4. A simulation of audit process using Adult data set

6.3 Comparing the Audit Performance of NST and UST

We apply UST to audit a few well-known classifiers built from sample data set. NST (introduced in the introduction) is used for the comparison since it is the only method for assessing the fairness of a classifier without accessing the training data set. We use precision and recall for the comparison. The ground truth for each audit case is calculated by using the population and the causes of Y. Audit cases are k% records randomly selected from the population. For each k, we resample audit cases 10 times and report the average precision and recall.

UST outperforms NST in both precision and recall as shown in Table 4. With the increasing number of audit cases, the deviations of both methods decrease. From the gaps between the precision and recall of NST and UST, we see that the collider bias deteriorates the detection performance of NST significantly.

Table 4. The audit performance comparison of NST and UST. The higher values are highlighted. The standard error is shown in brackets.

		k = 0.1%		k = 0.5%		k = 1%	
		NST	UST	NST	UST	NST	UST
DT	Recall	59.6%(0.96)	**79.8%(0.11)**	56.7%(0.27)	**73.3%(0.05)**	56.3%(0.05)	**71.7%(0.10)**
	Precision	84.4%(0.32)	**98.1%(0.16)**	78.9%(0.06)	**99.1%(0.01)**	80.3%(0.02)	**98.8%(0.01)**
SVM	Recall	77.9%(1.18)	**87.8%(0.26)**	75.7%(0.17)	**89.9%(0.03)**	74.1%(0.06)	**89.1%(0.02)**
	Precision	68.1%(0.73)	**83.4%(0.14)**	65.1%(0.11)	**79.9%(0.06)**	64.6%(0.13)	**81.0%(0.05)**
RF	Recall	56.1%(1.91)	**73.8%(0.32)**	58.2%(0.23)	**66.8%(0.03)**	57.7%(0.09)	**65.2%(0.14)**
	Precision	88.6%(0.17)	**96.4%(0.25)**	86.8%(0.02)	**97.8%(0.01)**	86.9%(0.04)	**98.4%(0.01)**
NN	Recall	65.9%(1.17)	**74.9%(0.14)**	67.6%(0.16)	**71.5%(0.10)**	67.9%(0.06)	**69.2%(0.08)**
	Precision	85.9%(0.24)	**96.7%(0.21)**	81.7%(0.04)	**97.1%(0.01)**	82.8%(0.02)	**97.2%(0.01)**

6.4 Data Based Audit May Be Biased

Removing the collider from the data can be used as an alternative method to UST. However, a data-based audit (DBA) relies on the representativeness of the audit cases for the population. The representativeness is difficult to be ensured because individuals who receive unfair treatments likely require the authority to audit their results. An audit agency does not have a resource to collect a representative sample for auditing. We simulate the unrepresentative audit cases by over (under) sampling discriminatory cases in the population. In the Adult data set, about 15% of the individuals are discriminatory and this ratio is the baseline. We vary discriminatory ratios of 1% data set.

The performance of DBA deteriorates significantly when a discriminatory ratio deviates from the baseline as shown in Table 5. Note that all discrimination detection methods based on data have the same limitation. In contrast, UST maintains similar performance.

Table 5. The audit performance comparison of DBA and UST with discriminatory sample. The higher values are highlighted. The standard error is shown in brackets.

		Discriminatory Ratio=0%		Discriminatory Ratio=10%		Discriminatory Ratio=20%	
		DBA	UST	DBA	UST	DBA	UST
DT	Recall	60.1%(1.18)	**72.2%(0.06)**	57.8%(1.72)	**71.7%(0.07)**	60.9%(0.91)	**71.8%(0.10)**
	Precision	84.6%(0.23)	**97.5%(0.01)**	78.2%(0.35)	**96.2%(0.01)**	70.9%(0.11)	**95.1%(0.01)**
SVM	Recall	39.8%(1.64)	**89.0%(0.01)**	40.0%(1.51)	**89.5%(0.01)**	35.6%(2.57)	**89.8%(0.01)**
	Precision	77.8%(0.19)	**82.5%(0.03)**	71.2%(0.31)	**76.0%(0.05)**	56.4%(1.02)	**67.7%(0.06)**
RF	Recall	39.9%(0.18)	**65.6%(0.06)**	40.4%(0.14)	**65.2%(0.06)**	39.5%(1.13)	**65.2%(0.05)**
	Precision	78.9%(0.14)	**97.3%(0.02)**	72.6%(0.16)	**96.4%(0.02)**	61.2%(0.26)	**94.2%(0.05)**
NN	Recall	46.1%(2.02)	**69.8%(0.07)**	45.7%(2.11)	**69.8%(0.06)**	39.4%(2.26)	**69.4%(0.09)**
	Precision	80.6%(0.11)	**95.0%(0.03)**	73.6%(0.29)	**93.0%(0.03)**	60.0%(0.51)	**90.1%(0.07)**

6.5 Rank Models Based on Fairness

We show that UST can be used for comparing the fairness of different models. We first discuss the metrics for the comparison. After discrimination detection on a model using audit cases with ground truths, we obtain True Positive (TF), False Positive (FP), True Negative (TN), and False Negative (FN). FN indicates the cases that are unfair but are corrected to be fair by the model. They are favourable for fair predictions, and we use correction rate, $CR = \frac{FN}{TP+FN}$, to represent the proportion of true unfair cases being corrected by a model. In contrast, FP represents that the cases that are fair become unfair after model predictions. These cases are called reversed discrimination and are unfavourite for predictions. We use the reversion rate, $RR = \frac{FP}{FP+TN}$, to represent the proportion of fair cases being reversed by a model. We wish the revision rate is as small as possible.

The CR and RR of four classifiers are shown in Table 6. Random Forest is the fairest model based on the two measures. Random Forest has corrected 35.24% of unfair cases, and only reversed 1.07% of fair cases to unfair. Note that, their prediction accuracies are very similar, but their CR and RR are different. This assessment shows that some errors made by a model are better than others in terms of the fairness.

Table 6. The audit results of various models.

	CR(↑)	RR(↓)
DT	29.09% ± 0.079	1.69% ± 0.001
SVM	10.05% ± 0.024	39.37% ± 0.049
RF	35.24% ± 0.147	1.07% ± 0.001
NN	31.48% ± 0.064	2.76% ± 0.001

7 Related Works

The work belongs to discrimination detection. Detection methods are divided into the group, and individual-based. Another division of the methods is association or causal based.

At the group level, a number of metrics have been defined to detect discrimination. Demographic parity, a well-known fairness measurement, is defined by [7]. Other measurements including equalised odds [9], predictive rate parity [23]. However, these group-based fairness does not necessarily mean individual fairness. Many algorithms focus on detecting discrimination at the individual level. Authors in [19] use existing inequality indices from economics to measure individual level fairness. Speicher et al. [14] propose an individual level discrimination detector, which is used to prioritise data samples and aims to improve the subgroup fairness measure of disparate impact.

Under the causal framework, Li et al. [13] use the (conditional) average causal effect to quantify fairness for (sub)group level discrimination detection. Counterfactual fairness [12] is an attractive definition of individual level fairness measurements by causality. It means that a decision is fair towards an individual if it is the same in both the actual world and a counterfactual world (when a value of a protected variable is changed). However, it needs strong assumptions. Zhang et al. [26] use nature direct effect and nature indirect effect to quantify fairness. The path-specific causal effect [4,21] have been used to quantify fairness when the regulatory policy recognises some causal paths involving a protected variable fair. Nature direct (indirect) effect and path-specific causal effect all need counterfactual reasoning and are difficult to implement in practice since the strong assumptions are related to counterfactual reasoning.

Situation test related work has been discussed in the introduction. The above-mentioned related work only introduces some main influential contributions. For more related work, please refer to the literature review [3,6,16,24].

8 Conclusions

In this paper, we have discussed collider bias in fairness assessment. We have presented theoretical results based on the graphical causal model to avoid collider biases in fairness assessment. The results are useful for discrimination detection and also for feature selection for building fair classifiers. We have proposed an Unbiased Situation Test (UST) algorithm for the fairness assessment of a classifier without accessing the training data set or a sample of the population. Experimental results show that UST effectively reduces collider biases and can be used to assess the fairness of a classifier without accessing to data. The UST is promising for an audit agency to audit machine learning models by private companies and organisations.

A Additional Definition and Theorem

Definition 8 (*d*-separation [17]). *A path π in a DAG is said to be d-separated (or blocked) by a set of nodes $\mathbf{Z}$ iff (1) π contains a chain $V_i \rightarrow V_k \rightarrow V_j$ and a fork $V_i \leftarrow V_k \rightarrow V_j$ node such that the middle node V_k is in $\mathbf{Z}$, or (2) π contains a collider V_k such that V_k is not in $\mathbf{Z}$ and no descendant of V_k is in $\mathbf{Z}$.*

Theorem 2 (Rules of *do*-Calculus [17]). *Let $\mathbf{X}, \mathbf{Y}, \mathbf{Z}, \mathbf{W}$ be arbitrary disjoint sets of variables in a causal DAG $\mathcal{G}$. The following rules hold, where $\mathbf{x}, \mathbf{y}, \mathbf{z}, \mathbf{w}$ are the shorthands of $\mathbf{X} = \mathbf{x}, \mathbf{Y} = \mathbf{y}, \mathbf{Z} = \mathbf{z}$ and $\mathbf{W} = \mathbf{w}$ respectively.*

Rule 1. (Insertion/deletion of observations):
$P(\mathbf{y} \mid do(\mathbf{x}), \mathbf{z}, \mathbf{w}) = P(\mathbf{y} \mid do(\mathbf{x}), \mathbf{w})$, *if* $(\mathbf{Y} \perp\!\!\!\perp \mathbf{Z} \mid \mathbf{X}, \mathbf{W})$ *in* $\mathcal{G}_{\overline{\mathbf{X}}}$.
Rule 2. (Action/observation exchange):
$P(\mathbf{y} \mid do(\mathbf{x}), do(\mathbf{z}), \mathbf{w}) = P(\mathbf{y} \mid do(\mathbf{x}), \mathbf{z}, \mathbf{w})$, *if* $(\mathbf{Y} \perp\!\!\!\perp \mathbf{Z} \mid \mathbf{X}, \mathbf{W})$ *in* $\mathcal{G}_{\overline{\mathbf{X}}\underline{\mathbf{Z}}}$.
Rule 3. (Insertion/deletion of actions):
$P(\mathbf{y} \mid do(\mathbf{x}), do(\mathbf{z}), \mathbf{w}) = P(\mathbf{y} \mid do(\mathbf{x}), \mathbf{w})$, *if* $(\mathbf{Y} \perp\!\!\!\perp \mathbf{Z} \mid \mathbf{X}, \mathbf{W})$ *in* $\mathcal{G}_{\overline{\mathbf{X}\mathbf{Z}(\mathbf{W})}}$, *where* $\mathbf{Z}(\mathbf{W})$ *is the nodes in* $\mathbf{Z}$ *that are not ancestors of any node in* $\mathbf{W}$ *in* $\mathcal{G}_{\overline{\mathbf{X}}}$.

B Proofs

B.1 Proof of Theorem 1

Theorem 1. DAG $\mathcal{G}$ contains variables A and Y, and variable set $\mathbf{X}$ where $(A \cup Y) \cap \mathbf{X} = \emptyset$. The causal sufficiency is satisfied. $P(y \mid do(A = a), do(\mathbf{X} = \mathbf{x})) = P(y \mid A = a, Pa'(Y) = \mathbf{pa})$ where $Pa'(Y)$ is the set of all parents of Y in $\mathcal{G}$ excluding A.

Proof. Firstly, let $\mathbf{X} = \{\mathbf{C} \cup \mathbf{Q}\}$ where $\mathbf{C}$ contains descendant nodes of Y, and $\mathbf{Q}$ contains non-descent nodes of Y. We have $P(y \mid do(A = a), do(\mathbf{C} = \mathbf{c}), do(\mathbf{Q} = \mathbf{q})) = P(y \mid do(A = a), do(\mathbf{Q} = \mathbf{q}))$. This is achieved by repeatedly using Rule 3 of Theorem 2. We show this by an example where $C \in \mathbf{C}$, $P(y \mid do(A = a), do(C = c), do(\mathbf{Q} = \mathbf{q})) = P(y \mid do(A = a), do(\mathbf{Q} = \mathbf{q}))$ because $Y \perp\!\!\!\perp C$ in DAG $\mathcal{G}_{\overline{A}, \overline{C}}$ where the incoming edges to A and to C have been removed.

Secondly, we consider $P(y \mid do(A = a), do(\mathbf{Q} = \mathbf{q}))$ only. Based on the Markov condition 1, Y is independent of all its non-descendant nodes given its parents. Therefore, $P(y \mid do(A = a), do(\mathbf{Q} = \mathbf{q})) = P(y \mid do(A = a), do(Pa'(Y) = \mathbf{pa}))$.

Thirdly, we will prove $P(y|do(A = a), do(Pa'(Y) = \mathbf{pa})) = P(y \mid A = a, Pa'(Y) = \mathbf{pa})$. This can be achieved by repeatedly applying Rule 2 of Theorem 2.

Let $Pa(Y) = \{A, X_1, X_2, \ldots, X_k\}$.

$$
\begin{aligned}
&P(y \mid do(A = a), do(X_1 = x_1), do(X_2 = x_2), \ldots, do(X_k = x_k)) \\
&= P(y \mid A = a, do(X_1 = x_1) do(X_2 = x_2), \ldots, do(X_k = x_k)) \\
&\textit{Since } Y \perp\!\!\!\perp A | X_1, X_2, \ldots, X_k \textit{ in } \mathcal{G}_{\overline{X_1}, \overline{X_2}, \ldots, \overline{X_k}, \underline{A}} \\
&= P(y \mid A = a, X_1 = x_1, do(X_2 = x_2), \ldots, do(X_k = x_k)) \\
&\textit{Since } Y \perp\!\!\!\perp X_1 | A, X_2, \ldots, X_k) \textit{ in } \mathcal{G}_{\overline{X_2}, \ldots, \overline{X_k}, \underline{X_1}} \\
&\textit{Repeat } k - 1 \textit{ times} \\
&= P(y \mid A = a, X_1 = x_1, X_2 = x_2, \ldots, X_k = x_k) \\
&= P(y \mid A = a, Pa'(Y) = \mathbf{pa})
\end{aligned}
$$

Now, we get,

$$P(y \mid do(A = a), do(\mathbf{X} = \mathbf{x})) = P(y \mid A = a, Pa'(Y) = \mathbf{pa})$$

B.2 Proof of Corollary 1

Corollary 1. Let $Ca(Y)$ include all the direct causes and only direct causes of Y except A. $P(y \mid do(A = a), do(\mathbf{X} = \mathbf{x})) = P(y \mid A = a, Ca(Y) = \mathbf{ca})$.

Proof. Direct causes of Y will be parent nodes of Y in any DAG even when the unobserved common causes are included, i.e. the causal sufficiency is unsatisfied. Since $Pa'(Y) = Ca(Y)$ and there is not an unobserved variable in between a direct cause and Y, $P(y \mid do(A = a), do(\mathbf{X} = \mathbf{x})) = P(y \mid A = a, Ca(Y) = \mathbf{ca})$ can be derived following the same procedure in Theorem 1.

Since other variables apart from $Pa'(Y)$ are not used in reducing $P(y \mid do(A = a), do(\mathbf{X} = \mathbf{x}))$, the unobserved common casues between these variables are irrelevant to the deduction and do not affect the above conclusion.

B.3 Proof of Corollary 2

Corollary 2. Let $\mathbf{B}$ include all the direct causes and some (or all) indirect causes of Y. We have $P(y \mid do(A = a), do(\mathbf{X} = \mathbf{x})) = P(y \mid A = a, \mathbf{B} = \mathbf{b})$.

Proof. Let $\mathbf{B} = Ca'(Y) \cup \mathbf{R}$, and $\mathbf{R}$ includes indirect causes of Y. Following Corollary 1, $P(y \mid do(A = a), do(\mathbf{X} = \mathbf{x})) = P(y \mid A = a, Ca'(Y) = \mathbf{ca})$. Based on the Markov condition 1, Y is independent of $\mathbf{R}$ given $A \cup Ca'(Y)$. Hence, $P(y \mid A = a, Ca'(Y) = \mathbf{ca}) = P(y \mid A = a, \mathbf{B})$.

References

1. Asuncion, A., Newman, D.: UCI Machine Learning Repository (2007)
2. Bendick, M.: Situation testing for employment discrimination in the United States of America. Horizons stratégiques **3**, 17–39 (2007)
3. Caton, S., Haas, C.: Fairness in machine learning: a survey (2020). arXiv preprint arXiv:2010.04053
4. Chiappa, S.: Path-specific counterfactual fairness. In: AAAI, pp. 7801–7808 (2019)
5. Cole, S.R., et al.: Illustrating BIAS due to conditioning on a collider. Int. J. Epidemiol. **39**(2), 417–20 (2010)
6. Corbett-Davies, S., Goel, S.: The measure and mismeasure of fairness: a critical review of fair machine learning (2018). arXiv preprint arXiv:1808.00023
7. Dwork, C., Hardt, M., Pitassi, T., Reingold, O., Zemel, R.: Fairness through awareness. In: Proceedings of the 3rd Innovations in Theoretical Computer Science Conference, pp. 214–226 (2012)
8. Feldman, M., Friedler, S.A., Moeller, J., Scheidegger, C., Venkatasubramanian, S.: Certifying and removing disparate impact. In: KDD, pp. 259–268 (2015)
9. Hardt, M., Price, E., Srebro, N.: Equality of opportunity in supervised learning. In: NeurIPS, pp. 3315–3323 (2016)
10. Hernán, M.A., Robins, J.M.: Causal Inference: What If. Chapman & Hall/CRC, Boca Raton (2020)
11. Kilbertus, N., Rojas-Carulla, M., Parascandolo, G., Hardt, M., Janzing, D., Schölkopf, B.: Avoiding discrimination through causal reasoning. In: NeurIPS, pp. 656–666 (2017)
12. Kusner, M., Loftus, J., Russell, C., Silva, R.: Counterfactual fairness. In: NeurIPS, pp. 4069–4079 (2017)
13. Li, J., Liu, J., Liu, L., Le, T.D., Ma, S., Han, Y.: Discrimination detection by causal effect estimation. In: BigData, pp. 1087–1094. IEEE (2017)
14. Lohia, P.K., Ramamurthy, K.N., Bhide, M., Saha, D., Varshney, K.R., Puri, R.: Bias mitigation post-processing for individual and group fairness. In: ICASSP, pp. 2847–2851. IEEE (2019)
15. Luong, B.T., Ruggieri, S., Turini, F.: K-NN as an implementation of situation testing for discrimination discovery and prevention. In: KDD, pp. 502–510 (2011)
16. Mehrabi, N., Morstatter, F., Saxena, N., Lerman, K., Galstyan, A.: A survey on BIAS and fairness in machine learning. ACM Comput. Surv. **54**(6), 1–35 (2021)
17. Pearl, J.: Causality. Cambridge University Press (2009)
18. Pearl, J., Mackenzie, D.: The Book of Why. Basic Books, New York (2018)
19. Speicher, T., et al.: A unified approach to quantifying algorithmic unfairness: measuring individual & group unfairness via inequality indices. In: KDD, pp. 2239–2248 (2018)
20. Spirtes, P., Glymour, C.N., Scheines, R., Heckerman, D.: Causation, Prediction, And Search. MIT Press (2000)
21. Wu, Y., Zhang, L., Wu, X., Tong, H.: Pc-fairness: a unified framework for measuring causality-based fairness. In: NeurIPS, vol. 32 (2019)
22. Xu, Z., et al.: Assessing Classifier Fairness With Collider Bias (2022). arXiv preprint arXiv:2010.03933
23. Zafar, M.B., Valera, I., Gomez Rodriguez, M., Gummadi, K.P.: Fairness beyond disparate treatment & disparate impact: Learning classification without disparate mistreatment. In: WWW, pp. 1171–1180 (2017)

24. Zhang, L., Wu, X.: Anti-discrimination learning: a causal modeling-based framework. Int. J. Data Sci. Anal. **4**(1), 1–16 (2017). https://doi.org/10.1007/s41060-017-0058-x
25. Zhang, L., Wu, Y., Wu, X.: Situation testing-based discrimination discovery: a causal inference approach. In: IJCAI, pp. 2718–2724 (2016)
26. Zhang, L., Wu, Y., Wu, X.: A causal framework for discovering and removing direct and indirect discrimination. In: IJCAI, pp. 3929–3935 (2017)

Hard Negative Sample Mining for Contrastive Representation in Reinforcement Learning

Qihang Chen, Dayang Liang, and Yunlong Liu(✉)

Department of Automation, Xiamen University, Xiamen, China
ylliu@xmu.edu.cn

Abstract. In recent years, contrastive learning has become an important technology of self-supervised representation learning and achieved SOTA performances in many fields, which has also gained increasing attention in the reinforcement learning (RL) literature. For example, by simply regarding samples augmented from the same image as positive examples and those from different images as negative examples, instance contrastive learning combined with RL has achieved considerable improvements in terms of sample efficiency. However, in the contrastive learning-related RL literature, the source images used for contrastive learning are sampled in a completely random manner, and the feedback of downstream RL task is not considered, which may severely limit the sample efficiency of the RL agent and lead to sample bias. To leverage the reward feedback of RL and alleviate sample bias, by using gaussian random projection to compress high-dimensional image into a low-dimensional space and the Q value as a guidance for sampling the hard negative pairs, i.e. samples with similar representation but diverse semantics that can be used to learn a better contrastive representation, we propose a new negative sample method, namely Q value-based Hard Mining (QHM). We conduct experiments on the DeepMind Control Suite and show that compared to the random sample manner in vanilla instance-based contrastive method, our method can effectively utilize the reward feedback in RL and improve the performance of the agent in terms of both sample efficiency and final scores, on 5 of 7 tasks.

Keywords: Contrastive learning · Reinforcement learning · Random projection · Hard negative sample mining

1 Introduction

With the development of deep neural network and by combining its feature extraction power with the decision ability of reinforcement learning (RL), Deep RL (DRL) has been widely and successfully applied to dozens of tasks with high dimensional inputs [6]. However, how to endow an agent with the ability to quickly master the task with less interactions is still a challenge nowadays. While

J. Gama et al. (Eds.): PAKDD 2022, LNAI 13281, pp. 277–288, 2022.
https://doi.org/10.1007/978-3-031-05936-0_22

model-based RL algorithms try to build and maintain an environment model which will help agent planning and making full use of precious interaction data, they usually suffer from enormous computations for planning and fragile model accuracy [7]. On the other side, model free RL algorithms directly learn from raw observation space, but plenty of training data is indispensable for good performance [8]. It's widely believed that a policy directly trained from the real state data will act better than those with raw and high dimensional inputs [9]. Hence a key to efficiency promotion for model free RL agents is to acquire good representations [1]. Lots of auxiliary tasks have been incorporated in traditional RL algorithms to accelerate better representation learning, such as auto-encoder [10], prediction [17,18,37], prototypes [11], goals [5] and so on. Recently, contrastive learning has make significant progress in natural language processing and computer vision areas [12,32], which has also been incorporated into RL algorithms. Srinivas et al. proposed an instance contrastive method called Contrastive Unsupervised Representations for Reinforcement Learning (CURL), which is the first contrastive based algorithm that beats model-based methods like PlaNet [15] and Dreamer [16] on several tasks of DeepMind Control Suite (DMC) [9].

As a self-supervised learning method, contrastive learning tries to define and contrast semantically similar (positive) pairs and semantically dissimilar (negative) pairs in the embedding space [18]. The success of contrastive methods mainly depends on the design of correct positive and negative pairs [19]. In RL setting, naively we can regard the real states as the label of observations and let contrastive learning gathers samples of the same state and pushes away those with different states. However, due to the limitation of perception, agents generally can not access to the real states. Under such circumstances, existing contrastive RL methods design positive or negative pairs in a random or unsupervised way, where the feedback of RL tasks is completely ignored and positive samples may sneak into negative samples. Such false-negative phenomenon is known as *sampling bias*. It may empirically induce to significant performance deterioration in some fileds [20].

Moreover, a plenty of work in metric learning believe that *hard negative samples* dominate the quality and efficiency of the representation learning [22, 36], where *hard negative samples* are the true negative samples that mapped nearby the anchor sample in the embedding space [21]. To mine hard negative samples and improve the sample efficiency of RL agents, by observing that hard negative samples are samples that look similar but with different semantics, in this paper, we propose a new hard negative samples mining method, namely Q-value based Hard Mining (QHM).

In the detail, in order to seek for samples embedded similarly, gaussian random projection and KD-Tree are firstly applied in QHM for dimensionality reduction and search. Then to further filtering semantically different pairs among these similar samples, as real states are inaccessible, we can take the advantage of cumulative reward Q value as a guidance for mining since it is the most key feedback of any RL tasks. Consequently, with the assistance of a handy K-means cluster method, QHM approximately treats those similar observation-action pairs but with different Q value in recent trajectories as hard negative samples pairs. Equipped with these unsupervised techniques, QHM is able to latch the RL task feedback and efficiently solve two key problems: (1) How to design task-relevant positive-negative pairs for contrastive representation learning in RL? (2) How to mine and exploit hard negative samples?

We conduct experiments on DMC and show that compared to contrastive learning with vanilla random sample method, our sample method combined with instance-based contrastive learning in RL can achieve better data efficiency and even better score performance on several tasks.

2 Related Work

2.1 Improving Sample Efficiency in RL

It is well known that learning policy directly from high dimensional data such as raw pixel images is inefficient [2]. Model-based RL agent builds an environment model and generates virtual rollouts to help better decisions, which is usually more data efficient than model-free agent. The related methods like SimPLe [2], PlaNet [15] and Dreamer [16] have successfully improve the data efficiency in Atari Games [13] and DMC [9], and even make a breakthrough on some challenging tasks such as MONTEZUMA'S REVENG. For the model-free approaches, to improve performance and efficiency, the agent mainly focus on constructing and adopting various auxiliary tasks, such as predicting future [17,18,37], prototypes cluster [11], particle-based entropy maximization [4] or multi-goals [5].

In recent years, contrastive learning is also incorporated into RL as an auxiliary task. Typical works include instance contrastive learning based CURL [1] and CPC [18] that leverages prediction information. Subsequent works also tried to use contrastive learning to force agents to learn temporal features [23]. Although these approaches have achieved some successes in various domains, the pairs in these contrastive learning methods are sampled in a random or unsupervised manner, and the possible signals that may help representation learning are not considered. In the work of Guoqing et al. [25], a return-based contrastive representation for RL (RCRL) method is introduced, where observation-action samples with similar cumulative rewards are regarded as positive pairs and vice versa. While the cumulative reward is used for the sampling in both RCRL and our approach, QHM further considers about the *hard* property of negative samples and uses a more adaptive manner to partition the experience buffer, where the explicit model structure or learning objective does not need to be changed.

Besides using auxiliary tasks for learning a better representation, recent work such as RAD [26], DrQ [24], DrQ-v2 [27] also show that simple combination of image augmentation is conducive to the improvement of data efficiency.

2.2 Sample Strategy in Unsupervised Contrastive Learning

Contrastive learning encourages semantically similar pairs (x, x^+) to be close and semantically dissimilar pairs (x, x^-) to be more distant in embedding space

$f(*)$ [34]. Since the labels of data are unknown under unsupervised conditions, the main differences among these methods are their strategies of obtaining positive and negative pairs [20]. In the literature, strategies including random crop, jittering in images [26,33] and random dropout in text missions [31] are commonly used to select positive samples, while less attention has been paid in the sampling of negative pairs and they are only simply sampled uniformly from the training data [19]. There exists two problems in randomly picking negative pairs. First, false negative samples will give rise to *sample bias* which is impossible to completely dismiss under unsupervised situations [20]. Second, we cannot ensure how informative the negative samples will be when they serve the downstream tasks [19]. The key to address the mentioned issues is hard negative mining, which in metric learning is well elucidated and proved to be most helpful for efficient representation learning [22]. But how to mine such hard negative samples for unsupervised contrastive learning? Based on Debiased Contrastive Loss (DCL) [20], Robinson et al. proposed to define the priority of a sample proportional to its similarity with the anchor to acquire hard samples, which has made a certain progress in images and sentences representations [19]. Wang et al. found that the choice of temperature τ in contrastive loss controls the granularity of penalties on hard negative samples [35].

3 Background

3.1 Instance Contrastive Learning in RL

In general, considering an embedding space $f(*)$, contrastive learning tries to gather the representations of positive pairs (x, x^+) but push away the representations of negative pairs (x, x^-):

$$\mathbb{E}_{x,x^+,\{x_i^-\}_{i=1}^N}[-log\frac{e^{f(x)^T f(x^+)}}{e^{f(x)^T f(x^+)} + \sum_{i=1}^N e^{f(x)^T f(x^-)}}] \tag{1}$$

Given an anchor x, a corresponding positive sample x^+ and N negative samples x^- will be used for contrast. For *Instance Discrimination* [14], x and x^+ are different views generated from the same sample whilst x and x^- are from different samples. CURL is the first method that combines *Instance Discrimination* with RL where views of different images are accomplished by random crop. In detail, given a batch of randomly sampled K raw-pixel images, for each of them $x_i (1 \leq i \leq K)$, we have:

$$x_{i1} = aug(x_i) \quad x_{i2} = aug(x_i) \tag{2}$$

where $aug(*)$ represents a fixed random method of data augmentation such as random crop. CURL simply takes samples (x_{i1}, x_{i2}) as positive pairs and $(x_{i*}, x_{j*})(j \neq i)$ as negative pairs according to whether they are generated from the same image. All these generated $2K$ samples will be used for the InfoNCE loss [18]:

$$\mathcal{L}_{CURL} = -log \frac{e^{z_q^T W z_k^+}}{e^{z_q^T W z_k^+} + \sum_{i=1}^{K} e^{z_q^T W z_{ki}^-}} \tag{3}$$

In Eq. (3), z_q are the encoded low-dimentional representations of cropped images x_{i1} through the query encoder $f_{\theta q}$ of the RL agent while z_k are from key encoder $f_{\theta k}$. Query and key encoders share the same neural framework but have different parameters weights. Similar to Moco [12], CURL detaches the gradient of the key encoder $f_{\theta k}$ whose parameters θ_k can be only updated by exponentially moving average (EMA) method as follow:

$$\theta_k = m\theta_k + (1-m)\theta_q \tag{4}$$

where $m \in [0,1]$ is a factor of trading off and such an update method has been proved to be helpful in improving agent's performance and avoiding model collapse [1,12].

In DMC tasks, CURL takes a SAC agent as the base policy learner by using Eq. (3) to learn contrastive representations. Our work will build upon CURL and aim to improve the completely random sample method to aquire more task-relevant negative samples for the calculation of Eq. (3).

3.2 Gaussian Random Projection

Gaussian random projection is a simple and convenient projection method to reduce high-dimensional space to low dimension. It defines a mapping function $\phi : x \rightarrow Px \in \mathbb{R}^F$ where $x \in \mathbb{R}^D$ is the original data with dimension D and will be multiplied by a random initialized matrix $P \in \mathbb{R}^{F \times D}$ to be transformed to a space of dimension F. Generally we have $F \ll D$ and each element P_{ij} in P are sampled independently from a predefined gaussian distribution $N(\mu, \sigma^2)$. According to Johnson-Lindenstrauss lemma [28], for arbitrary x_i and x_j, there exists a $\epsilon(0 \leq \epsilon \leq 1)$ and a map function ϕ that satisfy:

$$(1-\epsilon) \parallel x_i - x_j \parallel \leq \parallel \phi(x_i) - \phi(x_j) \parallel \leq (1+\epsilon) \parallel x_i + x_j \parallel \tag{5}$$

As showed in Eq. (5), the distance relationship can be well preserved in the mapped low-dimensional space even with a random initialized matrix as long as D is sufficient [29]. In our method, in order to search for hard negative samples, a computation efficient search method is urgently-needed and as it is time consuming and computation expensive by directly searching in the raw pixel space, gaussian random projection are adopted for reducing the high-dimensional input to a low-dimensional space.

4 Q Value Based Hard Mining

In this section, we will introduce QHM, a contrastive learning sampling method based on cumulative rewards. Our intention is to improve the unreasonable random sampling method of contrastive learning in RL such as CURL and try to use the reward feedback of a specific task to guide the sampling strategy of positive-negative pairs. So that the samples eventually used for contrastive training are semantically *mutually exclusive* in RL setting, which will also contribute to efficiency promotion. The most ideal result is that the samples divided into positive and negative pairs will not belong to the same real state.

4.1 Construct Task-Relevant Positive and Negative Pairs in RL

In contrastive learning, given an arbitrary anchor sample x, a sample x^+ is positive when it is semantically similar to anchor x and takes x^- as negative vice versa. In RL, considering an one-step observation o_t composed of successive images, through a given data augmentation method, accurate positive pairs can be guaranteed since we can simply generate two views of o_t and they do actually semantically matched. However, it is common to regard the augmentations of any two different observation o_{t1} and o_{t2} as negative pairs in existing contrastive method, where positive samples may be misdiagnosed as negative ones and that will inevitably lead to the *sample bias* problem and probably further, sample efficiency decline as mentioned before.

It is natural for an RL agent to distinguish different observations by their real states s, however, it is notoriously known that the real states is unavailable due to perceptual limitations in real world. When only high-dimensional observation o_t is available, we can turn to the most important feedback of RL tasks, i.e. Q value. Q value is the expected discount cumulative rewards after agent taking action a_t at observation o_t: $Q(o_t, a_t) = \mathbb{E}(\sum_{\tau=t}^{T} \gamma^{\tau-t} r_\tau(o_\tau, a_\tau))$, where $\tau \in (0, 1]$ is the discount factor. Hence, to define pos-neg samples in RL, intuitively we have:

Assumption 1. *In RL, given a policy $\pi_\psi : \mathbb{O} \rightarrow \mathbb{A}$ and* ***arbitrary*** *observation-action samples (augmented or not) $(o_{t1}, \pi(a_{t1} \mid o_{t1}))$, $(o_{t2}, \pi(a_{t2} \mid o_{t2}))$, if they* ***share the same*** *Q value, we can approximately regard them as a positive pair.*

However, strict conditions are required for the establishment of this hypothesis including a perfect reward function of environment to disambiguate Q value. But on the contrary, we can define the negative samples:

Assumption 2. *In RL, given a policy $\pi_\psi : \mathbb{O} \rightarrow \mathbb{A}$ and* ***representations similar*** *observation-action samples (augmented or not) $(o_{t1}, \pi(a_{t1} \mid o_{t1}))$, $(o_{t2}, \pi(a_{t2} \mid o_{t2}))$, if they have* ***quite different*** *Q values, we can approximately regard them as a negative pair.*

Please note that a sample mentioned above is composed of both observation and action. Such a definition for the negative pair is not perfect because of the uncertainty of the environment and the policy divergence, but it is still more reasonable than the random sampling method that is widely adopted for contrastive representation learning in RL. Since it is quite difficult to get the exact value of any $Q(o_t, a_t)$, in practice we simply use the cumulative rewards in historical trajectories to approximate Q.

4.2 Mine and Utilize Hard Negative Samples in RL

As mentioned, hard negative samples, i.e., the pairs with similar representation but different semantics are the key to efficient contrastive learning [21]. However, how to mine such samples from the data is still a challenging problem in the literature. In the following, how to mine and make use of hard negative samples in RL by using QHM will be introduced.

Given an anchor sample, it is infeasible to search the similar samples directly from the raw-pixel space due to heavy computational burden. We also should not search for the samples in agent's encoder embedding space since frequent forward propagation in model may deteriorate the overall running time. In QHM, we just take the advantage of gaussian random projection to map raw-pixel images to a far-less dimensional space and subsequently a KD-Tree is utilized to execute k-nearest searching on the projection space. Specifically, KD-Tree is a table-like buffer which is independent of the agent's replay buffer. Considering a gaussian projection function $\phi(*)$, QHM simply stores tuples $<[\phi(o), a], o, Q(o, a)>$ encountered in recent trajectories into the KD-Tree. Once the tree capacity hits the peak, samples visited most infrequently during training will be replaced. The specific process above is illustrated in Fig. 1.

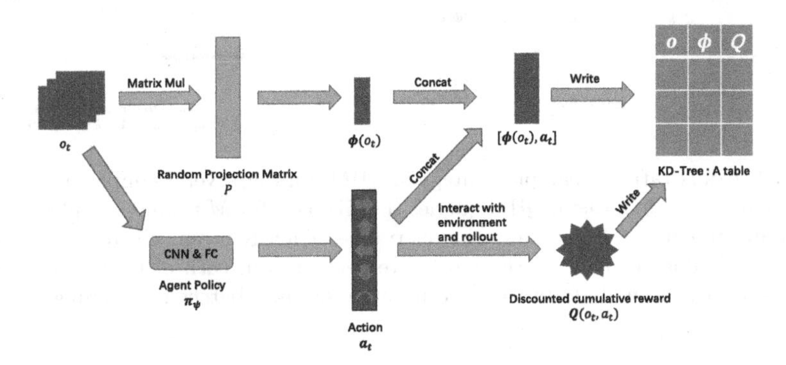

Fig. 1. KD tree storage process in QHM. For each observation o, QHM summarizes its cumulative reward Q in the trajectory after rollout and writes o, z and Q into the KD tree where z is the concatenation of projection $\phi(o)$ and corresponding action a.

In order to further screen hard negative samples, a simple K-means method is applied in QHM to cluster all these similar samples according to their Q value, and as all samples have been well scattered, QHM will eventually pick one sample at random from each cluster respectively. Then, all these left samples will share similar representations but with different Q values, which should be the hard negative samples we are seeking for. Please note that our QHM method mainly focus on the selection of the source images for negative samples generation. As for postive pairs, we adopt the same scheme as CURL, i.e. two views generating from a same image will be regard as positive to each other.

The implementation process is shown in Fig. 2: firstly QHM samples a batch of N samples $x_i(1 \leq i \leq N) = [\phi(o_i), a_i, o_i, Q_i]$ at random from the KD-Tree. For each x_i of these N samples, QHM queries the KD-Tree for M nearest samples $x_{ij}(1 \leq j \leq M)$ on $\phi(*)$ to form a similar batch $B_i^{in} = \{x_i, x_{i1}, , x_{i2}, ..., x_{iM}\}$ after absorbing the query sample x_i. Then K-Means cluster is applied based on their Q value to get K clusters $C_k(1 \leq k \leq K)$. Excluding the cluster C_{k^i} which the query sample x_i belongs to, we randomly pick one sample from each clusters to finally make up the hard negative batch $B_i^{out} = \{x_i, x_{ik}, ...\}(1 \leq k \leq K \,\&\, k \neq k^i)$ of x_i. Sequentially, $N \times K$ samples will be acquired and cropped randomly to generate totally $N \times K \times 2$ samples $\{x_{ik1}, x_{ik2}, ...\}(1 \leq i \leq N, 1 \leq k \leq K)$, which will be used for contrastive loss Eq. (3) as following:

$$\mathcal{L}_{QHM-CURL} = \sum\nolimits_{i=1}^{N} \mathcal{L}_{CURL}(x_{ik1}, x_{ik2}, ..., x_{iK1}, x_{iK2}) \tag{6}$$

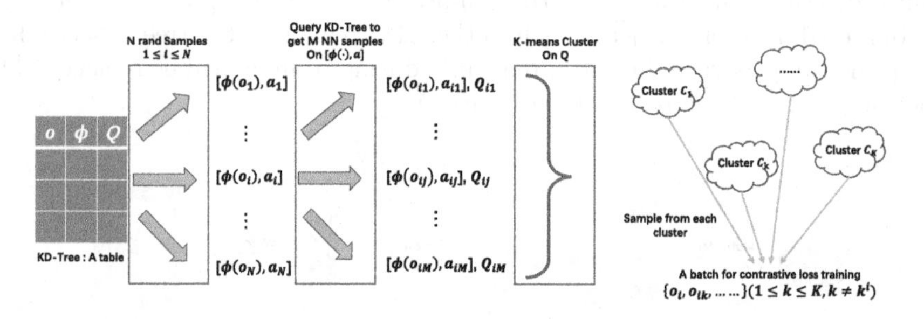

Fig. 2. The illustration of sample strategy in QHM. Firstly, several samples are sampled at random. For each of them, QHM queries the KD-Tree for M nearest samples to make up a state-similar batch including the query one. Then K-means cluster will conduct based on Q values to pick up the most divergent ones in each of these state-similar batchs, which will finally form the hard negative samples batchs for training.

5 Experiments

5.1 Environments

7 challenging tasks of DMC [9] are selected for evaluation. At every time step, the input of the agent is an 8-bits, 100×100, RGB image from the environment and 3 successive frames will be stacked as the observation to alleviate partially observable problems. And to accelerate training, the agent's action repeat numbers is set as 8 for cartpole swingup, 2 for walker run and 4 for the rest respectively. Corresponding task policy step can be calculated by $total\ frames/action\ reapeat$, which will be the abscissa of our experimental plot results.

5.2 Setting

QHM will be carried on CURL for experiments with the same neural network structure and hyperparameters. Specifically, encoders $f_{\theta q}$ and $f_{\theta k}$ composed of successive convolution layers and full connection layers that are in charge of a mapping from raw-pixel images to embedding space with a dimensionality of 50. The capacity of replay buffer is $100k$ and a batch of 512 samples will be randomly

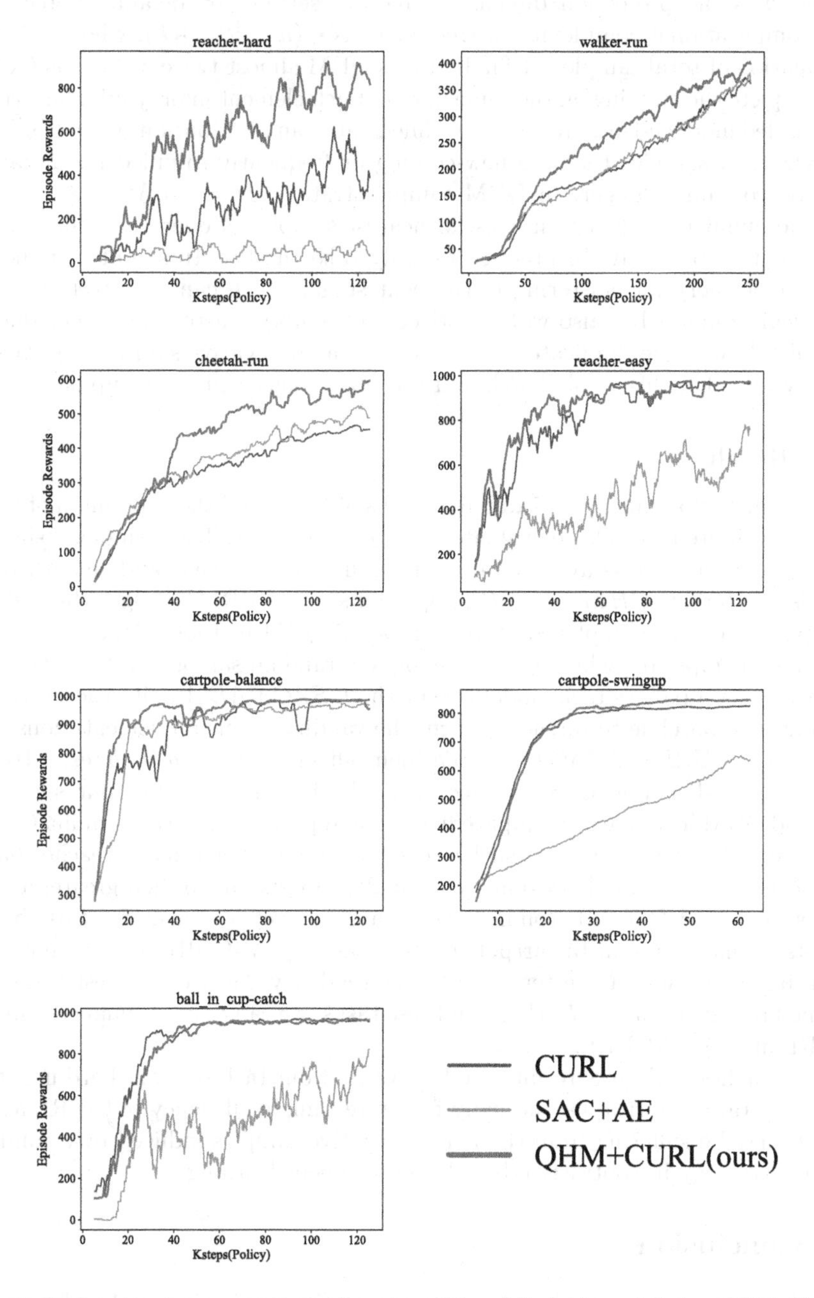

Fig. 3. Evaluation scores results on 7 tasks of DeepMind Control Suite. QHM-CURL indicates CURL equipped with our method and vanilla CURL is our main competitor. SAC+AE is also included as a competitive method.

selected for SAC [3] updating. Adam [30] optimizer for update and 84×84 random crop for image preprocessing. Specific QHM settings are shared across tasks: the dimensionality of the gaussian projection matrix P is $[h, 9 \times 100 \times 100]$ where h is the projection dimension that we set 128 by default. Notes that the computational complexity of tree query is $O(n^{(1-1/h)} + k)$, where k is the population of total samples, it finally takes QHM almost twice as long as CURL to complete the training on the same device. Each element in projection matrix P is sampled independently from a predefined gaussian distribution $N(0, (1/\sqrt{h})^2)$. In KD-Tree, samples that are new-coming or frequently sampled for updating will be contained longer. In QHM sample strategy, $N = 16, M = 31, K = 4$ are the numbers of query samples, M-nearest samples of query and clusters for K-Means, respectively. In practice we only take the last 16 samples of the M nearest of query for clustering in the next stage, which can effectively prevent identical samples. It's also worth noticed that empty cluster may occur due to completely Q value duplicate. To get rid of biased negative samples, QHM will simply abandon this whole batch and the subsequent contrastive update.

5.3 Results

We conduct experiments on 7 different tasks of DMC and the supremum of tasks frames is limited to $500k$ to evaluate the agent efficiency. The results are showed in Fig. 3. Every line is averaged over different random seeds and smoothed on abscissa interval. $QHM - CURL$ represents the CURL [1] algorithm whose sampling strategy is replaced by our proposed QHM method. Hence, $CURL$ is our main competitor which uses a completely random sample strategy. We also take another auxiliary task model-free method, $SAC+AE$ [10], into account as a competitive baseline to further confirm the validity of our implementations. We implement CURL and SAC+AE from their official codes on *github* respectively.

As showed in Fig. 3, we can see that CURL combined with our sampling method QHM has superior sample efficiency and performance than vanilla CURL and SAC+AE on several tasks such as cheetah-run, walker-run and reacher-hard. All of these 3 tasks are defined in *medium* [27] subsets due to their greater action dimensions and 500k environment steps are not yet sufficient for SAC-based agents to master them. In cartpole balance task, QHM-CURL acts more robust than baselines and has better convergence tendency. Most of the rest tasks are defined relatively *easy* [27]. Hence in these tasks, we can only see nuances among CURL and QHM-CURL.

Throughout all tasks results, we believe that our task-relevant hard negative mining strategy, QHM, can actually facilitate sample efficiency of the RL agent which may be suffering from the biased negative samples induced by a random sample strategy in contrastive-based reinforcement learning.

6 Conclusion

In this paper, we proposed QHM, a hard negative mining method dedicated to improving data-efficiency of RL agents. With the assistance of light components such as KD-Tree, K-Means Cluster and Random Projection, when compared to vanilla instance-based contrastive sampling method, QHM can achieve further efficiency and even performance improvements on a certain number of tasks from DeepMind Control Suite. However, as in general we have no access to the real state of the environments, differentiating samples by their Q value stored would still be biased. There is a long way for us to acquire a near-real hard negative distribution and we leave this for future work. We believe that in the forthcoming future, better hard sampling strategies for contrastive learning in RL will be discovered and make significant contribution to representation learning.

Acknowledgements. This work was supported by the National Natural Science Foundation of China (No. 61772438 and No. 61375077). This work was also supported by the Innovation Strategy Research Program of Fujian Province, China (No. 2021R0012).

References

1. Srinivas, A., Laskin, M., Abbeel, P.: CURL: contrastive unsupervised representations for reinforcement learning (2020)
2. Kaiser, L., et al.: Model-based reinforcement learning for atari. arXiv preprint arXiv:1903.00374 (2019)
3. Haarnoja, T., et al.: Soft actor-critic algorithms and applications. arXiv preprint arXiv:1812.05905 (2018)
4. Mutti, M., Pratissoli, L., Restelli, M.: A policy gradient method for task-agnostic exploration (2020)
5. Veeriah, V., Oh, J., Singh, S.: Many-goals reinforcement learning. arXiv preprint arXiv:1806.09605 (2018)
6. Mnih, V., et al.: Playing atari with deep reinforcement learning. arXiv preprint arXiv:1312.5602 (2013)
7. Talvitie, E.: Model regularization for stable sample rollouts. In: UAI (2014)
8. Racanière, S., et al.: Imagination-augmented agents for deep reinforcement learning. In: Proceedings of the 31st International Conference on Neural Information Processing Systems (2017)
9. Tassa, Y., et al.: Deepmind control suite. arXiv preprint arXiv:1801.00690 (2018)
10. Yarats, D., et al.: Improving sample efficiency in model-free reinforcement learning from images. arXiv preprint arXiv:1910.01741 (2019)
11. Yarats, D., et al.: Reinforcement learning with prototypical representations. arXiv preprint arXiv:2102.11271 (2021)
12. Chen, X., et al.: Improved baselines with momentum contrastive learning. arXiv preprint arXiv:2003.04297 (2020)
13. Bellemare, M.G., et al.: The arcade learning environment: an evaluation platform for general agents. J. Artif. Intell. Res. **47**, 253–279 (2013)
14. Wu, Z., et al.: Unsupervised feature learning via non-parametric instance-level discrimination. arXiv preprint arXiv:1805.01978 (2018)
15. Hafner, D., et al.: Learning latent dynamics for planning from pixels. In: International Conference on Machine Learning. PMLR (2019)
16. Hafner, D., et al.: Dream to control: learning behaviors by latent imagination. arXiv preprint arXiv:1912.01603 (2019)

17. Lee, K.-H., et al.: Predictive information accelerates learning in RL. arXiv preprint arXiv:2007.12401 (2020)
18. van den Oord, A., Li, Y., Vinyals, O.: Representation learning with contrastive predictive coding. arXiv preprint arXiv:1807.03748 (2018)
19. Robinson, J., et al.: Contrastive learning with hard negative samples. arXiv preprint arXiv:2010.04592 (2020)
20. Chuang, C.-Y., et al.: Debiased contrastive learning. arXiv preprint arXiv:2007.00224 (2020)
21. Le-Khac, P.H., Healy, G., Smeaton, A.F.: Contrastive representation learning: a framework and review. IEEE Access **8**, 193907–193934(2020)
22. Suh, Y., et al.: Stochastic class-based hard example mining for deep metric learning. In: Proceedings of the IEEE/CVF Conference on Computer Vision and Pattern Recognition (2019)
23. Zhu, J., et al.: Masked contrastive representation learning for reinforcement learning. arXiv preprint arXiv:2010.07470 (2020)
24. Kostrikov, I., Yarats, D., Fergus, R.: Image augmentation is all you need: regularizing deep reinforcement learning from pixels. arXiv preprint arXiv:2004.13649 (2020)
25. Liu, G., et al.: Return-based contrastive representation learning for reinforcement learning. arXiv preprint arXiv:2102.10960 (2021)
26. Laskin, M., et al.: Reinforcement learning with augmented data. arXiv preprint arXiv:2004.14990 (2020)
27. Yarats, D., et al.: Mastering visual continuous control: improved data-augmented reinforcement learning. arXiv preprint arXiv:2107.09645 (2021)
28. Johnson, W.B., Lindenstrauss, J.: Extensions of Lipschitz mappings into a Hilbert space 26. Contemp. Math. **26**, 28 (1984)
29. Blundell, C., et al.: Model-free episodic control. arXiv preprint arXiv:1606.04460 (2016)
30. Kingma, D.P., Ba, J.: Adam: a method for stochastic optimization. arXiv preprint arXiv:1412.6980 (2014)
31. Gao, T., Yao, X., Chen, D.: SimCSE: simple contrastive learning of sentence embeddings. arXiv preprint arXiv:2104.08821 (2021)
32. He, K., et al.: Momentum contrast for unsupervised visual representation learning. In: Proceedings of the IEEE/CVF Conference on Computer Vision and Pattern Recognition (2020)
33. Bachman, P., Devon Hjelm, R., Buchwalter, W.: Learning representations by maximizing mutual information across views. arXiv preprint arXiv:1906.00910 (2019)
34. Gutmann, M., Hyvärinen, A.: Noise-contrastive estimation: a new estimation principle for unnormalized statistical models. In: Proceedings of the thirteenth international conference on artificial intelligence and statistics. JMLR Workshop and Conference Proceedings (2010)
35. Wang, F., Liu, H.: Understanding the behaviour of contrastive loss. In: Proceedings of the IEEE/CVF Conference on Computer Vision and Pattern Recognition (2021)
36. Wu, C.-Y., et al.: Sampling matters in deep embedding learning. In: Proceedings of the IEEE International Conference on Computer Vision (2017)
37. Yan, W., et al.: Learning predictive representations for deformable objects using contrastive estimation. arXiv preprint arXiv:2003.05436 (2020)

Reduction of the Position Bias via Multi-level Learning for Activity Recognition

Aomar Osmani and Massinissa Hamidi(✉)

LIPN-UMR CNRS 7030, Université Sorbonne Paris Nord, Villetaneuse, France
{ao,hamidi}@lipn.univ-paris13.fr

Abstract. The relative position of sensors placed on specific body parts generates two types of data related to (1) the movement of the body part w.r.t. the body and (2) the whole body w.r.t. the environment. These two data provide orthogonal and complementary components contributing differently to the activity recognition process. In this paper, we introduce an original approach that separates these data and abstracts away the sensors' exact on-body position from the considered activities. We learn for these two totally orthogonal components (i) the bias that stems from the position and (ii) the actual patterns of the activities abstracted from these positional biases. We perform a thorough empirical evaluation of our approach on the various datasets featuring on-body sensor deployment in real-life settings. Obtained results show substantial improvements in performances measured by the f1-score and pave the way for developing models that are agnostic to both the position of the data generators and the target users.

Keywords: Meta-learning · Decentralized machine learning · Federated learning · Internet of Things · Human activity recognition

1 Introduction

The selection of the sensors' positions in moving targets is a constraint that is encountered in many fields, such as human activity recognition from on-body sensor deployments [4,5,11,12,30]. The movements of the area of the target on which the sensors are positioned generate data of two different but complementary natures (see Fig. 1). The first concerns the movement of the position relative to the target itself, and the second concerns the movement of the target relative to its surroundings. In the case of human activity recognition, we notice for example that the kinetics of the hand movements during a race can be decomposed into a circular movement (CM) of the hand relative to the shoulder and a translation movement (TM) associated with the whole body [23].

At least three practical implications can be devised from this: (i) CM data are enough to learn some target concepts, e.g., the hand kinetics movement is

J. Gama et al. (Eds.): PAKDD 2022, LNAI 13281, pp. 289–302, 2022.
https://doi.org/10.1007/978-3-031-05936-0_23

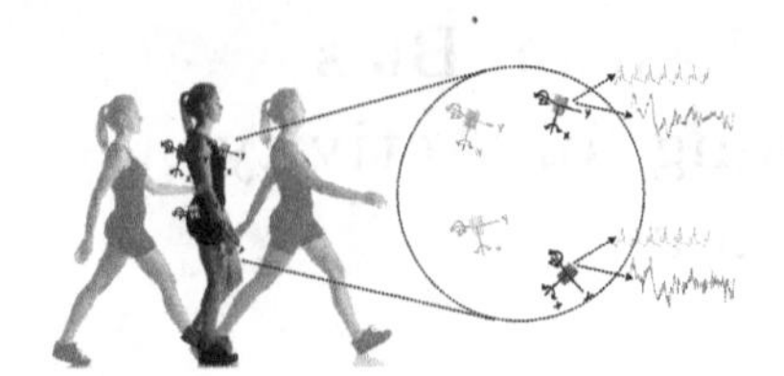

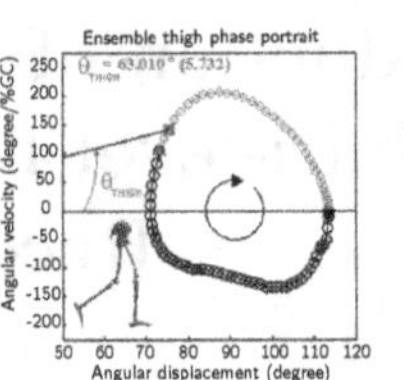

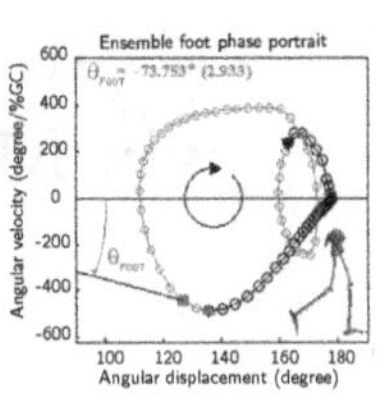

Fig. 1. (left) The hand sensor undergoes two types of movements. One is of the same nature as the torso and linked to the translational movement of the body. The other is linked to the movement of the hand locally relative to the body. (right) Phase plan showing the dynamics of the thigh and foot during gait cycle (GC) (-◆- 1%GC) extracted from the biomechanics works of [6].

enough to determine if a person is at rest or running; (ii) CM data from different positions, e.g., hand and torso, cannot be shared and mixed together. Otherwise, this generates noise and confusion during the learning process; (iii) only TM data can be shared among the different positions as these data are of the exact same nature but taken from different points of view (positions or perspectives).

In this paper, we leverage the data decomposition into universal and position-specific components to improve activity recognition models. These components have distinctive contributions concerning the target concepts to learn. This brings an interesting property that allows us to fuse the universal components as seen from different points of view (positions) while identifying the position-specific components, which could serve as additional knowledge in situations when the position-specific components are not sufficient to recognize an activity. Without this data decomposition process, the local part of the data adds position noise challenging to manage with centralized approaches, e.g., federated learning [22,34]. Indeed, to integrate data from different positions (or clients), it is necessary to separate the data of the same nature (shareable) from the pure local ones linked to the specific kinetics of the position. Similar data can and should be shared to improve recognition rates. However, the specific data must be processed locally, otherwise impacting the learning process.

Traditional HAR approaches [5,24,38] often consider the sensory inputs to be flattened therefore disregarding the significant impact of the various positional biases. Some approaches consider these problems from the perspective of deployment optimization, mainly focusing on the study of the optimal on-body sensors placement and its impact on the recognition of target activities [3,4,30].

There are also rare approaches offering pipelines which include recognition of the position of the data generator followed by the activity recognition [37] or including an explicit model of the context [2,8]. Other approaches, e.g., [19], try to develop heuristics to improve the robustness of activity recognition models to sensors displacements. Regardless of the devised techniques, these approaches rely on centralized processing of the data, which does not match the intrinsic complementary nature of the data, thus limiting their potential capacities.

To deal with these complementary data sources, we propose an original multi-level model of abstraction of the data generator position encompassing a central learner (or set of local generic learners) and a set of specific local learners. The local learners (one for each position) use only specific local data concerning the local relativity. They are responsible for learning (i) the position-dependent patterns of activities and (ii) the movements that link them to the individual. The aim is to abstract the learning examples from the bias arising from the position from which they are generated. The central learner (or set of local generic learners) uses the aggregated universal components from the local learners via a conciliation step based on the efficient federated learning (FL) setting. Extensive experiments on three representative datasets featuring real-world sensor deployment settings show the effectiveness of abstracting the impact of the data generator's position. We noticeably get substantial improvements in terms of the recognition performances of individual activities and robustness to the evolution of the sensor deployments. We perform a comprehensive comparative analysis of our proposed approach via ablation studies which shows the contribution of the dual interplay between the local and central learners.

2 Problem Formulation

In this section, we briefly characterize the problem of abstracting the exact position of a given sensor. We consider settings where a collection $\mathcal{S}$ of M sensors (also called data generators or data sources), denoted $\{s_1, \ldots, s_M\}$, are positioned respectively at positions $\{p_1, \ldots, p_M\}$ in the object of interest, e.g., human body. Each sensor s_i generates a stream $\mathbf{x}^i = (x_1^i, x_2^i, \ldots)$ of observations of a certain modality like *acceleration* or *gravity*, distributed according to an unknown generative process. Furthermore, each observation is composed of channels, e.g. three axes of an accelerometer. The goal is to continuously recognize a set of target concepts $\mathcal{Y}$ like *running* or *biking* in the case of the human activity recognition according to all sensor's positions. In the case of the SHL dataset, the sensors deployment features data generated from 4 smartphones, carried simultaneously at typical body locations (*hand*, *torso*, *hips*, and *bag*).

2.1 Abstraction of the Position

As described in the previous section, each sensor produces two types of orthogonal data. This problem can be formally defined as the construction, for the data generated by each sensor s_i, of a factorized representations z_i being a composition of (i) position-invariant (abstract or universal) components vector z_{iA}, and (ii) a position-specific (local) components vector z_{iP}. The position-invariant components vector captures the features that are shared across all positions. On the other hand, the position-specific components vector captures specific and complementary insights concerning the target concepts. The first problem to solve in our model is to build automatically this data decomposition process for each sensor automatically. Thanks to this process, each sensor $s \in \mathcal{S}$ will disentangle the data interlaced between the local and universal component $\mathbf{x}$ by projecting them into two separate representations z_A and z_P. Components z_P will be used only in a local learner, and z_A can be used in the local learner or shared with the same data coming from all other sensors in a global learner. This process allows us to have fine-grained control on the inference process where one can leverage different configurations in order to get optimal performances, while traditional HAR approaches often consider the inputs to be flattened and disregard the bias related to the position. We notice that in certain situations the position-specific component alone is enough to recognize the activity, e.g., the circular movements of the hand are sufficient to distinguish between running and walking. In addition, since only position-independent data is shared, this process considerably reduces data heterogeneity. It, therefore, improves data aggregation techniques or learners such as federated learning [36] by sharing only the position-invariant data. When the data are not decomposed, the position-specific part of the data represents noise for the global system.

To deal with these two challenging complementary representations, we propose a model based on multi-level processing to abstract the position as described below. In this model, we suppose that the position-invariant components share the data with a central learner.

3 Source Position Multi-level Abstraction

Here, we propose an instantiation of the proposed problem formulation composed of local and central learners. To perform the separation of the position-specific components from the universal ones, we use a family of models based on variational autoencoders (VAEs) [18] (Sect. 3.1). The proposed conciliation step is based on the federated learning (FL)-based aggregation setting where the position-specific learners in our formulation are assimilated to the decentralized clients in FL (Sect. 3.2). This instantiation is described in the following. Figure 2 summarizes the proposed instantiation. Algorithm 1 outlines the complete learning process.

3.1 Position-Specific (or Local) Learners

The position-specific learners L_p pursue their own learning steps locally using their own generated data. Their goal is to decompose the contents of the data into different factors of variations, particularly those related to the position itself. The objective of the local learner L_p can be formalized as the expected loss over the data distribution of the position p, $f_p(w_p) = \mathbb{E}_{\xi_p}[\tilde{f}_p(w_p;\xi_p)]$, where ξ_p is a random data sample drawn according to the distribution of position p and w_p the set of the learner's weights. In particular, the distributions from which are drawn the samples ξ_{p_i} and ξ_{p_j}, $p_i \neq p_j$, can be distinct. At the step t of communication round, each local learner independently runs τ_p iterations of the local solver, e.g., stochastic gradient descent, starting from the current global model $L_p^{(t,0)}$ until the step $L_p^{(t,\tau_p)}$ to optimize its own local objective (see the black arrows depicted in Fig. 2).

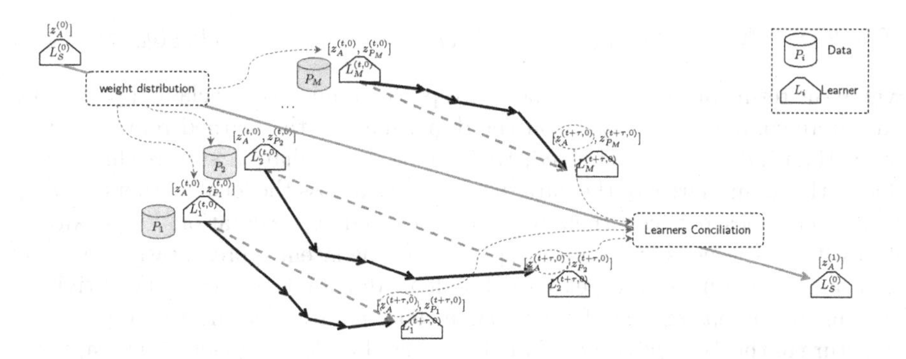

Fig. 2. Framework of the proposed multi-level abstraction architecture. The global learner L_S starts with an initial set of weights which are distributed to the local learners. The local learners L_P, one for each position p, learn the two vector components z_A and z_P, by performing independently a set of gradient steps which allows to get newer versions. These new versions are used during the conciliation step which results in a new version of the global learner, and subsequently a more robust position-independent representation. (Color figure online)

The objective function $f_p(w_p)$ is constructed using a family of models based on VAEs for their ability to deal with entangled representations. The task here is to learn these factors of variation, commonly referred to as learning a disentangled representation. It corresponds to finding a representation where each of its dimensions is sensitive to the variations of exactly one precise underlying factor and not the others.

Depending on the availability of explicit knowledge about the underlying factors of variation, different strategies are pursued to learn the disentangled representation. For example, in video prediction [7,16], temporal-invariance is often leveraged with a content representation which captures structure that is shared across all video frames and a pose representation capturing content that varies over time. These strategies require devising complex architectures and intricate loss functions to enforce prior knowledge. Alternatively, the disentanglement can be performed using separate representations for each factor of variation, which are jointly learned by different encoders, e.g. [28,29]. Although the representations are explicitly separated and learned by different encoders, getting exact correspondence with the factors of variation, i.e., non-overlapping dimensions, is not ensured and can lead to identical representations. Recent advances in unsupervised disentangling based on VAEs demonstrated noticeable successes in many fields using the β-VAE, which leads to improved disentanglement [15]. It uses a unique representation vector and assigns an additional parameter ($\beta > 1$) to the VAE objective, precisely, on the Kullback Leibler (KL) divergence between the variational posterior and the prior, which is intended to put implicit independence pressure on the learned posterior. The improved objective becomes:

$$\mathcal{L}(x;\theta,\varphi) = \mathbb{E}_{q_\varphi(z|x)}[\log p_\theta(x|z)] - \beta D_{KL}(q_\varphi(z|x)||p(z)) - \alpha D_{KL}(q_\varphi(z)||p(z)),$$

where the term controlled by α allows to specify a much richer class of properties and more complex constraints on the dimensions of the learned representation other than independence. Indeed, the proposed conciliation step is challenging due to the dissimilarity of the data distributions across the local learners, leading to discrepancies between their respective learned representations. One way to deal with this issue is by imposing sparsity on the latent representation in a way that only a few dimensions get activated depending on the learner and activities. We ensure the emergence of such sparse representations using the appropriate structure in the prior $p(z)$ such that the targeted underlying factors are captured by precise and homogeneous dimensions of the latent representation. We set the sparse prior as $p(z) = \prod_d (1-\gamma)\mathcal{N}(z_d; 0, 1) + \gamma\mathcal{N}(z_d; 0, \sigma_0^2)$ with $\sigma_0^2 = 0.05$. This distribution can be interpreted as a mixture of samples being either activated or not, whose proportion is controlled by the weight parameter γ [21].

3.2 Referential (or Central) Learner

Each local learner pursues its own "version" of the universal representation z_{pA} but has not to diverge from the *referential* universal representation z_A, which constitutes a consensus among all local learners. In our setting, we build the *referential* universal representation by making every learner contributes to it via a weighted aggregation defined as follow: given the objectives $f_p(w)$ of the local learners L_p, the referential learner objective function is formulated as:

$$\min_{w\in\mathbb{R}^d} \left\{ F(w) := \sum_{p=1}^{M} \alpha_p \times f_p(w_p) \right\} \text{ with } \sum_{p=1}^{M} \alpha_p = 1, \quad (1)$$

where α_p is used to weigh the contribution of every learner to the universal representation. After a predefined number of local update steps, we conduct a conciliation step (see the dotted red arrows in Fig. 2). Each conciliation step t produces a new version of the referential learner $L_S^{(t)}$ and, a new version of the referential universal representation $z_A^{(t)}$. The conciliation step has to be performed on the learned representations $z_{pA}^{(t)}$ via regularization, for example. In our approach, the conciliation step is performed via representation alignment, e.g., correlation-based alignment [1]. More formally, we instrument the objective function of the local learners with an additional term derived from the representation alignment [33].

$$\min_{w\in\mathbb{R}^d} \left\{ F(w) = \frac{1}{M}\sum_{p=1}^{M} F_p(w_p) \right\}, \; F_p(w_p) = \min_{w\in\mathbb{R}^d} \left\{ f_p(w_p) + \lambda R(z_{pA}, z_A^{(t)}) \right\}, \quad (2)$$

where R is a regularization term responsible for aligning the locally learned universal components with the ones learned by the referential learner and $\lambda \in [0,1]$ is a regularization parameter that balances between the local objective and the regularization term. Note that in this setting, it is required that, at conciliation step t, a copy of the referential learner's weights be available locally to perform the generative step. Position-specific and universal components will still be learned separately but locally. Then, the conciliation can be performed via the standard FL setting, where the weights of the local universal components learners are aggregated and used to update the referential learner. In this regard, the conciliation step can be implemented with any federated learning algorithm, e.g., federated averaging [22], federated normalized averaging [34]. The shared global model is updated based on the federated averaging as follows:

$$w^{(t+1,0)} - w^{(t,0)} = \sum_{p=1}^{M} \alpha_p \Delta_p^{(t)} = -\sum_{p=1}^{M} \alpha_p \cdot \eta \sum_{k=0}^{\tau_p - 1} g_p(w_p^{(t,k)}), \tag{3}$$

where $w_p^{(t,k)}$ denotes client p's model after the k-th local update in the tth communication round and $\Delta_p^{(t)} = w_p^{(t,\tau_p)} - w_p^{(t,0)}$ denotes the cumulative local progress made by client p at round t. η is the client learning rate and g_p represents the stochastic gradient over a mini-batch of samples.

Algorithm 1: Multi-level abstraction of sensor position

Input : $\{\mathbf{x}^p\}_{p=1}^{M}$ streams of annotated observations from the sensors
1 $w \leftarrow$ initWeights() ; % *Init. referential learner weights*
2 distributeWeights(w, $\mathcal{S}$) ; % *Weights distribution*
3 **while** *not converged* **do**
  ; % *Local updates*
4   **foreach** *position* $p \in \mathcal{S}$ **do**
5   **for** $t \in \tau_p$ *steps* **do**
6   Sample mini-batch $\{x_i^p\}_{i=1}^{n_p}$ from the stream of data $\mathbf{x}^p$
7   Evaluate $\nabla_{w_p}\mathcal{L}(w_p)$ with respect to the mini-batch
8   Compute adapted parameters: $w_p^{(t)} \leftarrow w_p^{(t-1)} - \eta\nabla_{w_p}\mathcal{L}(w_p)$
9   **end**
10   **end**
  ; % *Central updates*
11   Update central model's weights $L_{\mathcal{S}}$ by aggregating the incoming weights from the local models $L_p, p \in \{1,\ldots,M\}$ using Eq. 3
12 **end**
Result: $L_{\mathcal{S}}$ and $L_p, p \in \{1,\ldots,M\}$, the trained referential and local learners

4 Experiments and Results

We perform an empirical evaluation of the proposed approach, consisting of two major stages: (1) we evaluate the quality of the data separation into position-specific and universal components which is performed by the local learners and how each of these components contributes individually, with and without the conciliation process, to the recognition performances (Sect. 4.2); (2) we then evaluate various inference configurations where the position-specific and universal components are combined to improve the performances. We also provide a comparative analysis against baselines (Sect. 4.3). Code and supplementary material can be found in https://github.com/hamidimassinissa/positionAbstraction.

4.1 Experimental Setup

We evaluate our proposed approach on three large-scale real-world wearable benchmark datasets featuring multi-location and heterogeneous sensors: SHL [10], HHAR [32], and Fusion [31] datasets (see § A.1 for a detailed description). Implementation details can be found in § A.2. We compare our approach with the following closely related baselines.

- **DeepConvLSTM** [24]: a model encompassing 4 convolutional layers responsible of extracting features from the sensory inputs and 2 long short-term memory (LSTM) cells used to capture their temporal dependence.
- **DeepSense** [38]: a variant of the DeepConvLSTM model combining convolutional and a Gated Recurrent Units (GRU) in place of the LSTM cells.
- **AttnSense** [20]: features an additional attention mechanism on top of the DeepSense model forcing it to capture the most prominent sensory inputs both in the space and time domains to make the final predictions.

For the ablation study, we compare our approach with two baselines which do not perform the separation nor conciliation steps. These models consist of convolution-based circuits for each position which are then fused together and trained jointly. We implemented two types of fusion schemes [13]: concatenation-based and alignment-based fusion (see § A.3). To make these baselines comparable with our models, we make sure to get the same complexity, i.e., comparable number of parameters. We use the f1-score in order to assess performances of the architectures. We compute this metric following the method recommended in [9,25] to alleviate bias that could stem from unbalanced class distribution (see § C). In addition, to alleviate performance overestimation problem, we rely in our experiments on the meta-segmented partitioning proposed in [14] (see § D).

4.2 Evaluation of the Data Decomposition Process

In this part, we evaluate the ability of the local learners to decompose the sensor data into the position-specific components and the universal ones. We evaluate this process with and without the conciliation phase, then we show the impact of this step on the recognition performances. We measure the sparsity of a given representation using the *Hoyer* extrinsic metric [17] which is formally defined for a vector $\mathbf{y} \in \mathbb{R}^d$ to be $Hoyer(\mathbf{y}) = \frac{\sqrt{d} - \|\mathbf{y}\|_1 / \|\mathbf{y}\|_2}{\sqrt{d}-1} \in [0, 1]$ yielding 0 for a fully dense vector and 1 for a fully sparse one. Table 1 summarizes the average normalized sparsity of the obtained representations. Figure 3 illustrates the average latent magnitude computed for each dimension of the learned representations.

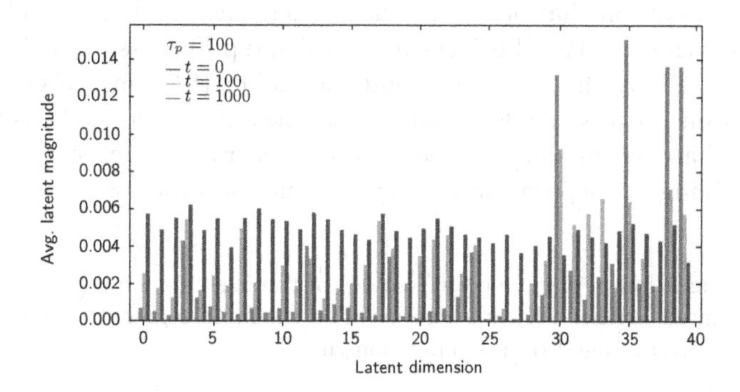

Fig. 3. Average latent encoding magnitude computed over different steps of the conciliation process.

Table 1. Summary of the per-position average normalized sparsity measured using the *Hoyer* extrinsic metric. Results w/ and w/o conciliation are shown.

Config.	Average normalized sparsity±std.			
	Bag	*Hand*	*Hips*	*Torso*
w/o concil.	0.42±.072	0.77±.002	0.71±.029	0.68±.024
w/ concil.	0.44±.0145	0.91±.0521	0.87±.038	0.727±.033

From Table 1, we can observe, as expected, that the representations learned by the local learners of the *hand* and *hips* have high sparsity compared to *bag* and *torso*. Sparsity increases further when the conciliation is performed as the dimensions that are less important are being pushed more and more towards zero. Regarding the latent magnitudes, we can observe that during conciliation some dimensions of the central learner's latent representation are getting more activated (e.g., dimensions 30, 35, 39, and 40 with an average magnitude of 0.0134, 0.146, 0.0138, and 0.138, resp.) corresponding to the universal components, while the remaining dimensions having low activation and some noticeable picks (e.g., at 3, 12, 18, and 24) corresponding to the position-specific components.

As demonstrated above, the dimensions of the learned representations have meaningful interpretation with regards to the activities that we seek to recognize. To further assess the usefulness of the separated components per se (without a conciliation step), we leverage them in a traditional discriminative setting. In other words, we take the learned representation and add, on top of it, a simple dense layer. This additional layer is trained to minimize classification loss while the rest of the circuit is kept frozen. To alleviate any effect that could be attributed to the model's complexity, the additional dense layer has low VC-dimension so that we ensure it has no capacity to improve the representation by itself. Table 2 compares the obtained performances with the baseline models on the considered representative datasets. Furthermore, to better understand how the process of conciliation among the learners, attached to the different positions, impacts the quality of both the universal and position-specific components, we leverage similarly the separated components but this time, after performing the conciliation process. Table 3, summarizes obtained results. We compare the results with baseline models trained on data generated from specific positions without applying the separation nor the conciliation processes.

Table 2. Recognition performances (f1-score) of the baseline models on different representative related datasets. Evaluation based on the meta-segmented cross-validation. Experiments were averaged over 7 repetition runs.

Model	HHAR	Fusion	SHL
DeepConvLSTM	$70.1 \pm .0018$	$68.5 \pm .002$	$65.3 \pm .0206$
DeepSense	$72.0 \pm .0022$	$69.1 \pm .0017$	$66.5 \pm .006$
AttnSense	$76.2 \pm .0074$	$70.3 \pm .0027$	$68.4 \pm .03$
Feature fusion	$72.9 \pm .004$	$68.7 \pm .001$	$66.8 \pm .009$
Corr. align.	$75.8 \pm .0014$	$70.2 \pm .04$	$69.1 \pm .015$
Proposed	$78.3 \pm .0045$	$72.8 \pm .002$	$74.5 \pm .0133$

We observe from Table 3 that, overall, the obtained performances using the position-specific and universal components are better than those obtained using the baseline (without separation nor conciliation). In theory, with the conciliation step, optimal representations would emerge in particular for the universal components. Indeed, this is achieved by the additional alignment term in Eq. 2 which should make them interchangeable regardless of the position from which they have been generated. This should nevertheless be harder in the case of the position-specific components which may activate very diverse dimensions of the learned representation (as described in the experimental results above). Surprisingly, this has a mild impact on the performances which stay comparable. This could potentially be explained by the importance of the position-specific components for the recognition of many of the activities that are considered in the SHL dataset. It is worth noticing though that the universal components achieve remarkable improvements in the case of *bag* and *torso*.

4.3 Inference Configurations

Here we evaluate the robustness of the proposed approach to the evolution of the sensors deployments via the flexibility that it offers for the inference step. Depending on the activity, the right prediction can be achieved by using either

Table 3. Performances obtained using either the universal or the position-specific components.

Config.	*Bag*	*Hand*	*Hips*	*Torso*
No sep.	63±.0089	63±.0014	·65±.0126	60±.0072
Universal				
w/o concil.	66±.0224	65±.0147	66±.0035	62±.013
w/ concil.	66±.016	67±.0015	67±.0354	63±.01
Pos.-specific				
w/o concil.	64±.3	66±.007	67±.0026	61±.087
w/ concil.	65±.029	68±.03	70±.07	61±.029

Table 4. Per-class performances obtained using various inference configurations.

Class	Best config.	Overall
Still	$z_{hi}; z_t$(85.77)	83.26±0.7
Walk	$z_A; z_{ha}$(88.54)	86.74±0.058
Run	z_{ha}(90.51)	89.46±0.03
Bike	$z_A; z_{hi}$(85.62)	83.22±0.086
Car	$z_A; z_{ha}$(78.24)	77.14±0.2
Bus	z_{ha}(78.08)	75.17±0.004
Train	$z_{hi}; z_{hi}$(76.13)	74.88±0.08
Subway	$z_A; z_{ha}; z_t$(75.89)	74.07±0.006

components z_A or z_{iP} taken individually, or a combination of the universal component z_A and the most appropriate position-specific component. In this part, we take a fine-grained look at the previously obtained performances by assessing the optimal configuration which allows the correct prediction of each of the individual activities. For this, we evaluate the predictions obtained using basic inference configurations, i.e., the combination of the universal components with *torso* $[z_A; z_t]$; *hand* $[z_A; z_{ha}]$; *bag* $[z_A; z_b]$; and *hips*-specific $[z_A; z_{hi}]$ components. Compared to the baseline models, the evaluated inference configurations yield better performances in general. For example, the combination of the universal and most of the position-specific components help discriminate efficiently activities like *walk*, *run*, and *bike*. On the other hand, some activities like *car*, *bus*, or *train* suffer from confusion and do no show significant improvements over the baseline (approx. 2% on avg.). Also, activity *subway* exhibits the same behavior with less proportion suggesting that this "on-wheels" group of activities need elaborate combination of points of views as demonstrated in [13,26,27]. This issue could potentially be circumvented by using more featured configurations where other position-specific representations, rather than a single one, can be leveraged to infer these problematic or hard-to-infer activities.

Table 4 summarizes the evaluation results of the inference configurations featuring the combination of various position-specific components. We observe an increase in the correct predictions for most of the activities compared to the previous setting. In particular, the "on-wheels" group of activities, i.e., *car*, *bus*, *train*, and *subway*, get improved substantially. At the same time, as expected, we see now that the configurations, which yield the highest performances for these activities, use genuine combinations like z_A alone in the case of *bus* or a combination of z_A, z_{ha}, and z_t in the case of *subway*. On the other hand, *still* gets the least improvement compared to the previous setting while the best configuration to infer it is a combination of z_{ha} and z_t (85.77±0.016). It is worth noticing that activities like *walk* and *bike* still achieve competitive performances (88.54±0.07 and 85.62±0.2, resp.) while using the same inference configuration, i.e., a combination of z_A and z_{ha} for *walk* and z_{hi} for *bike*, as in the previous setting. For *run*, the highest scores are achieved using only z_{ha}, which supports the observations presented in Sect. 1.

5 Summary and Future Work

This paper proposes an original approach for abstracting the impact of the specific position of the sensory data generators. Our approach is based on multi-level processing, starting with the disentanglement of the position-specific and universal components at a local level and the conciliation of the universal components at a global level. Experimental results show that the proposed approach improves recognition rates and has many advantages, including reducing the data sources' heterogeneity impact. The decomposition process allows a better recognition rate in several ways: (i) by reducing the noise induced by the data linked to the position itself, e.g., the local component of the movement of the hand constitutes noise for the local component of the movement of the feet; (ii) by aggregating only data of the same nature presenting different points of view and; (iii) for certain activities, the local component alone is sufficient to ensure recognition, e.g., hand movement during run. Future work follows two axes. (1) Improving the quality of the model, in particular, having a fine-grained control on the data decomposition process using additional domain knowledge, e.g., expliciting the dynamics of the body movements in the latent space like in [6,35]. (2) Improving federated multi-source approaches where the sources are entangled with local components. Sharing only mutualisable components has a promising potential.

References

1. Andrew, G., et al.: Deep canonical correlation analysis. In: ICML (2013)
2. Asim, Y., et al.: Context-aware human activity recognition (CAHAR) in-the-wild using smartphone accelerometer. IEEE Sens. J. **20**(8), 4361–4371 (2020)
3. Banos, O., Toth, M.A., Damas, M., et al.: Dealing with the effects of sensor displacement in wearable activity recognition. Sensors **14**(6), 9995–10023 (2014)
4. Barshan, B., Yurtman, A.: Classifying daily and sports activities invariantly to the positioning of wearable motion sensor units. IEEE Internet Things J. **7**, 4801–4815 (2020)
5. Bulling, A., Blanke, U., Schiele, B.: A tutorial on human activity recognition using body-worn inertial sensors. ACM Comput. Surv. (CSUR) **46**(3), 1–33 (2014)
6. Carollo, J.J., Worster, K., Pan, Z., Ma, J., et al.: Relative phase measures of intersegmental coordination describe motor control impairments in children with cerebral palsy who exhibit stiff-knee gait. Clin. Biomech. **59**, 40–46 (2018)
7. Denton, E.L., Birodkar, V.: Unsupervised learning of disentangled representations from video. In: NIPS (2017)
8. Ehatisham-Ul-Haq, M., et al.: Coarse-to-fine human activity recognition with behavioral context modeling using smart inertial sensors. IEEE Access **8**, 7731–7747 (2020)
9. Forman, G., Scholz, M.: Apples-to-apples in cross-validation studies. ACM SIGKDD **12**(1), 49–57 (2010)

10. Gjoreski, H., et al.: The university of Sussex-Huawei locomotion and transportation dataset for multimodal analytics with mobile devices. IEEE Access **6**, 42592–42604 (2018)
11. Hamidi, M., Osmani, A.: Data generation process modeling for activity recognition. In: Dong, Y., Mladenić, D., Saunders, C. (eds.) ECML PKDD 2020. LNCS (LNAI), vol. 12460, pp. 374–390. Springer, Cham (2021). https://doi.org/10.1007/978-3-030-67667-4_23
12. Hamidi, M., Osmani, A.: Human activity recognition: a dynamic inductive bias selection perspective. Sensors **21**(21), 7278 (2021)
13. Hamidi, M., Osmani, A., Alizadeh, P.: A multi-view architecture for the SHL challenge. In: UbiComp-ISWC 2020, p. 317–322. ACM (2020)
14. Hammerla, N.Y., Plötz, T.: Let's (not) stick together: pairwise similarity biases cross-validation in activity recognition. In: UbiComp 2015, pp. 1041–1051 (2015)
15. Higgins, I., et al.: Beta-VAE: learning basic visual concepts with a constrained variational framework. In: ICLR (2017)
16. Hsieh, J.T., et al.: Learning to decompose and disentangle representations for video prediction. arXiv preprint arXiv:1806.04166 (2018)
17. Hurley, N., Rickard, S.: Comparing measures of sparsity. IEEE Trans. Inf. Theory **55**(10), 4723–4741 (2009)
18. Kingma, D., Welling, M.: Auto-encoding variational Bayes. arXiv:1312.6114 (2013)
19. Kunze, K., Lukowicz, P.: Dealing with sensor displacement in motion-based on body activity recognition systems. In: UbiComp, pp. 20–29 (2008)
20. Ma, H., Li, W., Zhang, X., Gao, S., Lu, S.: AttnSense: multi-level attention mechanism for multimodal human activity recognition. In: IJCAI, pp. 3109–3115 (2019)
21. Mathieu, E., Rainforth, T., Siddharth, N., Teh, Y.W.: Disentangling disentanglement in variational autoencoders. In: ICML, pp. 4402–4412 (2019)
22. McMahan, B., Moore, E., Ramage, D., et al.: Communication-efficient learning of deep networks from decentralized data. In: AISTATS, pp. 1273–1282 (2017)
23. Melendez-Calderon, A., Shirota, C., Balasubramanian, S.: Estimating movement smoothness from inertial measurement units. bioRxiv (2020)
24. Ordóñez, F.J., Roggen, D.: Deep convolutional and LSTM recurrent neural networks for multimodal wearable activity recognition. Sensors **16**(1), 115 (2016)
25. Osmani, A., Hamidi, M.: Hybrid and convolutional neural networks for locomotion recognition. In: UbiComp-ISWC 2018, pp. 1531–1540. ACM (2018)
26. Osmani, A., Hamidi, M., Alizadeh, P.: Hierarchical learning of dependent concepts for human activity recognition. In: Karlapalem, K., Cheng, H., Ramakrishnan, N., Agrawal, R.K., Reddy, P.K., Srivastava, J., Chakraborty, T. (eds.) PAKDD 2021. LNCS (LNAI), vol. 12713, pp. 79–92. Springer, Cham (2021). https://doi.org/10.1007/978-3-030-75765-6_7
27. Osmani, A., Hamidi, M., Alizadeh, P.: Clustering approach to solve hierarchical classification problem complexity. In: AAAI, vol. 36 (2022)
28. Qian, H., et al.: Latent independent excitation for generalizable sensor-based cross-person activity recognition. In: AAAI, vol. 35, pp. 11921–11929 (2021)
29. Sadeghi, M., et al.: Audio-visual speech enhancement using conditional variational auto-encoders. IEEE/ACM TASLP **28**, 1788–1800 (2020)
30. Shi, J., Zuo, D., Zhang, Z., Luo, D.: Sensor-based activity recognition independent of device placement and orientation. Trans. ETT **31**(4), e3823 (2020)
31. Shoaib, M., Bosch, S., et al.: Fusion of smartphone motion sensors for physical activity recognition. Sensors **14**(6), 10146–10176 (2014)
32. Stisen, A., et al.: Smart devices are different: assessing and mitigating mobile sensing heterogeneities for activity recognition. In: SenSys 2015, pp. 127–140 (2015)

33. T Dinh, C., Tran, N., Nguyen, T.D.: Personalized federated learning with Moreau envelopes. In: NeurIPS, vol. 33 (2020)
34. Wang, J., Liu, Q., Liang, H., Joshi, G., Poor, H.V.: Tackling the objective inconsistency problem in heterogeneous federated optimization. In: NeurIPS, vol. 33 (2020)
35. Watter, M., Springenberg, J.T., et al.: Embed to control: a locally linear latent dynamics model for control from raw images. In: NeurIPS, pp. 2746–2754 (2015)
36. Woodworth, B.E., Patel, K.K., Srebro, N.: Minibatch vs local SGD for heterogeneous distributed learning. In: NeurIPS, vol. 33, pp. 6281–6292 (2020)
37. Yang, R., Wang, B.: PACP: a position-independent activity recognition method using smartphone sensors. Information **7**(4), 72 (2016)
38. Yao, S., et al.: DeepSense: a unified deep learning framework for time-series mobile sensing data processing. In: WWW 2017, pp. 351–360 (2017)

Modeling IsA Relations via Box Structure for Knowledge Graph Embedding

Yao Dong[1,2], Lei Wang[1(✉)], Ji Xiang[1], and Kai Liu[1]

[1] Institute of Information Engineering, Chinese Academy of Sciences, Beijing, China
{dongyao,wanglei,xiangji,liukai}@iie.ac.cn
[2] School of Cyber Security, University of Chinese Academy of Sciences, Beijing, China

Abstract. Knowledge graph completion (KGC) aims to predict missing connections by mining information already present in a knowledge graph (KG). Predicting such connections is heavily dependent on the inference patterns. *IsA* relations (i.e., `instanceOf` and `subclassOf`) play an essential part in inferencing the composition pattern. Some existing methods already exploit *isA* relations. However, most of them learn insufficient representations, which may limit the performance. To address this issue, we propose a box-based knowledge graph embedding model called **IBKE**, in which concepts are embedded as boxes, and instances are represented by vectors in the same semantic space. According to the relative positions of elements, IBKE can naturally formulate *isA* relations. In addition, we introduce a random update strategy (RUS) for optimizing training, which updates embeddings in a probability pattern. Experimental results on benchmark datasets show that IBKE outperforms most existing state-of-the-art methods, and demonstrate the effectiveness of RUS.

Keywords: Knowledge graph embedding · Link prediction · Box

1 Introduction

Knowledge graphs (KGs) are structured facts of the real world, where nodes represents entities and edges between nodes represents relations. Large-scale KGs such as WordNet [14], YAGO [20] and Freebase [3] find applications in a variety of downstream tasks including machine translation [31], relation extraction [24], question answering [9] and recommender systems [29]. Although KGs may contain millions of triples, most existing KGs are incomplete. Therefore, much research work has been devoted to link prediction task, which is also known as knowledge graph completion (KGC). The target of link prediction is to predict missing facts in KG based on the existing links. An effective solution for KGC is knowledge graph embedding (KGE), which learns embeddings in a continuous low-dimensional vector space, and predicts missing links by evaluating the similarity of facts.

J. Gama et al. (Eds.): PAKDD 2022, LNAI 13281, pp. 303–315, 2022.
https://doi.org/10.1007/978-3-031-05936-0_24

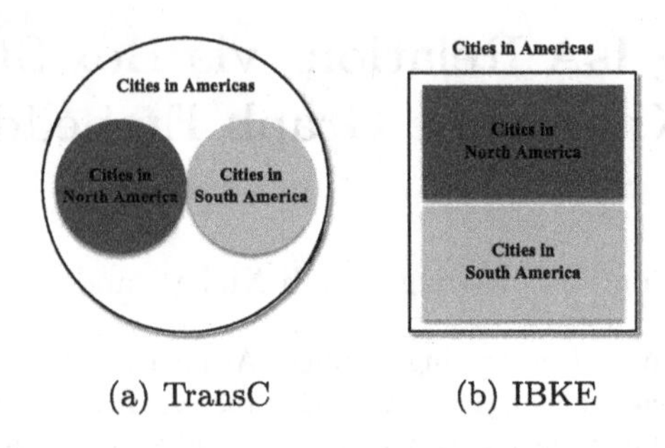

(a) TransC (b) IBKE

Fig. 1. Space utilization.

State-of-the-art KGE models can be broadly categorised as translational models [4,10,12,26], semantic matching models [11,23,28] and deep learning models [6,16,25]. Most approaches focus on translational models in early times, which provide competitive performance with fewer parameters. Afterwards, several methods turn to semantic matching models, which achieve better performance by matching latent semantics of entities and relations. Recently, deep learning models for KGC have received increasing research attention. Such models generally achieve more outstanding performance on account of the larger parameters.

Despite achieving remarkable performance, most existing methods still regard both instances and concepts as entities to make a simplification, which leads to the following two drawbacks: **insufficient concept representation** and **lacking transitivity** of ***isA*** relations. To address these issues, TransC [13] is proposed as the first KGC model for differentiating concepts and instances, which encodes each concept as a hypersphere and each instance as a vector. Although modeling concepts via hyperspheres can building the transitivity of *isA* relations, it still result in the **insufficient concept representation**. A typical case is shown in Fig. 1(a). Commonly, parent class *Cities in Americas* can be exactly divided into two disjoint subclasses: *Cities in North America* and *Cities in South America*. However, TransC cannot take full advantage of space in parent class *Cities in Americas* under any circumstances, which means the representations of subclass concepts are insufficient. Furthermore, the blank space in the hypersphere of *Cities in Americas* lacks practical significance, which may lead to weak interpretability.

The problem of insufficient representation gives rise to the box structure [19]. Boxes can be regarded as the extended hyperspheres, which have different radii in each dimension. Similarly, boxes can easily deal with *isA* relations. Due to the flexibility of hyper-rectangles, boxes need only a slight effort to fill the gaps between the parent class and subclasses. Thus, boxes not only have more promising representation power but also reserve the superiority of hyperspheres.

In this paper, we propose a new method called **IBKE** for knowledge graph embedding. IBKE encodes each concept as a box (hyper-rectangle), while instances and relations are encoded as vectors. Further, we utilize relative positions between instances and concepts to model *isA* relations. Specifically, IBKE represents `instanceOf` relation by checking whether an instance vector is inside the box. For `subclassOf` relation, we enumerate four relative positions and define different score functions for three non-target cases: **disjoint**, **intersect** and **inverse**. Moreover, we introduce a new parameter update method called Random update strategy (RUS) for optimizing, which randomly updates embeddings according to two update thresholds. Note that RUS has good generalization ability for closed-region models.

In summary, **our contributions** are listed as follows:

- We propose IBKE, to the best of our knowledge, the first method using box structure to distinguish instances and concepts for modeling *isA* relations.
- We present a random update strategy, which enhances the representation power by updating parameters in a probability pattern.
- Through extensive experiments on two datasets, we show that IBKE achieves state-of-the-art performance in most cases. Besides, we analyze the random update strategy in detail and prove its effectiveness.

2 Related Work

In this section, we give an overview of KGE models for link prediction, and divide previous methods into four categories.

Translational Models. TransE [4] is the first translational model, which encodes entities and relations as vectors based on the principle $\mathbf{h} + \mathbf{r} = \mathbf{t}$, where $\mathbf{h}$, $\mathbf{r}$, $\mathbf{t}$ denotes head entity, relation and tail entity, respectively. Then, several variants are proposed to solve the drawbacks of TransE, including TransH [26], TransR [12] and TransD [10]. By introducing manifold-wise modeling, ManifoldE [27] remedys the N-N problem in TransE. TorusE [7] expands the embedding space to a Non-Euclidean space, i.e., torus. RotatE [21] first regards translations as rotations from head entity to tail entity in complex plane.

Semantic Matching Models. RESCAL [18] is the first bilinear model that can perform collective learning, which is prone to overfitting. Hence, DistMult [28] simplifies RESCAL by using a diagonal matrix. ComplEx [23] extends DistMult to the complex domain for modeling antisymmetric relations. HolE [17] combines the quintessence in DistMult and ComplEx. Recently, SimplE [11] presents a simple enhancement of Canonical Polyadic (CP) decomposition, and TuckER [2] is based on Tucker decomposition. QuatE [30] first models relations as rotations in quaternion space to enable rich and expressive semantic matching.

Deep Learning Models. ConvE [6], ConvKB [16] and InteractE [25] use convolutional neural network to capture the interactions between entities and relations. In addition, KBGAT [15] learns graph attention-based embeddings by a *generalized* graph attention model.

Region-Based Models. Generally, region-based models encode elements by explicitly defining the regions. These elements can be both entities and relations.

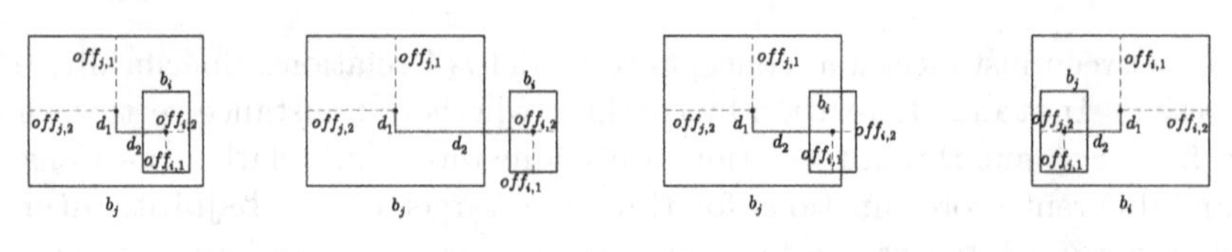

(a) $d_n \leq off_{j,n} - off_{i,n}$ (b) $d_n \geq off_{i,n} + off_{j,n}$ (c) $-off_{i,n} \leq d_n - off_{j,n} < off_{i,n}$ (d) $off_{i,n} > off_{j,n}$

Fig. 2. Four relative positions between box b_i and b_j.

Using a hypersphere to encode each concept, TransC [13] first differentiates concepts and instances. BoxE [1] that provides a solution to multi-arity KGC, encodes each relation as a box, while encodes each entity as a point and the corresponding *translational bump.*

Our proposed model IBKE belongs to the translational models. IBKE shares similarities with TransC, in which both models can deal with *isA* relations by differentiating concepts and instances. However, there are two major differences between TransC and IBKE:

- **Modeling.** IBKE encodes each concept as a box instead of a hypersphere, which is used in TransC.
- **Training.** Compared to TransC, we propose the random update strategy, which randomly learns parameters.

Note that we provide a comprehensive analysis about the computational complexity of several representative models in the supplemental material.[1]

3 Methodology

In this section, we propose a novel embedding method IBKE and present a new algorithm random update strategy (RUS).

3.1 IBKE

Formally, a knowledge graph is denoted by $\mathcal{G} = \{\mathcal{E}, \mathcal{R}, \mathcal{S}\}$. Entity set $\mathcal{E}$ consists of instance set $\mathcal{I}$ and concept set $\mathcal{C}$, i.e., $\mathcal{E} = \mathcal{I} \cup \mathcal{C}$. Relation set $\mathcal{R} = \{r_i, r_c\} \cup \mathcal{R}_r$, where r_i represents `instanceOf` relation, r_c represents `subclassOf` relation and $\mathcal{R}_r$ denotes the set of other relations. Therefore, the triple set $\mathcal{S}$ can be divided into three disjoint subsets according to the relation type: relational triple set $\mathcal{S}_r$, `instanceOf` triple set $\mathcal{S}_i$ and `subclassOf` triple set $\mathcal{S}_c$.

Given a knowledge graph $\mathcal{G}$, KGC aims at predicting the missing links in $\mathcal{G}$ by learning embeddings for instances, concepts and relations in the same vector space $\mathbb{R}^k$, where k denotes the dimension of vector space. In IBKE, for each instance $i \in \mathcal{I}$ and relation $r \in \mathcal{R}_r$, we learn a k-dimensional vector $\mathbf{i} \in \mathbb{R}^k$ and

[1] The supplemental material of our paper is available online: https://github.com/JensenDong/IBKE.

$\mathbf{r} \in \mathbb{R}^k$, respectively. For each concept $c \in \mathcal{C}$, we learn a box $b(\mathbf{cen}, \mathbf{off})$ with $\mathbf{cen}, \mathbf{off} \in \mathbb{R}^k$ denoting the box center and offsets of all dimensions, respectively.

Box structure is more flexible, but it also brings the challenge that it is difficult to measure nested boxes. Thus, we define different dimensional-wise score functions for `instanceOf`, `subclassOf` and relational triples.

Relational Triples. A relational triple denoted as (h, r, t) consists of one relation and two instances. IBKE learns k-dimensional vectors for instances and relations. Hence, we define the score function just like TransE as follows:

$$f_r(h, t) = \|\mathbf{h} + \mathbf{r} - \mathbf{t}\|_2^2. \tag{1}$$

InstanceOf Triples. For an `instanceOf` triple (i, r_i, c), when it holds, the instance i should be inside the **box** b. However, there is another relative position which i is out of the **box** b. Therefore, we define the following score function for optimizing:

$$f_i(i, c) = \sum_{n=1}^{k}(|i_n - cen_n| - off_n), \tag{2}$$

where i_n, cen_n and off_n represent the n-th element of $\mathbf{i}$, $\mathbf{cen}$ and $\mathbf{off}$, respectively.

SubclassOf Triples. For a `subclassOf` triple (c_i, r_c, c_j), when it holds, the **box** b_i should be inside the **box** b_j (as shown in Fig. 2(a)). However, there are three other relative positions between **box** b_i and b_j, i.e., **disjoint**, **intersect**, and **inverse**. Distance between the centers of b_i and b_j in n-th dimension is defined as follows:

$$d_n = |cen_{i,n} - cen_{j,n}|, \tag{3}$$

where $cen_{i,n}$ and $cen_{j,n}$ denote the n-th dimension of cen_i and cen_j, respectively. Further, we define a specific score function for each condition.

- **Disjoint.** b_i is disjoint from b_j (as shown in Fig. 2(b)). The two boxes should be closer in optimization. Therefore, the score function is defined as follows:

$$f_c(c_i, c_j) = \sum_{n=1}^{k}(d_n + off_{i,n} - off_{j,n}), \tag{4}$$

  where $off_{i,n}$ and $off_{j,n}$ denote the n-th dimension of $\mathbf{off_i}$ and $\mathbf{off_j}$, respectively.
- **Intersect.** b_i intersects with b_j (as shown in Fig. 2(c)). Similarly, we define the score fuction like the first condition as follows:

$$f_c(c_i, c_j) = \sum_{n=1}^{k}(d_n + off_{i,n} - off_{j,n}). \tag{5}$$

- **Inverse.** b_i is inside b_j (as shown in Fig. 2(d)). This condition is exactly the opposite of our optimization objective, so we define the following score function to reduce $off_{j,n}$ and increase $off_{i,n}$:

$$f_c(c_i, c_j) = \sum_{n=1}^{k}(off_{i,n} - off_{j,n}). \tag{6}$$

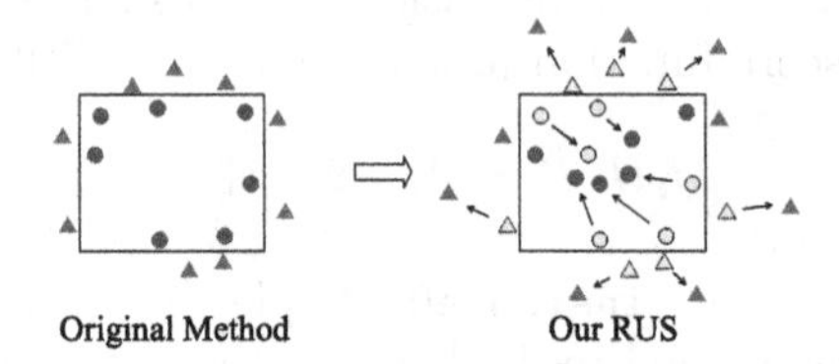

Fig. 3. Traditional method VS. RUS. Best view in colors. Red triangles represent negative instances and blue circles represent positive instances. Dotted triangles and circles represent their original positions. (Color figure online)

In experiments, we enforce constraints on embeddings, i.e., $\|\mathbf{h}\|_2 \leq 1$, $\|\mathbf{r}\|_2 \leq 1$, $\|\mathbf{t}\|_2 \leq 1$, $\|\mathbf{i}\|_2 \leq 1$, $\|\mathbf{cen}\|_2 \leq 1$ and $\forall n \in \{1, \ldots, k\}$, $off_n \leq 1$.

Optimization. We define a margin-based ranking loss function for relational triples as follows:

$$\mathcal{L}_r = \sum_{\xi \in \mathcal{S}_r} \sum_{\xi' \in \mathcal{S}'_r} [\gamma_r + f_r(\xi) - f_r(\xi')]_+, \tag{7}$$

where$[x]_+ \triangleq \max(0, x)$, ξ denotes a positive triple, ξ' denotes a negative triple and γ_r is the margin between positive triples and negative triples. Similarly, we define the loss function for `instanceOf` triples and `subclassOf` triples as follows:

$$\mathcal{L}_i = \sum_{\xi \in \mathcal{S}_i} \sum_{\xi' \in \mathcal{S}'_i} [\gamma_i + f_i(\xi) - f_i(\xi')]_+, \tag{8}$$

$$\mathcal{L}_c = \sum_{\xi \in \mathcal{S}_c} \sum_{\xi' \in \mathcal{S}'_c} [\gamma_c + f_c(\xi) - f_c(\xi')]_+. \tag{9}$$

We adopt stochastic gradient descent (SGD) to minimize the above loss functions, and use random update strategy (RUS) to randomly update embeddings.

Negative Sampling. Following Lv et al. [13], we randomly replace h or t to construct a negative triple (h', r, t) or (h, r, t'). (See details in supplemental material)

3.2 Random Update Strategy

During training `instanceOf` triples, the traditional method will stop updating parameters when score function $f_i(\xi) < 0$ for positive triples or $f_i(\xi') > 0$ for negative triples, which means that positive instances are inside the boxes or negative instances are outside the boxes, respectively. While the positive instances and negative instances are separated, they still gather around near the boundary of the boxes, i.e., the surface of hyper-rectangles. According to empirical regularity, instances and concepts should randomly distribute in the embedding space. Intuitively, we present a **R**andom **U**pdate **S**trategy (RUS) in place of the traditional algorithm. The comparison of these two algorithms is shown in Fig. 3. Note that we demonstrate this case in 2D for convenience.

In RUS, both positive triples and negative triples randomly update parameters depending on the score function f_i. We set two update thresholds $\phi_{pos}, \phi_{neg} \in [0, 1]$, where ϕ_{pos}, ϕ_{neg} denote positive update threshold and negative update threshold, respectively. For each positive training triple ξ, RUS updates parameters when $f_i > 0$ and updates parameters with probability ϕ_{pos} when $f_i \leq 0$. Similarly, for each negative training triple ξ', RUS updates parameters when $f_i < 0$ and updates parameters with probability ϕ_{neg} when $f_i \geq 0$. Details of RUS are summarized in supplemental material.

RUS is also applied to `subclassOf` triples. This strategy assists our model to separate positive and negative triples. Moreover, RUS can be generalized to a method that uses a structure of closed region such as hypersphere and box.

4 Theoretical Analyses

In this section, we provide some theoretical analyses of IBKE and Box structure. Note that all proofs for theorems can be found in the **supplemental material**.

4.1 Representation Power

Definition 1. *(**Filling Mode**) A filling mode is the way to stuff a box (hypersphere) with smaller ones. We define three types of filling modes.*

- ***Align** mode aligns the centers of the smaller boxes (hyperspheres) along each axis.*
- ***Compact** mode aligns the centers of the interlaced boxes (hyperspheres) to get a more compact spatial distribution.*
- ***Hybrid** mode is the mixture of **Align** mode and **Compact** mode.*

Theorem 1. *A hypersphere cannot be filled with several identical smaller hyperspheres without a single gap by any filling mode.*

Theorem 2. *A box can be filled with several identical smaller boxes without a single gap by a certain filling mode.*

Definition 2. *(**Representation Power**) The representation power of box (hypersphere) structure is the space utilization of embedding space.*

Theorem 3. *The representation power of box is superior to hypersphere.*

4.2 Inference Patterns

Knowledge graphs mainly consist of three relation patterns. We give their formal definitions here:

Definition 3. *A relation r is* ***symmetric(antisymmetric)*** *if*

$$\forall x, y \in \mathcal{E}, r(x, y) \Rightarrow r(y, x) \ (r(x, y) \Rightarrow \neg r(y, x))$$

A relation with such form is a ***symmetry(antisymmetry)*** *pattern.*

Definition 4. *Relation r_1 is* ***inverse*** *to r_2 if*

$$\forall x, y \in \mathcal{E}, r_2(x, y) \Rightarrow r_1(y, x)$$

Relations with such form is an ***inversion*** *pattern.*

Definition 5. *Relation r_1 is* ***composed*** *of relation r_2 and relation r_3 if*

$$\forall x, y, z \in \mathcal{E}, r_2(x, y) \wedge r_3(y, z) \Rightarrow r_1(x, z)$$

Relations with such form is a ***composition*** *pattern.*

According to the above definitions, we provide a comprehensive analysis on IBKE in supplemental material and come to the following theorem:

Theorem 4. *IBKE can infer the antisymmetric, inversion and composition patterns.*

5 Experiments

In this section, we evaluate IBKE and RUS on link prediction [4]. In addition, we conduct a series of ablation experiments for RUS.

5.1 Experimental Setup

Datasets. Most previous models are evaluated on FB15k [4] and WN18 [4]. To address the test leakage problem in FB15k and WN18, FB15k-237 [22] and WN18RR [6] are constructed, which are subsets of FB15k and WN18, respectively. However, FB15k and FB15k-237 mainly consist of instances; WN18 and WN18RR mainly consist of concepts. The imbalance in the number of instances and concepts makes these four datasets inappropriate for testing the ability of distinguishing instances and concepts. Besides, *isA* relations are not explicitly given on these datasets. Even YAGO26K-906 [8] and DB111K-174 [8], which have explicitly given the *isA* relations, are not applicable. Both YAGO26K-906 and DB111K-174 suffer from the severe imbalance of instances and concepts, either. Moreover, these two datasets have test leakage problem and contain a large number of repeating triples. Hence in experiments, following TransC, we evaluate IBKE on benchmark dataset YAGO39K [13], which is constructed from another popular knowledge graph YAGO [20], and contains a number of instances and concepts. The statistics of YAGO39K are listed in supplemental material. In addition, we also evaluate IBKE on Countries dataset [5,21] to explicitly test the ability of inferring the composition pattern. It consists of three sub-tasks which increase in difficulty in a step-wise fashion. For more details about Countries, please see supplemental material.

Evaluation Protocol. Following Bordes et al. [4], the link prediction performance is reported on the standard evaluation metrics: Mean Reciprocal Rank (MRR) and Hits@N for $N = 1, 3, 10$. MRR is the mean reciprocal rank of correct triples. Hits@N is the proportion of correct triples whose rank is not larger than N. Note that an excellent embedding model should achieve a higher MRR and a higher Hits@N. We report the filtered results to avoid possibly flawed evaluation.

Table 1. Link prediction results on YAGO39K with $k = 100$ and $k = 200$. Best results are in **bold** and second best results are underlined.

Model	$k = 100$				$k = 200$			
	MRR	H@1	H@3	H@10	MRR	H@1	H@3	H@10
TransE [4]	.248	.123	.287	.511	–	–	–	–
TransH [26]	.215	.104	.240	.451	–	–	–	–
TransR [12]	.289	.158	.338	.567	–	–	–	–
TransD [10]	.176	.089	.190	.354	–	–	–	–
HolE [17]	.198	.110	.230	.384	–	–	–	–
DistMult [28]	.362	.221	.436	.660	–	–	–	–
ComplEx [23]	.362	.292	.407	.481	–	–	–	–
SimplE [11]	.392	.283	.456	.590	.465	.367	.523	.644
TorusE [7]	.351	.295	.388	.449	–	–	–	–
TuckER [2]	.270	.187	.290	.428	.427	.315	.477	.653
KBGAT [15]	.469	.351	.539	.692	.475	.357	.543	.699
QuatE [30]	.399	.273	.452	.659	–	–	–	–
RotatE [21]	.504	.413	.560	.668	.552	.458	.611	.721
BoxE [1]	**.546**	**.462**	.598	.697	.566	.475	.626	<u>.726</u>
TransC [13]	.437	.299	.521	.700	.520	.406	.597	.720
IBKE (ours)	.522	.404	<u>.605</u>	**.731**	<u>.578</u>	<u>.487</u>	<u>.640</u>	**.729**
TransC-RUS	.448	.311	.534	<u>.704</u>	.531	.423	.602	.721
IBKE-RUS (ours)	<u>.532</u>	<u>.418</u>	**.613**	**.731**	**.582**	**.497**	**.641**	.725

Implementation. We select learning rate λ for SGD among {0.1, 0.01, 0.001}, the dimensionality of embedding space k among {20, 50, 100}, the three margins

γ_r, γ_i and γ_c among {0.1, 0.3, 0.5, 1, 2}, the two update thresholds ϕ_{pos} and ϕ_{neg} among {0.1, 0.2, 0.3, 0.4, 0.5, 0.6, 0.7, 0.8, 0.9}. The optimal configurations on YAGO39K and Countries are listed in the supplemental material. To maintain comparison fairness, we train each model for 1000 epoches.

5.2 Results and Analysis

Evaluation results for relational triples are shown in Table 1. Note that we use publicly available source codes to reproduce results of comparison models, i.e., SimplE, TorusE, TuckER, KBGAT, QuatE, RotatE, BoxE and TransC. Other results are taken from [13]. From Table 1, we conclude that: (1) IBKE outperforms all baseline models in terms of Hits@3 and Hits@10. Results indicate that IBKE can get better performance by explicitly modeling *isA* relations. Distinguishing instances and concepts play a crucial part in learning embeddings. (2) The trend of performance with $k = 200$ is basically consistent with the performance with $k = 100$. We can see that IBKE outperforms all baseline models on all metrics when $k = 200$. The reason is that the representation power of box is more significant with a larger dimension. (3) The RUS works well for both IBKE and TransC, which implies a good scalability.

Comparison with TransC. IBKE achieves significant performance improvement. In specific, the improvement is 0.522–0.437 = 0.085 on MRR and +8.1% on Hits@1 over TransC when $k = 100$, which indicates that with a higher space utilization, the box structure is superior to hypersphere.

Comparison with BoxE. IBKE is only less competitive than BoxE in MRR and Hits@1 with $k = 100$, but outperforms BoxE on all metrics when $k = 200$. The reason is that IBKE encodes concepts as boxes and BoxE encodes relations as boxes. The number of concepts is larger than relations. Therefore, IBKE can capture more information with a larger dimension.

Random Update Strategy. To verify the effectiveness of RUS, we conduct a series of ablation experiments as shown in Table 1, Fig. 4 and Fig. 5(a). From Table 1, TransC-RUS achieves relative improvement of 0.448–0.437 = 0.011 on MRR and +1.7% on Hits@1 over TransC. Compared to IBKE, IBKE-RUS achieves relative improvement of 0.532–0.522 = 0.010 on MRR and +1.4% on Hits@1. Figure 4(a) shows that despite the epoch, models with RUS always outperform the corresponding ones without RUS. Moreover, as shown in Fig. 5, RUS can achieve better performance by using specific update thresholds.

Results on Countries S1/S2/S3. To further investigate the ability of inferring composition pattern, we evaluate our model on Countries dataset. In Table 2, we report the results with respect to the AUC-PR metric, which is commonly used in the literature. We can see that IBKE outperforms all the baseline models on S1 and S3, and obtains competitive performance on S2. Note that S3 is the most difficult task.

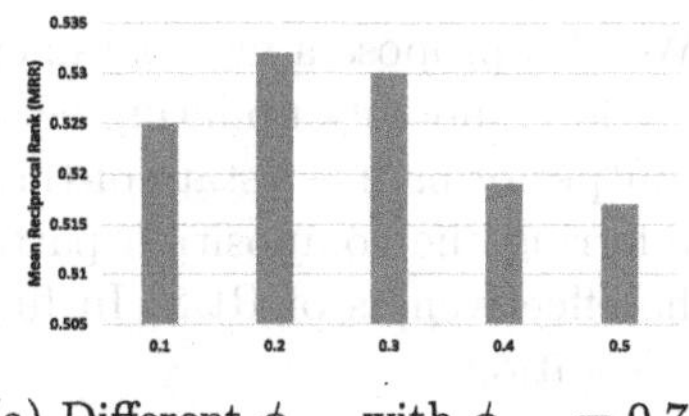

(a) Different ϕ_{pos} with $\phi_{neg} = 0.7$.

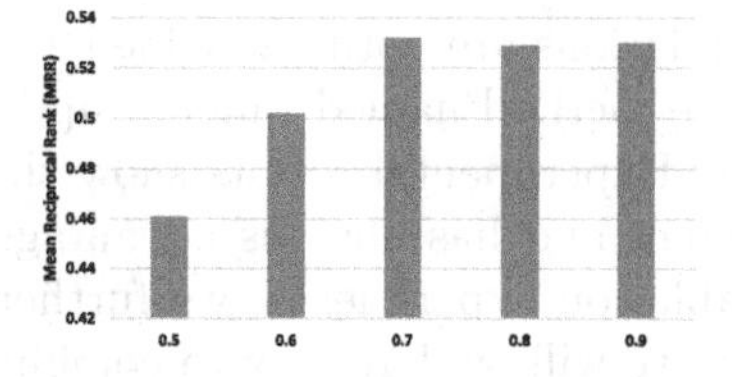

(b) Different ϕ_{neg} with $\phi_{pos} = 0.2$.

Fig. 4. Performance with different update thresholds on MRR.

Table 2. Link prediction results of Countries datasets. Best results are in **bold**.

Model	Countries (AUC-PR)		
	S1	S2	S3
DistMult	**1.00**	0.72	0.52
ComplEx	0.97	0.57	0.43
ConvE	**1.00**	0.99	0.86
RotatE	**1.00**	**1.00**	0.95
IBKE	**1.00**	0.99	**0.96**

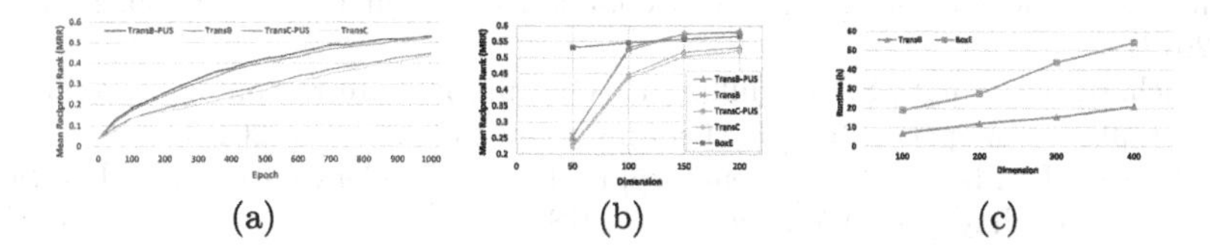

(a) (b) (c)

Fig. 5. (a) Learning curves of IBKE, IBKE-RUS, TransC, and TransC-RUS; (b) Performance versus dimensionality; (c) Runtime analysis.

Robustness Experiment. We evaluate the dependence of IBKE on dimensionality. Experimental results are shown in Fig. 4(b), in which we can conclude that: (1) Compared to IBKE, BoxE can obtain competitive performance with a smaller dimension. (2) When $k \geq 150$, IBKE can achieve state-of the-art performance relative to most models.

Running Time Analysis. We train IBKE-RUS on the CPU and BoxE on a single Tesla V100 GPU. Results are shown in Fig. 4(c), in which we can see that each runtime of IBKE is less than BoxE with same embedding dimension. Furthermore, as the dimension grows, so does the runtime gap between IBKE and BoxE. Hence, IBKE is more efficient than BoxE.

6 Conclusion

In this paper, we propose IBKE, which introduces a new use of box for knowledge graph completion. IBKE applies box structure to model concepts. Instances and relations are both embedded as vectors. We also propose a new parameters update method named random update strategy for randomly updating embeddings. Experimental results show that IBKE outperforms most state-of-the-art baselines and has obvious advantages when inferring the composition pattern. By ablation experiments, we further prove the effectiveness of RUS. In future work, we will explore how to combine box and rotation.

Acknowledgments. This work was supported by the National Key Research and Development Program of China.

References

1. Abboud, R., Ceylan, I., Lukasiewicz, T., Salvatori, T.: BoxE: a box embedding model for knowledge base completion. In: NeurIPS, vol. 33 (2020)
2. Balazevic, I., Allen, C., Hospedales, T.: TuckER: tensor factorization for knowledge graph completion. In: EMNLP, pp. 5185–5194 (2019)
3. Bollacker, K., Evans, C., Paritosh, P., Sturge, T., Taylor, J.: Freebase: a collaboratively created graph database for structuring human knowledge. In: SIGMOD, pp. 1247–1250 (2008)
4. Bordes, A., Usunier, N., Garcia-Duran, A., Weston, J., Yakhnenko, O.: Translating embeddings for modeling multi-relational data. In: NeurIPS, vol. 26, pp. 2787–2795 (2013)
5. Bouchard, G., Singh, S., Trouillon, T.: On approximate reasoning capabilities of low-rank vector spaces. In: AAAI Spring Symposium Series (2015)
6. Dettmers, T., Minervini, P., Stenetorp, P., Riedel, S.: Convolutional 2D knowledge graph embeddings. In: AAAI, vol. 32 (2018)
7. Ebisu, T., Ichise, R.: TorusE: knowledge graph embedding on a lie group. In: AAAI, vol. 32 (2018)
8. Hao, J., Chen, M., Yu, W., Sun, Y., Wang, W.: Universal representation learning of knowledge bases by jointly embedding instances and ontological concepts. In: SIGKDD, pp. 1709–1719 (2019)
9. Hao, Y., et al.: An end-to-end model for question answering over knowledge base with cross-attention combining global knowledge. In: ACL, pp. 221–231 (2017)
10. Ji, G., He, S., Xu, L., Liu, K., Zhao, J.: Knowledge graph embedding via dynamic mapping matrix. In: ACL, pp. 687–696 (2015)
11. Kazemi, S.M., Poole, D.: Simple embedding for link prediction in knowledge graphs. In: NeurIPS, pp. 4284–4295 (2018)
12. Lin, Y., Liu, Z., Sun, M., Liu, Y., Zhu, X.: Learning entity and relation embeddings for knowledge graph completion. In: AAAI, vol. 29 (2015)
13. Lv, X., Hou, L., Li, J., Liu, Z.: Differentiating concepts and instances for knowledge graph embedding. In: EMNLP, pp. 1971–1979 (2018)
14. Miller, G.A.: Wordnet: a lexical database for English. Commun. ACM **38**(11), 39–41 (1995)
15. Nathani, D., Chauhan, J., Sharma, C., Kaul, M.: Learning attention-based embeddings for relation prediction in knowledge graphs. In: ACL, pp. 4710–4723 (2019)
16. Nguyen, T.D., Nguyen, D.Q., Phung, D., et al.: A novel embedding model for knowledge base completion based on convolutional neural network. In: NAACL, pp. 327–333 (2018)

17. Nickel, M., Rosasco, L., Poggio, T.: Holographic embeddings of knowledge graphs. In: AAAI, vol. 30 (2016)
18. Nickel, M., Tresp, V., Kriegel, H.P.: A three-way model for collective learning on multi-relational data. In: ICML, vol. 11, pp. 809–816 (2011)
19. Ren, H., Hu, W., Leskovec, J.: Query2box: reasoning over knowledge graphs in vector space using box embeddings. In: ICLR (2019)
20. Suchanek, F.M., Kasneci, G., Weikum, G.: Yago: a core of semantic knowledge. In: WWW, pp. 697–706 (2007)
21. Sun, Z., Deng, Z., Nie, J., Tang, J.: Rotate: knowledge graph embedding by relational rotation in complex space. In: ICLR (2019)
22. Toutanova, K., Chen, D.: Observed versus latent features for knowledge base and text inference. In: Proceedings of the 3rd Workshop on Continuous Vector Space Models and their Compositionality, pp. 57–66 (2015)
23. Trouillon, T., Welbl, J., Riedel, S., Gaussier, É., Bouchard, G.: Complex embeddings for simple link prediction. In: ICML (2016)
24. Vashishth, S., Joshi, R., Prayaga, S.S., Bhattacharyya, C., Talukdar, P.: Reside: improving distantly-supervised neural relation extraction using side information. In: EMNLP, pp. 1257–1266 (2018)
25. Vashishth, S., Sanyal, S., Nitin, V., Agrawal, N., Talukdar, P.P.: Interacte: improving convolution-based knowledge graph embeddings by increasing feature interactions. In: AAAI, pp. 3009–3016 (2020)
26. Wang, Z., Zhang, J., Feng, J., Chen, Z.: Knowledge graph embedding by translating on hyperplanes. In: AAAI, vol. 28 (2014)
27. Xiao, H., Huang, M., Zhu, X.: From one point to a manifold: knowledge graph embedding for precise link prediction. In: IJCAI, pp. 1315–1321 (2016)
28. Yang, B., tau Yih, W., He, X., Gao, J., Deng, L.: Embedding entities and relations for learning and inference in knowledge bases (2015)
29. Zhang, F., Yuan, N.J., Lian, D., Xie, X., Ma, W.Y.: Collaborative knowledge base embedding for recommender systems. In: SIGKDD, pp. 353–362 (2016)
30. Zhang, S., Tay, Y., Yao, L., Liu, Q.: Quaternion knowledge graph embeddings. In: NeurIPS, pp. 2735–2745 (2019)
31. Zhao, Y., Zhang, J., Zhou, Y., Zong, C.: Knowledge graphs enhanced neural machine translation. In: IJCAI, pp. 4039–4045 (2020)

A GNN-Enhanced Game Bot Detection Model for MMORPGs

Xianyang Qi[1], Jiashu Pu[2], Shiwei Zhao[2], Runze Wu[2](✉), and Jianrong Tao[2]

[1] Zhejiang University, Hangzhou, China
22060015@zju.edu.cn
[2] Fuxi AI Lab, NetEase Games, Hangzhou, China
{pujiashu,zhaoshiwei,wurenze1,hztaojianrong}@corp.netease.com

Abstract. Game bots are automated programs that assist cheating players in obtaining huge superiority in Massively Multiplayer Online Role-Playing Games (MMORPGs), which has led to an imbalance in the gaming ecosystem and a collapse of interest among normal players. Game bot detection aims to identify cheating behaviors to ensure fair competition for MMORPGs. Due to the high practical value, there is much research on game bot detection at present. One main existing method is conventional machine learning algorithms, which require extensive feature engineering and get limited performance. The other main existing method is the recurrent neural network, but it fails to capture the complex behavioral patterns of players. To tackle the above problems, we propose a novel graph neural network-enhanced game bot detection model, namely GB-GNN. In the proposed model, we model players' trajectories as graph-structured data to capture the player's complex behavioral patterns that are difficult to reveal by traditional sequential methods. Extensive experiments on three real-world datasets show that GB-GNN outperforms the previous methods.

Keywords: Game bot detection · MMORPGs · Graph neural networks

1 Introduction

The COVID-19 pandemic has reshaped the world in many aspects, anecdotal reports claim that many people have turned to play video games during the pandemic; people play games to search for cognitive stimulation, opportunities to socialize, and stress reduction [2]. MMORPG is a popular and relaxing game genre in which players can take on various roles in a virtual world, complete main/side storylines, upgrade, socialize, and adventure [6]. Some MMORPGs allow players to use fiat money to buy virtual items or exchange virtual currency. The association of fiat and virtual currencies gives rise to game bot developers as the business is profitable [10]. Game bots can replace humans with scripts to complete tedious tasks, or complete a series of difficult operations through

X. Qi—Work was done during an internship at NetEase.

J. Gama et al. (Eds.): PAKDD 2022, LNAI 13281, pp. 316–327, 2022.
https://doi.org/10.1007/978-3-031-05936-0_25

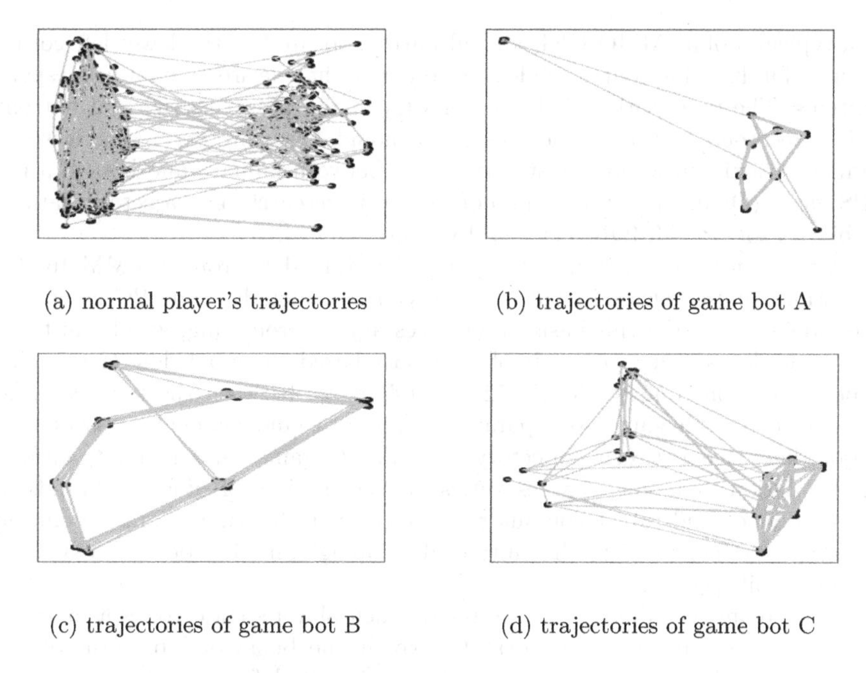

(a) normal player's trajectories (b) trajectories of game bot A

(c) trajectories of game bot B (d) trajectories of game bot C

Fig. 1. Some typical trajectories.

simulated clicks, to help players gain an unfair advantage in the game, such as quickly accumulating experience values, obtaining legendary items or equipment. This unfair game ecology will significantly reduce the game experience of honest players, leading to player loss and is detrimental to the healthy development of the entire game industry [11]. To mitigate the negative effects of game plug-ins and automated scripts, some previous work [7,8,19] performed game bot detection by designing domain-specific features and models. [18,22] resort to the integration of supervised and unsupervised learning and propose an auto-iteration mechanism to quickly acquire the ability to detect bots in new games or domains. Besides, the disparities between social network data constructed by humans and game bots can also be used for game bot detection [14,20].

However, The above studies revolve around data from the PC side. As the mobile game market is continuously expanding[1], the trajectory data inside the game is also rapidly accumulating. A survey from Adjust[2] finds that 41% of mobile gamers have used bots in 2020, indicating a real and urgent need to develop an effective game bot detection system for mobile games. While playing mobile games, the mobile device may generate a large amount of trajectory data.

[1] https://www.mordorintelligence.com/industry-reports/mobile-games-market.
[2] https://www.adjust.com.

In a typical mobile MMORPG, virtual buttons located at the lower-left corner are used for direction control while those on the right side are used for using spells or items. The trajectories of the player's finger touching these virtual buttons will be recorded by the mobile sensors. Although there are two research works aiming to distinguish movement trajectories between humans and bots of a PC FPS game (Quake2) [3,15], game bot detection research on trajectory data of mobile sensors for MMORPGs is still vacant.

We collect a large amount of original trajectory data from two MMORPGs in distinct genres released from NetEase Games[3], including one PC game and one mobile game. On this basis, after necessary preprocessing, we obtain three real-world datasets for the study of trajectory-based game bot detection in this paper. Note that due to the significant differences between the two scenes in our selected mobile game, we separate them into two independent datasets. One major difference between trajectory data in PC games and mobile games is the resolution of mobile devices varies, as screen sizes of different brands of mobile phones and tablet computers are often not the same[4]. This resolution inconsistency increases the difficulty of developing game bot detection systems for the mobile sensor data.

To overcome the limitations, we regard each player's trajectories as a graph and employ graph neural networks to exploit the behavioral patterns of the game bot as shown in Fig. 1 and avoid hand-crafted features. In this paper, we propose a novel graph neural network-enhanced game bot detection model, namely GB-GNN. Concretely, for capturing the patterns of the game bot, each trajectory should be discretized and built as a directed line graph. GB-GNN utilizes graph neural networks to pass messages and obtain the representation of nodes of the graph. Finally, we employ Gated Recurrent Units (GRUs) to get the vector of a trajectory and identify whether the trajectory is a game bot. The major contributions of this paper are summarized as follows:

- We model players' trajectories into graph-structured data and use graph neural networks to capture complex patterns of the game bots. To the best of our knowledge, it presents a novel perspective for modeling game bot detection in practical scenarios.
- Based on trajectory graph, we propose a novel model to unify Graph Neural Networks and Gated Recurrent Units for game bot detection across MMORPGs in distinct genres.
- By implementing extensive experiments on three commonly used real-world datasets, we demonstrate that our proposed model outperforms state-of-the-art non-graph-based methods.

[3] http://game.163.com/.

[4] https://en.wikipedia.org/wiki/Comparison_of_high-definition_smartphone_displays.

2 Related Work

In this section, we briefly review the work related to this paper from game bot detection and integration of GNN and RNN.

2.1 Game Bot Detection

Game bot detection is generally divided into three categories: client-side, network-side, and server-side. We only discuss the widely adopted server-side approaches as such approaches are prioritized to ensure the best gaming experience for players [18]. Game developers can collect many types of data on the server-side, including domain-specific features, behavior sequences, social network data, and trajectory data. [7,8,19] mine logs to distinguish differences between humans and robots in behavioral characteristics and network communication timing and sizes. The main drawback of these works is that they all rely on domain-specific features, leading to the inability to directly transfer such methods to new games. To address this problem, [18,22] integrate supervised and unsupervised methods for game bot detection, and its proposed auto-iteration mechanism approach can be quickly adapted to new games for plugin detection. Social network data can also be exploited to distinguish humans and bots. [14] assume that humans and game bots are different in their construction of social networks. [20] perform social activity detection by analyzing users' social interactions.

In addition, there is a large amount of trajectory data in games, such as mouse trajectories and movement trajectories in the game world. [3,15] propose a manifold method (Isomap) to perform bot detection on movement trajectories. However, the above work was only validated on quake2, a PC-based FPS game, and only three types of mainstream FPS plug-ins were validated. Nowadays, plug-ins and automation scripts are also proliferating in various MMORPGs, and players are also slowly moving from the PC platform to the mobile side. However, very little work has been done specifically on the trajectory data of mobile sensors. Game developers may record many touch sensor data but detecting bots on these specific data still remain challenging. To fill this void, we carefully investigate the characteristics of sensor trajectories and propose a model architecture combining GNN and GRU.

2.2 Integration of GNN and RNN

The integration of GRU and GNN is not uncommon, but the function of GRU and GNN may vary according to different tasks. [17] integrate GNN and LSTMs to predict COVID-19 new cases by exploiting spatial and temporal information. To facilitate prescription, a hybrid method of RNN and GNN are proposed to represent patient status sequences and temporal medical event graphs respectively [12]. [13] design Graph Long Short Term Memory (GLSTM) model for traffic speed prediction, using GNN to capture spatial-temporal dependencies and LSTM to capture long-term dependencies.

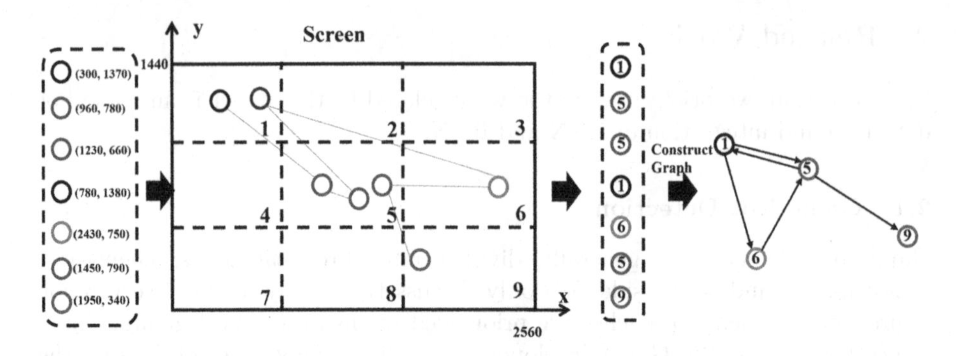

Fig. 2. The workflow for graph construction.

3 Preliminaries

Problem Formulation: In this paper, we intend to use the players' trajectories to detect game bots, where a player's trajectory T can be expressed as:

$$T = \{(x_0, y_0, t_0), (x_1, y_1, t_1), ..., (x_n, y_n, t_n)\}, \quad t_0 < t_1 < ... < t_n. \tag{1}$$

Here n is the length of the trajectory T, t_n is the timestamp, and (x_n, y_n) is the player's n-th coordinate. We use T_i to denote the i-th coordinate of T, and $T_{i,0}$ and $T_{i,1}$ to denote the corresponding x and y of T_i, respectively.

Graph Construction: We construct the graph as shown in Fig. 2. We first discretize the original trajectory T into $\hat{T}$ as follows:

$$\hat{T} = \{(w_0, t_0), (w_1, t_1), ..., (w_n, t_n)\}, \quad t_0 < t_1 < ... < t_n. \tag{2}$$

Here element w_i could be calculated as the following way:

$$w_i = [mT_{i,0}/L] + m[nT_{i,1}/W]. \tag{3}$$

Here $[\cdot]$ denotes rounding up. L and W are the length and width of the trajectory coordinate plane, respectively. And m and n mean that the length L is equally divided into m parts and the width W is equally divided into n parts.

On this basis, the discretized trajectory $\hat{T}$ can be modeled as a directed graph $\mathcal{G} = (\mathcal{V}, \mathcal{E})$. After the trajectory T is discretized, sequence $\hat{T}$ can be modeled as a directed graph $\mathcal{G} = (\mathcal{V}, \mathcal{E})$, where each node represents an element and each edge $(w_{i-1}, w_i) \in \mathcal{E}$ means that element w_i appear after w_{i-1}. We embed each element w_i into a unified embedding space and learn the node vector through a graph neural network. Finally, each trajectory can be represented as an embedding vector, denoted as T for simplicity.

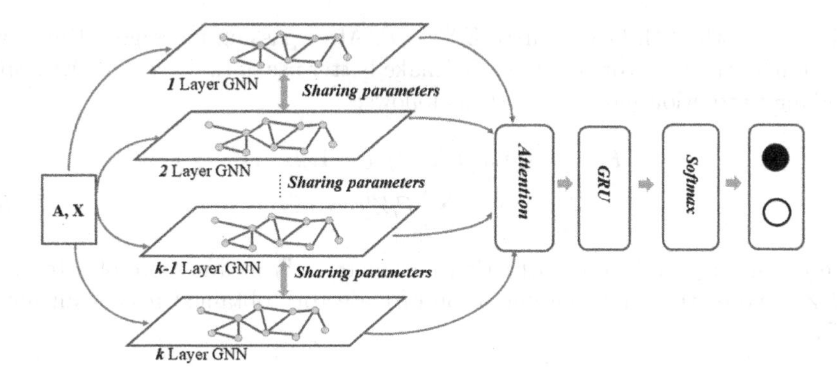

Fig. 3. The workflow of our proposed model.

4 The Proposed Model

In this section, we introduce the workflow of our proposed model as shown in Fig 3. The input of our proposed model is discrete trajectories while the output is the probability distribution of the game bots. Graph convolution networks (GCN) are employed in the model to capture repetitive patterns of the game bots. And we utilize Gated Recurrent Units (GRU) to explore normal players' intentions such as "attack" and "move". After obtaining the latent representation of trajectories, the probability distribution of the game bots could be obtained by softmax of the representation vector of the trajectories.

4.1 Capturing Repetitive Patterns

Because the trajectories of the game bots contain more repetitive patterns, the trajectories of the game bots are more predictable than that of normal players. The vanilla graph neural network is proposed by Scarselli et al. [16], extending neural network methods for processing the graph-structured data. Graph neural network is widely applied to natural language process, computer vision, and data mining due to their capability of capturing patterns. Multi-layer GCN is employed in the model to capture different repetitive patterns, we consider a multi-layer GCN with the following layer-wise propagation rule:

$$Z^{(l+1)} = \sigma(\widetilde{D}^{-\frac{1}{2}}\widetilde{A}\widetilde{D}^{-\frac{1}{2}}Z^{(l)}W^{(l)}). \tag{4}$$

Here $\widetilde{A} = A + E$, E is the identity matrix, A is the adjacency matrix of the directed graph $\mathcal{G}$ with added self-connections. $\widetilde{D}_{ii} = \sum_j \widetilde{A}_{ij}$ and W is trainable parameters. $\sigma(\cdot)$ means an activation function. And $Z^{(l)}$ is the matrix of activation in the l-th layer, where $Z^{(0)} = T$. After passing message k times, we concatenate the k vectors together. To make better use of the k vectors, we apply Bahdanau attention mechanism [1] as follows:

$$F_i = v_a^T tanh(w_a Z^{(i)}), i = 1, 2, ..., k. \tag{5}$$

$$\widetilde{Z} = \sum F_i Z^{(i)}. \tag{6}$$

Here v_a and w_a are learned attention parameters, F_i means score of attention, and $\widetilde{Z}$ denotes the final representation of trajectory obtained from multi-layer GCN.

4.2 Exploring Players' Intentions

In MMORPGs, some fragments of normal players' trajectories indicate the players' intentions such as "attack enemies" and "move to some places". Exploring the players' intentions from players' trajectories, we exploit recurrent neural networks. Gated Recurrent Unit (GRU) introduced by Kyunghyun Cho et al. [5] has a good performance on the task of natural language processing. The update functions are written as follows:

$$z_i = \sigma(W_z \widetilde{Z}_i + U_z h_{i-1} + b_z), \tag{7}$$

$$r_i = \sigma(W_r \widetilde{Z}_i + U_r h_{i-1} + b_r), \tag{8}$$

$$\hat{h}_i = \tanh(W_h \widetilde{Z}_i + U_h(r_i \odot h_{i-1}) + b_h), \tag{9}$$

$$h_i = (1 - z_i) \odot h_{i-1} + z_t \odot \hat{h}_i. \tag{10}$$

Here $\widetilde{Z}_i$ is the item embedding of i-th item of timestep t_i in the trajectory, W, U, b are the learnable parameters. $\sigma(\cdot)$ means the sigmoid function, $\odot$ represents the element-wise multiplication operations, $\hat{h}_i$ denotes the candidate hidden state, and h_i, z_i and r_i are respectively the hidden state, update gate vector, and reset gate vector.

4.3 Training and Inference

To generate the trajectory vector, we use the mean of hidden state h_i of GRU as trajectory vector:

$$\widetilde{h} = \sum_i^n h_i / n. \tag{11}$$

After getting the trajectory vector $\widetilde{h}$, we apply a softmax to get the output vector of the model $\hat{\mathbf{y}}$:

$$\hat{\mathbf{y}} = \text{softmax}(W_s \widetilde{h}), \tag{12}$$

where $\hat{\mathbf{y}}$ denotes the probabilities distribution of being a game bot and W_s is the learned parameters. The loss function is defined as cross-entropy of the prediction and the ground truth as follows:

$$\mathcal{L}(\mathbf{y}, \hat{\mathbf{y}}) = \sum_{i=1}^{N} \mathbf{y}_i \log(\hat{\mathbf{y}}_i) + (1 - \mathbf{y}_i) \log(1 - \hat{\mathbf{y}}_i). \tag{13}$$

Here $\mathbf{y}$ denotes the one-hot encoding vector of the ground truth. After calculating the above formula, we use the backpropagation algorithm to train our model.

5 Experiments

In this section, we first introduce the datasets, baseline, and evaluation metrics. Then we compare the proposed model with the baselines and analyze the results.

5.1 Experiment Settings

Datasets. We use three real-world datasets to evaluate the performance of our proposed model. We collect a large amount of original trajectory data from two MMORPGs in distinct genres released from NetEase Games, including one PC game and one mobile game. To be specific, the PC game is Justice[5], a martial game based on Wen Rui'an's novel The Four Great Constables (Si Da Ming Bu). And the mobile game contains two significantly different scenes. Hence, we specifically separate them as two independent datasets here, named rookie and battle. All the original trajectory data is generated from players' logs while labels are obtained by manual annotation from multiple game experts. On this basis, for each dataset, we select 5000 players' trajectories as a train set and 1000 players' trajectories as a validation set, respectively.

Evaluation Metrics. Due to category imbalance in the datasets, we utilize Area Under the Precision-Recall Curve (PR-AUC) to evaluate all methods. A high PR-AUC suggests that the evaluating model has a strong ability to distinguish positive and negative samples.

Parameter Setup. For the fair competition, we set up the same parameters for all models. We find that the performance of models gets better as l, m and the layers of GB-GNN k become larger. Due to the limitation of memory of GPU, we set the parameters l and m to 20, k to 12, dimension of hidden vector d to 64. All parameters are initialized by Gaussian distribution. We use a mini-batch Adam optimizer to optimize these parameters, setting the initial learning rate to 0.003, with a batch size of 32.

[5] https://leihuo.163.com/en/games.html?g=game-1nsh.

Table 1. The performance of the methods.

Method	Justice	rookie	battle
Metrics	PR-AUC	PR-AUC	PR-AUC
GRU	0.6854	0.8632	0.6050
Transformer	0.7228	0.8477	0.6503
XGBoost	0.7362	0.7702	0.4442
GCN	0.7669	0.6854	0.6012
maLstmFcn	0.7894	0.8542	0.7264
GB-GNN	**0.8558**	**0.8937**	**0.7350**

5.2 Baselines

In this work, we use players' trajectories to detect game bots, however, there are a lot of work detecting game bots by trading networks of virtual items or players' profiling [18]. The previous methods are unsuitable for detecting game bots using players' trajectories. To demonstrate the performance of our model, we choose several sequential models and some data mining methods for comparison:

- GRU [5]: GRU is an RNN-based model to exploit temporal information.
- Transformer [21]: A transformer is based solely on attention mechanisms.
- XGBoost [4]: XGBoost is a scalable tree boosting system, which is used widely by data scientists to achieve state-of-the-art results.
- GCN: GCN is a powerful neural network architecture on graph-structured data. We employ a multi-layers (7-layer) GCN as a baseline.
- maLstmFcn [9]: The model integrates CNN, LSTM, and attention mechanisms to exploit spatial and temporal information.

5.3 Comparison with Baseline Methods

The performance in terms of PR-AUC metrics on three datasets is shown in Table 1, and we highlight the best results in boldface. Our proposed method achieves the best performance compared with the other state-of-the-arts on all datasets, which verifies the effectiveness of GB-GNN.

In the Justice dataset, GCN has a better performance than GRU, which indicates that the trajectories of the game bots in the Justice dataset have more repetitive patterns. The maLstmFcn obtains an excellent result, which suggests CNN has a promising ability to capture repetitive patterns of game bots. Our proposed method GB-GNN has the most excellent results, which suggests the simple repetitive patterns of trajectories mainly consist of a certain number of nodes as shown in 1 and a specific layer of GB-GNN is capable of exploring the repetitive patterns. In the battle dataset, the performance of GCN is similar to the GRU, however, GCN exceeds GRU in the Justice dataset, which suggests that GCN with a specific layer is capable of specific repetitive patterns but

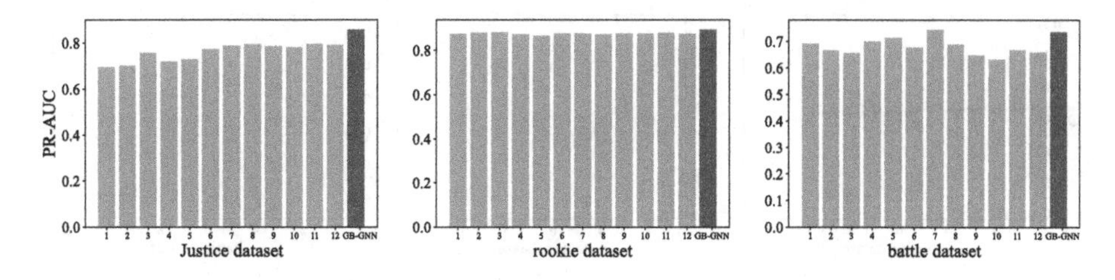

Fig. 4. Performance of each layer of GB-GNN.

fails to capture multiple patterns. As a comparison, GB-GNN still has the best performance, which indicates that the attention mechanism employed in GB-GNN can make use of the information extracted by each layer of GB-GNN. In the rookie dataset, GB-GNN has an absolute advantage over the other methods but GCN has the worst performance, which suggests that GNN is capable of exploring the repetitive patterns but the sequential model is also critical to capture the patterns of game bots.

5.4 Comparison with Each Layer of GB-GNN

GB-GNN aims at utilizing different layers to capture different repetitive patterns as shown in Fig. 1. We evaluate the idea by removing the attention mechanism of GB-GNN and using the output of each layer of GCN to detect game bots. More specifically, when layers of GB-GNN is 12, we will extract the output of i-th $(i = 1, ..., 12)$ layer of GNN and detect game bot respectively. The performance of each single-layer model is shown in Fig. 4.

We find GB-GNN outperforms all the single-layer models in the three real-world datasets. There are two differences between these models on the datasets. One is that the improvement in terms of PR-AUC of GB-GNN over single-layer GNN is different on the different datasets, the improvement is the largest on the Justice dataset, while on the rookie dataset, the improvement is not very obvious. The other difference is that layer of the best single-layer GNN is different on the different datasets. The layers of the best performance of the single-layer GNN in the Justice dataset are 11, while the best performance is obtained in the other two datasets when layers are 7 and 3, respectively.

The first difference is caused by the different motion patterns of the game bots on the different datasets, as shown in Fig. 1, which shows the typical motion patterns of game bots on the Justice, battle, and rookie datasets, respectively. As a result of the difference in game bot patterns, the improvement of GB-GNN on the three datasets is different. Secondly, we take a motion trajectory of a game bot in the battle dataset shown in Fig. 1c as an example to explain why layers of the best single-layer GCN are different in the datasets. The motion trajectory falls in seven fixed regions of a screen. The best-performing single-layer GNN is 7-th layer GCN. Combined with the propagation mechanism of GNN, we can speculate that the information of seven specific regions will be shared when layers of GCN are equal to 7, which helps the model to detect game bots better.

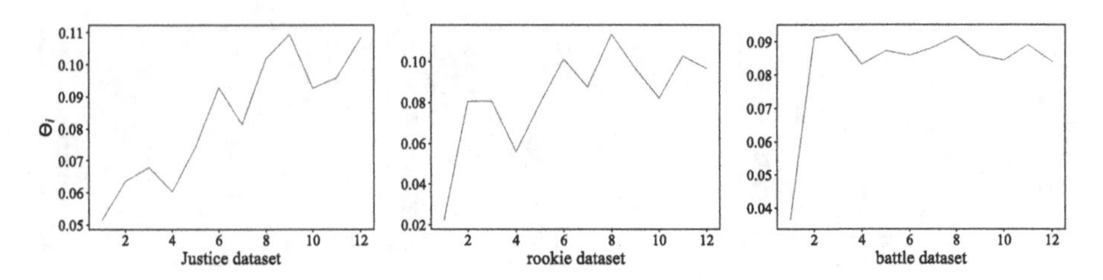

Fig. 5. The coefficients of each layer of GB-GNN.

To explore whether GB-GNN chooses the most effective layer, we obtain the coefficients θ_i in the GB-GNN model, as shown in Fig. 5. On the battle dataset, θ_i of 1-th layer is the smallest and θ_i of 7-th layer is the largest, which indicates that GB-GNN almost makes no use of information of 1-th layer and utilizes information of 7-th layer mostly. And game bot detection not only relies on the information obtained from a single layer of GNN but also the information of other layers.

6 Conclusion

In this paper, we propose GB-GNN to detect various motion patterns of game bots. We evaluate GB-GNN on three real-world datasets, our proposed method achieves state-of-the-art results over other baselines. We also analyze the structure of the model and find that our proposed model can effectively capture the motion patterns of the game-bot and extract the complex graph features for the game-bot detection. Our method can be utilized to detect game bots and can be also used to detect trajectory with local patterns, such as detecting human behavior based on data collected from mobile phone sensors.

References

1. Bahdanau, D., Cho, K., Bengio, Y.: Neural machine translation by jointly learning to align and translate. arXiv preprint arXiv:1409.0473 (2014)
2. Barr, M., Copeland-Stewart, A.: Playing video games during the COVID-19 pandemic and effects on players' well-being. Games Cult. **17**, 122–139 (2021). https://doi.org/10.1177/15554120211017036
3. Chen, K.T., Pao, H.K.K., Chang, H.C.: Game bot identification based on manifold learning. In: Proceedings of the 7th ACM SIGCOMM Workshop on Network and System Support for Games, pp. 21–26 (2008)
4. Chen, T., Guestrin, C.: XGBoost: a scalable tree boosting system. In: Proceedings of the 22nd ACM SIGKDD International Conference on Knowledge Discovery and Data Mining, pp. 785–794 (2016)
5. Chung, J., Gulcehre, C., Cho, K., Bengio, Y.: Empirical evaluation of gated recurrent neural networks on sequence modeling. arXiv preprint arXiv:1412.3555 (2014)

6. Ducheneaut, N., Moore, R.J.: The social side of gaming: a study of interaction patterns in a massively multiplayer online game. In: Proceedings of the 2004 ACM Conference on Computer Supported Cooperative Work, pp. 360–369 (2004)
7. Hilaire, S., Kim, H.c., Kim, C.k.: How to deal with bot scum in MMORPGs? In: 2010 IEEE International Workshop Technical Committee on Communications Quality and Reliability (CQR 2010), pp. 1–6. IEEE (2010)
8. Kang, A.R., Jeong, S.H., Mohaisen, A., Kim, H.K.: Multimodal game bot detection using user behavioral characteristics. SpringerPlus **5**(1), 1–19 (2016). https://doi.org/10.1186/s40064-016-2122-8
9. Karim, F., Majumdar, S., Darabi, H., Harford, S.: Multivariate LSTM-FCNs for time series classification. Neural Netw. **116**, 237–245 (2019)
10. Kwon, H., et al.: Crime scene reconstruction: online gold farming network analysis. IEEE Trans. Inf. Forensics Secur. **12**(3), 544–556 (2016)
11. Lee, E., Woo, J., Kim, H., Mohaisen, A., Kim, H.K.: You are a game bot!: uncovering game bots in MMORPGs via self-similarity in the wild. In: Ndss (2016)
12. Liu, S., et al.: A hybrid method of recurrent neural network and graph neural network for next-period prescription prediction. Int. J. Mach. Learn. Cybern. **11**(12), 2849–2856 (2020). https://doi.org/10.1007/s13042-020-01155-x
13. Lu, Z., Lv, W., Xie, Z., Du, B., Huang, R.: Leveraging graph neural network with lstm for traffic speed prediction. In: 2019 IEEE SmartWorld, Ubiquitous Intelligence & Computing, Advanced & Trusted Computing, Scalable Computing & Communications, Cloud & Big Data Computing, Internet of People and Smart City Innovation (SmartWorld/SCALCOM/UIC/ATC/CBDCom/IOP/SCI), pp. 74–81. IEEE (2019)
14. Oh, J., Borbora, Z.H., Sharma, D., Srivastava, J.: Bot detection based on social interactions in MMORPGs. In: 2013 International Conference on Social Computing, pp. 536–543. IEEE (2013)
15. Pao, H.K., Chen, K.T., Chang, H.C.: Game bot detection via avatar trajectory analysis. IEEE Trans. Comput. Intell. AI Games **2**(3), 162–175 (2010)
16. Scarselli, F., Gori, M., Tsoi, A.C., Hagenbuchner, M., Monfardini, G.: The graph neural network model. IEEE Trans. Neural Netw. **20**(1), 61–80 (2008)
17. Sesti, N., Garau-Luis, J.J., Crawley, E., Cameron, B.: Integrating LSTMs and GNNs for COVID-19 forecasting. arXiv preprint arXiv:2108.10052 (2021)
18. Tao, J., Xu, J., Gong, L., Li, Y., Fan, C., Zhao, Z.: NGUARD: a game bot detection framework for NetEase MMORPGs. In: Proceedings of the 24th ACM SIGKDD International Conference on Knowledge Discovery & Data Mining, pp. 811–820 (2018)
19. Thawonmas, R., Kashifuji, Y., Chen, K.T.: Detection of MMORPG bots based on behavior analysis. In: Proceedings of the 2008 International Conference on Advances in Computer Entertainment Technology, pp. 91–94 (2008)
20. Varvello, M., Voelker, G.M.: Second life: a social network of humans and bots. In: Proceedings of the 20th International Workshop on Network and Operating Systems Support for Digital Audio and Video, pp. 9–14 (2010)
21. Vaswani, A., et al.: Attention is all you need. In: Advances in Neural Information Processing Systems 30 (2017)
22. Xu, J., et al.: NGUARD+ an attention-based game bot detection framework via player behavior sequences. ACM Trans. Knowl. Discov. Data (TKDD) **14**(6), 1–24 (2020)

Online Learning with Regularized Knowledge Gradients

Donghun Lee[1(✉)] and Warren B. Powell[2]

[1] Korea University, Seoul 02841, South Korea
holy@korea.ac.kr
[2] Princeton University, Princeton, NJ 08544, USA
wbpowell328@gmail.com

Abstract. We introduce a simple and effective regularization of knowledge gradient (KG) and use it to present the first sublinear regret bound result for KG-based algorithms. We construct online learning with regularized knowledge gradients (ORKG) algorithm with independent Gaussian belief model, and prove that ORKG algorithm achieves sublinear regret upper bound with high probability facing bounded independent Gaussian multi-armed bandit (MAB) problems. The theoretical properties of regularized KG and ORKG algorithm are analyzed, and the empirical characteristics of ORKG algorithm are empirically validated with MAB benchmark simulations. ORKG algorithm shows top-tier performance comparable to select MAB algorithms with provable regret bounds.

Keywords: Knowledge gradient · Online learning · Regret analysis

1 Introduction

This paper considers the problem of making best possible decisions facing uncertainty, in which a decision-making agent repeatedly chooses from a set of decisions and then observes an outcome from which a bounded quantifiable reward can be derived. We assume that the agent knows the set of possible decisions, which is finite and remains the same over the time horizon in which the agent choose and learns. If such an agent is evaluated on how well it finds out which choice incurs the best reward, disregarding the rewards incurred by its choices while learning, the agent is facing a ranking and selection (R&S) problem.

Knowledge gradient (KG) is an algorithm proposed to solve R&S problems with independent Gaussian model assumption [4], and later with different assumptions such as correlated Gaussian model [5], Gaussian process model [15],

D. Lee—This work is supported by the National Research Foundation of Korea (NRF) grant funded by the Korea government (MSIT) (No. 2020R1G1A1102828).

Supplementary Information The online version contains supplementary material available at https://doi.org/10.1007/978-3-031-05936-0_26.

J. Gama et al. (Eds.): PAKDD 2022, LNAI 13281, pp. 328–339, 2022.
https://doi.org/10.1007/978-3-031-05936-0_26

binary cost function [22], locally nonlinear parametric models [9], and repeated noisy measurements [8]. Empirical effectiveness of KG-based algorithms has been demonstrated in diverse fields where R&S problems can be applied: for example, drug discovery [12], chemical engineering [3], fleet management [10,21], COVID responses [19], and clinical trials [20].

However, R&S problem disregards the reward incurred by the choices made by the agent while it is learning. As such, R&S problem is ill-suited to model online learning problems, in which every single reward incurred by the agent counts, and 2) the remaining number of choices the agent must make may be unknown. Little work has been done to utilize KG in online learning, where a most notable approach assuming the agent knows the remaining number of choices [13,14]. In this paper, we present a new approach to utilize KG to solve online learning problem *with unknown time horizon.*

Novel contribution of this manuscript is summarized as follow. We present Online learning with Regularized Knowledge Gradients (ORKG) algorithm with independent Gaussian belief, a novel online learning algorithm that uses knowledge gradient. We provide theoretical analysis of ORKG, including the proof of ORKG's regret upper bound of $O(\sqrt{KT\ln(KT)})$ in stochastic MAB problems with K bounded independent Gaussian arms, which is the first sublinear regret bound for knowledge gradient based algorithms. We also perform empirical validation of the theoretical properties of ORKG and empirical sensitivity analysis of the key hyperparameters of ORKG. Lastly, we verify empirical performance of ORKG in Gaussian stochastic MAB problems against other well-known MAB algorithms with provable regret bounds.

2 Problem Setting

We consider "online" sequential decision problem in which a decision-making agent faces partial information stochastic MAB problem, in particular with bounded Gaussian stochastic arms and unknown total number of decisions to make. For each time index $t \in \{0, 1, \cdots, T-1\}$ with unknown finite time horizon T, the agent must make a decision, denoted by x, among $K < \infty$ mutually independent arms that can be indexed by $i \in \{1, 2, \cdots, K\}$, and then observe a bounded random reward/contribution C_t from respective arm's distribution with mean μ^i and standard deviation σ^i that are unknown to the agent. We use x_t for decision made at time t, and $\mathcal{X}$ as the set containing all possible decisions. Hence, $\forall t : x_t \in \mathcal{X}$, and $|\mathcal{X}| = K$.

The goal of the agent is twofold: 1) to learn the best arm i^* whose reward distribution has the largest mean (i.e. $\mu^{i^*} = \max_i \{\mu^i\} =: \mu^*$ using the observations incurred by past decisions, and 2) to control the impact of inevitable suboptimality caused by choosing arms that are not the best arm without knowing the best arm *a priori*. Note that the term "online" is *not* the same as in online convex optimization, but instead is related to the second aspect of the goal of the learning agent – that the performance of the agent while it is learning (i.e. "online") is important, as opposed to batch learning such as R&S problems where only final performance matters.

The belief state B_t at time t, for KG-based algorithms with independent Gaussian belief model, is defined as the sufficient information to model Gaussian rewards incurred by each action $x \in \mathcal{X}$. Hence, we define $B_t := \{(\bar{\mu}_t^x, \bar{\sigma}_t^x) | \, x \in \mathcal{X}\}$, as the set of mean parameter estimates $\bar{\mu}_t^x$ and standard deviation parameter estimates $\bar{\sigma}_t^x$ for all $x \in \mathcal{X}$.

Under independent Gaussian belief model, KG of choosing x at time t can be efficiently computed [6] using the following closed form formula:

$$\nu_t^{KG,x} := \tilde{\sigma}_t^x \left(\xi_t^x \Phi\left(\xi_t^x\right) + \phi\left(\xi_t^x\right)\right), \tag{1}$$

where $\Phi(\cdot)$ and $\phi(\cdot)$ are the cumulative distribution function and the probability density function of standard Gaussian distribution, respectively. ξ_t^x is defined as:

$$\xi_t^x := -\frac{\left|\bar{\mu}_t^x - \max_{x' \neq x} \bar{\mu}_t^{x'}\right|}{\tilde{\sigma}_t^x}, \tag{2}$$

where $\tilde{\sigma}_t^x := \bar{\sigma}_t^x / \sqrt{1 + (\sigma^\epsilon / \bar{\sigma}_t^x)^2}$. σ^ϵ is the standard deviation of the zero-mean Gaussian measurement noise assumed to be found on all observed reward $C(x)$ for all $x \in \mathcal{X}$. Most KG-based algorithms have σ^ϵ as a hyperparameter.

Using KG as-is to solve online learning problems is expected to fail, because R&S problem disregards the rewards caused by a fixed, known number of choices which it considers as the learning process. From this perspective, KG algorithm for online learning problems (OKG) is proposed [13]. OKG algorithm chooses action x_t at time t as:

$$x_t = \begin{cases} \arg\max_{x \in \mathcal{X}} \left\{\bar{\mu}_t^x + (T - t)\, \nu_t^{KG,x}\right\} & (t < T) \\ \arg\max_{x \in \mathcal{X}} \left\{\bar{\mu}_t^x\right\} & (t \geq T) \end{cases}, \tag{3}$$

where T is the total number of choices to make in the online learning problem. Naturally, OKG algorithm requires knowing the true time horizon T, after which it exploits learned information and choose the action with best expected mean reward.

3 Online Learning with Regularized KG

We present Online learning with Regularized KG (ORKG) with independent Gaussian belief, a novel online learning algorithm with knowledge gradient, in Algorithm 1. Compared to OKG algorithm [13], ORKG introduces two key innovations: 1) standardizing and regularizing knowledge gradient; 2) adaptively learning exploration parameter ρ_t. ORKG contains two key hyperparameters $\kappa_R > 0$ and $0 < \delta < 1$, and we use $\kappa_R = 0.01$, $\delta = 0.01$ as their default values. These hyperparameters are explained in theoretical analysis of ORKG (Sect. 4) and their default values are justified in empirical sensitivity analysis of ORKG (Sect. 5.2).

Algorithm 1. ORKG with Independent Gaussian Belief

1: Initialize belief state: $\{\bar{\mu}_0^x, \bar{\sigma}_0^x\}_{x\in\mathcal{X}}$
2: **for** $t = 0, 1, 2, \cdots$ **do**
3: Compute standardized KG: $\kappa_t^x \leftarrow \frac{\nu_t^{KG,x}}{\bar{\sigma}_t^x}$ ▷ Compute $\nu_t^{KG,x}$ by (1)
4: Compute regularized KG: $\nu_t^{RKG,x} \leftarrow \bar{\sigma}_t^x \max(\kappa_R, \kappa_t^x)$
5: Compute coefficient $\rho_t \leftarrow \sqrt{2\ln\left(\frac{2|\mathcal{X}|}{\delta\pi_t}\right)\frac{1}{\max\{\kappa_R, \min_{x\in\mathcal{X}}\kappa_t^x\}}}$
6: Choose action: $x_t \leftarrow \arg\max_{x\in\mathcal{X}}\left\{\bar{\mu}_t^x + \rho_t\nu_t^{RKG,x}\right\}$
7: Observe $C_{t+1} \sim C(x_t)$
8: Update $\bar{\mu}_{t+1}^x, \bar{\sigma}_{t+1}^x$ for $x = x_t$ using observation C_{t+1} ▷ Use update rules in [6]

As in step 6 of Algorithm 1, ORKG algorithm chooses action at time t as:

$$x_t = \arg\max_{x\in\mathcal{X}}\left\{\bar{\mu}_t^x + \rho_t\nu_t^{RKG,x}\right\}, \tag{4}$$

where $\rho_t := \sqrt{2\ln\left(\frac{2|\mathcal{X}|}{\delta\pi_t}\right)\frac{1}{\max\{\kappa_R, \min_{x\in\mathcal{X}}\kappa_t^x\}}}$, in which $\delta \in (0,1)$ and π_t is a sequence satisfying $\sum_t^\infty \pi_t = 1$, for example, $\pi_t := \frac{1}{(t+1)^2}\frac{6}{\pi^2}$. With this ρ_t, ORKG balances the exploitation action to maximize $\bar{\mu}_t^x$, the current estimate of mean reward incurred by action x and the exploration action to maximize $\nu_t^{RKG,x}$, the regularized knowledge gradient of action x at time t. It is notable that ORKG does not need to know the time horizon T; whereas OKG algorithm explicitly requires knowing the true T as shown in its decision rule (3). This property allows ORKG to be easily applied to online learning problems where explicit end-of-horizon is unknown or changes over time.

With carefully constructed decision rule, ORKG controls the exploration-exploitation dilemma in online learning problem with unknown horizon, and achieves sublinear regret upper bound as shown in Theorem 1.

Theorem 1. *In stochastic MAB problems with bounded independent Gaussian arms, ORKG algorithm with independent Gaussian belief has regret upper bound:*

$$R_T \leq_p \sqrt{8|\mathcal{X}|T\ln\left(\frac{2|\mathcal{X}|T}{\delta\pi_{T-1}}\right)}L^{RKG}\sigma^\epsilon,$$

with probability $1-\delta$*, where* $0<\delta<1$*, and* $L^{RKG} < \infty$ *is a constant uniformly bounding smoothness of regularized KG surface.*

Our proof strategy, which is inspired from GP-UCB algorithm [17], is as follow: first, the deviations of Gaussian rewards are taken with union bounds to bound squared one-step regret with high probability, given δ and κ_R, and then we sum up one-step regrets and bound the regret $R(T)$ and derive ρ_t shown in Algorithm 1. The smoothness constant L^{RKG} is analyzed in greater detail in Sect. 4.2, and complete proof of Theorem 1 is given in appendix A.7.

Therefore, ORKG algorithm with independent Gaussian belief has a sublinear regret upper bound of $O\left(\sqrt{|\mathcal{X}|T\ln|\mathcal{X}|T}\right)$ with probability $1-\delta$, when its modeling assumption matches the problem specification.

4 Theoretical Analysis

4.1 Regularization of Knowledge Gradient in ORKG

In this section, we define the regularization of KG used in ORKG algorithm, and analyze the theoretical property of the regularized KG on which the sublinear regret bound of ORKG depends.

Conceptual summary of the regularization of KG in ORKG algorithm is as follows: 1) "standardize" KG into a unitless value, 2) force it to have a fixed uniform bound from below, 3) then give back its unit to match KG. Step 1 is achieved by computing standardized KG, and steps 2 and 3 are done in computing regularized KG from standardized KG.

Definition 1. *κ_t^x, standardized knowledge gradient of an action $x \in \mathcal{X}$ at time t is defined for all $x \in \mathcal{X}$ as:*

$$\kappa_t^x := \frac{\nu_t^{KG,x}}{\bar{\sigma}_t^x}, \tag{5}$$

where knowledge gradient $\nu_t^{KG,x}$ is computed from belief state B_t.

κ_t^x is "standardized" KG, in a sense that it has the same unit as ξ_t^x:

$$\kappa_t^x = \underbrace{\frac{\bar{\sigma}_t^x}{\sqrt{\left(\bar{\sigma}_t^x\right)^2 + \left(\sigma^\epsilon\right)^2}}}_{\text{unitless}} \underbrace{\left(\xi_t^x \Phi\left(\xi_t^x\right) + \phi\left(\xi_t^x\right)\right)}_{\text{same unit as } \xi_t^x}, \tag{6}$$

where ξ_t^x is as defined in (2), Φ is the cumulative distribution function, and ϕ is the probability density function of standard normal distribution.

We introduce the following regularization method, designed to achieve a needed property for a sublinear upper bound of the regret of ORKG, and at the same time easy to interpret.

Definition 2. *$\nu_t^{RKG,x}$, the regularized KG for making a decision x at time t given belief state B_t, is defined as*

$$\nu_t^{RKG,x} := \bar{\sigma}_t^x \max\left\{\kappa_R, \kappa_t^x\right\}, \tag{7}$$

where $\kappa_R > 0$ is the regularizing parameter, which is a small arbitrary constant uniform lower bound on κ_t^x for all x, t, and κ_t^x is standardized KG computed at time t given belief state B_t according to Definition 1.

Note that from this regularization originates κ_R, one of the two hyperparameters of ORKG algorithm. κ_R stands for the uniform lower bound on how small κ_t^x can get for all x, t.

4.2 Smoothness of Regularized KG Surface

In ORKG algorithm facing stochastic MAB with finite number of bounded Gaussian independent arms, $\nu_t^{KG,x}$ can be efficiently computed for all $x \in \mathcal{X}$ given B_t. To represent the "surface" of KG with respect to x at t, we consider $\nu_t^{KG} = \left[\nu_t^{KG,1}, \nu_t^{KG,2}, \cdots, \nu_t^{KG,K}\right]$ as a piecewise linear function measured at $x = 1, 2, \cdots, K$. We define a smoothness constant for the surface of KG as:

Definition 3. $L_t^{KG,x}$*, the smoothness constant of KG for action* x *at time* t*, is defined as:*

$$L_t^{KG,x} := \frac{\nu_t^{KG,x}}{\min_{x' \in \mathcal{X}} \nu_t^{KG,x'}} . \tag{8}$$

$L_t^{KG,x}$ represents the worst case relative difference between KG of x at t and smallest KG across all x at t, up to permutation of $\mathcal{X}$, in the unit of the value of smallest KG at t. It has trivial lower bound of 1, and upper bound of ∞ at $t \to \infty$, suggesting that the KG "surface" may have a very sharp point.

On the other hand, the surface of regularized KG, whose smoothness constant is shown in Definition 4, has a smoothness bound as shown in Lemma 1.

Definition 4. $L_t^{RKG,x}$*, the smoothness constant of regularized KG for action* x *at time* t*, is defined, analogous to that of KG (Definition 3), as:*

$$L_t^{RKG,x} := \frac{\max\left\{\kappa_R, \kappa_t^x\right\}}{\max\left\{\kappa_R, \min_{x' \in \mathcal{X}} \kappa_t^{x'}\right\}} . \tag{9}$$

Lemma 1. *There exists a finite constant* L^{RKG} *such that*

$$L_t^{RKG,x} \leq L^{RKG} < \infty \quad \forall x \in \mathcal{X}, \forall t \in \{0, 1, \cdots\} . \tag{10}$$

Existence of a constant L^{RKG} is needed to establish the sublinear regret upper bound of ORKG, as the constant appears in the regret bound in Theorem 1. We provide the proof of Lemma 1 in appendix A.3.

5 Empirical Verification

In this section, we present multifaceted empirical verification of the performance of ORKG algorithm in online learning. We use Python package `smpybandit` [1] to implement all stochastic multi-armed bandit (MAB) benchmarks, on an AMD Ryzen 3900x CPU with 64GB of RAM. Benchmarks are randomized and repeated 100 times, and the sample mean and standard deviation from all repeats are reported. For each benchmark scenario, the best result in sample mean and all runner-up results within 1 standard deviation of the best result are boldfaced.

5.1 ORKG Compared to Other KG Algorithms

First, we demonstrate how the theoretical improvements of ORKG is realized, by comparing empirical performance of KG based algorithms in bounded Gaussian stochastic MAB problems. We compare ORKG algorithm against KG with independent Gaussian belief algorithm (KG) [6] and KG for general class of online learning problems algorithm (OKG) [13]. We also test ϵ-greedy algorithm with constant $\epsilon(t) = 0.01$ as a widely known benchmark algorithm frequently seen in applications. The key differences of the algorithms are outlined in Table 1. We use $\sigma_\epsilon = 0.1$ as the value of the common hyperparameter among the KG algorithms for fair comparison.

Table 1. Comparison of algorithms used in Sect. 5.1

	Decision rule	Belief state	Hyperparameters	Regret bound
ϵ-greedy	$\bar{\mu}_t^x$ w.p. $1-\epsilon$	$\bar{\mu}_t^x$	$\epsilon(t)$	N/A
KG	$\nu_t^{KG,x}$	$\bar{\mu}_t^x, \bar{\sigma}_t^x$	σ_ϵ	N/A
OKG	$\bar{\mu}_t^x + (T-t)\nu_t^{KG,x}$	$\bar{\mu}_t^x, \bar{\sigma}_t^x$	σ_ϵ, T	N/A
ORKG	$\bar{\mu}_t^x + \rho_t\nu_t^{RKG,x}$	$\bar{\mu}_t^x, \bar{\sigma}_t^x$	$\sigma_\epsilon, \delta, \kappa_R$	$O\left(\sqrt{\lvert\mathcal{X}\rvert T \ln \lvert\mathcal{X}\rvert T}\right)$

We test the algorithms on the stochastic MAB benchmark problems with 5, 10, and 20 arms generating Gaussian rewards, whose mean parameter μ_x sampled equally distanced in $[-5, 5]$, with low variance scenario of $\sigma_x^2 = 0.1$ and high variance scenario $\sigma_x^2 = 1$ for all actions x. For each algorithm, we sum up observed regrets from $t = 1, \cdots, 10000$, and report their mean and standard deviations from 100 independent repeats in Table 2. ORKG shows expected behavior of controlling the cumulative regret throughout all tested settings, whereas other KG algorithms without sublinear regret bounds mostly show large regret. OKG, even when provided with additional information on the true time horizon $T = 10000$, achieves results comparable to ORKG only in 5 arms setting, not in the harder settings with 10 and 20 arms. KG, intended to solve R&S problem, shows worst performance in terms or regrets as expected. Note that ϵ-greedy,

Table 2. Cumulative Regrets in Gaussian Stochastic MAB. Lower is Better.

MAB setting		Algorithms			
Arms	Variance	ORKG	OKG	KG	ϵ-greedy
5	High	$\mathbf{215 \pm 102}$	$\mathbf{204 \pm 96}$	33100 ± 256	8830 ± 8200
5	Low	$\mathbf{17 \pm 9}$	$\mathbf{12 \pm 12}$	33200 ± 235	6570 ± 8110
10	High	$\mathbf{1060 \pm 85}$	2580 ± 3210	39600 ± 355	14700 ± 11100
10	Low	$\mathbf{40 \pm 9}$	1020 ± 2840	40600 ± 241	17400 ± 11900
20	High	$\mathbf{2210 \pm 105}$	5950 ± 3900	39900 ± 774	19600 ± 10500
20	Low	$\mathbf{96 \pm 10}$	6210 ± 4690	44400 ± 264	21100 ± 14500

a widely used algorithm in practice, shows extremely large standard deviation, suggesting hit-or-miss performance in online learning.

5.2 Sensitivity Analysis of ORKG

ORKG introduces new hyperparameters δ and κ_R compared to other KG algorithms as shown in Table 1. Since those hyperparameters play critical role in the sublinear regret bound of ORKG, we analyze empirical sensitivity of ORKG to δ and κ_R, one by one, tested on Gaussian MAB benchmarks. In the main paper, we present results with 10 arms and high variance only, and full results are found in appendix (Figs. B.4 and B.5).

First, we vary $\kappa_R \in \{0.0001, 0.001, 0.01, 0.1, 1\}$ while fixing $\delta = 0.01$, and report the time evolution of cumulative regret against t, averaged over 100 repeats, as trajectories shown in Fig. 1.

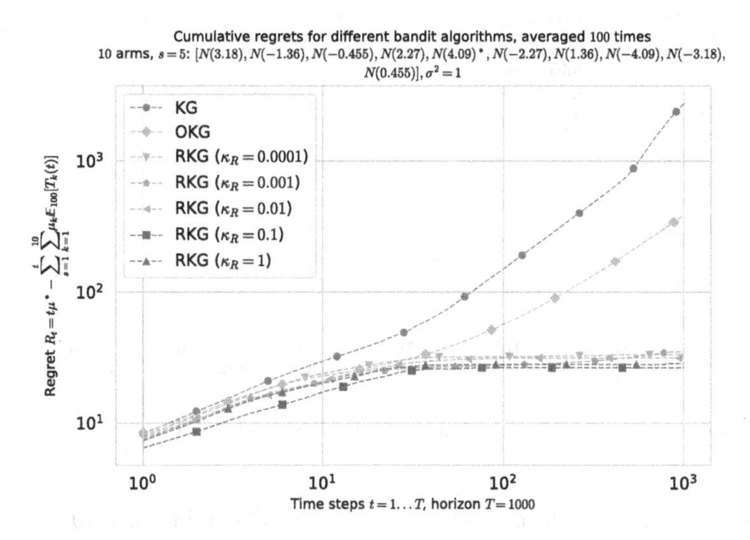

Fig. 1. Sensitivity of ORKG to κ_R in Gaussian MAB ($K = 10, \sigma^2 = 1, \delta = 0.01$).

It is evident that ORKG shows robust regret controls regardless of wide range of κ_R, and retains robust advantage over KG and OKG. Considering the intuitive role of κ_R in ORKG to enforce the lower bound of KG and regularizes the smoothness of the KG surface, the subtle sensitivity to κ_R is theoretically expected, and can be interpreted as follows: changing κ_R can change the values of the ORKG decision rule (4) transiently when exploration happens, visualized as minor difference in early-stage trajectories ($T < 10^3$) of ORKGs with different κ_R values in Fig. 1.

We recommend the default value of $\kappa_R = 0.01$, based on theoretical understanding of the value should be small enough to become the lower bound of KG, as the intuitive meaning of KG is the expected improvement from a single reward. Also, κ_R can be tuned with *a priori* information or at problem formulation stage: if the gap between the largest mean and the smallest mean of the rewards are known or can be enforced by clipping rewards, then κ_R can be set to be at least sufficiently smaller than the gap.

Next, we vary $\delta \in \{0.0001, 0.001, 0.01, 0.1, 0.9\}$ while fixing $\kappa_R = 0.01$, and report the time evolution of cumulative regret against t, averaged over 100 repeats, as the trajectories shown in Fig. 2.

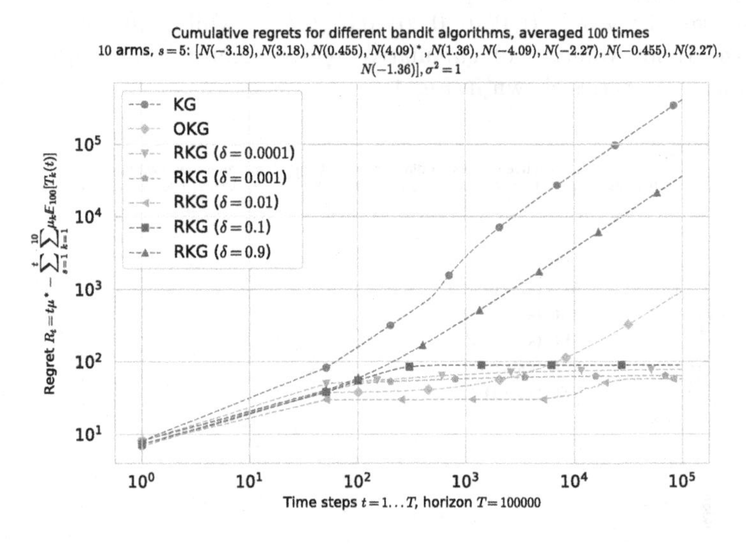

Fig. 2. Sensitivity of ORKG to δ in Gaussian MAB ($K = 10, \sigma^2 = 1$, $\kappa_R = 0.01$).

It is notable that δ affects the behavior of ORKG in mid-range $10^2 < T < 10^4$ to vary, and in most cases, the impact appears to be transient as the regret is controlled for $\delta \leq 0.1$ cases. Drastically different behavior of ORKG is observed for $\delta = 0.9$ case, and this is expected according to the role of δ in ORKG: the probability δ of encountering a reward deviates more than the estimated mean plus exploration bonus term scaled by ρ_t (as given in (4)). Intuitively, larger δ makes ORKG more cautious before greedily exploiting, since δ is the probability of a rare event of facing unexpected rewards after choosing the action according to ORKG decision rule (4), and this is empirically shown by ORKG with $\delta = 0.9$ case in Fig. 2. Therefore, it is reasonable to set δ in ORKG as a relatively small value even if $\delta \in (0, 1)$ is theoretically allowed, as δ adjusts how much ORKG should expect the rare events would happen. We recommend the default value of $\delta = 0.01$, as 1% appears to be a good reference point for encountering "rare" events; if more frequent surprises are expected, larger δ is recommended.

5.3 ORKG Performance Validation Against Other MAB Algorithms

We validate empirical performance of ORKG against other MAB algorithms with provable regret bounds, on stochastic Gaussian MAB benchmark problems set up the same way as described in Sect. 5.1. Both classic algorithms and cutting-edge algorithms for MAB are compared against ORKG in this validation, with abbreviated names as follow: UCB [11], kl-UCB [7], EXP3++ [16], TS [18], and BG [2]. Detailed rationale of choosing these algorithms are given in appendix Sect. B.1. For each algorithm, we sum up observed regrets from $t = 1, \cdots, 10000$, and report their mean and standard deviations from 100 independent repeats in Table 3.

Table 3. Cumulative regrets in Gaussian stochastic MAB. Lower is better.

MAB Setting		Algorithms					
Arms	Variance	ORKG	UCB	kl-UCB	TS (G)	EXP3++	BG
5	High	**215 ± 102**	**247 ± 90**	573 ± 2320	**246 ± 94**	1090 ± 113	**235 ± 105**
5	Low	**17 ± 9**	**30 ± 10**	**15 ± 10**	**41 ± 37**	919 ± 67	**36 ± 12**
10	High	**1060 ± 85**	**1060 ± 88**	1920 ± 2590	1420 ± 698	2920 ± 198	**1070 ± 99**
10	Low	**40 ± 9**	75 ± 11	**41 ± 10**	644 ± 1240	1920 ± 138	85 ± 12
20	High	**2210 ± 105**	**2260 ± 68**	3240 ± 1930	4590 ± 2210	5480 ± 212	**2240 ± 72**
20	Low	**96 ± 10**	182 ± 9	**91 ± 10**	3010 ± 2490	3470 ± 226	181 ± 12

As shown by boldfaced results across all scenarios, ORKG reliably performs well in all tested Gaussian MAB benchmark scenarios, with the cumulative regret of ORKG is on par with the top-performing algorithm within each scenario; whereas other algorithms show some scenario preferences in which they perform well. Both UCB, a classic algorithm, and Boltzmann-Gumbel (BG), a cutting edge algorithm are the runner-ups, closely followed by kl-UCB, an improved UCB with tighter bound that shows scenario preference different from UCB. We conjecture that the unexpectedly poor performance of EXP3++ may be an unwanted artifact of general-purposing EXP3 algorithm that is originally designed for adversarial MAB problems to have sublinear regrets for stochastic MAB problems as well. Thompson sampling (TS) also show unexpectedly poor performance in many-arms scenario, and we think that 10000 samples, although they are sufficiently many for 5 arms case, are not sufficient enough for 10 and 20 arms case, as there are more Bayesian estimates for TS to learn as the number of arms grow. All algorithms tested have regret bounds for Gaussian MAB problems tighter than the bound of ORKG we present in Theorem 1, and this empirical validation suggests existence of tighter regret bounds for ORKG.

6 Discussion

The simple regularization method for KG used in ORKG algorithm allows the first KG-based algorithm with sublinear regret bounds, yet this approach may be too simple to tighten regret bounds of ORKG on par with other stochastic MAB algorithms. Despite the theoretical gap in regret bounds, we witness impressive empirical performance of ORKG in MAB benchmarks with correct model specification. Notably, the empirical validations is performed with relatively few samples from MAB perspective, which suggests ORKG can perform well in real world applications where the number of samples are limited. Also, ORKG gives new insight to a long-standing question in KG literature on how to trade off exploration-exploitation correctly in online learning, and at the same time, ORKG allows interdisciplinary discussion between KG and MAB literature by providing the first regret bound result of KG-based algorithm in MAB problems.

7 Conclusion

We present a simple and effective method to regularize knowledge gradient (KG) that allows novel asymptotic regret analysis of KG-based algorithms with independent Gaussian belief model. Using regularized knowledge gradients, we construct ORKG, a KG-based online learning algorithm, and present its sublinear regret bound in partial information Gaussian MAB problem. We provide empirical validation of ORKG, and verify that ORKG algorithm performs comparable to select MAB algorithms with tighter regret bounds in Gaussian MAB benchmarks. Our result opens up an interesting stage for further research in KG from the perspective of MAB literature.

References

1. Besson, L.: SMPyBandits: an open-source research framework for single and multi-players multi-arms bandits (MAB) algorithms in python. GitHub.com/SMPyBandits/SMPyBandits (2018)
2. Cesa-Bianchi, N., Gentile, C., Lugosi, G., Neu, G.: Boltzmann exploration done right. Adv. Neural Inf. Process. Syst. **30** (2017)
3. Chen, S., Reyes, K.R.G., Gupta, M.K., McAlpine, M.C., Powell, W.B.: Optimal learning in experimental design using the knowledge gradient policy with application to characterizing nanoemulsion stability. SIAM/ASA J. Uncertain. Quant. **3**(1), 320–345 (2015)
4. Frazier, P., Powell, W.: The Knowledge Gradient Policy for Offline Learning with Independent Normal Rewards (2007)
5. Frazier, P., Powell, W., Dayanik, S.: The knowledge-gradient policy for correlated normal beliefs. INFORMS J. Comput. **21**(4), 599–613 (2009)
6. Frazier, P.I., Powell, W.B., Dayanik, S.: A knowledge-gradient policy for sequential information collection. SIAM J. Control Opt. **47**(5), 2410–2439 (2008)

7. Garivier, A., Cappé, O.: The KL-UCB algorithm for bounded stochastic bandits and beyond. In: Kakade, S.M., von Luxburg, U. (eds.) Proceedings of the 24th Annual Conference on Learning Theory. Proceedings of Machine Learning Research, vol. 19, pp. 359–376. PMLR, Budapest (2011)
8. Han, W., Powell, W.B.: Optimal online learning for nonlinear belief models using discrete priors. Oper. Res. **68**(5), 1538–1556 (2020)
9. He, X., Reyes, K.G., Powell, W.B.: Optimal learning with local nonlinear parametric models over continuous designs. SIAM J. Sci. Comput. **42**(4), A2134–A2157 (2020)
10. Huang, Y., Zhao, L., Powell, W.B., Tong, Y., Ryzhov, I.O.: Optimal learning for urban delivery fleet allocation. Transp. Sci. **53**(3), 623–641 (2019)
11. Lai, T.L., Robbins, H.: Asymptotically efficient adaptive allocation rules. Adv. Appl. Math. **6**(1), 4–22 (1985)
12. Negoescu, D.M., Frazier, P.I., Powell, W.B.: The Knowledge-Gradient Algorithm for Sequencing Experiments in Drug Discovery (2011)
13. Ryzhov, I.O., Powell, W.: The knowledge gradient algorithm for online subset selection. In: 2009 IEEE Symposium on Adaptive Dynamic Programming and Reinforcement Learning, March 2009, pp. 137–144 (2009)
14. Ryzhov, I.O., Powell, W.B., Frazier, P.I.: The knowledge gradient algorithm for a general class of online learning problems. Oper. Res. **60**(1), 180–195 (2012)
15. Scott, W., Frazier, P., Powell, W.: The correlated knowledge gradient for simulation optimization of continuous parameters using Gaussian process regression. SIAM J. Opt. Publ. Soc. Indust. Appl. Math. **21**(3), 996–1026 (2011)
16. Seldin, Y., Slivkins, A.: One practical algorithm for both stochastic and adversarial bandits. In: Xing, E.P., Jebara, T. (eds.) Proceedings of the 31st International Conference on Machine Learning. Proceedings of Machine Learning Research, vol. 32, pp. 1287–1295. PMLR, Bejing (2014)
17. Srinivas, N., Krause, A., Kakade, S.M., Seeger, M.: Gaussian Process Optimization in the Bandit Setting: No Regret and Experimental Design (2009)
18. Thompson, W.R.: On the likelihood that one unknown probability exceeds another in view of the evidence of two samples. Biometrika **25**(3/4), 285–294 (1933)
19. Thul, L., Powell, W.: Stochastic Optimization for Vaccine and Testing Kit Allocation for the COVID-19 Pandemic (2021)
20. Tian, Z., Han, W., Powell, W.B.: Adaptive learning of drug quality and optimization of patient recruitment for clinical trials with dropouts. Manuf. Serv. Oper. Manag. (2021)
21. Wang, Y., Do Nascimento, J.M., Powell, W.: Reinforcement Learning for Dynamic Bidding in Truckload Markets: An Application to Large-Scale Fleet Management with Advance Commitments (2018)
22. Wang, Y., Wang, C., Powell, W.: The knowledge gradient for sequential decision making with stochastic binary feedbacks. In: Balcan, M.F., Weinberger, K.Q. (eds.) Proceedings of The 33rd ICML. Proceedings of Machine Learning Research, vol. 48, pp. 1138–1147. PMLR, New York (2016)

Fact Aware Multi-task Learning for Text Coherence Modeling

Tushar Abhishek(✉), Daksh Rawat, Manish Gupta, and Vasudeva Varma

Information Retrieval and Extraction Lab, IIIT, Hyderabad, India
tushar.abhishek@research.iiit.ac.in,
daksh.rawat@students.iiit.ac.in, {manish.gupta,vv}@iiit.ac.in

Abstract. Coherence is an important aspect of text quality and is crucial for ensuring its readability. It is essential for outputs from text generation systems like summarization, question answering, machine translation, question generation, table-to-text, etc. An automated coherence scoring model is also helpful in essay scoring or providing writing feedback. A large body of previous work has leveraged entity-based methods, syntactic patterns, discourse relations, and traditional deep learning architectures for text coherence assessment. However, these approaches do not consider factual information present in the documents. The transitions of facts associated with entities across sentences could help capture the essence of textual coherence better. We hypothesize that coherence assessment is a cognitively complex task that requires deeper fact-aware models and can benefit from other related tasks. In this work, we propose a novel deep learning model that fuses document-level information with factual information to improve coherence modeling. We further enhance the model efficacy by training it simultaneously with Natural Language Inference task in multi-task learning setting, taking advantage of inductive transfer between the two tasks. Our experiments with popular benchmark datasets across multiple domains demonstrate that the proposed model achieves state-of-the-art results on a synthetic coherence evaluation task and two real-world tasks involving prediction of varying degrees of coherence.

1 Introduction

Coherence is a crucial metric for text quality analysis. It assimilates how well the sentences are connected and how well the document is organized. Coherent documents have clear topic transitions that are discussed throughout the text with a smooth flow of concepts, typically in an increasing order of complexity. Ideas are first introduced in preceding sentences and are referred to later in document. Connectives are often used to assist the structure and for smooth transitions within the document. Overall, coherence leads to better clarity.

M. Gupta—The author is also a Principal Applied Scientist at Microsoft.

J. Gama et al. (Eds.): PAKDD 2022, LNAI 13281, pp. 340–353, 2022.
https://doi.org/10.1007/978-3-031-05936-0_27

Coherence is vital for multiple Natural Language Processing (NLP) applications like summarization [3,44], question answering [51], machine translation [38,55], question generation [10], language assessment for essay scoring [8,16,46], story generation [34], readability assessment [41,45] and other text generation [22,26,43].

Many formal theories of coherence [2,19,33] have been proposed leading to further development of various coherence models. Based on such theories, multiple text coherence models like entity-grid [4] and its extensions have been proposed. Other linguistic approaches for text coherence include coreference resolution, discourse relations, lexical cohesion, and syntactic features. However, feature engineering is decoupled from the prediction task thus limiting model performance. Recently, various models have been proposed which leverage deep learning architectures like convolutional neural networks (CNNs), recurrent neural networks (RNNs), long short-term memory networks (LSTMs). Transformer [50] based approaches [23–25] have also been proposed that achieve better results on coherence modeling and its downstream tasks.

However, these approaches do not consider the factual information present in the document. Recent work has demonstrated usefulness of fact triples ⟨subject, verb, object⟩ for improving result on various NLP tasks, such as summarization [20], Question Answering (QA) [47], Natural Language Inference (NLI) [1] and language modeling [53]. In this work, we propose a novel architecture that fuses document-level information with factual information to improve coherence modeling. Further, we enhance the accuracy of coherence prediction by jointly modeling coherence and Natural Language Inference (NLI) in a multi-task learning (MTL) setting.

Overall, in this paper, we make the following main contributions. (1) We investigate the effectiveness of novel fact-aware MTL architecture. (2) We assess the extent to which the information encoded in the network generalizes to multiple domains and demonstrate the effectiveness of our approach not only on popular sentence order discrimination task but also on more realistic task like predicting coherence of varying degrees in people's everyday writings. (3) Experiments on popular benchmark datasets (GCDC and WSJ) indicate that our proposed methods establish SOTA across multiple (task, dataset) combinations. (4) On an automated essay scoring (AES) task, we demonstrate that addition of coherence signal from our model significantly improves AES accuracy.

2 Related Work

Entity-Grid Based Methods: Discourse coherence has been studied widely using both deep learning as well as non-deep learning models. Barzilay et al. [4] proposed the entity grid model, which is based on Centering Theory [19]. It captures the distribution of discourse entities and transition of grammatical roles (subject, object, neither) across the sentences. Several extensions were proposed by utilising entity specific features [13], modifying ranking scheme [17] or transforming problem into bipartite graph [35]. The entity grid method as well as extensions suffer from two main drawbacks: (1) they use discrete representation for grammatical roles and features, which prevents the model from considering sufficiently long transitions due to the curse of dimensionality problem. (2) Feature engineering is decoupled from the prediction task, which limits the model's capacity to learn task-specific features.

Other Feature Engineering Methods: Besides entity grid, other linguistic approaches for text coherence include coreference resolution, discourse relations, lexical cohesion, and syntactic features. Elsner et al. [13] proposed a maximum-entropy based discourse-new classifier that classifies mentions of all referring expression as first mention (discourse-new) or subsequent (discourse-old) mentions. Louis et al. [32] proposed a coherence model based on syntactic patterns by assuming that sentences in a coherent discourse should share the same structural syntactic patterns. Other approaches have used syntactic patterns [32], lexical cohesion [40,46] or capture topic shifts via HMMs [5].

Deep Learning Methods: Recently, multiple deep learning approaches have been proposed. Li et al. [29] propose a neural framework to compute the coherence score of a document by estimating a coherence probability for each clique of L sentences. Li et al. [30] propose generative methods to capture global topic information. Nguyen et al. [42] and Mohiuddin et al. [37] transform entity-grid based methods into deep learning versions that obtain better results than traditional counterparts. Farag et al. [15] propose a hierarchical attention model with multi-task learning objective. Xu et al. [56] and Moon et al. [39] show that modeling local coherence with discriminative models could capture both the local and the global contexts of coherence. Guz et al. [21] propose an RST-Recursive model, which takes advantage of the text's RST features. Farag et al. [14] extend some of the previous discriminative models using BERT (Bidirectional Encoder Representations from Transformers) [11] embeddings. Recently, Transformer [50] based approaches [23–25] have been proposed that achieve better results.

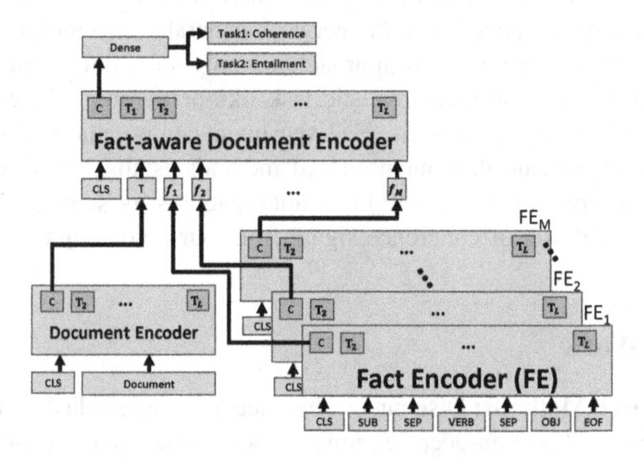

Fig. 1. An overview of our proposed fact-aware multi-task learning architecture. M distinct facts extracted from the document are fed to Fact Encoder individually to get permutation invariant representation. Fact-aware document encoder combines the document representation with M factual representation to obtain the fact-aware document representation.

3 Proposed Model

Given a document D, our goal is to assess its coherence according to the downstream task (binary classification, multi-class classification or regression task). Figure 1 provides an overview of our novel fact-aware multi-task learning model. It consist of three components: (i) Fact extractor to extract facts from textual content, (ii) Fact-aware document encoder that fuses the textual information with factual information, and (iii) Multi-task learning (MTL) framework that add auxiliary objective of textual entailment prediction to coherence objective. We discuss these components in detail in the following.

3.1 Fact Extractor

We leverage MinIE, an Open Information Extraction (IE) system [18] to generate a set of facts for each sentence. Open IE systems aim to exploit linguistic information including dependency relations in sentences to extract facts in a knowledge-agnostic manner. A fact is essentially an ordered 3-tuple ⟨subject, verb, object⟩ extracted from a particular sentence. A single sentence can produce multiple facts. Consider the sentence "They are trying to determine whether it was used to attack Steenkamp, if she used the bat in self-defense." Two facts that can be extracted from this sentence are ("it", "was used to attack", "Steenkamp") and ("she", "used bat in", "self-defense"). Each of the three components of a fact triple can contain multiple words.

For a given document D we pass the textual content through fact extractor (MinIE) to extract in-domain facts. Let M be the number of distinct facts obtained from the document D using MinIE.

3.2 Fact-Aware Document Encoder

This module follows a hierarchical structure with the following two encoders at the bottom level: (i) document encoder, and (ii) fact encoder. Each encoder uses a transformer model. Document encoder and fact encoder share weights. For i^{th} fact triple obtained from fact extractor for given document D, we create linear fact string by concatenating the subject, predicate and object delimited by separator token (SEP). The linear fact string is then fed to fact encoder FE_i individually to produce permutation invariant fact representation f_i. The document encoder encodes the document expressed using standard sub-word tokens to obtain document-level representation T. These fact and the document representations, T and f_i respectively, form the input for the fact-aware document encoder. Finally, we obtain fact-aware document representation as the CLS token vector from the last layer of the fact-aware document encoder. This is then fed to a fully-connected layer with ReLU, and then to a task specific output layer.

3.3 MTL Framework

When multiple related prediction tasks need to be performed, multi-task learning (MTL) has been found to be very effective. We experimented with various Natural

Language Understanding (NLU) tasks as auxiliary task and empirically found MTL combination of textual entailment and text coherence task provides better generalization and robustness. For a given a pair of sentences, the textual entailment task aims to predict whether the second sentence (hypothesis) is an entailment with respect to the first one (premise) or not. We share the fact-aware document encoder weights across the two tasks. Task specific layers for each task are conditioned on the shared fact-aware document encoder. For the sentence entailment task, we form input by concatenating the hypothesis and premise with sentence separator token SEP placed between them. For both the tasks (coherence and entailment), we use a fully-connected layer with ReLU, and then a softmax output layer. The final loss is computed as a sum of the individual losses for the two tasks. In the multi-task learning, we use mini-batch based stochastic gradient descent (SGD) to learn the parameters of our model (i.e., the parameters of all shared layers and task-specific layers) as shown in Algorithm 1.

Algorithm 1: Training a fact-aware MTL model

```
Model trainable parameters : θ (initialized to pretrained weights)
Set the max number of epochs: epoch_max.
for epoch in 1, 2, ..., epoch_max do
    Merge coherence and entailment dataset: D_global = D_coh ∪ D_entail
    Shuffle D_global
    for batch in D_global do
        Initialize losses: L_coh = 0, L_entail = 0
        if batch_coh ∈ batch then
            L_coh = Compute text coherence loss on batch_coh
        if batch_entail ∈ batch then
            L_entail = Compute text entailment loss on batch_entail
        Combine loss: L_total = L_coh + L_entail.
        Update the gradients and θ
```

4 Evaluation Tasks and Datasets

We experiment with two popular benchmark datasets: Wall Street Journal (WSJ) and Grammarly Corpus of Discourse Coherence (GCDC). GCDC is a real dataset while WSJ is a synthetic dataset. We use the Recognizing Textual Entailment (RTE) dataset [52] for training the auxiliary task head for our MTL model (2490 train and 277 validation instances) for experiments on GCDC. For WSJ, we found MTL to perform better when we use the Multi-Genre Natural Language Inference (MNLI) dataset [54] (21560 train and 6692 validation instances) for training the auxiliary task. We also evaluated the efficiency of proposed architecture on one downstream task: Automated Essay Scoring (AES). For AES task we use Automated Student Assessment Prize (ASAP) dataset. We make the code and dataset publicly available[1].

WSJ Sentence Order Discrimination Task. The WSJ portion of the Penn Treebank [13,42] is one of the most popular datasets for the sentence order discrimination task. It contains long articles without any constraint on style. Following previous

[1] https://www.dropbox.com/s/wolrmesgr4k1lf8/fact-aware-mtl-text-coh.zip.

work [4,42], we also use the sections 00–13 for training and 14–24 for testing (documents consisting of only one sentence are removed). We create 20 permutations per document, making sure to exclude duplicates or versions that happen to have the same ordering of sentences as the original article. We labeled these permuted documents as negative samples. The dataset is created by pairing the original document and the permuted document. The task is to rank the original document higher than the permuted one in terms of coherence. We present the basic statistics of the dataset in Table 1.

We evaluate model performance on this dataset using pairwise ranking accuracy (PRA) between original text and its 20 permuted counterparts, similar to previous work. PRA calculates the fraction of correct pairwise rankings in the test data (i.e., the original coherent text should be ranked higher than its permuted non-coherent counterpart).

For this task, the coherent and incoherent document representations are obtained by using proposed fact-aware document encoder using the architecture shown in Fig. 1. Further, on top of these representations, we apply Siamese network [7] as illustrated in Fig. 2. The document encoder for the coherent as well as the incoherent document, share weights. Both the document representations are separately connected to a dense layer with shared weights. Outputs of the dense layers are used to calculate margin ranking loss.

Table 1. Basic statistics of the WSJ dataset. #Docs represents the number of original articles and #Synthetic Docs represents the number of original articles and their permuted versions.

	#Docs	#Syn. Docs	Avg #Sents	Avg #Words
Train	1376	29720	21.0	529.8
Test	1090	21800	21.9	564.3

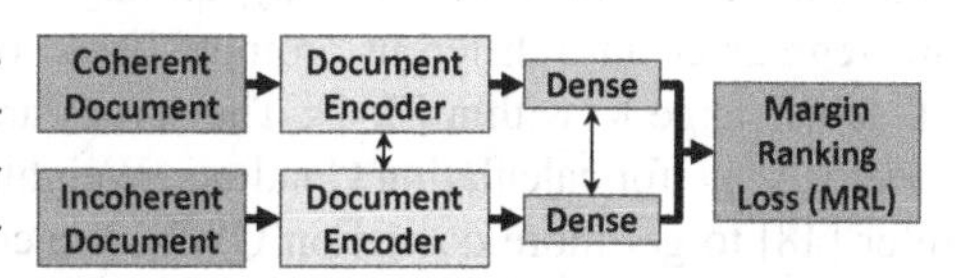

Fig. 2. Overview of Siamese neural approach applied for sentence order discrimination task. Document encoder weights are shared. Dense layer weights are also shared.

GCDC 3-Way Classification. The GCDC dataset contains emails and reviews written with varying degrees of proficiency and care [28]. The WSJ dataset contains documents that have been professionally written and extensively edited. In contrast to WSJ, the GCDC dataset contains writing from non-professional writers in everyday contexts. Rather than using permuted or machine generated texts as examples of low coherence, GCDC has real sentences in which people try but fail to write coherently. GCDC is a corpus that contains texts from four domains, covering a range of coherence, each annotated with a document-level coherence score. Specifically, the dataset contains texts from four domains: Yahoo online forum posts, emails from Hillary Clinton's office, emails from Enron and Yelp business reviews. We present the basic statistics of the dataset in Table 2.

Given a document, the task is to classify it into one of the three different labels (high, medium and low) which denotes the textual coherence level of the given document. For each of these domains, a fixed split of 1000 and 200 was used for train and test respectively as specified in [28]. Of the 1000 documents, we use 200 documents for validation and remaining 800 for training. For our experiments, we use the consensus rating of the expert scores as calculated by [28], and train our models for all the four domains. To evaluate model performance, we use 3-way classification accuracy.

Table 2. Basic statistics of the GCDC dataset. For each of these domains, a fixed split of 1000 and 200 was used for train and test respectively as specified in [28]

	#Docs	Avg #Words	Avg #Sents	Low, medium, high instances (%)
Yahoo	1200	162.1	7.5	46.6,17.4,37.0
Clinton	1200	189.0	6.6	28.2,20.6,51.2
Enron	1200	196.2	7.7	29.9,19.4,50.7
Yelp	1200	183.1	7.5	27.1,21.8,51.1

Table 3. Statistics of ASAP dataset.

Prompt	#Essays	Genre	#Avg length	Range of scores
1	1783	Argumentative	350	2–12
2	1800	Argumentative	350	2–12
3	1726	Response	150	0–3
4	1772	Response	150	0–3
5	1805	Response	150	0–4
6	1800	Response	150	0–4
7	1569	Narrative	250	0–30
8	723	Narrative	650	0–60

ASAP Automated Essay Scoring. Automated Student Assessment Prize (ASAP) dataset is taken from the Kaggle competition[2] which was organized and sponsored by the William and Flora Hewlett Foundation and ran on Kaggle from 10-Feb-12 to 30-Apr-12. The essays are associated with scores given by humans and categorized in eight prompts. Table 3 summarizes some properties of this dataset. The task is to assign an automatic score for a given essay, aiming to replicate human scoring results. Essays are segregated into different prompts based on essay topic and genre. We normalize all score range to within [0, 1]. The scores are re-scaled back to the original prompt-specific scale for calculating Quadratic Weighted Kappa (QWK) scores. The reader can refer [48] to get more details on QWK. We conduct the evaluation in prompt-specific fashion as done in [48].

For this task, we follow previous studies [36,57]. First, we obtain the essay's feature vector v_1 by training a Longformer model for AES task, and take CLS token representation from the last layer. Next, without any AES-task-specific finetuning, we obtain a coherence vector v_2 produced by our model finetuned on WSJ task. The concatenation of v_1 and v_2 is now "coherence augmented representation" of the essay. This representation is passed to a linear layer with sigmoid activation for final essay scoring. We hope that augmentation by v_2 obtained from our model will improve AES scoring accuracy.

5 Experiments

5.1 Baselines

For WSJ and GCDC Related Tasks. We perform extensive comparisons with the following baselines. While Flesch-Kincaid grade level (FKGL) [27] is a readability measure, previous work has treated readability and text coherence as overlapping tasks [4,35]. For coherence classification, Mesgar et al. [35] search over the grade level scores on the training data and select thresholds that result in the highest accuracy. Entity grid (EGRID) [4] builds an entity grid which is a matrix that tracks entity mentions over sentences. Random forest classifier is trained over features extracted from entity grid. CNN-Egrid [42] is a local coherence model that employs a CNN that operates over the entity grid representation. LCNN-Egrid [37] extends CNN-Egrid with lexical information about the entities. In Local Coherence Model (LC) [29], sentences are

[2] https://www.kaggle.com/c/asap-aes/.

encoded with a recurrent or recursive layer and a filter of weights is applied over each window of sentence vectors to extract scores that are aggregated to calculate overall document coherence score. Paragraph sequence (PARSEQ) [28] contains three stacked LSTMs to represent sentence, paragraph and document. Hierachical LSTM [15] is very similar to PARSEQ, but with attention and uses BiLSTMs. Coh+GR [15] extends Hierachical LSTM by training it to predict word-level labels indicating the predicted grammatical role (GR) type at the bottom layers of the network, along with the document-level coherence score. Coh+SOX [15] is same as Coh+GR where, for each word, we only predict subject (S), object (O) and 'other' (X) roles. Seq2Seq [30] consists of two LSTM generative language models and uses the difference between conditional log likelihood of a sentence given its preceding/succeeding context, and the marginal log likelihood of the current/next sentence to assess coherence. Local Coherence Discriminator (LCD-L) [56] uses max-pooling on the hidden state of the language model to get the sentence representation. A representation for two consecutive sentences is then computed by concatenating the output of a set of linear transformations applied to the two sentences. This is fed to a dense layer and used to predict a local coherence score. Coh+GR_BERT [14] is similar to Coh+GR, except that BERT embeddings are used instead of GloVe embeddings as input to BiLSTMs. LCD_BERT [14] is similar to LCD-L but uses averaged BERT (instead of GloVe) embeddings as the sentence representations. We also included LCD_RoBERTa which similar to LCD_BERT but uses RoBERTa embeddings instead of BERT. Unified [39] uses a combination of LSTMs and CNNs. Inc-lex-Coh [24] extracts sentence representations using a pretrained language model and combines the semantic centroid vector with semantic similarity vector to obtain coherence output. They also created another variant Avg-XLNET-Doc that encodes an text content at the document level and averages the encoded representations. We created RoBERTa variant of this model Avg-RoBERTa-Doc where we used RoBERTa embedding instead of XLNET.

For AES/ASAP Task. We perform extensive comparisons with the following baselines. EASE is publicly available, open-source[3] software which ranked third amongst 154 participants in the ASAP competition. It uses manual feature engineering with Support Vector Regression (SVR) and Bayesian Linear Ridge Regression (BLRR). EASE+cohLSTM [36] combines the feature vector computed by EASE, and the coherence vector produced by LSTM-based coherence model to obtain a more reliable representation of an essay. Constraint MTL [9] uses a constrained multi-task pairwise preference learning approach that enables the data from multiple tasks to be combined effectively. Attention based RCNN [12] uses hierarchical sentence-document model to represent essays, using the attention mechanism to learn the relative importance of words and sentences. SkipFlow [49] models coherence using the similarity between multiple states of an LSTM over time with a bounded window.

5.2 Experimental Settings and Reproducibility Information

All experiments were run on a machine equipped with four 32GB V100 GPUs. For all our models, we use 12-layer models, and embedding layer was frozen except for

[3] https://github.com/edx/ease.

Table 4. Sentence order discrimination task Pairwise Ranking Accuracy (PRA) results on WSJ

	Model	PRA
Baselines	LC	74.10
	PARSEQ	74.10
	Seq2Seq	86.95
	CNN-Egrid	88.69
	Unified (ELMo)	93.19
	Coh+GR	93.20
	LCD-L	95.49
	Coh+GR_BERT	96.10
	LCD_RoBERTa	96.45
	LCD_BERT	97.10
Ours	Vanilla Transformer	97.34
	Fact-aware Transformer	97.81
	Fact-aware MTL Trans	**98.22**

Table 5. 3-way classification accuracy results on GCDC.

	Models	Yahoo	Clinton	Enron	Yelp	Average
Baselines	EGRID+coref	41.5	48.0	47.0	49.0	46.4
	EGRAPH+coref	42.5	55.0	44.0	54.0	48.9
	LCNN-Egrid+coref	51.0	56.6	44.7	54.0	51.6
	FKGL	43.5	56.0	52.5	55.0	51.8
	Coh+SOX	50.5	58.5	51.0	–	53.3
	Hierachical LSTM	55.0	59.0	50.5	–	54.8
	PARSEQ	54.9	60.2	53.2	54.4	55.7
	LC	53.5	61.0	54.4	–	56.3
	PARSEQ (A)	58.5	61.0	53.9	56.5	57.5
	Coh+GR	56.0	62.0	56.0	–	58.0
	Inc-lex-Coh	57.3	61.7	54.5	**59.0**	58.1
	Avg-RoBERTa-Doc	60.0	65.3	55.0	58.8	59.8
	Avg-XLNet-Doc	60.5	65.9	**56.9**	**59.0**	60.6
Ours	Vanilla Trans.	58.1	63.9	55.3	57.6	58.7
	Fact-aware Trans.	59.2	67.2	56.3	58.5	60.3
	Fact-aware MTL Trans.	**60.7**	**67.4**	56.4	**59.0**	**60.8**

Table 6. Experimental results on ASAP dataset of our approach versus the baseline methods. Results are reported in terms of the quadratic weighted kappa (QWK) measure, using 5-fold cross-validation. Best QWK for each prompt is highlighted in bold.

	Models	Prompts								
		1	2	3	4	5	6	7	8	Average
Baselines	CohLSTM	0.669	0.634	0.591	0.710	0.639	0.716	0.729	0.641	0.666
	EASE (SVR)	0.781	0.630	0.621	0.749	0.782	0.771	0.727	0.534	0.699
	EASE (BLRR)	0.761	0.606	0.621	0.742	0.784	0.775	0.730	0.617	0.705
	EASE+CohLSTM	0.784	0.654	0.663	0.788	0.793	0.794	0.756	0.646	0.735
	Constraint MTL	0.816	0.667	0.654	0.783	0.801	0.778	0.787	0.692	0.747
	Attention based RCNN	0.822	0.682	0.672	0.814	0.803	0.811	0.801	**0.705**	0.764
	SkipFlow	**0.832**	**0.684**	0.695	0.788	**0.815**	0.810	0.800	0.697	0.765
Ours	Longformer	0.824	0.660	0.693	0.820	0.795	0.810	0.817	0.701	0.765
	Longformer+Fact aware MTL Trans.	0.822	0.674	**0.696**	**0.821**	0.798	**0.812**	**0.822**	0.699	**0.768**

the sentence order discrimination task on WSJ. For fact-aware document encoder, we used pretrained model for the fact encoders and document encoder, and a randomly initialized RoBERTa for fact-aware document encoder. For all experiments we cap the maximum number of facts to 100.

For all experiments, we run 10 epochs except ASAP where we use 5-fold cross validation, weight decay of 0.01 and use a dropout of 0.1. We use Adam optimizer for experiments on GCDC, and use AdamW for WSJ and ASAP experiments. For all the baseline models, we report results from their original papers. For all of our models, the reported results on WSJ and GCDC dataset, are the mean of 10 runs with different random seeds. Margin for the margin ranking loss is set to 1. For MTL framework, categorical cross entropy loss was used for the auxiliary task. We use Longformer based models for WSJ and ASAP dataset to handle the long input documents. For Longformer, we fixed max sequence length to 2048. For RoBERTa, we fixed it to 512. We use learning rate of 2e−5 for all experiments. We use batch size of 2 for all the models on all the tasks.

For model proposed for Automated Essay Scoring (ASAP), we use 5-fold cross validation to evaluate all systems with a 60/20/20 split for train, dev and test sets. We use the splits provided by [48] and closely follow the same experimental procedure. We train our models on ASAP using mean square error (MSE) for 10 epochs and select the best model based on the performance on the validation set.

5.3 Results

Tables 4 and 5 show the results for the two text coherence tasks for WSJ and GCDC datasets respectively. Broadly we observe that our proposed approach significantly outperforms baselines, establishing a new SOTA across all tasks. Across all tasks, the results using our method are statistically significantly better compared to the best baseline with $p \leq 10^{-3}$ at 95% confidence.

Sentence Order Discrimination Results: Table 4 shows results for the sentence order discrimination task for WSJ dataset. We make the following observations: (1) Fact aware transformer outperforms vanilla transformer model as it can incorporate the factual information flow (subject in discourse) in addition to textual information which helps it to correctly determine the coherent sentences. (2) fact-aware MTL model outperforms other variants as the auxiliary task helps in better generalization over test set.

3-Way Classification Results: Table 5 shows 3-way classification results on GCDC. We make the following observations: (1) The Fact-aware model performs better than the vanilla model across all the domains, demonstrating that transitions of facts associated with entities across sentences benefit the model in capturing textual coherence signals. (2) Out of the three gold coherence labels (low, medium, high) all the models have difficulty in correctly classifying documents of medium level coherence, which can be attributed to the smaller number of training examples for that particular class.

AES Results: From Table 6 we observe that Vanilla Longformer finetuned on ASAP dataset performs better than or comparable to previous baseline approaches. Among our models, the "coherence augmented representation" from Fact aware MTL obtains the best result. To understand this a little better we computed the correlation between the coherence score predicted by the Fact aware MTL Transformer and the essay scores in ASAP dataset. We found it to be 0.48 and 0.53 for Longformer and Longformer with fact aware MTL respectively, thereby explaining why our model outperforms vanilla Longformer model.

Qualitative Analysis: We also explore our model qualitatively, examining the coherence scores assigned to some artificial miniature discourses that exhibit various kinds of coherence. The score varies from 0 to 3 and higher score denotes higher level of textual coherence. (1) Case 1: Lexical Coherence. The examples in Table 7 (type = LC) suggest that the models handle lexical coherence, correctly favoring the first over the second, and the third over the fourth and fifth examples (for all our models except the

Table 7. Qualitative analysis: Lexical Coherence (LC), Temporal Order (TO), Centering/Referential Coherence (CRC) examples. Ours = Fact-aware MTL.

Type	S. No	Text	Vanilla	Ours
LC	1	Pinochet was arrested. His arrest was unexpected	1.81	2.76
	2	Pinochet was arrested. His death was unexpected	1.67	1.56
	3	Mary ate some apples. She likes apples	1.45	2.30
	4	Mary ate some apples. She likes pears	1.47	1.45
	5	Mary ate some apples. She likes Paris	1.36	1.27
TO	1	Washington was unanimously elected president in the first two national elections. He oversaw the creation of a strong, well financed national government	1.93	2.79
	2	Washington oversaw the creation of a strong, well-financed national government. He was unanimously elected president in the first two national elections	1.88	2.36
CRC	1	Mary ate some apples. She likes apples	1.45	2.52
	2	She ate some apples. Mary likes apples	1.31	2.49
	3	John went to his favorite music store to buy a piano. He had frequented the store for many years. He was excited that he could finally buy a piano. He arrived just as the store was closing for the day	2.38	2.86
	4	John went to his favorite music store to buy a piano. It was a store John had frequented for many years. He was excited that he could finally buy a piano. It was closing just as John arrived	2.45	2.67

fact-aware one). (2) Case 2: Temporal Order. We show an example of temporal order in Table 7 (type = TO). (3) Case 3: Centering/Referential Coherence. We show a few examples of Centering/Referential Coherence in Table 7 (type = CRC). We observe that our model provides intuitive results while the Vanilla Transformer does not. This suggests that straight-forward adaptation of Transformer models for coherence assessment may not be the best approach.

6 Conclusion

In this paper, we proposed a fact-aware MTL model for text coherence assessment. The proposed model incorporates factual information with document-level information to capture transitions of facts associated with entities across sentences. We observe that our Fact aware approaches outperform existing models on synthetic data (WSJ) as well as real-world data (GCDC). Our work also demonstrates that inductive transfer between tasks: textual coherence assessment and textual entailment, provides better generalization and robustness. Coherence vector obtained from our proposed coherence models also improves the effectiveness of simple models on the automated essay scoring downstream task. In the future, we plan to extend this work to evaluate the text coherence in an open domain setting.

References

1. Annervaz, K., Chowdhury, S.B.R., Dukkipati, A.: Learning beyond datasets: Knowledge graph augmented neural networks for natural language processing. arXiv:1802.05930 (2018)
2. Asher, N., Asher, N.M., Lascarides, A.: Logics of Conversation. Cambridge University Press, Cambridge (2003)
3. Barzilay, R., Elhadad, N.: Inferring strategies for sentence ordering in multidocument news summarization. JAIR **17**, 35–55 (2002)
4. Barzilay, R., Lapata, M.: Modeling local coherence: an entity-based approach. COLING **34**(1), 1–34 (2008)
5. Barzilay, R., Lee, L.: Catching the drift: probabilistic content models, with applications to generation and summarization. In: NAACL-HLT, pp. 113–120 (2004)
6. Beltagy, I., Peters, M.E., Cohan, A.: Longformer: the long-document transformer. arXiv:2004.05150 (2020)
7. Bromley, J., et al.: Signature verification using a "siamese" time delay neural network. Int. J. Pattern Recognit. Artif. Intell. **7**(04), 669–688 (1993)
8. Burstein, J., Tetreault, J., Andreyev, S.: Using entity-based features to model coherence in student essays. In: NAACL, pp. 681–684 (2010)
9. Cummins, R., Zhang, M., Briscoe, T.: Constrained multi-task learning for automated essay scoring. In: ACL, pp. 789–799 (2016)
10. Desai, T., Dakle, P., Moldovan, D.: Generating questions for reading comprehension using coherence relations. In: Workshop on NLP Techniques for Educational Applications, pp. 1–10 (2018)
11. Devlin, J., Chang, M.W., Lee, K., Toutanova, K.: BERT: pre-training of deep bidirectional transformers for language understanding. arXiv:1810.04805 (2018)
12. Dong, F., Zhang, Y., Yang, J.: Attention-based recurrent convolutional neural network for automatic essay scoring. In: CoNLL, pp. 153–162 (2017)
13. Elsner, M., Charniak, E.: Coreference-inspired coherence modeling. In: ACL, pp. 41–44 (2008)
14. Farag, Y., Valvoda, J., Yannakoudakis, H., Briscoe, T.: Analyzing neural discourse coherence models. arXiv:2011.06306 (2020)
15. Farag, Y., Yannakoudakis, H.: Multi-task learning for coherence modeling. arXiv:1907.02427 (2019)
16. Farag, Y., Yannakoudakis, H., Briscoe, T.: Neural automated essay scoring and coherence modeling for adversarially crafted input. arXiv:1804.06898 (2018)
17. Feng, V.W., Hirst, G.: Extending the entity-based coherence model with multiple ranks. In: EACL, p. 315–324 (2012)
18. Gashteovski, K., Gemulla, R., Del Corro, L.: MinIE: minimizing facts in open information extraction. In: EMNLP, pp. 2620–2630. ACL (2017)
19. Grosz, B.J., Weinstein, S., Joshi, A.K.: Centering: a framework for modeling the local coherence of discourse. COLING **21**(2), 203–225 (1995)
20. Gunel, B., Zhu, C., Zeng, M., Huang, X.: Mind the facts: knowledge-boosted coherent abstractive text summarization. arXiv:2006.15435 (2020)
21. Guz, G., Bateni, P., Muglich, D., Carenini, G.: Neural RST-based evaluation of discourse coherence. arXiv:2009.14463 (2020)
22. Holtzman, A., Buys, J., Forbes, M., Bosselut, A., Golub, D., Choi, Y.: Learning to write with cooperative discriminators. arXiv:1805.06087 (2018)
23. Jeon, S., Strube, M.: Centering-based neural coherence modeling with hierarchical discourse segments. In: EMNLP (1), pp. 7458–7472 (2020)

24. Jeon, S., Strube, M.: Incremental neural lexical coherence modeling. In: Proceedings of the 28th International Conference on Computational Linguistics, pp. 6752–6758 (2020)
25. Jeon, S., Strube, M.: Countering the influence of essay length in neural essay scoring. In: Second Workshop on Simple and Efficient Natural Language Processing, pp. 32–38 (2021)
26. Kiddon, C., Zettlemoyer, L., Choi, Y.: Globally coherent text generation with neural checklist models. In: EMNLP, pp. 329–339 (2016)
27. Kincaid, J.P., Fishburne Jr., R.P., Rogers, R.L., Chissom, B.S.: Derivation of new readability formulas (automated readability index, fog count and flesch reading ease formula) for Navy enlisted personnel. Technical report, Naval Technical Training Command Millington TN Research Branch (1975)
28. Lai, A., Tetreault, J.: Discourse coherence in the wild: a dataset, evaluation and methods. arXiv:1805.04993 (2018)
29. Li, J., Hovy, E.: A model of coherence based on distributed sentence representation. In: EMNLP, pp. 2039–2048 (2014)
30. Li, J., Jurafsky, D.: Neural net models of open-domain discourse coherence. In: EMNLP, pp. 198–209 (2017)
31. Liu, Y., et al.: RoBERTa: a robustly optimized BERT pretraining approach. arXiv:1907.11692 (2019)
32. Louis, A., Nenkova, A.: A coherence model based on syntactic patterns. In: EMNLP, pp. 1157–1168 (2012)
33. Mann, W.C., Thompson, S.A.: Rhetorical structure theory: toward a functional theory of text organization. Text **8**(3), 243–281 (1988)
34. McIntyre, N., Lapata, M.: Plot induction and evolutionary search for story generation. In: ACL, pp. 1562–1572 (2010)
35. Mesgar, M., Strube, M.: Graph-based coherence modeling for assessing readability. In: Joint Conference on Lexical and Computational Semantics, pp. 309–318 (2015)
36. Mesgar, M., Strube, M.: A neural local coherence model for text quality assessment. In: EMNLP, pp. 4328–4339 (2018)
37. Mohiuddin, T., Joty, S., Nguyen, D.T.: Coherence modeling of asynchronous conversations: a neural entity grid approach. In: ACL, pp. 558–568 (2018)
38. Mohiuddin, T., Jwalapuram, P., Lin, X., Joty, S.: CohEval: benchmarking coherence models. arXiv:2004.14626 (2020)
39. Moon, H.C., Mohiuddin, M.T., Joty, S., Xu, C.: A unified neural coherence model. In: EMNLP-IJCNLP, pp. 2262–2272 (2019)
40. Morris, J., Hirst, G.: Lexical cohesion computed by thesaural relations as an indicator of the structure of text. COLING **17**(1), 21–48 (1991)
41. Muangkammuen, P., Xu, S., Fukumoto, F., Saikaew, K.R., Li, J.: A neural local coherence analysis model for clarity text scoring. In: COLING, pp. 2138–2143 (2020)
42. Nguyen, D.T., Joty, S.: A neural local coherence model. In: ACL, pp. 1320–1330 (2017)
43. Park, C.C., Kim, G.: Expressing an image stream with a sequence of natural sentences. NIPS **28**, 73–81 (2015)
44. Parveen, D., Mesgar, M., Strube, M.: Generating coherent summaries of scientific articles using coherence patterns. In: EMNLP, pp. 772–783 (2016)
45. Pitler, E., Louis, A., Nenkova, A.: Automatic evaluation of linguistic quality in multi-document summarization. In: ACL, pp. 544–554 (2010)
46. Somasundaran, S., Burstein, J., Chodorow, M.: Lexical chaining for measuring discourse coherence quality in test-taker essays. In: COLING, pp. 950–961 (2014)
47. Sorokin, D., Gurevych, I.: Modeling semantics with gated graph neural networks for knowledge base question answering. arXiv:1808.04126 (2018)
48. Taghipour, K., Ng, H.T.: A neural approach to automated essay scoring. In: EMNLP, pp. 1882–1891 (2016)

49. Tay, Y., Phan, M.C., Tuan, L.A., Hui, S.C.: SkipFlow: incorporating neural coherence features for end-to-end automatic text scoring. In: AAAI (2018)
50. Vaswani, A., et al.: Attention is all you need. In: NIPS, pp. 5998–6008 (2017)
51. Verberne, S., Boves, L., Oostdijk, N., Coppen, P.A.: Evaluating discourse-based answer extraction for why-question answering. In: SIGIR, pp. 735–736 (2007)
52. Wang, A., et al.: SuperGLUE: a stickier benchmark for general-purpose language understanding systems. In: NIPS, pp. 3266–3280 (2019)
53. Wang, X., et al.: KEPLER: a unified model for knowledge embedding and pre-trained language representation. TACL **9**, 176–194 (2021)
54. Williams, A., Nangia, N., Bowman, S.: A broad-coverage challenge corpus for sentence understanding through inference. In: NAACL-HLT, pp. 1112–1122 (2018)
55. Xiong, H., He, Z., Wu, H., Wang, H.: Modeling coherence for discourse neural machine translation. In: AAAI, vol. 33, pp. 7338–7345 (2019)
56. Xu, P., et al.: A cross-domain transferable neural coherence model. arXiv:1905.11912 (2019)
57. Zesch, T., Wojatzki, M., Scholten-Akoun, D.: Task-independent features for automated essay grading. In: 10th Workshop on Innovative Use of NLP for Building Educational Applications, pp. 224–232 (2015)

Open Set Recognition for Time Series Classification

Tolga Akar[2(✉)], Thorben Werner[1], Vijaya Krishna Yalavarthi[1], and Lars Schmidt-Thieme[1]

[1] Information Systems and Machine Learning Lab, University of Hildesheim, Universitatsplätz 1, 31141 Hildesheim, Germany
{werner,yalavarthi,schmidt-thieme}@ismll.de

[2] Distributed Artificial Intelligence Laboratory, Technical University of Berlin, Straße des 17. Juni 135, 10623 Berlin, Germany
tolga.akar@tu-berlin.de

Abstract. Traditional classification problems often assume that the number of classes present in the data is finite. This may hold true for the training data, but in real life, the risk of encountering unknown samples is ubiquitous. Classifying these unknown samples into one of the target classes can have drastic effects in some situations like security systems or body sensors. To address this problem, recently, open set recognition models that can correctly classify the known samples and detect the unknowns simultaneously, are proposed. In contrast to the existing models where unknown detection depends on the classification model, we propose, to the best of our knowledge, an open set recognition model for time series classification that works independent of the classifier by employing class-specific barycenters. Specifically, DTW distance, and the cross-correlation between the class-specific barycenters, and the input are used for detecting the unknown classes during testing. Our extensive experimental evaluation on the UEA multivariate time series archive with 30 datasets shows that the proposed open set recognition architecture deployed on top of the InceptionTime outperforms the state-of-the-art open set recognition models by an average of 22% in terms of macro F1 score.

Keywords: Open set recognition · Time series classification · Machine learning

1 Introduction

The success of machine learning based solutions for various classification problems is undeniable. Most of the time, the number of target classes is assumed to be finite, and solutions for these problems are derived in such a way. However, in real-life applications, there is always a risk of encountering samples from

T. Akar—This work is done while the main author was doing his master's thesis at ISMLL, University of Hildesheim.

J. Gama et al. (Eds.): PAKDD 2022, LNAI 13281, pp. 354–366, 2022.
https://doi.org/10.1007/978-3-031-05936-0_28

unknown classes that are not seen during the training. This will, inevitably, lead to a situation where the classifier will classify those unknown samples as one of the target classes, which is of course wrong (e.g. Fig. 1b). Such wrong predictions can have drastic effects in certain situations, e.g. security systems, body sensors, machinery maintenance. To address this problem, open set recognition (OSR) models that can correctly classify the known samples and detect the unknowns at the same time are proposed in the past decade, starting with [15]. Even though OSR has received a lot of attention in recent years, the majority of the studies in this field focuses on computer vision problems, and best of our knowledge, there is no other work for open set recognition that focuses on the time series classification (TSC) task.

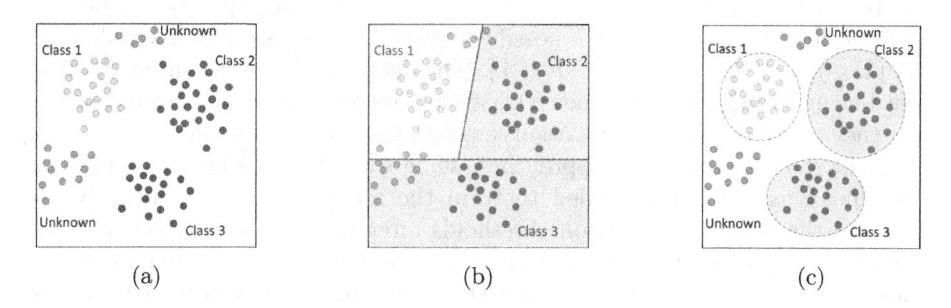

Fig. 1. Comparison between traditional classification (Fig. 1b) and open set recognition (Fig. 1c). Figure 1a shows the distribution of the original dataset.

This study focuses on proposing a methodology for achieving a solution for the open set recognition problem regarding the TSC tasks, that is generally applicable to multiple datasets and that can ideally work with different classifiers. The proposed method uses class-specific time series barycenters, i.e. the centroids representing a cluster of time series, for unknown detection. The DTW distance and the cross-correlation between a class-specific barycenter and an input determine whether the input belongs to the given class or not. As for the classifier, the proposed method benefits the state-of-the-art time series classification model InceptionTime [7]. Thus, the proposed method is referred to as Open Set InceptionTime (OS-InceptionTime or OS-IT). As the unknown detection methodology is independent of the classifier, it can be used alongside any other time series classification algorithm.

The scientific contribution of this study thesis is threefold:

- The first-ever open set recognition method that is specifically designed for time series is introduced. This answers the research question: *How can the open set problem be solved for the time series classification tasks?*
- It is shown that artificially created unknown samples can simulate the actual unknowns to some extent, eliminating the need for handpicking and thus, fully automating the training process. This answers the research question: *Can the training procedure be fully automated for the proposed method?*

- The proposed model is evaluated on 30 datasets to demonstrate that it is a generally applicable solution for open set problems regarding the various type of TSC datasets and can be used with any classifier. This answers the research question: *Is the proposed method generic enough to be applied for different types of TSC datasets?*

2 Background and Problem Definition

Let f_c be a traditional closed set classifier that takes a time series sequence x with L number of time steps and M number of dimensions (also referred as channels or variables) and assigns this input x a label y, deciding among the K number of known classes, i.e. $f_c : \mathbb{R}^{L \times M} \rightarrow \{1, ..., K\}$. An open set model f_o, on the other hand, has an extra possible class $K+1$ to assign for the unknown samples, $f_o : \mathbb{R}^{L \times M} \rightarrow \{1, ..., K, K+1\}$. The data that are seen during training that belong to the one of the known classes are referred to as the known samples and they are denoted with $\mathcal{D}_k$. A subset of $\mathcal{D}_k$ is used for training.

In most of the open set approaches, some sort of data that can represent the unknowns are either needed to train the model, or more commonly, to optimize the unknown detection thresholds after the training. These types of unknown samples that are used during the training will be referred to as the known unknowns. They are denoted with $\mathcal{D}_a$, having a vector of K+1 values as their labels. The last type of samples is the unknowns that are not a part of the training. They are referred to as the unknown unknowns. They may or may not appear during the testing in real-life applications. They are denoted with $\mathcal{D}_u$. Since all the samples that do not belong to the $\mathcal{D}_k$ are treated as unknowns, $\mathcal{D}_a$ can be considered as a subset of the $\mathcal{D}_u$ as well. In short, an ideal open set classifier f for an input x should predict the correct class label k for known samples, and $K+1$ for the unknowns as follows:

$$f(x) = \hat{y} = \begin{cases} k & \text{for } x \in \mathcal{D}_k,\ k \in \{1, ..., K\} \\ K+1 & \text{for } x \in \mathcal{D}_u \end{cases} \tag{1}$$

3 Related Work

Barycenters. There are multiple ways to compute barycenters. The first one, Euclidean, is simply the arithmetic mean of each point in time. It is much faster to compute than the others, however, it does not provide a meaningful representation enough since it does not take shifts in time into account like the DTW (Dynamic time warping) based methods. The second one, originally proposed in [13], is an iterative averaging method to compute the barycenters under DTW. The aim is to minimize the DTW distance between the center (average sequence) and the actual sequences in the given class or dataset. Expectation-maximization or stochastic subgradient methods are used to find optimal solutions with this method. Unlike the DBA (DTW Barycenter Averaging) approach, soft-DTW, introduced in [3], uses a differentiable loss function to solve this minimization problem, which makes it much more easier to obtain the optimal result. In other words, the soft-DTW method is able to find more accurate and smoother barycenters for sequential data. Thus, for this study, barycenters are computed using the soft-DTW geometry. An example result of this calculation can be seen in Fig. 2.

Time Series Classification. After their proven success [10], convolutional neural networks (CNNs) attracted a great amount of attention from the TSC community who were looking for scalable alternatives to traditional ensemble classifiers such as HIVE-COTE [11] or BOSS [14]. In [6], authors experimented on several CNN based deep learning solutions for TSC and reported that Fully Connected Neural Networks (FCNs) and deep Residual Networks (ResNet) achieved the best performances overall. More recently, following the footsteps of [6], InceptionTime method is proposed in [7] and shown to achieve state-of-the-art performance, on par with HIVE-COTE.

Open Set Recognition. In the last decade, the popularity of the open set recognition (OSR) domain has grown significantly after [15] revealed that unknown samples can generate high activation scores for some of the known classes in closed set classifiers. The first application of OSR on deep networks was [2], where authors introduced a novel OpenMax layer. During testing, this OpenMax layer replaces the final softmax layer, which enables the classifier to have a probability distribution with an extra class probability for the unknown class. Directly extending the OpenMax paper, [8] proposes G-OpenMax, which utilizes a generative adversarial network (GANs) to generate samples of unknown classes. The trend of using generative models for OSR tasks continued with Class Conditioned Auto-Encoder for Open-set Recognition (C2AE) [12] and Classification-Reconstruction Learning for Open-Set Recognition (CROSR) [18], both using slightly different auto-encoder networks, and are similar in the way that both of them uses EVT to decide on the reconstruction thresholds. CGDL [16] can be considered the state-of-the-art OSR method with a generative network. It uses a variational auto-encoder (VAE) which is forced to approximate different Gaussian models for different known classes. Another alternative approach, [5] introduces two novel loss functions for unknown detection, that maximize the entropy of the unknown samples, namely Entropic Open-Set Loss and Objectosphere Loss. The only OSR paper for time series is [9], however, they are using a specific dataset for combustion engines rather than one of the popular TSC benchmark datasets, and they perform open set recognition to label each time step, but not the time series as a whole sequence.

4 Methodology

The unknown detector of the proposed model consists of two separate criteria to minimize the chance of missing unknown data. The first one is the DTW

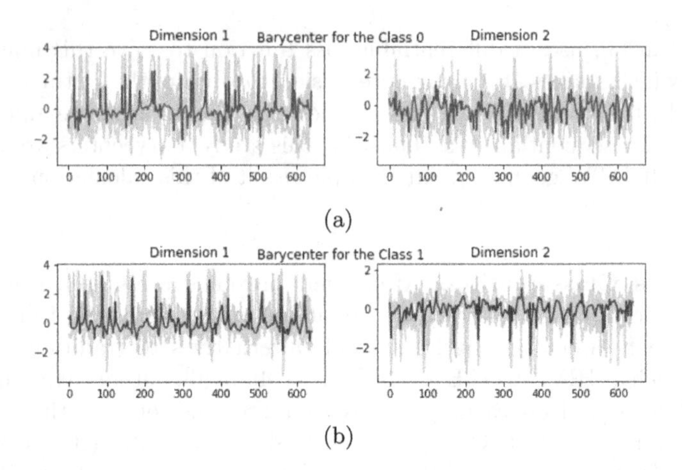

Fig. 2. Barycenters for the AtrialFibrillation dataset for the classes 0 and 1. The samples that belong to that class are in the background in gray.

similarity. It is basically the sum of squared distances between a barycenter and a sample, computed after aligning both time series using DTW. If the distance between a sample and the barycenter is above a certain threshold for all the known classes, that sample is considered unknown. In cases where the intraclass variation is high, barycenters are usually not able to represent meaningful patterns regarding that class. In such cases, an out-of-class sample that looks like a horizontal line along the mean can have a smaller distance, especially in low dimensional time series, than a sample that actually belongs to that class. For this reason, a second criterion is added to the unknown detector.

The second criterion is the cross-correlation (a.k.a sliding dot product) of a sample and a barycenter. Similar to the convolution operation, cross-correlation is mainly used for searching an input sequence for a given filter (usually a shorter filter representing a feature). In this case, the barycenter functions as the filter and slid through the input sample to calculate the cross-correlation. The idea behind this approach is that out-of-class samples should generate much lower cross-correlation values. Cross-correlations are computed for each dimension of the data separately. If a sample generates a lower cross-correlation value for at least one of the dimensions for a specific class, then, it is rejected by that class for extra safety.

Compared in isolation, the cross-correlation criterion works usually better than the distance criterion. However, there are some datasets where the cross-correlation threshold does not work well. Hence, combining the two yields better results in most of the datasets used in this study. The formulas for defining the thresholds are quite straightforward. In Eqs. 2 and 3, τ_k^{dist} and τ_k^{cc} represent the distance and the cross-correlation thresholds for a known class k. For each k, median distance to the barycenter of the class $\tilde{\mu}_k^{dist}$, median maximum cross-correlation $\tilde{\mu}_k^{cc}$ with the barycenter, and their standard deviations σ_k^{dist} and σ_k^{cc} are computed using the train samples belonging to the class k. Median values are chosen here over the means in order to reduce the effect of the outliers within the class.

$$\tau_k^{dist} = \tilde{\mu}_k^{dist} + \alpha \cdot \sigma_k^{dist}, \ \ for\ k \in \{1, ..., K\} \tag{2}$$

$$\tau_k^{cc} = \tilde{\mu}_k^{cc} - \beta \cdot \sigma_k^{cc}, \ \ for\ k \in \{1, ..., K\} \tag{3}$$

The crucial values in these equations are hyper-parameters α and β, since they determine the magnitude of the thresholds. A grid search among the combinations of possible values (ranging from 0 to 5) is performed using the whole train set in order to find the optimal values. The full outline of the inference procedure can be seen in Algorithm 1.

Since this study aims to propose a generic solution that is applicable to multiple datasets, the actual unknown samples are not used in this stage to prevent cherry-picking. Instead, an artificial set of unknown data (known unknowns) are generated for each dataset (see Algorithm 2) and used to evaluate the open set performance of the model. The aim of the grid search is to find optimal hyper-parameters that can help detect the unknowns while maintaining high accuracy for the known samples. To do this a simple formula (Eq. 4) is used to assess the performance after every iteration. Then, the combination of hyper-parameters with the highest score $s(\alpha, \beta)$ is chosen.

$$s(\alpha, \beta) = \lambda^4 \cdot \frac{acc_X \cdot acc_A}{acc_X + acc_A} \tag{4}$$

$$\lambda = \frac{acc_X}{acc_{closed\ set}} \tag{5}$$

Given a train set X and artificially created known unknown data matrix A, acc_X stands for the accuracy of the model with the given hyper-parameters for the original train samples (known classes) X. Similarly, acc_A is the accuracy for detecting the known unknown samples, i.e. the recall for the $(K+1)$th (the label for unknowns) class. The λ functions as a penalization parameter to prevent the model from sacrificing too much from acc_X to increase acc_A. This is undesirable for most cases since detecting unknowns will not worth it if the classification accuracy drops dramatically. In other words, λ puts more importance on the classification accuracy than the unknown detection in this trade-off. It is calculated by simply dividing acc_X by $acc_{closed\ set}$, the accuracy of the closed set model.

The proposed method employs the InceptionTime model, state-of-the-art deep learning ensemble of five CNNs with Inception modules [17] (see [7] for an in-depth explanation) as the classifier. Since the unknown detector is independent of the classifier, InceptionTime can easily be replaced with any other classification model. This also means that the training procedure of the classifier is also separate from the unknown detector. InceptionTime is trained the same way as in [7].

Algorithm 1: Testing procedure for the OS-InceptionTime.

Input: Test sample x
Input: Classifier $f()$
Input: Barycenters for each known class $B = \{b_1, ..., b_K\}$
Input: Unknown detection thresholds τ_k^{dist} and τ_k^{cc}
1 Predict an initial label: $\hat{y} = f(x_r)$
// **Unknown detection part**
2 Calculate distances and cross-correlations **for** $k \in \{1, ..., K\}$ **do**
3 | Calculate the DTW distance: $d_k = DTW(x, b_k)$
4 | Calculate the cross-correlation: $c_k = max(correlate(x, b_k))$
5 **end**
6 **if** $d_k > \tau_k^{dist}$ *or* $c_k < \tau_k^{cc}$, $\forall k \in \{1, ..., K\}$ **then**
7 | Modify the predicted label to be the unknown class: $\hat{y} = K + 1$
8 **end**
9 **return** $\hat{y}$

Algorithm 2: Known unknowns generation algorithm.

Input: Train samples X
Input: A mean μ, and a standard deviation σ for the random noise
1 Define augmented data matrix $A = X$
2 **for** $i \in N$ **do**
3 | Generate a random noise: $noise \sim \mathcal{N}(\mu, \sigma^2)$
4 | Add the noise to the original sample: A_i += $noise$
5 **end**
6 Define splitting index $cut\_idx = L/2$
7 Define $temp1 = A_{1:N,1:cut\_idx}$ to store the first halves of every sample
8 Define $temp2 = A_{1:N,cut\_idx:L}$ to store the second halves
9 Switch the places of the first and second halves:
$A = concatenate(temp2, temp1)$
10 Reverse the order of the time steps and dimensions: $A = flip(A)$
11 **return** $\mathcal{D}_a = \{A, \overrightarrow{K+1}\}$

5 Experiments

5.1 Datasets

The 30 multivariate time series classification datasets from the UEA archive are used for all the experiments in this work. Background information about these datasets can be seen in [1]. The unknown datasets are also chosen from the archive. They are presented in Table 1 alongside the *Openness* score for each test scenario. *Openness* takes percentage values between 0% and 100%, where 0% represents a completely closed set problem. For each known dataset, two other datasets from the archive were used as the unknowns. In order to avoid cherry-picking, datasets were picked according to their sizes and shapes. The most similar ones have been used to keep the integrity as much as possible after resampling to match the shape of the original known dataset.

Table 1. The chosen unknown datasets for each train set and the openness score of each open set problem. The hybrid dataset refers to the artificially created dataset forged by concatenating PEMS-SF, InsectWingbeat, FaceDetection, FingerMovements, and HandMovementDirection along their last axes.

Training dataset	Unknown dataset 1	Openness	Unknown dataset 2	Openness
ArticularyWordRecognition	PEMS-SF	7.15%	SpokenArabicDigits	9.46%
AtrialFibrillation	HandMovementDirection	26.15%	Heartbeat	18.35%
BasicMotions	SpokenArabicDigits	35.11%	InsectWingbeat	35.11%
CharacterTrajectories	Handwriting	22.73%	PhonemeSpectra	29.29%
Cricket	SelfRegulationSCP1	5.72%	SelfRegulationSCP2	5.72%
DuckDuckGeese	Hybrid dataset*	8.71%		
EigenWorms	MotorImagery	12.29%	Cricket	34.06%
Epilepsy	CharacterTrajectories	47.48%	PhonemeSpectra	59.18%
EthanolConcentration	SelfRegulationSCP2	14.72%	Cricket	38.28%
ERing	NATOPS	20.53%	FaceDetection	10.56%
FaceDetection	InsectWingbeat	48.36%	PEMS-SF	42.26%
FingerMovements	FaceDetection	24.41%	InsectWingbeat	48.36%
HandMovementDirection	Heartbeat	14.72%	DuckDuckGeese	24.41%
Handwriting	Epilepsy	4.49%	CharacterTrajectories	15.60%
Heartbeat	PEMS-SF	42.26%	DuckDuckGeese	36.75%
JapaneseVowels	NATOPS	15.15%	FingerMovements	7.42%
Libras	NATOPS	9.95%	LSST	14.46%
LSST	FaceDetection	4.96%	SpokenArabicDigits	15.27%
InsectWingbeat	DuckDuckGeese	12.29%	PEMS-SF	15.48%
MotorImagery	DuckDuckGeese	36.75%	PEMS-SF	42.26%
NATOPS	FingerMovements	10.56%	FaceDetection	10.56%
PenDigits	LSST	19.68%	FaceDetection	6.75%
PEMS-SF	DuckDuckGeese	16.33%		
PhonemeSpectra	FaceDetection	1.87%	SpokenArabicDigits	6.38%
RacketSports	JapaneseVowels	35.11%	LSST	35.11%
SelfRegulationSCP1	SelfRegulationSCP2	24.41%	Cricket	51.49%
SelfRegulationSCP2	Heartbeat	24.41%	MotorImagery	24.41%
SpokenArabicDigits	InsectWingbeat	19.68%	FaceDetection	6.75%
StandWalkJump	MotorImagery	18.35%	Cricket	43.80%
UWaveGestureLibrary	HandMovementDirection	12.71%	PhonemeSpectra	46.55%

5.2 Experimental Results

3 baselines are considered to compare and evaluate the results of the proposed method for open set recognition. The first baseline is the most primitive one among all. It is an ensemble of small binary CNN models for each known class in the dataset (One-vs-All), with two convolutional layers followed by max pooling and two fully connected layers.

The second baseline replaces the softmax layer of the vanilla InceptionTime network with the OpenMax layer introduced in [2].

The last baseline is the class conditional VAE with the probabilistic ladder net architecture, proposed in the CGDL paper [16]. Unlike the original model, which was designed for images, 1D convolutions are used for this case.

The performance measure for the closed set classification (without the involvement of the unknown data) is the classification accuracy. The values for the performance metrics are obtained after running the algorithm three times for each testing scenario and then averaging the results. Macro F1 score, on the other hand, comes in handy when evaluating the open set performance of the models with unknown samples included in the test set, and it is the standard metric for open set papers. It will be used to evaluate the overall performance of the open set algorithms. Table 2 presents the open set performances of the algorithms for each dataset. For almost two thirds of the datasets, the proposed algortihm achieves better results than the other baselines. Detailed results for the OS-InceptionTime are given in Table 3 alongside with the optimal hyper-parameter values.

Table 2. Comparison of the open set macro F1 scores for each dataset using the unknowns from Table 1

Dataset	OvA-CNNs	OM-IT	LCVAE	OS-IT
ArticularyWordRecognition	**0.98**	0.57	0.85	0.96
AtrialFibrillation	0.19	**0.70**	0.39	0.18
BasicMotions	0.77	0.81	0.52	**0.82**
CharacterTrajectories	0.91	0.88	**0.98**	0.96
Cricket	0.83	0.75	**0.90**	0.68
DuckDuckGeese	0.33	0.25	0.35	**0.64**
EigenWorms	0.08	0.45	0.40	**0.85**
Epilepsy	0.58	0.79	0.60	**0.82**
EthanolConcentration	0.16	0.36	0.24	**0.38**
ERing	**0.90**	0.32	0.58	0.86
FaceDetection	0.40	0.44	0.15	**0.54**
FingerMovements	0.24	0.35	0.00	**0.62**
HandMovementDirection	0.20	0.13	0.30	**0.42**
Handwriting	0.18	0.16	0.20	**0.43**
Heartbeat	0.28	0.42	0.16	**0.53**
JapaneseVowels	0.87	0.90	**0.95**	**0.95**
Libras	0.60	0.66	0.74	**0.80**
LSST	0.40	**0.45**	0.08	0.36
InsectWingbeat	0.55	0.64	0.03	**0.65**
MotorImagery	0.18	0.23	0.50	**0.53**
NATOPS	0.85	0.69	**0.92**	0.89
PenDigits	0.79	0.94	0.10	**0.95**
PEMS-SF	0.67	0.76	0.55	**0.87**
PhonemeSpectra	0.10	0.34	0.13	**0.37**
RacketSports	0.51	0.63	0.67	**0.85**
SelfRegulationSCP1	0.44	0.39	**0.48**	0.46
SelfRegulationSCP2	0.27	0.19	0.30	**0.54**
SpokenArabicDigits	0.74	0.92	0.67	**0.98**
StandWalkJump	0.31	0.25	**0.45**	0.17
UWaveGestureLibrary	0.77	0.67	0.73	**0.79**
Average Results	0.50	0.53	0.46	**0.66**

Table 3. The results for the OS-InceptionTime algorithm. Open set results are averaged for the unknown datasets given in Table 1.

Dataset	Closed set classification				Open set classification	
	InceptionTime accuracy	OS-InceptionTime accuracy	Performance decrease (%)	Hyper-parameters (α, β)	Open Set Macro F1 score	Recall for the Unknowns
ArticularyWordRecognition	0.99	0.92	−7.07	2.75, 3.25	0.96	1.00
AtrialFibrillation	0.27	0.00	−100.00	4, 0	0.18	1.00
BasicMotions	1.00	0.80	−20.00	1.75, 2.5	0.82	0.80
CharacterTrajectories	1.00	0.94	−6.00	2, 3	0.96	1.00
Cricket	0.99	0.58	−41.41	2.75, 2.75	0.68	1.00
DuckDuckGeese	0.62	0.50	−19.35	2.75, 4	0.64	1.00
EigenWorms	0.93	0.82	−11.83	2, 1.75	0.85	1.00
Epilepsy	0.97	0.79	−18.56	1.25, 1	0.82	0.81
EthanolConcentration	0.28	0.26	−7.14	1, 1	0.38	1.00
ERing	0.89	0.82	−7.87	2, 2	0.86	0.98
FaceDetection	0.67	0.39	−41.79	4, 2	0.54	1.00
FingerMovements	0.50	0.31	−38.00	4, 1.5	0.62	1.00
HandMovementDirection	0.32	0.27	−15.63	1.5, 1.5	0.42	1.00
Handwriting	0.50	0.38	−24.00	1, 1.25	0.43	1.00
Heartbeat	0.78	0.60	−23.08	1, 1.5	0.54	0.54
JapaneseVowels	0.98	0.93	−5.10	1.75, 2	0.95	1.00
Libras	0.88	0.75	−14.77	1, 1	0.8	1.00
LSST	0.45	0.45	0.00	0, 0.1	0.36	0.00
InsectWingbeat	0.71	0.67	−5.63	1, 1	0.65	0.92
MotorImagery	0.51	0.39	−23.53	1, 1.5	0.53	0.54
NATOPS	0.95	0.82	−13.68	2, 2	0.89	1.00
PenDigits	0.99	0.92	−7.07	1.5, 1.5	0.95	1.00
PEMS-SF	0.86	0.86	0.00	2.5, 3	0.87	1.00
PhonemeSpectra	0.37	0.34	−8.11	1, 1	0.37	1.00
RacketSports	0.89	0.76	−14.61	1, 1.5	0.85	1.00
SelfRegulationSCP1	0.78	0.41	−47.44	0.75, 0.5	0.46	0.61
SelfRegulationSCP2	0.51	0.37	−27.45	0.75, 1.5	0.54	0.86
SpokenArabicDigits	1.00	0.95	−5.00	1.5, 2.75	0.98	1.00
StandWalkJump	0.47	0.00	−100.00	2, 1.5	0.17	1.00
UWaveGestureLibrary	0.84	0.70	−16.67	1.5, 3	0.79	1.00

On average, the OS-InceptionTime sacrifices around 20% of the closed set classification accuracy compared to the vanilla version. In return, however, it achieves an outstanding performance for detecting the unknowns. The average recall for detecting the unknowns is 0.926. In 46 test cases out of 58 (79.3%), the proposed algorithm is able to detect all the unknowns with a perfect recall value of 1.00. In 51 cases (88%), it can detect at least half of the unknowns, and only in 7 cases (12%), it achieves 0.35 or less recall for the unknowns.

5.3 Discussion

The critical difference diagrams regarding the methods used in this work are presented in Fig. 3 separately for each evaluation metric. The ranks are calculated using the Wilcoxon signed-rank test, which is used to compare repeated measurements on the same samples (in this case, test datasets). Then Holm test is used to reject the null hypothesis, i.e. the mean ranks for each pair of algorithms are not significantly different from each other. According to the Fig. 3b, the proposed Open Set InceptionTime model has the highest ranking by a significant margin, clearly separating itself from the others. However, it lacks behind the

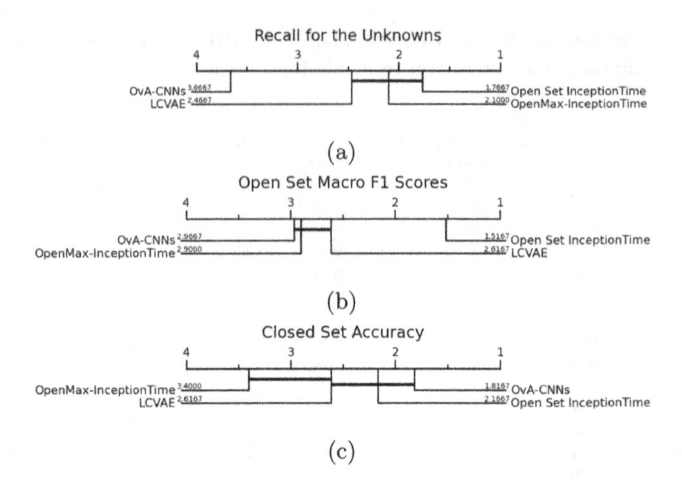

Fig. 3. Difference diagrams of the mean ranks of the algorithms by each metric.

OvA-CNNs algorithms in terms of closed set accuracy, which is understandable because it trades-off nothing to detect unknowns.

Future Work. Since all the datasets used in this work were multivariate, the proposed method can be tested and validated on the UCR time series archive with 128 univariate TSC datasets [4]. Moreover, to trade off less closed set accuracy, better alternatives/additions to the distance and cross-correlation thresholds can be incorporated into the OS-InceptionTime, such as the difference between the forecasting errors of known and unknown samples. Finally, parallelization can be introduced to speed up the grid search for hyper-parameter optimization, as it takes the longest time to compute during the training phase with the computational complexity of $O(N^2)$.

6 Conclusion

This study presents the first ever open set model for time series classification, Open Set InceptionTime. The proposed method makes use of the class-specific barycenters of the time series to detect unknowns, and combines it with a state-of-the-art classifier. Moreover, an automated algorithm for creating the known unknown data that is required to determine the unknown detection thresholds is also presented in this work.

The experiments show that OS-InceptionTime achieves near-perfect results for unknown detection, but it trades off closed set classification accuracy while doing so. Thus, it can be considered as more suitable in situations where detecting the unknowns are more vital than the classification accuracy of the known samples. OS-InceptionTime is able to outperform all the other baselines that are adapted from computer vision to the time series classification domain. The results are validated on 30 different datasets, which proves that the proposed method is generic and applicable to various time series classification datasets.

Being the first work that develops a generic method regarding the open set recognition for time series classification, this master thesis shall act as a baseline for the future research in this field. The full implementation of the Open Set InceptionTime algorithm in Python can be found publicly on the web[1].

References

1. Bagnall, A., et al.: The UEA multivariate time series classification archive 2018. arXiv preprint arXiv:1811.00075 (2018)
2. Bendale, A., Boult, T.E.: Towards open set deep networks. In: Proceedings of the IEEE conference on computer vision and pattern recognition. pp. 1563–1572 (2016)
3. Cuturi, M., Blondel, M.: Soft-DTW: a differentiable loss function for time-series. In: International Conference on Machine Learning, pp. 894–903. PMLR (2017)
4. Dau, H.A., et al.: The UCR time series archive. IEEE/CAA J. Automatica Sinica **6**(6), 1293–1305 (2019)
5. Dhamija, A.R., Günther, M., Boult, T.E.: Reducing network agnostophobia. arXiv preprint arXiv:1811.04110 (2018)
6. Ismail Fawaz, H., Forestier, G., Weber, J., Idoumghar, L., Muller, P.-A.: Deep learning for time series classification: a review. Data Min. Knowl. Disc. **33**(4), 917–963 (2019). https://doi.org/10.1007/s10618-019-00619-1
7. Fawaz, H.I., et al.: InceptionTime: finding AlexNet for time series classification. Data Min. Knowl. Disc. **34**(6), 1936–1962 (2020)
8. Ge, Z., Demyanov, S., Chen, Z., Garnavi, R.: Generative openmax for multi-class open set classification. arXiv preprint arXiv:1707.07418 (2017)
9. Jung, D.: Data-driven open-set fault classification of residual data using Bayesian filtering. IEEE Trans. Control Syst. Technol. **28**(5), 2045–2052 (2020)
10. Krizhevsky, A., Sutskever, I., Hinton, G.E.: ImageNet classification with deep convolutional neural networks. Adv. Neural Inf. Process. Syst. **25**, 1097–1105 (2012)
11. Lines, J., Taylor, S., Bagnall, A.: HIVE-COTE: the hierarchical vote collective of transformation-based ensembles for time series classification. In: 2016 IEEE 16th International Conference on Data Mining (ICDM), pp. 1041–1046. IEEE (2016)
12. Oza, P., Patel, V.M.: C2ae: Class conditioned auto-encoder for open-set recognition. In: Proceedings of the IEEE/CVF Conference on Computer Vision and Pattern Recognition, pp. 2307–2316 (2019)
13. Petitjean, F., Ketterlin, A., Gançarski, P.: A global averaging method for dynamic time warping, with applications to clustering. Pattern Recogn. **44**(3), 678–693 (2011)
14. Schäfer, P.: The boss is concerned with time series classification in the presence of noise. Data Min. Knowl. Disc. **29**(6), 1505–1530 (2015)
15. Scheirer, W.J., de Rezende Rocha, A., Sapkota, A., Boult, T.E.: Toward open set recognition. IEEE Trans. Pattern Anal. Mach. Intell. **35**(7), 1757–1772 (2012)
16. Sun, X., Yang, Z., Zhang, C., Ling, K.V., Peng, G.: Conditional gaussian distribution learning for open set recognition. In: Proceedings of the IEEE/CVF Conference on Computer Vision and Pattern Recognition, pp. 13480–13489 (2020)

[1] https://github.com/tolgaakar/Open-Set-Recognition-for-Time-Series-Classification.

17. Szegedy, C., Ioffe, S., Vanhoucke, V., Alemi, A.: Inception-v4, inception-ResNet and the impact of residual connections on learning. In: Proceedings of the AAAI Conference on Artificial Intelligence, vol. 31 (2017)
18. Yoshihashi, R., Shao, W., Kawakami, R., You, S., Iida, M., Naemura, T.: Classification-reconstruction learning for open-set recognition. In: Proceedings of the IEEE/CVF Conference on Computer Vision and Pattern Recognition, pp. 4016–4025 (2019)

Discretization Inspired Defence Algorithm Against Adversarial Attacks on Tabular Data

Jiahui Zhou[1], Nayyar Zaidi[2], Yishuo Zhang[2], and Gang Li[2(✉)]

[1] College of Computer Science, Xi'an Shiyou University, Shaanxi 710065, China
[2] School of I.T., Deakin University, Burwood, VIC 3216, Australia
{nayyar.zaidi,zhangyis,gang.li}@deakin.edu.au

Abstract. Deep learning methods are usually trained via a gradient-descent based procedure, which can be efficient as it is not only end-to-end but also suitable for large quantities of data. However, gradient-based learning is vulnerable to adversarial attacks – which account for unperceivable changes in the input data to misguide a trained model. Though a plethora of work explored the adversarial learning (attacks and defences) in image datasets, the exploration of adversarial learning in tabular datasets has seen little attention. In this work, we study adversarial learning in tabular datasets. We investigate the role of discretization and demonstrate that discretizing numeric attributes offers a strong defence mechanism. The main contribution of this work is the proposition of two new defence algorithms for numeric tabular datasets, that utilize cut-points obtained from discretization, to forge a defence against various forms of adversarial attacks. We evaluate the effectiveness of our proposed method on a wide range of machine learning datasets and demonstrate that the proposed algorithms lead to a state-of-the-art defence strategy on tabular datasets.

1 Introduction

At the heart of deep learning is a parametric model in the form of `Artificial Neural Network` (`ANN`), which is trained by optimizing a differentiable objective function. The error is propagated back through the network, and each weight of the model is updated in an iterative gradient-descent optimization manner. This end-to-end training process, as it is known, is efficient as it can process notably large quantities of data in a strictly online or in some batch processing manner. However, this gradient-based learning has a fundamental weakness – it opens the door to adversarial attacks. The idea behind adversarial learning is that any malicious entity, if, has access to model weights/parameters and can obtain the respective gradients, then it can modify the input in a way, such that the desired output can be obtained from the model [7]. E.g., for an input $\mathbf{x}$ to a given model

J. Zhou and N. Zaidi—Equal Contribution.

J. Gama et al. (Eds.): PAKDD 2022, LNAI 13281, pp. 367–379, 2022.
https://doi.org/10.1007/978-3-031-05936-0_29

$f(\mathbf{x})$, $\mathbf{r}$ is an adversarial noise if $f(\mathbf{x}+\mathbf{r}) \neq f(\mathbf{x})$, where $|\mathbf{r}| \leq \epsilon$. It can be seen that the only challenge for the attacker is that the noise ($\mathbf{r}$) should be unperceivable. Well, for high-resolution images, one can easily make unnoticeable changes in the input data to fool the model. Therefore, in computer vision, adversarial attacks are considered a serious threat, and a lot of research has focused on building effective defence mechanisms [15].

Tabular dataset has some characteristics that challenge the plain application of adversarial attacks. They can have categorical features, leading to values that are unique and distinct, that is `Y` or `N` or `High` and `Low`, etc. Note, these values are represented as whole numbers: $0, 1$ or 2, etc. – i.e., either encoded as bin numbers, or used in the one-hot-encoding format. Therefore, any changes to these values can easily be detected. Let us formalize adversarial attacks on tabular data in the following. We can denote the original unadulterated data as: $\mathcal{S}_{\text{data}} = [(\mathbf{x}^1, \mathbf{y}^1), \cdots, (\mathbf{x}^m, \mathbf{y}^m)]$, where $\mathbf{x} \in \mathcal{R}^n$ and $\mathbf{y} \in \mathcal{R}$. They are used to train the model: $f(\cdot)_{\boldsymbol{\theta}_s, \nabla_s}$. When an adversarial attack occurs, the adversarial sample $\mathcal{S}_{\text{adv}}$ is generated based on $f(\cdot)_{\boldsymbol{\theta}_s, \nabla_s}$, such that for each adversarial sample $\tilde{\mathbf{x}} \in \mathcal{S}_{\text{adv}}$, we have $f(\mathbf{x})_{\boldsymbol{\theta}_s, \nabla_s} = \mathbf{y} \neq \mathbf{f}(\mathbf{x} + \mathbf{r})_{\boldsymbol{\theta}_s, \nabla_s}$. The goal of the adversarial attack is defined as maximizing the following objective function: $\mathcal{L}_{\text{adv}}(f(\mathbf{x} + \mathbf{r})_{\boldsymbol{\theta}_s, \nabla_s}, \mathbf{y})$, where $|\mathbf{r}| \leq \epsilon$. Here, $\mathcal{L}_{\text{adv}}(f(\mathbf{x} + \mathbf{r})_{\boldsymbol{\theta}_s, \nabla_s}, \mathbf{y})$ is known as the adversarial risk on $\mathbf{x}$. We define adversarial risk over model $f(\cdot)$ as: $\mathbf{E}_{\mathbf{x} \in \mathcal{S}_{\text{data}}}(\mathcal{L}_{\text{adv}}(f(\mathbf{x} + \mathbf{r})_{\boldsymbol{\theta}_s, \nabla_s}, \mathbf{y}))$.

Let us suppose that every feature in our dataset is categorical. As discussed earlier, the final representation that we get for the features will only consist of well-round numbers. Any adversary aiming to choose a small value of ϵ, can easily be spotted[1]. Therefore, a simple strategy to defend against the adversarial samples for datasets with categorical data only is:

- Perform `Ceil` or `Floor` operations as advocated in [2]. E.g., if x_i is 3 and $\epsilon = 0.15$, the adversarial sample will have a value of 3.15, which will be converted back to 3 with `Floor` operation. Note, an adversary can also set $\epsilon = -0.15$, leading to $x_i = 2.85$. Now, if we perform the `Floor` operation, we obtain a value of 2. Note, if the value 2 is allowed, it is fine; otherwise, if the feature can only take values ≥ 3, the value 2 will violate the validity constraint of the feature and can be detected easily.

What if a tabular dataset has continuous features? In our previous example, if an adversarial sample has the values 3.15 or 2.85 – there is no way we can determine if it is not a legitimate value. And, therefore, adversarial attacks on numeric data can easily evade a manual inspection. Clearly, there is a need to determine whether a numeric value is adversarial or not. How about discretizing the feature and representing it as a categorical feature, or determining whether it is adversarial based on its distance from the discretization cut-point? These two questions will form the basis of our two proposed algorithms in this work. Generally, discretization is only employed to convert continuous features into categorical if a model can not handle continuous features. However, it has been

[1] This is one reason, why adversarial attacks against tabular data are not prevalent as compared to against image datasets.

shown recently [11], that discretization can lead to significantly better performance as well. In this paper, we show that it can be equally useful as a defence mechanism for adversarial attacks. Though related techniques such as quantisation have been used in adversarial defence on image datasets [2], their efficacy on tabular datasets is not well studied. Exploring what is the role, discretization can play in warding-off adversarial attack on tabular datasets has been the main motivation of this work.

In this work, we will devise defence strategies under two scenarios. The first scenario is where we are allowed to modify the input data while training the defence model. Here, we discretize the input continuous data – $\mathcal{S}_{\text{data}}$, and then adversarially train the model on this new discretized data – $\mathcal{S}_{\text{d-data}}$, as well as adversarially generated data – $\mathcal{S}_{\text{adv}}$. We demonstrate the efficacy of this strategy by formalizing it in form of our first proposed algorithm named `D2A3` – ***D****iscretized-based* ***D****efence* ***A****gainst* ***A****dversarial* ***A****ttacks*. The second scenario is where we are not allowed to modify the input features, and the input to the model has to be original continuous features. For this case, we believe, again discretization can offer an excellent defence mechanism, but rather implicitly. In this work, we have proposed a new defence algorithm named `D2A3N` – ***D****iscretized-based* ***D****efence* ***A****gainst* ***A****dversarial* ***A****ttacks with* ***N****umeric Input* – which leverages the cut-points (boundaries) definition obtained from discretization on original data and exploit the distance to cut-points to determine if a data point is adversarial or not. We summarize the main contributions of this work as follows:

- We highlight the importance of discretization as a defence mechanism for attacks on tabular datasets, and demonstrate that a simple discretization of continuous features can be very effective towards multiple forms of adversarial attacks.
- We propose two algorithms – `D2A3` and `D2A3N`, which utilize discretization to develop defence strategies for continuous features in tabular datasets. We evaluate the effectiveness of our proposed algorithms on a wide range of datasets, and against various forms of attacks.

The rest of this paper is organized as follows. We will discuss the related work in Sect. 2. Our proposed algorithms are presented and discussed in Sect. 3. We do an empirical evaluation of our proposed algorithms in Sect. 4, and conclude in Sect. 5, with some pointers to future works.

2 Background and Related Work

2.1 Tabular Data Adversarial Defence and Attack

There are three kinds of adversarial attacks that are common for tabular datasets. `LowProFool` [1] is a white-box attack method in the tabular domain for generating imperceptible adversarial examples. It is based on minimizing the addition of (imperceptible) adversarial noise on the features via the gradient descent approach. The gradients of the adversarial noise are used to guide the updates towards the opposite target class of the clean sample. At the same time, the penalty of the perturbation is set proportionally to the feature importance for confirming the minimal perceptibility of the adversarial noise. The benefit of the `LowProFool` is that the success rate is largely guaranteed, even the adversarial noise is imperceptible compared to many other white-box attack methods. It is the state-of-the-art white-box attack method on tabular data [1]. `DeepFool` [6] is another white-box attack method, and it works by adding adversarial noise to the clean sample by finding the distance between the sample and the model decision boundary. In `DeepFool`'s formulation, the smallest adversarial noise can be considered as the orthogonal projection between the sample and the class decision boundary (affine hyperplane). The advantage of the `DeepFool` compared to other classical gradient-based white box attacks, such as `FGSM`, is that the adversarial noise is more reliable and efficient as `DeepFool` always finds the generated adversarial sample close to the decision boundary and therefore the target class can be changed. The limitation of `DeepFool` is that the adversarial noise can be large when the sample is far away from the model decision boundary [1]. `FGSM` [3] is a classical white box attack method for image and typical numerical tabular datasets. The idea behind `FGSM` is quite simple and straightforward. It relies on adding the gradients into the original sample to create the adversarial sample.

Defence methods against tabular data adversarial attacks are still limited in the current literature. A commonly used defence method for adversarial attack on continuous data is `Madry` [5]. It leverages the adversarial training to minimize the adversarial risk of the model. `Trade` [12] is another commonly used defence method for continuous data which minimizes the regularized surrogate loss instead of directly training adversarially.

Finally, `Thermometer` [2] is another defence mechanism that relies on idea similar to discretization. The `Thermometer` discretization for tabular data on $\mathbf{x}_i$ with k cut-points can be expressed as: $t(\alpha = \phi_{\text{ew}}(f_{\text{scale}}(\mathbf{x}_i)))_l = 1, \quad l \geq \alpha$. The $f_{\text{scale}}(\cdot)$ is the min-max scaler to scale the value of $\mathbf{x}_i$ into the range $[0, 1]$. The $\phi_{\text{ew}}(f_{\text{scale}}(\mathbf{x}_i))$ is the quantization function that uses the `equal-frequency` to obtain bin α on k cut-points. The array $t(\alpha)_l$ has k dimensions and l is the l-th dimension of the array. It can be seen that the `Thermometer` discretization is similar to one-hot-encoding after `equal-width` discretization. However, in contrast to one-hot encoding, `Thermometer` discretization can ensure that the order remains the same after discretization.

2.2 Discretization as Defence and Discretization Methods

Discretization is a commonly used and well-studied technique in machine learning [13]. It is to convert a continuous feature into a set of discrete values, which is usually done by sorting the data and then identifying some cut-points (also known as boundaries), and placing the continuous data point based on which bin does it fall into [14]. Each bin is labelled a number in range: $\{0, 1, \ldots, k\}$, where k is the total number of bins.

A simple illustration of how discretization can lead to a defence is shown in Fig. 1, where one of the data points (in red) is maliciously tampered (Fig. 1a). It

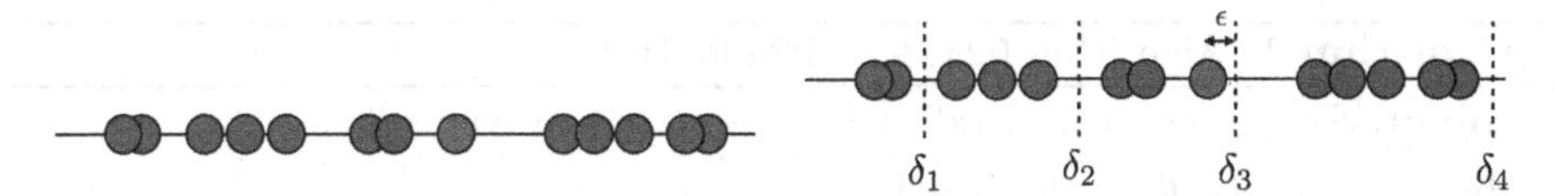

(a) Numeric feature, adversarial data in red.

(b) Cut-points after discretization, with adversarial sample closer to cut-points.

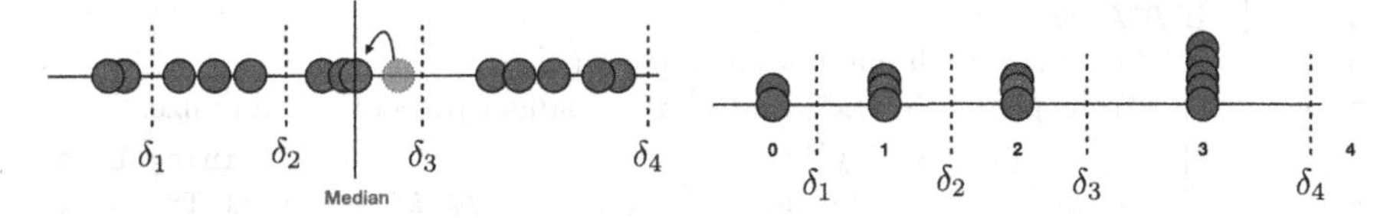

(c) Replacing the adversarial sample with the median of the bin.

(d) Data after discretization.

Fig. 1. Illustration of discretization as a defence against adversarial attack. (Color figure online)

can be seen that if we have the feature in a continuous format, there is no way for us to differentiate between these red data points and the others. However, let us suppose that we have obtained some boundaries $(\delta_1, \ldots, \delta_4)$ after running a discretization algorithm. We conjecture that the proximity of the data points to the boundaries can be an indication of an adversarial example. This can be seen in Fig. 1b, where this proximity is measured as a distance, ϵ. In practice, once the boundaries are identified, the data is discretized as the bin-number or represented with one-hot-encoding. In our example, the malicious red data will be assigned a value of 2, as shown in Fig. 1d. Now, if we just use bin-number (and train the original model) or do a one-hot-encoding (and train a modified model that takes in many more features as input) – the adversarial data is neutralized, which means that whatever the malicious intent of the attacker was, we have scaled it back to a value that our model expects (in our example, that is $\{0, 1, 2, 3\}$). Additionally, instead of using the bin-number, one can only replace the value of the adversarial data to the median of the bin, and keep all other data points the same – as depicted in Fig. 1c. This will help us in training a model that still takes as input the data in original format.

In Support of Discretization as Defence. We know that in `ANN` models, with all linear activation functions, the loss function tends to be linear with respect to the inputs as well. In such case, when the input $\mathbf{x}$ with the model $f(\mathbf{x}) = \sigma(\mathbf{w}^\top \mathbf{x})$ is under adversarial attack with $\mathbf{x} + \mathbf{r}$, we have:

$$f(\mathbf{x} + \mathbf{r}) = \sigma(\mathbf{w}^\top(\mathbf{x} + \mathbf{r})) = \sigma(\mathbf{w}^\top \mathbf{x} + \mathbf{w}^\top \mathbf{r})$$

It can be seen that $\mathbf{w}^\top \mathbf{r}$ determines the success of adversarial attack. Even though, non-linear functions are typically used in deep `ANN` models, such as `Relu`, they are only piece-wise linear. Much of the work in designing a defence against

Algorithm 1: Algorithm D2A3 and D2A3N Training

Input: $\mathcal{S}_{\text{data}} = [(\mathbf{x}^1, \mathbf{y}^1), ..., (\mathbf{x}^m, \mathbf{y}^m)]$, Discretization type – $\mathcal{C}$

```
1  Initial parameter θ_s of the model f(·)
2  Run discretization method C to obtain cut-points for each feature with Φ(·)
3  for iteration q ⊂ Q in training model f(·) do
4      for sample x^t in batch X ⊂ S_data and y^t do
5          if D2A3 then
6              Discretize with one-hot encoding Φ(x^t)
7              Train f(·) with Φ(x^t) and y^t via gradient descent to minimize:
                 L^{θ_s,∇_s}(f(Φ(x^t)), y^t) ;                    // Training Loss
8              Obtain adversarial sample x̃^t ∈ S_adv ;     // Adversarial Training
9              Discretize with one-hot encoding Φ(x̃^t)
10             Train f(·) with Φ(x̃^t) via gradient descent to minimize:
                 L^{θ_s,∇_s}(f(Φ(x̃^t)), y^t)
11         else
12             Train f(·) with x^t and y^t via gradient descent to minimize:
                 L^{θ_s,∇_s}(f(x^t), y^t)
13             Obtain adversarial sample x̃^t ∈ S_adv ;     // Adversarial Training
14             Obtain data transformation: M(Φ(x̃^t))
15             Train f(·) with M(Φ(x̃^t)) via gradient descent to minimize:
                 L^{θ_s,∇_s}(f(M(Φ(x̃^t))), y^t) ;      // Adversarial Training Loss
16 return f(·), Φ(·), M(·)
```

adversarial attacks focus on *how to break the linearity between inputs and the output?* [2]. Well, discretization followed by one-hot-encoding leads to a non-linear model, and we claim that it can break the linearity between the input and the output, and hence, can provide an effective defence mechanism against adversarial attacks.

3 Methodology

3.1 D2A3 – Model Discretization as Defence

Our proposed algorithm D2A3, relies on discretizing the continuous features to categorical features as input, and therefore, instead of using a model with continuous input, the discretized model is trained and used in D2A3. One can utilize any discretization method. The detailed pseudocode of D2A3 is given in Algorithm 1, where it takes as input the training data $\mathcal{S}_{\text{data}}$, as well as the discretization method $\mathcal{C}$ – `equal-frequency, equal-width, MDL`[2]. Of course, changing the input data format can be considered as a limitation. We will discuss D2A3N in the next section, which addresses this issue.

[2] Note, we have combined the two proposed algorithms into one due to space constraints.

Algorithm 2: Algorithm D2A3 and D2A3N Defence

Input: $\mathcal{S}_{\text{adv}}$, Algorithm 1 Output: $f(\cdot)$, $\Phi(\cdot)$

1 , $\mathcal{M}(\cdot)$ **while** In defence **do**
2 Load $f(\cdot)$
3 **for** *sample* $\tilde{\mathbf{x}} \sim \mathcal{S}_{adv}$ **do**
4 Discretize $\Phi(\tilde{\mathbf{x}})$
5 **if** *D2A3* **then**
6 One-hot encoding on $\Phi(\tilde{\mathbf{x}})$
7 **return** $f(\Phi(\tilde{\mathbf{x}}))_{\boldsymbol{\theta}_s,\nabla_s}$
8 **else**
9 Obtain data transformation: $\mathcal{M}(\Phi(\tilde{\mathbf{x}}^t))$
10 **return** $f(M(\Phi(\tilde{\mathbf{x}}^t))_{\boldsymbol{\theta}_s,\nabla_s}$

Let us discuss another salient feature of our D2A3, i.e., it relies on exploiting adversarial learning to enhance its defence capability, by optimizing the following objective function:

$$\mathcal{L}_{\text{D2A3}}^{\boldsymbol{\theta}_s,\nabla_s} = \arg\min_{\boldsymbol{\theta}_s,\nabla_s} \overbrace{\mathbf{E}_{\mathbf{x}^t \sim \mathcal{S}_{\text{data}}} \mathcal{L}^{\boldsymbol{\theta}_s,\nabla_s}(f(\Phi(\mathbf{x}^t)), \mathbf{y}^t)}^{\text{Training Loss}} + \overbrace{\mathbf{E}_{\tilde{\mathbf{x}}^t \sim \mathcal{S}_{\text{adv}}} \mathcal{L}^{\boldsymbol{\theta}_s,\nabla_s}(f(\Phi(\tilde{\mathbf{x}}^t)), \mathbf{y}^t)}^{\text{Adversarial Training Loss}}. \quad (1)$$

It can be seen that our proposed objective function is composed of two parts: $\min \mathcal{L}^{\boldsymbol{\theta}_s,\nabla_s}(f(\Phi(\mathbf{x}^t)), \mathbf{y}^t)$ and $\min \mathcal{L}^{\boldsymbol{\theta}_s,\nabla_s}(f(\Phi(\tilde{\mathbf{x}}^t), \mathbf{y}^t)$. The motivation for Equation 1 is to achieve better performance for minimizing the empirical loss and also add robustness to the model as recommended by the work of [4,9].[3] As we discussed earlier, minimization of training loss – $\mathcal{L}^{\boldsymbol{\theta}_s,\nabla_s}(f(\Phi(\mathbf{x}^t)), \mathbf{y}^t)$, can leverage the non-linearity of the discretization process and obtain a more accurate model $f(\cdot)$ [10]. However, only adding discretization into model $f(\cdot)$ is not enough as advocated in [4]. The adversarial training loss – $\min \mathcal{L}^{\boldsymbol{\theta}_s,\nabla_s}(f(\Phi(\tilde{\mathbf{x}}^t), \mathbf{y}^t)$, is further added to add robustness to the model. It is based on the optimization problem which is to minimize the adversarial training loss given by the inner attack type (i.e., maximizing the adversarial loss by finding the adversarial version of input) [5]. During the defence, the input samples $\tilde{\mathbf{x}}^t$ will be firstly discretized and format with one-hot encoding as $\Phi(\tilde{\mathbf{x}}^t)$. The outline of D2A3 in defence mode is given in Algorithm 2.

3.2 D2A3N – Input Discretization as Defence

Unlike D2A3, where the input of the model $f(\cdot)$ is discretized, in D2A3N, the input to the model is the same as the original data format (continuous or categorical). It still obtains the cut-point $\phi^\alpha \in \Phi(\cdot)$ for each feature, using discretization method $\mathcal{C}$ on original data $\mathcal{S}_{\text{data}}$. After discretization, we use $\Phi(\cdot)$ to build a

[3] The effectiveness of adversarial training loss component is also studied in ablation study in Sect. 4.4.

defence strategy. The strategy revolves around finding the closest data points to the cut-points. These data points are replaced with the median of the bin. This is achieved by implementing the transformation function: $\mathcal{M}(.)$ as:

$$\mathcal{M}(\tilde{\mathbf{x}}^t) = \begin{cases} \arg\min_{\alpha} \|\tilde{\mathbf{x}}^t - \mu(\Phi(\cdot))^{\alpha}\|, & |\tilde{\mathbf{x}}^t - \phi^{\alpha}| < \epsilon, \\ \tilde{\mathbf{x}}^t, & |\tilde{\mathbf{x}}^t - \phi^{\alpha}| \geq \epsilon. \end{cases} \tag{2}$$

Here, $\mu(\Phi(\cdot))^{\alpha}$ denotes the median value of bin α and ϕ_i^{α} is its cut-points. In addition, we maintain a constraint, i.e., the absolute value between the data and the corresponding cut-point has to be smaller than the tiny constant value ϵ (minimum threshold value to change the bin and fixed as constant during the defence). Additionally, just like D2A3, the adversarial training is also augmented here to add the robustness to the model to minimize the adversarial risk as:

$$\mathcal{L}_{\mathtt{D2A3N}}^{\boldsymbol{\theta}_s,\nabla_s} = \arg\min_{\boldsymbol{\theta}_s,\nabla_s} \overbrace{\mathbf{E}_{\mathbf{x}^t \sim \mathcal{S}_{\text{data}}} \mathcal{L}^{\boldsymbol{\theta}_s,\nabla_s}(f(\Phi(\mathbf{x}^t)), \mathbf{y}^t)}^{\text{Training Loss}} + \overbrace{\mathbf{E}_{\tilde{\mathbf{x}}^t \sim \mathcal{S}_{\text{adv}}} \mathcal{L}^{\boldsymbol{\theta}_s,\nabla_s}(f(\mathcal{M}(\Phi(\tilde{\mathbf{x}^t}))), \mathbf{y}^t)}^{\text{Adversarial Training Loss}} \tag{3}$$

When the D2A3N is used in defence (Algorithm 2), the input is discretized first and then processed by the transformation function $\mathcal{M}(\Phi(\tilde{\mathbf{x}^t}))$ to map it back to a numeric value.

4 Experiments

In this section, let us empirically verify the effectiveness of D2A3 and D2A3N. We will first implement white-box attacks to compare the robustness of both D2A3 and D2A3N with state-of-the-art methods. Later, we will conduct an ablation study to determine the effect of adding adversarial training to D2A3 and D2A3N.

4.1 Experimental Set-up

Datasets. We have used a total of 12 UCI classification datasets in our experiment. Note, all datasets are numeric in our experiments. Out of 12 datasets, 5 datasets have more than 10K samples and are denoted as Large, whereas 3 datasets have between $5-10$K samples and are denoted as Medium. The remaining 4 datasets have less than 5K samples and are denoted as Small. The statistics of the data are summarized in Sect. 4.1.

Adversarial Attack Setting and Evaluation Metric. We have made use of 3 commonly used white-box attack methods, which are common for tabular datasets, i.e., – FGSM, Deepfool and LowProfool. For each attack method, the architecture and the parameters of the target model are available, and the adversarial samples are directly generated. The step size of the FGSM is 0.1, and the maximum iteration for Deepfool and LowProFool is 50, which is the same as the default setting in their original implementation.

Table 1. Description of datasets.

Dataset	m	n	c	size	Dataset	m	n	c	size
SkinSegmentation	245057	4	2	Large	page-blocks	5473	11	5	Medium
connect-4	67557	43	3	Large	wall-following	5456	25	4	Medium
letter-recog	20000	17	26	Large	spambase	4601	58	2	Small
magic	19020	11	2	Large	diabetes	768	9	2	Small
sign	12546	9	3	Large	pid	768	9	2	Small
satellite	6435	37	6	Medium	credit-a	690	16	2	Small

Regarding the evaluation metrics, we have made use of **Accuracy** which generally determines the level of resistance of a defence mechanism against an adversarial attack. Notably, the **Standard Accuracy** for normal manner (no attack has occurred) and the **Robust Accuracy** for attack manner are used separately [9]. Moreover, the **Success Rate** of an adversarial attack is also used to measure the efficacy of the attack. According to the work [1], we define the **Success Rate** as:

$$\mathrm{SR} = \frac{\sum_{i=1}^{\mathcal{Z}} |f(\mathbf{x}_i + \mathbf{r}) \neq f(\mathbf{x}_i)|}{\mathcal{Z}} \tag{4}$$

which is a ratio of successfully crafted adversarial samples (e.g., the samples are able to fool the model) and the total targeted samples ($\mathcal{Z}$).

Experiment Configuration and Baselines. Each dataset is split into train and test via 3-fold cross-validation. The models in D2A3 and D2A3N are trained with 300 epochs and consist of a fully connected Artificial Neural Network (ANN) with Relu activation on 5 hidden layers and 1 Softmax for the output layer. We have used Thermometer encoding as a baseline for D2A3. Also, D2A3 is implemented with three commonly used discretization methods i.e., Equal-Width (D2A3-EW), Equal-Frequency (D2A3-EF), and the MDL (D2A3-MDL).

For D2A3N, as discussed in Sect. 3.2, we have used two commonly used defence methods as baselines – Madry [5] and Trades [12]. Again, just like D2A3. we have used three commonly used discretization methods for D2A3N, leading to D2A3N-EW, D2A3N-EF and D2A3N-MDL respectively.

Other than baselines for D2A3 and D2A3N, we have presented results with Clean model, which is the model without any defence mechanism. our code will be released in https://github.com/tulip-lab/open-code.

4.2 Model Discretization (D2A3) Results

Let us compare the performance of D2A3 method with baseline methods by considering the three commonly used white-box attack methods. Of course, these attack methods have the full access to the architecture and parameters of the model $f(\cdot)$, and can directly generate the adversarial samples depending on any

Table 2. Performance comparison for D2A3.

Models	Standard Accuracy (Avg)	Robust Accuracy (Avg)			Success Rate(SR) (Avg)		
		FGSM	DeepFool	LowProFool	FGSM	DeepFool	LowProFool
Clean	0.895	0.813	0.272	0.282	0.135	0.793	0.769
Thermometer	0.836	0.759	0.761	0.726	0.196	0.160	0.194
D2A3-EF	0.870	0.792	0.768	0.722	0.153	0.165	0.217
D2A3-EW	0.854	0.793	0.788	0.753	0.136	0.143	0.161
D2A3-MDL	**0.918**	**0.897**	**0.884**	**0.869**	**0.000**	**0.025**	**0.030**

attack method. This is one reason why the **robust accuracy** of the clean model can deteriorate easily.

Table 2 shows the averaged comparison results on all datasets between D2A3 and other baseline methods, including the Clean model. Our experiment suggests that FGSM as an attack method is not particularly effective for tabular datasets. On the other hand, DeepFool and LowProFool are quite effective attack methods which result in lowering the accuracy of the model quite significantly.

It can be seen that Thermometer method as a defence is quite effective against the three forms of attacks. However, the difference between **standard accuracy** and **robust accuracy** is quite large.

It can be seen that three variants of D2A3 lead to an effective defence mechanism against three forms of attacks. It is important to note that since we are discretizing differently, hence the quality of model in terms of Standard Accuracy is different. As expected, discretization based on mdl leads to the best results in terms of **standard accuracy** (91.8%). Note, the more accurate the trained model is, the higher the quality of the adversarial samples it can generate. Now, it is encouraging to see that D2A3-MDL not only achieve a higher performance on **robust accuracy** but also has the smallest difference between **standard accuracy** and **robust accuracy**. A similar pattern can be seen in terms of the **success rate**, where the success rates of the three attacks are 0%, 2.5% and 3% respectively. These results are extremely encouraging, as they demonstrate that the model discretization algorithm D2A3 is an effective defence method against white-box attacks for tabular datasets. In Sect. 4.4, we will discuss the advantage of incorporating adversarial training loss within D2A3.

4.3 Input Discretization (D2A3N) Results

It can be seen from Table 3 that the overall performances of D2A3N trained with three forms of discretization – D2A3N-EF, D2A3N-EW, and D2A3N-MDL is better than the two standard baselines namely Madry and Trades. It is encouraging to see that without changing the dimensionality of the model $f(\cdot)$, D2A3N can largely help to resist the 3 kinds of adversarial attack. Particularly, the D2A3N-EW has shown higher **robust accuracy** and lower **success rate** compared to D2A3N-EF and D2A3N-MDL.

Table 3. Performance comparison for D2A3N.

Models	Standard Accuracy (Avg)	Robust Accuracy (Avg)			Success Rate(SR) (Avg)		
		FGSM	DeepFool	LowProFool	FGSM	DeepFool	LowProFool
Clean	**0.895**	0.813	0.272	0.282	0.135	0.793	0.769
Madry	0.892	0.808	0.459	0.496	0.129	0.553	0.500
Trades	0.895	0.817	0.456	0.307	0.127	0.549	0.739
D2A3N-EW	0.881	0.839	**0.683**	**0.536**	**0.103**	**0.291**	**0.458**
D2A3N-EF	0.871	0.807	0.558	0.532	0.164	0.449	0.487
D2A3N-MDL	0.874	**0.860**	0.588	0.488	0.138	0.416	0.514

Table 4. Ablation study on adversarial training.

Model		Robust Accuracy (Avg)			Success Rate(SR) (Avg)		
		FGSM	DeepFool	LowProFool	FGSM	DeepFool	LowProFool
D2A3	Adv. Train	**0.897**	**0.884**	**0.869**	**0.000**	**0.025**	**0.030**
	With Out Adv. Train	0.862	0.883	0.847	0.039	0.025	0.064
D2A3N	Adv. Train	**0.835**	**0.609**	**0.518**	**0.115**	**0.385**	**0.486**
	With Out Adv. Train	0.753	0.579	0.495	0.218	0.429	0.519

The better performance of D2A3N-EW discretization is surprising but can be explained. It is well known that EW discretization is more robust to the skewness of the data [8]. Since adversarial training leads to the original data being skewed, EW discretization based D2A3N is more robust to the adversarial attack.

4.4 Ablation Study on Adversarial Training

In this section, we conduct an ablation study to verify the effectiveness of the adversarial training step.

In Table 4, both D2A3 and D2A3N are tested to obtain the **robust accuracy** and **success rate** with and without the adversarial training, denoted as Adv.Train and With Out Adv. Train. It can be seen that without the adversarial training, both D2A3 and D2A3N leads to a slightly worse **robust accuracy** and **success rate** than with the adversarial training. This highlights the importance and necessity of the adversarial training within both D2A3 and D2A3N. It is also interesting to see that even without adversarial training, the performance of D2A3 and D2A3N is mostly better than the corresponding baselines in Tables 2 and 3.

5 Conclusion

In this paper, we studied the role of discretization in devising a defence strategy against adversarial attacks on tabular datasets. We showed that not only discretization can be effective, but it can also lead to better performance than existing baselines. We proposed two algorithms namely `D2A3` and `D2A3N`, which leverages the cut-points obtained from discretization to devise a defence strategy. We evaluated the effectiveness of our proposed methods on 12 standard datasets and compared them against standard baselines of `Thermometer`, `Madry` and `Trades`. The effectiveness of `D2A3` and `D2A3N` clearly demonstrate the importance of discretization in warding-off adversarial attacks on tabular datasets. As future work, we are keen to explore theoretical justification over why `mdl` discretization is effective for `D2A3` and why `EW` is better than the other two forms of discretization for `D2A3N`. We are keen to extend our proposed methods to image datasets as well.

Acknowledgement. This research is partially supported by the National Natural Science Fund of China (Project No. 71871090).

References

1. Ballet, V., Renard, X., Aigrain, J., Laugel, T., Frossard, P., Detyniecki, M.: Imperceptible adversarial attacks on tabular data. arXiv preprint arXiv:1911.03274 (2019)
2. Buckman, J., Roy, A., Raffel, C., Goodfellow, I.: Thermometer encoding: one hot way to resist adversarial examples. In: International Conference on Learning Representations (2018)
3. Goodfellow, I.J., Shlens, J., Szegedy, C.: Explaining and harnessing adversarial examples. arXiv preprint arXiv:1412.6572 (2014)
4. Kurakin, A., Goodfellow, I., Bengio, S.: Adversarial machine learning at scale. arXiv preprint arXiv:1611.01236 (2016)
5. Madry, A., Makelov, A., Schmidt, L., Tsipras, D., Vladu, A.: Towards deep learning models resistant to adversarial attacks. arXiv preprint arXiv:1706.06083 (2017)
6. Moosavi-Dezfooli, S.M., Fawzi, A., Frossard, P.: DeepFool: a simple and accurate method to fool deep neural networks. In: CVPR (2016)
7. Qiu, S., Liu, Q., Zhou, S., Wu, C.: Review of artificial intelligence adversarial attack and defense technologies. Appl. Sci. **9**(5), 909 (2019)
8. Sulewski, P.: Equal-bin-width histogram versus equal-bin-count histogram. J. Appl. Stat. **48**(12), 2092–2111 (2021)
9. Yang, S., Guo, T., Wang, Y., Xu, C.: Adversarial robustness through disentangled representations. In: Proceedings of the AAAI Conference on Artificial Intelligence, vol. 35, pp. 3145–3153 (2021)
10. Yang, Y., Webb, G.I.: Discretization for Naive-Bayes learning: managing discretization bias and variance. Mach. Learn. **74**(1), 39–74 (2009)
11. Zaidi, N.A., Du, Y., Webb, G.I.: On the effectiveness of discretizing quantitative attributes in linear classifiers. IEEE Access **8**, 198856–198871 (2020). https://doi.org/10.1109/ACCESS.2020.3034955
12. Zhang, H., Yu, Y., Jiao, J., Xing, E., El Ghaoui, L., Jordan, M.: Theoretically principled trade-off between robustness and accuracy. In: ICML (2019)
13. Zhang, Y., Zaidi, N.A., Zhou, J., Li, G.: GANBLR: a tabular data generation model. In: 2021 IEEE International Conference on Data Mining (ICDM), pp. 181–190. IEEE (2021)

14. Zhang, Y., Zaidi, N.A., Zhou, J., Li, G.: GANBLR++: incorporating capacity to generate numeric attributes and leveraging unrestricted Bayesian networks. In: Proceedings of the 2022 SIAM International Conference on Data Mining (2022)
15. Zhou, M., Wu, J., Liu, Y., Liu, S., Zhu, C.: DAST: data-free substitute training for adversarial attacks. In: Proceedings of the IEEE/CVF Conference on Computer Vision and Pattern Recognition, pp. 234–243 (2020)

Safe Offline Reinforcement Learning Through Hierarchical Policies

Shaofan Liu and Shiliang Sun(✉)

School of Computer Science and Technology, East China Normal University, 3663 North Zhongshan Road, Shanghai 200062, People's Republic of China
51194506021@stu.ecnu.edu.cn, slsun@cs.ecnu.edu.cn

Abstract. Recently, offline reinforcement learning has gained increasing attention. However, the safety of offline reinforcement learning has been ignored. It poses a significant challenge to learn a safe and high-performance policy from a fixed dataset that contains unsafe or unexpected state-action pairs without interacting with the environment. Since the unsafe state-action pairs are usually sparse in the behavior data collected by humans, it is difficult to effectively model information about unsafe behaviors. This paper utilized the hierarchical reinforcement learning framework to alleviate the sparsity issue by modeling unsafe behaviors with hierarchical policies. Specifically, a high-level policy determines a prospective state, and a low-level policy takes action to reach the specified goal state. The training objective of the high-level policy is to improve the expected reward that the low-level policy collects when it moves toward the goal state and reduce the number of unsafe actions. We further develop data processing methods to provide training data for the high-level policy and the low-level policy. Evaluation experiments about performance and safety are conducted in simulation environments that return the rewards and unsafe costs obtained by agents during the interaction. Experimental results demonstrate that the proposed algorithm can choose safe actions while maintaining high performance.

Keywords: Safe reinforcement learning · Offline training · Hierarchical policies

1 Introduction

Deep Reinforcement Learning (DRL) has made significant progress on a series of complex control tasks [16]. DRL algorithms interact with an online environment or simulator and learn from their own collected experience [2]. However, collecting data online is difficult, risky, or costly in many real-world applications such as automatic driving, healthcare, and recommendation systems [15]. Offline Reinforcement Learning (Offline RL) is a promising method for learning a practical decision-making policy from a fixed historical dataset without direct interactions with the environment [14]. Thus, offline RL has excellent potential to play a role in the application scenarios mentioned above.

Recently, several offline RL algorithms have been proposed [2,7,9,10] which achieve competitive performance against online DRL baselines [2]. However, in real-world applications, historical behavior data collected by human operators or other

J. Gama et al. (Eds.): PAKDD 2022, LNAI 13281, pp. 380–391, 2022.
https://doi.org/10.1007/978-3-031-05936-0_30

behavioral policies may contain some unsafe or unexpected state-action pairs. For instance, drivers occasionally violate some trivial traffic rules when driving. Existing offline RL methods will not partially sacrifice the performance to improve decision-making safety when the dataset includes unsafe behavior.

A direct method to prevent the agent from taking unsafe actions or entering dangerous states is utilizing the value-based off-policy RL algorithm and setting negative rewards for unsafe actions [6,21]. Consequently, the agent will automatically learn to avoid those unsafe actions with negative rewards to maximize the value function. However, the trained agents will face the problem of "sparse rewards" in the offline setting since there are normally few unsafe state-action pairs in the dataset, and collecting dangerous behavior data is often expensive. Therefore, efficient utilization of hazardous behavior information in the offline setting becomes particularly meaningful.

In recent years, hierarchical RL has gained popularity in designing RL algorithms that converge in complex or potentially sparsely rewarded environments [18,19,24]. Drawing inspiration from the recent hierarchical reinforcement learning (hierarchical RL) literature, we propose an offline hierarchical RL framework to solve the problem of the scarcity of unsafe or unexpected state-action pairs. In our proposed method, the high-level policy sets sub-goals for the low-level policy while the low-level policy is responsible for reaching the sub-goals set by the high-level policy. A decision-making process of the high-level policy includes multiple interactions with the environment, and the sparse reward signal of each interaction is accumulated as the reward of the high-level policy. The input of high-level policy is a sequence of contiguous states. For each unsafe state-action pair, we will build multiple sequences containing it by adjusting the starting point of the sequence. In this way, the high-level policy can more effectively model the information of unsafe behaviors.

Our key contributions are summarized as follows: (1) We propose an offline RL algorithm that can learn to avoid the unsafe behaviors in the dataset. To effectively model the information of unsafe state-action pairs, we utilize hierarchical policies to predict the long-term safety risk of a state-action pair; (2) An adaptive weight is used to regulate the training objectives of safety and performance to achieve a better trade-off. (3) A data pre-processing method is proposed to obtain the datasets for the proposed offline hierarchical policies; (4) Extensive experiments show the effectiveness of the proposed method. We show that the proposed method achieves a competitive trade-off between safety and performance compared with the online safe RL algorithm. Ablation experiments demonstrate the effectiveness of various design choices in our method.

The rest of this paper is organized as follows. In Sect. 2, we first introduce the necessary background and core ideas of hierarchical RL. We present the proposed method in detail in Sect. 3. Experimental results are given in Sect. 4. Finally, we conclude this paper in Sect. 5.

2 Preliminaries

2.1 Reinforcement Learning

In RL, the agent interacts with the environment and receives the reward signal. Through careful exploration and trial and error, the agent can finally learn a decision-making policy that can maximize the received rewards. The interaction process is typically formulated by a Markov decision process (MDP) (S, A, R, T, γ), where $s \in S$ is the state space, $a \in A$ is the action space, $r \in R$ is the reward, $T(s_{t+1}|s_t, a_t)$ is the conditional probability distribution that describes the transition dynamics of the environment, and γ is a scalar discount factor. The essential goal of RL is to find an optimal policy $\pi(s_t)$ to maximize the expectation of discounted rewards $V_\pi(s_t) = \mathbf{E}_\pi[\sum_{i=t}^{\infty} \gamma^{i-t} r_i]$. For a given policy $\pi(s_t)$, we define the state-action value function as $Q_\pi(s_t, a_t) = \mathbf{E}_\pi[V(s_t)|s_t, a_t]$. Thus the greedy optimal policy $\pi^*(s_t) = \mathbf{argmax}_a Q(s_t, a_t)$. Given transition tuples (s_t, a_t, r_t, s_{t+1}) collected from interactions with the environment, the training objective is to minimize the Bellman error which can be expressed as the following:

$$(Q_\theta(s_t, a_t) - (r_t + \gamma \cdot Q_\theta(s_{t+1}, \pi^*(s_{t+1}))))^2 . \tag{1}$$

Offline Reinforcement Learning. In the offline RL setting, the agent can not interact with the environment to collect data. The goal of offline RL is to train the agent to maximize the cumulative discounted rewards with historical data from human operators or logs [2,10]. Off-policy RL algorithms learn from a data buffer that consists of transition tuples collected during training. A simple offline RL method is inserting the existing dataset into the buffer of the off-policy algorithms such as deep Q-learning [7,9]. The difference between offline RL and off-policy algorithms lies in the incapable of the offline RL to use the current policy to collect additional data during training.

Hierarchical Reinforcement Learning. Hierarchical RL has gained popularity in recent years in designing RL algorithms that converge in complex environments [4]. Hierarchical RL is proposed to solve the complex control tasks with sparse reward. In hierarchical RL, the high-level policy chooses the most suitable candidate from several low-level policies [3,22] or sets sub-goals for a specific low-level policy [12,19,24]. Goal-conditioned hierarchical designs have emerged as a practical paradigm for hierarchical RL [18,19]. The low-level policy receives intrinsic rewards from the high-level policy in every time step. The high-level policy makes decisions every c time step, and the reward of high-level policy comes from a potentially sparsely rewarded environment. The problem of sparse reward can be mitigated since a reward of high-level policy is gathered among several time steps. In our methods, we utilize the idea of hierarchical policies to alleviate the sparsity of unsafe state-action pairs.

3 The Proposed Method

In this section, we first introduce our data pre-processing method. Then we describe our hierarchical training framework in detail. The pseudo-code for the algorithm is outlined in Algorithm 1.

3.1 Data Pre-processing of Hierarchical Policies

Here we first define the form of state, action, and reward of the high-level and low-level policies, for a behavior trajectory $(t_1, t_2, t_3, ..., t_n)$ where $t_n = (s_n, a_n, r_n)$, we use $(s_n^h, a_n^h, r_n^h, s_{n+1}^h)$ and $(s_n^l, a_n^l, r_n^l, s_{n+1}^l)$ to represent the transition tuples of the high-level and low-level policies.

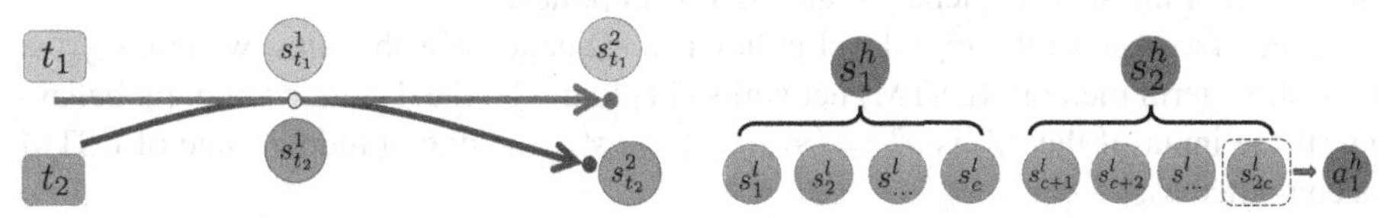

Fig. 1. t_1 and t_2 are two continuous trajectories. They go through the same state ($s_{t1}^1 = s_{t1}^1$), but they have different subsequent states ($s_{t2}^2 \neq s_{t2}^2$).

Fig. 2. The demonstration of the high-level policy state-action pair.

The high-level policy needs to predict a future goal according to the current state. In this paper, the goal means an expected future state. However, the state of one time step may have different subsequent states (see Fig. 1). Thus, we employ behavior sequences as states of the high-level policy: $s_n^h = (s_{cn+1}, s_{cn+2}, ..., s_{cn+c})$, where c is a hyper-parameter representing the length of the sequence and the next state $s_{n+1}^h = (s_{c(n+1)+1}, s_{c(n+1)+2}, ..., s_{c(n+1)+c})$. The final state s_{cn+c} in the sequence is the sub-goal of the other states, which is also the action of high-level policy. The reward of the high-level policy is the total reward of the sequence: $r_n^h = \Sigma_{i=1}^c r_{cn+i}$. To distinguish unsafe behaviors in the dataset, we use a binary variable **cost** to represent whether a transition tuple contains an unexpected state-action pair. That is, we use $\mathbf{cost}^h = \Sigma_{i=1}^c \mathbf{cost}_{cn+i}$ to represent the number of unsafe actions or unexpected state-action pairs contained in the sequence. We demonstrate the data composition of the high-level policy state-action pair in Fig. 2.

For the low-level policy, $s_n^l = [s_n, g_n]$ where the sub-goal g_n is the state of the future. For instance, in Fig. 2, the low-level state $s_1^l, s_2^l, ..., s_c^l$ all have the sub-goal of s_c^l. The low-level policy interacts directly with the environment, and the action of the low-level policy is the same as those in the original behavior data: $a_n^l = a_n$. The low-level policy aims to achieve the given goal state, which is different from the goal expected by the environment. Therefore, the reward from environmental feedback cannot be directly used as the low-level policy's reward. The reward of low-level policy $r_n^l = \mathbf{sim}(s_n, g_n)$, where **sim** measures the similarity between the two states. Here we choose the negative Euclidean distance as our similarity measure. r_n^l is called the intrinsic reward in the literature of hierarchical RL, [24] which guides the low-level policy to achieve the given sub-goal.

3.2 Safe Offline Reinforcement Learning Through Hierarchical Policies

We utilize hierarchical policies to model the information of sparse unsafe state-action pairs. Specifically, the high-level policy π^h predicts a target state as a sub-goal for the low-level policy, and the low-level policy π^l takes actions to reach the given sub-goal.

High-Level Policy. The design of the high-level policy is critical to safety and performance. We use the actor-critic framework of TD3 to build our high-level policy. The actor $\pi^h_{\theta_2}(s^h)$ outputs a sub-goal for the low-level policy every c time steps where c is the length of the state sequence of the high-level policy.

Since the state of the high-level policy is a sequence of sub-states, we employ the long short-term memory (LSTM) networks [11] to model the dynamic temporal behavior. If the input of the $\pi^h_{\theta_2}$ is $s^h = (s^l_1, ..., s^l_t, ..., s^l_c)$, the output hidden state of LSTM is computed as:

$$\begin{aligned} i_{(t)} &= \sigma\left(W_{is}s^l_{(t)} + W_{ih}h_{(t-1)} + W_{ic}c_{(t-1)}\right) \\ f_{(t)} &= \sigma\left(W_{fs}s^l_{(t)} + W_{fh}h_{(t-1)} + W_{fc}c_{(t-1)}\right) \\ c_{(t)} &= f_{(t)} \odot c_{(t-1)} + i_{(t)} \odot \tanh\left(W_{cs}s^l_{(t)} + W_{ch}h_{(t-1)}\right) \\ o_{(t)} &= \sigma\left(W_{os}s^l_{(t)} + W_{oh}h_{(t-1)} + W_{oc}c_{(t-1)}\right) \\ h_{(t)} &= o_{(t)} \odot \tanh\left(c_{(t)}\right), \end{aligned} \tag{2}$$

where W_* represents the network parameters, $\odot$ is the element-wise product, σ is the sigmoid function, $i_{(t)}$ is the input gate, $f_{(t)}$ is the forget gate, $c_{(t)}$ is the memory cell, and $o_{(t)}$ is the output gate. In the subsequent calculation process, we only use the last hidden state $h_{(c)}$.

For the critic network, $h_{(c)}$ is concentrated with a^h and be fed to a linear layer f_{q^h} to get $Q^h_{\phi_2}(s^h, a^h)$. To make the agent learn the safe decision-making policy, we further define the parameterized cost function $C^h_{\phi_3}(s^h, a^h)$ to estimate the cumulative discounted cost. Thus, a safe and efficient agent should take actions that can get a high $Q^h_{\phi_2}$ value and a low $C^h_{\phi_3}$ value. The action with maximum $Q^h_{\phi_2} - C^h_{\phi_3}$ value will be chosen in the decision-making stage, according to the greedy strategy. Since the tasks of predicting $Q^h_{\phi_2}$ and $C^h_{\phi_3}$ are similar, we formulate them as a multi-task learning task and utilize the shared-bottom model [17] to reduce redundant model parameters. Specifically, the expert networks of $Q^h_{\phi_2}$ and $C^h_{\phi_3}$ share the bottom of $[h_{(c)}, a^h]$.

In order to prevent the actor from being dominated by the value of $Q^h_{\phi_2}$ or $C^h_{\phi_3}$, we add an adaptive weight λ to $C^h_{\phi_3}$. When the $Q^h_{\phi_2}$ value is small and the $C^h_{\phi_3}$ value is large, we can set a large λ to punish this situation. Conversely, if $Q^h_{\phi_2}$ value is large, we set a relatively small λ to reduce the penalty of the $C^h_{\phi_3}$ value. Formally, λ is defined as:

$$\lambda = \begin{cases} \frac{1}{Q^h_{\phi_2}}, & Q^h_{\phi_2} > 0 \\ 1 \quad, & Q^h_{\phi_2} \leq 0 \end{cases}, \tag{3}$$

where we choose the action with the largest weighted value $Q^h_{\phi_2}(s^h, a^h) - \lambda \cdot C^h_{\phi_3}(s^h, a^h)$ as the final action.

We define $Q^h_{\phi_2,\phi_3} = Q^h_{\phi_2}(s^h, a^h) - \lambda \cdot C^h_{\phi_3}(s^h, a^h)$, and $r^\lambda = r^h - \lambda \cdot c^h$. During the training stage, $\pi^h_{\theta_2}$ is trained to maximize the estimated value of $Q^h_{\phi_2,\phi_3}$ with ϕ_2 and ϕ_3 fixed:

$$\min_{\theta_2} - Q^h_{\phi_2,\phi_3}(s^h, \pi^h_{\theta_2}(s^h)), \tag{4}$$

Algorithm 1. Hierarchical Safe Offline Reinforcement Learning

Input: Collected trajectory behavior data D.
Parameter: Parameters of actor and critic in high-level policy and low-level policy: θ^h_{actor}, θ^h_{critic}, θ^l_{actor}, θ^l_{critic}.
Output: Trained parameters.

1: pre-processing D with methods mentioned in Sect. 3.1 and get D_l and D_h.
2: Train high-level policy with D_h.
3: **while** no convergence **do**
4: Update θ^h_{actor} and θ^h_{critic} alternately to minimize actor loss Eq. 4 and critic loss Eq. 5.
5: **end while**
6: Use trained high-level policy to generate and replace half of sub-goals in the D_l.
7: **while** no convergence **do**
8: Update θ^l_{actor} and θ^l_{critic} to minimize actor loss and critic loss of the low-policy.
9: **end while**
10: **return** θ^l_{actor}, θ^l_{critic}, θ^h_{actor}, θ^h_{critic}.

and $Q^h_{\phi_2,\phi_3}$ is trained through temporal difference learning:

$$\min_{\phi_2,\phi_3} \left(Q^h_{\phi_2,\phi_3}(s^h_t, a^h_t) - (r^\lambda_t + \gamma \cdot Q^h_{\phi_2,\phi_3}(s^h_{t+1}, \pi^h_{\theta_2}(s^h_{t+1}))\right)^2. \tag{5}$$

Low-Level Policy. The low-level policy is built upon the continuous control algorithm of TD3 [8]. Compared with recent off-policy algorithms that aim to alleviate the distributional shift, TD3 has stable and powerful performance although it is not tailored for offline RL [2]. The actor $\pi^l_{\theta_1}$ and critic $Q^l_{\phi_1}$ are all building with deep neural networks. The actor $\pi^l_{\theta_1}(s^l_t)$ maps the low-level state s^l_t to a specific action a^l_t, and the critic $Q^l_{\phi_1}(s^l_t, a^l_t)$ estimates the cumulative discounted reward of the state-action pair (s^l_t, a^l_t). It is noted that the sub-goal is included in the low-level policy state. During the training stage, $\pi^l_{\theta_1}$ maximizes the estimated value of $Q^l_{\phi_1}(s^l_t, \pi^l_{\theta_1}(s^l_t))$ with ϕ_1 fixed. $Q^l_{\phi_1}(s^l, a^l)$ is updated through temporal difference learning like Eq. 5.

3.3 Parameter Combination of Hierarchical Policies

Since our high-level policy and low-level policy are trained separately with different datasets, we will save the network parameters of different training rounds during the training process. After the training, we need to test each combination of the parameters of the high-level policy and the low-level policy obtained to determine the best parameter combination.

The effects of two different parameter combinations are demonstrated in Fig. 3. They have the same high-level policy and different low-level policies. Although the final states of the two low-level policies have the same distance d_3 from the sub-goal given by the high-level policy, the distance between the final state of the two low-level policies and the optimal state may be very different ($d_2 > d_1$). However, it is time-consuming to evaluate each parameter combination. Therefore, we use a more

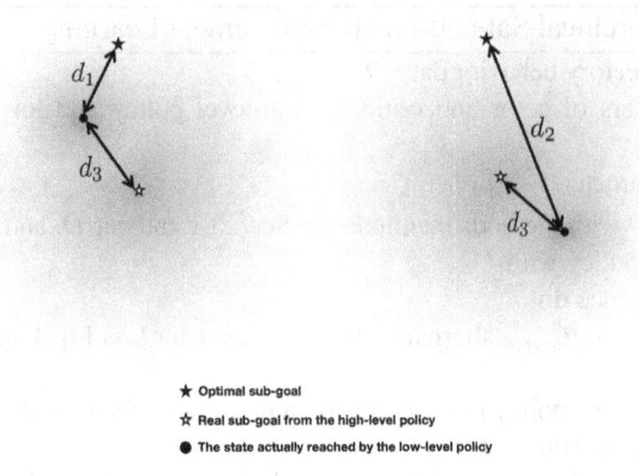

Fig. 3. The effects of two different parameter combinations. We assume that the distributions involved all obey Gaussian distribution, yellow region represents the distribution of the sub-goal given by the high-level policy, green region represents the distribution of the state reached after c steps of the low-level policy. d_* represents the distance between the two states. (Color figure online)

straightforward method that uses the trained high-level policy to meddle the low-level data to alleviate the mismatch between hierarchical policies.

Specifically, we first train the high-level policy until the loss function converges and the Q^h tends to be stable. Then, we use the learned high-level policy to generate and replace part of the sub-goals in the low-level policy data. The proportion of replaced sub-goals is a hyper-parameter. This method can enhance the robustness of the low-level policy to the distribution mismatch between the optimal sub-goal and the predicted sub-goal from the high-level policy. We call this method hierarchical policies adaptation (HPA). To assess the effectiveness of HPA, we conduct ablation experiments in Sect. 4.3.

4 Experiments

4.1 Experiment Settings

Safety Gym environments is a set of tools for facilitating research about safe RL [20], which uses the OpenAI Gym [5] interface and MuJoCo [23] physics simulator to construct environments with extensive layout randomization. In the environment of Safety Gym, there exists a goal and some traps. An agent interacts with a Safety Gym environment and receives the value of reward and cost. The reward represents the quality of the agent's action, and the cost represents whether the agent has entered a trap. Safety Gym can select different components to build the environment. Users only need to formulate the robot, task, and difficulty level. In this paper, we choose the robots of Point and Car, tasks of Goal and Button with the difficulty of 0,1 to construct our environments. Examples of these environments and other implementation details are introduced in the Appendix.

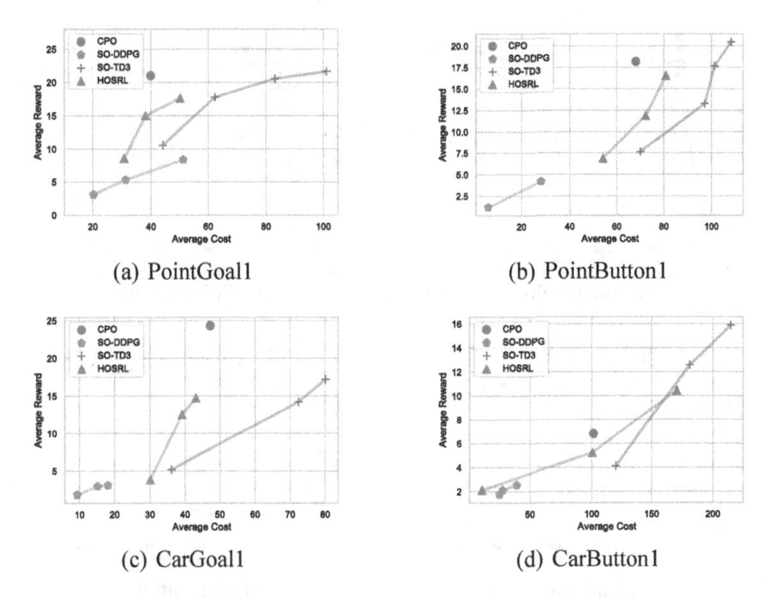

(a) PointGoal1 (b) PointButton1

(c) CarGoal1 (d) CarButton1

Fig. 4. Reward-cost trade-offs for CPO, SO-DDPG, SO-TD3 and the proposed HSORL. Upper-left corner (high reward, low cost) is preferable.

Our experiments use datasets collected by a partially-trained DDPG [16]. However, the goal in Safe Gym only appears randomly in a small area, and thus the trajectory of a trained agent will also always appear in this small area. Therefore, it is difficult to generalize the model to the region far away from the goal with offline training. Thus, we add the Gaussian noise to the goal's position when the behavior policy collects the data. We do not make any additional modifications to the environments.

4.2 Comparative Analysis

Experiments of Safety. In this subsection, we construct experiments to evaluate the safety of the proposed method in environments of difficulty 1. When the agent hits an obstacle or enters a trap, the environment will return an unsafety cost 1. It is expected that the agent can get fewer costs and more rewards from the environment.

To demonstrate the effectiveness of the proposed hierarchical training framework for modeling information of unsafe state-action pairs, we compared HSORL with the safe offline RL methods without hierarchical policies. To be specific, we add the cost function to DDPG and TD3 and train their actor and critic models following Eqs. 4 and 5. In this way, we get safe offline DDPG (SO-DDPG) and safe offline TD3 (SO-TD3) as our baselines. We choose the same hyper-parameters with experiments in Sect. 4.2. All the cost functions have the same network structure as the Q-function. We also consider the online safe exploration algorithms of CPO [1] which can be used as a dominant baseline [2].

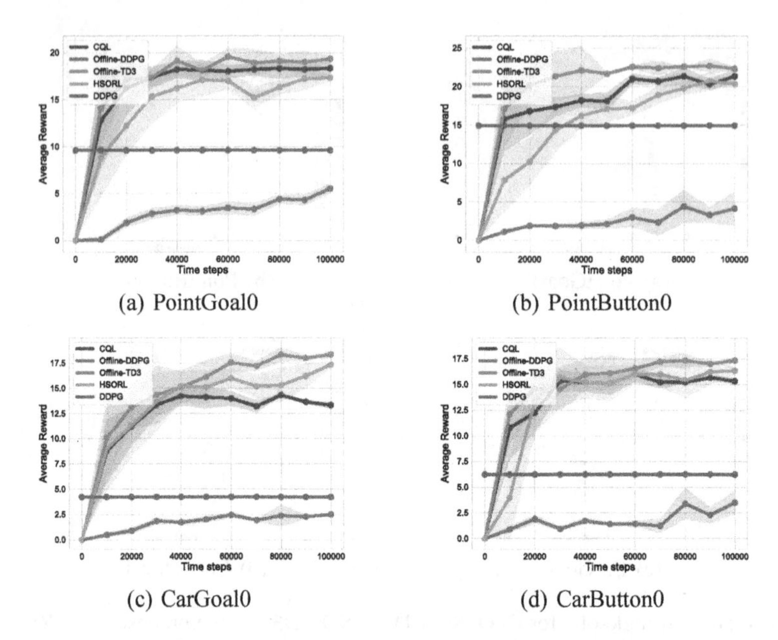

(a) PointGoal0 (b) PointButton0

(c) CarGoal0 (d) CarButton0

Fig. 5. The horizontal axis represents the number of training batches and the vertical axis represents the average value received from the environment. For HSORL, the horizontal axis denotes the low-level policy training batches.

In Fig. 4, we demonstrate the average values of rewards and costs that each algorithm obtained from the environments. As expected, CPO achieves the best trade-off between rewards and costs in most environments. However, CPO must interact with the environment to learn the safe decision-making policy, which is not feasible in safety-critical applications. Our proposed offline method can achieve a similar cost in every environment with a slightly lower reward compared with CPO. Compared with SO-TD3, HSORL can obtain more rewards when the cost is the same as CPOs, which validates that HSORL is capable of effectively modeling information of unsafe behaviors. Since SO-DDPG can not learn a practical policy in the environments of difficulty 1, the trained agent can only do some meaningless actions near the initial state. Thus both the reward and the cost of SO-DDPG are very few. The empirical results prove that the proposed hierarchical offline training framework can achieve an appropriate performance while effectively avoiding unsafe actions when learning from a fixed dataset that contains unsafe behavior.

Experiments of Performance. Although the proposed method mainly focuses on the safety of offline RL, it still shows strong competitiveness in maintaining qualified performance. To verify the performance of the proposed method, we choose baselines of continuous control agents of offline DDPG, and offline TD3, which proved to be comparable to algorithms designed specifically to learn from offline data [2], and recently proposed SOTA offline RL algorithms of CQL [13]. We build actor-critic CQL based on the details given in paper [13]. To demonstrate the performance of the proposed

Table 1. Ablation experiment results.

Method	Average cost	Average reward
SO	107.21	20.09
HSO	96.37	18.86
HSO+AW	62.19	17.87
HSO+HPA	79.10	19.12
HSO+LSTM	102.24	**20.53**
HSO+AW+HPA+LSTM	**57.22**	19.98

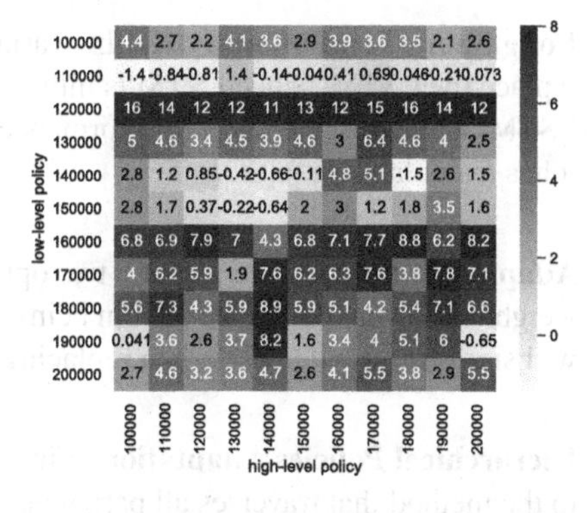

Fig. 6. The parameters of different high-level policies and low-level policies in different training epochs are used for online performance test. The number in the box is the mean reward of 20 online experiments with different random seeds.

methods trained on a fixed dataset without interactions with the environment, we collect data from environments of difficulty 0 that do not contain traps and dangerous states. Then we train our HSORL agent with $\lambda = 0$ and other baseline algorithms with the same datasets.

We demonstrate the average rewards among 10 episodes received from the environments during training in Fig. 5. Every episode includes 10000 time steps. As illustrated, offline TD3 outperforms CQL and achieves the best performance in every environment. HSORL achieves comparable performance with offline TD3 and CQL while outperforming offline DDPG in most environments. The empirical results show that our method has powerful offline learning performance.

4.3 Ablative Analysis

In this section, we conduct ablation experiments to understand the importance of various design choices of HSORL. The basic model is named SO, which has the same structure as SO-TD3. Then, by adding hierarchical policies into consideration, this becomes HSO. After that, we add external components to HSO. AW stands for the adaptive weight, HPA stands for the hierarchical policies adaptation, and LSTM stands for long short-term memory. Table 1 displays the results of variations of our model.

Hierarchical Policies. Firstly, we evaluate the benefit of hierarchical policies to model information about unsafe behaviors. As seen in Table 1, HSO incurs fewer safety violations than the SO, with a slight sacrifice in reward. It means that HSO can make a trade-off between safety and performance.

Long Short-Term Memory. In this variant, we use 2-layers feed-forward networks to replace the LSTM. Since LSTM is influential in modeling sequence information, HSO-LSTM achieves the choicest performance compared with the other variants. However, it has a similar cost to SO.

Adaptive Weight. Since we have two optimizer objectives in the critic, we use adaptive weight to prevent one of them from being dominant. We notice that the safety validation will significantly increase when replacing the adaptive weight with constant 1.

Hierarchical Policies Adaptation. Finally, we access the advantage of HPA compared to the method that traverses all parameter combinations. We randomly take 10 parameter combinations and choose the best for no-HPA variants. HSO-HPA achieves a better trade-off between safety and performance compared with HSO. We also drop HPA from HSORL and traverse all parameter combinations (Fig. 6). Compared with the inefficient traversing-based method, our HPA-based method achieves a reward of 16.37, similar to the best performance of the no-HPA method.

5 Conclusion

Recently, significant efforts have been dedicated to promoting the performance of offline reinforcement learning. However, behavior data scene in real-world scenarios inevitably contains sparse unsafe or unexpected state-action pairs, which may mislead the agents into undesirable states. To address this issue, we propose a novel offline training framework in conjunction with hierarchical policies to model the sparse unsafe behaviors effectively. Extensive experiments have been conducted to verify the effectiveness of the proposed framework on avoiding unsafe actions. Moreover, the results suggest that our method can achieve the best trade-off between safety and performance compared to a wide range of online and offline RL baselines.

Acknowledgments. This work was supported by the NSFC Projects 62076096 and 62006078, the Shanghai Municipal Project 20511100900, Shanghai Knowledge Service Platform Project ZF1213, the Shanghai Chenguang Program under Grant 19CG25, the Open Research Fund of KLATASDS-MOE and the Fundamental Research Funds for the Central Universities.

References

1. Achiam, J., Held, D., Tamar, A., Abbeel, P.: Constrained policy optimization. In: International Conference on Machine Learning, pp. 22–31. PMLR (2017)
2. Agarwal, R., Schuurmans, D., Norouzi, M.: An optimistic perspective on offline reinforcement learning. In: International Conference on Machine Learning, pp. 104–114. PMLR (2020)
3. Bacon, P.L., Harb, J., Precup, D.: The option-critic architecture. In: Proceedings of the AAAI Conference on Artificial Intelligence, pp. 1726–1734 (2017)
4. Bi, J., Dhiman, V., Xiao, T., Xu, C.: Learning from interventions using hierarchical policies for safe learning. In: Proceedings of the AAAI Conference on Artificial Intelligence, pp. 10352–10360 (2020)

5. Brockman, G., et al.: OpenAI Gym. arXiv preprint arXiv:1606.01540 (2016)
6. Dalal, G., Dvijotham, K., Vecerik, M., Hester, T., Paduraru, C., Tassa, Y.: Safe exploration in continuous action spaces. arXiv preprint arXiv:1801.08757 (2018)
7. Fujimoto, S., Conti, E., Ghavamzadeh, M., Pineau, J.: Benchmarking batch deep reinforcement learning algorithms. arXiv preprint arXiv:1910.01708 (2019)
8. Fujimoto, S., Hoof, H., Meger, D.: Addressing function approximation error in actor-critic methods. In: International Conference on Machine Learning, pp. 1587–1596. PMLR (2018)
9. Fujimoto, S., Meger, D., Precup, D.: Off-policy deep reinforcement learning without exploration. In: International Conference on Machine Learning, pp. 2052–2062 (2019)
10. Gulcehre, C., et al.: RL unplugged: a suite of benchmarks for offline reinforcement learning. arXiv preprint arXiv:2006.13888 (2020)
11. Hochreiter, S., Schmidhuber, J.: Long short-term memory. Neural Comput. **9**(8), 1735–1780 (1997)
12. Kulkarni, T.D., Narasimhan, K., Saeedi, A., Tenenbaum, J.: Hierarchical deep reinforcement learning: integrating temporal abstraction and intrinsic motivation. In: Advances in Neural Information Processing Systems, pp. 3675–3683 (2016)
13. Kumar, A., Zhou, A., Tucker, G., Levine, S.: Conservative q-learning for offline reinforcement learning. arXiv preprint arXiv:2006.04779 (2020)
14. Lange, S., Gabel, T., Riedmiller, M.: Batch reinforcement learning. In: Wiering, M., van Otterlo, M. (eds.) Reinforcement Learning. Adaptation, Learning, and Optimization, vol. 12, pp. 45–73. Springer, Heidelberg (2012). https://doi.org/10.1007/978-3-642-27645-3_2
15. Levine, S., Kumar, A., Tucker, G., Fu, J.: Offline reinforcement learning: tutorial, review, and perspectives on open problems. arXiv preprint arXiv:2005.01643 (2020)
16. Lillicrap, T.P., et al.: Continuous control with deep reinforcement learning. arXiv preprint arXiv:1509.02971 (2015)
17. Ma, J., Zhao, Z., Yi, X., Chen, J., Hong, L., Chi, E.H.: Modeling task relationships in multitask learning with multi-gate mixture-of-experts. In: International Conference on Knowledge Discovery & Data Mining, pp. 1930–1939 (2018)
18. Nachum, O., Gu, S., Lee, H., Levine, S.: Near-optimal representation learning for hierarchical reinforcement learning. In: International Conference on Learning Representations, pp. 1–18 (2018)
19. Nachum, O., Gu, S.S., Lee, H., Levine, S.: Data-efficient hierarchical reinforcement learning. In: Advances in Neural Information Processing Systems, pp. 3303–3313 (2018)
20. Ray, A., Achiam, J., Amodei, D.: Benchmarking safe exploration in deep reinforcement learning. arXiv preprint arXiv:1910.01708 (2019)
21. Srinivasan, K., Eysenbach, B., Ha, S., Tan, J., Finn, C.: Learning to be safe: deep reinforcement learning with a safety critic. arXiv preprint arXiv:2010.14603 (2020)
22. Sutton, R.S., Precup, D., Singh, S.: Between MDPS and semi-MDPS: a framework for temporal abstraction in reinforcement learning. Artif. Intell. **112**(1), 181–211 (1999)
23. Todorov, E., Erez, T., Tassa, Y.: MUJOCO: a physics engine for model-based control. In: International Conference on Intelligent Robots and Systems, pp. 5026–5033. IEEE (2012)
24. Vezhnevets, A.S., et al.: Feudal networks for hierarchical reinforcement learning. In: International Conference on Machine Learning, pp. 3540–3549. PMLR (2017)

Leveraged Mel Spectrograms Using Harmonic and Percussive Components in Speech Emotion Recognition

David Hason Rudd[1(✉)], Huan Huo[1], and Guandong Xu[1,2]

[1] The University of Technology Sydney, 15 Broadway, Ultimo, Australia
david.hasonrudd@student.uts.edu.au, {huan.huo,guandong.xu}@uts.edu.au
[2] Data Science Institute, 15 Broadway, Ultimo, Australia

Abstract. Speech Emotion Recognition (SER) affective technology enables the intelligent embedded devices to interact with sensitivity. Similarly, call centre employees recognise customers' emotions from their pitch, energy, and tone of voice so as to modify their speech for a high-quality interaction with customers. This work explores, for the first time, the effects of the harmonic and percussive components of Mel spectrograms in SER. We attempt to leverage the Mel spectrogram by decomposing distinguishable acoustic features for exploitation in our proposed architecture, which includes a novel feature map generator algorithm, a CNN-based network feature extractor and a multi-layer perceptron (MLP) classifier. This study specifically focuses on effective data augmentation techniques for building an enriched hybrid-based feature map. This process results in a function that outputs a 2D image so that it can be used as input data for a pre-trained CNN-VGG16 feature extractor. Furthermore, we also investigate other acoustic features such as MFCCs, chromagram, spectral contrast, and the tonnetz to assess our proposed framework. A test accuracy of 92.79% on the Berlin EMO-DB database is achieved. Our result is higher than previous works using CNN-VGG16.

Keywords: Speech Emotion Recognition (SER) · Mel spectrogram · Convolutional Neural Network (CNN) · Voice signal processing · Acoustic features

1 Introduction

The general motivation of SER systems is to recognize specific features of a speaker's voice in different emotional situations to provide a more personal and often superior user experience [6]. For example, a Customer Relationship Management (CRM) team can use SER to determine a customer's satisfaction by their voice during a call. Emotions are universal, although their understandings, interpretations and reflections are particular and partially associated with culture [1]. Unlike speech recognition, there is no standard or integrated approach for recognising emotions and analysing them through human voices [30].

J. Gama et al. (Eds.): PAKDD 2022, LNAI 13281, pp. 392–404, 2022.
https://doi.org/10.1007/978-3-031-05936-0_31

The fundamental challenge of SER is the extraction of discriminative and robust features from speech signals. Features used for SER are generally categorized as prosodic, acoustic, and linguistic features. The prosodic features include pitch, energy, and zero-crossings of the speech signal [16,19,28]. The acoustic features describe speech wave properties including linear predictor coefficients (LPC), mel-scaled power spectrograms (Mel), linear predictor cepstral coefficients (LPCC), power spectral analysis (FFT), power spectrogram chroma (Chroma), and mel-frequency cepstral coefficients (MFCC) [5]. In SER, the Mel spectrogram, MFCC, and chromagram are the most effective in decoding emotion from a signal [22].

Among the most common speech feature extraction techniques, this paper addresses a principal question in Emotion Recognition (ER): How can we maximise the advantage of the Mel spectrogram feature to improve SER? This study presents a novel implementation of emotion detection from speech signals by processing harmonic and percussive components of Mel spectrograms and combining the result with the log Mel spectrogram feature. Our primary contribution is the introduction of an effective hybrid acoustic feature map technique that improves SER. First, we employ CNN-VGG16 as a feature extractor of emotion identifier, then utilise the MLP networks for classification task. Furthermore, we tune the MLP network parameters using the random search model hyperparameter technique to obtain the best model. Based on empirical experiments, we assert that a data augmentation strategy using an efficient prosodic and acoustic feature combination analysis is the key to obtaining state-of-the-art results since input data represents more diversity with enriched features; these characteristics lead to better model generalisation.

2 Related Works

Early traditional SER models relied on modification and optimisation of Support Vector Machine (SVM) classifiers to predict emotions such as anger, happiness, and sadness, among others [15,23,25]. Wu et al. [31] implemented a traditional machine learning method based on EMO-DB [3] database. The authors proposed novel sound features named Modulation Spectral Features (MSFs) that combined prosodic features, and they ultimately obtained 85.8% validation accuracy for speaker-independent classification using a multi-class Linear Discriminant Analysis (LDA) classifier. Similarly, Milton et al. [21] proposed another classical machine learning method for SER by using a combination of three SVMs to classify emotions in the Berlin EMO-DB. Furthermore, Huang et al. [13] introduced a hybrid model called a semi-CNN, which used a deep CNN to learn feature maps and a classic machine learning SVM to classify seven emotions from EMO-DB. The authors utilised spectrograms as the input for their proposed model and achieved 88.3% and 85.2% test accuracy for speaker-dependent and speaker-independent classification, respectively.

The idea of exploiting pre-trained CNN image classifiers [7] for other tasks involves leveraging transfer learning methods in SER. Surprisingly, using speech-based spectrograms as the input images for pre-trained image classifiers produced competitive results when compared with other well-known traditional methods. Badshah et al. [2] extracted spectrogram speech features, which were then visualised in 2D images and passed to a CNN; this approach achieved a 52% test accuracy on EMO-DB. Demircan and Kahramanli [8] developed several different classifiers and obtained test accuracies 92.86%, 92.86%, and 90%, respectively on SVM, KNN and ANN. Additionally, Wang et al. [29] worked on MFCCs feature and proposed an acoustic feature called the Fourier Parameter (FP), which obtained 73.3% average accuracy with an SVM classifier. Furthermore, many similar studies were conducted on different databases. Popova et al. [24] used a fine-tuned DNN and CNN-VGG16 classifier to extract the Mel spectrogram features in the RAVDESS dataset [17] and obtained an accuracy of 71% [24]. Satt et al. [27] presented another multi-modal LSTM-CNN and proposed a novel feature extraction method based on the paralingual data from spectrograms. The authors obtained 68% accuracy on the IMOCAP [4] database.

In recent years, some works proposed the use of hybrid feature map techniques as input data for CNN-based networks. Meng et al. [20] proposed a feature extraction strategy for Log-Mel spectrograms that extracted a 3D voice feature representation map by combining log Mel spectrograms with the first and second derivatives of the log MelSpec of the raw speech signal. The authors proposed a CNN with a multimodal dilated architecture that used a residual block and BiLSTM (ADRNN) to improve the classifier accuracy. In addition, the ADRNN further enhanced the extraction of speech features using the proposed attention mechanism approach. The model achieved a remarkable performance of 74.96% and the 90.78% accuracy of the IEMOCAP and EMO-DB databases. On the other hand, Hajarolasvadi et al. introduced a 3D feature frame technique for use as input data to the network by extracting an 88-dimensional vector of voice features including MFCCs, intensity, and pitch. The model can reduce speech signal feature frames by applying k-means clustering on the extracted features and selecting the k most discriminant frames as keyframes. Then, the feature data placed in the keyframe sequence were encapsulated in a 3D tensor, which produced a final extracted feature map for use as input data for a 3D-CNN-based classifier that used the 10-fold cross-validation method. The authors achieved a weighted accuracy of 72.21% on EMO-DB. Zhao et al. [33] proposed a multimodal 2D CNN-LSTM network and extracted the log of the Mel-spectrograms from the speech signals for use as input data. The outcome of their work is state-of-the-art with the accuracy of 95.89% for speaker-independent classification on the Berlin EMO-DB.

3 Methodology

This section explains the work procedures used to build the hybrid feature map representation in our model. We compute the average of the signal's harmonic and percussive components and combine the result with the log Mel spectrogram feature. The proposed hybrid feature map method can be generalised with other supervised classifiers to obtain better prediction accuracy.

3.1 Proposed Hybrid Features: Harmonic and Percussive Components of Mel Spectrogram

Essential features in speech signal processing are the spectrograms on the Mel scale, chromograms [12], spectral contrast, the tonnetz [12] and MFCCs [32]. Since the average length of the recorded voice samples are four seconds, we digitise each original utterance signal at an 88 KHz sample rate using the Hanning window function [11] shown in (1) to provide sufficient frequency resolution and spectral leakage protection. Next, we apply Mel filter banks to the spectrogram by shifting 0.4 ms in a window time of 23 ms so that the output is a group of FFTs located next to one another. The Hanning window is described in (1),

$$H_m(n) = 0.5[1 - cos(\frac{2\pi.n}{M-1})] = sin^2(\frac{\pi.n}{M-1}) \qquad 0 \leq n < M-1 \quad (1)$$

where M represents the number of points in the output window, which is set to 128 and n denote the number of specific sample point from the signal. Finally, we construct the Mel spectrogram by multiplying the obtained energy matrix of the Mel scaled static with the STFT results formulated in (2),

$$LMS(m) = \sum_{k=f(m-1)}^{f(m+1)} log(H_m(k) \,.\, |X(k)|^2) \quad (2)$$

where $|X(k)|^2$ represents the energy spectrum in the kth energy block, $H(k)$ is a Mel-spaced filter bank function, m represents the number of filter banks, and k points to the number of FFT coefficients. LMS represents the log Mel spectrogram. To perform Mel spectrogram feature extraction, we use Librosa tools [18] to set the size of Mel filterbanks as 128, the window size as 2048 and hop length as 512. Figure 1 shows the Mel spectrogram of sample voices exhibiting five emotions from the EMO-DB dataset. It is clear that the amplitude and frequency of each emotion image have a high distinction from other samples.

The first feature map is built by applying a decomposition process to the Mel spectrum using the popular method in [9]. The decomposition method can be formulated such that the harmonic s_h and percussive s_p components are separated from the input signal s by applying a STFT on the frames to obtain spectrogram S of signal s as shown in definitions (3) and (4),

$$s \;=\; S_h \;+\; S_p \quad (3)$$

$$S(n,\; k) \;:=\; \sum_{r=0}^{N-1} s(r+nH) \,.\, \omega(r) \,.\, e^{(\frac{-j2\pi.k.n}{N})} \quad (4)$$

where S denotes a spectrum of signal s in k^{th} Fourier coefficient on the mth time frame, $\omega : [0 : N-1] := \{0, 1, ..., N-1\}$ is a sine windowing function that represents the window length N, H represents the hop size value, n indicates current frame number and N is the length of the discrete Fourier transform. We

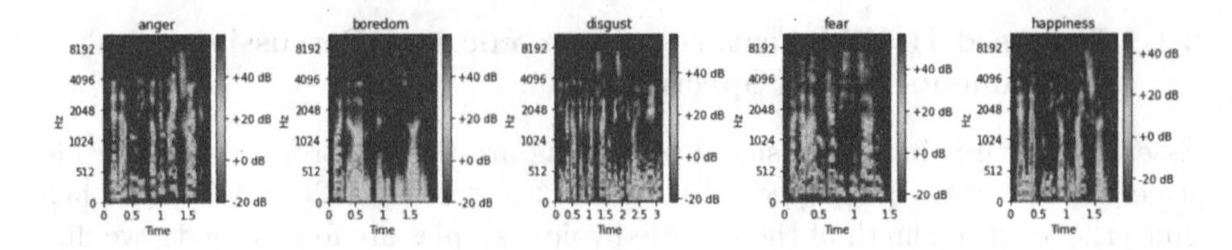

Fig. 1. The above sample Mel spectrograms clearly illustrate the distinction between amplitude and frequency in each emotion. The red colours represent frequencies that contribute more than orange and white colours. (Color figure online)

can obtain the harmonic and percussive components of the spectrum by applying a median filter in the horizontal (time-domain) and vertical (frequency-domain) direction on spectrum S. Finally, we extract the first feature map by obtaining the mean of both components as shown in the following summarised formulas in (5), (6) and (7)

$$\widehat{H} = \widehat{S} \bigotimes M_H \tag{5}$$

and

$$\widehat{P} = \widehat{S} \bigotimes M_P \tag{6}$$

obtained by

$$F_{2(LMS)} = \frac{(\widehat{H} + \widehat{P})}{2} \tag{7}$$

where $\bigotimes$ denotes the multiplication element of the median filter in M_H, which is the horizontal direction filtering used to obtain the $\widehat{H}$ harmonic components of the original spectrogram $\widehat{S}$. Subsequently, M_P represents the vertical median filtering results M_P, which is the percussive component of the original spectrogram, $\widehat{S}$ shown in (4). Figure 2 shows the harmonic and percussive components as two distinctive spectrograms in the 128 Mel filterbank.

The second feature map is extracted by applying the log of the Mel spectrogram obtained in (2) to measure the sensitivity of the Mel spectrogram output value fluctuation concerning changes in the voice signal amplitude. A sample 2D hybrid feature representation in our work is visualised in Fig. 3, which clearly shows that each sample feature map is combined in a two-dimensional image. This specific feature combination improves the prediction accuracy in a simple full contact neural network classifier based on our empirical experiments.

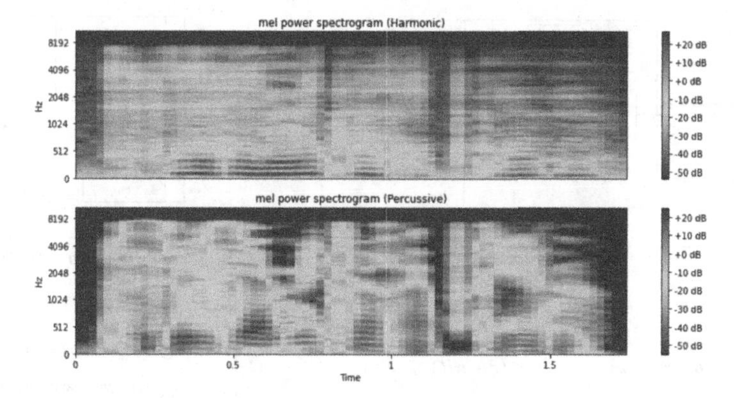

Fig. 2. The harmonic and percussive components of the Mel spectrograms for a sample neutral emotion

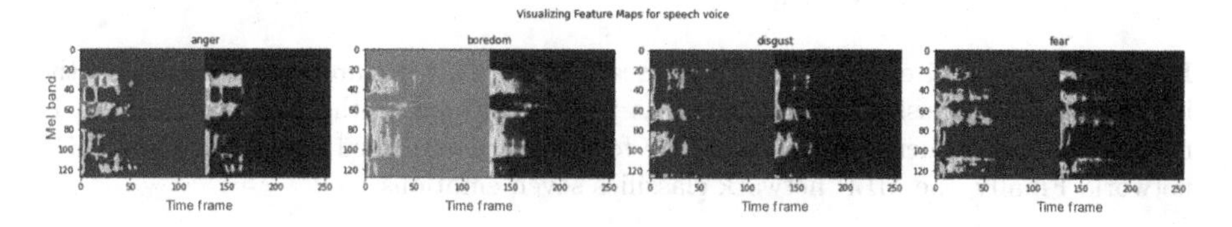

Fig. 3. Visualising an achieved 2D hybrid feature maps from the Berlin EMO-DB

3.2 Model Architecture and Training

We use the CNN-VGG16 [26] as a feature extractor to learn from high dimensional feature maps since the network can learn from small variations that occur in the extracted features maps. However, the high-capacity memory storage requirements for a simple classification task can be considered a partial limitation of VGG16 applications.

The details of the proposed architecture are shown in Fig. 4; the architecture consists of an VGG16 and MLP network, which serve as an feature extractor and emotion classifier, respectively. First, the subsamples are extracted from a fixed window size and then feature maps are built using the proposed feature map function. Therefore, the input to the VGG16 feature extractor is a 2-D feature map in the dimension of $(128 \times 128 \times 2)$. The input to the MLP classifier is a 2048 one-dimensional vector generated by VGG16. The MLP classifier includes four fully connected layers with the ReLU activation function and softmax in the output layer. Dense 1 and 2 have a 1024 input with a 0.5 dropout value, and dense 3 and 4 are set to 512 input with 0.3 dropouts. The ADAM optimiser with a learning rate of 0.0001 is selected for our architecture design.

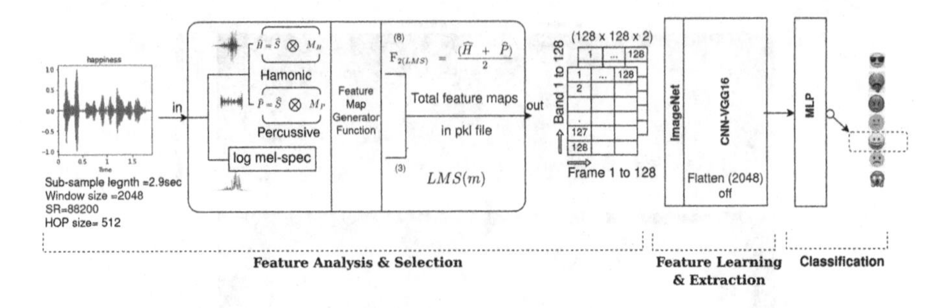

First Feature: $$S(n,\ k) \ := \sum_{r=0}^{N-1} s(r+nH) \ . \ \omega(r) \ . \ e^{(\frac{-2\pi.kn}{N})} : \begin{cases} \hat{H} = \hat{S} \otimes M_H & \text{horizontal median filtering} \\ \hat{P} = \hat{S} \otimes M_P & \text{vertical median filtering} \end{cases}$$

Second Feature: $$LMS(m) = \sum_{k=f(m-1)}^{f(m+1)} log(H_m(k) \ . \ |X(k)|^2)$$

Fig. 4. Model architecture, which includes a 2D hybrid feature map built using the harmonic and percussive components, as well as the log of the Mel spectrogram in the feature map generator function. The features are extracted using a CNN-VGG16 network. Finally, the MLP network classifies seven emotions.

4 Experimental Analysis

This section analyses the experimental configuration and the result of the feature extractor and MLP classifier on EMO-DB [3]. The sample voices are randomly partitioned and 80% are used for the training set and 10% for the validation and test set for the speaker-independent classification task. We apply an oversampling strategy to compensate the minority classes and increase the voice samples before feeding them to the feature extractor network during the pre-processing phase. The classifier is trained on 128 epochs with a batch size of 128 and used an Nvidia GPU. The window size is set to 2048 with (128×128) bands and frames to obtain each subsample length $= 2.9$ s. Then, the subsamples are created in each defined data frame. Finally, 167426 signal subsamples and 9717 feature maps are obtained from a sample rate of 88 KHz. Based on the time-frequency trade-off, large frame size is chosen to obtain high-frequency resolution rather than time resolution since analysing the frequency of speech signal enables us to decode emotion. Several time-consuming experiments are conducted to assess the effectiveness of the proposed hybrid feature, which aims to find the best data augmentation through feature combination.

4.1 Results Analysis

To assess our enriched feature representation method in the MLP classifier, the result of evaluation metrics such as the confusion matrix and test accuracy are observed on different sample rates and feature map dimensions (bands and frames). We also evaluate our model output based on the setting of various parameters in the feature map function. For example, the prediction accuracy results based on some different parameters setting are shown in Table 1. These results indicate that the superior result is achieved on feature map dimensions of 128 × 128 with a sample rate of 88200, Since the highest subsample length of 2.9 s is achieved and more sample points can contribute in each subsample.

Table 1. The emotion classifier accuracy based on different feature map representation dimensions and signal sampling ratios

Band	Frame	Sample rate	Accuracy
32	32	88200	71.92%
128	128	44100	87.04%
128	128	22050	88.54%
32	32	22050	89.54%
64	64	44100	92.02%
128	128	88200	**92.79%**

We examine the effect of the different number of subsamples from the signal by increasing the window size and sample rate on ten different feature map representations, including 1D, 2D, and 3D maps, and we then compare their results with our hybrid feature extraction method. With respect to the primary research question, it is found that we can take maximum advantage of the powerful Mel spectrogram feature through harmonic and percussive components in emotion recognition.

As shown in Table 2, the proposed hybrid feature map representation achieves better results than other well-known feature combinations techniques. Furthermore, the results in Table 2 indicate that the accuracy increases in the high range of the sample rate and window size in most represented methods since the feature map generator function handles more data points via a higher overlapping between frames. Consequently, for most feature extraction methods, the VGG16 network can learn from better-enriched features when the sample rate is higher. In contrast, an increased number of data points in the subsamples requires a memory capacity in the gigabyte range to store the base, train, validation, test feature map files in the pkl format. For instance, in our model, a signal sampling rate of 88 KHz and a window size of 2048 occupy an approximately 3-GB memory space to store the pkl files for analysing the whole voice files in the EMO-DB; this requirement can limit its application.

The fluctuation in the prediction accuracy per emotion class is illustrated for various feature representation methods in Fig. 5. The boxplot graph shows that the model output is more reliable and stable when predicting seven emotions using our proposed hybrid feature extraction "2D-log-MSS+Avg.HP" and two more feature representations built by combining the delta of the Mel spectrogram (MSS) and log Mel spectrogram or MFCCs features.

Table 2. Evaluation of the prediction accuracy based on the different feature extraction methods, sample rate and window size

Window size	512	1024	2048
Sample rate	22050	44100	88200
Feature extraction methods	Acc %	Acc %	Acc %
1D-MFCCs	65.81	68.39	69.03
1D-Mel spectrogram	75.48	75.48	82.71
1D-Chromagram	80.01	80.13	81.29
1D-Tonnets	56.77	63.08	56.81
1D-Spectral	54.84	50.93	47.10
2D-MFCCs+Chromagram	83.87	83.23	91.59
2D-Mel spectrogram+MFCCs	88.39	85.16	85.81
2D-Mel spectrogram+Spectral	82.01	85.13	80.65
3D-Mel spectrogram+MFCCs+Chromagram	83.87	88.39	81.94
2D-log-MSS+Avg.HP (proposed)	92.02	89.54	**92.79**

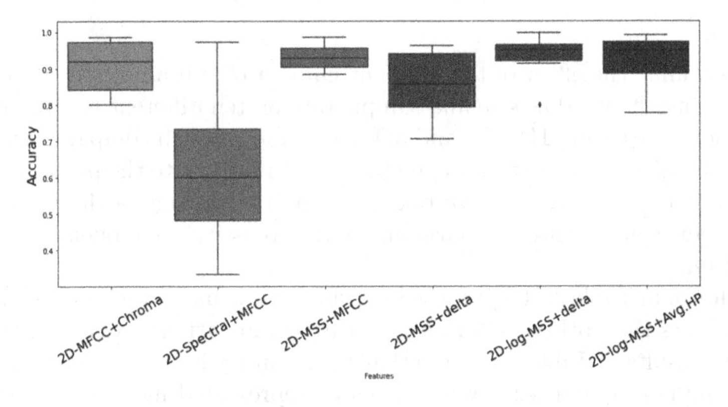

Fig. 5. Variation in the prediction accuracy per emotion class for different feature representation methods

The model's confusion matrix in Table 3 shows that the network performs better when recognising specific emotions (anger, sadness, happiness, and fear) while its performance is comparatively poor when predicting emotions such as neutral and boredom. Many experiments are conducted and the highest test accuracy of 92.79% is achieved. The Python Keras based network implementation for the proposed model and more experimental results and visualisations are available in our GitHub repositories[1].

[1] https://github.com/DavidHason/ser.

Table 3. Confusion matrix (%) of the model with an average accuracy of 92.71% on the EMO-DB dataset

Emotion	Anger	Boredom	Disgust	Fear	Happiness	Neutral	Sadness
Anger	**94.92**	0	0	0	5.12	0	0
Boredom	0	**78.77**	0	0	0	9.9	11.54
Disgust	0	0	**89.47**	0	9.8	0	0
Fear	0	0	0	**96**	0	0	3.85
Happiness	0	0	0	0	**100**	0	0
Neutral	0	12.81	0	0	0	**88.87**	0
Sadness	0	0	0	0	0	0	**100**

4.2 Model Comparison with Previous Works on EMO-DB

As shown in Table 4, our method achieves superior results compared with most previous studies except for two works in terms of accuracy that are not significantly higher than our results. However, their work frame is more sophisticated than our proposed model. Zhao et al. [33] combined two 1-D and 2-D LSTM CNN networks in the feature learning process. Demirican et al. [8] used a model with three classifiers KNN, SVM and ANN to improve the prediction accuracy. Nevertheless, the major advantage of our architecture comes from its simplicity and generality, which can be employed for other acoustic features, as shown in Table 2. Another advantage of the architecture is the capability of storing the feature maps into cloud storage in pkl format that enables us to share them for simultaneous analysis with other networks.

Table 4. Model comparison with previous works on EMO-DB

Previous works	Learner	Feature extraction method	Accuracy
Badshah et al. [2]	CNN	log Mel spectrogram	52%
Popova et al. [24]	VGG16	Mel spectrograms	71%
Hajarol et al. [10]	CNN	Mel spectrograms+MFCCs	72.21%
Wang et al. [29]	SVM	Fourier Parameter+MFCCs	73.3%
Huang et al. [13]	CNN	Spectrogram	85.2%
Issa et al. [14]	VGG16	MFCCs+Chroma.+Mel spec.+Contrast+Tonnetz	86.10%
Meng et al. [20]	CNN-LSTM	log Mel spec.+1st & 2nd delta(log Mel spec.)	90.78%
Wu et al. [31]	SVM	Modulation Spectral Features (MSFs)	91.60%
Our model	**VGG16-MLP**	**Harmonic-Percussive (HP)+log Mel spec.**	**92.79%**
Demircan et al. [8]	SVM	LPC+MFCCs	92.86%
Zhao et al. [33]	CNN-LSTM	log Mel spectrogram	95.89%

5 Conclusion

The key research question in this study focuses on leveraging Mel spectrogram components in a hybrid-based feature engineering technique as well as proposing a novel acoustic feature extraction method to improve emotion recognition. The proposed feature map generator function extracts the harmonic and percussive components by applying a median filter on the horizontal (time-domain) and vertical (frequency-domain) directions of the spectrum, and is implemented with a four-layer MLP classifier to predict emotions in the human voice. The performance of the proposed hybrid feature technique is tested on the Berlin EMO-DB and compared with other 1D, 2D, and 3D feature extraction methods. To the best of our knowledge, this is the first study on speech emotion recognition that combines this specific component of the spectrogram. The results show that our work significantly outperforms most previous works due to its achievement of a 92.79% test accuracy which is also a superior result in VGG16 feature learning methods. In future investigations, facial expression analysis and linguistic features can be embedded into the framework to improve the emotion recognition as an acoustic-only method is not constant across different languages and cultures.

Acknowledgement. This work is partially supported by Australian Research Council under grant number: DP22010371, LE220100078, DP200101374 and LP170100891.

References

1. Alu, D., Zoltan, E., Stoica, I.C.: Voice based emotion recognition with convolutional neural networks for companion robots. Sci. Technol. **20**, 222–240 (2017)
2. Badshah, A.M., Ahmad, J., Rahim, N., Baik, S.W.: Speech emotion recognition from spectrograms with deep convolutional neural network. In: 2017 International Conference on Platform Technology and Service (PlatCon), pp. 1–5 (2017)
3. Burkhardt, F., Paeschke, A., Rolfes, M., Sendlmeier, W.F., Weiss, B., et al.: A database of German emotional speech. In: Interspeech, vol. 5, pp. 1517–1520 (2005)
4. Busso, C.: IEMOCAP: interactive emotional dyadic motion capture database. Lang. Resour. Eval. **42**(4), 335–359 (2008). https://doi.org/10.1007/s10579-008-9076-6
5. Chu, S., Narayanan, S., Kuo, C.C.J.: Environmental sound recognition with time-frequency audio features. IEEE Trans. Audio Speech Lang. Process. **17**(6), 1142–1158 (2009)
6. Cowie, R.: Emotion recognition in human-computer interaction. IEEE Sig. Process. Mag. **18**(1), 32–80 (2001)
7. Cummins, N., Amiriparian, S., Hagerer, G., Batliner, A., Steidl, S., Schuller, B.W.: An image-based deep spectrum feature representation for the recognition of emotional speech. In: Proceedings of the 25th ACM International Conference on Multimedia, pp. 478–484 (2017)
8. Demircan, S., Kahramanli, H.: Application of fuzzy c-means clustering algorithm to spectral features for emotion classification from speech. Neural Comput. Appl. **29**(8), 59–66 (2018)

9. Fitzgerald, D.: Harmonic/percussive separation using median filtering. In: Proceedings of the International Conference on Digital Audio Effects (DAFx), vol. 13, pp. 1–4 (2010)
10. Hajarolasvadi, N., Demirel, H.: 3D CNN-based speech emotion recognition using k-means clustering and spectrograms. Entropy **21**(5), 479–495 (2019)
11. Harris, F.J.: On the use of windows for harmonic analysis with the discrete Fourier transform. Proc. IEEE **66**(1), 51–83 (1978)
12. Harte, C., Sandler, M., Gasser, M.: Detecting harmonic change in musical audio. In: Proceedings of the 1st ACM Workshop on Audio and Music Computing Multimedia, pp. 21–26 (2006)
13. Huang, Z., Dong, M., Mao, Q., Zhan, Y.: Speech emotion recognition using CNN. In: Proceedings of the 22nd ACM International Conference Media, pp. 801–804 (2014)
14. Issa, D., Demirci, M.F., Yazici, A.: Speech emotion recognition with deep convolutional neural networks. Biomed. Sig. Process. Control **59**, 101894–101904 (2020)
15. Jin, Q., Li, C., Chen, S., Wu, H.: Speech emotion recognition with acoustic and lexical features. In: 2015 IEEE International Conference on Acoustics, Speech and Signal Processing (ICASSP), pp. 4749–4753 (2015)
16. Li, M., Han, K.J., Narayanan, S.: Automatic speaker age and gender recognition using acoustic and prosodic level information fusion. Comput. Speech Lang. **27**(1), 151–167 (2013)
17. Livingstone, S.R., Russo, F.A.: The Ryerson audio-visual database of emotional speech and song (RAVDESS): a dynamic, multimodal set of facial and vocal expressions in North American English. PLoS ONE **13**(5), 1–35 (2018)
18. McFee, B., et al.: librosa: audio and music signal analysis in Python. In: Proceedings of the 14th Python in Science Conference, vol. 8, pp. 18–25 (2015)
19. Meinedo, H., Trancoso, I.: Age and gender classification using fusion of acoustic and prosodic features. In: 11th Annual Conference of the International Speech Communication Association, pp. 1–4 (2010)
20. Meng, H., Yan, T., Yuan, F., Wei, H.: Speech emotion recognition from 3D Log-Mel spectrograms with deep learning network. IEEE Access **7**, 125868–125881 (2019)
21. Milton, A., Sharmy Roy, S., Tamil Selvi, S.: SVM scheme for speech emotion recognition using MFCC feature. Int. J. Comput. Appl. **69**(9), 34–39 (2013). https://doi.org/10.5120/11872-7667
22. Motlıcek, P.: Feature extraction in speech coding and recognition. Technical Report of Ph.D. research internship in ASP Group, pp. 1–50 (2002)
23. Pérez-Rosas, V., Mihalcea, R., Morency, L.P.: Utterance-level multimodal sentiment analysis. In: Proceedings of the 51st Annual Meeting of the Association for Computational Linguistics (Volume 1: Long Papers), pp. 973–982 (2013)
24. Popova, A.S., Rassadin, A.G., Ponomarenko, A.A.: Emotion recognition in sound. In: International Conference on Neuroinformatics, pp. 117–124 (2017)
25. Rozgić, V., Ananthakrishnan, S., Saleem, S., Kumar, R., Prasad, R.: Ensemble of SVM trees for multimodal emotion recognition. In: Proceedings of the 2012 Asia Pacific Signal and Information Processing Association Annual Summit and Conference, pp. 1–4 (2012)
26. Russakovsky, O.: ImageNet large scale visual recognition challenge. Int. J. Comput. Vis. **115**(3), 211–252 (2015)
27. Satt, A., Rozenberg, S., Hoory, R.: Efficient emotion recognition from speech using deep learning on spectrograms. In: Interspeech, pp. 1089–1093 (2017)

28. Shriberg, E., Ferrer, L., Kajarekar, S., Venkataraman, A., Stolcke, A.: Modeling prosodic feature sequences for speaker recognition. Speech Commun. **46**(3–4), 455–472 (2005)
29. Wang, K., An, N., Li, B.N., Zhang, Y., Li, L.: Speech emotion recognition using Fourier parameters. IEEE Trans. Affect. Comput. **6**(1), 69–75 (2015)
30. Weninger, F., Wöllmer, M., Schuller, B.: Emotion recognition in naturalistic speech and language-a survey. In: Emotion Recognition: A Pattern Analysis Approach, pp. 237–267 (2015)
31. Wu, S., Falk, T.H., Chan, W.Y.: Automatic speech emotion recognition using modulation spectral features. Speech Commun. **53**(5), 768–785 (2011)
32. Xu, M., Duan, L.-Y., Cai, J., Chia, L.-T., Xu, C., Tian, Q.: HMM-based audio keyword generation. In: Aizawa, K., Nakamura, Y., Satoh, S. (eds.) PCM 2004. LNCS, vol. 3333, pp. 566–574. Springer, Heidelberg (2004). https://doi.org/10.1007/978-3-540-30543-9_71
33. Zhao, J., Mao, X., Chen, L.: Speech emotion recognition using deep 1D & 2D CNN LSTM networks. Biomed. Sig. Process. Control **47**, 312–323 (2019)

SelectAug: A Data Augmentation Method for Distracted Driving Detection

Yuan Li[1,2], Wei Mi[1(✉)], Jingguo Ge[1,2], Jingyuan Hu[1], Hui Li[1], Daoqing Zhang[1], and Tong Li[1]

[1] Institute of Information Engineering, Chinese Academy of Sciences, Beijing, China
miwei@iie.ac.cn

[2] School of Cyber Security, University of Chinese Academy of Sciences, Beijing, China

Abstract. With distracted driving becoming one of the main causes of traffic accidents, deep learning technology has been widely used in distracted driving detection, which achieves high accuracy when the training and test data are identically distributed. However, this assumption cannot correspond to the real-world situation. In case of a small sample size, we usually utilize open datasets as training dataset. Thus the training data distribution and test data distribution are different, which may induce accuracy plummets. Concentrating on the unforeseen data shifts encountered under different data distributions in distracted driving detection application, it is extremely desired to develop the detection technique with high robustness. In order to alleviate the issue about data shifts encountered under different data distributions, we propose an innovative method, SelectAug, to enhance images by applying the selected important features of the images. The experimental evaluations on the StateFarm dataset show that our method outperforms prior methods, demonstrating its efficacy in detecting distracted driving behaviors scenes. Furthermore, our method also improves generalization performance under different data distributions for distracted driving detection, which allows open datasets to be applied to real-world scenarios.

Keywords: Deep learning · Data augmentation · Distracted driving detection · Different data distributions

1 Introduction

Drivers are generally required to stay focused during driving, otherwise they will face the risk of the traffic accidents. According to the National Highway Traffic Safety Administration (NHTSA) [1], 3142 people died in traffic collisions on United States due to the driver distraction in 2019. Most states in US now have enacted laws against distracted driving behaviors such as texting and talking on the phone. The distractions of driver are usually classified as visual, auditory, manual and cognitive distractions. Visual distractions are considered as a behavior that driver moves his eyes off the road, auditory distractions mean that the

J. Gama et al. (Eds.): PAKDD 2022, LNAI 13281, pp. 405–416, 2022.
https://doi.org/10.1007/978-3-031-05936-0_32

driver is busy answering a cell phone or engaged in a talk. Manual distractions refer to that the hands of the driver are not on the wheels. Cognitive distraction is that the mind of the driver is occupied by other thoughts rather than driving.

Deep learning has attracted extensive attention from researchers in the field of artificial intelligence during recent years. More and more deep learning methods have been proposed in the distracted driving detection. Deep learning requires a large amount of reliable data to achieve better performance. However, deep learning often faces the situation of the lack of data in the real-world scenarios. In order to solve this problem, existing methods usually introduce open datasets as the initial datasets to train the deep neural networks. Currently the distributions between training data from open datasets and test data in real-world scenarios are generally different, yet there are few studies focusing on this problem. In computer vision field, most researches pay more attention to the structural design of deep neural network [2–4] using open datasets. It has been proved that when training and test data are from the same dataset, the model trained by using deep learning will obtain high accuracy [5]. However, when a mismatch occurs between the training data and the test data, the performance of the model will decrease dramatically.

In this work, we propose a data augmentation method called SelectAug, banding with distracted driving detection models to promote the accuracy for different data distributions. Our method first extracts the important features of the image by utilizing image segmentation technology, then the extracted features are used to enhance the image in order to obtain the prominent regions of the image. The augmentation method is mainly applied to improve the accuracy under different data distributions. It decreases the influence of background information and enables models to focus more on the most important features in the driving scenario. The main contributions of this paper are as follows:

1. We propose a data augmentation method in distracted driving detection to solve the data insufficient problem in real-world scenarios. Compared with the baseline methods and other augmentation methods, experiment results show great improvement on the accuracy and generalization performance using SelectAug in these cases.
2. Our method is designed for distracted driving detection, which implements the Yolact++ [17] model trained as the upstream task and can automatically segment important features in complex driving scenes thus avoiding a lot of manual annotation work.
3. Extensive experimental evaluations on the StateFarm dataset show that the models using SelectAug outperforms prior methods by a considerable margin for distracted driving detection.

2 Related Work

2.1 Distracted Driving Detection

With distracted driving becoming one of the main causes of traffic accidents, distracted driving detection has gradually been paid more attention. In the early days, researchers determined whether a driver was distracted by the movement of the eyes and head. [20,23,24] based on the movement of the eyes and head detected behaviors of distracted driving. Currently, deep learning for action recognition has become mainstream [14,15,19,21,22]. There are several open datasets (SUE-DP dataset [27], StateFarm dataset [18] and AUC dataset [25]) that are exploited for the distracted driving detection. SUE-DP dataset includes four types of distracted driving behaviors "grasping the steering wheel", "operating the shift lever", "eating," and "talking on a cellular phone". [26] used machine learning to detect distracted driver behaviors in SUE-DP dataset. The distracted driver detection dataset of the StateFarm is comprised of images labeled with one of the following ten actions: safe driving, texting using the right hand, talking on the phone using the right hand, texting using the left hand, talking on the phone using the left hand, operating the radio, drinking, reaching behind, hair and makeup, and talking to passengers. In [14–16], they used deep learning technology to detect drivers' behaviors. American University in Cairo (AUC) distracted driving dataset is similar to StateFarm. In 2018, Eraqi and Abouelnaga etc. [25] introduced the AUC dataset which combined three of the most advanced methods in deep learning, namely inception module with residual blocks and hierarchical recursive neural network, in order to improve the performance of driver distraction detection.

Although at present a lot of attention has been paid to the distracted driving detection field, most of these tasks [13–15,19–22,25] only focus on the accuracy improvement under the same data distributions, rather than the different data distributions. Due to the different data distribution (resolution of the camera on different datasets, installation location, wide Angle, installation angle different, or seat, steering wheel position is different), it causes a large decrease in the accuracy of the model trained by open dataset in the real-world scenarios with fewer samples.

2.2 Data Augmentation

Currently, data augmentation [5–7,11] has proven to be a crucial method for solving various challenges in the deep learning tasks, including image classification, natural language understanding and semi-supervised learning. Data augmentation mainly is employed to reduce the over-fitting phenomenon of the network. By transforming the training images, a network with the stronger generalization ability will be obtained, which can better adapt to the application scenarios. Random Erasing [7] and Cutmix [11] mask or modify randomly selected rectangular regions of images. Mixup [6] combines two images to generate an unseen training sample. Augmix [5] utilizes reinforcement learning to find the best strategy for selecting and combining label-invariant transforms. Although these methods [5–7,11] increase the effective data size and promote diversity in training examples, the randomness and uncertainty of these data augmentation methods may destroy the key features of images that is critical for detecting driving behaviors. Thus, a new data augmentation method is proposed for distracted driving

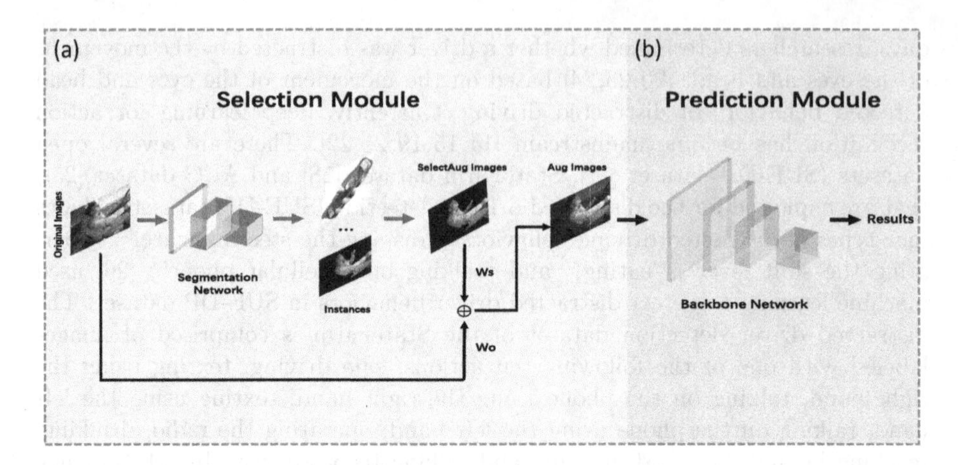

Fig. 1. The architecture of Network. The overall operation of the network is as follows: (a) In the selection module, the image segmentation network is used to segment instances of an image and select the most important instances of the image according to the rules we set, such as drivers, water cups and cell phones. These important instances are mixed according to the position of the original image to obtain a SelectAug image. Then the original image and the SelectAug image are weighted combined to obtain the Aug image. (b) In the prediction module, the Aug images are input into the deep neural network to obtain the results of image classification.

detection, which aims to enhance the selected key features of the images and make other unimportant information be weakened.

3 Preliminaries

In the Section, we first introduce an augmentation method "SelectAug", and then illustrate the architecture of the network. We use the term "SelectAug" to refer to a data augmentation method that enhances images by using selected important regions which are obtained by image segmentation technology.

3.1 SelectAug

SelectAug is a kind of data augmentation method, which can improve the robustness of the model and the uncertainty estimation of migration when training and test data are mismatched.

The SelectAug aims to make the regions including important features enhanced, while other regions will be weakened. In order to get the regions containing important features, we apply the image segmentation model to segment an image and obtain all instances of the image. However, many instances are unimportant for distracted driving detection, we only need to segment the instances that are critical for identifying distracted driving behaviors. Thus SelectAug algorithm is applied to solve the problem. Figure 1 shows that SelectAug consists of two main steps. Firstly, in the selection module, the image segmentation network is used to segment key instances of images and select the most important instances according to the stated rules. These important instances are mixed according to the position where it is in the original image to obtain SelectAug image. Then, we weighted it with the original image to get the augmentation image. The implementation details of SelectAug algorithm are as follows.

Algorithm 1. SelectAug

Require: Image segmentation model M, Image I, Classes C_n, Thresholds T_n;
Ensure: SelectAug Image S_I;
1: Initialize the selected objects O_s;
2: Getting all objects O_{all} and their confidence values Con_{all} by the model M;
3: k = len(O_{all})
4: **for** i from 1 to k **do**
5: **if** $C_{O_i} \in C_n$ & $Con_{O_i} \geq T_c$ **then**
6: $O_s \leftarrow O_i$
7: **end if**
8: **end for**
9: Getting S_I by mixing O_s in an image according to the position of the image itself.
10: **return** S_I

We input image segmentation model M which has been trained, an image I, classes C_n we need in the distracted driving scene and the thresholds T_n. Each class corresponds to a threshold T_c in the T_n. We segment the image I by using the model M in order to get all objects O_{all} and their confidence values T_{all}. It is assumed that the object O_i belongs to the class C_{O_i} and its confidence value is Con_{O_i}. If C_{O_i} is included in the classes C_n and its confidence value Con_{O_i} is greater than the threshold T_c of the class C_{O_i}, the object O_i will be selected. O_i will be added to O_s, where all the selected objects O_s are combined according to their positions in the original image, the SelectAug image S_I is obtained. All images are processed by this algorithm.

In order to decrease the error caused by the inaccurate image segmentation model, we introduce the original image I and S_I for weighted combination to get the augmentation image Aug. The combination of S_I and I is as follows:

$$I_A = w_s \times I_s + w_o \times I_o \tag{1}$$

where I_s, I_o and I_A represent the SelectAug image, the original image and the Aug image, w_s and w_o represent the weights of SelectAug image and the original image. Because $I_s \in I_o$, $I_s \in I_A$ when $w_o + w_s = 1$. In other words, it's certain that the regions that we select will be retained, while the other regions become weaker with a decrease of w_o when $w_o + w_s = 1$. Due to I_A contains all key instances of the original image, the influence of the inaccurate image segmentation model will also be reduced.

3.2 Architecture

The architecture of our method is shown in Fig. 1, including selection module and prediction module.

Selection Module. Selection module enhances the original images so as to obtain the Aug images. In this module, we select the Yolact++ [17] model which has been trained on the COCO [10] dataset to segment images. We get the Aug images by using the SelectAug method, which retains the key features of the original image. The non-critical background information in original images will be weakened, while the critical features will be retained. Thus the model will focus on the regions including key features and model's ability of identifying critical features will be strengthened. Besides, it also reduces unnecessary noise interference from background information.

Figure 3 shows the enhanced effect with different weighting schemes.

Prediction Module. After image processing, the convolutional neural network is used as the backbone network, which can extract the features of images and classify the images. ResNet50 [12] is applied as the backbone networks to predict the driving behaviors of the drivers. Besides, we consider integrating VGG16, ResNet50 and Xception [9,12,13] to achieve optimal prediction results.

4 Experiments

In this section, we first describe the experimental setting and implementation details. Then we evaluate our models with quantitative comparisons to other methods. Finally, we show the sensitivity analysis of the parameter to validate the enhanced effect in different weights of the selectAug image.

4.1 Experimental Setting

Datasets. Our datasets mainly include two parts, the StataFarm dataset [18] and Self-Collection dataset. In order to verify the feasibility of our work under different data distributions, we need to collect Self-Collection dataset from the real-world scenes which is mismatched from the StateFarm dataset. The State-Farm dataset contains 22714 images (18277 images in the training set and 4,437 in the test set). The dataset consists of 10 classes (C0–C9, safe driving, texting using the right hand, talking on the phone using the right hand, texting using the left hand, talking on the phone using the left hand, operating the radio, drinking, reaching behind, hair and makeup, and talking to passengers) as shown in Fig. 2. The Self-Collection dataset includes 3,200 images, which are taken from 15 persons. This dataset is similar to the StataFarm dataset and it is only used as a test dataset in order to ensure that the training and test datasets are distributed differently. Figure 2 shows some samples of our datasets. Pedestrians

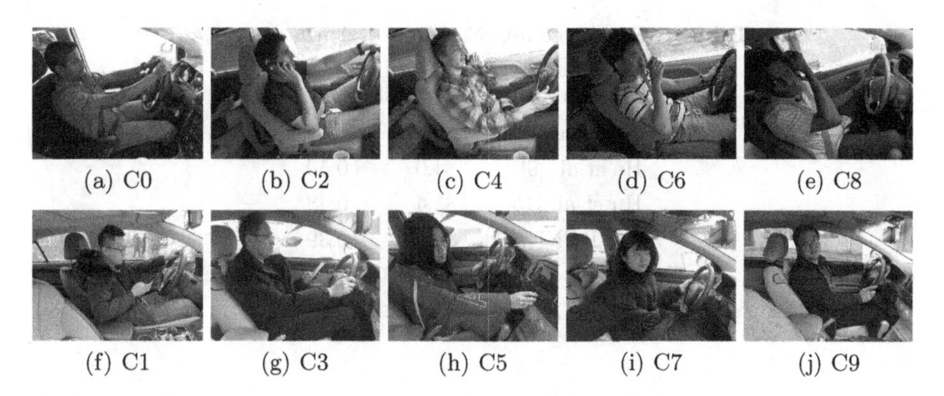

Fig. 2. Samples from StateFarm and Self-Collection datasets represent the ten classes (C0–C9). The five samples (a–e) are from StateFarm dataset, and the others (f–j) are from Self-Collection dataset. Pedestrians and other objects outside the window are not filtered out in the Self-Collection dataset, and the images of the Self-Collection dataset are taken from different cameras and angles. The Self-Collection dataset is more complex and diverse.

and other objects outside the window were not filtered out in the Self-Collection dataset, and the angle we took was also different from the StateFarm dataset. Due to the images of the Self-Collection dataset were taken from different cameras and angles, it was more complex and diverse. And we need to process the StataFarm dataset by the SelectAug in order to obtain an enhanced dataset Driver_IMGS-PBP including all Aug Images.

Evaluation Metric. We use accuracy as an evaluation metric.

$$Accuracy = \frac{TP + TF}{TP + TF + FP + FN} \tag{2}$$

Accuracy is closeness of the measurements to a specific value. This metric generally describes the ability of the classifier to predict the labels. Formula 2 is the formula of accuracy, where TP = True positive, FP = False positive; TN = True negative, FN = False negative.

$$-\sum_{c=1}^{M} y_{o,c} \log(p_{o,c}) \tag{3}$$

Our experiment applies cross-entropy cost function as loss function as shown in Formula 3. The purpose of the models is to minimize this value. In Formula 3, M, log and y respectively refer to the number of classes, the natural log and binary indicator (0 or 1) if class label c is the correct classification for observation o. And p refers to predicted probability that observation o belongs to class c.

Table 1. Results comparison on StateFarm dataset.

Method	Acc. (%)	Loss
Yan et al. [16]	70.6	0.72
He et al. [9]	84.0	0.50
Hu et al. [15]	85.4	0.48
DD-RCNN [14]	86.0	0.39
Ours	90.1	0.33
Ours+Ensemble	95.1	0.15

4.2 Implementation Details

We use Yolact++ [17] model trained on the COCO [10] dataset to segment image avoiding a lot of manual annotation work, and we apply VGG16 [9], ResNet50 [12] and Xception [13] as image classification network, which are pre-trained on the ImageNet [8] dataset. We set person, water cup and mobile phone as selected classes C_n. And we set T_n including the highest confidence value of the person class and the top three in the water cup and mobile phone classes. Besides, we set $w_o = 0.7$, w_o is the weight of selectAug image combined with the original image. In order to ensure that the experiment is only affected by data augmentation, we use the same hyperparameters for the StataFarm and Driver_IMGS-PBP datasets in the image classification network. The ratio of training dataset to test dataset is 8: 2. We set $lrate = 0.001$, the $drop = 0.5$, and $epochs\_drop = 40$. Besides, the early stopping and SGD strategies are used in all experiments.

4.3 Results

We experimented with the StateFarm dataset and the Self-Collection dataset. We train models on the StateFarm dataset and Driver_IMGS-PBP dataset. And we test these models on the StateFarm dataset to evaluate the results under the same data distribution. In order to obtain the results under different data distributions, we test these models on the Self-Collection dataset.

Results Under the Same Data Distribution. In order to verify the effectiveness of SelectAug on distracted driving detection, we designed relevant experiments and compared them with previous methods on the StateFarm dataset. Our method uses ResNet50 as backbone to detect driving behaviors. It should be noted that Ensemble model uses hard voting of the VGG16, ResNet50 and Xception models for classification. Compared with other methods, the accuracy of our method is greatly improved. From Table 1, our method achieves state-of-the-art accuracy under the same data distribution. Our method pays more attention to the foreground information including key features, no longer pays attention to the background information in the scene of distracted driving detection. Table 1 shows that the SelectAug has an excellent effect on improving the accuracy of distracted driving detection.

Table 2. Results under different data distributions.

Method	Acc. (%)	Loss
Vgg16	43.8	2.43
ResNet50	48.0	2.43
Xception	40.6	2.11
Ensemble	50.4	2.23
Ours+Vgg16	77.6	0.99
Ours+ResNet50	71.8	1.20
Ours+Xception	79.0	0.76
Ours+Ensemble	85.2	0.64

Table 3. SelectAug vs. other data augmentation methods

Model	Acc. (%)	Loss
RandomErasing [7]	35.2	4.83
MIXUP [6]	53.1	1.68
CUTMIX [11]	55.1	1.63
AUGMIX [5]	56.0	1.59
Ours	71.8	1.20
Ours+Ensemble	85.2	0.64

Results Under Different Data Distributions. In order to verify the accuracy and loss function of our method under different data distributions, we test models on Self-Collection dataset. Since there is almost no method for different data distributions of distracted driving detection scenarios in the past, we implemented several recent classic image classification [9,12,13] models as the baseline methods. Table 2 shows all baseline methods do not work well under different data distributions. The experiment shows that all results using the SelectAug are better than the baseline methods under different data distributions. SelectAug retains the regions including key characteristics and weaken other regions, which enables the model to learn the knowledge related to distracted driving detection. Therefore, the models using our method have better generalization performance and robustness under different data distributions.

SelectAug vs. Other Data Augmentation Methods. To make the comparison fair, we carried out experiments using the same network and datasets. We trained ResNet50 models for all methods on the StateFarm dataset and tested the models on the Self-Collection dataset. From Table 3, our method achieves state-of-the-art accuracy on Self-Collection dataset. Compared with other augmentation methods, it shows that SelectAug augmentation is more suitable for distracted driving scenarios under different data distribution. Most previous data augmentation methods [5–7,11] may destroy the key features of original images,

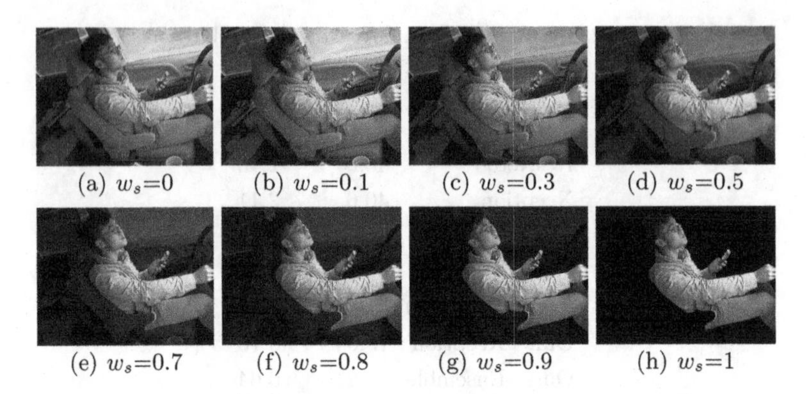

(a) w_s=0 (b) w_s=0.1 (c) w_s=0.3 (d) w_s=0.5

(e) w_s=0.7 (f) w_s=0.8 (g) w_s=0.9 (h) w_s=1

Fig. 3. The enhanced effects with different weight schemes. w_s indicates the weight of the SelectAug image in Aug Image.

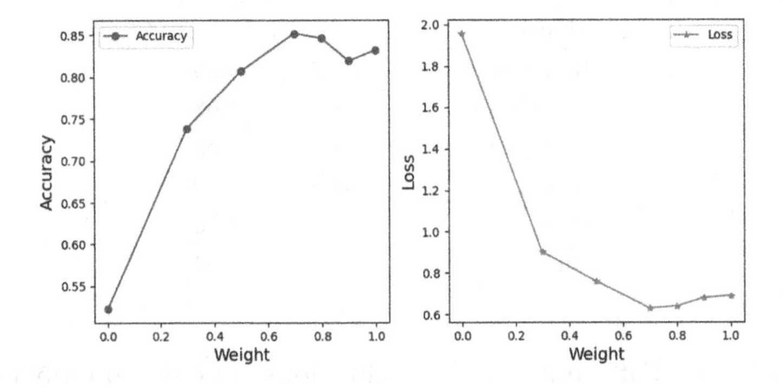

Fig. 4. Accuracy and Loss of SelectAug in the different weights.

which create augmented images with wrong or ambiguous labels. The difference of our method to theirs is that our method tries to retain key regions directly and weaken others regions of original images whereas their methods just try to make sure the augmented images which are enhanced randomly are similar to the current training images. The method can increase the data volume and promote the diversity of data. At the same time, this method can preserve the key areas of the image so that the newly generated images contain the key features, so as to ensure that the newly generated images are effective data rather than noise. Therefore, our method has significant improvement under different data distribution in distracted driving detection scenarios.

4.4 Sensitivity Analysis of the Parameter

In the following experiments we show the influence about different weights of SelectAug images under different data distributions. The weight represents the proportion of the selectAug images when we weighted combine it with the original image. Figure 3 shows the enhanced effect to images with different weighting schemes. Figure 4 shows the effects of our method under different weights of the SelectAug image. When the weight = 0, it means that we input the original images into the prediction module, the model is almost unusable under different data distributions. When SelectAug is applied in the scene, the accuracy is greatly improved. And our method works best when the weight equals to 0.7. Thus we choose 0.7 as the weight of the selectAug image.

5 Conclusion

In this paper, we propose a novel augmentation method called SelectAug which weighted combines the original images and the regions including important features according to the spatial position, in order to solve the problem about data shifts encountered under different data distributions. Throughout an extensive evaluation, we have demonstrated that not only the accuracy of the model is significantly improved under the same data distribution, the generalization performance of the model also becomes better when the training data and the test data are distributed differently by applying our method. The experimental results on the StateFarm dataset show that our method outperforms prior methods by a considerable margin. Furthermore, our method achieves good robustness under different data distributions for distracted driving detection. Our method enables models trained in open datasets to be used in real-world scenarios. In the future work, we will consider to improve the generality of our method in other vision tasks such as visual scene description and human object interaction detection.

References

1. National Highway Traffic Safety Administration: Traffic safety facts 2019 data: Distracted Driving 2019 (DOT HS 813 111). Washington, DC (2021)
2. Zoph, B., Vasudevan, V., Shlens, J., Le, Q.V.: Learning transferable architectures for scalable image recognition. In: Proceedings of IEEE Conference on Computer Vision and Pattern Recognition (2017)
3. Hu, J., Shen, L., Sun, G.: Squeeze-and-excitation networks. arXiv preprint arXiv:1709.01507 (2017)
4. Mohamed, A., Qian, K., Elhoseiny, M., et al.: Social-STGCNN: a social spatio-temporal graph convolutional neural network for human trajectory prediction. In: 2020 IEEE/CVF Conference on Computer Vision and Pattern Recognition (CVPR). IEEE (2020)
5. Hendrycks, D., et al.: AugMix: a simple data processing method to improve robustness and uncertainty. In: ICLR (2020)
6. Zhang, H., Cisse, M., Dauphin, Y.N., et al.: mixup: Beyond empirical risk minimization. In: ICLR (2018)
7. Zhong, Z., Zheng, L., Kang, G., Li, S., Yang, Y.: Random erasing data augmentation. arXiv preprint arXiv:1708.04896 (2017)
8. Deng, J., Dong, W., Socher, R., et al.: ImageNet: a large-scale hierarchical image database. In: 2009 IEEE Computer Society Conference on Computer Vision and Pattern Recognition (CVPR 2009), Miami, Florida, USA, June 20–25 2009. IEEE (2009)

9. Simonyan, K., Zisserman, A.: Very deep convolutional networks for large-scale image recognition. arXiv preprint arXiv:1409.1556 (2014)
10. Lin, T.-Y., et al.: Microsoft COCO: common objects in context. In: Fleet, D., Pajdla, T., Schiele, B., Tuytelaars, T. (eds.) ECCV 2014. LNCS, vol. 8693, pp. 740–755. Springer, Cham (2014). https://doi.org/10.1007/978-3-319-10602-1_48
11. Yun, S., Han, D., Chun, S., et al.: CutMix: regularization strategy to train strong classifiers with localizable features. In: International Conference on Computer Vision, 0. ICLR (2019)
12. He, K., Zhang, X., Ren, S., Sun, J.: Deep residual learning for image recognition. In: 2016 IEEE Conference on Computer Vision and Pattern Recognition (CVPR). IEEE (2016)
13. Chollet, F.: Xception: deep learning with depthwise separable convolutions. In: 2017 IEEE Conference on Computer Vision and Pattern Recognition (CVPR). IEEE (2017)
14. Lu, M., Hu, Y., Lu, X.: Driver action recognition using deformable and dilated faster R-CNN with optimized region proposals. Appl. Intell. **50**(4), 1100–1111 (2019). https://doi.org/10.1007/s10489-019-01603-4
15. Hu, Y., Lu, M., Lu, X.: Driving behaviour recognition from still images by using multi-stream fusion CNN. Mach. Vis. Appl. **30**(5), 851–865 (2018). https://doi.org/10.1007/s00138-018-0994-z
16. Yan, C., Coenen, F., Zhang, B.: Driving posture recognition by convolutional neural networks. IET Comput. Vis. **10**(2), 103–114 (2016)
17. Bolya, D., Zhou, C., Xiao, F., et al.: YOLACT++: better real-time instance segmentation. IEEE Trans. Pattern Anal. Mach. Intell. **PP**(99), 1 (2020)
18. Kaggle Competition: State Farm Distracted Driver Detection. https://www.kaggle.com/c/state-farm-distracted-driver-detection. Accessed 12 Apr 2017
19. Baheti, B., Gajre, S., Talbar, S.: Detection of distracted driver using convolutional neural network. In: 2018 IEEE/CVF Conference on Computer Vision and Pattern Recognition Workshops (CVPRW). IEEE (2018)
20. Pohl, J., Birk, W., Westervall, L.: A driver-distraction-based lane-keeping assistance system. Proc. Inst. Mech. Eng. Part I J. Syst. Control Eng. **221**, 541–552 (2007)
21. Alotaibi, M., Alotaibi, B.: Distracted driver classification using deep learning. Signal Image Video Process. **14**, 617–624 (2019)
22. Hu, J., Xu, L., He, X., et al.: Abnormal driving detection based on normalised driving behaviour. IEEE Trans. Veh. Technol. **66**(8), 6645–6652 (2017)
23. Eren, H., Celik, U., Poyraz, M.: Stereo vision and statistical based behaviour prediction of driver. In: Proceedings of the 2007 IEEE Intelligent Vehicles Symposium, Istanbul, Turkey, June 13–15 2007, pp. 657–662 (2007)
24. Murphy-Chutorian, E., Doshi, A., Trivedi, M.M.: Head pose estimation for driver assistance systems: a robust algorithm and experimental evaluation. In: Intelligent Transportation Systems Conference, pp. 709–714. IEEE (2007)
25. Eraqi, H.M., Abouelnaga, Y., Saad, M.H., Moustafa, M.N.: Driver distraction identification with an ensemble of convolutional neural networks. J. Adv. Transp. (2019)
26. Berri, R.A., Silva, A.G., Parpinelli, R.S., Girardi, E., Arthur, R.: A pattern recognition system for detecting use of mobile phones while driving. In: International Conference on Computer Vision Theory and Applications, VISAPP, pp. 411–418. IEEE (2014)
27. Zhao, C.H., Zhang, B.L., He, J., Lian, J.: Recognition of driving postures by contourlet transform and random forests. IET Intell. Transp. Syst. **6**, 161–168 (2012)

Neural Topic Modeling with Gaussian Mixture Model and Householder Flow

Cangqi Zhou[1,2], Sunyue Xu[1], Hao Ban[3], and Jing Zhang[1(✉)]

[1] Nanjing University of Science and Technology, Nanjing, China
{cqzhou,sue,jzhang}@njust.edu.cn
[2] SenseDeal Intelligent Technology Co., Ltd., Beijing, China
[3] Southeast University, Nanjing, China
banhao@seu.edu.cn

Abstract. To overcome high computational cost suffered by statistical inference algorithms used in traditional topic models, recently, Variational AutoEncoder (VAE) frameworks have been proposed for topic modeling. However, the vanilla VAE model is originally introduced for unsupervised learning, which cannot meet more precise and customized requirements. In addition, the approximate posterior distribution in VAE is often selected as a Gaussian with a diagonal covariance matrix. This unimodal choice may hinder the ability of representation of latent space. In view of these limitations, in this paper, we propose to use Gaussian mixture model and Householder Flows for topic modeling under semi-supervised settings. We assume a document is associated with a mixture of classes, and a class is modeled as a multivariate Gaussian over latent topics. Specifically, an input document is encoded by a network into a discrete distribution, which not only serves the classifier for prediction, but also acts as mixing weights of Gaussian components. Another network is adopted to learn the parameters of Gaussian components. Additionally, Gaussian mixture is transformed by a Householder Flow to produce a more general posterior distribution. The effectiveness of the proposed model has been validated by the experiments performed on several standard datasets.

Keywords: Topic model · Variational autoencoder · Semi-supervised learning · Gaussian mixture model · Householder flow

1 Introduction

Topic models are proposed to summarize and cluster documents by unveiling the patterns of words and phrases [6]. A wide range of real-world applications could benefit greatly from topic modeling [2,27]. Arguably the most well-known topic model is Latent Dirichlet Allocation (LDA) [7]. There are also numerous variants that extend LDA [3,21]. However, the inference approaches adopted by LDA and its variants suffer from heavy computational cost and burdensome mathematical derivations, which reduces the scalability to large datasets [4].

J. Gama et al. (Eds.): PAKDD 2022, LNAI 13281, pp. 417–428, 2022.
https://doi.org/10.1007/978-3-031-05936-0_33

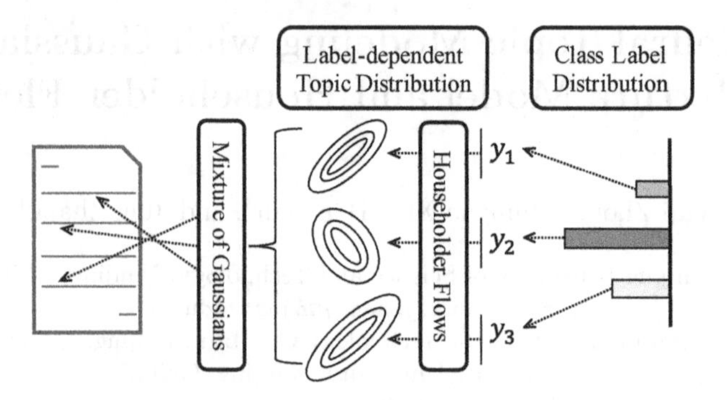

Fig. 1. A schematic of the basic assumption. First, a distribution over classes is associated with each document. Under each class, latent variables are modeled as a multivariate Gaussian with a diagonal covariance matrix at first. After Householder transformations, these Gaussians become more complex distributions with non-diagonal covariance matrices. The class weights are used to combine Gaussian mixture components. They are also used to predict class labels.

In view of these limitations, neural networks (NNs) are introduced to undertake the task of inference [22]. One typical example is Variational AutoEncoder (VAE) [11]. It utilizes the neural network structure to encode continuous latent variables, and the Stochastic Gradient Variational Bayes algorithm (SGVB) to approximate true posterior distributions [11]. VAE uses an inference network to approximate latent topic variables conditioned on input text [19,25]. However, these VAE-based topic modeling approaches have two main limitations.

The first limitation is that the vanilla VAE is mainly proposed for learning unlabeled data. However, an increasing number of demands for more precise, customized and controllable topic models require researchers to utilize partially labeled text data to build topic models under semi-supervised settings [18]. Partially labeled data are not only valuable for performance [31], but also feasible and practicable to obtain. Unfortunately, extending VAE-based topic models to semi-supervised learning is often non-trivial, since the vanilla VAE is not applicable to model both input labels and data simultaneously [11,24].

The second limitation is that the selected approximate posterior distribution in VAE, usually a Gaussian with diagonal covariance matrix, is often not sophisticated enough to model the true posterior [8]. This Gaussian distribution for hidden variables in topic models is a rigid choice to represent the latent space, since it cannot model multi-mode factors and correlated latent variables. A number of approaches have been proposed to increase the flexibility of variational distributions to accommodate real-world applications, such as Structured VAE [10], Auxiliary VAE [17], etc. Among them, the normalizing flow approach [26] is proved to be intuitive, effective, and compatible to VAE framework.

To address the above limitations, in this paper, we propose `SVAE-GH`, a VAE-based semi-supervised topic modeling framework with the combination of Gaussian mixture model (GMM) and Householder Flows (HF), for a more flexible modeling of the latent space. SVAE-GH extends our previous model [30] by introducing Householder Flows for a better topic correlation modeling. The basic assumption is that each document is endowed with two types of latent variables, class labels and latent topics. A document is associated with a mixture of classes with different weights. A class-dependent latent topic is modeled as a multivariate Gaussian distribution. By now, each Gaussian component is associated with a diagonal covariance matrix, which means topics are uncorrelated between each other. To improve the flexibility and to model topic correlations, a Householder Flow is further applied to the Gaussian mixture distribution to produce more complex approximate posteriors. The main idea is illustrated in Fig. 1. The efficacy of the proposed model has been validated by the experiments performed on three benchmark datasets, with comparisons to four baseline topic models.

2 Preliminaries

2.1 Variational Autoencoder

Given a dataset $\boldsymbol{X} = \{\boldsymbol{x}_1, \ldots, \boldsymbol{x}_N\}$, generally, we intend to maximize the log-likelihood $\ln p(\boldsymbol{X}) = \sum_{i=1}^{N} \ln p(\boldsymbol{x}_i)$. By introducing some latent variable $\boldsymbol{z}$, the likelihood can be marginalized as $p(\boldsymbol{x}) = \int p_\theta(\boldsymbol{x}|\boldsymbol{z})p(\boldsymbol{z})\,\mathrm{d}\boldsymbol{z}$. However, the integral of the marginal likelihood is often computational intractable. To overcome this issue, the vanilla VAE proposes an inference procedure (an encoder) $q(\boldsymbol{z}|\boldsymbol{x})$ to approximate the true posterior. And the so-called Evidence Lower BOund (ELBO) will be maximized instead,

$$\ln p(\boldsymbol{x}) \geq \mathbb{E}_{q(\boldsymbol{z}|\boldsymbol{x})}[\ln p(\boldsymbol{x}|\boldsymbol{z})] - \mathbb{D}_{\mathrm{KL}}[q(\boldsymbol{z}|\boldsymbol{x})||p(\boldsymbol{z})] \tag{1}$$

where $p(\boldsymbol{x}|\boldsymbol{z})$ is the decoder and $p(\boldsymbol{z})$ is the prior, and KL denotes the Kullback-Leibler divergence (KLD). During implementation, q is often assumed to be a Gaussian with a diagonal covariance matrix, parameterized by neural networks.

2.2 Normalizing Flow

To construct more flexible latent variables, a series of invertible transformations can be applied to the original variables. The invertible transformation is denoted as an invertible function f. Suppose the probability density function (PDF) of a continuous random vector $\boldsymbol{X}$ is $p(\boldsymbol{x})$, and $\boldsymbol{Y} = f(\boldsymbol{X})$. The PDF of $\boldsymbol{Y}$ is

$$\boldsymbol{Y} \sim p(\boldsymbol{y}) = p(\boldsymbol{x})\left|\det \frac{\mathrm{d}f}{\mathrm{d}\boldsymbol{x}}\right|^{-1} \tag{2}$$

where $|\det J|$ denotes the absolute value of the determinant of the Jacobian.

If a series of invertible functions f_t, $t = 1, \ldots, T$ are applied, we obtain a normalizing flow. Suppose the initial random vector is $\boldsymbol{z}_0$, and its distribution is denoted as $q_0(\boldsymbol{z}_0)$. Then $\boldsymbol{z}_T$ can be obtained by $\boldsymbol{z}_T = f_T \circ \cdots \circ f_1(\boldsymbol{z}_0)$, and the distribution of the final random vector $\boldsymbol{z}_T$ is

$$\boldsymbol{z}_T \sim p(\boldsymbol{z}_T) = q_0(\boldsymbol{z}_0) \prod_{t=1}^{T} \left| \det \frac{\mathrm{d} f_t}{\mathrm{d} \boldsymbol{z}_{t-1}} \right|^{-1} \tag{3}$$

According to [23], any expectation $\mathbb{E}_{q_T}[h(\boldsymbol{z}_T)]$ over q_T can be represented as the expectation over q_0

$$\mathbb{E}_{q_T}[h(\boldsymbol{z}_T)] = \mathbb{E}_{q_0}[h(f_t \circ \cdots \circ f_1(\boldsymbol{z}_0))] \tag{4}$$

Equipped with normalizing flow, the ELBO can be re-written as

$$\ln p(\boldsymbol{x}) \geq \mathbb{E}_{q_0(\boldsymbol{z}_0|\boldsymbol{x})}[\ln p(\boldsymbol{x}|\boldsymbol{z}_T)] - \mathbb{D}_{\mathrm{KL}}[q_0(\boldsymbol{z}_0|\boldsymbol{x}) \| p(\boldsymbol{z}_T)] + \sum_{t=1}^{T} \ln \left| \det \frac{\partial f_t}{\partial \boldsymbol{z}_{t-1}} \right| \tag{5}$$

The first term on the RHS of Eq.(5) is the reconstruction error. However, different from the ELBO in vanilla VAE, the encoder models $\boldsymbol{z}_0$ given $\boldsymbol{x}$, and the decoder models $\boldsymbol{x}$ given $\boldsymbol{z}_T$. The second term acts as a regularization term. The third term comes from the normalizing flow. In this paper, we adopt the volume-preserving Householder Flow (HF) to model the transformation.

3 Methodology

Now we describe the principle of SVAE-GH. Some details can be found in our previous work [30]. Here we highlight the differences brought by HF.

3.1 Problem Formalization

We denote a corpus, i.e., a collection of documents as $\boldsymbol{X} = \{\boldsymbol{x}_1, \boldsymbol{x}_2, \ldots, \boldsymbol{x}_N\}$, where $\boldsymbol{x}_i \in \mathbb{R}^{|\mathcal{V}|}$ denotes the vector that corresponds to the representation of the i-th document, and $\mathcal{V}$ denotes the vocabulary. By introducing a latent variable $\boldsymbol{z} \in \mathbb{R}^K$, where the dimension of latent space K is consistent with the number of topics, $\boldsymbol{x}_i$ is allowed to be parameterized and then sampled from the conditional distribution $p(\boldsymbol{x}_i|\boldsymbol{z})$. The dataset can be divided into a labeled set $\boldsymbol{X}_l$ and an unlabeled set $\boldsymbol{X}_u$. $\boldsymbol{X}_l$ is associated with its labels as $\boldsymbol{X}_l = \{(\boldsymbol{x}_i, y_i)\}$, where $y_i \in \{y_1, y_2, \ldots, y_M\}$ is the label of corresponding $\boldsymbol{x}_i$. We intend to use VAE to train a probabilistic model, to mimic the generation of $\boldsymbol{X}$. Therefore, a) the posterior $p(y_j|\boldsymbol{x}_i)$ will be used to predict class labels; b) the posterior $p(z_k|\boldsymbol{x}_i)$ will be used to recover the latent topics of documents.

3.2 Variational Objective

Under the semi-supervised setting, the marginalization of $p(\boldsymbol{x})$ can be factorized by the chain rule of conditional probability as $p(\boldsymbol{x}) = \iint p_\theta(\boldsymbol{x}|y, \boldsymbol{z}) p(y, \boldsymbol{z}) \, \mathrm{d}y \, \mathrm{d}\boldsymbol{z}$

by introducing both variable y and $\boldsymbol{z}$. The likelihood $p_\theta(\boldsymbol{x}|y, \boldsymbol{z})$ will be learned by an decoder network. And the posterior $p(y, \boldsymbol{z}|\boldsymbol{x})$ of prior $p(y, \boldsymbol{z})$ will be approximated by another distribution $q_\phi(y, \boldsymbol{z}|\boldsymbol{x})$, which will be modeled as an encoder parameterized by ϕ. The KLD between the true posterior and q_ϕ will be minimized to get the objective.

If normalizing flows are not taking into consideration, one can refer to Kingma's work [12] to obtain the ELBO for labeled and unlabeled data. When normalizing flow is considered, we need to slightly change the forms of objectives.

Specifically, we assume the factorization as $p(y, \boldsymbol{z}_T) = p(y)p(\boldsymbol{z}_T)$. We also factorize q_ϕ as the product of the encoder of class labels $q_{\phi_1}(y|\boldsymbol{x})$ and the encoder of latent variables $q_{\phi_2}^{(0)}(\boldsymbol{z}_0|\boldsymbol{x}, y)$, where $q_{\phi_2}^{(0)}$ represents the initial distribution of latent variable $\boldsymbol{z}_0$, which is a Gaussian with diagonal covariance matrix. ϕ_1, ϕ_2 and θ are the parameters for the encoder of the latent label variable, the encoder of the latent topic variable and the decoder, respectively.

Since y can be observed in $\boldsymbol{X}_l$, the ELBO of a single labeled sample is only conditioned on $\boldsymbol{z}$

$$\begin{aligned}\log p(\boldsymbol{x}, y) \geq & -\mathcal{L}(\boldsymbol{x}, y) \\ & = \mathbb{E}_{q_0(\boldsymbol{z}_0|\boldsymbol{x},y)}\left[\ln p_\theta(\boldsymbol{x}|y, \boldsymbol{z}_T)\right] - \mathbb{D}_{\mathrm{KL}}\left[q_0(\boldsymbol{z}_0|\boldsymbol{x}, y) \mid\mid p(y)p(\boldsymbol{z}_T)\right] \\ & + \sum_{t=1}^{T} \ln \left| \det \frac{\partial f_t}{\partial \boldsymbol{z}_{t-1}} \right|\end{aligned} \tag{6}$$

Following [12], a cross entropy objective is introduced to empower the model to classify unlabeled data with only limited labeled data. Then the objective for all $(\boldsymbol{x}, y) \in \boldsymbol{X}_l$ is

$$\mathcal{J}_l = \mathbb{E}_{\tilde{p}_l}[\mathcal{L}(\boldsymbol{x}, y)] + \mathbb{E}_{\tilde{p}_l}[-\log q_{\phi_1}(y|\boldsymbol{x})] = \mathbb{E}_{\tilde{p}_l}[\mathcal{L}(\boldsymbol{x}, y)] + \mathcal{H}[\tilde{p}_l, q_{\phi_1}] \tag{7}$$

where $\tilde{p}_l$ is the empirical distribution of labeled data, $\mathcal{H}$ is the cross entropy.

The ELBO for one sample in $\boldsymbol{X}_u$ is

$$\begin{aligned}\log p(\boldsymbol{x}) \geq & \mathbb{E}_{q_\phi^{(0)}}\left[\log p_\theta(\boldsymbol{x}|y, \boldsymbol{z}_T)\right] - \mathbb{D}_{\mathrm{KL}}\left[q_\phi^{(0)}(y, \boldsymbol{z}_0|\boldsymbol{x}) \mid\mid p(y, \boldsymbol{z}_T)\right] \\ & + \sum_{t=1}^{T} \ln \left| \det \frac{\partial f_t}{\partial \boldsymbol{z}_{t-1}} \right|\end{aligned} \tag{8}$$

Here y cannot be observed and it is considered as a hidden variable. Equation(8) can then be factorized and re-written as $\log p(\boldsymbol{x}) \geq \mathbb{E}_{q_{\phi_1}}\left[-\mathcal{L}(\boldsymbol{x}, y) - \log q_{\phi_1}(y|\boldsymbol{x})\right]$. The objective for the whole dataset $\boldsymbol{X}_u$ is

$$\begin{aligned}\mathcal{J}_u & = \mathbb{E}_{\tilde{p}_u(\boldsymbol{x})}\left[\mathbb{E}_{q_{\phi_1}}[\mathcal{L}(\boldsymbol{x}, y)] - \mathbb{E}_{q_{\phi_1}}[-\log q_{\phi_1}]\right] \\ & = \mathbb{E}_{\tilde{p}_u(\boldsymbol{x})}\mathbb{E}_{q_{\phi_1}}\mathcal{L}(\boldsymbol{x}, y) - \mathcal{H}[y|\boldsymbol{x}]\end{aligned} \tag{9}$$

where $\mathcal{H}[y|\boldsymbol{x}]$ denotes the conditional entropy of y given $\boldsymbol{x}$.

Finally, the objective for the whole set $\boldsymbol{X}$ is $\mathcal{J} = \mathcal{J}_l + \mathcal{J}_u$.

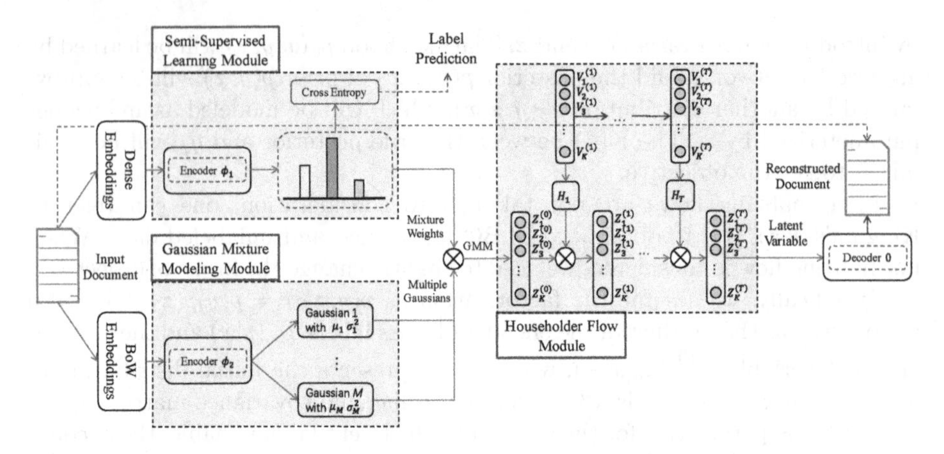

Fig. 2. An overview of our modeling framework. Two encoders, ϕ_1 and ϕ_2, are implemented separately, one for class probabilities and the other for the parameters of Gaussian components. Latent variables are first sampled from the GMM, and then transformed by a series of Householder matrices. After that, the decoder θ is used to reconstruct documents.

3.3 Model Framework

An overview of our method is illustrated in Fig. 2. The framework consists of basically three modules, a semi-supervised learning module, a Gaussian mixture module and a Householder Flow module, which will be detailed below.

The basic assumptions about the generation procedure of a corpus of documents are given as follows.

- Each class corresponds to a Gaussian distribution on latent variables, which represent the topics that guide the generation of documents.
- A document is assumed to be a mixture of several classes. And the weight for each class corresponds to the weight of each Gaussian.
- The approximate posterior of the initial latent variable $\boldsymbol{z}_0$ is modeled as a mixture of Gaussians with diagonal covariance matrices. And the final latent variable $\boldsymbol{z}_T$ is transformed by a Householder Flow from $\boldsymbol{z}_0$.

Encoder. As mentioned before, two encoders q_{ϕ_1} and q_{ϕ_2} will be implemented for latent variable y and $\boldsymbol{z}_0$ separately. Specifically, multi-layer perceptron (MLP) will be used for both encoders. Parameter $\boldsymbol{\pi}_{\phi_1}$, which is parameterized by the first network, produces label y after a softmax function. It also acts as the mixing weights for the generation of GMM variable $\boldsymbol{z}_0$,

$$q_{\phi_2}^{(0)}(\boldsymbol{z}_0|\boldsymbol{x}, y) = \sum_{j=1}^{M} \pi_j(\boldsymbol{x};\phi_1)\mathcal{N}\big(\boldsymbol{z}_0|\boldsymbol{\mu}_j(\boldsymbol{x}, y;\phi_2), \text{diag}(\boldsymbol{\sigma}_j^2(\boldsymbol{x};\phi_2))\big) \tag{10}$$

where $\boldsymbol{\mu}_j$ and $\boldsymbol{\sigma}_j^2$ are produced by the second inference network parameterized by ϕ_2. And the re-parametrization trick will be adopted for the sampling of $\boldsymbol{z}_0$.

Then $\boldsymbol{z}_0$ will be transformed by $H_1, H_2, \ldots, H_T$ to obtain a more complex latent variable $\boldsymbol{z}_T$ for the decoder to reconstruct the data.

Decoder. The decoder network $p_\theta(\boldsymbol{x}|\boldsymbol{z}_T)$ will adopt the implementation used in NVDM [19], in which a document will be decomposed as D words $\boldsymbol{w}_d$, where $\boldsymbol{w}_d$ is a one-hot representation of the d-th word. Then decoding a document will be transformed as decoding D words by

$$p_\theta(\boldsymbol{x}|\boldsymbol{z}_T) = \prod_{d=1}^{D} p_\theta(\boldsymbol{w}_d|\boldsymbol{z}_T) \tag{11}$$

NVDM introduces a $K \times |\mathcal{V}|$ matrix R to represent the correlations between topics and the vocabulary. It uses a softmax regression model to evaluate the score of a word given a topic. Details can be found in [19].

3.4 KLD Between Two GMMs

Equation (6) requires the computation of the KLD between a GMM with a diagonal covariance matrix and a standard Gaussian. The details can be found in [14]. We only show the final results here. The upper bound of $L(\boldsymbol{x}, y)$ is

$$\begin{aligned}\tilde{L}(\boldsymbol{x}, y) = &- \mathbb{E}_{q_0(\boldsymbol{z}_0|\boldsymbol{x},y)}\left[\ln p_\theta(\boldsymbol{x}|y, \boldsymbol{z}_T)\right] \\ &+ \sum_{j=1}^{M} \pi_j \left[\log \frac{\pi_j}{\tilde{\pi}_j} - \frac{1}{2}\sum_{k=1}^{K}\left(1 + \log \sigma_{jk}^2 - \sigma_{jk}^2 - \mu_{jk}^2\right)\right]\end{aligned} \tag{12}$$

where π_j is the mixture weight in Eq.(10), $\tilde{\pi}_j$ is any mixture weight of standard Gaussians, μ_{jk} and σ_{jk}^2 are the parameters of the k-th coordinate of the j-th mixture component of the approximate posterior GMM of the initial latent variable $\boldsymbol{z}_0$, i.e., $\boldsymbol{\mu}_j(\boldsymbol{x}, y; \phi_2)$ and $\boldsymbol{\sigma}_j^2(\boldsymbol{x}; \phi_2)$ in Eq.(10), respectively.

Then the final objective becomes

$$\mathcal{J} = \mathbb{E}_{\tilde{p}_l}\tilde{\mathcal{L}}(\boldsymbol{x}, y) + \mathbb{E}_{\tilde{p}_u}\mathbb{E}_{q_{\phi_1}}\tilde{\mathcal{L}}(\boldsymbol{x}, y) + \mathcal{H}[y|\boldsymbol{x}] - \mathcal{H}[\tilde{p}_l, q_{\phi_1}] \tag{13}$$

To minimize the objective, some instantiations of the stochastic gradient descent algorithm, such as ADAM will be used to obtain optimized network parameters.

4 Experiment

4.1 Datasets

Three public available benchmark datasets are used to perform our experiments, including 20NewsGroup[1], a news dataset covering 20 newsgroups with different topics; IMDB[2], containing the positive and negative polarized movie reviews; and

---

[1] http://qwone.com/~jason/20Newsgroups.

[2] https://ai.stanford.edu/~amaas/data/sentiment/.

Table 1. Dataset Statistics. *Training* and *Testing* show how the datasets are split. $|\mathcal{V}|$ is the vocabulary size, and $|\boldsymbol{Y}|$ is the number of labels.

Dataset	Total	Training	Testing	$\|\mathcal{V}\|$	$\|\boldsymbol{Y}\|$
20NewsGroup	18,846	11,314	7,532	2,000	20
IMDB	50,000	25,000	25,000	10,000	2
AGNews	127,600	120,000	7,600	5,000	4

AGNews [29][3], another dataset containing news articles from 4 classes. Details of the descriptions and the statistics of these datasets can be found in corresponding links and Table 1. To comply the convention of semi-supervised learning, the labels of only 20% of training samples are kept.

4.2 Evaluation Metrics

To evaluate the performance of our model, we adopt two commonly-used metrics, perplexity and Pointwise Mutual Information (PMI), to validate the performance of topic models. Perplexity measures to what degree the predicted topic distribution is consistent with test data. It is defined as a function of cross entropy

$$Perplexity = \exp\big(-\frac{1}{N}\sum_i \frac{1}{D_i}\log p(\boldsymbol{x}_i)\big) \tag{14}$$

where N is the number of documents, D_i is the number of words in the i-th document. This value of perplexity is often approximated by its upper bound [19,20]. PMI [1] evaluates the co-occurrence probabilities of topic words as

$$PMI(\omega_i, \omega_j) = \log \frac{p(\omega_i, \omega_j)}{p(\omega_i)p(\omega_j)} \tag{15}$$

where $p(\omega)$ is the probability of seeing the topic word ω in a random document in the test set. And the PMI of a topic is evaluated by summing the PMI of the top ordered pairs of topic words, $\sum_{i<j} PMI(\omega_i, \omega_j)$. For the whole dataset, PMI is evaluated by averaging the PMI of each topic.

4.3 Comparison Methods

Four well-known topic models are chosen as our baselines. 1) LDA [7], the most famous and well-known topic model based on statistical inference [9,16]. 2) CTM [13], an extension of LDA by introducing topic correlations using logistic normal distribution for topic mixtures [5,28]. 3) NVDM [19], a pioneer study that reformulates topic models as deep generative neural networks. NVDM significantly improves model performance over statistical inference based models. 4) CHTM [15] adopts HF to extend NVDM. The difference between CHTM and our method is that we use GMM to model the latent space based on class latent variables, while CHTM just uses a single Gaussian for topic modeling.

[3] http://groups.di.unipi.it/~gulli/AG_corpus_of_news_articles.html.

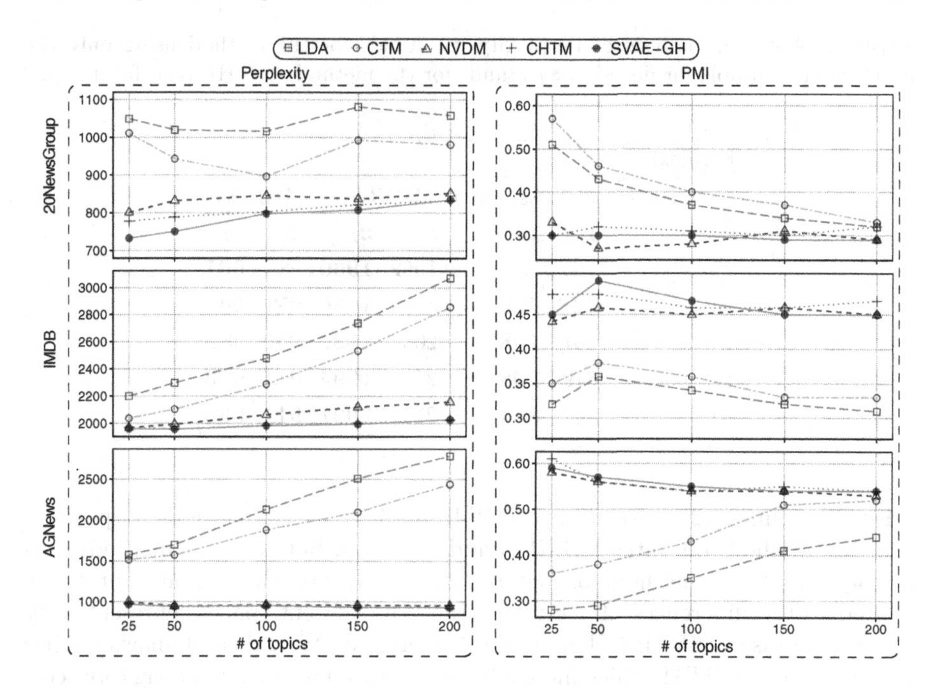

Fig. 3. Comparison results of Perplexity (the lower the better) and NMI (the higher the better) with respect to the variations of the numbers of latent topics.

4.4 Parameter Settings

Encoder ϕ_1 takes dense representations of documents as its inputs, which are obtained by summing the pre-trained Glove vectors of words. On the contrary, encoder ϕ_2 takes sparse bag-of-words (BoW) representations as its inputs. Hence, the dimensions of these two representations are different. For dense embeddings, it is set to 100. For BoW, it is set to the vocabulary size of the dataset. The sizes of the hidden layers of the MLP for both ϕ_1 and ϕ_2 are set to 500. The number of nodes in the output layers of the MLP are consistent with the number of labels. Then HF is applied. The length of HF is set to 10. The settings of the decoder is the same as those used in NVDM, except that we use the transformed latent variables to reconstruct document words. During training, learning rate, batch size and the maximum epochs are set to 10^{-4}, 64, and 1,000, respectively. Other settings are identical to our previous work [30].

4.5 Comparison Results

According to the results shown in Fig. 3, we find that, (1) The proposed method, SVAE-GH, achieves the lowest (best) perplexities under almost all settings. For dataset 20NewsGroup and IMDB, SVAE-GH outperforms other 3 baselines under all 5 settings of topic numbers (dimension of latent variable). For datasets AGNews, SVAE-GH and NVDM achieve similar performance, and outperform

Table 2. Ablation analysis of Perplexity. z_0 stands for the method using only the initial latent variable for decoder. z_T stands for the method with HF transformations.

Perplexity						
# of Topics	*20NewsGroup*		*IMDB*		*AGNews*	
	z_0	z_T	z_0	z_T	z_0	z_T
25	**732**	733	1962	**1960**	988	**967**
50	**748**	751	1984	**1956**	959	**942**
100	801	**798**	2064	**1982**	**943**	952
150	819	**808**	2114	**1994**	949	**933**
200	842	**834**	2153	**2027**	942	**932**

other 2 baselines. (2) With regard to PMI, all 3 NN-based methods outperform Non-NN methods on datasets IMDB and AGNews, but Non-NN methods outperform NN-based methods on dataset 20NewsGroup. The reason is probably related to the differences of the true structure of latent spaces represented by different datasets. SVAE-GH performs better than NVDM, and shows similar performance as CHTM, which indicates the effectiveness of introducing topic correlations. (3) Non-NN-based methods, such as LDA and CTM, can not achieve comparable results as NN-based methods, such as NVDM and our method. Further, as the number of topics increases, non-NN-based methods show obvious deviation of performance. On the contrary, NN-based methods, including SVAE-GH, are robust to this parameter.

In conclusion, the proposed SVAE-GH exhibits effectiveness in topic modeling, especially under relatively small number of latent topics.

4.6 Ablation Study

Adopting GMM has been validated in our previous work [30]. In this paper, we perform ablation experiments to examine whether there are gains when HF is adopted. We compare the performance between using the initial latent variable z_0 and the transformed latent variable z_T for decoding. The corresponding perplexities are shown in Table 2, from which we discover that, by using Householder Flow, the performance of our model with respect to perplexity improves, especially with IMDB and AGNews. With dataset 20NewsGroup, perplexities are improved as the number of latent topics increases. In general, introducing HF to latent space can remarkably improve the model's performance under a majority of conditions.

5 Conclusion

We propose SVAE-GH, a novel semi-supervised framework based on variational autoencoder for topic modeling. We introduce Gaussian mixture model and

Householder Flow to learn a more general and more flexible distribution for latent space modeling. Specifically, we use two MLP neural networks for encoding the parameters of the Gaussian components and the mixing weights of these components separately. After learning the initial Gaussian mixture model with diagonal covariance matrices, Householder Flows are adopted to transform the distribution into a more general and flexible one. Transformed variables are then sampled from the learned distribution for decoding and reconstructing original documents. This model has been validated by the experiments performed on three benchmark datasets, comparing to four baselines, including both none-neural network-based and neural network-based methods. We hope that this study could shed some light on neural topic modeling and related applications.

Acknowledgements. This work was supported by the National Key Research and Development Program of China (No. 2018AAA0102002), and the National Natural Science Foundation of China (No. 61902186, 62076130 and 91846104).

References

1. Aletras, N., Stevenson, M.: Evaluating topic coherence using distributional semantics. In: Proceedings of the 10th International Conference on Computational Semantics (IWCS 2013), pp. 13–22 (2013)
2. Arnold, C., Speier, W.: A topic model of clinical reports. In: Proceedings of the 35th International ACM SIGIR Conference on Research and Development in Information Retrieval, pp. 1031–1032 (2012)
3. Blei, D., Lafferty, J.: Correlated topic models. Adv. Neural. Inf. Process. Syst. **18**, 147 (2006)
4. Blei, D.M., Kucukelbir, A., McAuliffe, J.D.: Variational inference: a review for statisticians. J. Am. Stat. Assoc. **112**(518), 859–877 (2017)
5. Blei, D.M., Lafferty, J.D.: A correlated topic model of science. Ann. Appl. Statist. **1**(1), 17–35 (2007)
6. Blei, D.M., Lafferty, J.D.: Topic models. In: Text Mining, pp. 101–124. Chapman and Hall/CRC, Boca Raton (2009)
7. Blei, D.M., Ng, A.Y., Jordan, M.I.: Latent Dirichlet allocation. J. Mach. Learn. Res. **3**, 993–1022 (2003)
8. Burda, Y., Grosse, R., Salakhutdinov, R.: Importance weighted autoencoders. In: 4th International Conference on Learning Representations (ICLR) (2016)
9. Jelodar, H., et al.: Latent Dirichlet Allocation (LDA) and topic modeling: models, applications, a survey. Multimedia Tools Applicat. **78**(11), 15169–15211 (2019)
10. Johnson, M.J., Duvenaud, D.K., Wiltschko, A., Adams, R.P., Datta, S.R.: Composing graphical models with neural networks for structured representations and fast inference. Adv. Neural. Inf. Process. Syst. **29**, 2946–2954 (2016)
11. Kingma, D.P., Welling, M.: Auto-encoding variational Bayes. arXiv preprint arXiv:1312.6114 (2013)
12. Kingma, D.P., Mohamed, S., Rezende, D.J., Welling, M.: Semi-supervised learning with deep generative models. In: Advances in Neural Information Processing Systems, pp. 3581–3589 (2014)
13. Lafferty, J.D., Blei, D.M.: Correlated topic models. In: Advances in Neural Information Processing Systems, pp. 147–154 (2006)

14. Liu, G., Liu, Y., Guo, M., Li, P., Li, M.: Variational inference with Gaussian mixture model and Householder flow. Neural Netw. **109**, 43–55 (2019)
15. Liu, L., Huang, H., Gao, Y.: Correlated topic modeling via Householder flow. In: Proceedings of the 1st International Workshop on Explainable Recommendation and Search (2018)
16. Liu, Z., Li, M., Liu, Y., Ponraj, M.: Performance evaluation of latent Dirichlet allocation in text mining. In: 2011 Eighth International Conference on Fuzzy Systems and Knowledge Discovery (FSKD), vol. 4, pp. 2695–2698. IEEE (2011)
17. Maaløe, L., Sønderby, C.K., Sønderby, S.K., Winther, O.: Auxiliary deep generative models. In: Proceedings of the 33rd International Conference on Machine Learning, pp. 1445–1454 (2016)
18. Mcauliffe, J., Blei, D.: Supervised topic models. Adv. Neural. Inf. Process. Syst. **20**, 121–128 (2007)
19. Miao, Y., Yu, L., Blunsom, P.: Neural variational inference for text processing. In: Proceedings of the 33rd International Conference on Machine Learning, pp. 1727–1736 (2016)
20. Mnih, A., Gregor, K.: Neural variational inference and learning in belief networks. In: Proceedings of the 31st International Conference on Machine Learning, pp. II-1791 (2014)
21. Perotte, A.J., Wood, F., Elhadad, N., Bartlett, N.: Hierarchically supervised latent Dirichlet allocation. In: Advances in Neural Information Processing Systems, pp. 2609–2617 (2011)
22. Ranganath, R., Gerrish, S., Blei, D.M.: Black box variational inference. In: Proceedings of the 17th International Conference on Artificial Intelligence and Statistics (2014)
23. Rezende, D.J., Mohamed, S.: Variational inference with normalizing flows. arXiv preprint arXiv:1505.05770 (2015)
24. Rezende, D.J., Mohamed, S., Wierstra, D.: Stochastic backpropagation and approximate inference in deep generative models. arXiv preprint arXiv:1401.4082 (2014)
25. Srivastava, A., Sutton, C.: Autoencoding variational inference for topic models. arXiv preprint arXiv:1703.01488 (2017)
26. Tomczak, J.M., Welling, M.: Improving variational auto-encoders using Householder flow. arXiv preprint arXiv:1611.09630 (2016)
27. Xia, X., Lo, D., Ding, Y., Al-Kofahi, J.M., Nguyen, T.N., Wang, X.: Improving automated bug triaging with specialized topic model. IEEE Trans. Software Eng. **43**(3), 272–297 (2016)
28. Xun, G., Li, Y., Zhao, W.X., Gao, J., Zhang, A.: A correlated topic model using word embeddings. In: Proceedings of the 26th International Joint Conference on Artificial Intelligence, pp. 4207–4213 (2017)
29. Zhang, X., Zhao, J., LeCun, Y.: Character-level convolutional networks for text classification. In: Advances in Neural Information Processing Systems, pp. 649–657 (2015)
30. Zhou, C., Ban, H., Zhang, J., Li, Q., Zhang, Y.: Gaussian mixture variational autoencoder for semi-supervised topic modeling. IEEE Access **8**, 106843–106854 (2020)
31. Zhu, X., Goldberg, A.B.: Introduction to semi-supervised learning. Synth. Lect. Artif. Intell. Mach. Learn. **3**(1), 1–130 (2009)

Dynamic Topic-Noise Models for Social Media

Rob Churchill(✉) and Lisa Singh

Georgetown University, Washington DC, USA
rjc111@georgetown.edu

Abstract. Temporal topic models often cannot effectively approximate topics on social media data sets due to the noise levels inherent in these types of data. Topic-noise models are important for modeling the short, sparse, noisy posts that we see throughout social media platforms. We propose using topic-noise models for temporal topic modeling, specifically D-TND (dynamic topic-noise discriminator). It enables topic and noise distributions to be generated together, modeling both the relationships between words in documents and the evolution of words and noise. We also propose Dynamic Noiseless Latent Dirichlet Allocation (D-NLDA), which integrates D-TND's time-dependent noise distribution with the topic distributions of Dynamic LDA, and show its propensity for improving dynamic topic models by effectively separating noise and topics on two large Twitter data sets.

1 Introduction

Topic models are important unsupervised tools for quickly understanding large textual data sets. They can be particularly useful when attempting to understand the discussion surrounding a large number of social media posts [4,22,26]. A number of topic models have been designed specifically to more accurately model social media data [5,16,21,27]. More recently, a class of topic models called topic-noise models was proposed to jointly model topic and noise distributions on social media data [8]. None of these models incorporate a temporal dimension. Temporal topic models enable researchers to not only identify the relevant underlying topics in a data set, but also to track the evolution of these topics through time. Recently, there has been a renewed interest in temporal topic models, with the publication of a graph-based dynamic topic model [12], and an embedding-based dynamic topic model [10]. Even though these and other dynamic topic models have been proposed, they do not explicitly model noise.

In this paper, we adapt a topic-noise model to a temporal social media setting, with the goal of improving topic coherence by successfully removing noise from evolving topics. We accomplish this by adapting a topic-noise model, Topic-Noise Discriminator (TND) [8] to a temporal setting, Dynamic Topic-Noise Discriminator (D-TND). D-TND takes advantage of the joint topic-noise distribution generation of TND, while at the same time enabling the tracking of topics and

J. Gama et al. (Eds.): PAKDD 2022, LNAI 13281, pp. 429–443, 2022.
https://doi.org/10.1007/978-3-031-05936-0_34

noise through time by passing topic and noise distributions from one time period to the next. We then propose a new temporal model, Dynamic Noiseless Latent Dirichlet Allocation (D-NLDA), a temporal version of Noiseless Latent Dirichlet Allocation (NLDA) [8] that integrates the proposed D-TND with Dynamic LDA (D-LDA). The advantage of this approach is that the noise distribution of D-TND and topic distribution of D-LDA evolve together, allowing for more accurate filtering of noise, and better-trained topic-word distributions at each time period. As we will see, D-NLDA is much greater than the sum of its parts.

The contributions of this paper are as follows. 1) We propose a new *temporal topic-noise model* that models noise and topics over time. 2) We propose a new *temporal topic model* that accounts for noise and generates higher quality topic sets. 3) To improve scalability, we introduce a vocabulary limiting function that reduces the vocabulary size of temporal data sets while maintaining topic quality. 4) We conduct an empirical analysis, both quantitative and qualitative, using two large Twitter data sets, that demonstrates the abilities of D-TND and D-NLDA to scale to accommodate such data sets, and to successfully identify high-quality topics. 5) We publish our model and evaluation code for others to use to continue advancing research in the field of temporal topic modeling.[1]

The paper is organized as follows: Sect. 2 presents related literature. Section 3 defines the notation used throughout the paper, details the models that were used in creating our proposed models, and presents our proposed models. Section 4 presents our quantitative and qualitative empirical analyses of our models. Finally, Sect. 5 presents our conclusions.

2 Related Literature

2.1 Static and Social Media Topic Models

The most well-known topic model is Latent Dirichlet Allocation (LDA) [3]. The basis upon which many topic models are built today, LDA is a bag-of-words model that approximates topics by maximizing the likelihood of documents in a k-dimensional Dirichlet distribution, where k is the number of topics. As documents are observed, words are probabilistically placed into topics and the probability distribution of each document over the topic set slowly changes to reflect the co-occurrence of words within the data set. After the model is trained, words that occur together in the same documents are more likely to be in the same topic. The result is topics containing words that are related according to the observed documents.

It has been apparent for some time that social media data sets require specially-constructed topic models to deal with the noise levels, short length, and sparsity of the data at hand. Biterm Topic Model detects topics, not from unigrams, but from bigrams generated from text [27], decreasing the vocabulary to improve quality. Self-Aggregating Topic Model (SATM) follows a similar

[1] Our code can be found at https://github.com/GU-DataLab/gdtm.

vein, aggregating related short posts into longer pseudo-documents and generating topics from the pseudo-documents. GPUDMM [16] improves the coherence of topics by sampling related words from an embedding space. Percolation-based Topic Models (PTM) [5] detects topics in social media data using a graph structure. A word co-occurrence graph is broken down into small communities and then built back up into small but coherent topics. For a more complete survey of unsupervised topic models, including ones designed for social media, see [6].

Topic-Noise Models [8] jointly model topics and noise distributions in order to more effectively remove noise from topics. Churchill and Singh propose a topic-noise model called Topic-Noise Discriminator (TND), which adds a noise distribution to LDA [3]. They use TND in an ensemble with LDA, called Noiseless Latent Dirichlet Allocation (NLDA), to create low-noise topics in domain-specific social media data sets. In this paper, we use the static TND and NLDA as a starting point to build two dynamic models, D-TND and D-NLDA.

2.2 Temporal Topic Models

Topics over Time (TOT) [25] jointly models time and topics, allowing for a continuous timeline of topics as opposed to discretized time periods like in D-LDA. Other early temporal topic models include MTTM [18], continuous DTM [24], Topic Tracking Model [13], and MDTM [14]. Dynamic Topic Models (D-LDA) [2] is a direct temporal adaptation of LDA [3]. Approximated probability distributions from a given time period are passed into the subsequent time period, in order to track the evolution of topics over time. We will draw on this temporal structure to create our dynamic topic-noise model, which incorporates a noise distribution into the model.

Bhadury et al. optimize D-LDA using multithreading and an optimized inference algorithm [1]. Topic Flow Model (TFM) [9] models temporal social media data using a graph structure. It runs a directed depth-first search from selected seed words to connected words and back to confirm mutual association. Dynamic Embedded Topic Model (D-ETM) [10] takes the Embedded Topic Model (ETM) [11], and adds a time-varying aspect. D-ETM runs ETM for each time period in the data set, passing parameters into the next time period like in D-LDA. The graph-based Dynamic Topic Model (GDTM) [12] is a scalable dynamic topic model for social media. The model assigns documents to topics based on the overlap of documents' graph representations, and partitions the documents based on graph density. One issue with GDTM is that it does not output the most probable words per topic, instead opting to output partitioned documents. Because of this, it is not directly comparable to models such as DTM, D-ETM, and our proposed models. In our experiments, we test our models against D-LDA, TFM, ToT, and D-ETM. The largest difference between our models and this previous work is that we explicitly model noise as a separate distribution. None of these other dynamic models do that.

3 Approach

In this section, we define our notation (Sect. 3.1) and review D-LDA, TND, and NLDA (Sect. 3.2). We then describe how we adapt the topic-noise models TND and NLDA to a dynamic setting to produce D-TND (Sect. 3.3) and D-NLDA (Sect. 3.4). We then propose a method for improving the scalability of dynamic topic models, with the goal of producing dynamic models capable of handling large social media data sets (Sect. 3.5).

3.1 Notation

Let D represent a *dataset* consisting of M documents, where $D = \{d_0, d_1, \ldots, d_{M-1}\}$. A *document* d is a group of N words, where $d = \{w_0, w_1, \ldots, w_{N-1}\}$. A vocabulary V is the set of unique words in D. In our context, a word is a unigram. However, words can be replaced by phrases without loss of generality.

A topic z consists of ℓ related words, $z = \{w_0, w_1, \ldots, w_{\ell-1}\}$. The words in z should be coherent and interpretable by a human. A topic set Z contains k topics, $Z = \{z_0, z_1, \ldots, z_{k-1}\}$, that represent a summary of D. A noise distribution Ω is a probability distribution over V, where each word has a non-zero probability of being a noise word. In the case of temporal models, we use discretized time. We refer to a data set as consisting of T time periods, $\{t_0, t_1, \ldots, t_T\}$, where topics within a time period are constructed together.

3.2 Dynamic LDA, Topic-Noise Discriminator, and Noiseless LDA

Dynamic LDA (D-LDA) was designed to approximate topics over time, but does not take into account the noise inherent in social media data [2]. Topic-Noise Discriminator (TND) was designed to simultaneously approximate noise and topic distributions in social media data sets. While it can be used as a standalone topic model, it is best used in an ensemble, like Noiseless LDA (NLDA)[8]. NLDA leverages the noise distribution of TND and topic-word distribution of LDA to produce more coherent, high quality topics in social media data sets [8]. We briefly describe these core components of our dynamic topic-noise models here.

D-LDA. D-LDA defines a topic-word distribution $\beta_{t,k}$, where t is the time period, and k is the number of topics. For a document d, its document-topic distribution $\alpha_{t,d}$ is a probability distribution over β_t. When generating a word for document d on time slice t, a topic z is chosen from β_t conditioned on $\alpha_{t,d}$. The word $w_{t,d}$ is drawn from $\beta_{t,z}$. This results in topics that are generated relative to time, as well as the observed documents.

TND. Topic-Noise Discriminator is a generative model that assumes that documents are a mixture of topics and noise. Words are drawn from a mixture of the topic-word distribution and noise distribution to generate documents. Each word in an observed document is assigned to the noise or topic-word distribution, based on its prior probabilities of being in each. A Beta distribution (Eq. 1) is used to determine whether a word belongs to the noise or topic distribution. β_z^i is the frequency of word i in topic z, and Ω_i is the frequency of word i in the noise distribution. The γ parameter can be increased to weight the Beta distribution toward assigning a word to the chosen topic over the noise distribution.

$$\lambda = Beta(\sqrt{\beta_z^i + \gamma}, \sqrt{\Omega_i}) \quad (1)$$

NLDA. While TND effectively models noise, it does not always independently find the strongest topics. Noiseless LDA [8] joins the noise distribution from TND with the topic distribution of LDA [3] to produce more coherent topics than those generated by TND or LDA. Assuming that we have a noise distribution Ω from TND and topic-word distribution β_z from LDA, NLDA integrates them based on each word's probability of being in the noise distribution and topic-word distribution for a given topic, in a process similar to Eq. 1.

3.3 Constructing a Dynamic Topic-Noise Model

We now describe how D-TND is constructed. $\beta_{t,k}$ is the topic-word distribution for t over k topics. The document-topic distribution α_t is a probability distribution over β_t. α and β are initially group Dirichlet priors (document-topic and topic-word distributions, respectively) in the first time period, but once trained, are passed to future time periods as individual priors. α_t and β_t are initialized from their $t-1$ counterparts.

We define Ω_t to be the noise distribution at time t. Like α_t and β_t, Ω_t is conditioned on Ω_{t-1}. This inherently assumes that words that were noise in $t-1$ are still noise in t. While this will make it harder for noise words from $t-1$ to be included in topics, it does not make it impossible, merely less likely.

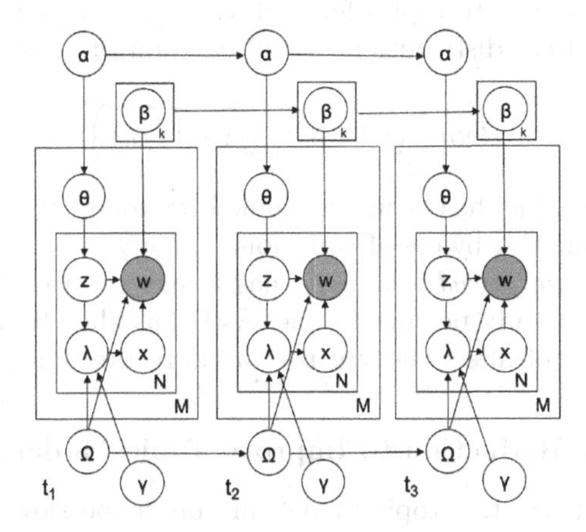

Fig. 1. Plate notation for D-TND, for three time periods.

Figure 1 shows plate notation for D-TND. Observed words are designated as noise or topic words within a time period based on a Beta distribution conditioned on the word's probability of being in the chosen topic or in noise. This process is represented by λ in Fig. 1, and is tuned by γ. The designation is indicated by the switching variable x. For a given time period t, we generate a document d as follows:

1. Draw the number of words N for d.
2. Draw the topic distribution $\theta_{t,d}$ from the Dirichlet distribution, conditioned on α_t.
3. For each word w_i, $0 \leq i < N$:
   (a) Draw a topic z_i from the topic distribution $\theta_{t,d}$.
   (b) Draw a word from either z_i or the noise distribution Ω_t, according to λ_t, indicated by switching variable x.
   (c) If drawing from z_i, draw w_i from β_{t,z_i}.
   (d) If drawing from Ω_t, draw w_i from Ω_t.

3.4 Constructing Dynamic NLDA

D-TND's most versatile feature is its noise distribution, which is trained along-side topics. Like TND for static models, D-TND can be easily integrated into generative temporal topic models. This makes D-TND particularly useful because researchers can use it in concert with whichever model they prefer.

Just as NLDA integrates TND's noise distribution and LDA's topic-word distribution, D-NLDA integrates D-TND's noise distribution and D-LDA's topic-word distribution. To create D-NLDA, we train a noise distribution Ω on D for each time period $t \in T$ using D-TND. Our assumption that noise and topics both evolve over time and in relation to each other allows us to track and integrate topics and noise in the style of NLDA, with a temporal aspect. We generate topics on D using D-LDA, and combine D-LDA's topic-word distribution β_t, k with D-TND's Ω_t to create topics for each time period. A word is removed or retained using the Beta distribution, conditioned on $\beta^i_{t,z}$ and $\Omega_{t,i}$ (Equation 2).

$$Beta\left(\sqrt{\beta^i_{t,z} + \gamma}, \sqrt{\Omega_{t,i}(\phi/k)}\right) \tag{2}$$

After the status of w_i has been determined, we follow the same guidelines as NLDA, incrementing $\Omega_{t,i}$ by one if w_i is noise, or by $\beta^i_{t,z}$ if w_i belongs to z. This ensures that, *for time period* t *only*, w_i has a high chance of not being put in another topic if it already belongs to one. As Ω has already been computed for all $t \in T$, this does not affect the status of w_i in future time periods.

3.5 Vocabulary Reduction to Improve Topic Model Performance

As we mentioned in Sect. 1, topic models are often too slow to infer topics on large data sets in a temporal setting. The original D-LDA [2], D-ETM [10], and ToT [25] only show results on data sets of tens of thousands of documents. In order to facilitate better scaling for topic models, we propose reducing the vocabulary size of data sets.

We define a *vocabulary limiting function* (VLF) to be a function that removes words from the vocabulary V, resulting in a smaller vocabulary V'. We define the frequency of a word $w_i \in V$ to be f_w. Given a threshold f_{min}, we compute the VLF as follows:

$$V' = V' \cup \{w_i\} \; \forall w_i \in V | f_{w_i} > f_{min} \tag{3}$$

In practice, we set f_{min} such that $|V'|$ is approximately equal to some target vocabulary size. It is also worth pointing out that this approach indirectly reduces the size and possibly the number of documents. Instead of removing documents that may have important words, we remove words from documents that are less likely to be high probability words in a topic model. We evaluate the effects of VLF in Sect. 4. We note that the performance impact of this may be small in cases where the lowest frequency words are much less frequent than the average word.

4 Empirical Evaluation

In this section, we present our empirical evaluation of D-TND and D-NLDA using quantitative and qualitative approaches. We begin by describing our experimental setup, including data sets, preprocessing, and model parameters (Sect. 4.1). We then present a quantitative evaluation (Sect. 4.2), and a qualitative evaluation of our models' performance (Sect. 4.3).

4.1 Experiment Setup

Baseline Algorithms. In our experiments, we tested against four state-of-the-art temporal topic models: D-LDA [2], ToT [25], TFM [9], and D-ETM [10]. They are each described in Sect. 2.

Data Sets. In our analysis, we use two Twitter data sets. The first data set contains posts about the 2020 United States Presidential Election from August 1 to November 14, with weekly time periods. We refer to this data set as *Election 2020.* The second data set, *Covid-19*, contains posts about the Covid-19 pandemic, collected between March 2020 and February 2021, with monthly time periods. We collected these documents using hashtags related to the election and Covid-19, respectively, via the Twitter Streaming API, and randomly sampled 200,000 posts per time period.[2]

We use our vocabulary limiting function (VLF) to create different versions of each data set. The large version is the full vocabulary, ($f_{min} = 0$). We set f_{min} such that $|V'| \approx 10,000$ for each time period to get medium-size data sets.[3] f_{min} was set such that $|V'| \approx 5,000$ for each time period for small-size data sets.[4] Table 1 shows the exact effects of the VLF for each data set. While there is a

[2] Leaving data sets in their original form, with a large skew in data set size from time period to time period, reinforces the skews in more pronounced ways in the probability distributions, leaving effects on future time periods.

[3] $f_{min} = 15, 20$ in Election2020 and Covid-19 for the medium-size data sets.

[4] $f_{min} = 40, 50$ for Election2020 and Covid-19 for the small-size data sets.

significant reduction in vocabulary, the number of documents remains high. In Covid-19, just over 100,000 documents, or about 4%, are lost, while in Election 2020, about 200,000 documents, or about 6.67% are lost. As we will see, this loss in documents has very little effect on the quality of topics.

Table 1. Data Set Qualities for different size variants of vocabulary. $|D|/t$ and $|V|/t$ are average data set size and vocabulary size within a time period.

		$\|D\|$	$\|D\|/t$	$\|V\|$	$\|V\|/t$
Covid-19	Large	2,400,103	200,008	1,041,552	172,116
	Medium	2,326,370	193,864	28,198	10,483
	Small	2,292,266	191,022	13,890	5,645
Election 2020	Large	3,000,042	200,002	648,193	96,671
	Medium	2,836,549	189,103	33,391	9,484
	Small	2,800,209	186,680	18,010	5,153

Text Preprocessing. Text processing can have a positive impact on topic model performance [7]. For our data sets, we tokenize on whitespace, remove lowercase text, remove URLs, punctuation (including hashtags), and stopwords. We also remove deleted posts and user tags.

Model Parameters. We conduct a sensitivity analysis for D-TND and D-NLDA, testing each model with an array of different parameter settings. Due to space limitations, we present the results for the best-performing settings. For D-TND, we found the best parameter settings to be $\alpha = 1$, $\beta = 0.01$, $\gamma = 25$, and $k = 30$. The best settings for D-NLDA were the same settings as D-TND, with $\phi = 10$.[5] For D-LDA, the best parameter settings were $\alpha = 1$, $\beta = 0.01$, and $k = 30$. The chosen α and β parameters consistently resulted in better topic quality than other options. The γ parameter is less sensitive than α and β, but $\gamma = 16$ was also a reasonable choice. We found that $\gamma = 0$ or 36 were too extreme of settings for our data sets, designating too few and too many words as noise, respectively. Changing the ϕ parameter can lead to far more coherent topic sets. We found that $\phi = 5$ resulted in too few noise words being filtered from topics, but that $\phi > 15$ resulted in some quality words being removed from topics. We note that it is straightforward to quickly iterate through ϕ values, since the filtering of noise is the fastest part of the model. For D-ETM and ToT, the parameters suggested in the papers were used, with $k = 30$ to match the parameters of the other models. While a sensitivity analysis was conducted, It is possible that with more extensive hyperparameter tuning, performance could be improved.

[5] Parameters for sensitivity analysis across our models: $k = \{10, 20, 30, 50, 100\}$, $\alpha, \beta = \{0.01, 0.1, 1.0\}$, $\gamma = \{0, 16, 25, 36\}$, $\phi = \{5, 10, 15, 20, 25, 30\}$.

Table 2. Time per iteration on each data set (s=seconds, m=minutes).

Model	Covid-19			Election 2020		
	Large	Medium	Small	Large	Medium	Small
D-TND	19.0 s	18.2 s	14.6 s	21.9 s	15.2 s	13.8 s
D-LDA	**1.5 s**	**1.4 s**	**1.32 s**	**2.0 s**	**1.2 s**	**0.9 s**
D-NLDA	20.6 s	19.7 s	16.0 s	23.6 s	16.4 s	14.8 s
D-ETM	480 m	117 m	87 m	360 m	177 m	138 m

4.2 Quantitative Analysis

Evaluation Metrics. Similar to previous work, we assess a model's ability to detect coherent, interpretable topics using a normalized point-wise mutual information score (NPMI) [15]. NPMI attempts to quantify the relatedness of two words within a topic, given their cofrequency, and is a commonly used evaluation metric [8,10,11,16,20,21]. For a pair of words (x, y), we define the probability of them appearing in the same document as $P(x, y)$. We define the probability of any word w appearing in a document as $P(w)$. Using these probabilities, we compute the NPMI of a topic $z \in Z$:

$$NPMI(z) = \frac{\sum_{x,y \in z} \frac{\log(\frac{P(x,y)}{P(x)P(y)})}{-\log(P(x,y))}}{\binom{|z|}{2}}$$

A higher NPMI indicates high topic coherence and lower noise penetration, or that a topic model is creating meaningful topics. We refer to the topic-wise NPMI score as *topic coherence*.

Unfortunately, a model can, in theory, find ten variants of the same meaningful topic. We care about the ability of a topic model to detect unique topics from the data,. *Topic diversity* is the fraction of unique words in the top 20 words of all topics in a topic set [11]. A model with high topic diversity is able to find almost entirely unique topics, while a model with low diversity is not able to successfully delineate between unique topics. *Topic quality*, proposed by Dieng et al. [10], is the product of the coherence and diversity scores. As we care about both metrics, a product of the two gives a good overall score for a topic set.

Given the size of our data sets, we are concerned about efficiency. For our experiments, models were run on a machine with twelve 2.2GHz virtual cores, with 50GB of memory. D-TND, D-NLDA, D-LDA, and D-ETM take advantage of parallelization or multi-threading (Mallet for D-LDA and D-TND [17], PyTorch for D-ETM [19]). ToT did not scale to the size of our data sets. It was allowed to run for three days, and did not complete an iteration for either data set. TFM ran for three days and did not finish constructing topics. The topics found contained only a single word, meaning its topic coherence would be zero. As a result, we do not include TFM and ToT in the analysis that follows.

Efficiency. To analyze efficiency, we compute time per iteration for the other methods (see Table 2). As we can see, D-LDA is the most efficient model.

Because it is only computing a topic distribution and not a noise model as well, this result is not surprising. D-TND and D-NLDA are the next most efficient models and are comparable to each other. Our models are between 300 and 1500 times faster than D-ETM, the most recent temporal topic model in our study.

D-ETM is implemented using PyTorch, a highly optimized Python framework for neural networks [19]. It is run for ten iterations on the small and medium data sets and five iterations on the large data set given that each iteration took approximately 8 h to run. D-LDA and our models were run for 500 iterations. Highlighting its ability to work on larger data sets. Part of the differential in computation time between D-ETM and the other models is likely due to the fact that D-ETM is implemented in Python, whereas the other models are implemented in Java. It is possible that a Java implementation of D-ETM would be faster than its Python implementation, but given the complexities of the underlying model, it is unlikely that a Java version of D-ETM would be faster than D-TND or D-NLDA.

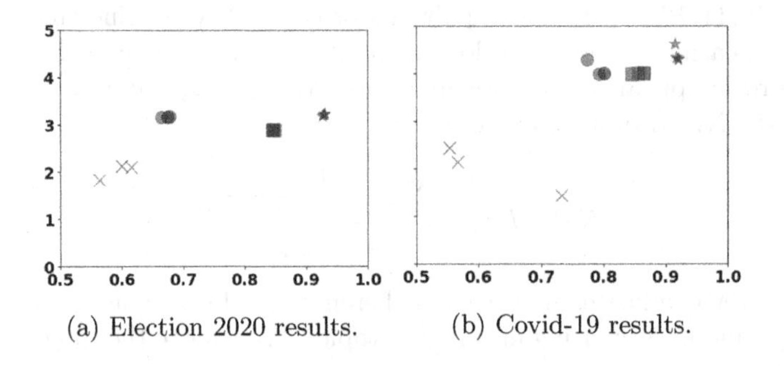

(a) Election 2020 results. (b) Covid-19 results.

Fig. 2. Coherence (y-axis) and Diversity (x-axis)

Coherence and Diversity. This section focuses on the quality of the models. Figure 2 plots the mean coherence and diversity score for D-TND (circles) and D-NLDA (stars) alongside D-LDA (squares) and D-ETM (X) for each data set. Coherence is plotted on the Y-axis, and diversity is plotted on the X-axis. The results for the large-size data set are colored red, the medium-size blue, and the small-size green. The closer to the top-right corner of the plot a model is, the better. D-NLDA performs the best out of any model on both data sets. In the Election 2020 data set, there's little difference in D-NLDA's performance across the different-size data sets. In the Covid-19 data set, we see a slight deterioration in terms of coherence when we use VLF to remove words from the vocabulary. Table 1 shows the difference between the Election 2020 and Covid-19 data sets in terms of how many words are removed from each vocabulary. By aiming to retain approximately 10,000 and 5,000 words per time period in the medium and small data sets, far more words were removed from the Covid-19 vocabulary than the Election 2020 vocabulary. It seems that in the case of Covid-19, we removed too many words from the vocabulary. This adversely affected the topic quality. In the Election 2020 data set, we retained all or most of the topic words, which is reflected in the maintained high topic quality across data set sizes. While we can certainly improve the scalability of topic models by reducing vocabulary size, removing too many words can sacrifice topic quality.

D-NLDA has a slightly higher (1.5%) coherence than D-TND, but a 25% higher diversity score. Compared to D-LDA, its diversity is 7% higher, and its coherence is 8% higher. D-NLDA once again is the best model on all data set sizes, beating D-TND by 0.35 in coherence and 14% in diversity, and beating D-LDA by 0.64 in coherence and 7% in diversity. D-ETM's poor performance is likely due to its inability to finish enough iterations in a reasonable amount of time to detect high-quality topics. Finally, for all models except D-ETM, the vocabulary size had little effect on the overall topic quality. The low variance in performance on D-TND, D-LDA, and D-NLDA reflects our theoretical assertion that removing the lowest frequency words should have very little affect on topic model performance. For D-ETM, the coherence of topics increases with the use of VLF, indicating that D-ETM benefits from smaller vocabularies.

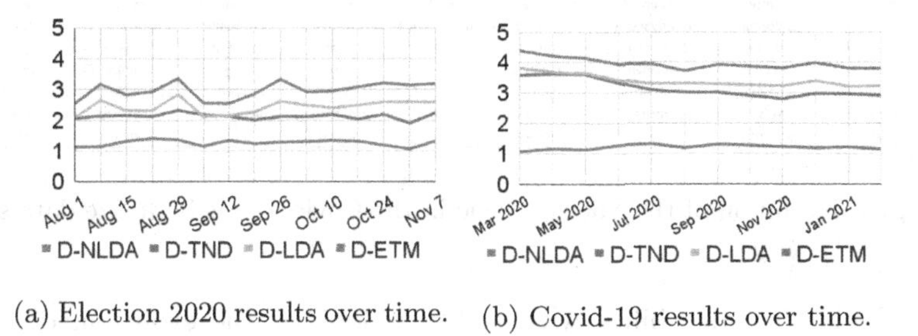

(a) Election 2020 results over time. (b) Covid-19 results over time.

Fig. 3. Topic Quality Plot for Election2020 and Covid-19 medium-size data sets.

In order to understand how models perform over time in relation to one another, we plot topic quality, the product of the coherence and diversity scores, for each model in each time period in Fig. 3. Plotting topic quality over time highlights the similarity of D-NLDA and D-LDA, but also highlights the clear improvement of D-NLDA with the addition of D-TND's noise distribution.

Table 3. Percent judge agreement on Covid-19 temporal topics.

Topic	Vaccines	Lockdowns	Cases	Testing	Schools	Masks	Global impact	Economy	India	China
Agreed %	100	100	100	100	100	100	100	100	80	100

4.3 Qualitative Analysis

Our qualitative analysis shows D-NLDA's ability to track topics through time. For the Covid-19 data set, we asked five human judges to individually label ten topics generated by D-NLDA that persisted throughout every time period. In Table 3, we show agreement between the judges, with the agreed upon label for each topic. For all topics but one, every judge independently concluded the same topic label. These findings indicate that D-NLDA is able to generate high quality topics that humans can easily comprehend.

March 2020	April	May	June	July	August	September	October	November
medical	vaccine	vaccine	vaccine	vaccine	vaccine	vaccine	vaccine	vaccine
workers	family	scientists	china	sarscov2	research	trials	emergency	vaccines
back	doctors	hydroxychloroquine	world	human	sarscov2	fauci	vaccines	effective
healthcare	nurses	research	virus	study	study	trial	public	trial
rate	treatment	treatment	lives	results	effective	experts	political	family
made	dying	study	black	research	scientists	find	health	results
night	heart	evidence	united	made	vaccines	clinical	data	pfizer
full	science	prevent	matter	early	immunity	trust	mental	missed
vaccine	policy	hands	human	shows	emergency	company	information	friends
action	scientists	spread	sarscov2	trials	governments	hold	control	spent
national	didnt	effective	chinese	complete	told	ready	billion	months

December	Dec (2)	January 2021	Jan. (2)	Jan. (3)	Jan. (4)	February	Feb. (2)	Feb. (3)
vaccine	vaccines	vaccine	vaccination	people	vaccines	vaccine	vaccines	vaccination
pfizer	effective	pfizer	india	video	doses	dose	countries	house
days	data	effective	minister	sick	healthcare	received	global	biden
vaccination	hospitals	rollout	working	vaccinated	workers	covidvaccine	india	president
heres	county	received	time	work	million	vaccinated	world	years
doses	california	covidvaccine	narendramodi	laurie_garrett	countries	pfizer	make	team
vaccinated	developed	started	prime	taking	county	rollout	access	past
years	christmas	receive	families	paid	iran	single	communities	economy
covidvaccine	shows	moderna	global	weeks	high	shot	safe	vaccinations
drleanawen	kids	mrna	response	america	frontline	distribution	country	mass
reach	tomorrow	dose	corona	bill	worker	leading	ensure	white

Fig. 4. Evolution of the Vaccine topic in the Covid-19 medium-size data set.

We highlight this ability with a deeper look at the evolution of the vaccine topic through time periods, seeing it evolve and grow (see Fig. 4). Words highlighted green appeared in the topic in the previous time period, and words highlighted yellow appeared in the topic in any previous time period. The topic starts out with a wish for a vaccine and with concern for healthcare workers. It evolves into a reality in the middle of 2020 and goes through drug trials. Finally, the vaccine is approved in late 2020 and rolled out at the beginning of 2021. D-NLDA allows us to see a very detailed evolution of the Vaccine topic that contains limited noise throughout the entirety of the year, showing the promise of topic-noise models within a temporal setting.

In the Election 2020 medium data set, we show the ability of D-NLDA to accurately track multiple relevant topics through time. To produce topic labels for this data set, we relied on a manually-generated topic set, curated by political scientists who closely studied the 2020 Election on social media platforms [23]. Figure 5 shows how the topic proportions of selected topics change throughout the election. New topics emerge and disappear throughout the campaign. We can see the large impact of the party conventions in time periods two and three (late August 2020), and how quickly talk about conventions ceases after they are over. The same happens with topics about Presidential and Vice Presidential Debates in time periods eight to ten (early October).

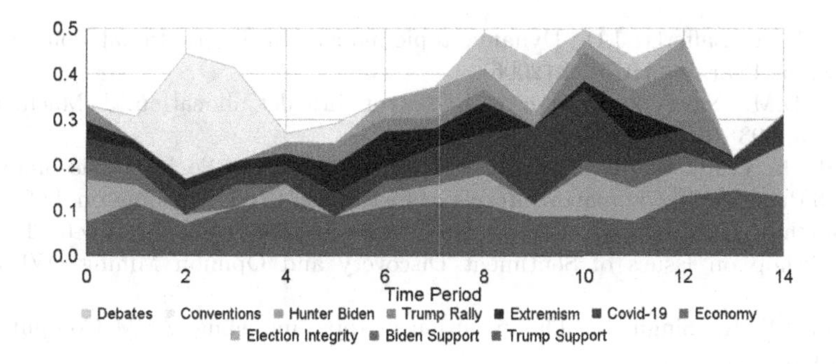

Fig. 5. Election 2020 Topic proportions (y-axis) over time periods (x-axis).

The Conventions and Debates topics represent bursty topics, which appear out of nowhere and disappear quickly as attention turns away from them. These bursty topics could be missed or muddled with other topics by static models. In general, this topic flow visualization highlights the ability of a dynamic topic-noise model like D-TND or D-NLDA to produce highly relevant and easily understandable topics with a temporal aspect. The ability to understand how topics evolve over an election is important both for voters and candidates.

5 Conclusions

In this paper we create a dynamic temporal-noise model that incorporates a noise distribution into a temporal topic model for the first time (D-TND), and weave together D-TND with the well-known D-LDA model to create D-NLDA. These approaches bring to temporal topic models the noise-filtering benefits of topic-noise models that are so necessary for social media data sets.

We demonstrate the ability of our proposed methods to both scale to large temporal data sets, and produce high quality topics on the data sets through time periods spanning weeks and months. We show how using a vocabulary limiting function (VLF) can speed up topic models, and in some cases, produce better topics. Finally, we share our code on GitHub for others to use.[6]

Acknowledgement. This work was supported by the National Science Foundation grant numbers #1934925 and #1934494 and by the Massive Data Institute (MDI) and McCourt Impacts at Georgetown University. We would also like to thank the Mosaic Project and SSRS for access to the Covid-19 Survey data set.

References

1. Bhadury, A., Chen, J., Zhu, J., Liu, S.: Scaling up dynamic topic models. In: The Web Conference (WWW) (2016)

[6] Our code can be found at https://github.com/GU-DataLab/gdtm.

2. Blei, D.M., Lafferty, J.D.: Dynamic topic models. In: International Conference on Machine Learning (ICML) (2006)
3. Blei, D.M., Ng, A.Y., Jordan, M.I.: Latent dirichlet allocation. J. Mach. Learn. Res. **3**, 993–1022 (2003)
4. Bode, L., et al.: Words that Matter: How the News and Social Media Shaped the 2016 Presidential Campaign. Brookings Institution Press, Washington, D.C. (2020)
5. Churchill, R., Singh, L.: Percolation-based topic modeling for tweets. In: KDD Workshop on Issues of Sentiment Discovery and Opinion Mining (WISDOM) (2020)
6. Churchill, R., Singh, L.: The evolution of topic modeling. ACM Comput. Surv. (CSUR) (2021)
7. Churchill, R., Singh, L.: textprep: a text preprocessing toolkit for topic modeling on social media data. In: The DATA Conference (2021)
8. Churchill, R., Singh, L.: Topic-noise models: modeling topic and noise distributions in social media post collections. In: International Conference on Data Mining (ICDM) (2021)
9. Churchill, R., Singh, L., Kirov, C.: A temporal topic model for noisy mediums. In: Pacific-Asia Conference on Knowledge Discovery and Data Mining (PAKDD) (2018)
10. Dieng, A.B., Ruiz, F.J.R., Blei, D.M.: The dynamic embedded topic model. CoRR (2019)
11. Dieng, A.B., Ruiz, F.J., Blei, D.M.: Topic modeling in embedding spaces. arXiv preprint arXiv:1907.04907 (2019)
12. Ghoorchian, K., Sahlgren, M.: GDTM: graph-based dynamic topic models. Progress Artif. Intell. **9**, 195–207 (2020)
13. Iwata, T., Watanabe, S., Yamada, T., Ueda, N.: Topic tracking model for analyzing consumer purchase behavior. In: International Joint Conference on Artificial Intelligence (2009)
14. Iwata, T., Yamada, T., Sakurai, Y., Ueda, N.: Online multiscale dynamic topic models. In: Conference on Knowledge Discovery & Data Mining (KDD) (2010)
15. Lau, J.H., Newman, D., Baldwin, T.: Machine reading tea leaves: automatically evaluating topic coherence and topic model quality. In: Conference of the European Chapter of the Association for Computational Linguistics (EACL) (2014)
16. Li, C., Wang, H., Zhang, Z., Sun, A., Ma, Z.: Topic modeling for short texts with auxiliary word embeddings. In: SIGIR Conference on Research and Development in Information Retrieval (2016)
17. McCallum, A.K.: Mallet: A machine learning for language toolkit (2002)
18. Nallapati, R.M., Ditmore, S., Lafferty, J.D., Ung, K.: Multiscale topic tomography. In: Conference on Knowledge Discovery & Data Mining (KDD) (2007)
19. Paszke, A., et al.: Pytorch: an imperative style, high-performance deep learning library. In: Advances in Neural Information Processing Systems (NIPS) (2019)
20. Qiang, J., Chen, P., Wang, T., Wu, X.: Topic modeling over short texts by incorporating word embeddings. In: Pacific-Asia Conference on Knowledge Discovery and Data Mining (PAKDD) (2017)
21. Quan, X., Kit, C., Ge, Y., Pan, S.J.: Short and sparse text topic modeling via self-aggregation. In: International Joint Conference on Artificial Intelligence (2015)
22. Singh, L., Bode, L., Budak, C., Kawintiranon, K., Padden, C., Vraga, E.: Understanding high-and low-quality URL sharing on Covid-19 twitter streams. J. Comput. Soc. Sci. **3**(2), 343–366 (2020)
23. Singh, L., et al.: The breakthrough [polling project] (2020)

24. Wang, C., Blei, D., Heckerman, D.: Continuous time dynamic topic models. arXiv preprint arXiv:1206.3298 (2012)
25. Wang, X., McCallum, A.: Topics over time: a non-Markov continuous-time model of topical trends. In: Conference on Knowledge Discovery & Data Mining (KDD) (2006)
26. Williams, J.B., Singh, L., Mezey, N.: # metoo as catalyst: a glimpse into 21st century activism. University of Chicago Legal Forum, p. 371 (2019)
27. Yan, X., Guo, J., Lan, Y., Cheng, X.: A biterm topic model for short texts. In: The Web Conference (WWW) (2013)

Contrastive Attributed Network Anomaly Detection with Data Augmentation

Zhiming Xu[1], Xiao Huang[2], Yue Zhao[3], Yushun Dong[1], and Jundong Li[1(✉)]

[1] University of Virginia, Charlottesville, USA
{zx2rw,yd6eb,jundong}@virginia.edu
[2] The Hong Kong Polytechnic University, Hung Hom, Hong Kong
xiao.huang@polyu.edu.hk
[3] Carnegie Mellon University, Pittsburgh, USA
zhaoy@cmu.edu

Abstract. Attributed networks are a type of graph structured data used in many real-world scenarios. Detecting anomalies on attributed networks has a wide spectrum of applications such as spammer detection and fraud detection. Although this research area draws increasing attention in the last few years, previous works are mostly unsupervised because of expensive costs of labeling ground truth anomalies. Many recent studies have shown different types of anomalies are often mixed together on attributed networks and such invaluable human knowledge could provide complementary insights in advancing anomaly detection on attributed networks. To this end, we study the novel problem of modeling and integrating human knowledge of different anomaly types for attributed network anomaly detection. Specifically, we first model prior human knowledge through a novel data augmentation strategy. We then integrate the modeled knowledge in a Siamese graph neural network encoder through a well-designed contrastive loss. In the end, we train a decoder to reconstruct the original networks from the node representations learned by the encoder, and rank nodes according to its reconstruction error as the anomaly metric. Experiments on five real-world datasets demonstrate that the proposed framework outperforms the state-of-the-art anomaly detection algorithms.

Keywords: Anomaly detection · Graph neural networks · Self-supervised learning

1 Introduction

Attributed networks are a kind of graph structured data, which exists ubiquitously in many real-world scenarios, such as social networks, biological networks, and financial transaction networks [1,22]. Over the past few decades, many research efforts have been devoted to performing different learning tasks on attributed networks. Anomaly detection is one such task, which in the context of attributed networks aims to identify nodes with significantly different

J. Gama et al. (Eds.): PAKDD 2022, LNAI 13281, pp. 444–457, 2022.
https://doi.org/10.1007/978-3-031-05936-0_35

patterns from other nodes in terms of their attributes, communities, etc. [1,28]. It has become a critical research area that has broad applications in various real-world scenarios [4], such as spammer detection [1] and fraud detection [3].

Extensive progress has been made towards anomaly detection on attributed networks over the past few years [8–10,19,20,25,30,31]. Generally speaking, existing anomaly detection approaches can be mainly divided into two mainstreams, namely Non-deep Learning (Non-DL) methods and Deep Learning (DL) methods. Non-DL methods typically rely on various types of heuristic anomaly measurements [30,31,34,35] or employ matrix decomposition techniques [19,20,29] to detect anomalies while DL methods often resort to Graph Neural Networks (GNNs) for the detection of anomalies [8,10,21]. It should be noted that DL methods have shown superior performance over traditional Non-DL methods [19,25,30,31] due to the strong capability of GNNs for learning node representations. Specifically, DL methods usually follow an encoder-decoder learning scheme, where the encoder takes the given attributed network as input, while the decoder reconstructs the graph structure and node attributes and compares the reconstructed data with the original input for anomaly detection [8–10]. However, despite the superior performance, these approaches mainly detect anomalies in an unsupervised manner due to the expensive labeling cost of ground truth anomalies. Many recent studies have shown that there often exist mixed types of anomalies on attributed networks, w.r.t. graph structure and node attributes [19,44].

For example, we present two typical anomaly types, namely attribute anomaly and structure anomaly in Fig. 1. There are a community of CA software engineers and a community of MA salesperson in this network. For attribute anomaly, the attribute value of node 4 is significantly different from others, thus it is suspicious to be an attribute anomaly; for structure anomaly, node 6 belongs to the CA software engineer community by its attributes, however, it also connects to a remotely related community of MA salesperson, rendering it structurally abnormal. Beyond the anomalies in the above example, more types of commonly encountered anomalies, e.g., community anomalies, have also been identified and summarized by existing works [1,20]. As a summary, these studies equipped us with rich prior human knowledge of different anomaly types. In fact, many learning related problems have witnessed a significant performance improvement when human knowledge is considered [33,36,42]. Motivated by such success, in this paper, we study an important research problem: *whether the prior human knowledge of different anomaly types could be harnessed to advance anomaly detection on attributed networks.*

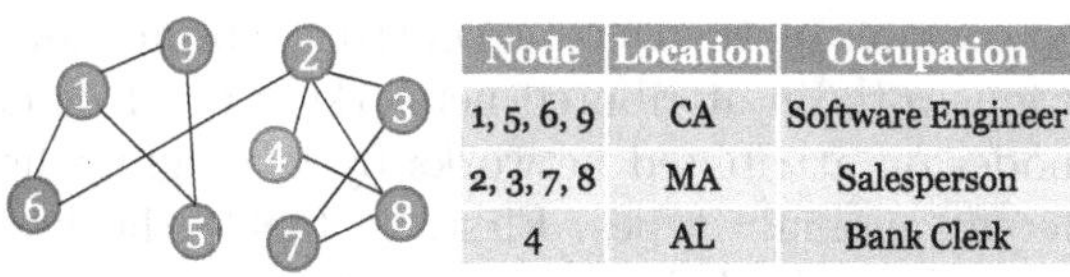

Node	Location	Occupation
1, 5, 6, 9	CA	Software Engineer
2, 3, 7, 8	MA	Salesperson
4	AL	Bank Clerk

Fig. 1. A toy example of attribute anomaly and structure anomaly on an attributed network.

Although leveraging prior human knowledge of different anomaly types could be potentially helpful for attributed network anomaly detection, how to properly model and utilize such knowledge remains a daunting task mainly because of the following two challenges: (1) *Knowledge Modeling Challenge.* How to properly model the prior human knowledge of different types of anomalies on attributed networks is the first challenge that needs to be tackled. The major problem here is that such knowledge only encodes human understanding of possible anomalous patterns on attributed networks, thus it does not have a concrete form and cannot be directly leveraged. While many existing studies proposed to model human knowledge as an invaluable data resource in addition to the original input data [18,33,42], it still remains unclear how to model human knowledge into concrete data resource that can be directly utilized in our case. (2) *Knowledge Integration Challenge.* The second challenge centers around integrating the prior human knowledge of anomaly types on attributed networks seamlessly into the detection model. Traditionally, many existing works regard human knowledge as an explicit supervision signal and integrate it into learning models by designing a specific loss term [6,26,36]. However, in our problem, existing human knowledge of anomaly types is not exhaustive, and an effective knowledge integration mechanism needs to be flexible enough to accommodate the available knowledge rather than design a flawed loss term informed only by partial observation.

To tackle the above challenges, in this paper, we propose CONTRASTIVE ANOMALY DETECTION (CONAD), a principled contrastive anomaly detection framework on attributed networks. CONAD is capable of identifying anomalous nodes on attributed networks by leveraging the prior human knowledge of different anomaly types. First, to tackle the knowledge modeling challenge, we propose a novel data augmentation strategy which explicitly models and formalizes the prior human knowledge of different anomaly types as contrastive samples (i.e., nodes whose patterns deviate significantly from existing nodes on the input attributed network) on the augmented attributed network. Second, to address the knowledge integration challenge, we propose to tightly integrate the contrastive samples on the augmented attributed network into the anomaly detection model with a well-designed contrastive loss. Methodologically, we first propose to generate an augmented attributed network to model known anomaly types. A Siamese GNN is employed as the encoder function to map both the input attributed network and the augmented attributed network into an embedding space. After that, a contrastive loss is designed based upon the normal nodes on the input attributed network and contrastive samples on the augmented attributed network, through which the human knowledge of different anomaly types can be well harnessed. The proposed contrastive loss is jointly considered with a graph reconstruction loss for end-to-end model training. During the detection phase, the suspicious score of each node is measured by the magnitude of the reconstruction error, which serves as the metric to identify anomalies, i.e., a larger error indicates the node has a higher chance of being abnormal.

The main contributions of this paper can be summarized as follows: (1) **Problem Formulation.** We study a novel problem of modeling and leveraging prior human knowledge of different anomaly types for anomaly detection on attributed networks. (2) **Algorithmic Design.** We propose a principled framework that models prior human knowledge of different anomaly types as contrastive samples in the augmented attributed network; and integrates the contrastive samples into the anomaly detection model with a well-designed contrastive loss. (3) **Experimental Evaluations.** We perform comprehensive experimental evaluations on real-world datasets to demonstrate the superiority of the proposed contrastive attributed network anomaly detection framework.

2 Problem Definition

Notations. We use bold uppercase letters (e.g. $\mathbf{A}$), bold lowercase letters (e.g. $\mathbf{x}$), and regular lowercase letters (e.g. a) to denote matrices, vectors, and scalars, respectively. Besides, for a matrix $\mathbf{A}$, we represent its (i, j)-th entry as $\mathbf{A}_{ij}$. Similarly, for a vector $\mathbf{y}$, its i-th element is denoted by y_i.

Let $\mathcal{G} = \{\mathbf{A}, \mathbf{X}\}$ be an input attributed network, where $\mathbf{A} \in \mathbb{R}^{n \times n}$ and $\mathbf{X} \in \mathbb{R}^{n \times d}$ denote the adjacency matrix and attribute matrix, respectively. The problem of *anomaly detection on attributed networks* aims to assign a suspicion score to each node that quantifies how likely it is to be abnormal. To utilize prior human knowledge of anomaly types in this process, we assume there is an additional human knowledge input ξ that consists of typical types of anomalies studied in previous works and observed in real-world scenarios [1,8,22], e.g., attribute and structure anomalies shown in Fig. 1 before. With the additional knowledge ξ, we hence formulate the following research problem.

Definition 1 ***Modeling and Leveraging Prior Human Knowledge of Anomaly Types for Attributed Network Anomaly Detection.*** *Given an attributed network $\mathcal{G} = \{\mathbf{A}, \mathbf{X}\}$, prior human knowledge ξ of anomaly types, our goal is to model and formalize the abstract human knowledge ξ into concrete data (denoted as $\mathrm{M}(\xi)$), and then integrate it into a principled detection model f that is capable of encoding both $\mathrm{M}(\xi)$ and $\mathcal{G}$ and ultimately detect anomalies in $\mathcal{G}$.*

3 The Proposed Framework

In this section, we introduce the proposed framework CONAD. It consists of three major components as shown in Fig. 2, namely, knowledge modeling module, knowledge integration module, and anomaly detection module. The overview of each module is listed below followed by detailed descriptions.

Knowledge Modeling Module. Given the prior human knowledge ξ of different anomaly types, we first use a novel data augmentation strategy to model and formalize it as concrete contrastive samples. We achieve this by introducing each known anomaly type encoded in ξ to the input attributed network $\mathcal{G}$ and generate the augmented attributed network $\mathcal{G}_{ano}$ accordingly.

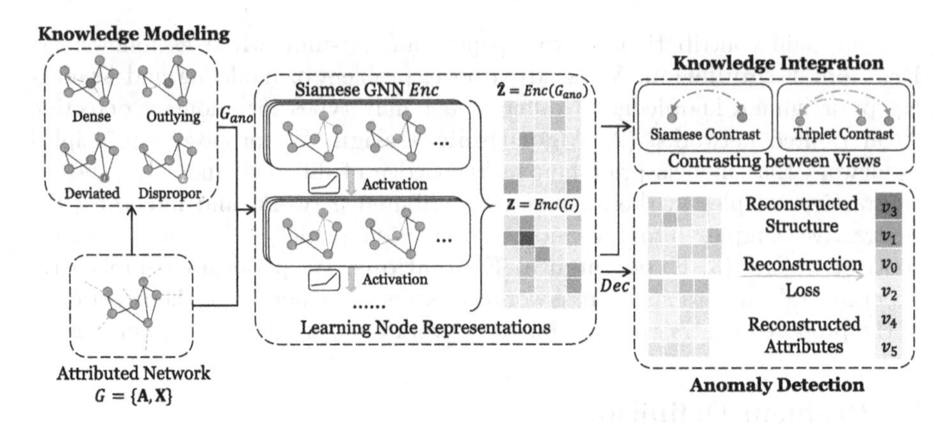

Fig. 2. Overview of CONAD. The lower-left box is the input attributed network for anomaly detection. The upper-left box shows the knowledge modeling module. Dense, Outlying, Deviated, and Disproportionate correspond to the prior human knowledge of anomaly types. The middle is the encoder built on Siamese GNN to learn node representations. The upper-right box presents two contrast strategies to integrate the prior human knowledge modeled in $\mathcal{G}_{ano}$. The lower-right part is the decoder that reconstructs both the structure and attributes of the input attributed network, which detects anomalies with the reconstruction error.

Knowledge Integration Module. After modeling prior human knowledge ξ, we feed both $\mathcal{G}$ and each $\mathcal{G}_{ano}$ into a graph encoding architecture in which a Siamese GNN acts as the encoder to learn representations of nodes. By using a Siamese network, both graphs will be encoded into the same latent space, making it possible to contrast between the node representations of $\mathcal{G}$ and $\mathcal{G}_{ano}$. After the encoding phase, to tightly integrate the human knowledge in $\mathcal{G}_{ano}$, we propose a well-designed contrastive loss. Specifically, the contrastive loss will guide the encoder to represent normal nodes on the input attributed network and contrastive samples on the augmented attributed network differently. Consequently, anomaly patterns of the augmented nodes can be captured.

Anomaly Detection Module. With the learned node representations, we aim to reconstruct the graph structure and node attributes of the input attributed network $\mathcal{G}$ with a decoder. The reconstruction errors produced by the reconstruction phase are leveraged as suspicion scores in detecting anomalies on $\mathcal{G}$.

3.1 Knowledge Modeling Module

We introduce the data augmentation strategy used to model the prior human knowledge of different anomaly types on attributed networks in this subsection. We consider four different types of anomalies on attributed networks (from both the structure side and the attribute side), and introduce a certain amount of anomalies belonging to each anomaly type to the input attributed network $\mathcal{G}$ to form an augmented attributed network $\mathcal{G}_{ano}$. Each of these four augmented anomaly types is illustrated in Fig. 3.

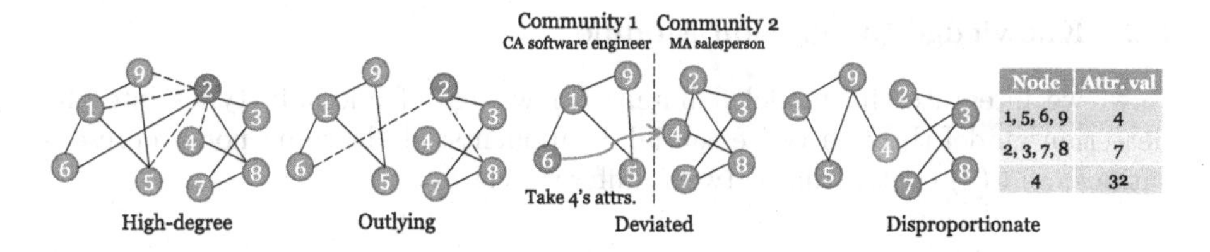

Fig. 3. An illustration of four different types of anomalies on attributed networks based on prior human knowledge.

Structure – high-degree. In social networks, spammers often follow and interact with excessively numerous users [14]. To simulate this anomaly type, we choose a certain amount of nodes with average degrees, and then connect them to many other random nodes. The chosen nodes thus have an unusually high degree and are considered structurally abnormal in the attributed network.

Structure – outlying. Another abnormal account type in social or e-commerce networks is created in large quantities to spam certain posts [38]. They behave like regular users but few users will follow them, and thus they do not belong to any communities, thus different from the majority of the whole network and deemed structurally abnormal. We simulate anomaly type by choosing a certain amount of nodes and drop most of their edges on the input attributed network.

Attribute – deviated. A common attribute anomaly on attributed network is a node with deviated attribute values from its neighbors [31]. In other words, the attribute value of this node could be rather different from others in the same community. To model this anomaly type, we first choose certain center nodes. For each center node, we randomly sample a number of other nodes from the entire network. We then calculate the similarity between the attribute vectors of this center node and the others, and then assign the attribute vector of the least similar one to the center node. In fact, through such generation process, we are introducing community anomalies to the input attributed network.

Attribute – disproportionate. In e-commerce websites, dishonest sellers might want to promote their products by setting unreasonably low prices or achieve high sale volumes by recruiting dishonest buyers [11]. Both of these sale frauds will result in unusually small or large numbers in certain node attributes. We hence largely scale up or scale down the values of certain node attributes with a preset probability to simulate this anomaly type of disproportionate numerical values in certain node attributes.

After applying the four augmentation strategies above, we obtain an augmented attributed network $\mathcal{G}_{ano}$, referred to as *anomalous view*. In the anomalous view $\mathcal{G}_{ano}$, we have a label vector $\mathbf{y}$, where $y_i = 1$ denotes that node i corresponds to one of those four known anomaly types, and $y_i = 0$ otherwise.

3.2 Knowledge Integration Module

Now, we integrate the modeled human knowledge of anomaly types into the detection model through two essential components: (1) learning node representations; and (2) contrasting between different views.

Learning Node Representations. We first discuss how to encode both $\mathcal{G}$ and $\mathcal{G}_{ano}$. In particular, we employ a Siamese GNN architecture as an encoder to learn embeddings for nodes in both $\mathcal{G}$ and $\mathcal{G}_{ano}$. Generally, various GNNs can be leveraged to learn node representations from attributed networks [40] based on the information aggregation mechanism: $\mathbf{h}_i^{(l+1)} = \text{AGG}(\{\mathbf{h}_i^{(l)}\} \cup \{\mathbf{h}_j^{(l)} : j \in \mathcal{N}_i\})$ where $\mathbf{h}_i^{(l)}$ denotes the representation of node i in the l-th layer, and $\mathbf{h}_i^{(0)}$ is the input attribute of node i. $\mathcal{N}_i$ is the set of all neighbors of node i. AGG$(\cdot)$ is an aggregation function that can be implemented by mean pooling, max pooling, and many other operations [24]. In this paper, we specify the information aggregation based on the self-attention mechanism in Graph Attention Networks (GAT) [39]. The reason is that GAT is able to account for different neighbors' contributions to the central node via assigning appropriate corresponding importance weights. Thus it is capable of capturing complicated relations in attributed networks. Each GAT layer follows the information propagation scheme of $\mathbf{h}_i^{(l+1)} = \sigma(\sum_{j \in \mathcal{N}_i} \alpha_{ij} \mathbf{W}^{(l)} \mathbf{h}_j^{(l)})$, where $\alpha_{ij} = \text{softmax}(e_{ij}) = \frac{e_{ij}}{\sum_{k \in \mathcal{N}_i} e_{ik}}$ and $e_{ij} = \sigma\left(\mathbf{a}^\top \left[\mathbf{W}\mathbf{h}_i \| \mathbf{W}\mathbf{h}_j\right]\right)$. Here α_{ij} is the attention weight between node i and j. $\mathbf{W}^{(l)}$ is a learnable parameter matrix for the l-th layer. $\mathbf{W}$ and $\mathbf{a}$ are learnable parameters that are shared by all GAT layers for learning the attention weights. $\|$ denotes the concatenation operation. In practice, we stack multiple GAT layers to form the encoder *Enc* for node representation learning.

Contrasting Between Views. To fully harness the power of human knowledge in $\mathcal{G}_{ano}$, we propose to make a contrast between $\mathcal{G}_{ano}$ and the given attributed network (normal view) $\mathcal{G}$. We expect the anomalous patterns on the attributed network can be well characterized through such contrastive process. Since the augmented anomalous nodes become different from both themselves and their neighbors, we consider two different contrast strategies in this paper, and we name them as Siamese contrast and Triplet contrast. The former contrast strategy is performed by comparing the embedding representation of each abnormal node in the anomalous view and its counterpart in the normal view. The latter contrastive strategy is performed for a connected node pair (i, j) where j is considered abnormal in the anomalous view and i remains intact. It is called "triplet" because three representations, i.e., the representation of i in the normal view and the representations of j in both the normal and anomalous views are involved. These two contrast strategies are described in detail below.

Strategy 1: Siamese Contrast. Suppose *Enc* encodes $\mathcal{G}$ and $\mathcal{G}_{ano}$ through stacked GAT layers into the final representations $\mathbf{Z}$ and $\hat{\mathbf{Z}}$. Siamese contrast is performed between $\mathbf{z}_i$ and $\hat{\mathbf{z}}_i$, i.e., the representations of each node i in the normal view and the anomalous view. The loss function of Siamese contrast is defined as follows:

$$\mathcal{L}^{\mathrm{sc}} = \frac{1}{n}\sum_{i=1}^{n}(\mathrm{I}_{y_i=0} \cdot d(\mathbf{z}_i, \hat{\mathbf{z}}_i) + \mathrm{I}_{y_i=1} \cdot \max\{0, m - d(\mathbf{z}_i, \hat{\mathbf{z}}_i)\}) \tag{1}$$

where I is the indicator function of the condition in its subscript. When applying the Siamese contrastive loss, if $\mathbf{y}_i = 1$, i.e., node i is considered abnormal in G_{ano}, the distance between its representation in the normal and the anomalous view, $d(\mathbf{z}_i, \hat{\mathbf{z}}_i)$ will be maximized with a margin no smaller than m. If $y_i = 0$, i.e., node i is not considered abnormal in $\mathcal{G}_{ano}$, then $d(\mathbf{z}_i, \hat{\mathbf{z}}_i)$ will be minimized.

Strategy 2: Triplet Contrast. In addition to the above strategy, we further propose Triplet contrast that works on a triplet of node representations. Specifically, we consider each connected node pair (i, j) where j is an augmented anomaly in $\mathcal{G}_{ano}$ while i remains intact. The triplet of representations consists of three representations $\mathbf{z}_i$, $\mathbf{z}_j$, and $\hat{\mathbf{z}}_j$, and the loss function is defined as:

$$\mathcal{L}^{\mathrm{tc}} = \sum_{\substack{\forall \mathbf{A}_{ij}=1,\\ y_i=0,\ y_j=1}} \max\{0, m - (d(\mathbf{z}_i, \hat{\mathbf{z}}_j) - d(\mathbf{z}_i, \mathbf{z}_j))\}. \tag{2}$$

Through minimizing this loss function, our model will increase the gap between two distances with a margin no smaller than m. Here $d(\mathbf{z}_i, \mathbf{z}_j)$ is the distance between the representations of i and its neighbor j in the normal view, and $d(\mathbf{z}_i, \hat{\mathbf{z}}_j)$ is the distance between the representation of node i in the normal view and that of its neighbor j in the anomalous view. Therefore, CONAD can enforce an augmented anomaly to be far away from its neighbors, and thus the human knowledge regarding this anomaly type can be harnessed.

3.3 Anomaly Detection Module

Besides learning from $\mathcal{G}_{ano}$ which models prior human knowledge of anomaly types, CONAD also needs to learn from the input attributed network $\mathcal{G}$ to detect anomalies in it. Towards this objective, we aim to reconstruct the graph structure and node attributes based on the learned node representations in normal view $\mathbf{Z}$. It has been proved in previous works [7,8,19] that reconstructing structures and attributes helps the model to learn the normal patterns of the input attributed networks, and since anomalies cannot be well reconstructed, they will therefore be detected. Specifically, our model uses a decoder function *Dec* on the encoder output $\mathbf{Z}$. *Dec* consists of a GAT layer to reconstruct the adjacency and attribute matrix from $\mathbf{Z}$. Frobenius norm of the difference between the input and the reconstructed matrix, i.e., reconstruction error, serves as the loss function:

$$\hat{\mathbf{A}} = \sigma\left(\mathbf{Z} \cdot \mathbf{Z}^{\top}\right), \hat{\mathbf{X}} = \mathrm{GATLayer}(\mathbf{A},\ \mathbf{Z}). \tag{3}$$

$$\mathcal{L}^{\mathrm{recon}} = \lambda \left\|\mathbf{A} - \hat{\mathbf{A}}\right\|_F + (1-\lambda) \cdot \left\|\mathbf{X} - \hat{\mathbf{X}}\right\|_F. \tag{4}$$

Here, $\sigma(\cdot)$ is a non-linear activation function, e.g., ReLU [27]. $(\cdot)^{\top}$ and $\|\cdot\|_2$ are the transpose and Frobenius norm on matrices. λ is a weighting factor to balance the scales of the two reconstruction errors on the structure and attributes.

3.4 Summary

We summarize the whole process of our proposed model CONAD in this subsection. Our input is an attributed network $\mathcal{G} = \{\mathbf{A}, \mathbf{X}\}$ and prior human knowledge of anomaly types ξ. We first model ξ through a novel data augmentation strategy described in Sect. 3.1. We then have two attributed networks $\mathcal{G}$ and $\mathcal{G}_{ano}$. A Siamese GNN encoder *Enc* is used to learn from prior human knowledge modeled in $\mathcal{G}_{ano}$ by contrasting between node representations in $\mathcal{G}$ and $\mathcal{G}_{ano}$ with the contrastive loss defined in Eq. (1) or Eq. (2). During this process, *Enc* learns to distinguish normal and abnormal representations in the latent space, and thus integrates the prior human knowledge. The node representations of $\mathcal{G}$ are further fed into an anomaly detection module described in Eq. (3) to learn the normal patterns in $\mathcal{G}$ with the reconstruction loss $\mathcal{L}^{\text{recon}}$. Hence CONAD learns from both knowledge ξ and attributed network $\mathcal{G}$ with $\mathcal{L}^{\text{cl}}$ and $\mathcal{L}^{\text{recon}}$, respectively. The total loss of CONAD becomes the summation of the contrastive and reconstruction loss (η is also a weighting factor to balance the two loss terms).

$$\mathcal{L}^{\text{CONAD}} = \eta \cdot \mathcal{L}^{\text{cl}} + (1-\eta)\mathcal{L}^{\text{recon}}, \ \mathcal{L}^{\text{cl}} \in \left\{\mathcal{L}^{\text{sc}}, \mathcal{L}^{\text{tc}}\right\}. \tag{5}$$

4 Experiments

4.1 Datasets

Five different real-world datasets, namely, Flickr [15], Amazon [35], Enron [25], Facebook [23], and Twitter [23], are used to evaluate the anomaly detection performance of CONAD. (1) *Flickr* dataset contains user following and follower relations on the eponymous photo-sharing website. There are 7,575 nodes (600 ground truth anomalies) and 23,938 edges in the entire network, and we follow the same settings as [8,10,21] to obtain ground truth anomalies. (2) *Amazon & Enron.* These two datasets contain ground truth anomalies. The Amazon dataset represents co-purchase relations between items. The anomalies here consist of erroneous categories or prices. There are 1,418 nodes (28 ground truth anomalies) and 3,695 edges. Enron is a corporate email network. The anomalies are employees who involve in the accounting fraud in this company. There are 13,533 nodes (5 ground truth anomalies) and 176,987 edges in total. (3) *Facebook & Twitter.* We also use social networks in Facebook and Twitter, where users form relations with others and share their "circles" of friends. We obtain ground truth anomalies by introducing nodes that connect to randomly selected circles or have abnormal attributes like [8]. There are 4,039 nodes (400 ground truth anomalies) and 88,234 edges in the Facebook dataset, and we use 4,865 nodes (500 ground truth anomalies) and 66,772 edges in the Twitter dataset[1].

[1] The anomaly labels in Flickr, Facebook, and Twitter datasets result from manual injection, and the injection rule coincides with two of our data augmentations.

4.2 Experimental Settings

We compare our proposed framework with the following four popular baseline methods, including LOF [2], DOMINANT [8], AEGIS [7], and AnomalyDAE [10]. Among them, the latter three are the state-of-the-art methods that employ GNNs and a comparison with them can validate the superiority of our proposed framework which harnesses the power of human knowledge.

For our proposed framework CONAD, the encoder *Enc* is initialized with two layers of GAT, where the hidden sizes are 128 and 64, respectively. For the reconstruction part, an additional GAT layer is applied for attribute reconstruction, while dot product and sigmoid activation are applied for structure reconstruction. Two attention heads and LeakyReLU [41] activation are used for all GAT layers. The margin m is set to 0.5 for both Siamese and Triplet losses, and the model is denoted by CONAD-S and CONAD-T corresponding to the specific contrastive loss used, i.e., **S**iamese and **T**riplet. Euclidean distance is used as the distance function $d(\cdot, \cdot)$. The ratio of augmented anomalies r is 10% for smaller networks, i.e., Amazon, Flickr, Facebook, and Twitter, and 20% for the larger one, i.e., Enron. The weighting factors λ and η are set to 0.9 and 0.7, respectively. We train the model with Adam [17]. The area under ROC (AUC) serves as the evaluation metric of anomaly detection performance.

Table 1. Anomaly detection performance (AUC scores) comparison. CONAD consistently performs the best across all three datasets (higher is better).

Dataset	Amazon	Enron	Flickr	Facebook	Twitter
LOF	0.510	0.581	0.661	0.522	0.511
DOMINANT	0.592	0.716	0.749	0.554	0.571
AEGIS	0.556	0.602	0.765	0.659	0.645
AnomalyDAE	0.610	0.552	0.694	0.741	0.688
CONAD-S	**0.635**	**0.731**	**0.782**	0.612	0.670
CONAD-T	0.620	**0.731**	0.759	**0.863**	**0.742**

4.3 Anomaly Detection Performance Comparison

Table 1 shows the anomaly detection performance of CONAD and baselines, where CONAD outperforms all others in all of the five real-world datasets used. Specifically, GNN-based models generally perform better than LOF, which does not consider structure information. By modeling and integrating prior human knowledge, CONAD achieves better performance than the other three GNN-based unsupervised anomaly detection models. Besides, for networks with explicit communities, i.e., Facebook and Twitter, CONAD-T, which contrasts between each pair of neighbors, performs better than CONAD-S, which only contrasts between the representations of each individual node in the normal and anomalous views.

4.4 Ablation Study

In this subsection, we conduct further experiments to study the improvements brought by each module individually in the proposed framework CONAD on Amazon dataset. The results are shown in Table 2, and similar observations can also be found in other datasets. We first study the influence of the types of contrasting between views, i.e., Siamese contrast and Triplet contrast. The performance of CONAD-T with Triplet contrast is slightly worse than CONAD-S with Siamese contrast. However, the performance on Facebook and Twitter datasets shown in the previous subsection demonstrates the opposite. We speculate that it is because the co-purchase relation in Amazon datasets does not have explicit communities, contrary to the friendship relation in the two social networks. Therefore, contrasting between neighbors is not very helpful. We then study how the amount of prior human knowledge modeled affects the performance of CONAD. Towards this goal, we change the data augmentation strategy in 3.1, where we solely model human knowledge of structure (w/o attribute anomalies) or attribute (w/o structure anomalies) anomalies. The performance of CONAD decrease with either of these two types removed, showing that the more knowledge of anomaly types is given, the more CONAD can harness it to facilitate anomaly detection. We also investigate the effectiveness of the knowledge integration module. Concretely, we remove this module which contrasts between normal and anomalous views entirely. The resulting model becomes almost identical to DOMINANT, and the corresponding performance drops drastically, which demonstrates that integrating prior human knowledge is crucial in the superior performance of CONAD. Lastly, we study the influence of the reconstruction. We remove the decoder used to reconstruct the structure and attributes of the input attributed network, and apply LOF instead to the nodes representations learned by the encoder. The performance shows that LOF fails to detect anomalies from only those node representations. It proves that the decoder and reconstruction also contribute a lot to anomaly detection.

Table 2. Ablation study on the Amazon dataset.

Variants of CONAD	AUC score
CONAD-S	**0.635**
CONAD-T	0.620
w/o attribute anomalies	0.621
w/o structure anomalies	0.628
w/o contrasting between views	0.592
w/o reconstruction	0.510

4.5 Robustness of CONAD W.r.t. Different Ratios of Anomalies

At last, we study the robustness of CONAD on Flickr dataset where the ground truth anomalies can be easily tuned. We omit the results on other datasets due to the observation of similar patterns. We vary the ratios of ground truth anomalies in Flickr among 2.5%, 5%, 7.5%, and 10% of the total number of nodes, and find that CONAD maintains steady performances with AUC scores of 0.760, 0.772, 0.781, and 0.778. It demonstrates that CONAD is very robust in detecting anomalies in attributed networks when the ratio of anomalies present varies.

5 Related Works

5.1 Attributed Network Anomaly Detection

Attributed networks are a kind of graph structured data that exist ubiquitously in many real-world scenarios. Detecting anomalies in attributed networks is of vital importance for anti-fraud, anti-money laundering, and other safety-critical applications [1,22]. Therefore, attributed network anomaly detection has attracted an increasingly amount of research attentions in recent years. Existing approaches can be broadly categorized as traditional machine learning (Non-DL) methods and deep learning (DL) methods. Non-DL methods are often developed based on certain heuristic anomaly metrics, e.g., ConSub [35], FocusCO [31], and AMEN [30], or matrix decomposition techniques, e.g., Radar [19], ANOMALOUS [29], and ALAD [20]. More recently, many DL anomaly detection methods have been proposed, which often resort to GNNs due to their superior representation learning capability. Typical methods along this line include DOMINANT [8], AEGIS [7], and AnomalyDAE [10]. Our proposed CONAD differs from the methods introduced above as the above methods are mainly unsupervised while ours explicitly models the human knowledge of different anomaly types on attributed networks and tightly incorporate such knowledge into the detection model.

5.2 Contrastive Learning

Supervised learning achieves great success in numerous machine learning areas, but one major disadvantage of it is that a large amount of labeled data is required to train a descent model. To ease the reliance on labeled data, contrastive learning (CL) has gained popularity as a novel self-supervised learning (SSL) paradigm. It often utilizes data augmentation techniques to obtain different views of the data, and leverages InfoMax principle [13] to maximize the similarity between pairs of positive views while minimize pairs of negative views. With contrastive learning, SSL models [5,16,37] achieve comparable performance in image classification against their supervised counterparts. CL frameworks also enjoys successes in graph representation learning [12,32,43] where techniques designed specifically for graph structured data, such as random walk and graph diffusion, can be used to generative positive views.

6 Conclusions

In this paper, we propose CONAD, a contrastive learning framework capable of leveraging human knowledge to detect anomalies on attributed networks. Specifically, we first model human knowledge of real-world anomalies through a data augmentation approach. We then train a Siamese graph neural network with a contrastive loss to encode both the modeled knowledge and the original attributed networks. Finally, we use reconstruction loss to obtain anomaly scores. Experiments on several datasets with different nature and characteristics show detection performance improvements compared to state-of-the-art models. Furthermore, we analyze the benefit brought about by each part in CONAD and show its robustness w.r.t. different anomaly ratios on the attributed network.

Acknowledgements. Yushun Dong and Jundong Li are partially supported by the National Science Foundation (NSF) under grants #2006844.

References

1. Akoglu, L., Tong, H., Koutra, D.: Graph based anomaly detection and description: a survey. Data Min. Knowl. Disc. **29**(3), 626–688 (2014). https://doi.org/10.1007/s10618-014-0365-y
2. Breunig, M.M., Kriegel, H., Ng, R.T., Sander, J.: LOF: identifying density-based local outliers. In: SIGMOD (2000)
3. Cao, S., Yang, X., Chen, C., Zhou, J., Li, X., Qi, Y.: Titant: online real-time transaction fraud detection in ant financial. Proc. VLDB Endow. **12**(12), 2082–2093 (2019)
4. Chandola, V., Banerjee, A., Kumar, V.: Anomaly detection: a survey. ACM Comput. Surv. **41**(3), 15:1–15:58 (2009)
5. Chen, T., Kornblith, S., Norouzi, M., Hinton, G.E.: A simple framework for contrastive learning of visual representations. In: ICML (2020)
6. Deng, C., Ji, X., Rainey, C., Zhang, J., Lu, W.: Integrating machine learning with human knowledge. iScience **23**(11), 101656 (2020)
7. Ding, K., Li, J., Agarwal, N., Liu, H.: Inductive anomaly detection on attributed networks. In: IJCAI (2020)
8. Ding, K., Li, J., Bhanushali, R., Liu, H.: Deep anomaly detection on attributed networks. In: SDM (2019)
9. Ding, K., Li, J., Liu, H.: Interactive anomaly detection on attributed networks. In: WSDM (2019)
10. Fan, H., Zhang, F., Li, Z.: Anomalydae: dual autoencoder for anomaly detection on attributed networks. In: ICASSP (2020)
11. Fei, G., Mukherjee, A., Liu, B., Hsu, M., Castellanos, M., Ghosh, R.: Exploiting burstiness in reviews for review spammer detection. In: ICWSM (2013)
12. Hassani, K., Ahmadi, A.H.K.: Contrastive multi-view representation learning on graphs. In: ICML (2020)
13. Hjelm, R.D., et al.: Learning deep representations by mutual information estimation and maximization. In: ICLR (2019)
14. Hu, X., Tang, J., Zhang, Y., Liu, H.: Social spammer detection in microblogging. In: IJCAI (2013)
15. Huang, X., Li, J., Hu, X.: Label informed attributed network embedding. In: WSDM 2017
16. Khosla, P., et al.: Supervised contrastive learning (2020)
17. Kingma, D.P., Ba, J.: Adam: a method for stochastic optimization. In: ICLR (2015)
18. Ladický, L., Jeong, S., Solenthaler, B., Pollefeys, M., Gross, M.: Data-driven fluid simulations using regression forests. ACM Trans. Graph. (TOG) **34**(6), 1–9 (2015)
19. Li, J., Dani, H., Hu, X., Liu, H.: Radar: residual analysis for anomaly detection in attributed networks. In: IJCAI (2017)
20. Liu, N., Huang, X., Hu, X.: Accelerated local anomaly detection via resolving attributed networks. In: IJCAI (2017)

21. Liu, Y., Li, Z., Pan, S., Gong, C., Zhou, C., Karypis, G.: Anomaly detection on attributed networks via contrastive self-supervised learning. CoRR abs/2103.00113 (2021)
22. Ma, X., et al.: A comprehensive survey on graph anomaly detection with deep learning (2021)
23. McAuley, J.J., Leskovec, J.: Learning to discover social circles in ego networks. In: NeurIPS (2012)
24. Mesquita, D.P.P., Jr., A.H.S., Kaski, S.: Rethinking pooling in graph neural networks. In: NeurIPS (2020)
25. Müller, E., Sánchez, P.I., Mülle, Y., Böhm, K.: Ranking outlier nodes in subspaces of attributed graphs. In: ICDE Workshop (2013)
26. Muralidhar, N., Islam, M.R., Marwah, M., Karpatne, A., Ramakrishnan, N.: Incorporating prior domain knowledge into deep neural networks. In: IEEE Big Data (2018)
27. Nair, V., Hinton, G.E.: Rectified linear units improve restricted Boltzmann machines. In: ICML (2010)
28. Pang, G., Shen, C., Cao, L., van den Hengel, A.: Deep learning for anomaly detection: a review. CoRR abs/2007.02500 (2020)
29. Peng, Z., Luo, M., Li, J., Liu, H., Zheng, Q.: Anomalous: a joint modeling approach for anomaly detection on attributed networks. In: IJCAI (2018)
30. Perozzi, B., Akoglu, L.: Scalable anomaly ranking of attributed neighborhoods. In: SDM (2016)
31. Perozzi, B., Akoglu, L., Sánchez, P.I., Müller, E.: Focused clustering and outlier detection in large attributed graphs. In: KDD (2014)
32. Qiu, J., et al.: GCC: graph contrastive coding for graph neural network pre-training. In: KDD (2020)
33. von Rueden, L., et al.: Informed machine learning-a taxonomy and survey of integrating knowledge into learning systems. arXiv preprint arXiv:1903.12394 (2019)
34. Sánchez, P.I., Müller, E., Irmler, O., Böhm, K.: Local context selection for outlier ranking in graphs with multiple numeric node attributes. In: SSDBM (2014)
35. Sánchez, P.I., Müller, E., Laforet, F., Keller, F., Böhm, K.: Statistical selection of congruent subspaces for mining attributed graphs. In: ICDM (2013)
36. Stewart, R., Ermon, S.: Label-free supervision of neural networks with physics and domain knowledge. In: AAAI (2017)
37. Tian, Y., Krishnan, D., Isola, P.: Contrastive multiview coding. In: ECCV (2020)
38. Varol, O., Ferrara, E., Davis, C., Menczer, F., Flammini, A.: Online human-bot interactions: detection, estimation, and characterization. In: ICWSM (2017)
39. Velickovic, P., Cucurull, G., Casanova, A., Romero, A., Liò, P., Bengio, Y.: Graph attention networks. In: ICLR (2018)
40. Wu, Z., Pan, S., Chen, F., Long, G., Zhang, C., Yu, P.S.: A comprehensive survey on graph neural networks. IEEE Trans. Neural Networks Learn. Syst. **32**(1), 4–24 (2021)
41. Xu, B., Wang, N., Chen, T., Li, M.: Empirical evaluation of rectified activations in convolutional network. CoRR abs/1505.00853 (2015)
42. Xu, J.G., Zhao, Y., Chen, J., Han, C.: A structure learning algorithm for bayesian network using prior knowledge. J. Comput. Sci. Technol. **30**(4), 713–724 (2015)
43. You, Y., Chen, T., Sui, Y., Chen, T., Wang, Z., Shen, Y.: Graph contrastive learning with augmentations. In: NeurIPS (2020)
44. Zhu, M., Zhu, H.: Mixedad: a scalable algorithm for detecting mixed anomalies in attributed graphs. In: AAAI (2020)

Cross-Lingual Product Retrieval in E-Commerce Search

Wenya Zhu[1(✉)], Xiaoyu Lv[1], Baosong Yang[1], Yinghua Zhang[2], Xu Yong[1], Linlong Xu[1], Yinfu Feng[1], Haibo Zhang[1], Qing Da[1], Anxiang Zeng[3], and Ronghua Chen[4]

[1] Alibaba (China) Technology Co., Ltd., Hangzhou, China
zhuwenya1991@gmail.com, {linlong.xll,yinfu.fyf,daqing.dq}@alibaba-inc.com
[2] Hong Kong University of Science and Technology, Clear Water Bay, Hong Kong
yzhangdx@cse.ust.hk
[3] Nanyang Technological University, Singapore, Singapore
zeng0118@ntu.edu.sg
[4] Fudan University, Shanghai, China
chenrh@fudan.edu.cn

Abstract. Cross-lingual product retrieval (CLPR) recalls semantically relevant products that match multilingual search queries. It plays a crucial role in E-commerce sites to serve cross-border customers. However, there exists no public large-scale dataset on CLPR, hindering the research on this topic. We present CLPR-9M (https://tianchi.aliyun.com/dataset/dataDetail?dataId=121505), the first large-scale CLPR dataset containing 9 million query-product pairs, covering 10 major commodity categories and 3 language pairs, mined from real-world user logs. We also release a test dataset, annotated by bilingual experts with fine-grained labels. We build our baselines upon the widely used cross-lingual embedding retrieval framework and improve it from a range of aspects, including the pretrain-finetune paradigm, negative sampling, as well as optimization objective. Benchmarks are assessed and reported using multiple evaluation metrics, and will be beneficial for future research in this area.

Keywords: Cross-lingual information retrieval · E-commerce search

1 Introduction

With the growth of international market, E-commerce websites have to cope with not only monolingual but also multilingual queries, in order to serve cross-border customers. For example, a seller from America can serve customers from Southeastern Asia. In this case, the product information is written in English, while the query may be in Thai, Filipino, or Bahasa Indonesia. The products remain monolingual for two reasons. Firstly, it requires non-trivial efforts for sellers to provide multilingual item descriptions; Secondly, building the multilingual item indexes with the machine translation is limited by the quality of the machine translation. We refer to the product retrieval [12,27] in this setting

J. Gama et al. (Eds.): PAKDD 2022, LNAI 13281, pp. 458–471, 2022.
https://doi.org/10.1007/978-3-031-05936-0_36

as cross-lingual product retrieval (CLPR), where the product descriptions and the user queries are in different languages. As more sellers are expanding their business in emerging markets, the CLPR setting is becoming popular.

However, few studies explored CLPR, due to the lack of in-domain dataset, especially for the state-of-the-art deep learning models which heavily depends on large-scale training samples. Although [28] and [17] paid their attention to out-of-domain cross-lingual retrieval tasks, these studies may fail to generalize to the E-commerce domain due to non-trivial domain discrepancy. Figure 1 provides the taxonomy of information retrieval datasets from the domain and the language aspects.

To fill the gap, we collect and release the first large-scale cross-lingual product retrieval dataset (CLPR-9M). We construct the training set by extracting query-product pairs from real-world user logs. Since labeling the negative samples requires non-trivial efforts, past studies obtained the negatives by sampling with the human-crafted strategy, which has achieved reasonable performance [6,7, 21,27]. In our dataset, we provide the irrelevant query-product pairs from two sampling strategies, including random sampling and category-based sampling. In total, the training set is composed of 9 million query-product relevant pairs that are from 10 categories. The queries are in Russian, Spanish, and English, while the product titles are in English only. To evaluate the generalization ability of the retrieval model, we provide the high-quality test set with three labels (relevant, weak relevant, and irrelevant) by carefully manual annotation. As shown in Fig. 2, we provide several samples from the proposed dataset.

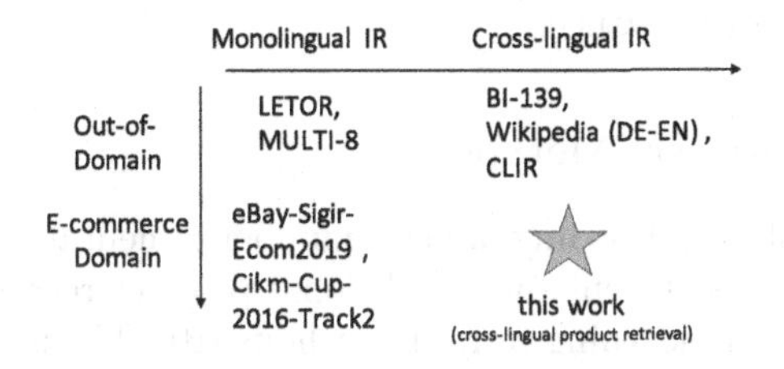

Fig. 1. Taxonomy of information retrieval (IR) datasets. We divided the information retrieval into monolingual and cross-lingual settings. Our work in this paper is to provide benchmarks for the cross-lingual IR in E-commerce domain. LETOR [15] and MULTI-8 [22] are the monolingual IR datasets in Wikipedia. Wikipedia (DE-EN) [19], CLIR [18] and BI-139 [22] are the datasets for cross-lingual IR in Wikipedia.

Building cross-lingual retrieval models has its unique challenges, such as how to bridge the lexical gap between languages [14]. Recently, the pretrained language models, such as multilingual BERT (M-BERT) [5] and XLM [11] can induce shared cross-lingual semantic space by learning the pretrained tasks based on sentence-aligned parallel data. We finetune the pretrained cross-lingual language model on the dataset, and provide extensive experiments to explore the loss function and negative sampling strategy. For the loss function, we propose a bi-log loss that maximizes the log-likelihood of positives from query and item directions. For the negative sampling, we compare random sampling, category-based sampling, and mixes of them. The experiments show that random negative sampling with bi-log loss can achieve a decent performance.

2 Related Work

2.1 Product Retrieval Datasets

The monolingual datasets (e.g., CIKM Cup 2016 Track 2 [1] and eBay Sigir Ecom 2019 [2]) have enabled the development of product retrieval for E-commerce search. However, to the best of our knowledge, there is no public large-scale dataset for the cross-lingual product retrieval task. A similar effort is cross-lingual information retrieval datasets for general domains, such as Wikipedia (DE-EN) [19], CLIR [18], and BI-139 [22]. They may not be applied to the E-commerce domain due to the non-trivial domain discrepancy. In addition, the multilingual queries of these datasets are extracted from the title or the first sentence of the document, rather than real-world user inputs. The relevance label is determined by various hand-crafted rules, such as smoothing out the BM25 score into discrete relevance labels in [22]. Our contribution differs from the above studies in three aspects: 1) CLPR-9M is the first large-scale dataset for cross-lingual information retrieval in E-Commerce search; 2) All multilingual search queries are from real users, and the dataset is closer to the real-world application; 3) CLPR-9M provides a high-quality benchmark by human annotation with finer-grained levels of relevance.

2.2 Product Retrieval Methods

With the success of deep learning, a large number of neural network based models have been proposed to enhance traditional product retrieval methods (e.g., BM25[16], LSI [4]) and learning to rank methods [10]. The neural retrieval models represent queries and products as dense vectors, which are further exploited to produce relevance scores. Particularly, DSSM [7] and its variant CDSSM [21] have pioneered the context of using deep neural networks for relevance scoring. Van Gysel et al. [23] proposed a latent vector space model (LSE) to learn the query and product representations with the entities as bridge. Zhang et al. [27] proposed two tower model to achieve the personalized and semantic retrieval goal. These methods have shown promising results on monolingual product retrieval tasks.

Nevertheless, due to the lack of large-scale public datasets, few studies in terms of deep model explore the cross-lingual scenario, especially for the E-commerce domain. Existing cross-lingual information retrieval (CLIR) systems usually adopt a translation-based approach that consists of three stages, including language identification, machine translation, as well as monolingual information retrieval [3,13,29]. However, the performance of the translation-based approaches is limited by the quality of the language identification and machine translation [29]. Recently, multiple pretrained language models have been developed, such as M-BERT [5] and XLM [11], that model the underlying data distribution and learn the linguistic patterns or features across languages and have been applied in cross-lingual information retrieval [9]. In this way, the cross-lingual information retrieval can be trained end-to-end, thus avoids the error propagation from language identification and machine translation.

3 Dataset

In this section, we introduce the construction of the CLPR-9M dataset. The dataset is composed of a set of query-product triplets $\{\mathbf{q}_n, \mathbf{i}_n, r_n\}_{n=1}^{N}$, where $\mathbf{q}_n$, $\mathbf{i}_n$, and r_n denote the query, product, and the semantic relevance between the query and the product, respectively. The query $\mathbf{q}_n$ is a sequence of m words, and there is $\mathbf{q}_n = \{q_{1,n}, q_{2,n}, \cdots, q_{m,n}\}$. Similarly, the product $\mathbf{i}_n$, which is also a sequence of k words, is denoted by $\mathbf{i}_n = \{i_{1,n}, i_{2,n}, \cdots, i_{k,n}\}$. There are 3 possible values for the relevance with 0 to represent irrelevant, 1 to represent weak relevant and 2 to represent relevant. In the following, we describe the collection of the training set and the annotation of the test set, followed by a brief summary of dataset statistics.

Query Language	Query	Item Title	Item Category	Relevance Label
En	army bag	Military Nylon Outdoor Hiking Backpack Waterproof Army Bag	Bags	Relevant
Es	bolsa del ejército	2020 Anti-theft Bag Women Men Fashion Travel Bags Schoolbag	Bags	Weak Relevant
Ru	армейская сумка	Infant Baby Swimwear Suspenders Bikini One-piece Swimwear	Swimwear	Irrelevant

Fig. 2. The samples of the dataset CLPR-9M. For each query-product pair, we provide the query language, the query content, the item title, the item category and relevance label. The terms in item title denoted what the product is are marked in red. (Color figure online)

3.1 Training Data Mining

The training set is mined from real-world user logs. Since it is difficult to determine whether a query-product pair is weakly related from the user logs, the relevance is a binary value in the training set. A semantically relevant query-product pair is considered as a positive sample, and similarly, an irrelevant pair is negative. Recent studies [6,27] suggest that using click results as positives and randomly sampling negatives can provide a reasonable model performance. Inspired by the previous studies, we randomly sample clicked pairs of 10 categories from online 1-month logs as the positives. The category of the query-product pair is determined by that of the product. There are several sampling strategies to obtain negative pairs. We provide negatives with two sampling strategies: **Random sampling** (for each query, we randomly sample products from all candidate products as negatives) and **Category-based sampling** (for a positive pair $\{\mathbf{q}_n, \mathbf{i}_n\}$, randomly sample products under the same domain of $\mathbf{i}_n$ as irrelevant products). Compared with random sampling, category-based sampling can produces hard negatives, since the products under the same domain tend to be similar.

3.2 Human Annotated Test Set

To build the test set, we first select the query-product pairs clicked by users as seed positive samples. Then, for a query in an arbitrary seed positive pair, we obtain the potential irrelevant products with three sampling strategies, and hence form potential negative pairs. To include more hard negatives, we added **Unclicked Impressed Sampling**: random sampling the products impressed to the user but not clicked, except for **Random sampling** and **Category-based sampling**. Notice that we do not utilize **Unclicked Impressed Sampling** to form the negatives in the training data, since the impressed products usually have some degree of relevance with search queries and the users do not click them may due to personal preference. Finally, each query-product pair is rated by two bilingual experts with three labels, namely "relevant", "weak relevant", and "irrelevant". The annotation instruction is provided in the Appendix. A pair with same labels from two bilingual experts is accepted; otherwise, the third language expert will make a decision.

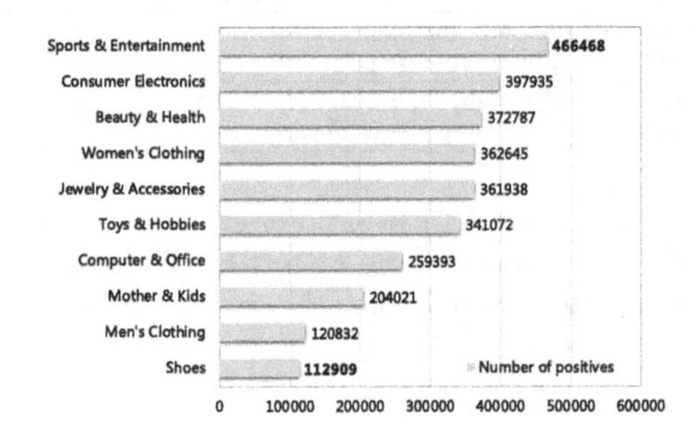

Fig. 3. The number of positive samples per category in the training set.

3.3 Dataset Statistics

The dataset contains 10 categories, and a total of 9 million query-product pairs for training, 21,700 pairs for testing. The training dataset contains 3 million relevant query-product pairs, 3 million irrelevant query-product pairs from random sampling, and 3 million irrelevant query-product pairs from category-based sampling. Figure 3 shows the number of query-product pairs for each category in the training set. In both the training and test set, the product title is in English, and the search queries have 3 languages, namely English(En), Russian(Ru) and Spanish(Es). Table 1 shows the statistics per language of the dataset. If all the tokens of a query appear in the product title, the query-product pair is considered as an "exact-match" pair; otherwise, it is an "inexact-match" pair. The inexact-match pairs cannot be handled by simple template matching methods, making the CLPR task challenging. The number of inexact-match pairs is 2.44 times larger than that of exact-match pairs, which indicates that CLPR on the CLPR-9M is indeed a demanding task.

Table 1. The statistics for the training and the test dataset. For each language X, we show the total number of queries (#Query) in language X, and the number of query product pairs (#QP pairs) where the query in language X. The number of products is shown in column #Product.

Dataset	#Query			#QP pairs			#Product
	English	Russian	Spanish	English	Russian	Spanish	
Training	1.14M	0.647M	0.648M	4.26M	2.36M	2.38M	1.70M
Test	0.76K	0.73K	0.68K	7.6K	7.35K	6.8K	26K

3.4 Human Evaluation on Dataset Quality

To evaluate the quality of the training data, we hire the bilingual expert to evaluate the relevance of 20,000 query product pairs randomly selected from the training data. For the positives in the training data, the accuracy of the labels is 80%. The error rate of the negatives is 0.02% and 2% for random sampling and category-based sampling, respectively. Besides, the agreement rate of two raters for test data annotation is 96.1%.

4 Baseline Approaches

In this section, we present a neural network retrieval model as a baseline model for the CLPR task on the CLPR-9M dataset. Motivated by the framework in [6], the model first converts the query and the product tokens into embeddings, and then generates a relevance score based on the extracted embeddings. An overview of the model is shown in Fig. 4. In the following, we first describe the method to extract embeddings and the scoring function to measure relevance. Then we explore two design choices, namely negative sampling strategies and loss functions.

4.1 Retrieval Model

As shown in 4, there are two major components in the retrieval model, namely the embedding model that encodes the query and product tokens into dense vectors and the scoring function that measures the relevance of the query-product pair. Both the query encoder and the product encoder adopt the same multi-layer transformer architecture, and the parameters are shared. With the self-attention mechanism [24], the transformer-based encoder outputs context-based token embeddings. The query embedding $\vec{q_n}$ and product embedding $\vec{i_n}$ are obtained by the average pooling of the token embeddings. The encoders are initialized with the pre-trained cross-lingual language model, such as M-BERT [5] and XLM [11]. However, existing public pretrained language models are trained on the Wikipedia corpus, which may not generalize to the E-commerce domain. To avoid the domain discrepancy, we utilize the E-commerce corpus to learn the cross-lingual language model, denoted as EXLM, with the pretrained task proposed in XLM [11]. In detail, the Translation Language Modeling is trained with translated query pairs, and the Masked Language Modeling is trained with monolingual queries and English item titles.

After obtaining the query and item embeddings, we choose cosine similarity as the score function $S(q_n, i_n)$ which is commonly used in the retrieval task [6]:

$$S(q_n, i_n) = \frac{\vec{q_n} \cdot \vec{i_n}}{\|\vec{q_n}\| \|\vec{i_n}\|}, \tag{1}$$

where $\cdot$ denotes the dot-product of two vectors and $\|\cdot\|$ is the l_2-norm of the vector.

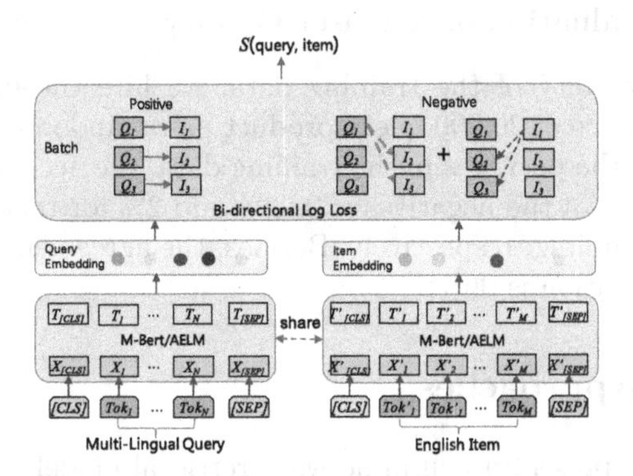

Fig. 4. The model architecture for CLPR with the batch negatives. The query embedding and item embedding is obtained by average pooling of outputs of multi-layer transformers. The transformers are initialized by the pre-trained cross-lingual model (M-BERT, XLM etc.). The solid line in the batch denotes the positive pairs. The dotted line denotes the batch negatives obtained in two ways. One is combining the irrelevant items with the query (the green dotted line), and the other is combining the irrelevant queries with the item (the red dotted line). (Color figure online)

4.2 Negative Sampling

Labeling negative samples for the retrieval task requires a large amount of labor and time cost. In the past studies, negatives are usually obtained by sampling based on human-crated rules. Here, we compared several sampling strategies, including random sampling, category-based sampling and a mixing strategy that combines random sampling and category-based sampling.

- **Random sampling:** for each query, we randomly sample items from all candidate items as negatives.
- **Category-based sampling:** for a positive pair $\{q_n, i_n\}$, randomly sample items under the same domain of i_n as irrelevant items.
- **Mixing strategies:** compared with random sampling, category-based sampling can produces hard negatives, since the items under the same domain tend to be similar. We explore two ways to combine the two sampling strategies. One is to train the model with random sampling first and then with category-based sampling (RANDOM -> CATEGORY). The other is to train the model with category-based sampling first and then train with random sampling (CATEGORY -> RANDOM).

Both random sampling and category-based sampling are computational expensive, with computational complexity $O(N_q \times N_i)$, where N_q is the number of queries and N_i is the candidate item pool size. To reduce computational complexity, the **batch negative** is adopted to approximate random sampling, where the irrelevant items for the query are the positive items from other queries in the same batch, as shown in Fig. 4. To implement category-based sampling with batch negatives, we organize the positive pair with the same category together.

4.3 Loss Function

We consider two popular loss functions for the retrieval task, namely the triplet loss [20,25] and the log-likelihood loss [21,26]. The triplet loss enforces a positive pair, denoted by $\{q_n, i_n^+\}$, to separate from a negative pair, denoted by $\{q_n, i_n^-\}$, by a distance margin m and is defined as:

$$\mathcal{L}_{triplet} = \sum_{n=1}^{N} max(0, D(q_n, i_n^+) - D(q_n, i_n^-) + m), \quad (2)$$

where $D(u, v)$ is a distance metric between vectors u and v, and is defined as $1 - S(u, v)$ in this paper.

The log-likelihood objective with the softmax function aims to place positives over the negatives. For the positive pair $\{q_n, i_n^+\}$, we can utilize the irrelevant item i_n^- for query q_n to compose the negative sample $\{q_n, i_n^-\}$ or the irrelevant query q_n^- for item i_n to compose the negative sample $\{q_n^-, i_n\}$. Thus, we can compute two log loss, one with $\{q_n, i_n^-\}$ as negatives (denotes as q-log loss) and the other with $\{q_n^-, i_n\}$ as negatives (denoted as i-log loss). The q-log loss and i-log loss are defined as:

$$\mathcal{L}_{q\_log} = -\frac{1}{N} \sum_{n=1}^{N} \log \frac{exp(S(q_n, i_n))}{exp(S(q_n, i_n)) + \sum_{i_k^- \in I_{q_n}} exp(S(q_n, i_k^-))}, \quad (3)$$

$$\mathcal{L}_{i\_log} = -\frac{1}{N}\sum_{n=1}^{N}\log\frac{exp(S(q_n, i_n))}{exp(S(q_n, i_n)) + \sum_{q_k^- \in I_{i_n}} exp(S(q_k^-, i_n))}, \tag{4}$$

where I_{q_n} is the set of irrelevant items for the query q_n, and I_{i_n} is the set of irrelevant queries for the item i_n. The sum of $\mathcal{L}_{q-log}$ and $\mathcal{L}_{i\_log}$ is denoted as bi-log loss $\mathcal{L}_{bi\_log}$. Figure 4 illustrates the bi-log loss computed with batch negatives.

$$\mathcal{L}_{bi\_log} = \mathcal{L}_{q\_log} + \mathcal{L}_{i\_log}. \tag{5}$$

5 Experiments

5.1 Evaluation Metrics

AUC is widely used to evaluate the product retrieval system. However, it cannot measure the effectiveness of an individual query, since it is computed over the whole test set. Inspired by the **Group AUC (GAUC)** proposed in [30], we define **GAUC** for the retrieval task as the mean of the AUC for each query. Both **AUC** and **GAUC** can only measure the ability to distinguish relevant and irrelevant pairs. To measure the ability to distinguish relevant, weak relevant, and irrelevant query-product pairs, we utilize **NDCG** [8], which is a popular metric for the ranking algorithms. In detail, the NDCG computes the similarity between the ranking results for each query and that based on relevance labels, and then is averaged over all test queries. Notice that the weak relevant label is used as irrelevant label when computing the **AUC** and **GAUC** metric.

Table 2. AUC of different negative sampling strategies on the CLPR task.

Model	Ru⇒En	Es⇒En	En⇒En	AVG
CATEGORY	81.47	77.36	82.76	80.53
RANDOM	82.12	76.84	82.51	80.49
CATEGORY -> RANDOM	82.05	77.30	82.49	80.61
RANDOM -> CATEGORY	**82.74**	**77.39**	**83.19**	**81.11**

5.2 Experimental Setting

The query encoder and item encoder are initialized with the pretrained cross-lingual language model, the 12-layer transformers with 768 hidden size. The max sequence length of the query encoder and the item encoder is 20 and 40, respectively. To finetune the pretrained cross-lingual language model, we use the Adam optimizer with $\beta_1 = 0.9$, $\beta_2 = 0.999$, L2 weight decay of 0.01, and learning rate of 3×10^{-5}. The margin value in the triplet loss and bi-log loss is set to 0.2, which leads to the best performance. We train all models by 10 epoches with a batch size of 512.

5.3 Effect of Negative Sampling

Table 2 shows the performance of various negative sampling strategies, including category-based sampling (Category), random sampling (Random), transferring random sampling to category-based sampling (Random -> Category), and transferring category-based sampling to random sampling (Category -> Random). All sampling strategies are implemented with *batch negatives*. Although category-based sampling may produce hard negatives, random sampling exhibits better performance than category-based sampling. This shows that the presence of easy negatives in training data is necessary. Besides, mixing easy and hard negatives in the training process is advantageous. Our experiment shows that transferring easy to hard achieves better performance than transferring hard to easy negatives. Consequently, transferring random sampling to category-based sampling (Random -> Category) is applied as the default in subsequent experiments.

5.4 Effectiveness of Bi-Directional Log Loss

We compare two loss functions, the triplet loss and our proposed bi-log loss. Since the bi-log loss is the sum of q-log loss and i-log loss, we further analyse the effectiveness of these two losses, respectively. Table 3 shows the results obtained with various loss functions. The i-log loss achieves better performance than the q-log loss, indicating forming the negatives by sampling the irrelevant multilingual queries given the item is more effective in the CLPR task. Although using q-log and i-log loss independently cannot achieve better performance than using the triplet loss, bi-log loss performs the best in terms of the AUC metric on all language pairs. This observation suggests that the q-log loss and i-log loss are complementary to each other.

Table 3. AUC of various loss functions on the CLPR task.

Model	Ru⇒En	Es⇒En	En⇒En	AVG
Triplet	79.28	76.88	78.71	78.29
Q-Log-Loss	76.85	73.87	77.6	76.11
I-Log-Loss	76.95	74.47	79.56	76.99
Bi-Log-Loss	**81.47**	**77.36**	**81.47**	**80.10**

Table 4. AUC, GAUC and NDCG of different models on CLPR task

Model	Ru⇒En			Es⇒En			En⇒En			AVG		
	AUC	GAUC	NDCG	AUC	GAUC	NDCG	AUC	GAUC	NDCG	AUC	GAUC	NDCG
DSSM	79.08	84.04	91.40	76.00	83.13	91.80	81.63	86.22	93.04	78.90	84.46	92.08
M-Bert+FT	82.74	87.81	93.05	77.39	84.37	91.90	83.19	88.63	94.21	81.11	86.94	93.05
EXLM+FT	**83.78**	**88.76**	**94.43**	**78.43**	**84.62**	**92.03**	**83.35**	**88.71**	**94.37**	**81.85**	**87.36**	**93.61**

5.5 Main Results

Given the best negative sampling strategy and loss function explored in the above sections, we explore the model architectures and the pretrained cross-lingual language models in this section. The best performance is reported as the baseline of the CLPR-9M dataset. Table 4 shows the results in terms of various evaluation metrics (AUC, GAUC and NDCG). **M-Bert+FT** and **EXLM+FT** finetune the pretrained models M-Bert and EXLM respectively. The overall performance of the models by finetuning pretrained language models achieves better performance than **DSSM**. The performance of **EXLM+FT** is better than that of **M-Bert+FT**, which indicates that the pretraining with parallel corpora and in-domain data can facilitate the CLPR learning. For all models, the English-English language direction achieves the best performance. This suggest that cross-lingual training is more challenging than monolingual training. The best performance is reported as the baseline of the CLPR-9M dataset.

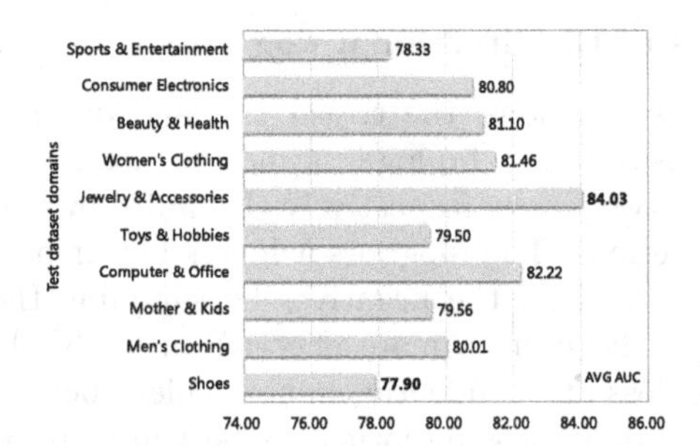

Fig. 5. AUC of XLM+Bilog on different categories.

5.6 Effect of the Category

Figure 5 illustrates the AUC on 10 categories for the best model. We find that the different categories have various levels of learning hardness. For example, the performance of the category **Sports & Entertainment** with the largest training data size ranked ninth place, while the category **Jewelry & Accessories** outperforms other categories with the medium training data size.

6 Conclusions

We construct CLPR-9M, a large-scale cross-lingual product retrieval benchmark. The CLPR-9M includes the training data by sampling online click logs, and the manually labelled test data. We conduct extensive experiments comparing different negative sampling strategies, and baseline models. Additionally, cross-lingual data facilitates the study of the cross-lingual language model.

Acknowledgement. We would like to thank for the support from the the National Key R&D Program of China under Grant 2018YFB1403200.

A Appendix

A.1 The annotation instructions for test dataset

The test set is obtained by the annotation of bilingual experts. We provide the detailed rating criteria to guarantee labeling Quality. For each label (relevant, weak relevant and irrelevant), we provide multiple criteria and the example to illustrate each criterion. The rating criteria and examples are shown in Table 5.

Table 5. The rating criteria and examples for human raters

Label	Criteria	Examples	
		Query	Item Title
Relevant	The item is consistent with the intention of query, and the title of the product exactly matches the literal or meaning of query	Wedding dress	**Wedding Dress** Long Sleeve Sheer Neck Appliques Bridal Gowns 2020 Spring
	The item is consistent with the intention of query, and item title does not match query literal, but it is synonym or abbreviation or original meaning	Mobile phone	Apple iPhone X 4G LTE **Mobile Cellphone** 3 GB RAM 64 GB 256 GB ROM 5.8"
	Brand and category have exactly the same intention as query	Apple iphone 12	**Apple iPhone 12** 5G LTE Mobile Phone 64 GB 256GB ROM 6.1"
	The item is consistent with the intention of query, but query is the hypernym	Lady shoes	Eilyken 2021 New Summer Fashion Design High heels **Ladies Sandals** Open Toe Shoes
Weak Relevant	The item is consistent with the intention of query, but query is the hyponym	Calf leather shoes	TUINANLE 2021 Autumn Winter **Shoes** Women Plush Snow Boot Heel Fashion Keep Warm Women's Boots Woman Size 36–42 Ankle Botas Pink
	The main product of title is consistent with the main product of query, but the attributes are different	64G usb driver	20pcs/lot Hot sale USB Flash Drive pendrive 8 GB 16 GB 32 GB
	The item is accessory of query	iPhone 11	Camera Lens Protection **Phone Case For iPhone 11** 12 Pro Max 8 7 6 6s
Irrelevant	The brand for item is different with query	Huawei phone case	Flower **Case For Samsung Galaxy** A50 A51 Plus Ultra S10E TPU
	The category for item is different with query	Apple iPhone	**Apple IPad Mini** 1st 7.9" 2012 16 Gb Silver Black 80% New Original Refurbish
	Item is related to intention of the query, and both belong to the same concept/category/industry, but not the same kind of products	Slippers	Eilyken 2021 New Summer Fashion Design Weave Women **Sandals**
	The item is totally different with query	Power cable	15 Pack LED S14 Replacement **Light Bulbs**, Warm White Edison Bulbs for Outdoor String Lights

References

1. CIKM Cup 2016 Track 2 (2016). https://competitions.codalab.org/competitions/
2. eBay SIGIR 2019 eCommerce search challenge (2019). https://sigir-ecom.github.io/ecom2019/data-task.html
3. Chen, A., Gey, F.C.: Combining query translation and document translation in cross-language retrieval. In: Peters, C., Gonzalo, J., Braschler, M., Kluck, M. (eds.) CLEF 2003. LNCS, vol. 3237, pp. 108–121. Springer, Heidelberg (2004). https://doi.org/10.1007/978-3-540-30222-3_10
4. Deerwester, S., Dumais, S.T., Furnas, G.W., Landauer, T.K., Harshman, R.: Indexing by latent semantic analysis. J. Am. Soc. Inf. Sci. **41**(6), 391–407 (1990)
5. Devlin, J., Chang, M.W., Lee, K., Toutanova, K.: Bert: pre-training of deep bidirectional transformers for language understanding. arXiv preprint arXiv:1810.04805 (2018)
6. Huang, J.T., et al.: Embedding-based retrieval in Facebook search. In: KDD, pp. 2553–2561 (2020)
7. Huang, P.S., He, X., Gao, J., Deng, L., Acero, A., Heck, L.: Learning deep structured semantic models for web search using clickthrough data. In: CIKM, pp. 2333–2338 (2013)
8. Järvelin, K., Kekäläinen, J.: Cumulated gain-based evaluation of IR techniques. TOIS **20**(4), 422–446 (2002)
9. Jiang, Z., El-Jaroudi, A., Hartmann, W., Karakos, D., Zhao, L.: Cross-lingual information retrieval with bert. arXiv preprint arXiv:2004.13005 (2020)
10. Karmaker Santu, S.K., Sondhi, P., Zhai, C.: On application of learning to rank for e-commerce search. In: SIGIR, pp. 475–484 (2017)
11. Lample, G., Conneau, A.: Cross-lingual language model pretraining. arXiv preprint arXiv:1901.07291 (2019)
12. Li, H., Xu, J.: Semantic matching in search. Found. Trends Inf. Retr. **7**(5), 343–469 (2014)
13. Monz, C., Dorr, B.J.: Iterative translation disambiguation for cross-language information retrieval. In: SIGIR, pp. 520–527 (2005)
14. Nie, J.Y.: Cross-language information retrieval. Synth. Lect. Hum. Lang. Technol. **3**(1), 1–125 (2010)
15. Qin, T., Liu, T.Y., Xu, J., Li, H.: Letor: a benchmark collection for research on learning to rank for information retrieval. Inf. Retrieval **13**(4), 346–374 (2010)
16. Robertson, S., Zaragoza, H.: The Probabilistic Relevance Framework: BM25 and Beyond. Now Publishers Inc., Delft (2009)
17. Sarvi, F., Voskarides, N., Mooiman, L., Schelter, S., de Rijke, M.: A comparison of supervised learning to match methods for product search. arXiv preprint arXiv:2007.10296 (2020)
18. Sasaki, S., Sun, S., Schamoni, S., Duh, K., Inui, K.: Cross-lingual learning-to-rank with shared representations. In: NAACL, pp. 458–463 (2018)
19. Schamoni, S., Hieber, F., Sokolov, A., Riezler, S.: Learning translational and knowledge-based similarities from relevance rankings for cross-language retrieval. In: ACL, pp. 488–494 (2014)
20. Schultz, M., Joachims, T.: Learning a distance metric from relative comparisons. NeurIPS **16**, 41–48 (2004)
21. Shen, Y., He, X., Gao, J., Deng, L., Mesnil, G.: Learning semantic representations using convolutional neural networks for web search. In: WWW, pp. 373–374 (2014)

22. Sun, S., Duh, K.: Clirmatrix: a massively large collection of bilingual and multilingual datasets for cross-lingual information retrieval. In: EMNLP, pp. 4160–4170 (2020)
23. Van Gysel, C., de Rijke, M., Kanoulas, E.: Learning latent vector spaces for product search. In: CIKM, pp. 165–174 (2016)
24. Vaswani, A., et al.: Attention is all you need. arXiv preprint arXiv:1706.03762 (2017)
25. Weinberger, K.Q., Saul, L.K.: Distance metric learning for large margin nearest neighbor classification. JMLR **10**(2), 1 (2009)
26. Yang, Y., et al.: Improving multilingual sentence embedding using bi-directional dual encoder with additive margin softmax. arXiv preprint arXiv:1902.08564 (2019)
27. Zhang, H., et al.: Towards personalized and semantic retrieval: an end-to-end solution for e-commerce search via embedding learning. In: SIGIR, pp. 2407–2416 (2020)
28. Zhang, Y., Wang, D., Zhang, Y.: Neural IR meets graph embedding: a ranking model for product search. In: WWW, pp. 2390–2400 (2019)
29. Zhou, D., Truran, M., Brailsford, T., Wade, V., Ashman, H.: Translation techniques in cross-language information retrieval. CSUR **45**(1), 1–44 (2012)
30. Zhu, H., et al.: Optimized cost per click in Taobao display advertising. In: CIKM, pp. 2191–2200 (2017)

An Adaptable Indexing Pipeline for Enriching Meta Information of Datasets from Heterogeneous Repositories

Siamak Farshidi(✉) and Zhiming Zhao(✉)

Multiscale Networked Systems, University of Amsterdam,
Amsterdam, The Netherlands
{s.farshidi,z.zhao}@uva.nl

Abstract. Dataset repositories publish a significant number of datasets continuously within the context of a variety of domains, such as biodiversity and oceanography. To conduct multidisciplinary research, scientists and practitioners must discover datasets from various disciplines unfamiliar with them. Well-known search engines, such as Google dataset and Mendeley data, try to support researchers with cross-domain dataset discovery based on their contents. However, as datasets typically contain scientific observations or collected data from service providers, their contextual information is limited. Accordingly, effective dataset indexing can be impossible to increase the Findability, Accessibility, Interoperability, and Reusability (FAIRness) based on their contextual information. This paper presents an indexing pipeline to extend contextual information of datasets based on their scientific domains by using topic modeling and a set of suggested rules and domain keywords (such as essential variables in environment science) based on domain experts' suggestions. The pipeline relies on an open ecosystem, where dataset providers publish semantically enhanced metadata on their data repositories. We aggregate, normalize, and reconcile such metadata, providing a dataset search engine that enables research communities to find, access, integrate, and reuse datasets. We evaluated our approach on a manually created gold standard and a user study.

Keywords: Dataset indexing · Dataset discovery · Inverted indexing · Metadata standard · Data repository

1 Introduction

Data are increasingly used in decision-making, such as establishing public policies and conducting scientific experiments [17], and are published by various organizations [7], such as scientific publishers, commercial or governmental data providers, research consortia, specialized data repositories, and data aggregators. The more data organizations publish, the more complicated the problem of

J. Gama et al. (Eds.): PAKDD 2022, LNAI 13281, pp. 472–484, 2022.
https://doi.org/10.1007/978-3-031-05936-0_37

data discovery becomes [6]. Datasets are typically offered by scientific repositories [1,25] or shared via open data portals [14,19,21,28,29,35]. Data regarding a set of relevant scientific or practical observations are collected, organized, and formatted for a particular purpose, called dataset [4,30]. Accordingly, a dataset can be a collection of alphanumeric data, such as entities, diagrams, graphs, design decisions, or textual documents. So that dataset search concerns the discovery, exploration, and retrieval of datasets based on search criteria of searchers [4,5].

Communities such as Wikidata or the Linked Open Data Cloud [35] offer open and general-purpose data resources that software practitioners can employ in various application domains [7], such as intelligent assistants, recommender systems, and search engine optimization [11–13]. The primary goal is to increase the findability, accessibility, interoperability, and reusability of zillions of publicly available datasets by enabling data discovery and sharing across organizations within various domains. This trend is reinforced by advances in machine learning and information retrieval, which rely on data to train, validate and enhance their algorithms [32]. To support these applications, we need to search for datasets, which have been researched for decades [8]. However, many characteristics of datasets are unique, with particular requirements and constraints, which have been recognized by well-known dataset search engines, such as Google [6]. According to the literature, we identified the following three challenges in dataset indexing that we are going to address in this study.

$Challenge_1$: General-purpose web search engines typically fail at finding datasets because of *lacking enough description on landing pages of datasets* [16]. In other words, data repositories do not create an individual webpage for each dataset that can be easily recognizable and crawlable by general-purpose web search engines. Data repositories are typically accessible through queries and encrypted web Application Programming Interfaces (Web APIs); this is a well-known phenomenon called *deep Web* [24]. Accordingly, general-purpose search engines index a limited set of datasets.

$Challenge_2$: In literature, various open standards are introduced for describing structured (including dataset metadata) [6]. For instance, https://Schema.org and the W3C Data Catalog Vocabulary (DCAT) [9] are well-known metadata standards for indexing datasets. Based on our observations (see Sect. 3), and Brickley et al. [6] there is *a limited agreement among dataset repositories in using such metadata*, and they typically define and employ their metadata features to index datasets. Thus, extracting metadata features of datasets from different data repositories automatically based on metadata standards is not possible.

$Challenge_3$: *Links between datasets are still rare, making identifying and using extra contextual information difficult* [6]. In order to offer cross-domain discovery, dataset search engines must improve their ingesting, indexing, and cataloging processes. So that incorporating external knowledge in the data handling process and better management and usage of dataset-intrinsic information can be considered two alternative solutions [7]. Incorporating external contextual information, whether through domain ontologies, tacit knowledge of domain experts, external quality indicators, domain keywords (e.g., essential variables [22]), or even unstructured information (e.g., in natural language) that describes the datasets, is a fundamental problem.

We introduce a novel dataset indexing pipeline to address these three challenges, incorporating information retrieval techniques including web crawling, metadata extraction, language models, human in the loop, and topic modeling to identify semantic similarities and generate indexing documents. The novelty of the proposed pipeline lies in (1) using domain experts' insights to collect *an extendable set of rules* for extracting and refining metadata of dataset records from heterogeneous repositories. Moreover, (2) it employs machine learning techniques, such as topic modeling and similarity approaches, as *replaceable components* to identify topic similarities. Furthermore, (3) the pipeline generates a mapping for each dataset record that adds *additional contextual information* to it. The final mappings can be used to generate effective indexing (e.g., inverted indexing). The proposed pipeline is adjusted based on extendable rules, domain keywords (e.g., essential variables), and domain experts observe and monitor their impacts on the mapping quality.

This paper is structured as follows. Section 2 elaborates on the proposed pipeline and its constituent components. Section 3 explains the experiment that we have conducted with four real-world dataset repositories to evaluate the pipeline. Section 3.4 analyzes the results of the experiment and assesses the performance of the pipeline on the selected dataset repositories. Section 4 discusses the lessons learned, the pipeline limitations, and feedback from the experts. Section 5 concludes this study and highlights our future research directions.

2 Dataset Indexing Pipeline

In this section, we elaborate on the constituent components of the proposed indexing pipeline. Figure 1 shows the components of the pipeline and its workflow.

Dataset Repositories refer to datasets isolated to be mined for data reporting and analysis. Data repositories are an extensive database of research infrastructures, such as ICOS and SeaDataNet, (see Sect. 3) that collect, manage, and store datasets for data analysis, sharing, and reporting.

Web crawling is the process of a spider bot that systematically browses dataset repositories and extracts dataset records in terms of RDF documents or their landing pages. It retrieves such contents in structured formats (e.g., JSON or key-values). The Web crawling process starts with a list of URLs to visit (seeds). The crawler identifies all the hyperlinks in the retrieved documents/landing pages and adds them to the list of its frontiers to visit them subsequently.

Metadata extraction is the process of retrieving any embedded metadata present in a document. It is responsible for extracting metadata features such as classes and properties inside an RDF document or textual contents of potential features mentioned on landing pages of datasets. The metadata of the retrieved documents will be extracted based on the rules that domain experts define them.

Language model employs various statistical and probabilistic methods to specify the probability of a given sequence of words occurring in a textual document.

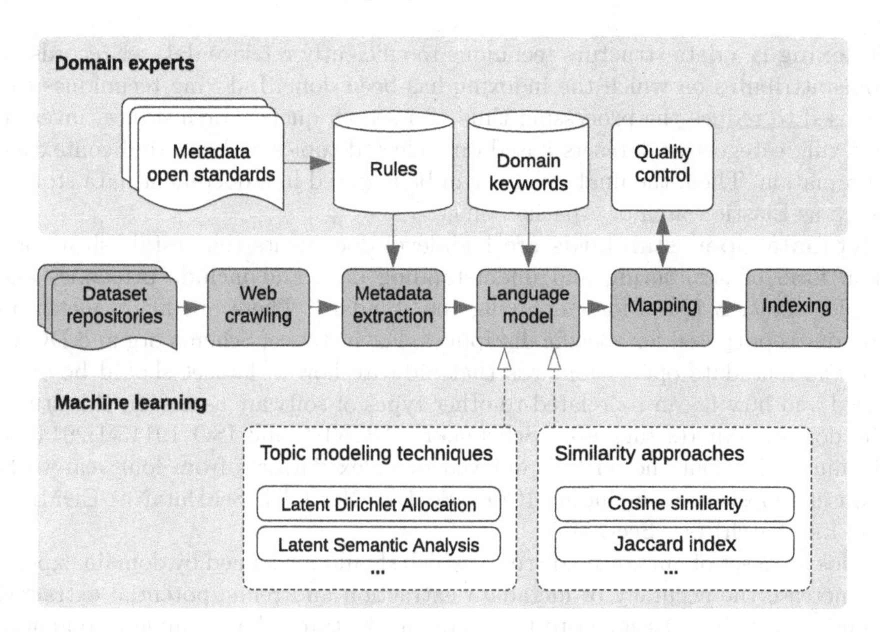

Fig. 1. Shows the constituent components of the pipeline and its workflow. The pipeline crawls dataset repositories and extracts metadata based on rules that domain experts define according to metadata open standards and domain knowledge. Then, the language model of the pipeline employs topic modeling techniques and similarity approaches to map the domain keywords and extra contextual information to the extracted metadata and create a mapping. The mapping quality will be checked frequently, and the hyperparameters will be adjusted accordingly. Finally, the mapping will be used to create indexes for the records of the dataset repositories.

It analyzes bodies of documents to convert qualitative information into quantitive information. In other words, the language model calculates similarities among metadata features of a particular dataset record and potential contextual information that could be assigned to it. Since contextual information, such as domain keywords, can be seen as vectors, we can use different similarity approaches, such as the cosine similarity or Jaccard index, to calculate the similarity of these vectors [34].

Mapping refers to the process of adding external contextual information, such as domain keywords, to extracted metadata features based on predefined rules by domain experts and language models' predictions. For instance, "sea surface salinity" and "sea surface temperature" as two domain keywords (essential variables[1]) can be mapped to a dataset record that the language model identified the following topics for it: *(water- temperature- dimension- dissolved- salinity- gas- oceanography- chemical- pigment- oceanographic- custodian- sea- geographical- coordinate- spatial)*.

[1] https://earthdata.nasa.gov/learn/backgrounders/essential-variables.

Indexing is a data structure technique to efficiently retrieve dataset records on some attributes on which the indexing has been done. Indexing techniques can be used to reduce the processing time of a search query. For instance, inverted indexing categorizes datasets based on collected topics and external contextual information. Then, the final indexes can be ingested in a document data storage (such as ElasticSearch or Apache Solr).
Metadata open standards are high-level documents that establish a common form of structuring and understanding data and include principles and implementation issues for employing the standard. There are many metadata standards purposed for specific disciplines. For instance, Schema.org and DCAT are two metadata open standards that indicate how a dataset should be organized and how it can be related to other types of software assets. In this study, the domain experts suggested Schema.org, DCAT3, and ISO 19115-1:2014 for defining rules that should be employed to index datasets from four real-world dataset repositories (including ICOS, SeaDataNet CDI, SeaDataNet EDMED, and LifeWatch) (see Sect. 3).
Rules are a set of human-made rules, which should be defined by domain experts to increase the accuracy of metadata extraction and refine potential extracted values that can be assigned to the metadata features. An example of potential rules in the rule base is presented as follows. The example shows a metadata feature called "identifier", which is a "unique identifier for this metadata record", and its data type is "PropertyValue/Text". The length of potential values for this metadata feature should be at least 15 characters. The *metadata extraction* component should look for metadata features such as "ISBN", "GTIN", or "UUID" to extract potential values that can be mapped to "identifier".

```
"identifier": [
  "datatype" : "PropertyValue/Text",
  "description": "unique identifier for this metadata record",
  "constraint": ["len(15)"],
  "suggested fields": ["ISBN", "GTIN", "UUID", "URI","URL","id","metadataIdentifier",
                      "gmd:fileIdentifier", "gco:CharacterString","pid"] ], ...
```

Domain keywords are content-related terms that are specific to a particular scientific domain. Terms in glossaries of social studies textbooks or essential variables in environmental sciences are examples of such vocabularies. Domain experts are the main source of knowledge for suggesting domain keywords.
Quality control is an essential phase of the pipeline as the quality of the mapping will be evaluated based on the number of values that mapped correctly to metadata features and the number of potential topics, the number of mapped values to the domain keywords. If the mapping quality is not acceptable, the number of topics of topic modeling algorithms, the threshold of the cosine similarity, and the rules should be revised to improve the mapping quality. This process can be considered as hyperparameter tuning, which is the process of choosing a set of optimal hyperparameters for the similarity approaches, such as cosine similarity and Jaccard index.

Topic modeling techniques are employed in the language model to identify the topics of dataset records. This study uses Latent Dirichlet Allocation (LDA) to find potential topics assigned to datasets. LDA is a generative model for the creation of natural language documents [3]. Note, a topic is a subject discussed in one or more documents. Examples of topics include dataset domains such as "Oceanography" entities such as "SeaDataNet" and long-standing subjects such as "climate change". Each topic is assumed to be represented by a multinomial distribution of words.
Similarity approaches are essential in solving many pattern recognition problems such as classification and clustering. Various similarity approaches, such as Cosine similarity and Jaccard index, are available in the literature to compare two text documents and determine how close their context or meaning are. Various text similarity approaches exist. Typically, similarity approaches have their specification to measure the similarity between two queries. For instance, cosine similarity measures the text-similarity between two documents irrespective of their size in Natural language Processing. The text documents are mainly represented in n-dimensional vector space.

3 Evaluation

One of the well-known issues in evaluating dataset indexing is the lack of benchmarks [7]. So, it is essential to identify a set of appropriate metrics to assess dataset indexing techniques and observe if they mimic information retrieval metrics, such as precision and recall. Such metrics should be employed [20] to evaluate the correctness of the indexing pipeline. In this study, we conducted an experiment in the context of four dataset repositories to assess the pipeline's impact on the quality of mappings and evaluate its effectiveness in addressing the dataset indexing challenges.

3.1 Dataset Repositories

The dataset repositories used for the evaluation are based on RDF datasets that have been published by four real-world dataset repositories, namely ICOS, SeaDataNet CDI, SeaDataNet EDMED, and LifeWatch.
(1) ICOS[2] (Integrated Carbon Observation System) is a European-wide greenhouse gas research infrastructure that produces standardized data on greenhouse gas concentrations in the atmosphere and carbon fluxes between the atmosphere, the earth, and oceans. The ICOS dataset repository contains more than *400K* dataset records.
(2) SeaDataNet CDI[3] (Common Data Index service) provides aggregated datasets (collections of all unrestricted SeaDataNet measurements of temperature and salinity by sea basins) and climatologies based on the aggregated

[2] https://data.icos-cp.eu/portal/.
[3] https://cdi.seadatanet.org/search.

datasets and data from external data sources such as the Coriolis Ocean Dataset for Reanalysis and the World Ocean Database for all the European sea basins and the Global Ocean. The CDI dataset repository contains more than *2,6M* dataset records.
(3) SeaDataNet EDMED[4] covers a wide range of disciplines, including marine meteorology; physical, chemical, and biological oceanography; sedimentology; marine biology and fisheries; environmental quality, coastal and estuarine studies, marine geology, and geophysics. Currently, EDMED contains more than *4K* dataset records, held at over 700 Data Holding Centres across Europe.
(4) LifeWatch[5] provides open data access and facilitates exploratory data analysis of data generated by the local marine-freshwater-terrestrial LifeWatch observatory. The LifeWatch dataset repository contains more than *1,1K* dataset records.

3.2 The Pipeline Configuration

Rule base - Ten domain experts within geology, oceanography, agriculture, environment, and biology research domains were selected based on their expertise and years of experience to participate in the research and assist us with building the rule base and evaluating the pipeline outcomes. Accordingly, we conducted a survey to identify the features and rules employed to extract metadata from the selected dataset repositories. The experts selected the features we need to use from three metadata standards, including DCAT 3, ISO 19115-1:2014, and Shema.org. It is interesting to highlight that almost less than half of the features that the research infrastructures have been employed in their own metadata were compatible with the metadata open standards[6].
Domain keywords - The domain experts suggested three sets of essential variables [10] based on the domains (atmosphere, oceanography, biodiversity) of the dataset repositories. Note, essential variables are variables known to be critical for observing and monitoring a given facet of the Earth system.
Topic modeling technique - We used Latent Dirichlet Allocation (LDA) as a topic modeling technique for generating potential topics of each dataset record based on its textual explanation.
Similarity approaches - We employed cosine similarity and Jaccard index in this study to estimate similarities among generated potential topics (by the LDA algorithm) of each dataset record and the essential variables. In cosine similarity, data objects in a dataset are treated as a vector. The Jaccard similarity index (sometimes called the Jaccard similarity coefficient) compares members for two sets to see which members are shared and which are distinct. It is a measure of similarity for the two sets of data, with a range from *0%* to *100%*. The higher the percentage, the more similar the two populations.

---
[4] https://edmed.seadatanet.org/search/.
[5] https://metadatacatalogue.lifewatch.eu.
[6] We published the results of our observations, analysis, script, and contextual information on Mendeley Data [10].

3.3 Experiment

First, we randomly selected 100 datasets from the dataset repositories to generate a training set. Two researchers independently determined the correctness of the mapped domain keywords and potential topics (generated by the LDA algorithms) to the selected datasets. To solve this task, they got the description of those datasets besides their extracted metadata feature, mapped domain keywords, potential topics, and the possibility of taking a deeper look inside the datasets themselves. Finally, we compared their responses, and in the case of inconsistencies, we asked both of them to recheck their responses to reach an agreement between their responses. The training set was used to adjust the hyperparameters, such as the similarity thresholds for both cosine similarity and the Jaccard index, and train the topic model. We altered the hyperparameters dynamically to reach their optimal values for the dataset. Then, we employed the Jaccard index and cosine similarity as the quality control approach to reject irrelevant topics.

One of the main weaknesses of information retrieval measures (including recall, accuracy, and F-measure) is the assumption of binary relevance, with human assessors asked to determine, for a set of documents, which members are relevant to the query and which are not. In other words, human experts are needed to judge the retrieved information and evaluate the effectiveness and efficiency of information retrieval methods [26]. The significant number of dataset records makes it impossible to ask human experts to evaluate the pipeline's outcomes thoroughly. We used a fitness function to assess the quality of the mapping automatically. The fitness function gets the mappings and datasets as its inputs, and then it uses the Jaccard index to assess their relevance. In other words, if a domain keyword is mapped correctly (based on the threshold) to a dataset, it will be highlighted as a true positive. Otherwise, it will be marked as a false positive. We calculated the res of the metrics accordingly.

Table 1 shows the results of the analysis on the dataset repositories (ICOS, SeaDataNet *CDI*, SeaDataNet *EDMED*, and LifeWatch) and 100 randomly selected dataset records form each of them to generate the validation set (contains 400 dataset records). Note, for the sake of validity of our evaluation, the intersection of the training set and the validation dataset records is the empty set.

3.4 Analysis

To evaluate the pipeline components and their impacts on the mapping quality, we perform the experiment incrementally. In each step of the experiment, we evaluate the impact of the absence of each component (Cosine similarity, Rules, Topic mining, and Jaccard index) on the quality of the mappings (involved pipeline components). Note that cosine similarity has been considered the baseline in our analysis to calculate similarity in the language model. Moreover, the fitness function cannot analyze the pipeline's impact on the potential topics

Table 1. Shows the results of the analysis on the dataset repositories (ICOS, SeaDataNet *CDI*, SeaDataNet *EDMED*, and LifeWatch)

Analysis / Dataset repositories		ICOS				CDI				EDMED				LifeWatch			
Involved pipeline components	Cosine similarity	Yes	Yes	Yes	Yes	Yes	Yes	Yes	Yes	Yes	Yes	Yes	Yes	Yes	Yes	Yes	Yes
	Rules	No	Yes	Yes	Yes	No	Yes	Yes	Yes	No	Yes	Yes	Yes	No	Yes	Yes	Yes
	Topic modeling	No	No	Yes	Yes	No	No	Yes	Yes	No	No	Yes	Yes	No	No	Yes	Yes
	Jaccard index	No	No	No	Yes	No	No	No	Yes	No	No	No	Yes	No	No	No	Yes
Domain keywords (essential variables)	Precision (EV)	0.01	0.00	0.01	0.00	0.58	0.58	0.61	1.00	0.48	0.48	0.29	0.66	0.26	0.26	0.09	0.36
	Recall (EV)	0.00	0.00	0.00	0.00	0.21	0.21	0.27	0.23	0.08	0.09	0.15	0.19	0.10	0.10	0.05	0.18
	Accuracy (EV)	0.99	1.00	1.00	1.00	0.59	0.59	0.66	0.63	0.79	0.79	0.72	0.81	0.86	0.86	0.83	0.88
	F (EV)	0.00	0.00	0.00	0.00	0.30	0.30	0.36	0.36	0.13	0.14	0.17	0.28	0.14	0.14	0.06	0.23
Potential topics	Precision (To)	N/A	N/A	0.20	1.00	N/A	N/A	0.41	1.00	N/A	N/A	0.31	1.00	N/A	N/A	0.42	1.00
	Recall (To)	N/A	N/A	0.41	1.00	N/A	N/A	0.71	0.74	N/A	N/A	0.74	0.85	N/A	N/A	1.00	1.00
	Accuracy (To)	N/A	N/A	0.38	1.00	N/A	N/A	0.40	0.78	N/A	N/A	0.29	0.85	N/A	N/A	0.46	1.00
	F (To)	N/A	N/A	0.26	1.00	N/A	N/A	0.51	0.85	N/A	N/A	0.43	0.92	N/A	N/A	0.57	1.00
Mapping	# Mapped values	0.38	0.52	0.53	0.45	0.34	0.42	1.00	0.42	0.48	0.51	0.52	0.42	0.35	0.54	0.55	0.46
Inverted Indexing	# key > 1	0.01	0.01	0.83	0.78	0.51	0.51	0.69	0.74	0.16	0.16	0.48	0.43	0.12	0.12	0.42	0.33
	# keys	101	101	23	18	441	441	320	189	251	251	504	334	858	858	602	401
	# singletone links	100	100	4	4	215	215	99	49	212	212	262	190	755	755	348	268

when the topic modeling is not applied. So, in such a scenario, the measures are equal to Not Applicable ("N/A").

The average F-measures of the domain keywords, $F(EV)$, and potential topics, $F(To)$, in Table 1, represent that the pipeline outperforms when all its components are involved.

It has already been shown that LDA does not perform well on short documents in which many different words rarely appear, e.g., messages of short messaging services [36]. It is essential to highlight that the datasets from ICOS typically have limited contextual information, so the $F(EV)$ values have not changed significantly by adding or removing a pipeline component. However, they increase the average F-measures of the rest of the datasets in the validation set. Note, to generate an almost stable list of topics using the LDA algorithm, and we repeated the topic modeling ten times. Increasing the number of iterations leads to higher accuracy and higher time consumption.

The number of assigned values to metadata features (# Mapped values) has been increased by applying the components. As the Jaccard index refines the irrelevant candidate values in the mapping, it reduces the number of mapped values and increases the mapping quality.

Keys are combinations of generated topics and successfully assigned domain keywords. In the last section of Table 1, the quality of the mapping has been evaluated. In the absence of topic modeling and Rules, the performance of the pipeline to generate high-quality mapping and keys decreases significantly. In this scenario, the number of identified keys (# keys) and for generating inverted indexed have increased. Most of the generated keys were singleton and meaningless quality values. In contrast, when the rules and topic modeling components have been applied, the number of keys decreased as the pipeline rejected low-quality values, and the number of the singleton keys decreased significantly as the pipeline aggregated more keys (# key > 1). For instance, the number of meaningful keys increased from 0.01 (1%) to 0.83 (83%) in the ICOS dataset records by adding topic modeling these two components. It is essential to highlight that applying the Jaccard index can lead to lower numbers of keys, as it further reduces the number of noisy or meaningless keys.

4 Discussion

In the literature, we observed that dataset search had been studied for decades by other researchers and practitioners and can be categorized into two types [7]: general-purpose and domain-specific dataset search. In *general-purpose dataset search approaches* such as Dataverse [1], Elsevier Data Search [25]), open data portals [14,19,21,28,29,35] and search engines such as DataMed [33], and Google Dataset Search [27], a collection of public and free datasets, in terms of scientific or practical observations, can be searched through their web portals. These dataset search engines are typically domain-independent, so they are not customized for a particular community. However, *domain-specific dataset search approaches* are designed for searching a set of related observations organized for a particular domain by searchers. This pattern of behavior is particularly marked in data lakes [15,31], data markets [2,18], and tabular search [23].

This study identified three challenges that general-purpose and domain-specific dataset search approaches face in their indexing phases. ($Challenge_1$) lack of enough description on landing pages of datasets [16] (deep Web [24]), ($Challenge_2$) a limited agreement among dataset repositories in using metadata standards [6], and ($Challenge_3$) complexity of identifying and using extra contextual information in dataset indexing [7]. To address $Challenge_1$, the pipeline contains an extendable set of domain keywords based on domain experts' insights on the dataset's domain. Domain keywords can improve the findability of dataset records by adding more contextual information to them. The proposed pipeline addresses $Challenge_2$ by suggesting an extendable set of rules based on open standards' definitions and properties. This pipeline component increases the quality of mapping and indexing significantly (see Sect. 3.4). The pipeline uses topic modeling (e.g., LDA) and similarity approaches to generate potential topics regarding a dataset record according to its contents. Then, it maps the most similar domain keywords to dataset records based on the generated topics.

Probabilistic topic modeling approaches such as LDA employ statistical reasoning to discover underlying patterns of data. As the model hyperparameters should be inferred from observations, the accuracy of statistical reasoning depends on the number of observations. LDA models a dataset as a mixture of topics, and then each word is drawn from one of its topics. Thus, the performance of LDA can be reduced dramatically in the case of short contextual documents (as happened with ICOS dataset records).

Similarity approaches are low sensitive to semantics. For instance, such methods do not consider the words "marine", "seawater", and "oceanic" as semantically similar. Additionally, they do not distinguish phrases based on their orders and conceptual meaning. For instance, "the ocean color is lighter than sky color" is similar to "the sky color is lighter than the ocean color". It is essential to highlight that similarity approaches, such as the Jaccard index, do not consider word frequency in a given document and count the number of common words in two documents. Accordingly, rare words that are mainly more informative in a document will be ignored. Moreover, the number of repetitions of similar words in two documents would not change the results of such similarity approaches. To sum up, before using a similarity approach in the language model, all its characteristics and behaviors should be investigated. Additionally, a performance testing analysis should be conducted beforehand to select the optimal solution for a particular usage.

Although the pipeline proposed in this study addresses three identified challenges in the literature, there are challenges in the literature regarding FAIRness of dataset discovery that requires profound attention. For instance, European Commission highlighted the following dataset discovery challenges: (1) lack of information that specific datasets exist and are available; (2) a lack of transparency of which public authority maintains datasets; (3) a lack of evidence concerning the terms of reuse; (4) datasets which are made available only in formats that are difficult or expensive to use; (5) complex licensing procedures or restrictive fees; (6) exclusive reuse agreements with one commercial third-party or reused restricted to a government-owned organization.

5 Conclusion and Future Work

Generating value from data needs the ability to find, access, and make sense of datasets. Many efforts are initiated to support dataset sharing and discovery. For instance, the Google dataset allows users to discover data stored in various online dataset repositories via keyword queries. This study highlighted three challenges that general-purpose and domain-specific dataset search approaches face in their indexing phases. ($Challenge_1$) lack of enough description on landing pages of datasets [16] (deep Web [24]), ($Challenge_2$) a limited agreement among dataset repositories in using metadata standards [6], and ($Challenge_3$) complexity of identifying and using extra contextual information in dataset indexing [7]. To address these challenges effectively, we proposed a novel dataset indexing pipeline based on information retrieval techniques. Next, we conducted an experiment incrementally on the pipeline components to evaluate their effectiveness in addressing the challenges. The results confirmed that the pipeline outperforms when all its components are involved.

Probing deeper, the pipeline presented in this paper also provides a foundation for future work in software asset discovery. We intend to conduct research to address*software asset recommendation* and *context aware search engines* as our (near) future work.

Acknowledgment. This work has been partially funded by the European Union's Horizon 2020 research and innovation programme, by the project of ARTICONF (825134), ENVRI-FAIR (824068) and BLUECLOUD (862409).

References

1. Altman, M., Castro, E., Crosas, M., Durbin, P., Garnett, A., Whitney, J.: Open journal systems and dataverse integration-helping journals to upgrade data publication for reusable research. Code4Lib J. **50**(30) (2015)
2. Balazinska, M., Howe, B., Koutris, P., Suciu, D., Upadhyaya, P.: A discussion on pricing relational data. In: Tannen, V., Wong, L., Libkin, L., Fan, W., Tan, W.-C., Fourman, M. (eds.) In Search of Elegance in the Theory and Practice of Computation. LNCS, vol. 8000, pp. 167–173. Springer, Heidelberg (2013). https://doi.org/10.1007/978-3-642-41660-6_7
3. Blei, D.M., Ng, A.Y., Jordan, M.I.: Latent Dirichlet allocation. J. Mach. Learn. Res. **3**, 993–1022 (2003)
4. Borgman, C.L.: The conundrum of sharing research data. J. Am. Soc. Inform. Sci. Technol. **63**(6), 1059–1078 (2012)
5. Borgman, C.L.: Big Data, Little Data, No Data: Scholarship in the Networked World. MIT Press, Cambridge (2016)
6. Brickley, D., Burgess, M., Noy, N.: Google dataset search: building a search engine for datasets in an open web ecosystem. In: The World Wide Web Conference, pp. 1365–1375 (2019)
7. Chapman, A., Simperl, E., Koesten, L., Konstantinidis, G., Ibáñez, L.-D., Kacprzak, E., Groth, P.: Dataset search: a survey. VLDB J. **29**(1), 251–272 (2019). https://doi.org/10.1007/s00778-019-00564-x
8. Codd, E.F., et al.: Relational completeness of data base sublanguages. IBM Corporation (1972)
9. Data Catalog Vocabulary (DCAT) - Version 3. https://www.w3.org/TR/vocab-dcat-3/. Accessed 30 Sept 2021
10. Farshidi, S.: The observations, analysis, script, and contextual information regarding this paper. Mendeley Data (2022). https://doi.org/10.17632/3yb7mhxtyf.1
11. Farshidi, S., Jansen, S.: A decision support system for pattern-driven software architecture. In: Muccini, H., Avgeriou, P., Buhnova, B., Camara, J., Caporuscio, M., Franzago, M., Koziolek, A., Scandurra, P., Trubiani, C., Weyns, D., Zdun, U. (eds.) ECSA 2020. CCIS, vol. 1269, pp. 68–81. Springer, Cham (2020). https://doi.org/10.1007/978-3-030-59155-7_6
12. Farshidi, S., Jansen, S., Deldar, M.: A decision model for programming language ecosystem selection: seven industry case studies. Inf. Softw. Technol. **139**, 106640 (2021)
13. Farshidi, S., Jansen, S., Fortuin, S.: Model-driven development platform selection: four industry case studies. Softw. Syst. Model. **20**(5), 1525–1551 (2021). https://doi.org/10.1007/s10270-020-00855-w
14. Find open data. https://data.gov.uk. Accessed 30 Sept 2021
15. Gao, Y., Huang, S., Parameswaran, A.: Navigating the data lake with datamaran: automatically extracting structure from log datasets. In: Proceedings of the 2018 International Conference on Management of Data, pp. 943–958 (2018)
16. Goel, S., Broder, A., Gabrilovich, E., Pang, B.: Anatomy of the long tail: ordinary people with extraordinary tastes. In: Proceedings of the Third ACM International Conference on Web Search and Data Mining, pp. 201–210 (2010)

17. Gohar, M., Muzammal, M., Rahman, A.U.: Smart TSS: Defining transportation system behavior using big data analytics in smart cities. Sustain. Urban Areas **41**, 114–119 (2018)
18. Grubenmann, T., Bernstein, A., Moor, D., Seuken, S.: Financing the web of data with delayed-answer auctions. In: Proceedings of the 2018 World Wide Web Conference, pp. 1033–1042 (2018)
19. Hendler, J., Holm, J., Musialek, C., Thomas, G.: US government linked open data: semantic. data. gov. IEEE Intell. Syst. **27**(03), 25–31 (2012)
20. Kacprzak, E., Koesten, L., Ibáñez, L.D., Blount, T., Tennison, J., Simperl, E.: Characterising dataset search-an analysis of search logs and data requests. J. Web Semant. **55**, 37–55 (2019). Article no. 106640
21. Kassen, M.: A promising phenomenon of open data: a case study of the Chicago open data project. Gov. Inf. Q. **30**(4), 508–513 (2013)
22. Lehmann, A., Masò, J., Nativi, S., Giuliani, G.: Towards integrated essential variables for sustainability (2020)
23. Lehmberg, O., Bizer, C.: Stitching web tables for improving matching quality. Proc. VLDB Endowment **10**(11), 1502–1513 (2017)
24. Madhavan, J., Ko, D., Kot, Ł, Ganapathy, V., Rasmussen, A., Halevy, A.: Google's deep web crawl. Proc. VLDB Endowment **1**(2), 1241–1252 (2008)
25. Mendeley data. https://data.mendeley.com/research-data/. Accessed 30 Sept 2021
26. Moffat, A., Zobel, J.: Rank-biased precision for measurement of retrieval effectiveness. ACM Trans. Inf. Syst. (TOIS) **27**(1), 1–27 (2008)
27. Nguyen, T.T., Nguyen, Q.V.H., Weidlich, M., Aberer, K.: Result selection and summarization for web table search. In: 2015 IEEE 31st International Conference on Data Engineering, pp. 231–242. IEEE (2015)
28. Open data monitor. https://www.opendatamonitor.eu/. Accessed 30 Sept 2021
29. Open knowledge foundation (CKAN). https://ckan.org/. Accessed 30 Sept 2021
30. Pasquetto, I.V., Randles, B.M., Borgman, C.L.: On the reuse of scientific data. Data Sci. J. **16**, 8 (2017)
31. Reynolds, P., Neuman, K.L., Officer, C.P.: DHS data framework. dhs.gov (2014)
32. Roh, Y., Heo, G., Whang, S.E.: A survey on data collection for machine learning: a big data-AI integration perspective. IEEE Trans. Knowl. Data Eng. (2019)
33. Sansone, S.A., et al.: Dats, the data tag suite to enable discoverability of datasets. Sci. Data **4**(1), 1–8 (2017)
34. Steyvers, M., Griffiths, T.: Probabilistic topic models. In: Handbook of Latent Semantic Analysis, pp. 439–460. Psychology Press, Hove (2007)
35. The linked open data cloud. https://www.lod-cloud.net/. Accessed 30 Sept 2021
36. Zhao, W.X., et al.: Comparing Twitter and traditional media using topic models. In: Clough, P., et al. (eds.) ECIR 2011. LNCS, vol. 6611, pp. 338–349. Springer, Heidelberg (2011). https://doi.org/10.1007/978-3-642-20161-5_34

FLiB: Fair Link Prediction in Bipartite Network

Piyush Kansal, Nitish Kumar(✉), Sangam Verma, Karamjit Singh, and Pranav Pouduval

Mastercard, Gurugram, India
{nitish.srivasatava,sangam.verma,pranav.poduval}@mastercard.com

Abstract. Graph neural networks have become a popular modeling choice in many real-world applications like social networks, recommender systems, molecular science. GNNs have been shown to exhibit greater bias compared to other ML models trained on i.i.d data, and as they are applied to many socially-consequential use-cases, it becomes imperative for the model results and learned representations to be fair. Real-world applications of GNNs involve learning over heterogeneous networks with several nodes and edge types. We show that various kinds of nodes in a heterogeneous network can pick bias from a particular node type and remain non-trivial to debias using standard fairness algorithms. We propose a novel framework- Fair Link Prediction in Bipartite Networks (FLiB) that ensures fair link prediction while learning fair representations for all types of nodes with respect to the sensitive attribute of one of the node type. We further propose S-FLiB, which effectively mitigates bias at the subgroup level by regularising model predictions for subgroups defined over problem-specific grouping criteria.

Keywords: Fairness · GNN · Link prediction · Bipartite graph

1 Introduction

GNNs [10,12,16] are being widely used for representation learning and modeling many real-world applications involving interaction between a variety of nodes via different relation types. For example, the song recommendation graph involves two types of nodes - Users-Songs, whereas, IMDB graph consists of three types of nodes - Movie-Actor-Director. While, machine learning algorithms are known to exhibit algorithmic biases such as women being discriminated [2] in a job-recommendation system or African-Americans being subjected to higher-interest credit cards [1], GNNs can further exacerbate algorithmic biases [6] on sensitive attributes like age, gender, or color. Such biases can significantly hamper the applicability of these models in real-world use-cases, limiting their adoption. Hence, it becomes essential to audit GNNs on Fairness. Much of the research on Fairness in GNNs is focused on homogeneous networks [6,15] and is limited for heterogeneous networks [4,19]. Due to the cross-interaction between various types of nodes, most of the existing work in fair graph learning cannot invariably be extended for heterogeneous graphs. FairGNN [6] proposes a

J. Gama et al. (Eds.): PAKDD 2022, LNAI 13281, pp. 485–498, 2022.
https://doi.org/10.1007/978-3-031-05936-0_38

framework using adversarial debiasing [20] for fair node classification in homogeneous networks. For heterogeneous graphs, Fair-HIN [19] uses metric-specific regularization loss functions to correct bias in model predictions. However, the learned representations can not be used for other downstream tasks like node regression, node classification, or link prediction since it does not necessarily learn **fair representations**. [4] aims to solve this challenge by using adversarial filters and achieves representational invariance for a node's embeddings with respect to its sensitive attributes in a heterogeneous graph. However, it assumes that in a **link prediction task** between two types of nodes, only the nodes with sensitive attribute gives rise to the bias in the predictions. Our work moves away from this assumption and shows that, bias from one node type can also propagate to the other types, making a task like link prediction involving different types of nodes unfair. In this work we study the effect of bias in a bipartite graph- A special case of heterogeneous graph, due to its wide applicability in real life applications. Without loss of generality, we assume that a bipartite network has user and item nodes with user nodes having a binary sensitive attribute gender where bias can propagate from user nodes to item nodes. This problem of removing bias from item nodes is challenging due to the absence of any label for them. It becomes imperative to ensure that bias from user nodes is blocked from propagating to other types of nodes for keeping the task of link prediction fair. Hence fairness in GNNs for bipartite graphs poses two challenges- a) Fair representation learning for downstream tasks and b) Fairness in link prediction task involving both type of nodes.

To solve these challenges, we introduce a novel framework, "Fair Link Prediction in Bipartite Networks" (FLiB), for fair link prediction and fair representation learning in bipartite networks. FLiB adopts a multi-layer adversarial debiasing approach that debiases user nodes and promotes fairness in item nodes by blocking bias from propagating to them. FLiB further determines and explicitly debiases item nodes that are highly susceptible to bias using a "*susceptibility factor*" (SF) calculated from the graph structure. The subset of nodes with high SF are assigned a pseudo-sensitive label which allows the integration of a parallel debiasing framework for item nodes as well. FLiB debiases these item nodes similar to user nodes via multi-layer adversarial debiasing using their pseudo-sensitive attributes. Finally, FLiB integrates metric-specific regularization losses to optimize for fairness in the node predictions. While group fairness metrics such as Demographic Parity (DP) and Equal Opportunity [11] (EO) are essential requirements for fairness, fair-model predictions at a sub-group level may still exhibit significant bias. For example, in a movie recommendation system, a model might perform fairly at an overall level but still be biased towards a particular gender when audited on movies of a particular genre. We propose an extended version S-FLiB to mitigate bias at the sub-group level while achieving low Demographic Parity and Equal Opportunity at the overall level by integrating sub-group level regularisation losses into our model. We perform thorough experiments and detailed ablation studies on two publicly available datasets, namely, MovieLens, and LastFM to demonstrate the effectiveness of FLiB. We show that FLiB outperforms other baseline algorithms in fairness while ensuring fairness in representations and minimal to no loss in accuracy and AUC-ROC. To summarise, key contributions of our paper are -

- We introduce a novel framework FLiB for fair representation learning and link prediction in bipartite networks.
- We show that bias from one node type having a sensitive attribute can propagate to other node types in a heterogeneous network.
- We develop a new pseudo-sensitive attribute based debiasing technique leveraging graph structure to identify and debias item nodes that are highly susceptible to bias with respect to the user's sensitive attribute.
- We extend FLiB to S-FLiB, which explicitly keeps the DP, EO for problem-specific sub-groups in check using Fairness Aware Regularization Losses.

2 Preliminaries

This section introduces the notations used in this paper, followed by details of the datasets used. We then describe fairness metrics used for evaluating model performance and formalize our problem statement.

2.1 Notations

We define a bipartite graph $G = (U, V, \xi)$ where U is a set of nodes with sensitive attributes s_v such as gender, also referred to as user nodes while V is the set of nodes which don't have the sensitive attributes, also referred to as item nodes. $|U|$ and $|V|$ are the sizes of the sets U and V respectively. ξ denotes the set of edges. For a user node $u \in U$ and an item node $v \in V$, h_u and h_v are the learnt representations by the GNN. An edge $e \in \xi$ between u and v is represented as $e : (h_u, h_v)$. For predicting if a link exists between $u \in U$ and $v \in V$, a binary classifier f_c is trained that outputs $(\hat{y_e})$.

2.2 Datasets

We use two open-source bipartite graph datasets for our experiments:

MovieLens[1] dataset contains a set of movies rated by users in the range 1–5. Every user is associated with age, gender, and occupation attributes, while every movie is tagged with at least one genre. A user likes a movie if the rating given to it is over 3. MovieLens consists of 6040 users, 3883 movies, and 1M ratings with 0.6M ratings > 3.

LastFM[2] dataset is a song recommendation dataset between users and songs. Users are associated with gender, age, and country attributes along with their listening history. Songs are tagged with their corresponding artists. A user is said to have liked a song if they have listened to it more than once. Only the users with all the known attributes are considered, and the final dataset contains 262 users and 84K songs across 8K artists with 369k liked ratings for the songs.

[1] https://grouplens.org/datasets/movielens/1m/.

[2] http://millionsongdataset.com/lastfm/.

2.3 Evaluation Metrics

We evaluate model fairness in terms of the two most widely used fairness metrics:

Demographic Parity (DP) quantifies the degree of independence between the model outputs and the sensitive attributes. For binary-valued target outcome and sensitive attributes, demographic parity seeks to achieve:

$$P(\hat{y} = 1|a = 0) = P(\hat{y} = 1|a = 1) \tag{1}$$

where $\hat{y}$ is the predicted label, y is the ground truth class label, and a is the sensitive attribute such as gender.

Equal Opportunity (EO) requires that the probability of an instance in a positive class being assigned to a positive outcome be independent of its sensitive attribute. It is defined as:

$$P(\hat{y} = 1|a = 0, y = 1) = P(\hat{y} = 1|a = 1, y = 1) \tag{2}$$

A model is said to have fair predictions if both DP and EO are minimal, and its learned representations are said to be fair if an external adversary/classifier can predict the sensitive attribute from them only with an accuracy close to 50.

2.4 Problem Definition

We consider the task of link prediction in bipartite networks. Our goal is twofold: 1.) To make this task between the two types of nodes fair (in terms of EO & DP) with respect to the binary sensitive attribute user nodes. 2.) Learn fair representations for both types of nodes such that those embeddings can be used for any further downstream task. This is measured by the accuracy of sensitive attribute prediction from these representations using an external adversary/classifier.

3 Proposed Approach

In this section, we provide the details of our proposed approach FLiB. The four major components of FLiB are 1) A GNN based link-predictor, 2) Multi-layer Adversarial Debiasing, 3) Multi-Nodal Adversarial Debiasing via Pseudo sensitive attributes, 4) Fairness Aware Regularization Loss (FARL). We later propose S-FLiB, which extends FLiB to mitigate bias at the subgroup level. An illustration of our proposed framework, FLiB, is shown in Fig. 1.

3.1 Graph Based Link-Predictor

The graph-based link predictor consists of a GNN that is used to learn the embeddings for the nodes in a bipartite graph followed by a binary classifier f_c which predicts if an edge exists between a user node $u \in U$ and an item node

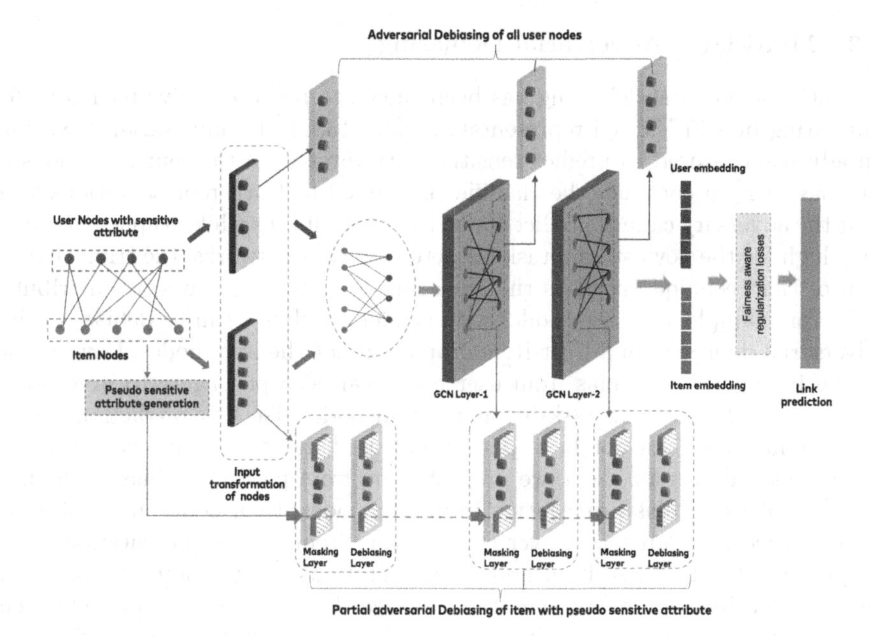

Fig. 1. An overview of the complete architecture of FLiB

$v \in V$. The link-predictor can flexibly use any feature aggregator like a GCN, GAT, RGCN. The classification loss for the link-predictor is given by

$$L_C = -\frac{1}{|\xi|} \cdot \sum_{e \in \xi} y_e \cdot \log(\hat{y_e}) + (1 - y_e) \cdot (1 - \log(\hat{y_e})) \tag{3}$$

where y_e are the true edge labels and $\hat{y_e} = f_c([h_u, h_v])$ are the predictions from the link-estimator. $|\xi|$ is the size of the set of nodes ξ.

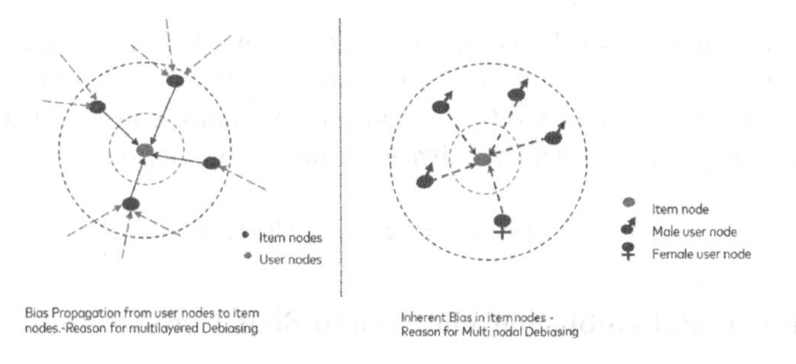

Fig. 2. a) Intuition behind Multi-layered (Left) and Multi-nodal (Right) Debiasing

3.2 Multi-layer Adversarial Debiasing

Recently, adversarial debiasing has been shown to be an effective technique for mitigating bias in learned representations [3,7,13,14]. In adversarial debiasing, an adversary is used to predict sensitive attributes from the representations of the classifier. In contrast, the classifier is trained to learn representations such that the adversary cannot predict the sensitive attributes while keeping the accuracy high for the downstream task. In heterogeneous networks, the trivial extension of this technique permits the debiasing of nodes with sensitive attributes only. For example, only user nodes with gender attributes can be debiased using adversarial debiasing in a User-Item graph and not the item nodes. However, as shown in experiments, bias from user nodes can also propagate to item nodes making any downstream task involving item nodes like link prediction unfair. Since item nodes can not be explicitly debiased due to the absence of actual sensitive attributes for them, we prevent them from picking up bias in the first place by blocking bias propagation from user nodes. Item nodes at graph layer l are learned by aggregation over the user and item node representations from graph layer $l-1$. These item node representations are promoted to be fair if the user and item node representations at layer $l-1$ are fair. Recursively, Item nodes at layer $l-1$ are promoted to learn fair representations by making user and item node representations fair at layer $l-2$. Hence, user and item node representations at all the graph layers should be fair. To achieve this, we deploy an adversary on the user node representations of every graph layer, which ensures user node representations remain fair throughout and, in turn, promote item node representations to remain fair till the last layer. Instead of directly applying an adversary on the input user features, we transform it using a feed-forward layer which is debiased and passed onto the graph structure. The total multi-layered adversarial loss L_A is given by the sum of cross-entropy loss for each adversary. For a graph with $L-1$ layers,

$$L_A = \sum_{l=1}^{L} -\frac{1}{|U|} \sum_{u \in U} w_1 \cdot s_u \cdot \log(\hat{s}_u^l) + w_0 \cdot (1 - s_u) \cdot \log(1 - \hat{s}_u^l) \tag{4}$$

where w_1, w_0 refers to the class weights for class 1 and 0 of the sensitive attributes respectively. s_u and $\hat{s}_u^l$ refers to the user sensitive attribute and predicted sensitive attribute respectively. We further add a covariance constraint [6,17,18] on the outputs of f_c to improve the fairness in model predictions.

$$L_{Cov} = |\, Cov(s, \hat{y}) \,| = |\, E[(s - E(s))(\hat{y} - E(\hat{y}))] \,| \tag{5}$$

3.3 Multi-nodal Debiasing via Pseudo Sensitive Attributes

As discussed in Sect. 3.2, item nodes cannot be explicitly debiased using adversarial debiasing since no sensitive labels exist for them. In this section we propose

Multi-nodal adversarial debiasing that explicitly debiases item nodes using pseudo-sensitive labels for them. As shown in Fig. 2 (right), an item node can have a large number of connections with user nodes belonging to a particular sensitive user group compared to the other. For example, in the MovieLens dataset, the movie "Sabotage" is highly liked by males, whereas "All over me" is much more liked by females. Such nodes will be largely aggregated over the members of one of the sensitive user groups during the learning phase of the GNN, making them prone to picking up bias against the other group since neighbors tend to learn representations that are closer to each other in the embedding space. We define "susceptibility factor" **SF**, which aims to quantify the bias inherent to each item node due to its connectivity. **SF** for every item node represents its association with each of the sensitive user groups. Specifically, SF quantifies the association with each group as a fraction of edges of an item node with the members of that group. For example, in the MovieLens dataset, SF for a movie is defined as the ratio of the percentage of females liking that movie to the percentage of males liking that movie. For any movie, a very high or a very low susceptibility factor indicates a expected bias in the graph learned representations towards one of the sensitive group (male/female in this case). To solve this, we assign pseudo-labels to item nodes with extreme values of the susceptibility factor and debias such item nodes using multi-layer adversarial debiasing. Since only a subset of item nodes would have the pseudo-label available, we filter out the non-susceptible nodes via a masking layer, as can be seen in Fig. 1.

3.4 Fairness Aware Regularization Losses

While fair representation learning is imperative to learn embeddings that can be used for many downstream tasks, ultimately, the fairness of each model is adjudicated via its outputs. Our extensive ablation studies below show that by integrating metric-specific loss functions, we can further optimize the model performance in terms of fairness with minimal to no loss in fairness in graph representations. We integrate the loss functions for DP and EO metrics [19] as regularisers in FLiB such that

$$L_{DP} = \frac{\sum\limits_{e \in \xi} I^s(e) \cdot f_c(e)}{\sum\limits_{e \in \xi} I^s(e)} - \frac{\sum\limits_{e \in \xi} (1 - I^s(e)) \cdot f_c(e)}{\sum\limits_{e \in \xi} 1 - I^s(e)} \tag{6}$$

$$L_{EO} = \frac{\sum\limits_{e \in \xi} I^s(e) \cdot I^y(e) \cdot f_c(e)}{\sum\limits_{e \in \xi} I^s(e) \cdot I^y(e)} - \frac{\sum\limits_{e \in \xi} (1 - I^s(e)) \cdot I^y(e) \cdot f_c(e)}{\sum\limits_{e \in \xi} (1 - I^s(e)) \cdot I^y(e)} \tag{7}$$

where $I^s(e)$ indicates if an user node in the edge has the sensitive attribute 1, and $I^y(e)$ indicates if an edge has a label 1.

3.5 Overall Objective Function

The overall objective function for FLiB is defined as

$$\min_{\theta_G, \theta_C} \max_{\theta_A} L_C + \alpha \cdot L_{Cov} - \beta \cdot L_A + \gamma \cdot (L_{DP} + L_{EO}) \tag{8}$$

where L_C, L_{Cov}, L_A, L_{DP}, L_{EO} refers to the losses for classification, covariance constraint, adversaries, DP and EO respectively. θ_G, θ_C refers to the parameters of the graph and the binary classifier while θ_A refers to the parameters of all the adversaries. The hyperparameters α, β, γ are used to control the contributions of Covariance, Adversarial and Regularization losses respectively.

3.6 S-FLiB: Sub-Group Fairness

Statistical fairness measures like DP and EO can result in a model that seems fair at an overall level but continue being unfair when looked at the subgroup level. For example, a movie recommendation system might satisfy fairness criteria on a population but remain biased towards recommending specific genres to a particular gender. We inspect the fairness at a sub-group level for a GCN-based movie recommendation system and compare it to FLiB. While FLiB is shown effective in mitigating bias even at the sub-group level, we further improve it using sub-group level regularization loss functions and propose an extended version, **S-FLiB**. For each sub-group, we calculate the regularization loss as the sum of DP and EO loss using Eqs. 6 and 7. S-FLiB final loss function is

$$L_{S-FLiB} = L_{FLiB} + \sum_{s \in S} L^s_{DP} + L^s_{EO} \tag{9}$$

where s indicates a sub-group and S is the set of sub-groups. L^s_{DP}, and L^s_{EO} refers to the DP and EO losses for each of the sub-groups $s \in S$.

4 Experimentation

This section describes the experiments performed to evaluate the efficacy of our proposed methods FLiB and S-FLiB, by reporting their performance on two open-source data sets and a comprehensive ablation study showing the importance of each component.

4.1 Baselines

We compare FLiB with two state-of-the-art fairness models in GNNs, FairGNN, and Fair-HIN on both datasets. Both are very recent and well-known methods that build over GNN to make the downstream tasks fair. FairGCN, although developed for homogeneous graphs, has been used for heterogeneous graphs by debiasing only the nodes with sensitive attributes. Fair-HIN uses fairness-aware regularization loss functions to make the model predictions fair with respect to some fairness metrics in both homogeneous and heterogeneous networks.

4.2 Implementation Details

For all our experiments, the datasets are split in 6:1:3 for training, validation, and testing respectively. For a fair comparison, we train each model up to 10000 epochs or convergence, whichever is earlier. The hyperparameters are determined via the fairness performance on validation set. The parameters α, β, and γ are varied over the values $\alpha = \{0.1, 1, 10, 20, 50\}$, $\beta = \{0.01, 0.01, 1, 10\}$, and $\gamma_1 = \{1, 10, 100, 1000\}$. For pseudo sensitive attribute labelling, item nodes with $\max(SF, \frac{1}{SF}) >= 2$ are assigned a label while others are masked. For Sub-group level fairness, sub-group regularization loss is weighted by $\gamma_2 = \{1, 10, 100, 1000\}$.

Table 1. Comparative results of FLiB on MovieLens, and LastFM datasets

	Dataset	Movie Lens					LastFM				
GNN	Model	Acc	AR	$\|DP\|$	$\|EO\|$	AAc	Acc	AR	$\|DP\|$	$\|EO\|$	AAc
GCN	GNN	83.57	0.915	5.79	4.91	62.58	67.45	0.723	8.53	6.03	56.60
	FAIR-HIN	83.64	0.914	0.85	0.79	72.45	67.55	0.726	**0.19**	**1.44**	90.57
	FairGNN	83.98	0.918	5.33	4.40	**51.40**	67.90	0.730	1.79	4.06	**51.14**
	FLiB	83.40	0.914	**0.32**	**0.12**	55.06	67.92	0.740	1.53	2.56	56.48
GAT	GNN	78.04	0.852	5.73	7.02	62.33	70.57	0.776	9.32	7.19	52.83
	FAIR-HIN	77.48	0.829	**0.52**	0.67	61.75	67.31	0.725	0.63	**0.76**	84.91
	FairGNN	77.00	0.869	0.71	2.61	53.55	66.93	0.730	1.29	4.19	**51.14**
	FLiB	70.84	0.802	1.15	**0.54**	**50.74**	67.19	0.745	**0.16**	0.83	51.52

4.3 Results and Analysis

Table 1 presents the results of FLiB and compares it against the state-of-the-art Fairness models on two fairness metrics, DP and EO. We can see that in a link prediction task on Bipartite graphs, GCN shows high bias as indicated by a (DP, EO) of $(5.79, 4.91)$, for MovieLens and $(8.53, 6.03)$ for LastFM. Compared to it, FairGCN is effective in learning fair representations for the user nodes as seen by the adversarial accuracy of 51.40% in MovieLens and 51.14% in LastFM. However, the bias still exists in the model outputs as indicated by a high (DP, EO) of (5.33, 4.40) for MovieLens and (1.79, 4.06) for LastFM. The existence of bias in model outputs despite learning fair representations for user nodes verifies our intuition that the bias can propagate to the item nodes, further discussed in Sect. 4.3.1. Fair-HIN is effective in making model predictions fair as the (DP, EO) has been reduced to (0.85, 0.79) for MovieLens and to (0.19, 1.44) for LastFM. However, for both the datasets, the bias in representations is even further exaggerated compared to the vanilla GCN, as seen by the sensitive attribute classification accuracies of 72.45% in MovieLens and 90.57% in LastFM. FLiB, however, can achieve both of our goals: 1) Reduce both DP and EO to achieve fair link predictions as seen by a (DP, EO) of (0.32, 0.12) for MovieLens and (1.53, 2.56) for LastFM. 2) Learn fair representations for both MovieLens and LastFM as seen by adversarial accuracies of 55.06% and 56.48%. In Table 1, we further show that a similar study can be done on other graph aggregator like GAT. **AAc** denotes classifier/adversarial accuracy and **AR** denotes AUC-ROC.

Table 2. Bias Percolation from user nodes to item nodes

Task	Model	Accuracy	AUC-ROC	$\|DP\|$	$\|EO\|$
Node-Classification	GCN	79.80	0.86	5.88	5.15
	GCN + Adversary	76.15	0.84	0.39	1.47
Link-Prediction	GCN	83.57	0.91	5.79	4.90
	GCN + Adversary	83.98	0.91	5.33	4.40

4.3.1 Bias Percolation from User Nodes to Item Nodes

We analyze the impact of single-layer adversarial debiasing and its effect on node classification and link prediction tasks. We show that while a single adversary can be effectively used to debias user nodes for a fair node classification task, it may not be good enough in a task that includes item nodes as well, like link-prediction. To show this, we create two tasks - a) node classification - a task involving user nodes only, b) link prediction - task involving both user and item nodes. For both tasks, we use the MovieLens dataset. The node classification task is to predict a user's age as a binary attribute indicating > 35 and $<= 35$ based on the user's movie choices keeping gender as the sensitive attribute, while the link prediction task is to recommend movies to the users. As seen in Table 2, the reduction in both DP and EO via a single adversary is (5.49%, 3.68%) in the node-classification task compared to the (0.34%, 0.50%) in the link-prediction task. Hence, bias in model predictions for a fair link prediction task via a single adversary can be attributed to bias in item nodes which is to say that the bias has propagated from user nodes to item nodes.

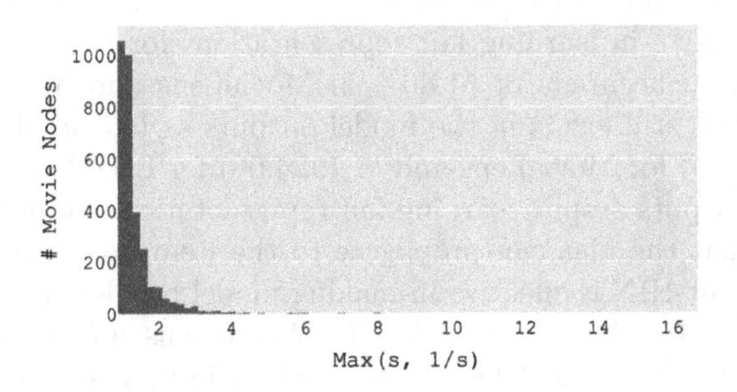

Fig. 3. max(s, 1/s), s: Susceptibility Factor

4.3.2 Inherent Bias in Item Nodes - Susceptibility Factor

Figure 3 shows a histogram for SF's of the movie nodes in the MovieLens dataset, which is used to identify a subset of item nodes with extreme SF values. In MovieLens, there are 818 movies out of 3706 with a SF > 2 and can be expected to contain bias in recommendations, therefore, a pseudo-label can be assigned to them. For example, the movie "Sabotage" has been liked by 71.7% of the males who watched it compared to only 17.42% females who watched it. This movie with an SF value of 4.11 is expected to be male-biased which indeed happens since a GCN based link-predictor recommends it to 20% males who had liked it compared to only 11% females, giving Δ of 9% whereas, on the application of FLiB, this Δ decreases to 4% thus improving the fairness for this movie recommendation.

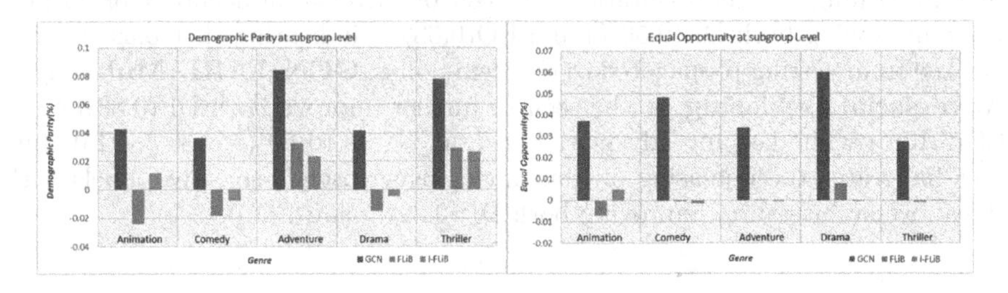

Fig. 4. DP and EO across different sub groups for the Movie dataset

4.3.3 Sub-Group Fairness

In Fig. 4 and Table 3, we evaluate the model fairness for 5 most biased movie genres in the GCN implementation on the MovieLens dataset. FLiB, based on statistical definitions of fairness, is mitigating bias even at the sub-group level. S-FLiB, however, explicitly keeps this bias at the sub-group levels in check and is shown to improve both DP and EO.

Table 3. Sub-group level fairness for GCN, FLiB, and S-FLiB

Metric	Accuracy			DP			EO		
Genre	GCN	FLiB	S-FLiB	GCN	FLiB	S-FLiB	GCN	FLiB	S-FLiB
Overall	83.58	83.41	73.54	5.79	0.33	0.79	4.91	0.12	−0.49
Animation	80.73	80.99	69.71	4.27	−2.43	1.16	3.72	−0.73	0.49
Comedy	83.03	83.4	75.20	3.63	−1.82	−0.77	4.84	−0.07	−0.15
Adventure	83.51	82.97	69.86	8.4	3.26	2.32	3.4	0.01	−0.05
Drama	84.28	84.36	73.65	4.19	−1.51	−0.46	6.02	0.78	−0.05
Thriller	82.61	81.84	68.52	7.76	2.92	2.65	2.74	−0.09	0.02

4.4 Ablation Study

In this section, we show the effectiveness of each component of FLiB by conducting experiments on the Movie Lens datasets. Results of the ablation are listed in Table 4. **GCN** based link prediction shows high DP(5.79%) and EO(4.91%), indicating the presence of high bias in the model predictions. **GCN+Fairness aware regularisation Loss(FARL)** makes the task of link prediction fair by decreasing the parity and EO to 0.85% and 0.79% but leads to high sensitive attribute classification accuracy of 72.45, indicating that the learned embeddings are not fair. **GCN+Single Adversary** makes the user representations fair by debiasing them at the last layer, as seen by an adversarial accuracy of 51.4% but leads to very high DP(5.33%) and EO(4.90%) due to the propagation of bias as discussed in Sec. 4.3.1. **GCN+Multi-layer Adversarial Debiasing** is effective in learning fair representations as seen by adversarial accuracy of 52.84% while improving both DP(0.56%) and EO(0.68%), indicating that bias is being blocked from getting propagated to the item nodes. **GCN+FARL+Multi-layer Adversarial Debiasing** can be seen to further improve both DP(0.53%) and EO(0.13%) while keeping adversarial accuracy close to 50 (55.13%). FLiB further integrates the debiasing of item nodes using pseudo-sensitive labels with the above architecture, improving both DP(0.32%) and EO(0.12%).

Table 4. Ablation results of FLiB on Movie Lens Dataset using GCN aggregator

Model architecture	Accuracy	AUC-ROC	DP	EO	Adv. Acc
GNN	83.57	0.915	5.79	4.91	62.58
GNN+Single Adv	83.98	0.918	5.33	4.40	51.40
GNN+FARL	83.64	0.914	0.85	0.79	72.45
GNN+Multi-layer Adv	83.40	0.912	0.56	0.68	52.84
GNN+FARL+Multi-layer Adv	83.56	0.913	0.53	0.13	55.13
FLiB	83.40	0.914	0.32	0.12	55.06

5 Related Work

In this section, we survey recent literature relevant to our work on fairness in GNN based methods. Graph Neural Networks [10,12,16] have evolved as one of the most promising techniques to learn node representation based on their neighborhood structure. Although these representations capture the neighborhood structure very well, they tend to become discriminative towards sensitive attributes. Fairwalk [15] uses the sensitive attribute and modifies the random walk of node2vec [9] to obtain a more diverse network neighborhood representation to adjust bias but is limited to homogeneous graphs only. Fair-HIN ([6]) discusses a range of debiasing algorithms to mitigate demographic biases in Heterogeneous Information Networks (HINs) and proposes the use of fairness-aware loss function to make the model predictions fair. Although it decreases the bias, it does not learn fair representations limiting the use of embeddings for any other downstream tasks. [4–6] propose the use of adversarial discriminators from GAN to learn graph embeddings that are agnostic to sensitive information. GAN [8] are deep learning frameworks where a generator generates the embeddings, and the discriminator tries to identify real and fake data. In the fairness domain, discriminators are trained to identify sensitive attribute, whereas generators are trained to fool the generator by generating sensitive attribute agnostic embedding. [6] has been shown effective in mitigating bias in homogeneous networks whereas [4] uses adversarial filters to achieve representational invariance for a node's embeddings with respect to its sensitive attributes in a heterogeneous graph but has a major assumption that the edge representation can only be biased via user node embeddings. In this work, we address the problem of bias in bipartite graphs and learning fair representations simultaneously.

6 Conclusion

This paper presents FLiB, a novel framework for fair link prediction and fair representation learning in Bipartite graphs. Our experimental results show that FLiB shows promising results on fairness metrics such as DP, EO without compromising fairness of learned representations. It is noteworthy that FLiB is generalizable to more general heterogeneous networks, where each node type can have 0 or 1 sensitive attribute. As a future research direction, the work can be extended to general heterogeneous graphs where each node type can have multiple and non-binary sensitive attributes.

References

1. https://www.bankingdive.com/news/artificial-intelligence-lending-bias-model-regulation-liability/561085/
2. https://www.reuters.com/article/us-amazon-com-jobs-automation-insight-iduskcn1mk08g
3. Beutel, A., Chen, J., Zhao, Z., Chi, E.H.: Data decisions and theoretical implications when adversarially learning fair representations. arXiv:1707.00075 (2017)
4. Bose, A., Hamilton, W.: Compositional fairness constraints for graph embeddings. In: International Conference on Machine Learning, PMLR, pp. 715–724 (2019)
5. Bozdag, V.: Bursting the Filter Bubble: Democracy, Design, and Ethics (2015)
6. Dai, E., Wang, S.: Say no to the Discrimination: Learning Fair Graph Neural Networks with Limited Sensitive Attribute Information (2021)
7. Edwards, H., Storkey, A.: Censoring representations with an adversary (2015). arXiv preprint, arXiv:1511.05897
8. Goodfellow, I., et al.: Generative adversarial networks. Commun. ACM **63**(11), 139–144 (2020)
9. Grover, A., Leskovec, J.: node2vec: scalable feature learning for networks. In: Proceedings of the 22nd ACM SIGKDD, pp. 855–864 (2016)
10. Hamilton, W.L.: Inductive representation learning on large graphs. In: Proceedings of the 31st International Conference on NIPS, pp. 1025–1035 (2017)

11. Hardt, M., Price, E., Srebro, N.: Equality of opportunity in supervised learning. Adv. Neural. Inf. Process. Syst. **29**, 3315–3323 (2016)
12. Kipf, T.N., Welling, M.: Semi-supervised classification with graph convolutional networks (2016). arXiv preprint arXiv:1609.02907
13. Liao, J., Huang, C., Kairouz, P., Sankar, L.: Learning generative adversarial representations (gap) under fairness and censoring constraints (2019). arXiv:1910.00411
14. Madras, D., Creager, E., Pitassi, T., Zemel, R.: Learning adversarially fair and transferable representations. In: International Conference on Machine Learning (2018)
15. Rahman, T., Surma, B., Backes, M., Zhang, Y.: Fairwalk: towards fair graph embedding (2019)
16. Veličković, P., Cucurull, G., Casanova, A., Romero, A., Lio, P., Bengio, Y.: Graph attention networks (2017). arXiv preprint. arXiv:1710.10903
17. Zafar, M.B., Valera, I., Gomez Rodriguez, M., Gummadi, K.P.: Fairness beyond disparate treatment & disparate impact: learning classification without disparate mistreatment. In: WWW 2017, pp. 1171–1180 (2017)
18. Zafar, M.B., Valera, I., Rogriguez, M.G., Gummadi, K.P.: Fairness constraints: mechanisms for fair classification. In: Artificial Intelligence and Statistics (2017)
19. Zeng, Z., Islam, R., Keya, K.N., Foulds, J., Song, Y., Pan, S.: Fair Representation Learning for Heterogeneous Information Networks (2021). arXiv:2104.08769
20. Zhang, B.H., Lemoine, B., Mitchell, M.: Mitigating unwanted biases with adversarial learning. In: ACM Conference on AI, Ethics, and Society, pp. 335–340 (2018)

SMITH: A Self-supervised Downstream-Aware Framework for Missing Testing Data Handling

Chih-Chun Yang[1], Cheng-Te Li[2(✉)], and Shou-De Lin[1]

[1] National Taiwan University, Taipei, Taiwan
{r08922050,sdlin}@csie.ntu.edu.tw
[2] National Cheng Kung University, Tainan, Taiwan
chengte@ncku.edu.tw

Abstract. Missing values in testing data has been a notorious problem in machine learning community since it can heavily deteriorate the performance of downstream model learned from complete data without any precaution. To better perform the prediction task with this kind of downstream model, we must impute the missing value first. Therefore, the imputation quality and how to utilize the knowledge provided by the pre-trained and fixed downstream model are the keys to address this problem. In this paper, we aim to address this problem and focus on models learned from tabular data. We present a novel Self-supervised downstream-aware framework for MIssing Testing data Handling (SMITH), which consists of a transformer-based imputation model and a downstream label estimation algorithm. The former can be replaced by any existing imputation model of interest with additional performance gain acquired in comparison with that of their original design. By advancing two self-supervised tasks and the knowledge from the prediction of the downstream model to guide the learning of our transformer-based imputation model, our SMITH framework performs favorably against state-of-the-art methods under several benchmarking datasets.

Keywords: Missing testing data · Downstream-aware · Transformer · Self-supervised learning · Tabular data

1 Introduction

Missing values in testing tabular data can heavily deteriorate the performance of downstream model learned from complete data. Despite meticulous control over the data collection pipeline, missing data arise under several circumstances such as the malfunctioning of storage device or the privacy concern that customers are reluctant to provide such information. Regarding this crucial issue raised by the missing values, previous methods addressing the missing testing data problem fall into two categories C1: Take precaution during the learning of downstream

J. Gama et al. (Eds.): PAKDD 2022, LNAI 13281, pp. 499–510, 2022.
https://doi.org/10.1007/978-3-031-05936-0_39

model [10,12]. C2: Impute the missing testing data then feed it into the downstream model as normal inference pipeline[1,4,8,11,12]. Although prevention is better than cure, one major drawback of C1 is that most existing downstream models do not consider the missing issue during designing their architecture, so they usually can not handle missing data. Therefore, C2 is usually preferred since it is applicable to most of existing downstream methods.

To perform imputation on missing testing data, previous work [8] exploits the instance correlation by statistically assuming that similar instances have similar feature. Other methods such as [1,11,12] additionally exploit the correlation among features to predict the missing value by learning a prediction model taking known features as input and estimating the missing value. However, the former suffers from high performance variance considering the wide variety of data distribution of different datasets, and the latter provides imputation of poor quality if the correlation between missing and known features is weak. Aside from the potential issues of imputation quality, previous imputation methods [1,4,8,11] focus on filling missing value via knowledge from the observed feature only. However, they neglect that the information provided by the prediction of pre-trained downstream model could be beneficial. To summarize, missing values in testing data is a challenging problem since the potential issues of the weak correlation among features and the over ideal statistic assumption can lead to poor imputation quality. The imputation methods can only learn from incomplete data, which further increase the difficulty of estimating missing value. Besides, previous methods fail to exploit the beneficial knowledge from the prediction of downstream model.

This paper presents a novel downstream-aware framework, **S**elf-supervised downstream-aware **MI**ssing **T**esting data **H**andling (SMITH), to provide better prediction with missing testing data. SMITH follows the pipeline of C2 to address the problem of missing data during testing phase. We present a transformer-based imputation model that exploits the feature correlation by co-relating input features with self-attention mechanism to provide accurate estimation about missing value. To effectively optimize our imputation model, we design two self-supervised tasks to guide the learning. The first is the masking and prediction task, i.e., we mask the known features then requires our model to predict the masked values. The second is the missing adversarial task, which serves as the regularization term to prevent our model from predicting missing values that is out of regular feature distribution. For leveraging the knowledge from the downstream prediction, we propose a downstream label estimation algorithm, which provides a more confident prediction by aggregating the predictions of similar neighbor instances. The downstream label estimation has two potential usages. One is to be performed on the imputation model optimized already. The other is to serve as extra aid during the learning of imputation model by maximizing the probability of the most confidential label. Besides, our SMITH framework also provides the flexibility of model integration, i.e., it can be applied to any of existing imputation methods of interest by simply replacing the imputation module. We highlight the contribution of this paper below:

- We present a novel downstream-aware framework, SMITH, which performs imputation on the missing testing data through exploiting the knowledge from the downstream model.
- The imputation module in SMITH is realized by a novel transformer-based learning, which is guided by two self-supervised tasks to perform imputation of high quality on the missing testing data.
- The downstream-aware nature of SMITH provides the flexibility of affording existing imputation methods to ameliorate their imputation performance.
- Experiment results show that SMITH outperforms state-of-the-art imputation methods on several benchmarking datasets for downstream predictions.

2 Related Work

Missing Data Imputation. Before the breakthrough of neural network, missing data imputation has already been a well-known research topics in different domain. In this work, we focus on handling imputation on tabular data. For the non-neural-based imputation methods, Mean imputation fill the missing value with global average while KNN imputation [8] fill it with the average of top-k similar instances. However, the performance of such imputation methods heavily relies on the distribution of input data. Other methods typically utilize the correlation among input features by learning a prediction model. For example, MICE [1] adopts multiple chained equations to predict the missing value by learning a regressor for each missing feature with the rest features served as the input for prediction. Nevertheless, the capability of previous non-neural based prediction models is typically not sufficient to give an accurate estimation for the missing value and thus could perform even worse than the non-learning based methods during the circumstances that the correlation among features is weak. With the progress of neural network nowadays, recent neural-based methods directly learn a black-box imputation model by exploiting the strong data fitting capability and robustness of neural network. Additionally, they do not rely on assumptions about the distribution of data. For example, GAIN [11] proposed a GAN-like [5] framework comprised of a generator predicting the missing value and a discriminator measuring the quality of the prediction. By backpropagating the goodness of prediction to the generator, this method outperforms previous traditional learning-based method on several datasets and demonstrate the possibility of bringing neural network into this research topic. However, the optimization of GAIN is more like learning an autoencoder with an extra discriminator, which could possibly result in learning an identity mapping from the input feature to the output. Another neural-based framework, GRAPE [12], adopts graph neural network to address the missing data problem, which constructs a bipartite graph with feature and each observation as nodes and the feature value as the edge. By formulating the imputation as the edge prediction task between feature and observation node, GRAPE demonstrates huge improvement over previous works on several regression task and is claimed to be the state-of-the-art method for missing data imputation.

3 The Proposed SMITH Framework

3.1 Notations and Problem Statement

Now we define the notations. A tabular dataset consists of N instances $\mathcal{D} = \{(x_i, m_i, y_i)\}_{i=1}^{N}$, in which $x_i \in R^d$ that denotes the features vector of each instance, m_i denotes the binary mask with values in $\{0,1\}^d$ to indicate whether a certain feature is missing, and y_i denotes the ground truth label that is only for evaluation. Besides, we consider the most general missing scenario, missing completely at random *(MCAR)*.

The goal of missing testing data problem is to maximize the performance of the downstream prediction task performed with a fixed classifier $\mathcal{C}$, which is trained on the complete training data while taking missing testing data as input during prediction. To have the fixed downstream classifier to perform inference on missing testing data, an imputation model $\mathcal{M}$ is required to fill the missing value first, taking (x_i, m_i) as input and outputting imputed data $\hat{x}_i$. Thus, the quality of the imputation by $\mathcal{M}$ plays the key role for the downstream performance.

3.2 SMITH Framework

As depicted in Fig. 1, SMITH consists of an imputation model $\mathcal{M}$ and a downstream label estimation algorithm. The former learns imputation from incomplete testing data, and the latter improves the performance further by leveraging knowledge from the downstream prediction task. While the overall framework is optimized, we can perform inference on the missing testing data without exploiting the ground-truth downstream label. With the above-mentioned framework, although testing data contains missing values and the downstream model can only take complete data as input, the downstream task can be achieved by performing imputation on the missing testing data (x_i, m_i), then feeding the imputed features $\hat{x}_i$ into the downstream model to acquire the downstream prediction $\hat{y}_i$. It is worth noting that the imputation model in our framework can be replaced by any existing imputation methods to meet the desire of different downstream tasks. We will present the downstream label estimation in Sect. 3.2, and leave the imputation model introduced in Sect. 4.

3.3 Downstream Label Estimation

Although the ground-truth labels during the testing phase is not available, we propose a downstream label estimation (DLE) algorithm that can exploit favorable knowledge from the downstream prediction to improve the performance. Our downstream label estimation algorithm follows the assumption that instances with similar feature values should be within the same class. For each instance, we perform voting within the prediction of top-k similar instances to give a more confident estimation of downstream label, denoted as $\tilde{y}_i$. Considering the incompleteness of testing data, we perform imputation with the imputation

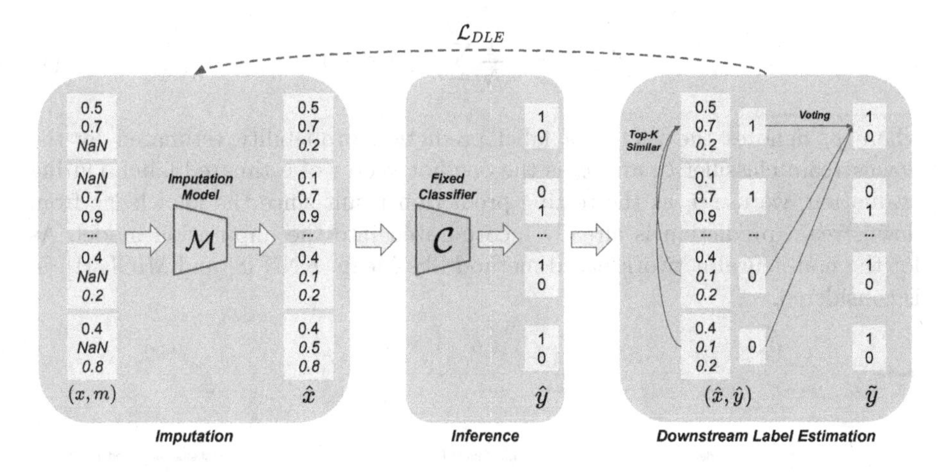

Fig. 1. The proposed SMITH framework.

Algorithm 1: Downstream Label Estimation

Input: Imputation Model $\mathcal{M}$, Downstream Classifier $\mathcal{C}$
Data: Testing data $\mathcal{D} = \{(x_i, m_i)\}_{i=1}^{N}$
Output: Estimated downstream label $\tilde{y}$
1 Perform imputation on missing data $\mathcal{D}$ with imputation model $\mathcal{M}$ and get imputed data $\hat{x}$.
2 Get downstream prediction $\hat{y}$ with downstream classifier $\mathcal{C}$.
3 **foreach** x_i **do**
4 Calculate the similarity between $\hat{x}_i$ and all the other instances $\hat{x}$.
5 Find the top-k similar instances according to similarity.
6 Perform voting within the prediction of top-k neighbors and acquire the estimated label $\tilde{y}_i$.
7 **end**

---

model $\mathcal{M}$ to acquire data with full feature values, $\hat{x}_i$, then calculate the instance similarity. The full DLE pipeline is detailed in Algorithm 1.

The downstream label estimation can be exploited in two ways, within and after the optimization of imputation model. The former "within imputation" is applicable to neural network-based methods only while the latter "after imputation" is for any supervised learning methods. More specifically, for neural network-based imputation methods like GAIN [11] and GRAPE [12], we can backpropagate the extra knowledge of the estimated label to help the learning of imputation model by maximizing the downstream probability of the corresponding label $\tilde{y}_i$, in which cross-entropy can be used as the objective. We denote this learning objective as $\mathcal{L}_{DLE}$, given by:

$$\mathcal{L}_{DLE} = -\frac{1}{N}\sum_{i=1}^{N} \tilde{y}_i^T \log(p_i) \quad (1)$$

where p_i denotes the vector of label prediction probability estimated by the downstream classifier $\mathcal{C}$, and $\tilde{y}_i$ is the one-hot vector of estimated label. During evaluation, we use $\hat{y}_i$ as the testing prediction result since the knowledge from downstream prediction is already incorporated into the imputation model. As for the non-neural network based method like Mean, KNN [8] and MICE [1], $\tilde{y}_i$ is considered.

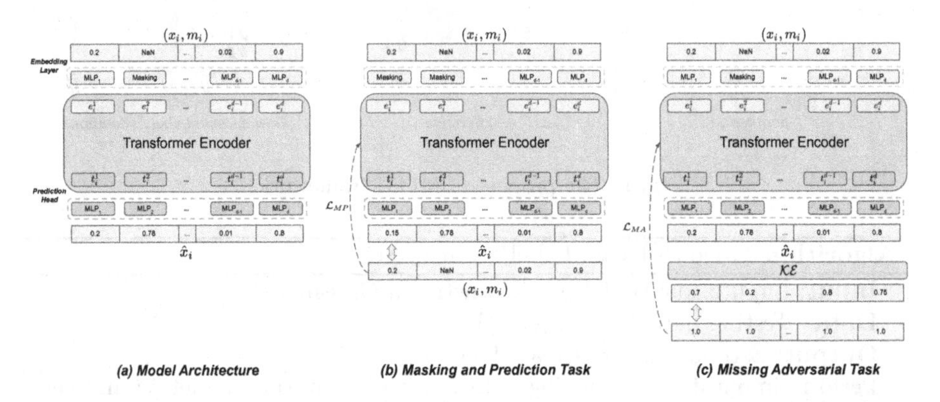

Fig. 2. Proposed transformer-based imputation model and its learning: (a) Model Architecture, (b) Masking and Prediction Task, and (c) Missing Adversarial Task

4 The Proposed Transformer-based Imputation Model

In this section, we introduce the learning of our transformer-based imputation model. The goal of the imputation model is to estimate the missing value based on the known input feature. As shown in Fig. 2(a), our imputation model consists of three modules, including the feature-specific embedding layer, transformer encoder, and feature-specific prediction head. The forward pass of our model is detailed as follows. First, we encode each known feature x_i^j in instance x_i into feature embedding $e_i^j \in R^k$, with feature-specific embedding layer respectively. For the numerical feature, we apply a two-layer MLP for each feature, which transforms the input scalar into a one-dimensional embedding. For the categorical feature, we simply use an embedding layer, similar to that of the word embedding [7]. As for the missing feature values, we adopt a special embedding as the feature embedding for each missing feature to indicate the missingness. After the above-mentioned encoding pipelines are completed, an extra embedding for each feature is added to emphasize the feature modality information, similar to the positional encoding [3]. As a result, we transform each instance from one-dimensional feature vector x_i into two-dimensional one, $e_i \in R^{d \times k}$. As the encoding process is completed, we feed the feature-specific representation e_i into the transformer encoder that leverages the correlation of each feature via self-attention [9] and output the representation t_i. The final prediction for imputation is acquired by feeding the representation t_i into the feature-specific prediction head, transforming the two-dimensional representation into a one-dimensional one, the same shape as the input feature. To optimize our imputation model, we introduce two self-supervised tasks, masking and prediction task and missing adversarial task, to guide its learning. The former enforces our model to estimate the missing values with the known features, and the latter prevents our model from predicting values which is out of normal data distribution.

4.1 Masking and Prediction Task

In this task, we aim at exploiting the correlation among the known feature values to guide the learning of our imputation model. Aside from the missing feature values which are unobserved at the testing stage, we perform extra masking on known feature values selected randomly, and require our model to recover them. Unlike the learning of autoencoder that the ground-truth feature value is already in its input, our masking operation reinforces the awareness between missing and known features, and prevents our model from identically mapping the input to the output. We illustrate this task in Fig. 2(b). For numerical features, we use mean square error as the learning objective. As for categorical feature, cross-entropy is considered. We denote this learning objective as $\mathcal{L}_{MP}$ and formally define it as follows:

$$\mathcal{L}_{MP} = \begin{cases} \frac{1}{N}\sum_{i=1}^{N}||\hat{x}_i^j - x_i^j||_2, & \text{if } x^j \text{ is numerical.} \\ -\frac{1}{N}\sum_{i=1}^{N}(x_i^j)^T \log(\hat{x}_i^j), & \text{if } x^j \text{ is categorical.} \end{cases} \tag{2}$$

4.2 Missing Adversarial Task

In addition to the concrete guidance provided by the known feature values, in this task, we prevent our model from predicting feature values that is out of their reasonable distribution with the help of an extra module, known estimator $\mathcal{KE}$. The $\mathcal{KE}$ estimates the goodness of the prediction by learning to distinguish known features from the predicted features generated from the imputation model. As depicted in Fig. 2(c), the known estimator $\mathcal{KE}$ takes the prediction from the imputation model $\mathcal{M}$, and outputs scores ranging from 0 to 1 to indicate the goodness of each feature value. The learning of $\mathcal{KE}$ is formulated as a feature-wise binary classification task. Formally, the learning objective for $\mathcal{KE}$ is as follows,

$$\mathcal{L}_{Est} = -\frac{1}{N}\sum_{i=1}^{N} m_i^T \log(\mathcal{KE}(\hat{x}_i)) \tag{3}$$

To backpropagate the knowledge of the prediction quality to the imputation model, we fix the model parameters of the known estimator $\mathcal{KE}$, and update those of the imputation model $\mathcal{M}$ to maximize the goodness score. We denote this learning objective as $\mathcal{L}_{MA}$, given by:

$$\mathcal{L}_{MA} = -\frac{1}{N}\sum_{i=1}^{N} \mathbf{1}^T \log(\mathcal{KE}(\hat{x}_i)). \tag{4}$$

Algorithm 2: Full Optimization Pipeline

Input: Imputation Model $\mathcal{M}$, Downstream Classifier $\mathcal{C}$, Known Estimator $\mathcal{KE}$

Data: Testing data $\mathcal{D} = \{(x_i, m_i)\}_{i=1}^{N}$ with missing values

1 **foreach** *epoch* **do**
2 Get imputation value with imputer $\mathcal{M}$.
3 Get known estimation with $\mathcal{KE}$.
4 Calculate known estimation loss $\mathcal{L}_{Est}$.
5 Update known estimator $\mathcal{KE}$.
6 Calculate mask prediction loss $\mathcal{L}_{MP}$.
7 Calculate missing adversarial loss $\mathcal{L}_{MA}$.
8 **if** *epoch* $\neq 0$ **then**
9   Calculate downstream label estimation loss $\mathcal{L}_{DLE}$.
10   Update imputer $\mathcal{M}$.
11   Update estimated downstream label $\hat{y}_i$ following Algorithm 1.
12 **else**
13   Update imputer $\mathcal{M}$.
14   Get estimated downstream label $\hat{y}_i$ following Algorithm 1.
15 **end**
16 **end**

Following the general learning pipeline of Generative Adversarial Network (GAN) [5], we train the known estimator and our imputation model iteratively with the objectives mentioned above.

4.3 The Overall Learning Objective

Now we summarize all the learning objectives introduced in the sections above. For the learning of our imputation model, we have $\mathcal{L}_{SST}$ sum up the two self-supervised objectives related to imputation as follows: $\mathcal{L}_{SST} = \mathcal{L}_{MP} + \lambda_a \mathcal{L}_{MA}$. After introducing the downstream label estimation into our imputation model, the full learning objective is formulated as below: $\mathcal{L} = \mathcal{L}_{SST} + \lambda_d \mathcal{L}_{DLE}$, where λ_a and λ_d denote the balancing factors of the missing adversarial task and the utilization of knowledge from downstream prediction during optimization. Since the guidance provided by these objectives is imprecise, in practice, we set relatively small learning weights, e.g., 0.01, to them. Note that all the existing neural network-based imputation methods can be incorporated into our framework by replacing the imputation model and the $\mathcal{L}_{SST}$ will be substituted with their learning objective accordingly. We detail the full optimization pipeline of the integration of our transformer-based imputation model into our SMITH framework in Algorithm 2.

5 Experiments

In this section, we conduct extensive experiments to quantitatively verify the effectiveness of SMITH on the task of missing testing data.

Table 1. Statistics of datasets.

Statistics	Breast	Spam	Electric	Letter	Plate	Bean	Wifi	Wine	Digit	Yeast
Abbr.	BR	SP	EL	LE	PL	BE	WIF	WIN	DI	YE
Instance	540	4601	10000	20000	1941	13611	2000	4898	10992	1484
Attribute	18	57	12	16	27	16	7	11	16	8
Class	2	2	2	26	7	7	4	7	10	10

5.1 Experimental Setup

We conduct experiments on 10 real world datasets from the UCI Machine Learning Repository [2], and the statistics is listed in Table 1. Since these datasets are originally complete, we drop features following the missing completely at random *(MCAR)* setting to simulate the missing testing data problem. In all experiments, we split the dataset into training/validation/testing set with 60/10/30% ratio. The downstream classifier is optimized on training set with complete instances and corresponding labels. The imputation model is optimized on the testing set with instances containing missing features only and validated with the validation set which is similar to the former but with the ground-truth label to select the best checkpoint. Following the same pipeline as previous works [11,12], we scale feature values to [0-1] with MinMax Scalar [6] as the feature preprocessing. We use a 3-layer MLP with *tanh* as the downstream classifier. For the analyses, we conduct experiments with the setting mentioned above 5 times with different random seeds and report the mean accuracy.

We compare SMITH with the following imputation methods: (a) **Mean**: Impute the missing value with the mean of observed feature value; (b) **KNN** [8]: Impute with the mean value among top-k most similar instances; (c) **MICE** [1]: Impute the missing value based on a simple prediction model conditioned on the other feature and optimized on instances with observed value; (d) **GAIN** [11]: State-of-the-art neural method with generative adversarial training; (e) **GRAPE** [12]: State-of-the-art neural method following the graph architecture.

5.2 Experimental Results

Comparison of Imputation Models. In this section, we compare the imputation capability of SMITH with competing methods on different levels of missing rate. To focus on the imputation capability, knowledge from the downstream model is not involved in these experiments. We found that though some neural-based methods are claimed to be the state-of-the-art, they could lose to non-neural based imputation methods on several datasets. Since there is no single

Table 2. Performance comparison of imputation models on different missing rates.

Method	BR	SP	EL	LE	BE	WIF	DI	PL	WIN	YE	Rank
30% Missing Rate											
Mean	94.5	87.96	75.4	45.19	76.16	85.17	70.02	57.39	46.45	43.73	5.5
KNN	94.69	88.16	74.75	60.45	90.01	86.47	91.06	65.57	46.99	43.06	4.2
MICE	**96.96**	88.22	76.79	52.46	90.69	91.87	87.29	65.4	46.88	44.0	3.1
GAIN	94.97	88.83	75.25	51.44	88.9	88.43	83.99	62.92	46.49	42.43	4.6
GRAPE	95.67	**89.45**	76.46	64.46	90.81	93.83	94.9	66.49	47.09	43.51	2.3
SMITH	96.26	88.72	**76.99**	**68.71**	**90.84**	**94.1**	**95.29**	**66.7**	**47.16**	**44.72**	**1.3**
50% Missing Rate											
Mean	92.28	81.93	71.36	28.77	55.78	69.37	50.07	44.95	46.26	40.76	5
KNN	93.68	82.06	68.23	32.59	76.62	77.3	59.39	57.94	46.05	39.69	4.85
MICE	**94.85**	84.23	71.59	36.77	87.25	80.6	69.77	58.59	46.14	40.4	3
GAIN	93.62	84.59	71.33	33.79	83.73	76.57	66.7	56.46	46.0	40.49	4.3
GRAPE	93.68	**86.09**	72.37	47.42	**89.47**	84.57	**87.8**	62.13	46.11	40.27	2.35
SMITH	94.27	84.33	**72.66**	**48.18**	89.35	**84.73**	86.77	**62.3**	**46.34**	**41.26**	**1.5**
70% Missing Rate											
Mean	88.77	73.19	67.85	16.41	35.57	50.8	30.48	38.01	45.78	36.31	4.85
KNN	88.07	73.74	63.95	18.08	63.55	61.83	41.87	48.73	45.78	34.7	4.45
MICE	**92.05**	77.68	68.03	19.4	69.2	63.93	39.16	47.73	**46.03**	35.19	2.9
GAIN	87.13	77.39	67.6	14.96	59.49	56.63	40.99	47.66	45.63	35.51	4.7
GRAPE	90.29	**80.62**	67.23	**25.9**	**85.47**	67.73	**65.3**	**54.47**	45.89	35.42	2.2
SMITH	91.46	74.78	**68.25**	24.6	85.2	**68.0**	62.57	52.78	45.97	**36.63**	**1.9**

method outperforming the others among all datasets, we compare them by the average ranking. Table 2 shows SMITH outperforms all competitors. However, as the missing rate increases, the ranking of our model gradually decreases.

Analysis on Downstream Label Estimation Algorithm. We investigate whether our downstream label estimation (DLE) algorithm can improve the performance of downstream task. Besides, we compare the two strategies of incorporating the algorithm into existing imputation methods, within or after the optimization of imputation, denoted as *single* (-s) and *iterative* (-i) respectively. The results are reported in Table 3. In comparison with the results in Table 2, we see that our algorithm improves the performance of all imputation methods. As for the comparison of different strategies, we observe that the option *iterative* always outperforms the option *single* on 30% and 50% missing rates. This is expected since the extra knowledge of downstream model is backpropagated to the learning of imputation. However, in the setting of 70% missing rate, the performance of *iterative* is worse than *single* in some datasets. We attribute this phenomenon to the uncertainty of the imputation learning on 70% missing rate. Incorporating such an objective could bring additional noise since the learning is already difficult. Consequently, our model still outperforms all competitors across all missing rates after we advance the downstream label estimation.

Ablation Study. To verify the effectiveness of each component in SMITH, we conduct ablation analyses on two self-supervised tasks and the downstream label estimation (DLE) algorithm on 10 datasets with 30% missing values. Table 4 shows the effectiveness of each learning objective is confirmed.

Table 3. Performance comparison between *iterative* and *single* learning strategies.

Method	BR	SP	EL	LE	BE	WIF	DI	PL	WIN	YE	Rank
30% Missing Rate											
Mean-s	95.02	88.01	75.62	51.42	81.13	87.97	76.51	58.16	46.55	43.92	5.4
KNN-s	94.91	88.23	74.98	64.31	90.45	88.79	91.37	65.82	47.03	43.23	4.5
MICE-s	97.01	88.31	77.01	54.57	90.81	92.76	88.81	65.77	46.91	44.03	3.5
GAIN-s	95.11	89.01	75.34	62.48	89.98	93.1	92.46	63.39	46.52	43.73	4.2
GAIN-i	95.32	89.11	75.43	63.12	90.03	93.99	93.67	63.57	46.6	44.05	3.9
GRAPE-s	95.92	89.52	76.52	65.83	90.96	93.87	95.16	66.57	47.12	44.02	2.2
GRAPE-i	96.07	**89.67**	76.66	67.59	91.02	94.47	95.34	66.71	47.19	44.22	2.1
SMITH-s	97.03	88.91	77.08	69.23	91.08	94.33	95.37	66.85	47.29	44.91	1.2
SMITH-i	**97.13**	88.98	**77.23**	**70.12**	**91.17**	**94.61**	**95.48**	**66.93**	**47.38**	**44.97**	**1.2**
50% Missing Rate											
Mean-s	93.15	82.07	71.45	31.88	70.66	73.48	61.73	48.12	46.47	41.23	5
KNN-s	94.03	82.15	69.76	39.92	80.89	80.21	65.88	58.23	46.33	40.11	4.8
MICE-s	95.12	84.45	71.88	41.98	88.56	83.11	72.19	59.44	46.41	40.72	3.2
GAIN-s	93.78	84.65	71.42	41.1	87.72	83.13	80.91	57.51	46.07	40.77	4
GAIN-i	93.98	84.68	71.48	41.59	88.62	83.22	81.27	57.98	46.31	41.03	3.7
GRAPE-s	93.89	86.23	72.44	50.46	89.51	85.07	88.64	62.31	46.19	40.58	2.5
GRAPE-i	94.04	**86.31**	72.53	50.77	**89.63**	85.18	**88.89**	62.69	46.2	40.95	2.4
SMITH-s	94.98	84.47	72.71	51.24	89.47	85.34	88.43	62.42	46.51	41.31	1.5
SMITH-i	**95.16**	84.51	**72.83**	**51.45**	89.52	**85.45**	88.63	**62.78**	**46.62**	**41.39**	**1.4**
70% Missing Rate											
Mean-s	89.11	73.22	68.05	18.12	50.23	53.41	40.77	43.38	45.8	36.57	5
KNN-s	88.93	73.78	65.59	19.2	67.88	62.78	45.91	49.22	45.91	35.11	4.7
MICE-s	92.66	77.72	68.13	21.01	71.37	65.77	42.83	48.55	**46.22**	35.56	2.9
GAIN-s	91.46	77.63	67.79	17.12	76.02	59.37	52.71	48.18	45.73	35.57	4.1
GAIN-i	88.77	76.78	67.19	16.01	74.94	56.83	47.21	48.32	45.54	35.73	4.5
GRAPE-s	91.35	**80.96**	67.44	**26.45**	85.63	68.6	**68.7**	**54.74**	45.96	35.51	2.4
GRAPE-i	90.76	79.32	67.63	26.15	**85.82**	68.63	65.65	54.37	45.97	35.06	2.3
SMITH-s	92.21	76.15	68.31	25.78	85.47	68.83	67.16	53.26	46.09	**36.81**	1.9
SMITH-i	**92.98**	76.57	**68.39**	25.81	85.67	**68.89**	65.51	53.78	46.13	36.74	**1.8**

Table 4. Results of ablation study.

Setting	BR	SP	EL	LE	BE	WIF	DI	PL	WIN	YE
$\mathcal{L}_{MP}$	95.74	88.01	76.51	68.23	90.75	93.88	95.01	66.31	46.89	44.39
$+\mathcal{L}_{MA}$	96.26	88.72	76.99	68.71	90.84	94.1	95.29	66.7	47.16	44.72
$+\mathcal{L}_{DLE}$	**97.13**	**88.98**	**77.23**	**70.12**	**91.17**	**94.61**	**95.48**	**66.93**	**47.38**	**44.97**

6 Conclusion

In this paper, we propose a novel downstream-aware framework, SMITH, comprised of a transformer-based imputation model learned from two self-supervised tasks and a downstream label estimation algorithm to handle missing data during prediction. SMITH is flexible to be applied to any existing imputation methods. By advancing the extra knowledge from the downstream model, we demonstrate improvement over 10 benchmark datasets. With SMITH, we outperform previous state-of-the-art methods regarding the overall average ranking. Extensive experiments are conducted to quantitatively verify the effectiveness of our method.

Acknowledgements. This material is based upon work supported by Taiwan Ministry of Science and Technology (MOST) under grant numbers 110-2634-F-002-050, 110-2221-E-006-136-MY3, 110-2221-E-006-001, and 110-2634-F-002-051.

References

1. Buuren, S., Groothuis-Oudshoorn, C.: Mice: Multivariate imputation by chained equations in R. J. Stat. Softw. **45**, 1–67 (2011)
2. Dua, D., Graff, C.: UCI machine learning repository (2017)
3. Gehring, J., Auli, M., Grangier, D., Yarats, D., Dauphin, Y.N.: Convolutional sequence to sequence learning. In: Proceedings of the 34th International Conference on Machine Learning, vol. 70. pp. 1243–1252. ICML 2017 (2017)
4. Gondara, L., Wang, K.: Mida: multiple imputation using denoising autoencoders. In: PAKDD (2018)
5. Goodfellow, I., et al.: Generative adversarial nets. In: Advances in Neural Information Processing Systems. vol. 27 (2014)
6. Leskovec, J., Rajaraman, A., Ullman, J.D.: Mining of Massive Datasets. Cambridge University Press, 3 edn. Cambridge 2020)
7. Mikolov, T., Chen, K., Corrado, G., Dean, J.: Efficient estimation of word representations in vector space. In: 1st International Conference on Learning Representations, ICLR 2013, Workshop Track Proceedings (2013)
8. Troyanskaya, O., et al.: Missing value estimation methods for DNA microarrays. Bioinformatics **17**(6), 520–525 (2001)
9. Vaswani, A., et al.: Attention is all you need. In: Advances in Neural Information Processing Systems, vol. 30 (2017)
10. Yi, J., Lee, J., Kim, K.J., Hwang, S.J., Yang, E.: Why not to use zero imputation? correcting sparsity bias in training neural networks. In: International Conference on Learning Representations (2020)
11. Yoon, J., Jordon, J., van der Schaar, M.: GAIN: Missing data imputation using generative adversarial nets. In: Proceedings of the 35th International Conference on Machine Learning. pp. 5689–5698 (2018)
12. You, J., Ma, X., Ding, Y., Kochenderfer, M.J., Leskovec, J.: Handling missing data with graph representation learning. In: Larochelle, H., Ranzato, M., Hadsell, R., Balcan, M.F., Lin, H. (eds.) Advances in Neural Information Processing Systems. pp. 19075–19087 (2020)

Tlife-GDN: Detecting and Forecasting Spatio-Temporal Anomalies via Persistent Homology and Geometric Deep Learning

Zhiwei Zhen[1(✉)], Yuzhou Chen[2,3], Ignacio Segovia-Dominguez[1,4], and Yulia R. Gel[1,5]

[1] The University of Texas at Dallas, Richardson, TX 75080, USA
{Zhiwei.Zhen,Ignacio.SegoviaDominguez,ygl}@utdallas.edu
[2] Princeton University, Princeton, NJ 08544, USA
yc0774@princeton.edu
[3] Lawrence Berkeley National Laboratory, Berkeley, CA 94720, USA
[4] Jet Propulsion Laboratory, Pasadena, CA 91109, USA
[5] National Science Foundation, Alexandria, VA 22314, Egypt

Abstract. Most recently, the tools of geometric deep learning (GDL) and, in particular, graph neural networks emerge as a promising new alternative in unsupervised anomaly detection problems where the data exhibit a sophisticated nonlinear dependence structure such as various geospatial surveillance systems. However, prevailing GDL-based methods for anomaly detection tend to exhibit limited capabilities to capture multiscale spatio-temporal variability which is ubiquitous in many applications, particularly, related to biosurveillance and biothreats. Motivated by the problem of assessing COVID-19 severity, we develop a novel approach to unsupervised anomaly detection in spatio-temporal data by fusing the notion of GDL with the emerging direction of persistent homologies and topological data analysis. In particular, our key idea is to bolster the GDL performance by leveraging the complementary insight on the intrinsic multiscale data organization which topological descriptors can provide. We also go one step further and show how our ideas at the interface of topological and geometric deep learning can be used not only for detection but for prediction of future anomalies. We show the utility of the new approach to detecting, forecasting and interpreting risks in COVID-19 clinical severity, measured in terms of hospitalization rates, in three U.S. states: California, Texas, and Pennsylvania.

Keywords: Anomaly detection · Geometric deep learning · Persistent homology · COVID-19

J. Gama et al. (Eds.): PAKDD 2022, LNAI 13281, pp. 511–525, 2022.
https://doi.org/10.1007/978-3-031-05936-0_40

1 Introduction

Efficient identification of data instances which differ noticeably from the expected baselines is the core behind such diverse tasks as combating money laundering on blockchain, river water-quality monitoring, and defending information systems against breaches of cybersecurity. With a long history in robust statistics and continually emerging new types of threats, anomaly detection remains one of the most actively developing fields at the nexus of machine learning and statistical sciences. Efficient detection of anomalies in dynamic settings such as biological and cyber threats is further exacerbated, first, by the limited or even non-existing records of labeled attack examples and, second, by a sophisticated dependence structure among entities of the underlying time-evolving object. For instance, transmission of many pathogens exhibit complex spatio-temporal interactions with atmospheric conditions, and moreover, pathogenicity of biothreats mat vary across spatial and temporal scales [28,29,32,41].

To address the first challenge, anomaly detection is often viewed as an unsupervised problem. Among some most widely used unsupervised tools for anomaly detection are Connectivity-based Outlier Factor (COF) [42] and Influenced Outlierness (INFLO) [23]. However, such approaches tend to focus on linear relationships among system entities, and as a result, show limited ability to account for early warning signals induced by nonlinear interactions exhibited by most complex real-world systems. Various deep learning (DL) tools such as variational autoencoders (VAE) [25], Long short-term memory (LSTM) [31], and Generative adversarial Networks (GANs) [26] partially mitigate this problem and are found to be promising approaches for anomaly detection in high-dimensional settings.

However, such DL methods are restricted in their ability to learn multiple types of interactions among system entities in dynamic settings, e.g., georeferencing. As such, in the last couple of years, there has been a spike of interest in bringing the tools of Graph Neural Networks (GNNs) [10] and other methods of geometric deep learning (GDL) to anomaly detection tasks [18]. Indeed, GDL offers a systematic framework for learning non-Euclidean objects such as graphs and manifolds, and hence, GDL allows us for more flexible modeling of complex interactions among entities in a broad range of complex data structures, including multivariate time series and dynamic networks.

Our goal here is to further enhance this emerging GDL direction in anomaly detection and to bolster its performance by leveraging the power of data topological (or shape) descriptors. By topological descriptors, we broadly understand data characteristics that are preserved under continuous transformations such as bending, twisting, and stretching. In turn, a few most recent studies show that integration of topological summaries of time-evolving structures such as spatio-temporal processes into DL, either in a form of a topological layer or as additional data attributes, can noticeably improve forecasting performance [12,13,47]. This phenomenon can be explained by the complementary information on the underlying intrinsic system organization at multiple scales which topological descriptors (or more precisely, tools of persistent homology) can deliver. Motivated by biothreat applications where variation of pathogenicity is ubiquitous across spatio-temporal scales, we believe that integration of topological summaries into GNNs may enhance not only anomaly detection performance but bring an invaluable insight about various hidden mechanisms behind anomaly formation. To investigate this hypothesis, we consider anomaly detection in COVID-19 clinical severity, measured in terms of hospitalization rates, in three U.S. states: California, Texas, and Pennsylvania, Moreover, we make a step forward in not only *detecting the existing anomalies* but *forecasting the future anomalies.* While assessing future anomalous patterns is the core behind proactive risk mitigation, especially, in healthcare analytics such as during COVID-19 pandemic, to the best of our knowledge, neither GDL nor any other DL tools have ever been used for spatio-temporal forecasting of anomalies.

The key novelty and contributions of this paper are summarized as follows:

- We are the first to integrate topological descriptors within GDL for anomaly detection tasks. Our Tlife-GDN model with a fully trainable topological layer within GNN shows competitive performance against existing state-of-the-art approaches and allows improving tractability of the latent mechanisms behind emergence of anomalies.
- This is the first paper to address the problem of *future* anomaly forecasting with GDL, which is the key behind developing proactive risk mitigation strategies.
- This is the first approach to assess evolution of existing and future spatio-temporal anomalies in COVID-19 clinical severity, measured in terms of hospitalization rates.

2 Related work

Anomaly Detection in Time-Evolving Processes. Traditional tools for this task include Principal Component Analysis [39] and K Nearest Neighbors (KNN) [5]. Most recently, there has been suggested a number of approaches that leverage topological descriptors for anomaly detection within statistical algorithms. For instance, [22] proposes to detect change points in topological summaries of the observed data instead of analyzing the observed data directly, as in prevailing tools. In turn, [43] considers topological summaries as a supplement to observed data as the input for arrhythmia detection. Finally, [27] and [33] propose anomaly detection in Ethereum blockchain graphs based on assessing similarity among the topological summaries of the data at adjacent time snapshots.

Most recently, DL tools emerge as powerful alternatives to address anomaly detection in spatio-temporal processes. Among such notable DL approaches are Autoencoders (AE) of [2] based on the idea of reconstruction errors; Deep Autoencoding Gaussian Model (DAGMM) of [48] which expands AE with the Gaussian Mixture Model, and Variational Autoencoders (VAE) of [25] with regularized encoding's distribution. Furthermore, inspired by the Support Vector

Data Description (SVDD) [34,35] proposes a Deep Support Vector Data Description (DEEP-SVDD) for anomaly detection tasks which is capable of learning the nodes' representation and hypersphere center of the data simultaneously.

Finally, in the last couple of years, there has been a spike of interest in bringing the power of GNNs to anomaly detection tasks on spatio-temporal data [30]. For instance, most recently [14] proposes a Graph Neural Network-Based Anomaly Detection tool based on the approach of graph attention mechanism with location embedding and structure learning. Although all those methods intend to discover the hidden relationships between system entities, to our best knowledge, there exists *no GNN which have explored the power of topological data descriptors for enhancing anomaly detection* in time-evolving processes.

Different from the anomaly detection task, anomaly prediction is the task of recognizing *future* abnormal instances relative to the currently recorded data patterns. The problem of anomaly prediction is noticeably more challenging due to elevated uncertainty of forecasting and, while playing a key role in efficient and proactive management of emergency preparedness, remains largely understudied. Previous works in this filed include applications of machine learning tools like Support Vector Machines (SVMs) [45] and epsilon-Support Vector Regression (ϵ-SVR)[6] in software programs [4], water pipeline [46]. However, to the best of our knowledge, neither the utility of GNNs nor DL tools, in general, has been explored before for anomaly prediction in conjunction with analysis of time-evolving processes.

COVID-19 Severity Prediction. Many recent studies have analyzed the risk factors for the severe acute respiratory syndrome coronavirus 2 (i.e., SARS-CoV-2, the virus which causes COVID-19). For example, [8] examines the possible correlation between obesity and COVID-19 clinical severity by surveying patients in a hospital, while [7] considers the linkage between anticancer therapy and COVID-19. More generally, [16] reviews the factors in demographics, comorbidities, hypoxia and radiographic features that might worse COVID-19 outcomes. However, the majority of the COVID-19 severity research focuses on the patients' clinical features rather than on the severity in a certain geographical area.

Two notable studies on spatio-temporal anomaly detection in conjunction with COVID-19 are [19] and [24] who consider topological data analysis (TDA) and the deep hybrid autoencoder networks for assessing daily new cases, respectively. Furthermore, [36–38] consider various GDL and LSTM models, coupled with topological descriptors for tracking COVID-19 hospitalizations and number of cases, but do not address the problem of spatio-temporal anomaly detection in COVID-19 clinical severity. As such, spatio-temporal anomalies in COVID-19 clinical severity and, particularly, anomalies in hospitalization rates remain largely under-explored. To the best of our knowledge, there exists *no current method assessing risk scoring in COVID-19 clinical severity using GNNs or TDA based on hospitalization data.* Our paper aims to take advantage of GNNs with topological descriptors to improve the performance and tractability of the unsupervised spatio-temporal anomaly detection and anomaly prediction for COVID-19 hospitalization rates.

3 Preliminaries on Persistent Homology

Persistent homology (PH) is a methodology under the framework of topological data analysis, which aims to study the most inherent shape characteristics of the observed data. The PH machinery is applicable to a broad range of data types, e.g., point clouds in Euclidean spaces, images, graphs, and more generally, objects in metric spaces. Here, we primarily focus on shape characteristics of the graph $\mathcal{G}$ generated from spatio-temporal time series[1] [9,11]. The approach consists of the three main steps. First, we convert $\mathcal{G}$ into a filtration of graphs $\mathcal{G}_1 \subseteq \mathcal{G}_2 \subseteq \ldots \subseteq \mathcal{G}_k = \mathcal{G}$. We can now track evolution of various patterns in this graph filtration, which ought to reveal the underlying structure of $\mathcal{G}$ at different scales. Second, to make the tracking process systematic and efficient, we build a simplicial complex $\mathscr{C}$ on top of $\mathcal{G}$ and, as such, our graph filtration is now associated with a nested sequence of complexes $\mathscr{C}(\mathcal{G}_1) \subseteq \mathscr{C}(\mathcal{G}_2) \subseteq \ldots \subseteq \mathscr{C}(\mathcal{G}_n)$. That is, we can now compute simplicial homologies and record which shape characteristics, for example, connected components, loops, and cavities, appear in the filtration of complexes. In particular, we say that a topological feature is born at i_b if $\mathscr{C}(\mathcal{G}_{i_b})$ is the complex where we first observe it. In turn, we record death of a topological feature at j_d if this feature is last seen in $\mathscr{C}(\mathcal{G}_{j_d})$. The longer the lifespan $j_d - i_b$ of the topological feature is, the likelier this feature contains important structural information on $\mathcal{G}$. Features with longer lifespans are also said to persist, while features with shorter lifespans are sometimes referred to as topological noise. Finally, in our third step, we summarize all the extracted topological features in a form of a multi-set $\mathcal{D} = \{(i_b, j_d) \in \mathbb{R}^2 | i_b < j_d\}$, called *persistence diagram* (PD). Since lifespan $j_d - i_b \geq 0$, all points in $\mathcal{D}$ are in the half-space on or above $y = x$. Finally, there exists multiple options to construct graph filtrations [20]). For instance, consider a continuous function $f : \mathcal{V} \to \mathbb{R}$ acting on nodes of $\mathcal{G}$ and a sequence of non-negative scales $\xi_1 < \xi_2 < \ldots < \xi_n$. Then, we can define the corresponding simplicial complex as $\mathscr{C}_i = \{\sigma \in \mathscr{C} : \max_{v \in \sigma} f(v) \leq \xi_i\}$. Similarly, filtration can be defined as $\mathcal{E}$ of $\mathcal{G}$. In this paper, we consider the weight rank clique filtration [40] and Vietoris-Rips abstract simplicial complexes [15], due to their computational benefits.

Since $\mathcal{D}$ is a multi-set, we cannot directly feed it into DL framework. As such, we use its vectorized representation, i.e., *persistence image (PI)* [1]. To construct PI, we first map $\mathcal{D}$ to an integrable function $\rho_{\mathcal{D}} : \mathbb{R} \to \mathbb{R}^2$, which is referred to as *the persistence surface* and which is given by sums of weighted Gaussian functions centered at each point in $\mathcal{D}$. We then integrate $\rho_{\mathcal{D}}$ over each grid box to obtain PI such that the value of each pixel z is given by

$$\mathrm{PI}(z) = \iint_z \sum_{\mu \in T(\mathcal{D})} \frac{g(\mu)}{2\pi\delta_x\delta_y} e^{-\left(\frac{(x-\mu_x)^2}{2\delta_x^2} + \frac{(y-\mu_y)^2}{2\delta_y^2}\right)} dydx. \quad (1)$$

Here $T(\mathcal{D})$ is the transformed PD $\mathcal{D}$ (i.e., $T(x, y) = (x, y-x)$), $g(\mu)$ is a weighting function, where $\mu = (\mu_x, \mu_y) \in \mathbb{R}^2$), while μ_x and δ_x and μ_y and δ_y are the mean and the standard deviation of the Gaussians in x and y direction, respectively.

[1] Generation details are available in Algorithm 1.

Graph Neural Network-Based Anomaly Detection. The GDN architecture addresses the structure learning process with graph neural networks and combines it with attention weights to detect anomaly. The GDN model learns the vector embedding for each location during the training process and uses the similarity between vectors to build the connection relationships. The observed data at time t is $\mathbf{s}^{(\mathbf{t})}$. When the size of the sliding window is w, the input $\mathbf{x}^{(t)}$ is $\mathbf{x}^{(t)} = \left[\mathbf{s}^{(\mathbf{t-w})}, \mathbf{s}^{(\mathbf{t-w+1})}, \cdots, \mathbf{s}^{(\mathbf{t-1})}\right]$. Based on the learned graph structure, the aggregated representation of node is computed as

$$\mathbf{z}_i^{(t)} = \text{ReLU}(\alpha_{i,i}\mathbf{W}\mathbf{x}_i^{(t)} + \sum_{j \in \mathcal{N}(i)} \alpha_{i,j}\mathbf{W}\mathbf{x}_j^{(t)}), \tag{2}$$

where $\mathbf{W}$ is a weighted matrix, $\mathbf{x}_i^{(t)}$ is the input feature of node i, $\mathcal{N}(i)$ denotes the neighbors of node i from structure learning, and $\alpha_{i,j}$ is attention coefficient.

Then the GDN model[14] utilizes the representation of the node i, i.e., $\mathbf{z}_i^{(t)}$ and embeds the corresponding vector $\mathbf{v}_i$ to predict the current value. Lastly, GDN generates the anomaly score and identify anomaly.

4 Topological Lifespan Graph Neural Network-Based Anomaly Detection Approach (Tlife-GDN)

Problem Statement. Mathematically, the anomaly detection problem can be formulated as follows. Let $\mathbf{s}^{(t)}$ be records (e.g., COVID-19 hospitalizations) from N locations, where $t = \{1, 2, \ldots, T\}$. Let $l^{(t)}$ be the binary anomaly status at time t, e.g., $l^{(t)} = 0$ represents a normal behaviour, whilst $l^{(t)} = 1$ when some abnormality occurs. Let $\mathcal{G}^{(t)} = (V, E, \omega^{(t)})$ be a weighted connectivity network among locations $\mathbf{s}^{(t)}$, with node set $V = \{v_1, v_2, \ldots, v_N\}$, i.e., each node represents a location, edge set $E \in V \times V$ and the non-negative symmetric edge-weight matrix $\omega^{(t)}$ with entries $\{\omega_{ij}^{(t)}\}_{1 \le i,j \le N}$. In this paper, we focus on two problems: 1) current anomaly prediction and 2) forecasting of future anomalies.

`Problem 1`: To learn a mapping function $\mathcal{H}(\{\mathbf{s}^{(t)}\}_{t=1}^{T-1}, \{\mathcal{G}^{(t)}\}_{t=1}^{T-1})$ which maps the records to a binary anomaly output $l^{(t)}$.

`Problem 2`: Given an ahead horizon h, our goal is to learn a mapping function $\mathcal{H}(\{\mathbf{s}^{(t)}\}_{t=1}^{T-1}, \{\mathcal{G}^{(t)}\}_{t=1}^{T-1})$ which maps the records to a binary anomaly output $l^{(t+h)}$.

In order to capture the complex topological features of the spatio-temporal data, we construct dynamic networks, and extract the n-dimensional features in the form of persistence diagram and vectorize the persistence diagram to obtain persistence image. Then, we integrate the persistence image into the GNN framework for detection of existing anomalies and prediction of future anomalies.

4.1 Topological Features of Dynamic Networks

Topological features provide a way to systematically describe the graph structure and track the evolution of hidden patterns of data. In this paper, we make

Algorithm 1. Topological Features from Dynamic Networks

1: **INPUT:** Location Records $\{\mathbf{s}_i^{(t)}\}_{i=1}^N, t = \{1, 2, \ldots, T\}$
2: **OUTPUT:** Topological summaries
3: **for** $t \leftarrow 1 : T$ **do**
4: **for** $i \leftarrow 1 : N - 1$ **do**
5: **for** $j \leftarrow i + 1 : N$ **do**
6: Compute $H_{ij}^{(t)} = |\mathbf{s}_i^{(t)} - \mathbf{s}_j^{(t)}|$
7: Keep only bottom-m values in $H^{(t)}$
8: Compute $\omega^{(t)}(e) = 1 - H^t / \max(H^{(t)})$
9: Generate $\mathcal{G}^{(t)}$ based on $\omega^{(t)}(e)$
10: Apply persistent homology on dynamic networks $\mathbb{G} = \{\mathcal{G}^{(t)}\}_{t=1}^T$ for different dimensions and generate persistence diagram(PD) for each timestamp t
11: Apply equation 1 in section 3 to generate persistence image (PI) from PD

use of lifespans of those topological features from different nodes in the dynamic network. Specially, with records $\{\mathbf{s}_i^{(t)}\}_{i=1}^N$, we calculate the L_1 distance matrix $H^{(t)}$ of record values $\{\mathbf{s}_i^{(t)}\}_{i=1}^N$ to build connections between locations (e.g., counties) as shown in Algorithm 1. The locations with close values are considered to have similar patterns. In our study, the counties with similar COVID-19 cases rate may have similar geometric structure information regarding the COVID-19 transmission, empty ICU beds and hospitalization severity. We take the lowest-m values in the connection matrix, where m is a predefined number based on the dataset. Then, we generate an edge weight matrix $\omega(e)$ by taking 1 minus the standardized $H^{(t)}$ and get its corresponding weighted graph $\mathcal{G}^{(t)}$. The next step is to use persistent homology to track the invariant structure features, and compute a persistence diagram (PD) for each network $\mathcal{G}^{(t)}$ and its corresponding lifespan information. Finally, we generate the vectorized represented persistent image (PI) defined in Sect. 3 as the topological features from the location's dynamic networks.

4.2 Tlife-GDN Architecture

With the spatio-temporal dataset $\mathbf{s}^{(t)}$ (where $t = \{1, 2, \ldots, T\}$), we capture the topology features PI defined in Section 4.1. Then we train our topology-based GDN to capture the hidden structure between different locations. Equation 3 shows the implementation of persistence image $\mathrm{PI}^{(t-1)}$ in the graph neural networks framework

$$\mathbf{z}_i^{(t)} = \mathrm{ReLU}\left((\alpha_{i,i}\mathbf{W}\mathbf{x}_i^{(t)} + \sum_{j \in \mathcal{N}(i)} \alpha_{i,j}\mathbf{W}\mathbf{x}_j^{(t)})\mathbf{Q}^{(t)}\right), \tag{3}$$

where $\mathbf{z}_i^{(t)}$ denotes the latent representation of the node i at timestamp t. $\mathbf{Q}^{(t)} \in \mathbb{R}^d$ is the topological representation from the CNN based model (where d is the length of embedding vector for each location), which is formulated as $\mathbf{Q}^{(t)} =$

$f_{cnn}(\mathrm{PI}^{(t-1)})$, where f_{cnn} is a CNN-based model and $\mathrm{PI}^{(t-1)}$ denotes the PI for the network at $(t-1)$ timestamp. Then, we add the latent nodes' representation into the graph detection network architecture to predict the location's value. For anomaly detection/prediction, we use the loss function and error score as

$$L_{\mathrm{MSE}} = \frac{1}{T-w}\sum_{t=w+1}^{T-h}\left\|\hat{\mathbf{s}}^{(\mathbf{t+h})} - \mathbf{s}^{(\mathbf{t})}\right\|_2^2, \quad \mathrm{Err}_i(t+h) = \left|\mathbf{s}_\mathbf{i}^{(\mathbf{t})} - \hat{\mathbf{s}}_\mathbf{i}^{(\mathbf{t}+h)}\right|, \quad (4)$$

where h is the prediction window and $h = 0$ correspond with the detection task. For both detection and prediction tasks, the anomalousness score at time t is the maximal score across locations $A(t) = \max_i a_i(t)$ where $a_i(t)$ is the standardized error score.

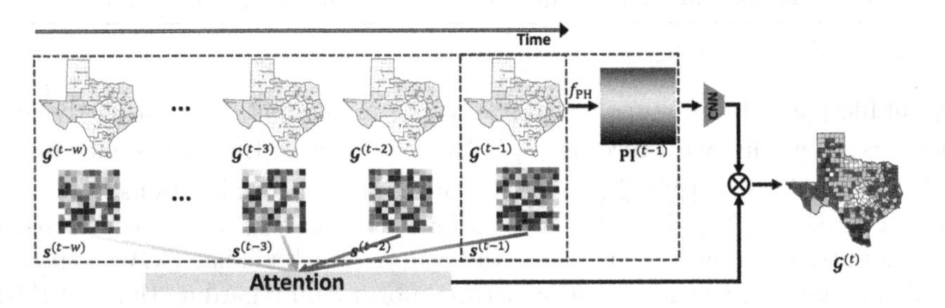

Fig. 1. Architecture overview of the Tlife-GDN model, where $(\mathcal{G}^{(t-w)}, \ldots, \mathcal{G}^{(t-2)}, \mathcal{G}^{(t-1)})$ and $(\mathbf{s}^{(t-w)}, \ldots, \mathbf{s}^{(t-2)}, \mathbf{s}^{(t-1)})^\top \in \mathbb{R}^{N\times F\times w}$ denote all graph structures and values of all features for each node over w time slices, respectively.

The overall architecture is shown in Figure 1. The intuition here is to combine the topological features along with the records (e.g., COVID-19 hospitalization rates) as the input for the topology-based GDN model. At timestamp t, we generate PIs for the latest timestamp $t-1$, as the topological summaries and use a CNN-based model to learn its representation. With the enriched input data, we use equation 3 to get the latent node's representation. Although different DL methods have been proposed to improve the anomaly detection accuracy, PIs have not been incorporated into this task. Furthermore, regrading COVID-19 spreading, topological summaries can help the learning grasp on the persistent hidden features behind the progression process caused by environmental or social-demographic variables. As a result, Tlife-GDN model extracts the complex spatio-temporal dependence properties which are inaccessible with other GDL tools.

5 Experiment

5.1 Datasets, Experiment Setup and Evaluation Metrics

We conduct experiments on 5 datasets: COVID-19 records in Texas (TX), California (CA) and Pennsylvania (PA), Curiosity Rover on Mars (MSL) and Water

Distribution (WADI). Table 1 summarize the properties of each dataset. The daily records for COVID-19 cases and hospitalizations come from CovidActNow project[2] and Johns Hopkins University[3]. These data sources contain COVID-19 time series from official state and county websites. We take 2 per thousand people as the anomaly threshold for hospitalization rate at state level. New cases rate at county level, which indicates the spread of COVID-19, is used for training and prediction. The Curiosity Rover on Mars (MSL) is an expert-labeled telemetry anomaly data which originally comes from Incident Surprise, Anomaly (ISA) [21]. The reports assists in reducing the risk of the unexpected events which influence the post lunch operations. In our study, we use a public available sub-set[4]. The anomaly ratio in the MSL test dataset is 78.13%, to make the data more balanced, we use the first 500 observations, which has anomaly ratio 20.24%. Water Distribution (WADI) is a sensor-based dataset derived from a distribution system comprising numerous pipelines[5] [17]. Here, a test with size 16 days is conducted, with 14 days under normal operation which are used as training data and 2 days under controlled attack scenarios which is our test set.

Table 1. Summary of the datasets. The anomaly rate is the ratio of true anomaly in the testing set.

Statistics	MSL	WADI	TX	CA	PA
Number of variables	28	128	252	56	61
Training size	1565	1784	200	200	200
Testing size	500	577	175	175	175
Anomaly rate	20.24%	5.55%	56.57%	30.86%	31.42%

We conduct our experiments using a Google colab sever with Intel(R) Xeon(R) CPU @ 2.20GHz, 52 GB RAM, K80, T4 and P100 graphic cards. All models are trained under ADAM optimizer with learning rate 1×10^{-6} and no decay rate. We perform 10 runs, train the models using 100 epochs, and use early stopping of 10. For GDN and Tlife-GDN, we use 128 as the length of embedding vectors and the number of neurons for all datasets. For COVID-19 anomaly prediction, we set similar setting of parameters as in the detection task and set the prediction window h to 7, and the validation ratio to 0.2.

To evaluate the performance of anomaly detection, we use the metrics: F1-Score (F1) and the area under the receiver operating characteristic curve (AUC). As the anomaly score range and the way to choose a suitable threshold is different from method to method, in order to keep the comparison fair for different detection baselines, we set the threshold to be the one which maximizes F1 score

[2] Available at https://covidactnow.org/?s=24821397.
[3] Available at https://github.com/CSSEGISandData/COVID-19.
[4] Available at https://github.com/d-ailin/GDN/tree/main/data/msl.
[5] Further details at https://itrust.sutd.edu.sg/testbeds/water-distribution-wadi/ [3].

for all baselines. The scores above the threshold are considered as anomaly. Our source codes are publicly available in Github[6].

5.2 Experimental Results

Are Persistent Images Really Helpful for COVID-19 Anomaly Detection and Prediction? The anomaly detection results for COVID-19 datasets in TX, CA, and PA are shown in Table 2. For all baselines, we take the average value for F1 score and AUC score in 10 runs, and the standard deviation is shown in parenthesis. From the result, we can see that Tlife-GDN outperforms all baselines across both F1 score and AUC on all 3 states. The topological features extracted from the counties tend to improve the detection performance through comparing Tlife-GDN with GDN model (which is the best baseline). In addition, Table 2 also indicates that integrating topological summaries into GDN model will not increase standard deviation of F1 score and AUC score. Furthermore, Fig. 2 shows the box-plot of AUC score for Tlife-GDN and GDN, from which we conclude that Tlife-GDN exhibits high stability.

Table 2. Average F1 and AUC scores on COVID-19 datasets in 10 runs. For each metric, the best result is highlighted in yellow.

Model	TX		CA		PA	
	F1	AUC	F1	AUC	F1	AUC
PCA [39]	0.570 (<0.0001)	0.739 (<0.0001)	0.550 (<0.0001)	0.498 (<0.0001)	0.536 (<0.0001)	0.498 (<0.0001)
KNN [5]	0.640 (<0.0001)	0.757 (<0.0001)	0.767 (<0.0001)	0.663 (<0.0001)	0.631 (<0.0001)	0.570 (<0.0001)
AE [2]	0.729 (0.0001)	0.739 (0.0002)	0.550 (0.0022)	0.498 (0.0010)	0.534 (0.0023)	0.495 (<0.0001)
DAGMM [48]	0.525 (0.0171)	0.710 (0.0422)	0.680 (0.0697)	0.6390 (0.0422)	0.875 (0.0443)	0.533 (0.0443)
VAE [25]	0.565 (0.0050)	0.519 (0.0016)	0.535 (0.0059)	0.484 (0.0026)	0.531 (0.0032)	0.516 (<0.0001)
DEEP-SVDD [34]	0.675 (0.0156)	0.739 (0.0122)	0.776 (0.0129)	0.436 (0.0189)	0.960 (0.0242)	0.492 (0.0112)
GDN [14]	0.754 (0.0352)	0.742 (0.0122)	0.928 (0.0015)	0.743 (0.0008)	0.994 (0.0013)	0.975 (0.0002)
Tlife-GDN	0.767 (0.0374)	0.759 (0.0092)	0.962 (0.0020)	0.754 (0.0006)	0.995 (0.0220)	0.976 (0.0001)

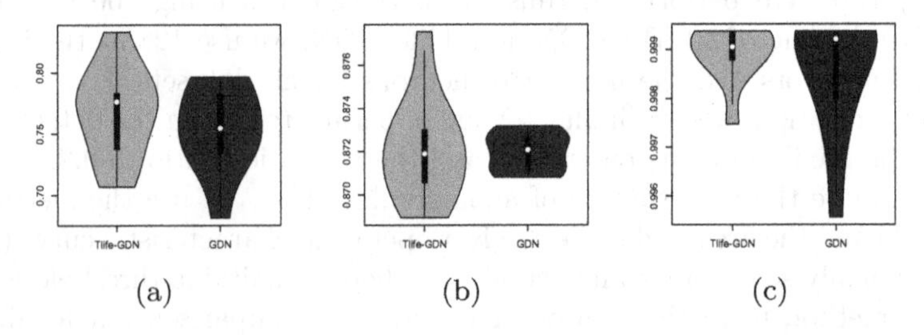

(a) (b) (c)

Fig. 2. Box plot of AUC scores in 10 runs from Tlife-GDN and GDN in (a) Texas (b) California (c) Pennsylvania.

[6] https://github.com/ZhiweiZhen/Tlife-GDN.

In addition, for the traditional anomaly detection problem, we utilize Tlife-GDN and GDN (i.e., the best baseline) to predict future anomalies and verify the significance of topological features. Figure 3 shows that Tlife-GDN achieves a better performance on TX and CA. On PA, both GDN and Tlife-GDN perform well. We can see that the complex hidden topological relationships between counties have a profound impact on future hospitalization anomalies as it may contain the information about the COVID-19 transmission at that moment.

Table 3. Average precision, recall, and F1 score on COVID-19 datasets for one-week ahead anomaly prediction in 10 runs based on GDN and Tlife-GDN.

Model	TX		CA		PA	
	F1	AUC	F1	AUC	F1	AUC
GDN	0.728 (0.0647)	0.762 (0.0521)	0.869 (0.0353)	0.940 (0.1028)	0.927 (0.0398)	0.994 (0.0315)
Tlife-GDN	0.741 (0.0321)	0.765 (0.0198)	0.894 (0.0265)	0.962 (0.0374)	0.927 (0.0379)	0.993 (0.0452)

What is the Performance of Tlife-GDN on MSL and WADI Datasets? To verify the value added by topological summaries for different types of anomaly detection problems, we also evaluate the performance of our Tlife-GDN model on MSL and WADI datasets. The results are shown in Table 4. We find that Tlife-GDN outperforms all baselines in terms of both F1 score and AUC score for WADI. For MSL, Tlife-GDN achieves the best result in F1 score and also competitive result in AUC.

Table 4. Average F1 and AUC scores on MSL and WADI datasets in 10 runs. For each metric, the best result is highlighted in yellow. The results from Tlife-GDN is highlighted in blue if there is improvement compared to GDN.

Model	MSL		WADI	
	F1	AUC	F1	AUC
PCA	0.151 (<0.0001)	0.533 (<.0001)	0.120 (<0.0001)	0.504 (<0.0001)
KNN	0.109 (<0.0001)	0.664 (<0.0001)	0.119 (<0.0001)	0.475 (<0.0001)
AE	0.152 (<0.0001)	0.553 (0.06187)	0.120 (<0.0001)	0.503 (0.02546)
DAGMM	0.361 (0.0549)	0.631 (0.0708)	0.289 (0.0250)	0.603 (0.0603)
VAE	0.120 (0.2210)	0.553 (0.2317)	0.148 (0.1376)	0.503 (0.0557)
DEEP-SVDD	0.337 (0.0555)	0.665 (0.1003)	0.100 (0.0483)	0.477 (0.0019)
GDN	0.407 (0.0125)	0.496 (0.0267)	0.356 (0.0745)	0.785 (0.0632)
Tlife-GDN	0.419 (0.0198)	0.563 (0.1054)	0.371 (00319)	0.797 (0.0480)

Possible Linkage Between Detection Results and Environment. In this study, we also explore the impact of topological features on the detection results. We investigate the timestamps where Tlife-GDN achieves the accurate anomaly detection performance compared with GDN. We believe that those timestamps may share some similarity in terms of environmental variables. Figure 3 shows the Aerosol Optical Depth (AOD) values, a measure of light extinction by aerosol in the atmospheric column above the earth's surface [44], in TX and CA whenever Tlife-GDN outperforms GDN at county level. In addition, Fig. 3 suggests that topological features can improve the ability of non-anomaly detection when AOD is low and help detect anomalies when AOD is high. Furthermore, we can find that the hospitalization rate can be well reflected by the AOD values, which can be used to define anomalies in the anomaly detection task.

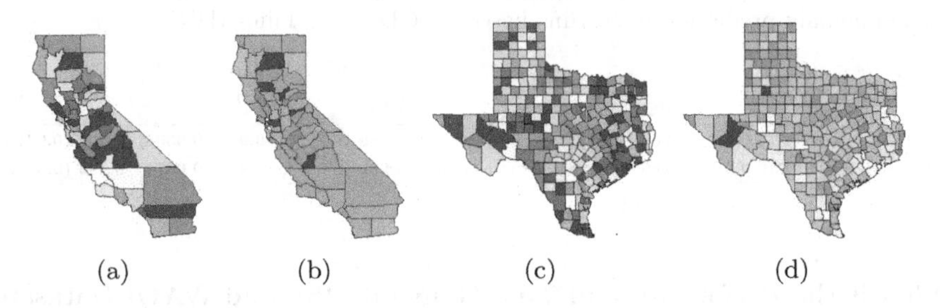

(a) (b) (c) (d)

Fig. 3. Aerosol Optical Depth (AOD) values in CA and TX. The color goes from red to green as AOD increase. (a) Non-anomaly CA. (b) Anomaly CA. (c) Non-anomaly TX. (d) Anomaly TX.

6 Conclusion

In this paper, we introduce a new topology-based graph neural network, i.e., Tlife-GDN to detect and predict anomaly. The experimental results show that Tlife-GDN provides more accurate detection and prediction for COVID-19 hospitalization anomalies in Texas, California, and Pennsylvania, which is critical to forecast pandemic trend, announce travel warnings and help local government prepare potential waves in advance. In the future, we can take pre-existing health conditions, distribution of medical resources and demographic variables into consideration and extend the application of Tlife-GDN to anomaly regarding network defense and national cyber security.

Acknowledgement. This work has been supported in part by grants NSF DMS 1925346, NSF ECCS 2039701, NASA 20-RRNES20-0021, and the Department of the Navy, Office of Naval Research under ONR award number N00014-21-1-2530. Part of this material is also based upon work supported by (while serving at) the National Science Foundation. Any opinions, findings, and conclusions or recommendations expressed in this material are those of the author(s) and do not necessarily reflect the views of the National Science Foundation and/or the Office of Naval Research. The authors are grateful to Huikyo Lee, NASA's Jet Propulsion Lab for the motivating discussion.

References

1. Adams, H., et al.: Persistence images: a stable vector representation of persistent homology. JMLR **18**, 1–35 (2017)
2. Aggarwal, C.C.: Data Mining: The Textbook. Springer, Cham (2015)
3. Ahmed, C.M., Palleti, V.R., Mathur, A.P.: WADI: a water distribution testbed for research in the design of secure cyber physical systems. In: CySWATER (2017)
4. Alonso, J., Belanche, L., Avresky, D.R.: Predicting software anomalies using machine learning techniques. In: IEEE NCA, pp. 163–170 (2011)
5. Angiulli, F., Pizzuti, C.: Fast outlier detection in high dimensional spaces. In: ECML PKDD (2002)
6. Basak, D., Pal, S., Patranabis, D.C.: Support vector regression. Neural Inf. Process. Lett. Rev. **11**(10), 203–224 (2007)
7. Brar, G., et al.: COVID-19 severity and outcomes in patients with cancer: a matched cohort study. J. Cl. Oncol. **38**(33), 3914–3924 (2020)
8. Cai, Q., et al.: Obesity and COVID-19 severity in a designated hospital in Shenzhen. China Diab. care **43**(7), 1392–1398 (2020)
9. Carlsson, G.: Topology and data. BAMS **46**(2), 255–308 (2009)
10. Chaudhary, A., Mittal, H., Arora, A.: Anomaly detection using graph neural networks. In: COMITCon, pp. 346–350. IEEE (2019)
11. Chazal, F., Michel, B.: An introduction to topological data analysis: fundamental and practical aspects for data scientists. Frontiers in Artificial Intelligence (2021)
12. Chen, Y., Segovia-Dominguez, I., Coskunuzer, B., Gel, Y.R.: TAMP-S2GCNets: coupling time-aware multipersistence knowledge representation with spatio-supra graph convolutional networks for time-series forecasting. In: ICLR (2022)
13. Chen, Y., Segovia-Dominguez, I., Gel, Y.R.: Z-GCNETs: time zigzags at graph convolutional networks for time series forecasting. In: ICML (2021)
14. Deng, A., Hooi, B.: Graph neural network-based anomaly detection in multivariate time series. In: AAAI (2021)
15. Dey, T.K., Wang, Y.: Computational Topology for Data Analysis. Cambridge University Press, Cambridge (2022)
16. Gallo Marin, B., et al.: Predictors of COVID-19 severity: a literature review. Rev. in Med. Virol. **31**(1), 1–10 (2021)
17. Goh, J., Adepu, S., Junejo, K.N., Mathur, A.P.: A dataset to support research in the design of secure water treatment systems. In: CRITIS (2016)
18. Golan, I., El-Yaniv, R.: Deep anomaly detection using geometric transformations. arXiv:1805.10917 (2018)
19. Hickok, A., Needell, D., Porter, M.A.: Analysis of spatiotemporal anomalies using persistent homology: case studies with COVID-19 data. arXiv:2107.09188 (2021)
20. Hofer, C.D., Graf, F., Rieck, B., Niethammer, M., Kwitt, R.: Graph filtration learning. In: ICML, vol. 119, pp. 4314–4323. PMLR (2020)
21. Hundman, K., Constantinou, V., Laporte, C., Colwell, I., Soderstrom, T.: Detecting spacecraft anomalies using LSTMs and nonparametric dynamic thresholding. arXiv:1802.04431 (2018)
22. Islambekov, U., Yuvaraj, M., Gel, Y.R.: Harnessing the power of topological data analysis to detect change points in time series. Environmetrics **31**(1), e2612 (2020)
23. Jin, W., Tung, A.K.H., Han, J., Wang, W.: Ranking outliers using symmetric neighborhood relationship. In: Ng, W.-K., Kitsuregawa, M., Li, J., Chang, K. (eds.) PAKDD 2006. LNCS (LNAI), vol. 3918, pp. 577–593. Springer, Heidelberg (2006). https://doi.org/10.1007/11731139_68

24. Karadayi, Y., Aydin, M.N., Öğrenci, A.S.: Unsupervised anomaly detection in multivariate spatio-temporal data using deep learning: early detection of COVID-19 outbreak in Italy. IEEE Access **8**, 164155–164177 (2020)
25. Kingma, D.P., Welling, M.: Auto-encoding variational bayes. arXiv:1312.6114 (2013)
26. Li, D., Chen, D., Goh, J., Ng, S.k.: Anomaly detection with generative adversarial networks for multivariate time series. arXiv:1809.04758 (2018)
27. Li, Y., Islambekov, U., Akcora, C., Smirnova, E., Gel, Y.R., Kantarcioglu, M.: Dissecting ethereum blockchain analytics: What we learn from topology and geometry of the ethereum graph? In: SDM, pp. 523–531. SIAM (2020)
28. Liang, L., Gong, P.: Climate change and human infectious diseases: a synthesis of research findings from global and spatio-temporal perspectives. Environ. Int. **103**, 99–108 (2017)
29. Liu, D., Veeramachaneni, K., Geiger, A., Li, V.O.K., Qu, H.: AQEyes: visual analytics for anomaly detection and examination of air quality data. arXiv:2103.12910 (2021)
30. Ma, X., Wu, J., Xue, S., Yang, J., Zhou, C., Sheng, Q.Z., Xiong, H., Akoglu, L.: A comprehensive survey on graph anomaly detection with deep learning. IEEE Trans. Knowl. Data Eng. (2021)
31. Malhotra, P., Vig, L., Shroff, G., Agarwal, P.: Long short term memory networks for anomaly detection in time series. In: ESANN, vol. 89, pp. 89–94 (2015)
32. Moore, M., Landree, E., Hottes, A.K., Shelton, S.R.: Environmental biodetection and human biosurveillance research and development for national security. Tech. rep, Homeland Security Operational Analysis Center, RAND Corp (2018)
33. Ofori-Boateng, D., Dominguez, I.S., Kantarcioglu, M., Akcora, C.G., Gel, Y.R.: Topological anomaly detection in dynamic multilayer blockchain networks. In: ECML (2021)
34. Ruff, L., et al.: Deep one-class classification. In: ICML, vol. 80, pp. 4393–4402 (2018)
35. Sanchez-Hernandez, C., Boyd, D.S., Foody, G.M.: One-class classification for mapping a specific land-cover class: SVDD classification of fenland. GRSS-IEEE **45**(4), 1061–1073 (2007)
36. Segovia Dominguez, I., Lee, H., Chen, Y., Garay, M., Gorski, K.M., Gel, Y.R.: Does air quality really impact COVID-19 clinical severity: coupling NASA satellite datasets with geometric deep learning. In: ACM SIGKDD, pp. 3540–3548 (2021)
37. Segovia-Dominguez, I., et al.: Using NASA satellite data sources and geometric deep learning to uncover hidden patterns in COVID-19 clinical severity. arXiv:2110.10849 (2021)
38. Segovia-Dominguez, I., Zhen, Z., Wagh, R., Lee, H., Gel, Y.R.: TLife-LSTM: Forecasting future COVID-19 progression with topological signatures of atmospheric conditions. In: PAKDD, pp. 201–212 (2021)
39. Shyu, M.L., Chen, S.C., Sarinnapakorn, K., Chang, L.: A novel anomaly detection scheme based on principal component classifier. Miami Univ Coral Gables Fl Dept of Electrical and Computer Engineering, Technical report (2003)
40. Stolz, B.J., Harrington, H.A., Porter, M.A.: Persistent homology of time-dependent functional networks constructed from coupled time series. Chaos **27**(4), 047410 (2017)
41. Tack, A.J., Thrall, P.H., Barrett, L.G., Burdon, J.J., Laine, A.L.: Variation in infectivity and aggressiveness in space and time in wild host-pathogen systems: causes and consequences. J. Evol. Biol. **25**(10), 1918–1936 (2012)

42. Tang, J., Chen, Z., Fu, A.W., Cheung, D.W.: Enhancing effectiveness of outlier detections for low density patterns. In: Chen, M.-S., Yu, P.S., Liu, B. (eds.) PAKDD 2002. LNCS (LNAI), vol. 2336, pp. 535–548. Springer, Heidelberg (2002). https://doi.org/10.1007/3-540-47887-6_53
43. Umeda, Y., Kaneko, J., Kikuchi, H.: Topological data analysis and its application to time-series data analysis. Fujitsu Sci. Tech. J. **55**(2), 65–71 (2019)
44. Van Donkelaar, A., Martin, R.V., Brauer, M., Kahn, R., Levy, R., Verduzco, C., Villeneuve, P.J.: Global estimates of ambient fine particulate matter concentrations from satellite-based aerosol optical depth: development and application. Environ. Health Perspectives **118**(6), 847–855 (2010)
45. Vapnik, V.N.: An overview of statistical learning theory. IEEE Trans. Neural Netw. **10**(5), 988–999 (1999)
46. Vries, D., Van Den Akker, B., Vonk, E., De Jong, W., Van Summeren, J.: Application of machine learning techniques to predict anomalies in water supply networks. Water Sci. Technol. **16**(6), 1528–1535 (2016)
47. Zeng, S., Graf, F., Hofer, C., Kwitt, R.: Topological attention for time series forecasting. In: NeurIPS (2021)
48. Zong, B., Song, Q., Min, M.R., Cheng, W., Lumezanu, C., Cho, D., Chen, H.: Deep autoencoding gaussian mixture model for unsupervised anomaly detection. In: ICLR (2018)

Layer Adaptive Deep Neural Networks for Out-of-Distribution Detection

Haoliang Wang(✉), Chen Zhao, Xujiang Zhao, and Feng Chen

University of Texas at Dallas, Richardson, USA
{haoliang.wang,chen.zhao,xujiang.zhao,feng.chen}@utdallas.edu

Abstract. During the forward pass of Deep Neural Networks (DNNs), inputs gradually transformed from low-level features to high-level conceptual labels. While features at different layers could summarize the important factors of the inputs at varying levels, modern out-of-distribution (OOD) detection methods mostly focus on utilizing their ending layer features. In this paper, we proposed a novel layer-adaptive OOD detection framework (LA-OOD) for DNNs that can fully utilize the intermediate layers' outputs. Specifically, instead of training a unified OOD detector at a fixed ending layer, we train multiple One-Class SVM OOD detectors simultaneously at the intermediate layers to exploit the full-spectrum characteristics encoded at varying depths of DNNs. We develop a simple yet effective layer-adaptive policy to identify the best layer for detecting each potential OOD example. LA-OOD can be applied to any existing DNNs and does not require access to OOD samples during the training. Using three DNNs of varying depth and architectures, our experiments demonstrate that LA-OOD is robust against OODs of varying complexity and can outperform state-of-the-art competitors by a large margin on some real-world datasets.

Keywords: OOD detection · Deep neural networks · One-Class SVM

1 Introduction

Recently, deep neural networks (DNNs) have demonstrated remarkable performance in classification problems. However, DNNs are often designed for a static and closed world, assuming the same data distribution during training and test times. In an open-world environment, it is important to detect examples from novel class distributions in safety-critical applications (*e.g.* detecting new categories of objects during autonomous driving and diagnoses of unknown diseases, such as COVID-19). It is hence necessary to develop DNNs that can identify OOD examples while at the same time classifying samples from known class distributions with high accuracy.

A number of recent methods have been proposed to detect OOD examples based on DNNs. The majority of these methods detect OOD examples based on predictive uncertainty measures of a softmax classifier, such as entropy [15],

J. Gama et al. (Eds.): PAKDD 2022, LNAI 13281, pp. 526–538, 2022.
https://doi.org/10.1007/978-3-031-05936-0_41

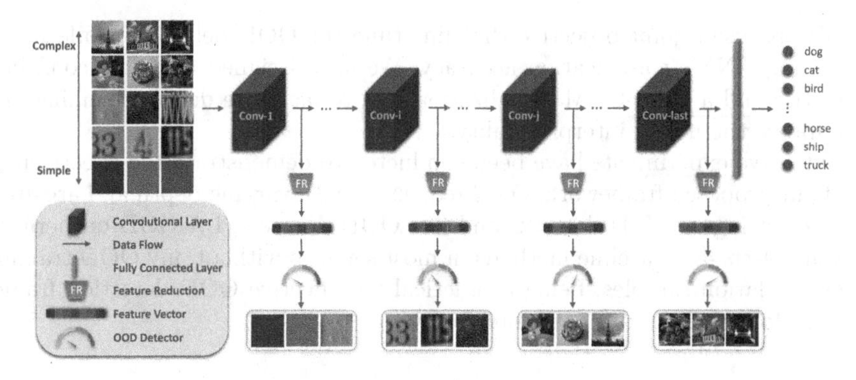

Fig. 1. An overview of our proposed Layer Adaptive Deep Neural Networks for OOD Detection (LA-OOD).

epistemic uncertainty [12], and others [4,11,18,19]. A more recent work presents the Deep-MCDD [7], that estimates a spherical decision boundary for each class based on support vector data description (SVDD), such boundaries will enclose the in-distribution (InD) samples and distinguish OODs based on their closest class-conditional distribution. Instead of using the last layer outputs, [1] proposed to find the best intermediate layer based on a holdout validation OOD dataset. However, all of the above methods detect the OOD examples at the same level of representation (*i.e.* outputs at one single layer) and they hence fail to account for the different representation complexities of OOD examples. Particularly, our empirical study indicates that different OODs may be better detected at their appropriate levels of representations (see Sect. 4.2).

This observation motivates us to propose a novel framework, namely Layer-Adaptive OOD detection (LA-OOD), a generic modification to off-the-shelf DNNs that introduces OOD detectors to intermediate layers. Specifically, we train separate One-Class SVM (OCSVM) OOD detectors using different layers' outputs and employ a simple yet effective layer-adaptive policy function to identify the best layer for detecting each potential OOD sample (see Fig. 1). We tune the OOD detectors through self-adaptive data shifting [16] to improve its accuracy and robustness against unseen OODs, and fine tune the framework using alternating optimization, in which the DNN classification error and the OOD detectors' training errors are minimized jointly.

The main contributions are stated as follows:

- We propose a novel layer-adaptive OOD detection framework (LA-OOD) that is practical for any off-the-shelf DNNs. Multiple OOD detectors are attached to the intermediate layers of a DNN, through a simple yet effective layer-adaptive policy, our proposed framework is able to fully utilize the intrinsic characteristics of inputs encoded in the intermediate latent space, hence, detect OODs with varying complexity.

- We propose a joint objective that fine-tune the OOD detectors while maintaining DNN's classification accuracy. We also designed an OOD confusion metric and a Grad-CAM visualization tool to facilitate decision making and improve the model interpretability.
- Extensive experiments have been conducted to demonstrate the effectiveness of our proposed framework. On three DNNs with varying depth and architectures, using two InD datasets and five OOD datasets, LA-OOD outperform state-of-the-art baseline methods in most settings without any OOD training or validation samples, being a practical yet effective OOD detection framework for OODs of different complexity.

2 Related Work

Dynamic Neural Networks with Early-Exit. Adaptive early-exist is a rising research topic in deep learning. By attaching early exits to a DNN, such methods allow "simple" samples to be output at early layers without "overthinking" [5,6]. For a given input, an early-exit could be determined by either a confidence metric [9] or a learned decision function [2]. However, these methods aim to improve DNN performance by focusing on InD sample evaluation without giving enough attention to OODs. *In this paper, we adopt the idea of early exits for the out-of-distribution detection problem and propose a novel framework in which each OOD sample is detected at its best layer.*

OOD Detection for Deep Neural Networks. In recent years, researchers have developed a number of OOD detection methods, where the majority of such techniques use the final outputs of a DNN to separate the OODs from the InD samples [15]. [4] proposes a baseline method that detects OODs based on the maximum softmax probabilities of a DNN's final outputs. ODIN [11] incorporates the temperature scaling and input perturbation into the maximum softmax probabilities to enhance the margin between InD and OOD samples. More recently, [7] extends Deep-SVDD to a multi-class setting and proposes the Deep-MCDD, It integrates multiple SVDDs into a single deep model where each SVDD is trained to surround one InD class sample. However, these works mainly focus on the high-level conceptual features outputted by the ending layers of DNNs while ignoring the low-level representations at the intermediate layers, hence, may "overthink" the problem and fail on OODs of relatively low complexity. *In contrast, LA-OOD not only considers the ending layers' outputs but also takes the intermediate layers into consideration to generate more accurate OOD predictions.*

Two existing methods [1,8] utilize intermediate outputs of a DNN for OOD detection. [8] defines the confidence score of input as a weighted average of the Mahalanobis distance to the closest class-conditional distribution at each layer, such weighting function is trained using an additional validation set. [1] proposes the OODL which decides an optimal discernment layer based on a holdout OOD dataset. Both methods require the OOD samples during the training, such OOD samples not only are hard to obtain in real-world applications, but also make the trained models susceptible to unseen OODs. *In this work, we tune the OOD detectors using pseudo OODs generated through self-adaptive data shifting* [16] *of the InD training samples, hence, does not require any OOD samples during the training.*

3 Adaptive One-Class Deep Neural Network

Since OOD samples are rarely available during the training, here we formulate the OOD detection as a one-class classification problem, in which OOD detectors only target to determine whether an input is in-distribution or not.

3.1 Problem Formulation

Let $\mathbf{x} \in \mathcal{X}$ be an input, $y \in \mathcal{Y} = \{1, \cdots, K\}$ being its label, given a deep neural network $\mathcal{M}$ with L layers, it tries to classify each input to K classes: $\hat{y} = \mathcal{M}(\mathbf{x}) \in \mathcal{Y}$. With the intermediate outputs $\mathbf{x}^{(\ell)}$ at layer $\ell \in \{1, \cdots, L\}$, its OOD score $s^{(\ell)} = C_\ell(\mathbf{x}^{(\ell)})$ is computed by a layer-specific OOD detector C_ℓ. Separate OOD detectors could be attached to different layers of $\mathcal{M}$, the final OOD score of $\mathbf{x}$ could be obtained by taking the maximum OOD scores outputted by all the OOD detectors: $s_{\text{final}} = \max\left[\{C_\ell(\mathbf{x}^{(\ell)})\}_{\ell=1}^{L}\right]$. Such OOD score then can be used to determine whether $\mathbf{x}$ is in-distribution or not based on a predefined threshold δ.

3.2 Framework Overview

In the context of one-class classification, there are many possible selections for the OOD detector (KDE, GMM, k-NN, *etc.*) In this paper, we use the One-Class Support Vector Machine (OCSVM) [13] which is one of the most commonly used one-class classifier in the literature. Note that, we could replace OCSVM with any other one-class classifiers as our framework design does not depend on a specific choice of one-class classifiers.

For the OCSVM, a feature mapping $\Phi : \mathcal{X} \subset \mathbb{R}^d \to \mathcal{F} \subset \mathbb{R}^h$ is defined, where $h > d$, it maps the input samples $\{\mathbf{x}_i\}_{i=1}^n \in \mathbb{R}^d$ into a high dimensional feature space $\mathcal{F}$. An OCSVM will try to find the best separating hyperplane that separates all the input samples from the origin such that the distance to the origin is maximized. Normally, the calculation of the feature mapping Φ is avoided by using the kernel trick $k(\mathbf{x}_i, \mathbf{x}_j) = (\Phi(\mathbf{x}_i) \cdot \Phi(\mathbf{x}_j))$. In this paper, we select the commonly used Gaussian Radial Base Function (RBF) kernel: $k(\mathbf{x}_i, \mathbf{x}_j) = \exp\left(-\gamma \|\mathbf{x}_i - \mathbf{x}_j\|^2\right)$, where γ is the kernel width.

Using Lagrange multipliers, optimizing the OCSVM C_ℓ at layer ℓ is equivalent to solving the following dual Quadratic Programming (QP) problem:

$$\min_{\boldsymbol{\alpha}^{(\ell)}} \quad \frac{1}{2}\sum_{i,j} \boldsymbol{\alpha}_i^{(\ell)} \boldsymbol{\alpha}_j^{(\ell)} k\left(\mathbf{x}_i^{(\ell)}, \mathbf{x}_j^{(\ell)}\right) \quad \text{s.t. } 0 \le \boldsymbol{\alpha}_i^{(\ell)} \le \frac{1}{\nu n}, \text{and } \sum_i \boldsymbol{\alpha}_i^{(\ell)} = 1 \quad (1)$$

where $\boldsymbol{\alpha}_i^{(\ell)}$ are the Lagrange multipliers, and $\nu \in (0, 1]$ is the upper bound of the training error.

Given an input sample $\mathbf{x}$ and its layer ℓ outputs $\mathbf{x}^{(\ell)}$, its OOD score at layer ℓ is calculated using the decision function:

$$C_\ell(\mathbf{x}) = -\sum_i \boldsymbol{\alpha}_i^{(\ell)} k\left(\mathbf{x}_i^{(\ell)}, \mathbf{x}^{(\ell)}\right) + \rho^{(\ell)} \tag{2}$$

where the offsets $\rho^{(\ell)}$ can be recovered by $\rho^{(\ell)} = \sum_j \boldsymbol{\alpha}_j^{(\ell)} k\left(\mathbf{x}_j^{(\ell)}, \mathbf{x}_i^{(\ell)}\right)$. Positive scores represent OODs, and negative scores represent InDs (assuming the default zero threshold is used, *i.e.*, $\delta = 0$).

3.3 Framework Training

Given a pre-trained DNN model $\mathcal{M}_{\boldsymbol{\theta}}$ parameterized by $\boldsymbol{\theta}$, using the OCSVMs as OOD detectors, we propose a joint objective for training both the backbone model and the OOD detectors:

$$\begin{aligned} \min_{\boldsymbol{\theta}} \min_{\boldsymbol{\alpha}^{(\ell)}{}_{\ell=1}^{L}} \quad & L(\boldsymbol{\theta}) + \frac{\lambda}{2} \cdot \sum_{\ell=1}^{L} \sum_{i,j} \boldsymbol{\alpha}_i^{(\ell)} \boldsymbol{\alpha}_j^{(\ell)} k\left(\mathbf{x}_i^{(\ell)}, \mathbf{x}_j^{(\ell)}\right) \\ \text{subject to} \quad & 0 \leq \boldsymbol{\alpha}_i^{(\ell)} \leq \frac{1}{\nu n}, \text{and } \sum_i \boldsymbol{\alpha}_i^{(\ell)} = 1 \end{aligned} \tag{3}$$

Here the first term $L(\boldsymbol{\theta})$ denotes the loss function of the backbone network, and the second term is the summation of losses for all the OOD detectors multiplied by a regularization parameter $\lambda > 0$. We aim to fine-tune the layer-dependent feature representations and the parameters of layer-dependent OCSVM jointly so that the training errors of the OOD detectors are minimized while maintaining DNN's classification accuracy.

To solve Eq.(3), an alternating optimization technique is applied in which the $\boldsymbol{\theta}$ and $\{\boldsymbol{\alpha}^{(\ell)}\}_{\ell=1}^{L}$ will be updated alternatively:

- Step I: Fix $\{\boldsymbol{\alpha}^{(\ell)}\}_{\ell=1}^{L}$ and re-estimate the model parameters $\boldsymbol{\theta}$ using a Eq. 4.
- Step II: Fix $\boldsymbol{\theta}$ and generate the updated intermediate outputs to re-estimate $\{\boldsymbol{\alpha}^{(\ell)}\}_{\ell=1}^{L}$ using Eq. 1.

In step I, we fix the estimated dual coefficients $\{\boldsymbol{\alpha}^{(\ell)}\}_{\ell=1}^{L}$ for all OCSVMs, then re-estimate the backbone model parameter $\boldsymbol{\theta}$:

$$\min_{\boldsymbol{\theta}} \quad L(\boldsymbol{\theta}) + \frac{\lambda}{2L} \sum_{\ell=1}^{L} \sum_{i,j} \boldsymbol{\alpha}_i^{(\ell)} \boldsymbol{\alpha}_j^{(\ell)} k(\mathbf{x}_i^{(\ell)}(\boldsymbol{\theta}), \mathbf{x}_j^{(\ell)}(\boldsymbol{\theta})) \tag{4}$$

In step II, we fix the backbone model to update the intermediate outputs for the training samples, then based on the newly generated outputs, we re-train all the OOD detectors using Eq. 1.

Algorithm 1. LA-OOD Training Procedure

Input: Pre-trained DNN model $\mathcal{M}_{\theta}$, InD sample set $\mathcal{X}$
Output: Jointly trained $\mathcal{M}_{\theta}$ and OOD detectors $\{C_\ell\}_{\ell=1}^{L}$
1: Generate the intermediate outputs $\{\mathcal{X}^{(\ell)}\}_{\ell=1}^{L}$
2: Generate pseudo-outliers
$\{\mathcal{X}^{(\ell)}_{\text{pseudo}}\}_{\ell=1}^{L} = \text{selfAdaptiveDataShifting}(\{\mathcal{X}^{(\ell)}\}_{\ell=1}^{L})$
3: Hyper-parameter tuning for $\{C_\ell\}_{\ell=1}^{L}$ using $\{\mathcal{X}^{(\ell)}\}_{\ell=1}^{L}$ and $\{\mathcal{X}^{(\ell)}_{\text{pseudo}}\}_{\ell=1}^{L}$
4: **while** not done **do**
5: Fix the $\{\boldsymbol{\alpha}^{(\ell)}\}_{\ell=1}^{L}$ and re-estimate $\boldsymbol{\theta}$ (Eq. 4)
6: Update the intermediate outputs $\{\mathcal{X}^{*(\ell)}\}_{1}^{L}$
7: Re-train $\{C_\ell\}_{\ell=1}^{L}$ using the updated intermediate outputs
$\{\mathcal{X}^{*(\ell)}\}_{1}^{L}$ (Eq. 1)
8: **return** trained $\mathcal{M}_{\theta}$ and $\{C_\ell\}_{\ell=1}^{L}$

Two important hyper-parameters for OCSVM training are the Gaussian kernel width γ and the training error upper bound ν. γ controls the smoothness of the decision boundary. The smaller the γ, the smoother the decision boundary will be. ν controls the error ratio, which is often tuned to reject the noisy samples in the training set and it also determines a lower bound on the fraction of support vectors. These two hyper-parameters are critical for OCSVM to achieve good performance. In general, these hyper-parameters are tuned using a held-out validation set that includes both InD and OOD samples. In this work, we adopt the self-adaptive data shifting [16] to generate pseudo-OODs for hyper-parameter tuning. Such pseudo-OODs are created purely using InD samples through edge pattern detection [10]. We summarized our LA-OOD training procedure in Algorithm 1.

3.4 Layer-Adaptive Policy Design

Having L OCSVM OOD detectors $\{C_\ell\}_{\ell=1}^{L}$ that each outputs an OOD score $s_i^{(\ell)}$ for input $\mathbf{x}_i$, we either need to define a threshold for each of these OOD detectors or design a decision policy that consolidates all the OOD scores into a final prediction. Empirically, we found that a layer-adaptive policy performs better than some fixed thresholds as it is very common that the predictions of OOD detectors diverge from each other (see Sect. 4.3). Here we choose a simple yet effective layer-adaptive policy that propagates the most confident opinion among all OOD detectors as the final prediction, specifically, the policy is design as $s_{i,\text{final}} = \max\left[\{C_\ell(\mathbf{x}_i^{(\ell)})\}_{\ell=1}^{L}\right]$. One challenge to such policy design is that OCSVMs trained on different features generally will have a different scale of scores, this effect could be alleviated by normalizing the training features for each OCSVM, here we simply use the standardization: $\mathbf{x}' = (\mathbf{x} - \bar{\mathbf{x}})/\sigma$, with $\bar{\mathbf{x}}$ being the sample mean and σ being its standard deviation.

4 Experimental Results

Empirical Settings[1]**.** (1) **Datasets.** Two InD datasets (CIFAR10 and CIFAR100) and five OOD datasets (LSUN, Tiny ImageNet, SVHN, DTD [3], and Pure Color) are considered in the experiments. The "Pure Color" dataset is a synthetic dataset that contains 10,000 randomly generated pure-color images. For each InD-OOD combination, we construct a training set using all the training images in the InD dataset and form a balanced test set using all the test images in both InD and OOD datasets, when the sizes of their test set mismatch, we randomly selected the same number of images from the larger dataset to match the test sample size of the smaller one. All images are down-sampled to 32×32 resolution using Lanczos interpolation. (2) **Backbone Models.** We evaluate our method using three popular CNNs in computer vision and machine learning studies. Particularly, we select the VGG-16, ResNet-34, and DenseNet-100 to demonstrate the effectiveness of our framework for DNN models of varying depth and architectures. (3) **Feature Reduction.** A feature reduction operation is applied to the intermediate outputs to maintain the scalability [1]. Among the pooling methods we have tested: max/average pooling with various sizes, global max/average pooling, the global average pooling performs the best. The pooled features are then standardized using the training set mean and deviation. (4) **Hyper-Parameters Tuning.** We fix ν to be 0.001 so that only a small number of InD samples will be considered as noise, the γ is tuned using pseudo-OODs generated by self-adaptive data shifting [16] of only the InD training samples. We search γ in $[0.001, 0.0025, 0.005, 0.01, 0.025, 0.05, 0.1, 0.25, 0.5, 1.0]$, for different InD-Backbone settings, we will shrink the value range to accommodate the differences in feature complexity and to reduce training time. (5) **Baseline Methods and Evaluation Metrics.** We compare our method with four state-of-the-art OOD detection baselines: MSP [4], ODIN [11] (both temperature scaling and input preprocessing are used to achieve optimum performance), OODL [1] (we use the iSUN [17] as an additional OOD dataset to find its optimal discernment layer), and Deep-MCDD [7]. Three commonly adopted OOD detection metrics are used: AUROC, AUPR, and FPR at 95% TPR.

4.1 Performance Evaluation

The experimental results are reported in Table 1, the mean values of the each evaluation metric are also reported to demonstrate the overall performance on OOD datasets with varying complexities. It is worth noting that previous works often choose to use linear interpolation for the down-sampling operation [1,7,8,11], however, we found that *using linear interpolation will create severe aliasing artifacts which make such OOD samples easily detectable*, therefore, to generate more genuine OOD samples, we down-sampled the OOD images using the Lanczos interpolation which is much more sophisticated than the linear interpolation.

[1] The source code and datasets are available at: https://github.com/haoliangwang86/LA-OOD.

From Table 1, it could be seen that OODs that of higher complexity will be harder to detect, such as the LSUN and Tiny ImageNet images that could contain complex backgrounds or multiple objects in a single image. OODs of lower complexity are easier to detect, such as the SVHN that contains cropped street view house numbers or DTD that contains images of different textures. The synthetic Pure Color dataset is of the lowest complexity as it contains limited information. Such dataset complexity could be easily verified using entropy or energy metrics.

Table 1. Performance evaluation. Metrics with "↑" indicate the bigger the better and "↓" indicate the smaller the better. Best performance are labeled in **bold**.

InD/Model	OOD	AUROC ↑	AUPR ↑	FPR at 95% TPR ↓
		MSP/ODIN/ Deep-MCDD/OODL/LA-OOD (Ours)		
Cifar10 VGG-16	LSUN	86.25/86.75/85.19/**88.03**/87.26	85.26/87.06/84.76/**88.01**/84.42	69.27/67.72/59.09/62.38/**54.88**
	Tiny	85.66/86.35/83.95/87.10/**88.39**	84.23/86.22/83.49/**86.98**/86.02	67.36/64.30/61.56/64.08/**44.00**
	SVHN	91.12/91.47/89.81/91.68/**97.27**	87.06/89.29/93.99/88.46/**97.15**	21.78/25.45/64.02/23.52/**14.25**
	DTD	87.73/90.26/88.33/92.16/**97.35**	87.05/89.58/80.60/90.82/**97.45**	66.24/46.33/53.56/25.04/**14.06**
	Pure Color	98.57/99.77/98.42/99.41/**99.93**	98.18/99.75/98.30/98.94/**99.84**	04.66/01.24/05.68/02.08/**00.21**
	Mean	89.87/90.92/89.14/91.68/**94.04**	88.36/90.38/88.23/90.64/**92.98**	45.86/41.01/48.78/35.42/**25.48**
Cifar100 VGG-16	LSUN	73.00/73.58/72.83/**75.10**/72.48	68.49/69.78/**69.92**/69.68/65.28	75.43/**74.92**/85.12/74.99/80.24
	Tiny	77.10/77.83/76.37/79.84/**80.57**	72.64/74.82/73.27/**75.20**/75.19	63.53/68.89/80.50/60.68/**56.22**
	SVHN	75.43/78.18/74.98/78.43/**87.07**	71.53/76.20/86.52/72.63/**85.82**	66.26/70.29/82.31/62.78/**48.94**
	DTD	75.75/76.81/73.80/77.76/**93.28**	70.20/72.94/58.84/70.63/**93.33**	62.13/64.66/82.20/57.82/**33.20**
	Pure Color	62.66/51.22/78.28/58.10/**96.71**	54.24/49.93/73.44/49.13/**95.24**	72.32/95.31/81.83/64.85/**30.08**
	Mean	72.79/71.52/75.25/73.85/**86.02**	67.42/68.73/72.40/67.45/**82.97**	67.93/74.81/82.39/64.22/**49.74**
Cifar10 ResNet-34	LSUN	90.16/90.26/88.02/**91.97**/89.06	87.62/90.19/86.74/**90.56**/84.48	33.24/50.28/55.75/**31.19**/37.35
	Tiny	86.53/85.46/83.34/88.81/**89.29**	84.79/86.46/83.25/**87.66**/86.47	58.26/74.41/61.28/46.15/**36.90**
	SVHN	84.33/81.22/88.08/87.74/**97.77**	81.88/81.89/93.97/85.13/**97.67**	66.58/81.16/57.06/42.84/**12.17**
	DTD	87.64/83.96/84.56/92.10/**97.91**	85.24/84.39/75.07/91.10/**98.06**	51.61/78.01/62.13/30.82/**11.84**
	Pure Color	94.59/96.84/96.11/95.52/**99.99**	93.48/96.93/93.81/94.35/**99.99**	17.84/15.54/36.80/19.50/**00.04**
	Mean	88.65/87.55/88.02/91.23/**94.80**	86.60/87.97/86.57/89.76/**93.33**	45.51/59.88/54.60/34.10/**19.66**
Cifar100 ResNet-34	LSUN	75.63/**77.52**/74.65/51.91/65.25	70.76/**72.81**/70.14/51.92/59.65	**62.63**/63.51/84.34/94.84/78.61
	Tiny	78.70/**81.28**/78.29/67.05/75.82	74.47/77.39/**78.26**/66.91/73.74	57.97/**57.47**/78.84/90.27/68.91
	SVHN	78.76/84.16/78.62/79.00/**84.61**	73.71/78.74/**88.50**/69.18/76.09	55.29/46.58/77.50/45.81/**36.85**
	DTD	75.32/78.94/77.11/86.25/**91.39**	70.07/74.52/84.85/83.45/**91.97**	62.59/60.60/81.49/**40.94**/41.19
	Pure Color	55.23/62.25/63.47/96.46/**99.80**	48.09/52.11/53.16/91.14/**99.78**	67.52/59.04/99.32/04.98/**01.04**
	Mean	72.73/76.83/74.43/76.13/**83.37**	67.42/71.11/74.98/72.52/**80.25**	61.20/57.44/84.30/55.37/**45.32**
Cifar10 DenseNet-100	LSUN	92.07/**94.01**/87.19/88.47/84.38	89.47/**93.12**/86.23/84.87/80.95	26.40/**23.71**/55.00/40.69/51.55
	Tiny	89.96/**91.95**/85.22/84.62/88.75	87.69/**91.32**/84.44/80.90/87.80	35.09/**34.04**/58.14/57.25/43.73
	SVHN	89.00/89.54/89.48/97.19/**97.79**	85.73/88.11/94.46/**97.54**/97.51	36.33/43.54/51.29/16.07/**09.41**
	DTD	88.65/85.42/86.93/95.10/**97.61**	86.06/84.75/77.33/96.14/**97.58**	39.61/60.98/59.57/33.07/**12.00**
	Pure Color	91.83/96.78/96.21/79.15/**99.97**	87.80/95.01/95.08/69.92/**99.97**	16.06/09.31/23.84/40.08/**00.17**
	Mean	90.30/91.54/89.01/88.91/**93.70**	87.35/90.46/87.51/85.87/**92.76**	30.70/34.32/49.57/37.43/**23.37**
Cifar100 DenseNet-100	LSUN	76.38/**77.41**/75.17/59.11/69.69	72.14/**73.19**/71.18/57.10/64.28	**62.62**/65.02/82.93/91.64/72.59
	Tiny	79.73/**84.27**/78.25/61.84/81.29	76.10/**81.66**/75.11/59.22/78.81	55.24/**50.97**/77.48/81.85/62.76
	SVHN	80.08/81.30/74.99/71.73/**86.99**	75.29/74.89/**86.25**/65.36/78.23	51.73/49.32/82.48/66.07/**32.89**
	DTD	73.18/70.29/79.34/84.69/**93.79**	69.03/67.93/66.09/84.72/**93.95**	73.09/91.60/75.11/56.15/**30.67**
	Pure Color	79.60/80.86/91.14/85.39/**99.47**	73.54/77.68/89.64/79.53/**99.41**	44.87/61.26/49.77/34.72/**02.84**
	Mean	77.79/78.83/79.78/72.55/**86.25**	73.22/75.07/77.65/69.19/**82.94**	57.51/63.63/73.55/66.09/**40.35**

The OOD detection methods that utilize the ending layers' features (MSP, ODIN, and Deep-MCDD) generally perform well on detecting OODs with higher complexity, such as the LSUN and the Tiny ImageNet datasets, however, they tend to give poor decisions for OODs of lower complexity such as the SVHN,

DTD, and the Pure Color datasets. The OODL baseline method could utilize the intermediate features, from the performance evaluation, we could see that OODL exhibit the same performance pattern as MSP, ODIN, and MCDD, however, it is due to that LSUN and Tiny ImageNet have similar complexity as the iSUN dataset, which is used to determine the optimal discernment layers for OODL, when the test OODs are of different complexity compare to iSUN, its performance could degrade significantly.

Through multiple intermediate OOD detectors and the layer-adaptive policy, LA-OOD can exploit the full-spectrum characteristics encoded in different intermediate layers. Specifically, by taking the early layers' outputs into consideration, LA-OOD outperforms the other four baseline methods by a large margin on OOD datasets of lower complexity (SVHN, DTD, and Pure Color). More importantly, LA-OOD achieves the best average AUROC/AUPR/FPR at 95% TPR for all InD-Backbone settings, which indicates our proposed method is robust against OODs of different complexity. Overall, LA-OOD achieves an 8.21% improvement margin on AUROC, 7.8% improvement margin on AUPR, and 29.98% improvement margin on FPR at 95% TPR compare to the second-best baseline method.

4.2 Understanding the Behaviors of Different Layers

As the layer of a DNN goes deeper, more complex features could be learned [20], by attaching OOD detectors to the intermediate layers, we could detect OODs based on features of different complexities. Figure 2 shows the number of OODs identified by different OOD detectors. For the LSUN and Tiny ImageNet OOD datasets which are of higher complexity, most of them are identified by the last two OOD detectors, while for the other three OOD datasets that have relatively lower complexity, they are mainly detected by the first seven detectors.

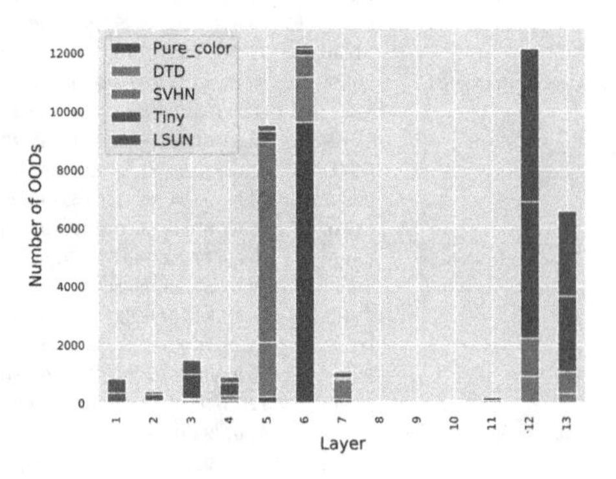

Fig. 2. Number of OODs detected by OOD detectors at different layers using VGG-16 and CIFAR10 InD.

In Fig. 3 we show the correctly identified Tiny ImageNet samples by different layer's OOD detectors using the VGG backbone and CIFAR10 as InD dataset. It could be seen that the OOD detectors at the initial layers are more sensitive to the image colors and textures which relate to the fine-scale details of the input images, while the OOD detectors at the ending layers tend to detect OODs based on objects or scenes. As the layer goes deeper, more and more complex OODs can be detected. Similar pattern could also be found on the DTD dataset, as shown in Fig. 4.

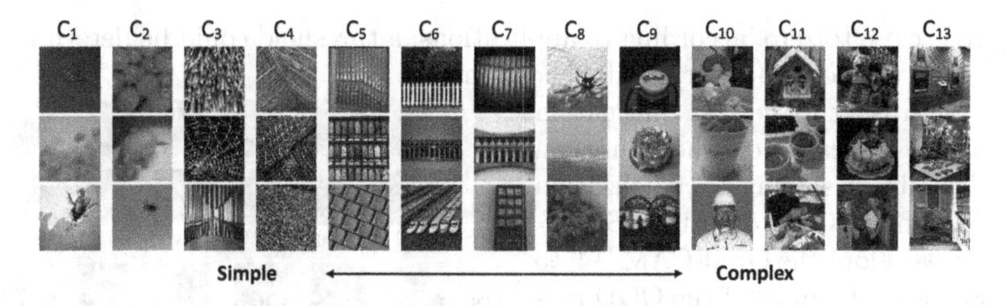

Fig. 3. Correctly identified Tiny ImageNet OODs by OOD detectors at different layers, using VGG backbone and CIFAR10 as InD dataset.

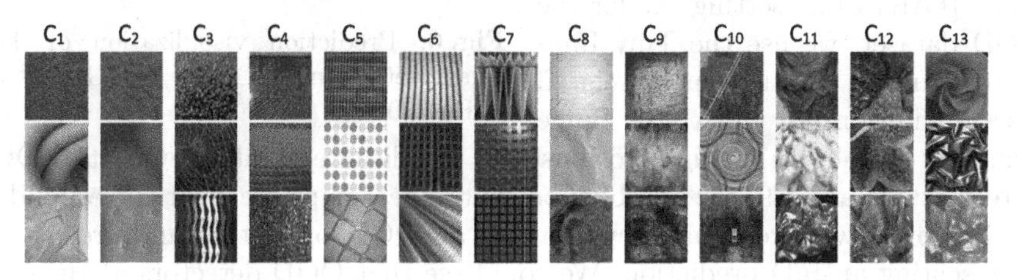

Fig. 4. Correctly identified DTD OODs by OOD detectors at different layers, using VGG backbone and CIFAR10 as InD dataset.

4.3 Framework Confusion Analysis

The disagreement between the OOD detectors indicates that their predictions are inconsistent and *confused.* Here we define a confusion score $D(\mathbf{x}) = \sum_1^L C_\ell(\mathbf{x}^{(\ell)})$ to measure the prediction divergence between the OOD detectors. For a good OOD detector, this confusion score should be negative for most of the InD test samples and positive for predicted OODs, the confusion occurs when the confusion score is close to 0.

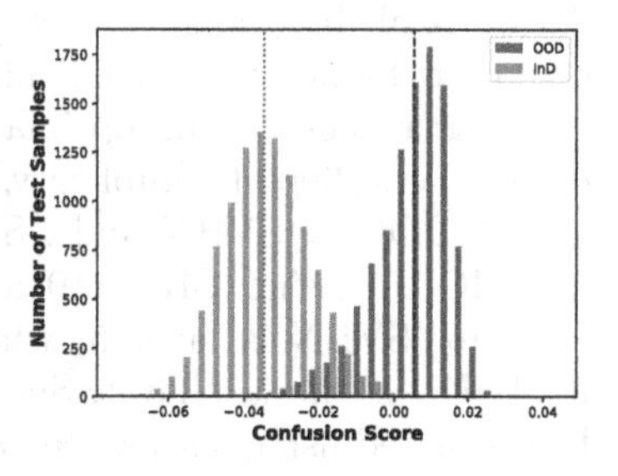

Fig. 5. Confusion score of SVHN vs. CIFAR10 on VGG-16.

We expect this confusion metric to be a reliable indicator in cases where the framework is unable to make a confident prediction and may have misclassified a test sample. Such an indicator has significant importance in handling errors due to the possible severe impact of false positives in real-world applications. We performed a confusion analysis on VGG backbone, using CIFAR10 as InD and SVHN as OOD, the confusion scores are shown in Fig. 5. While the InD samples tend to have small negative values (with an average of −0.16), the OOD samples are more concentrated on the positive side (with an average of 0.02). More importantly, the majority of the InD samples (99.78%) have negative confusion scores and this makes the confusion analysis highly reliable and less prone to false positives. The confusion happens when the confusion score is close to zero, according to applications, a threshold could be determined based on the tolerance for misclassification.

Towards this error mitigation problem, we carry on the confusion analysis by designing a visualization tool for image OOD detection. Specifically, we adopt the Grad-CAM [14] to show the root causes of the OOD predictions in the input space. The analysis is continued on the VGG backbone and CIFAR10 InD setting. As for the OOD dataset, we use the Tiny ImageNet since it has the most related class definition as CIFAR10. Some examples are shown in Fig. 6 to illustrate the disagreement between two OOD detectors: C4 and C9, the numbers below the heatmaps are their corresponding OOD scores, with red color representing an OOD prediction and green color representing an InD prediction. We could see that OOD detectors at the early layers are more sensitive to textures and colors, while OOD detectors at the ending layers are more focused on objects and scenes.

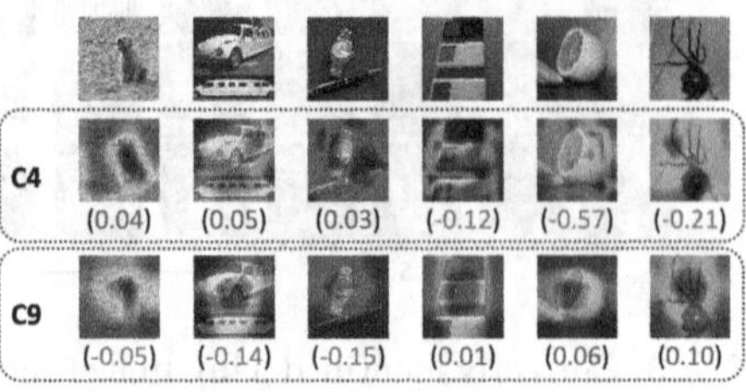

Fig. 6. Prediction visualization of Tiny ImageNet samples, on VGG-16 and CIFAR10 InD.

4.4 Advantages of Using Intermediate OOD Detectors

An optimal discernment layer [1] (or best layer) could be found for a particular OOD dataset, but it may not be the optimal choice for OOD datasets of different complexity. In Fig. 7 we show the AUROC of SVHN and LSUN at each layer of VGG-16 (using CIFAR10 as InD). The best layer for SVHN is layer 5, while the best layer for LSUN is the last layer. Such best layer could be estimated using a separate OOD dataset, however, as we could see from Table 1, OODL that estimates the best layer using the iSUN dataset could have its performance degrade significantly when OODs of different complexity are encountered. Therefore, instead of choosing the best layers for different OODs, LA-OOD propagates the most confident OOD prediction across all layers, and could effectively construct a good OOD confidence measurement for unseen OODs. For all five OOD datasets considered in this paper, LA-OOD can achieve competitive or even better accuracy compare to their corresponding best layers.

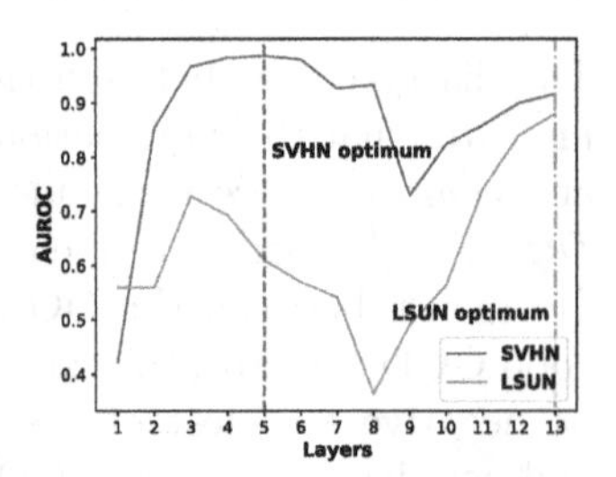

Fig. 7. The optimal discernment layers of SVHN and LSUN on VGG-16.

4.5 Ablation Studies

Table 2. Performance average on all the OOD datasets. Evaluation metrics with "↑" indicate the bigger the better and "↓" indicate the smaller the better. Best performance is labeled in **bold**.

InD/Model	Metric	C_1	C_2	C_3	C_4	C_5	C_6	C_7	C_8	C_9	C_{10}	C_{11}	C_{12}	C_{13}	LA-OOD
CIFAR10 VGG-16	AUROC ↑	60.20	78.14	89.55	89.00	83.92	81.13	77.99	70.45	65.96	71.58	83.57	89.20	91.62	**93.73**
	AUPR ↑	89.18	94.60	97.55	97.51	96.36	95.75	94.87	92.80	87.07	89.27	94.46	97.08	97.79	**98.48**
	FPR at 95% TPR ↓	95.47	84.39	61.76	64.55	77.97	85.61	89.12	95.19	88.47	82.01	56.04	63.22	37.16	**28.25**
CIFAR100 VGG-16	AUROC↑	51.87	70.09	83.71	82.61	79.72	77.08	76.17	69.55	72.43	70.38	62.66	37.80	73.46	**85.34**
	AUPR↑	85.56	91.89	95.93	95.85	95.19	94.51	93.84	91.90	91.11	90.27	86.70	76.10	90.42	**95.93**
	FPR at 95% TPR↓	94.75	89.32	76.22	80.75	83.96	86.80	86.00	90.39	81.74	86.29	85.16	96.38	65.37	**52.58**

Here we evaluate the effectiveness of the "early exits". We compare the results of the proposed LA-OOD with the average performance using each OOD detector solely on five OOD datasets mixed (LUSN + Tiny ImageNet + SVHN + DTD + Pure Color). Results are shown in Table 2. Using VGG-16 as an example, for both CIFAR10 and CIFAR100 InD settings, LA-OOD can achieve consistently better performance than any single OOD detector.

5 Conclusion

We proposed the LA-OOD, a layer-adaptive OOD detection framework for deep neural networks. By attaching multiple intermediate OOD detectors to the DNNs, LA-OOD can fully exploit the intrinsic characteristics of the intermediate latent space and reveal OODs with increasing complexity at deeper layers. Extensive experiments have been conducted to verify the effectiveness and interpretability of LA-OOD. On three DNNs with varying depth and architectures, our framework outperforms the state-of-the-art baselines without using any OOD training/validation data, being a reliable method for detecting unseen OODs.

References

1. Abdelzad, V., Czarnecki, K., Salay, R., Denounden, T., Vernekar, S., Phan, B.: Detecting out-of-distribution inputs in deep neural networks using an early-layer output. arXiv:1910.10307 (2019)
2. Bolukbasi, T., Wang, J., Dekel, O., Saligrama, V.: Adaptive neural networks for efficient inference. In: ICML, pp. 527–536. PMLR (2017)
3. Cimpoi, M., Maji, S., Kokkinos, I., Mohamed, S., Vedaldi, A.: Describing textures in the wild. In: CVPR, pp. 3606–3613 (2014)
4. Hendrycks, D., Gimpel, K.: A baseline for detecting misclassified and out-of-distribution examples in neural networks. arXiv:1610.02136 (2016)
5. Huang, G., Chen, D., Li, T., Wu, F., Van Der Maaten, L., Weinberger, K.Q.: Multi-scale dense convolutional networks for efficient prediction. arXiv:1703.09844 **2** (2017)
6. Kaya, Y., Hong, S., Dumitras, T.: Shallow-deep networks: understanding and mitigating network overthinking. In: ICML, pp. 3301–3310. PMLR (2019)

7. Lee, D., Yu, S., Yu, H.: Multi-class data description for out-of-distribution detection. In: KDD, pp. 1362–1370 (2020)
8. Lee, K., Lee, K., Lee, H., Shin, J.: A simple unified framework for detecting out-of-distribution samples and adversarial attacks. In: Advances in Neural Information Processing Systems, pp. 7167–7177 (2018)
9. Leroux, S., et al.: The cascading neural network: building the internet of smart things. Knowl. Inf. Syst. **52**(3), 791–814 (2017)
10. Li, Y., Maguire, L.: Selecting critical patterns based on local geometrical and statistical information. PAMI **33**(6), 1189–1201 (2010)
11. Liang, S., Li, Y., Srikant, R.: Enhancing the reliability of out-of-distribution image detection in neural networks. arXiv:1706.02690 (2017)
12. Malinin, A., Gales, M.J.F.: Predictive uncertainty estimation via prior networks. arxiv:1802.10501 (2018)
13. Schölkopf, B., Williamson, R.C., Smola, A.J., Shawe-Taylor, J., Platt, J.C.: Support vector method for novelty detection. In: Advances in Neural Information Processing Systems, pp. 582–588 (2000)
14. Selvaraju, R.R., Cogswell, M., Das, A., Vedantam, R., Parikh, D., Batra, D.: Grad-cam: visual explanations from deep networks via gradient-based localization. In: ICCV, pp. 618–626 (2017)
15. Vyas, A., Jammalamadaka, N., Zhu, X., Das, D., Kaul, B., Willke, T.L.: Out-of-distribution detection using an ensemble of self supervised leave-out classifiers. In: ECCV, pp. 550–564 (2018)
16. Wang, S., Liu, Q., Zhu, E., Porikli, F., Yin, J.: Hyperparameter selection of one-class support vector machine by self-adaptive data shifting. Pattern Recogn. **74**, 198–211 (2018)
17. Xu, P., Ehinger, K.A., Zhang, Y., Finkelstein, A., Kulkarni, S.R., Xiao, J.: Turkergaze: crowdsourcing saliency with webcam based eye tracking. arXiv:1504.06755 (2015)
18. Zhao, C., Chen, F.: Rank-based multi-task learning for fair regression. In: IEEE International Conference on Data Mining (ICDM) (2019)
19. Zhao, C., Chen, F., Thuraisingham, B.: Fairness-aware online meta-learning. In: ACM SIGKDD (2021)
20. Zhou, B., Bau, D., Oliva, A., Torralba, A.: Interpreting deep visual representations via network dissection. PAMI **41**(9), 2131–2145 (2018)

Improving Energy-Based Out-of-Distribution Detection by Sparsity Regularization

Qichao Chen[1], Wenjie Jiang[2], Kuan Li[1](✉), and Yi Wang[1]

[1] Dongguan University of Technology, Dongguan, China
{likuan,wangyi}@dgut.edu.cn
[2] Shenzhen University, Shenzhen, China
1910272058@email.szu.edu.cn

Abstract. Out-of-distribution (OOD) detection is critical for safely deploying machine learning models in the open world. Recently, an energy-score based OOD detector was proposed for any pre-trained classification models. The energy score, which is less susceptible to overconfidence, proves to be a better substitute for the conventional approaches leveraging the softmax confidence score. However, current energy-score based methods rely heavily on large-scale auxiliary datasets and introduce several dataset-dependent hyperparameters. In this paper, we propose a simple yet effective sparsity-regularized learning objective for deep neural networks so that the energy-based detector works better. Our learning objective is parameter-free and its key idea is to enlarge the differences between network outputs of in-distribution data and OOD data by regularizing the networks to generate high sparsity representations for in-distribution data. We also contribute to a tiny auxiliary outlier dataset to replace the previous one, which reduces the volume size significantly (230G vs. 40M). Besides, a new energy-score based OOD detector named **S**parsity-**R**egularized **O**utlier **E**xposure **(SROE)** is proposed to incorporate the proposed sparsity-regularized loss function into the traditional Outlier Exposure method. Experimental results show that the proposed sparsity-regularized loss strategy is effective, and the SROE OOD detector outperforms the other SOTA methods with a large margin. The source code and dataset are available at https://github.com/kuan-li/SparsityRegularization.

Keywords: OOD detection · Energy score · Sparsity regularization

1 Introduction

Deep neural networks (DNNs) have achieved high accuracy on many machine learning systems when the training and testing data are sampled from the same distribution, e.g., image recognition [22], speech recognition [20], person re-identification [16]. However, the real world is open and full of unknowns.

J. Gama et al. (Eds.): PAKDD 2022, LNAI 13281, pp. 539–551, 2022.
https://doi.org/10.1007/978-3-031-05936-0_42

Machine learning models in deployment often encounter testing data that have a large covariate shift from training data. Such distribution shift may lead to serious consequences since the model still attempts to classify test data into a certain training class, even though it may not belong to any training classes. The above issue gives rise to the importance of OOD detection, which is a great concern to AI Safety, especially in high-risk applications. OOD detection aims to determine whether the test data is in-distribution or OOD data.

Plenty of recent researches have emerged to address this problem [1,5,7,8,12–14]. The most straightforward methods directly utilized the output from the posterior distribution of DNNs. Hendrycks and Gimpel [5] observed that the predicted probability of OOD examples tend to be lower than that of the in-distribution data, establishing a common baseline for OOD detection named Maximum Softmax Probability (MSP). Follow-up works attempted to improve the OOD detection performances. ODIN [14] leveraged temperature scaling and input preprocessing to compute a more effective score using the max probability. Generalized ODIN [7] further improved the OOD detection performance by decomposing confidence and modifying input preprocessing. Maha [13] used pre-trained classification to model the representation of training data as class-conditional Gaussian distributions, then calculates Mahalanobis distance between test samples and the Gaussian distribution as the OOD score. ATOM [1] improves the robustness of the OOD detector by mining informative auxiliary OOD data and generalize to unseen adversarial attacks. GramNorm [8] utilized information extracted from the gradient space to compute the OOD score.

Although those methods are computationally simple, they only focus on improving inference procedures. Their performances are highly dependent on the pre-trained classification models. Recently, the outlier exposure was proposed to leverage diverse data as an auxiliary outlier dataset when training the classification model [6]. Within the outlier exposure method, an energy-score based OOD function was utilized, along with an energy-based learning objective using diverse data to fine-tune the network. The learning procedure trys to assign relatively lower energy values to the in-distribution data and higher energy values to OOD data [15]. However, current energy-based loss function introduces several dataset-dependent hyperparameters, and the cost of training (including time and space consumptions) is particularly high due to the large-scale auxiliary outlier dataset. Meanwhile, to prune redundant computation, SMSR [21] explores the sparsity in image Super-Resolution (SR) to improve the inference efficiency of SR networks. Inspired by SMSR, we firstly introduce the sparsity of neural networks to our OOD detection framework.

In this paper, we propose a sparsity-regularized learning objective for a pre-trained classification model in order to improve the performance of the energy-score based OOD detector. The intuitive motivation is that the energy-based OOD detection algorithm may work better if we further increase the sparsity of the feature vector generated by neural networks on in-distribution data. In contrast, our method adopts a parameter-free strategy that is easy to use and implement, without tuning any dataset-dependent hyperparameters. We also modify the outlier exposure [6] with our parameter-free strategy to propose a **S**parsity-**R**egularized **O**utlier **E**xposure (**SROE**) OOD detector, which further enlarges the energy score gap between in-distribution and OOD samples and leads to better performances. In addition, we build a tiny dataset as the auxiliary outlier dataset to replace the previous bulky one [6,15], which significantly reduces time and space consumptions in the learning process. Our contributions in this paper are summarized as follows:

1. We propose a parameter-free sparsity-regularized learning objective for fine-tuning the deep neural networks, our learning objective guides the networks to generate sparser feature vectors for in-distribution data while no limitations for OOD data. The energy-based detector works better even when no external (OOD) dataset used.
2. We propose the **S**parsity-**R**egularized **O**utlier **E**xposure **(SROE)** by incorporating the new parameter-free loss function into the original outlier exposure method. Our SROE learning objective guides the representation sparser for in-distribution data while more uniform for the predictive distribution on OOD examples. Experimental results show that our method achieves a new state-of-the-art performance on the OOD detection tasks.
3. The original outlier exposure method uses *80 Million Tiny Images* as the auxiliary outlier dataset, which is bulky and bloated, leads to more time and space consumptions in the learning procedure. Thus, we build a tiny dataset to replace the previous one that significantly reduces time and space consumptions without any detection performance loss.

2 Background and Related Work

2.1 OOD Detection Problem Statement

OOD detection can be formulated as a binary classification problem [5]. Let P_X denote a data distributions defined on the sample space $\mathcal{X}$. We consider a training dataset $\mathcal{D}_{in}$ drawn i.i.d from P_X (called the **in-distribution**), with label space $\mathcal{Y} = \{1, 2, \cdots, C\}$. **Out-of-distribution** (OOD) usually refers to the samples from an irrelevant distribution whose label set has no intersection with $\mathcal{Y}$ and therefore should not be predicted by the model.

We use $\mathcal{D}_{in}$ to train a neural network classification model $f(x)$. The goal of OOD detection is to design a binary function estimator $G(x; f)$ that evaluate whether a input data $x \in \mathcal{X}$ from in-distribution P_X or not. Specifically, the OOD detector $G(x; f)$ is defined as follows:

$$G(x; f) = \begin{cases} 0 & \text{if } S(x; f) \leq \delta \\ 1 & \text{if } S(x; f) > \delta \end{cases}$$

where $S(x; f)$ is the scoring function that captures uncertainty of input data, and δ is the threshold commonly chosen so that a high fraction (e.g., 95%) of in-distribution data is correctly classified. The detector $G(x; f)$ assigns label "1" for input data if the confidence score $S(x; f)$ is above δ, considered to be in-distribution data. Otherwise, it assigns label "0", considered to be out-of-distribution data.

2.2 Outlier Exposure

One of the difficulties within OOD detection is that we have no prior knowledge of OOD data. Thus, it is not possible to model OOD data directly. Most methods detect OOD examples by using representations from only in-distribution data. Based on this phenomenon, A heuristic approach named Outlier Exposure (OE) was proposed for OOD detection by exposing the model to OOD examples [6].

Specifically, OE utilized an auxiliary dataset D_{out}^{OE} entirely disjoint from test-time data and in-distribution data to teach the network learning better representations. OE introduced the OE loss on D_{out}^{OE} , and added it to the classification loss (e.g., cross-entropy) on training data. The OE loss is the KL divergence loss between posterior and uniform distribution on $\mathcal{D}_{out}^{OE}$, which forces the predictive distribution on OOD samples to be closer to the uniform distribution:

$$\mathcal{L}_{OE} = \mathbb{E}_{x \sim \mathcal{D}_{out}^{OE}}\left[KL\big(\mathcal{U}(y) \mid\mid P_\theta(y|x)\big)\right],$$

where KL denotes the Kullback-Leibler (KL) divergence, $\mathcal{U}(y)$ is the uniform distribution, and $P_\theta(y|x)$ is the output probability of samples from $\mathcal{D}_{out}^{OE}$ in the neural network.

2.3 Energy-Based Model for OOD Detection

Energy-based Models (EBMs) capture dependencies by associating scalar energy to each configuration of the variables [11]. A collection of energy values could be turned into a probability density $p_\theta(x)$ through th Gibbs distribution:

$$p_\theta(x) = \frac{exp\big(-E_\theta(x)\big)}{Z_\theta},$$

where $E_\theta(x) : \mathbb{R}^D \rightarrow \mathbb{R}$ is the energy function that map each point to a scalar, and $Z_\theta = \int_x exp\big(-E_\theta(x)\big)$ is the partition function that normalize constant with respect to x.

The essence of the EBMs is to build an energy function $E(x)$, and $E(x)$ is very flexible, it can be any function that takes x from sample space $\mathcal{X}$ as the input then returns a scalar. Recently, the energy-based OOD detection [15] was proposed to leverage the energy value differences between data to detect OOD examples. Examples with higher energy values are considered as OOD inputs, and the **energy score function** $E(x; f)$ is defined as follow:

$$E(x; f) = -\log \sum_i exp(f_i),$$

where score function $E(x; f)$ maps all logits to a non-probabilistic scalar called the **energy score** [15].

The energy score mitigates a critical problem of softmax confidence with arbitrarily high values for OOD examples [4]. It is less susceptible to overconfidence, proves to be a better substitute for the conventional approaches using the softmax confidence score. In this work, we focus on improving the performance of energy score for OOD detection.

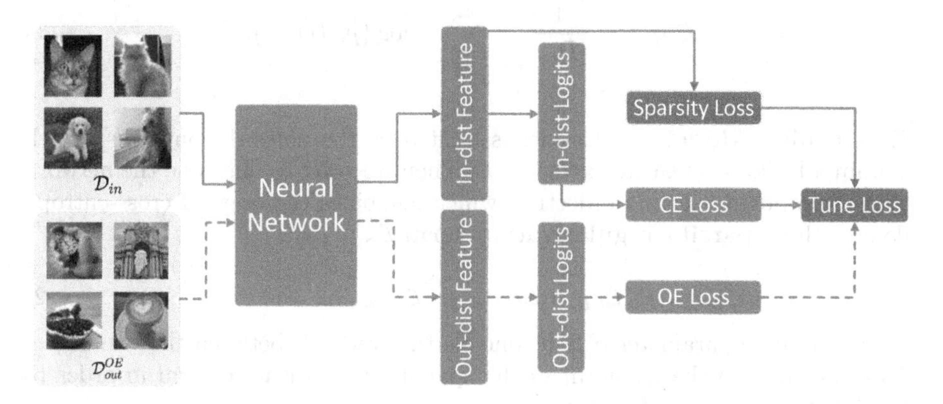

Fig. 1. Sparsity-Regularized OOD detection framework. We proposed the red modules as the picture shown above, the solid lines indicate the flow of $\mathcal{D}_{in}$ loss in neural network, and the dashed lines indicate the flow of $\mathcal{D}_{out}^{OE}$ loss in neural network. Note that the dashed part is optional. The total tune loss $\mathcal{L}_{Tune} = \mathcal{L}_{CE} + \alpha \cdot \mathcal{L}_{Sparsity}$ when there is no $\mathcal{D}_{out}^{OE}$ available; otherwise, $\mathcal{L}_{Tune} = \mathcal{L}_{CE} + \alpha \cdot \mathcal{L}_{Sparsity} + \beta \cdot \mathcal{L}_{OE}$ (Color figure online)

3 The Sparsity-Regularized Framework

In this section, we present our parameter-free sparsity-regularized (SR) learning method for fine-tuning the classification models in order to improve the performance of energy-based OOD detection [15]. Then we show how to incorporate the parameter-free learning strategy into the traditional Outlier Exposure [6] to form Sparsity-Regularized Outlier Exposure (SROE), which further improves the performance. Finally we describe how the new tiny auxiliary dataset $\mathcal{D}_{out}^{OE}$ is built to reduce time and space consumptions in the learning process. The framework of our work is shown in Fig. 1.

3.1 Sparsity-Regularized Loss Function

The intuitive motivation is that the energy-based OOD detection algorithm may work better if we further increase the sparsity of the feature vector generated by neural networks for in-distribution data, while giving no limitations for those of the out-of-distribution data. In this work we focus on fine-tuning a pre-trained classifier and then use energy-score based detector to measure its performance of OOD detection.

Pre-training Model. Firstly, we train a neural classifier under the supervision of labeled in-distribution training samples $\mathcal{D}_{in}$. Unless otherwise noted, we use the standard cross-entropy loss as follow:

$$\mathcal{L}_{CE} = -\frac{1}{|\mathcal{D}_{in}|} \sum_{(x_i, y_i) \in \mathcal{D}_{in}} \log \left(p_{y_i}(y|x_i) \right). \tag{1}$$

Fine-Tuning Model. A classifier is built once the network converges, which contain a backbone feature extractor F. Then we start to fine-tune the network using a proposed learning objective, which combines the original cross-entropy along with **a sparsity regularization term** $\mathcal{L}_{Sparsity}$:

$$\mathcal{L}_{SR} = \mathcal{L}_{CE} + \alpha \cdot \mathcal{L}_{Sparsity}, \tag{2}$$

where the hyperparameter $\alpha > 0$ controls the trade-off between the two objectives, and we use the L_1 norm as the sparsity regularization term in order to penalize the feature vector and make it sparser:

$$\mathcal{L}_{Sparsity} = \frac{1}{|\mathcal{D}_{in}|} \sum_{x_i \in \mathcal{D}_{in}} \left(\|F(x_i)\|_1 \right). \tag{3}$$

Unlike the energy-bounded loss function [15] that introduces two dataset-dependent hyperparameters, the $\mathcal{L}_{Sparsity}$ is a nearly parameter-free strategy. We believe that this strategy could guide the networks generating sparser feature vectors for in-distribution data, while not limiting/ignoring the out-of-distribution data. This corresponds to the solid lines in Fig. 1, which achieve better OOD detection performance without any auxiliary datasets, just using the in-distribution training dataset for tuning the network.

3.2 Sparsity-Regularized Outlier Exposure

Fine-Tuning Model with $\mathcal{D}_{out}^{OE}$. Same as Sect. 3.1, we construct a pre-trained neural network model. Unlike the previous sparsity regularization method, we further introduce an auxiliary dataset $\mathcal{D}_{out}^{OE}$ to it by OE loss [6], forming our **S**parsity-**R**egularized **O**utlier **E**xposure (**SROE**). The overall framework is shown in Fig. 1, composing by both the solid lines and the dashed lines:

$$\mathcal{L}_{SROE} = \mathcal{L}_{CE} + \alpha \cdot \mathcal{L}_{Sparsity} + \beta \cdot \mathcal{L}_{OE}, \tag{4}$$

where the hyperparameters $\alpha > 0$, and $\beta > 0$ control the trade-off between the three objectives.

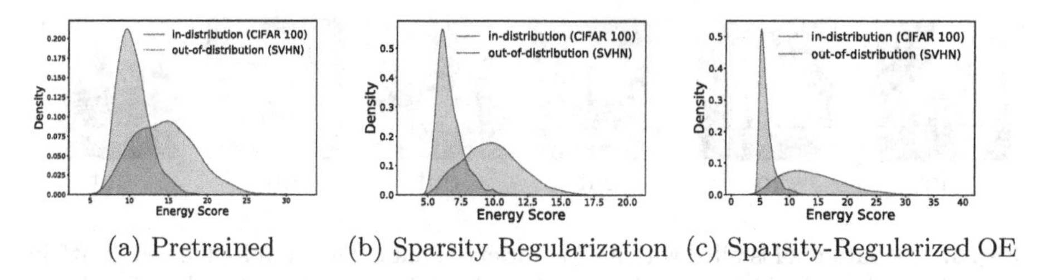

(a) Pretrained (b) Sparsity Regularization (c) Sparsity-Regularized OE

Fig. 2. Distribution of energy scores. We contrast the distribution of the models from Pretrained, Sparsity-Regularization, Sparsity-Regularized Outlier Exposure, respectively. We chose the CIFAR-100 as the $\mathcal{D}_{in}$ and the SVHN as the $\mathcal{D}_{out}^{test}$, which performed consistently on the other OOD test datasets. (Color figure online)

The cross-entropy loss keeps the performance of classification on in-distribution samples, the OE loss forces the predictive distribution on out-of-distribution samples to be closer to the uniform one, and our sparsity-regularized loss increases the sparsity of the feature vectors generated by neural networks for in-distribution samples. By combining the above three losses and adjusting their weights, the overall learning objective $\mathcal{L}_{SROE}$ maintains the classification performance while guiding the representation sparser on in-distribution samples, and more uniform for the predictive distribution on OOD samples.

The experimental results validate the effectiveness of our method. As shown in Fig. 2, the green curve is the distribution of energy scores on in-distribution examples, and the red curve is the distribution of energy scores on OOD examples. The intersecting region of the two curves indicates that in-distribution and OOD examples cannot be effectively distinguished by energy score. The larger area of the intersecting region is, the worse performance of OOD detection for the model will be, and vice versa. From Fig. 2, we see that using the proposed sparsity regularization could significantly reduce the area of the intersecting region, which demonstrates the performance of OOD detection has been improved. And the area of the intersecting region reduces more by combining the OE Loss with our sparsity regularization term.

3.3 A New Auxiliary Outlier Dataset

Previous works use the *80 Million Tiny Images* as the $\mathcal{D}_{out}^{OE}$ [6,15]. It is bulky and bloated, the volume size is bigger than 230G. Furthermore, when using as the auxiliary dataset for OOD researches, samples within 80 Million Tiny Images should be removed manually if it appears as (or similar as) the in-distribution samples. However, the dataset is too enormous and it is impossible to cleanly remove such samples manually or even by automated technique. This motivates us to reconstruct a dataset $\mathcal{D}_{out}^{OE}$ in a more economical and friendly way [19].

The auxiliary dataset needs to meet two constraints [6]:

1) Diverse enough to represent unknown out-of-distribution data in some way;
2) Completely disjoint with the in-distribution data and the data to be tested.

Fig. 3. The data pool is filled with a wide variety of data. The samples labeled "**out**" in the data pool can be kept as auxiliary outlier data. The samples labeled "**in**" indicate that those samples have similar semantic information as in-distribution data, and we need to filter it out from the data pool manually.

The Tiny ImagesNet dataset [10] is a 200-class subset of the ImageNet dataset [18], we use it as a data pool to build our auxiliary dataset. The data pool selected is shown in Fig. 3, which is filled with a wide variety of data. We spend a lot of time manually filtering out data with similar semantic information to in-distribution data from it. Once the filtering operation in the data pool is completed, we then resize all images to 32×32. The volume size of our dataset is small enough compared to the previous one (40M vs. 230G), we call it "**tiny**".

4 Experiments

In this section, we describe the experimental setup, including evaluation metrics, evaluation datasets, network architectures and training details. Then, we also demonstrate the effectiveness of our proposed method by comparing it with the SOTA ones.

4.1 Experimental Setup

At inference time, given one test image, if it comes from CIFAR-10 or CIFAR-100 datasets, it is viewed as an in-distribution example; otherwise, it will be viewed as an out-of-distribution example.

Evaluate Metrics. In order to measure the effectiveness of our method on distinguishing in-distribution with OOD examples, we followed the same metrics as previous methods [5,6,15]. **(1) FPR95** shows the false positive rate (FPR) of OOD examples when the true positive rate (TPR) of in-distribution examples is 95%; **(2) AUROC** is the area under the receiver operating characteristic (ROC) curve; and **(3) AUPR** is the area under the Precision-Recall (PR) curve. In our experimental results, all the values of metrics are percentages and the results are averaged over 10 runs. The larger values are better for AUROC and AUPR, while the lower value is better for FPR95.

In-Distribution Datasets. We use CIFAR-10 and CIFAR-100 as in-distribution datasets $\mathcal{D}_{in}$ where CIFAR-10 contains 10 classes; CIFAR-100 is more complicated and it contains 100 classes [9]. Both of them are consisted by 60,000 32 $\times$ 32 natural color images, with 50,000 for training and 10,000 for testing. CIFAR-10 and CIFAR-100 are disjoint but have similarities.

Out-of-Distribution Datasets. The OOD evaluation benchmarks are rigorously selected to ensure that the semantic knowledge in these datasets does not overlap with in-distribution data. We use six OOD common test datasets $\mathcal{D}_{out}^{test}$ to evaluate our method, including fine-grained images, scene images, and textural images, etc. All the datasets considered are listed below: **iSUN** [3], **SVHN** [17], **Texture** [2], **Places365** [25], **LSUN-Crop** and **LSUN-Resize** [23]. For the auxiliary outlier dataset D_{out}^{OE} that is required during the training procedure in the experiment, the proposed **tiny** dataset introduced in Sect. 3.3 is used.

Table 1. OOD detection performance improvements by sparsity-regularized strategy. ↑ indicates larger values are better, and ↓ indicates smaller values are better. **Boldface** values indicate the relatively better results.

$\mathcal{D}_{in}$	$\mathcal{D}_{out}^{test}$	**FPR95** ↓	**AUROC** ↑	**AUPR** ↑
		Baseline [15]/**Ours**		
CIFAR-10	iSUN	44.65/**20.55**	88.61/**96.31**	98.85/**99.22**
	SVHN	30.35/**29.70**	93.70/**94.07**	98.52/**98.64**
	Texture	53.10/**26.70**	85.04/**94.05**	95.48/**98.45**
	Place 365	42.75/ **27.50**	88.31/**93.51**	96.75/**98.36**
	LSUN-Crop	7.90/**3.20**	98.26/**99.25**	99.63/**99.84**
	LSUN-Resize	37.55/**16.10**	91.46/**98.88**	97.82/**99.32**
	Average	34.92/**20.26**	91.27/**95.70**	97.57/**98.97**
CIFAR-100	iSUN	77.15/**52.19**	79.54/**87.53**	94.91/**97.57**
	SVHN	82.41/**58.25**	81.59/**85.86**	95.92/**95.52**
	Texture	82.10/**58.90**	77.26/**85.86**	94.14/**96.52**
	Place 365	82.10/**58.90**	77.26/**85.86**	94.14/**96.52**
	LSUN-Crop	40.60/**33.30**	92.73/**93.93**	98.46/**98.71**
	LSUN-Resize	77.05/**66.40**	78.59/**84.36**	94.51/**96.32**
	Average	71.86/**56.59**	81.94/**87.49**	95.59/**97.10**

Network Architectures and Training Details. All the images from CIFAR-10 and CIFAR-100 are normalized using per-channel mean and standard deviation. WideResNet-40-2 is used as the classification network. When pre-training the model, a SGD optimizer with a weight decay of 0.0005 and a momentum of 0.9 is used; the initial learning rate is 0.1 with cosine decay and the dropout rate is 0.3; the batch size is 128 for 100 epochs. When fine-tuning the classification model, we use the same network configuration as the training phase. The learning rate reduces to 0.001, and set the number of epochs to 10. In the sparsity-regularized (SR) experiment, the α in Eq. (3) is 0.001. And in the sparsity-regularized outlier exposure (SROE) experiment, the batch size is 256 for axuiliary data $\mathcal{D}_{out}^{OE}$. the α and β in Eq. (4) is 0.001 and 0.5, respectively.

4.2 Experimental Results

OOD Performance Improved by Sparsity Regularization. At training time, we use the WideResNet-40–2 network architecture trained on in-distribution D_{in} as our energy-score based OOD detection baseline, and use neural networks fine-tuned by the proposed sparsity-regularized learning objective as the comparison.

Table 1 reports the details of OOD detection performance when using CIFAR-10 and CIFAR-100 as $\mathcal{D}_{in}$ against six OOD test datasets $\mathcal{D}_{out}^{test}$. The results show that the proposed parameter-free sparsity-regularized loss achieved better performance even without any external data. The average FPR95 reduced by 14.66% on CIFAR-10 and 15.27% on CIFAR-100 respectively, which demonstrate the effectiveness of the sparsity-regularized strategy.

Table 2. Comparisons between previous methods and ours on the D_{in}. $\uparrow$ indicates larger values are better, and $\downarrow$ indicates smaller values are better. **Boldface** values indicate the relatively better results. Our method obtains consistently better results on almost all metrics.

$\mathcal{D}_{in}$	Method	FPR95 $\downarrow$	AUROC $\uparrow$	AUPR $\uparrow$
CIFAR-10	MSP [5]	51.35	90.45	97.82
	ODIN [14]	35.59	90.96	97.64
	Maha [13]	37.08	93.27	98.49
	EBD [15]	34.92	91.27	97.57
	SR (ours)	**19.19**	**96.03**	**99.08**
CIFAR-100	MSP [5]	80.56	75.44	93.45
	ODIN [14]	76.64	77.43	94.23
	Maha [13]	62.03	80.33	95.02
	EBD [15]	71.86	81.94	95.59
	SR (ours)	**56.59**	**87.49**	**97.10**

Comparisons with Other Methods Which Need No Auxiliary Dataset. We compare our work with previous methods that do not require any auxiliary data. All experiments use WideResNet-40–2 [24] for fair comparisons. Specifically, we compare the proposed sparsity-regularized strategy with MSP [5],

ODIN [14], Mahalanobis [13], as well as energy-based detection [15]. Those methods do not require external data, only a pre-trained model needed.

Table 2 reports the OOD detection performance of various methods and all numbers reported are averaged over six OOD test datasets $\mathcal{D}_{out}^{test}$. The experiments show that the **energy score** (EBD) function is indeed a better replacement for the **softmax confident score** (MSP). And our sparsity-regularized learning objective could improve energy score OOD detection by fine-tuning the pre-trained model with an extra sparsity related loss term.

Performance of SROE. Using an auxiliary dataset could further improve OOD detection performance. We compare different methods which need fine-tuning the networks. WideResNet-40–2 network model is used as the baseline, and it is used as the pre-trained model before fine-tuning for all different methods. We reproduced the results of the Outlier Exposure and Energy based OOD, using the *80 Million Tiny Images* dataset as the auxiliary outlier dataset. As for SROE, we use the propsed **tiny** dataset instead, this greatly save a lot of disk space and training time consumptions.

Table 3 reports the performances of various fine-tuning OOD detection methods. The experimental results illustrate that the fine-tuning methods only affect the classification ability of the model on in-distribution data slightly. In contrast, SROE is dataset-dependent-parameter-free, while the energy-based method(EBD) introduces two dataset-dependent hyperparameters, which have to be fixed manually. More importantly, SROE greatly improve the performance of the outlier exposure by incorporating the proposed sparsity-regularized loss strategy.

Table 3. OOD detection performance comparisons when using auxiliary datasets to fine-tune the networks. The unit of "**Time**" is **second**, and all values except "**Time**" are percentages. Note that the "**Time**" in the table represents time required when fine-tuning the model, and the "**Time** = -" for the Baseline method means that the model needs not fine-tuning. We could see that the total fine-tuning time of the model is largely reduced when using our **tiny** dataset. ↑ indicates larger values are better, and ↓ indicates smaller values are better. Boldface values indicate relative the better results. SROE obtains consistently better results on almost all metrics.

$\mathcal{D}_{in}$	Method	FPR95 ↓	AUROC ↑	AUPR ↑	Time ↓	Err(D_{in}) ↓
CIFAR-10	Baseline[15]	34.92	91.27	97.57	–	5.24
	OE [6]	8.53	95.30	99.63	48	5.30
	Energy [15]	**3.32**	98.22	99.75	48	5.12
	SROE (ours)	4.15	**98.92**	**99.76**	**5**	**5.06**
CIFAR-100	Baseline [15]	71.86	81.94	95.59	–	24.04
	OE [6]	56.57	86.79	96.82	48	24.52
	Energy [15]	49.28	88.23	97.09	49	24.56
	SROE (ours)	**28.04**	**93.14**	**98.12**	**5**	**23.98**

5 Conclusion

In this paper, we proposed a novel sparsity-regularized learning objective for neural networks, making the energy-based OOD detector work better. Our method does not introduce dataset-dependent hyperparameters, nor require extra dataset for auxiliary neural network training. Furthermore, we also proposed a new SROE OOD detector, which means Sparsity-Regularized Outlier Exposure, by incoporating the sparsity-regularized learning objectives into the traditional OE. To speed up the fine-tuning process, we built a **tiny** auxiliary outlier dataset to replace the bulky and large-scale *80 Million Tiny Images* dataset.

Experimental results show that our method establishes a new state-of-the-art performance on the OOD detection task. Although we primarily focus on image classification in our experiments, our method can be applied to any other related field, e.g., audios and videos etc.

Acknowledgements. This work is supported in part by NSFC Grant 61876038, as well as Dongguan Science and Technology of Social Development Program under Grant 2020507140146, and Characteristic Innovation Projects of Guangdong Colleges and Universities (Grant No.2021KTSCX134).

References

1. Chen, J., Li, Y., Wu, X., Liang, Y., Jha, S.: Robust Out-of-Distribution Detection via Informative Outlier Mining, vol. 1(2), p. 7 (2020). arXiv preprint arXiv:2006.15207
2. Cimpoi, M., Maji, S., Kokkinos, I., Mohamed, S., Vedaldi, A.: Describing textures in the wild. In: CVPR (2014)
3. Girshick, R., Donahue, J., Darrell, T., Malik, J.: Rich feature hierarchies for accurate object detection and semantic segmentation. In: CVPR, pp. 580–587 (2014)
4. Hein, M., Andriushchenko, M., Bitterwolf, J.: Why RELU networks yield high-confidence predictions far away from the training data and how to mitigate the problem. In: CVPR, pp. 41–50 (2019)
5. Hendrycks, D., Gimpel, K.: A baseline for detecting misclassified and out-of-distribution examples in neural networks. In: ICLR (2017)
6. Hendrycks, D., Mazeika, M., Dietterich, T.: Deep anomaly detection with outlier exposure. In: ICLR (2019)
7. Hsu, Y.C., Shen, Y., Jin, H., Kira, Z.: Generalized ODIN: detecting out-of-distribution image without learning from out-of-distribution data. In: CVPR, pp. 10951–10960 (2020)
8. Huang, R., Geng, A., Li, Y.: On the importance of gradients for detecting distributional shifts in the wild. In: NeurIPS (2021)
9. Krizhevsky, A., Hinton, G., et al.: Learning Multiple Layers of Features from Tiny Images. Citeseer (2009)
10. Le, Y., Yang, X.: Tiny ImageNet Visual Recognition Challenge (2015). http://cs231n.stanford.edu/tiny-imagenet-200.zip
11. LeCun, Y., Chopra, S., Hadsell, R., Ranzato, M., Huang, F.: A tutorial on energy-based learning. In: Predicting Structured Data, vol. 1(0) (2006)

12. Lee, K., Lee, H., Lee, K., Shin, J.: Training confidence-calibrated classifiers for detecting out-of-distribution samples. In: ICLR (2018)
13. Lee, K., Lee, K., Lee, H., Shin, J.: A simple unified framework for detecting out-of-distribution samples and adversarial attacks. In: NeurIPS, vol. 31 (2018)
14. Liang, S., Li, Y., Srikant, R.: Enhancing the reliability of out-of-distribution image detection in neural networks. In: ICLR (2018)
15. Liu, W., Wang, X., Owens, J., Li, Y.: Energy-based out-of-distribution detection. In: NeurIPS (2020)
16. Meng, J., Zheng, W.S., Lai, J.H., Wang, L.: Deep graph metric learning for weakly supervised person re-identification. In: TPAMI (2021)
17. Netzer, Y., Wang, T., Coates, A., Bissacco, A., Wu, B., Ng, A.Y.: Reading digits in natural images with unsupervised feature learning. In: Proceedings of NIPS Workshop on Deep Learning and Unsupervised Feature Learning (2011)
18. Russakovsky, O., et al.: ImageNet large scale visual recognition challenge. In: IJCV (2015)
19. Schwartz, R., Dodge, J., Smith, N.A., Etzioni, O.: Green AI. Commun. ACM **63**(12), 54–63 (2020)
20. Takahashi, N., Singh, M.K., Basak, S., Sudarsanam, P., Ganapathy, S., Mitsufuji, Y.: Improving voice separation by incorporating end-to-end speech recognition. In: ICASSP, pp. 41–45. IEEE (2020)
21. Wang, L., Dong, X., Wang, Y., Ying, X., Lin, Z., An, W.: Exploring sparsity in image super-resolution for efficient inference. In: CVPR, pp. 4917–4926 (2021)
22. Xie, S., Kirillov, A., Girshick, R., He, K.: Exploring randomly wired neural networks for image recognition. In: ICCV, pp. 1284–1293 (2019)
23. Yu, F., Zhang, Y., Song, S., Seff, A., Xiao, J.: LSUN: Construction of a large-scale image dataset using deep learning with humans in the loop (2015). arXiv preprint arXiv:1506.03365
24. Zagoruyko, S., Komodakis, N.: Wide residual networks. In: British Machine Vision Conference 2016. British Machine Vision Association (2016)
25. Zhou, B., Lapedriza, A., Khosla, A., Oliva, A., Torralba, A.: Places: A 10 million image database for scene recognition. In: TPAMI (2017)

Author Index

Printed in the United States
by Baker & Taylor Publisher Services